Scottish Life and Society

Scotland's Buildings

Publications of the European Ethnological
Research Centre

Scottish Life and Society:
A Compendium of Scottish Ethnology
(14 Volumes).

Already published:
Volume 3 *Scotland's Buildings*
Volume 11 *Institutions of Scotland: Education*
Volume 14 *Bibliography*

GENERAL EDITOR:
Alexander Fenton

Scottish Life *and* Society

A COMPENDIUM OF SCOTTISH ETHNOLOGY

SCOTLAND'S BUILDINGS

Edited by Geoffrey Stell, John Shaw
and Susan Storrier

TUCKWELL PRESS

THE EUROPEAN ETHNOLOGICAL RESEARCH CENTRE

First published in Great Britain in 2003 by
Tuckwell Press
The Mill House, Phantassie
East Linton, East Lothian
EH40 3DG
Scotland

COPYRIGHT © The European Ethnological
Research Centre 2003

All rights reserved
ISBN 1 86232 123 X

We gratefully acknowedge the support of the following bodies:
The Scotland Inheritance Fund
The Russell Trust
The Binks Trust

Typeset by Carnegie Publishing Ltd, Chatsworth Road, Lancaster
Printed and bound by The Cromwell Press, Trowbridge, Wiltshire

MORAY COUNCIL LIBRARIES & INFO.SERVICES	
20 25 58 99	
Askews	
305.800941	

Contents

List of Figures	*vii*
List of Contributors	*xvii*
Foreword	*xix*
Abbreviations	*xxii*
Glossary	*xxiii*

PART ONE: INTRODUCTION

1. Buildings in Context   3
   Geoffrey Stell
2. Continuity and Change   9
   Alexander Fenton

PART TWO: DWELLINGS

3. Castles, Palaces and Fortified Houses   27
   Charles McKean
4. Country Houses, Mansions and Large Villas   48
   Kitty Cruft
5. Middle-Sized Detached Houses   67
   Deborah Mays
6. Small Houses and Cottages   90
   Annette Carruthers and John Frew
7. Tenements and Flats   108
   Miles Glendinning
8. Inns, Hotels and Related Building Types   127
   David Walker
9. Ancillary Estate Buildings   190
   Elizabeth Beaton
10. Garden Layouts and Garden Buildings   211
    Tim Buxbaum
11. Seasonal and Temporary Dwellings   224
    Roger Leitch
12. Housing for Seasonal Agricultural Workers   236
    Heather Holmes

PART THREE: COMMUNITY BUILDINGS

13. Buildings of Administration   249
    Caroline A MacGregor

14.	Worship and Commemoration *Neil Cameron*	276
15.	Education *Robert D Anderson*	294
16.	Health and Welfare *Harriet Richardson and Geoffrey Stell*	311
17.	Public Services *John R Hume*	341
18.	Buildings for Recreation *Charles McKean*	358
19.	Public Defences *Nigel A Ruckley*	381

PART FOUR: WORK PLACES

20.	Agricultural Buildings 1: Introduction; Equipment Storage and Traction *John Shaw*	423
21.	Agricultural Buildings 2: Storing and Processing Crops *John Shaw*	438
22.	Agricultural Buildings 3: Livestock Housing and Products *John Shaw*	465
23.	Workshops: Small-Scale Processing and Manufacturing Premises *John Shaw*	494
24.	Mills and Factories *Mark Watson*	510
25.	Mines, Quarries and Mineral Works *Miles K Oglethorpe*	551
26.	Harbours, Docks and Fisheries *John R Hume*	571
27.	Roads, Canals and Railways *John R Hume*	594
28.	Structures Associated with the Retail Trade † *Owen F Hand*	612
29.	Business and Commercial Buildings *David Walker*	624

Index 681

List of Figures

1.1	Map of Scotland showing pre-1975 counties and principal cities	2
1.2	Rural context: Ardchyle and Ben More, Perthshire, c. 1900	5
1.3	Urban context: East Church Street, Fordyce, Banffshire, 1937	5
1.4	High-density urban housing: 517 Lawnmarket and 10 Milne's Court, Edinburgh, 1910	6
2.1	Uishal, Lewis: plan and long section of shieling hut	11
2.2	Uishal, Lewis: rectangular shieling hut with gable chimney	12
2.3	Uishal, Lewis: plan of rectangular shieling hut	12
2.4	Uishal, Lewis: general view of shieling hut	13
2.5	Àiridh nan Sileag, Argyll: plan of shielings	14
2.6	Jura, Argyll: shieling huts	15
2.7	42 Arnol, Lewis: cross-section of blackhouse	17
2.8	Shawbost, Lewis: ruined blackhouse wall	18
3.1	Mingary Castle, Argyll	29
3.2	Coxton Tower, Moray	34
3.3	Pitsligo, Aberdeenshire	36
3.4	Castle Fraser, Aberdeenshire	38
3.5	Thirlestane, Berwickshire	39
3.6	Craignethan Castle, Lanarkshire: aerial view	41
3.7	Earl's Palace, Kirkwall, Orkney: view by R W Billings	44
4.1	Kinross House, Kinross-shire	50
4.2	Dalkeith House, Midlothian: late nineteenth-century view	50
4.3	Glendoick House, Perthshire	53
4.4	Taymouth Castle, Perthshire	55
4.5	Millearne, Perthshire	56
4.6	Castlemilk House, Dumfriesshire: main entrance	58
4.7	Cringletie House, Peeblesshire	60
4.8	Mar Lodge, Aberdeenshire, 1880s	62
4.9	Gribloch House, Stirlingshire	63
5.1	Haddington House, Sidegate, Haddington, East Lothian: a seventeenth-century urban L-plan house with stair in re-entrant angle, typical of domestic urban and rural work of this scale and period	66
5.2	Leaston House, Humbie, East Lothian: an early eighteenth-century two-storeyed, five-bay house (with early	

	nineteenth-century wings) which, typical of contemporary practice, mixes simple classical symmetry with traditional Scottish elements	68
5.3	*Rudiments of Architecture*, 1778: designs XX and XXI	72
5.4	Royal Circus, Edinburgh	74
5.5	Lagarie, Rhu, Dunbartonshire: design by A N Paterson	81
5.6	Lagarie, Rhu, Dunbartonshire	82
5.7	46a Dick Place, Edinburgh	85
6.1	Glen Dochart, Killin, Perthshire: long house, photograph probably late nineteenth century	91
6.2	Shandwick, Nigg, Easter Ross, mid nineteenth century (photographed 1975)	94
6.3	Campbeltown, Argyll: lodge house, c. 1900	96
6.4	Reid Terrace, Stockbridge, Edinburgh, 1861–2	98
6.5	34–6 Bayview Crescent, Methil, Fife	100
6.6	Winram Place, St Andrews, Fife	104
7.1	Milne's Court, Lawnmarket, Edinburgh	110
7.2	Walmer Crescent, Glasgow	116
7.3	Glenboig, Lanarkshire, inter-war settlement of two-storeyed flats built by Lanark County Council: aerial view	120
7.4	Piershill, Edinburgh: aerial view	121
7.5	Red Road development, Glasgow	123
8.1	The Black Bull, 12 Grassmarket, Edinburgh: early twentieth-century view	128
8.2	Great Inn, Inveraray, Argyll: elevation as completed, drawn by John Campbell, 1771	134
8.3	Great Inn, Inveraray, Argyll: ground plan and stables as intended, John Adam, 1750	134
8.4	Annandale Arms, Moffat, Dumfriesshire, c. 1880	136
8.5	Caledonian Hotel, Inverness, c. 1830	138
8.7	Fife Arms Hotel, Low Street, Banff, c. 1890	144
8.8	Duke of Gordon's Inn, Kingussie, Inverness-shire, c. 1890	145
8.9	The New Club, 85 Princes Street, Edinburgh, 1966	147
8.12	The New Club, 144–6 West George Street, Glasgow: sketch of dining room by James Sellars or assistant, c. 1877–8	151
8.10	The Western Club, 147 Buchanan Street, Glasgow	149
8.11	The Western Club, 147 Buchanan Street, Glasgow: interior	149
8.6	Star Hotel, Princes Street, Edinburgh, c. 1835	141
8.13	St Enoch's Hotel, Glasgow	161
8.14	North British Hotel (The Balmoral), Edinburgh, c. 1930	164
8.15	Trossachs Hotel ('Ardcheanochrochan Inn'): dining room, watercolour by Sarah Sherwood Clarke, 1854	167

8.16	Royal Hotel, Bridge of Allan, 1920s	169
8.17	Moffat Hydropathic Hotel, Dumfriesshire, c. 1875	171
8.18	Marine Hotel, Elie, Fife	174
8.19	Panmure Arms, Edzell, Angus, early twentieth century	175
8.20	Aviemore Station Hotel, Inverness-shire, c. 1930	176
8.21	Turnberry Hotel, Ayrshire	177
8.22	Orient House, 16 McPhater Street, Glasgow	180
9.1	Aberlour House, Banffshire: probable tea house, later gate lodge, by William Robertson, c. 1830–5, architect of Aberlour House in similar classical style, 1838–9	191
9.2	Penicuik House, Midlothian: stables	194
9.3	Orton House, Moray: 'washing house' design by A & W Reid, Elgin, 1855	196
9.4	Girnals, Portmahomack, Ross and Cromarty: measured survey drawings	198
9.5	The Whim, Peeblesshire: plan and section of ice house	199
9.6	Cadboll, Ross and Cromarty: rectangular 'lectern' dovecot	201
9.7	Gordonstoun, Moray: cylindrical dovecot converted from windmill	202
9.8	Carmylie Manse, Angus: beeboles	203
9.9	Balfour Village, Shapinsay, Orkney: tide-washed toilet	204
9.10	Darnaway Castle, Moray: kennels design by A & W Reid, Elgin, 1878	206
9.11	John Starforth, *The Architecture of the Park*, 1890: boathouse design, plate XCIX	208
10.1	Edzell Castle, Angus: garden	212
10.2	Hatton House, Midlothian: late seventeenth-century Slezer view, incorrectly labelled 'Argile House', showing high walls sheltering the terraced gardens immediately adjacent to the house	213
10.3	Glamis Castle, Angus: sundial	213
10.4	Pitfour, Aberdeenshire: bath house, 'Temple of Theseus'	216
10.5	Dunmore, Stirlingshire: The Pineapple	217
10.6	Dunmore, Stirlingshire: The Pineapple, cross-section	218
10.7	Dalkeith House, Midlothian: conservatory	219
10.8	Drummond Castle, Perthshire: garden	220
10.9	Earlshall, Fife, 1900	221
11.1	River North Esk, Angus/Kincardineshire: salmon netsmen in their bothy, c. 1918	225
11.2	Badentarbat, Achiltibuie, Ross-shire: former salmon-fishing bothy	227

11.3	The Lag, Monreith Bay, Wigtownshire: lobster fishermen's hut	231
11.4	Stenness, Eshaness, Shetland: fishing station lodges with haaf fishermen and sixareen boats, c. 1880	231
11.5	Ullapool, Ross-shire: herring gutters' hut	233
12.1	67 Eastside, Kirkintilloch, Dunbartonshire: interior of women's sleeping accommodation, 1937	241
13.1	Crail Tolbooth, Fife	250
13.2	Dingwall Tolbooth, Ross and Cromarty: reconstruction drawing	251
13.3	Stirling Town House	252
13.4	Strichen Town House, Aberdeenshire	254
13.5	Duns Town Hall, Berwickshire: perspective by James Gillespie Graham, c. 1816	255
13.6	Annan Town Hall, Dumfriesshire	256
13.7	Falkland Town House, Fife: bell	257
13.8	Kirkcudbright Tolbooth: clock (now in Stewartry Musuem)	259
13.9	Kilmaurs Town House, Ayrshire: jougs	260
13.10	Stirling Town House: panelled interior	264
13.11	Dunbar Tolbooth, East Lothian: council chamber, painted panel with royal arms of King James VII, 1686	265
13.12	Pirlie Pig, 1602	268
13.13	Greenlaw, Berwickshire: detail of Armstrong's map showing tolbooth steeple between county house and church, 1771	270
14.1	St Andrews Cathedral and St Rule's Church, Fife: aerial view	278
14.2	Dalmeny Parish Church, West Lothian: south doorway and sarcophagus	281
14.3	Seton Chapel, East Lothian: chancel interior	283
14.4	Dunblane Cathedral, Perthshire: choir stalls	284
14.5	St Columba's Parish Church, Burntisland, Fife	286
14.6	Collegiate Church, Castle Semple, Renfrewshire: monument to John, 1st Lord Semple (d 1513)	290
14.7	The Necropolis, Glasgow, 1870s	292
15.1	Clandeboye School, Dunlop, Ayrshire: a rare example of an early schoolhouse (1641) which, as was common at the time, is located adjacent to a churchyard	295
15.2	Gordon Schools, Huntly, Aberdeenshire	296
15.3	Scotland Street School, Glasgow	299
15.4	Dundee High School: view by Joseph Swan	301
15.5	Dundee High School, 1956	301
15.6	Dundee High School, 1956	302
15.7	George Heriot's Hospital, Edinburgh	303

15.8	Knightswood Secondary School, Glasgow; designed 1938, begun 1954, this was one of a number of similar commissions by Gillespie, Kidd & Coia, architects	305
15.9	University of Aberdeen, Marischal College, 1906	306
15.10	Gorbals District Library, Glasgow: general reading room, c. 1907	309
16.1	Royal Infirmary, Edinburgh: bird's-eye view of David Bryce's 1870–9 design	312
16.2	Millbank Pavilion, Astley Ainslie Hospital, Edinburgh: a typical unit of this convalescent home associated with Edinburgh Royal Infirmary, which was developed between 1925 and 1939 along similar lines to contemporary sanatoria	314
16.3	Craig Dunain Hospital (formerly Northern Counties District Lunatic Asylum), Inverness: specimen ward, c. 1902	322
16.4	New Craig House, Thomas Clouston Clinic, Craighouse, Edinburgh: designs by Sydney Mitchell & Wilson, 1889	323
16.5	Lennox Castle Hospital, Lennoxtown, Stirlingshire	324
16.6	Royal Hospital for Sick Children, Sciennes Road, Edinburgh: bird's-eye view	332
17.1	Loch Katrine Waterworks, Perthshire: sluice operating mechanisms	342
17.2	Loch Vennachar, Perthshire: sluice house	343
17.3	Eglinton Toll and St Andrews Electricity Generating Works, Glasgow, early twentieth century	346
17.4	Rothesay Gas Works, Bute, 1960s	351
18.1	St Cecilia's Hall, Cowgate, Edinburgh: interior of hall, 1960	359
18.2	Former Exchange Coffee House, 15 Shore Terrace, Dundee	360
18.3	Theatre Royal, 254–90 Hope Street, Glasgow: auditorium, c. 1930	363
18.4	The Royal Museum of Scotland, Chambers Street, Edinburgh: interior of main hall, 1960s	365
18.5	The Vine, 43 Magdalen Yard Road, Dundee: lithograph by David Walker, 1954–5	366
18.6	The National Gallery of Scotland and Royal Scottish Academy, The Mound, Edinburgh, c. 1860	367
18.7	The Kibble Palace, Botanic Gardens, 730 Great Western Road, Glasgow	369
18.8	The Kibble Palace, Botanic Gardens, 730 Great Western Road, Glasgow: interior	370
18.9	St Andrew's Hall, Granville Street, Glasgow, c. 1888	373
18.10	Playhouse Cinema, Murray Street, Perth	377
19.1	Eyemouth Fort, Berwickshire: aerial view	383

19.2	Blackness Castle, West Lothian: aerial view	384
19.3	Killiwhimen, Ruthven of Badenoch, Bernera (Inverness-shire) and Inversnaid (Stirlingshire) Barracks: design drawings c. 1718	392
19.4	Fort Augustus, Inverness-shire: as designed by John Romer and completed 1742	394
19.5	Fort George, Inverness-shire: aerial view	396
19.6	Fort, Lamer Island, Dunbar, East Lothian	399
19.7	Martello Tower, Hackness, Hoy, Orkney: cross-section, 1814–15	400
19.8	Broughty Ferry Castle and Battery, Angus: aerial view	403
20.1	Weem, Perthshire: glebe steading	424
20.2	Sidinish, South Uist, Inverness-shire: croft byre and hen house	424
20.3	Greendykes Farm, East Lothian: work-horse stables	429
20.4	Garth, near Sullom Voe, Shetland: pony pund	431
20.5	Allangrange Mains Farm, Ross-shire: house stables	432
20.6	Kilrie Farm, Fife: cart shed and granary	433
20.7	Greendykes Farm, East Lothian: implement shed and yard	434
21.1	Cairn Farm, Fife: horse-engine house	441
21.2	Baillieknowe Farm, Roxburghshire: grain range	447
21.3	Appleby Farm, Wigtownshire: bank barn	451
21.4	Thurston Home Farm, East Lothian: hay barn	452
21.5	Glenochar Farm, Lanarkshire: hay barn	453
21.6	Wellside Farm, Moray: tower silo	454
22.1	Ross Farm, Kirkcudbrightshire: cheese-making dairy and bower's house	472
22.2	Dercullich Home Farm, Perthshire: dairy	473
22.3	Blackrig Farm, Dumfriesshire: cattle shelter	474
22.4	Edenmouth Farm, Roxburghshire: cattle court	475
22.5	Low Gameshill Farm, Ayrshire: pigsties	479
22.6	Heylor, near Ronas Voe, Shetland: lamb house	484
22.7	Shepherdscleuch, Selkirkshire: keb house	487
23.1	Ballachulish quarry, Argyll: slate-trimmers' shelter	497
23.2	Cashlie, Loch Tay, Perthshire: lime kiln	500
23.3	Mill of Montgarrie, Aberdeenshire	501
23.4	Falkirk, Stirlingshire: bleachfield, late eighteenth century	503
23.5	Auchtermuchty, Fife: weaver's house and loom shop	504
23.6	Gullane, East Lothian: smithy	505
24.1	Kirkcaldy, Fife: aerial view of an urban industrial landscape: flax and flour mills, lineoleum and engineering works	511

24.2	Arthur Street Works, Greenock, Renfrewshire: interior of pattern store showing timber joists (cut through) on wrought-iron beams and cast-iron columns	515
24.3	Sentinel Works, Jessie Street, Glasgow: pattern shop in Hennebique ferroconcrete, 1904	517
24.4	Dangerfield Mills, Commercial Road, Hawick, Roxburghshire: lineshaft bearing in mule mill	522
24.5	Kilncraigs Mill, Alloa, Clackmannanshire: aerial view, c. 1940: The school, office and yarn stores are in the foreground; scouring and dye-works behind, beside the tower; power station with timber cooling tower; five-storeyed woollen mills and single-storeyed worsted spinning mills; wool stores on the left. The Alloa corn mill and Youngers brewery are also in this view	527
24.6	Spring Gardens Ironworks, Aberdeen: bird's-eye view, c. 1908–20	530
24.7	Eglinton Engine Works, Cook Street, Glasgow: The four-storeyed building is the pattern shop and store over the ground-floor weighing machine shop; engine house between it and the original (1855) erecting shop	531
24.8	Eglinton Engine Works, Cook Street, Glasgow: interior of heavy machine shop	531
24.9	Douglas Fraser Works, Orchard Street, Arbroath, Angus: bird's-eye view. Left to right: Belfast roofs, Inch Flax Mill, machine and erecting shops, Westgate Works (small jute mill and Aspargartas shoe factory), reinforced-concrete pattern shop foundry and jute warehouses. 1947 advertisement	533
24.10	Bonthrone Maltings and Brewery, Newton of Falkland, Fife	542
24.11	Slateford Maltings, Edinburgh: floor malting in progress	543
24.12	Craigmillar Brewery, Edinburgh: timber-clad cooling unit	544
24.13	Maclay's Brewery, Alloa, Clackmannanshire: plans and section through brewhouse	546
25.1	Preston Island, Culross, Fife: aerial view	552
25.2	Murrayshall Limeworks, Cambusbarron, Stirlingshire: measured survey drawings	553
25.3	Bonawe Ironworks, Argyll: furnace	555
25.4	Lady Victoria Colliery, Newtongrange, Midlothian	561
25.5	Barony Colliery, Auchinleck, Ayrshire	562
25.6	Kinneil Colliery, Bo'ness, West Lothian	564
25.7	Prestongrange Colliery, East Lothian: Cornish beam pumping engine and engine house	565
26.1	Banff Harbour, c. 1880	572

26.2	Keiss Harbour, Caithness: fish-curing house and stilling basin with vertically-set masonry wall	573
26.3	Princes Dock, Glasgow: containers being loaded onto cargo ship from floating crane, c. 1970	574
26.4	Stromness, Orkney: private quay with house and warehouse	576
26.5	Tugnet, Speymouth, Moray: salmon-fishing station with gear store and boiling house to left and fishermen's two-storeyed accommodation block to right	579
26.6	Troon Harbour, Ayrshire: steam crane and puffer	581
26.7	Port Errol, Aberdeenshire: salmon-fishery net-boiling tub	580
27.1	Bridge over the River Tay, Perth, by John Smeaton (1766–72 with later cantilevered footpath)	596
27.2	Bonar Bridge, Sutherland: bowstring steel arch bridge, the third on this site, by Crouch & Hogg (1973)	598
27.3	Forth and Clyde Canal between Dalmuir and Old Kilpatrick, Dunbartonshire: typical two-leaf bascule bridge and bridge-keeper's house	599
27.4	Port Dundas, Forth and Clyde Canal, Glasgow: view by Joseph Swan showing warehouses and masts of sailing ships, c. 1830	600
27.5	Guthrie Castle estate, Angus: castellated railway bridge and gate lodge (1839)	602
27.6	Connel Ferry Viaduct, Loch Etive, Argyll (1903)	603
27.7	Dunkeld and Birnam Railway Station, Perthshire (1856 and 1863)	605
28.1	The Barras and Paddy's Market, Glasgow	612
29.1	Leith Bourse, 7–9 Tolbooth Wynd, Leith: view of (Queen Street) rear of building, c. 1929	625
29.2	Royal Exchange, City Chambers, High Street, Edinburgh, c. 1800	629
29.3	Royal Bank of Scotland, Royal Exchange Square, Glasgow	633
29.4	Royal Bank of Scotland (formerly Head Office, Commercial Bank of Scotland), 14 George Street, Edinburgh: perspective by David Rhind, c. 1845	634
29.5	British Linen Bank, 37 St Andrew Square, Edinburgh, 1890s	635
29.6	British Linen Bank, 37 St Andrew Square, Edinburgh: interior of banking hall, 1890s	636
29.7	Royal Bank of Scotland (formerly Commercial Bank of Scotland), 8 Gordon Street, Glasgow: elevation as designed, 1854	637
29.8	Bank of Scotland, 2 St Vincent Place, Glasgow, c. 1890	638

29.9 High Street, Irvine, Ayrshire, showing Royal Bank of
    Scotland, c. 1890   639
29.10 Life Association of Scotland, 81–3 Princes Street, Edinburgh,
    1966   641
29.11 Royal Exchange, Dundee, c. 1880   643
29.12 Edinburgh Corn Exchange, 35 Constitution Street, Leith   644
29.13–15 Edinburgh Corn Exchange, 35 Constitution Street, Leith:
    sections of frieze   645
29.16 37–51 Miller Street, Glasgow   648
29.17 73–97 Commercial Street, Dundee, 1897   652
29.18 Lion Chambers, 170–2 Hope Street, Glasgow   661
29.19 Waterloo Chambers, 15–23 Waterloo Street, Glasgow:
    perspective view by J J Burnet   663

List of Contributors

PROFESSOR ROBERT D ANDERSON
Department of History, University of Edinburgh.

ELIZABETH BEATON
Independent architectural historian; formerly Assistant Inspector, Historic Scotland.

TIM BUXBAUM
Architect in private practice.

NEIL CAMERON
Listed Buildings Recording Programme, Royal Commission on the Ancient and Historical Monuments of Scotland.

ANNETTE CARRUTHERS
Senior Lecturer, School of Art History, University of St Andrews.

KITTY CRUFT
Formerly Curator, National Monuments Record of Scotland, Royal Commission on the Ancient and Historical Monuments of Scotland.

PROFESSOR ALEXANDER FENTON
Director, European Ethnological Research Centre.

DR JOHN FREW
Senior Lecturer, School of Art History, University of St Andrews.

DR MILES GLENDINNING
Threatened Buildings Survey, Royal Commission on the Ancient and Historical Monuments of Scotland.

† OWEN F HAND
Formerly Research Assistant, European Ethnological Research Centre.

DR HEATHER HOLMES
The Scottish Executive, and freelance writer.

JOHN R HUME
Formerly Chief Inspector of Historic Buildings, Historic Scotland, Honorary Professor in the Factulty of Arts, University of Glasgow and School of History University of St Andrews.

DR ROGER LEITCH
Freelance writer, Dundee.

DR DEBORAH MAYS
Principal Inspector, Head of Listing, Historic Scotland.

CAROLINE A MacGREGOR
Formerly Research Assistant, Royal Commission on the Ancient and Historical Monuments of Scotland.

PROFESSOR CHARLES McKEAN
Professor of Scottish Architectural History, Department of History, University of Dundee.

DR MILES K OGLETHORPE
Industrial Survey, Royal Commission on the Ancient and Historical Monuments of Scotland.

HARRIET RICHARDSON
Survey of London, English Heritage.

NIGEL A RUCKLEY
Formerly Geoarchaeology Co-ordinator, British Geological Survey (Scotland).

DR JOHN SHAW
Curator of Environmental Social History, National Museums of Scotland.

GEOFFREY STELL
Head of Architecture, Royal Commission on the Ancient and Historical Monuments of Scotland.

SUSAN STORRIER
Deputy Director, European Ethnological Research Centre.

PROFESSOR DAVID WALKER
Emeritus Professor of Art History, University of St Andrews; formerly Chief Inspector of Historic Buildings, Historic Scotland.

MARK WATSON
Principal Inspector of Historic Buildings, Historic Scotland.

Foreword

Scotland's Buildings is the second volume to appear in the planned fourteen volume series, *Scottish Life and Society: A Compendium of Scottish Ethnology*, currently under preparation by the European Ethnological Research Centre. The first volume to have been published was Volume 11, *Institutions of Scotland: Education*. The full range is as follows:

1. An Introduction to Scottish Ethnology
2. Farming and the Landscape
3. Scotland's Buildings (now published)
4. Boats and Fishing, Coast and Sea
5. The Food of the Scots
6. Scotland's Domestic Life
7. Working Life: Crafts, Trades and Professions
8. Transport and Communications
9. The Individual and Community Life in Scotland
10. Oral Literature and Performance Culture
11. Institutions of Scotland: Education (published 2000)
12. Institutions of Scotland: Religious Expression
13. Institutions of Scotland: the Law. General Index
14. Bibliography

The Bibliography for Scottish Ethnology not only covers the thematic territory of the *Compendium*, but will also be a useful research tool for related disciplines.

Scotland's Buildings is in many ways a summation of ideas about buildings in their environmental and functional contexts that have been developing and expanding over several years. In 1972, for example, the Scottish Vernacular Buildings Working Group (SVBWG) was formed, as a focal point not only for architects but also for all those whose work and interests related to buildings and their environment. This body has continued to act as a stimulant through its journal, *Vernacular Building*, its thematic publications and its annual conferences. An early emphasis on rural buildings resulted in, for example, Fenton, A, Walker, B. *The Rural Architecture of Scotland*, Edinburgh, 1981. In more recent times there has been a substantial flow of books on architectural themes, such as Glendinning, M, MacInnes, R, MacKechnie, R. *A History of Scottish Architecture*, Edinburgh, 1996, and Mays, D, ed. *The Architecture of Scottish Cities*, East Linton, 1997. Members of the SVBWG and authors of and contributors to such books are well represented in *Scotland's Buildings*.

Scotland's Buildings goes further than most of these. The concept of 'buildings' is taken to include structures built for any functional purpose. That is why harbours, mines, roads and canals, and gardens, for example, appear in this book. The perspective of the volume is thus very wide, which gives it a unique quality. In line with the editorial plan, a three-pronged approach was adopted, and implemented by leading experts on the subjects. The first considers dwellings at all social levels from castles and palaces to housing for seasonal agricultural workers. The second is concerned with community buildings across a broad range of social functions and public services, from religion and local government to recreation and defence. The third relates to work places associated with Scottish industry and business. The picture presented, therefore, has to do first and foremost with the living environment of human beings within their households and communities, as they seek to earn their daily bread and as they spend their later years.

Buildings in cities, towns and villages, and the scatter of farm buildings, big houses, and other structures in the countryside are the oldest and most prominent features of human intervention in the landscape. They constitute a network linked by road, rail and waterways. This concept should be kept in mind when this book is being used, for it is possible to get from the various chapters a feeling for the gradual increase in density up to our day of the settlement pattern. Ancient buildings commemorate the past, but many have been subjected to alteration in line with changing economic and social circumstances and new technologies. To study and interpret the sequence of changes in a building is to reveal layers of history, and the importance of learning to use this major three-dimensional historical source cannot be overestimated.

It is, of course, inevitable that many areas of research remain. Some relevant topics, such as the stone dykes that characterise so much of the farming and sheep-grazing landscape, will be picked up in other volumes of the *Compendium*, but *Scotland's Buildings* will also be serving a purpose when it brings into the awareness of readers subject areas that have not been treated, or that have been too briefly touched on. The kind of contextualisation of architectural phenomena in this volume is seen as a starting point for further investigations.

In the preparation of the book, there has been close collaboration with other bodies, especially the National Museums of Scotland (NMS) and the Royal Commission on the Ancient and Historical Monuments of Scotland (RCAHMS), a partnership which is reflected in the sources of illustration and in the editorial work. The range of illustrations is especially large, in view of the nature of the subject matter, and there has been much dependence on the Scottish Life Archive of the NMS and on the pictorial resources of the RCAHMS. One of the two guest editors, Geoffrey Stell of the RCAHMS, has contributed an introduction and organised the illustrative content, as well as maintaining a detailed overview of the whole volume. The second is Dr John Shaw of the NMS, who has in addition contributed four chapters. They were the two who organised the original plan of the volume, and at a later stage Susan Storrier, Deputy Director of the European Ethnological Research Centre, was given the task of working with the Guest Editors and pulling the volume together. Thanks are due to all three for the immense amount of meticulous work that preceded this successful outcome, and to the individual contributors for maintaining a high level of scholarship.

Grateful thanks are also due to the following bodies:

The Trustees of the National Museums of Scotland, who provide the EERC with premises and a core grant.

The Russell Trust, which covered the cost of the illustrations.

The Scotland Inheritance Fund, which continues to provide substantial support for the preparation and production costs of the *Compendium* volumes.

ALEXANDER FENTON
General Editor

Abbreviations

AHSS	Architectural Heritage Society of Scotland
CBA	Council for British Archaeology
EERC	European Ethnological Research Centre
FN&NBA	*Farming News and North British Agriculturalist*
JRASE	*Journal of the Royal Agricultural Society of England*
NAS	National Archives of Scotland
NBA	*North British Agriculturalist*
NCR	National Cash Registers
NLS	National Library of Scotland
NMRS	National Monuments Record of Scotland
NMS	National Museums of Scotland
NSA	*The New Statistical Account of Scotland, 1845*
OSA	*The Statistical Account of Scotland, 1791–99*
PRO	Public Record Office
PSAS	*Proceedings of the Society of Antiquaries of Scotland*
RCAHMS	Royal Commission on the Ancient and Historical Monuments of Scotland
RCHME	Royal Commission on Historical Monuments (England)
RIAS	Royal Incorporation of Architects in Scotland
RIBA	Royal Institute of British Architects
ROSC	*Review of Scottish Culture*
SBA	Scottish Brewing Archive
SBRS	Scottish Burgh Records Society
SCWS	Scottish Co-operative Wholesale Society
SFBS	Scottish Farm Buildings Survey
SLA	Scottish Life Archive, NMS
SVBWG	Scottish Vernacular Buildings Working Group
TCM	Town Council Minutes
TDGNHAS	*Transactions of the Dumfriesshire and Galloway Natural History and Antiquarian Society*
THASS	*Transactions of the Highland and Agricultural Society of Scotland*

Glossary

This selective list contains only those terms found in the text that may aid a non-specialist reader. Reflecting the broad scope of the book, the terms cover many areas, including agriculture, architecture, engineering, industry and military technology. Gaelic and Scots terms are distinguished by [G] and [S] respectively; local or regional usage in Scotland is also noted.

acanthus, classical formalised leaf ornament.
activated sludge, method of sewage treatment which involves adding aerobic bacteria and then blowing air through the mixture.
aedicule, opening framed by classical columns, **entablature** and pediment.
American reaction/Francis turbine, turbine in which the water enters radially inwards and leaves axially.
Anderson shelter, air-raid shelter of arched corrugated-iron construction, designed for family use.
anti-tank blocks, series of concrete and/or metal obstacles, to obstruct mobile attack.
arabesque, surface decoration consisting of lines and scrolls, based on geometrical patterns.
arch ring, innermost structural order of an arch, made up of arch stones or voussoirs.
architectonic, systemised knowledge of architecture.
architrave, lowest member of a classical **entablature**, generally applied to the moulded frame of a door or window of classical profile.
arles [S], earnest money.
ascititious, adscititious, supplementary or additional. In bridge arch construction, as devised by Thomas Telford (1757–1834), an external arch, adjacent and parallel to the main arch or arches.
ashlar, hewn masonry wrought to an even face and square edges, and bedded with a fine joint.
astragal, classical moulding of semi-circular profile, or, generally [S], a glazing bar.
astylar, classical façade without columns or **pilasters**.

atrium, open or top-lit inner court of a building.
awn [S], spike on the end of a growing barley grain.

bailie [S], officer of a barony; a town official, next in rank to a provost (mayor).
barrage balloon, balloon tethered by a wire cable to obstruct low-flying aircraft.
bartizan, corbelled turret at the top angle of a building.
bascule bridge, vertically-opening bridge hinged at one or both ends.
bastion, defence work of polygonal or semi-circular plan, projecting from the main wall or angle of a fortification.
bathing, method of protecting sheep from parasites, formerly used on lowland and arable farms. The fleece was parted, in lines about three inches apart, and a liquid poured or rubbed onto the exposed skin. Replaced by dipping from the later nineteenth century. Also **pouring**.
battery, unit of ordnance or searchlights.
beetling mill, mill where cotton or linen cloth is finished (given a sheen) by beating with powered wooden mallets.
Belfast roof, roof made up of trusses of curved profile and latticed construction.
bell-cast, of a roof or similar feature, having a bell-shaped profile.
belvedere, a **gazebo** located on the roof of a house.
bink [S], ledge, as at the side of an open fire.
blockship, vessel deliberately sunk to block or obstruct a navigation channel.
blowing-engine house, building housing a steam engine, linked to an air pump, to provide the blast of air to a metal-smelting furnace.
bombing decoy, system of lights, fires or dummy buildings, simulating a target for enemy aircraft.
boom, protective curtain across a waterway to prevent hostile incursions, particularly by submarines.
boscage, areas of trees or shrubs.
bothan àiridh [G], shieling hut.
bothy, hut for temporary accommodation; self-contained quarters for unmarried male farm servants, typical of east central Scotland.
bourse/burse [S], exchange building.
bowed truss, bridge of arched or bowed construction above the level of the carriageway.
bower, in south-west Scotland, a dairy herd who, under a renewable contract with a farmer, was provided with working and fixed capital to operate an on-farm dairying business.
bowes, bows [S], seed heads (of flax).
boyne [S], shallow, wide vessel, usually for separating cream from milk.

brattishing, ornamental crest, usually formed of leaves, flowers and battlements.
bucht [S], small, roofed structure for night-housing sheep; an enclosure for milking ewes. Also **cott**.
but and ben [S], two-roomed cottage, the **ben** being the inner or best room of a house.
byke [S], store of man's height for threshed grain made of coiled straw ropes and similar in shape to an old type of bee hive.

cap-house, small roofed superstructure of a stair leading to the parapet or upper floor of a building.
carding, disentangling wool or cotton fibres in preparation for spinning.
caryatid, a female figure used instead of a column to support an **entablature**.
casemate, bomb-proof vaulted chamber.
chaumer [S], farm servants' accommodation in a farm steading, chiefly in north-east Scotland.
cinquecento, Italian Renaissance style of the '(mil) cinque cento', fifteenth century.
citadel, inner fortress, adjacent to or within a larger fortified circuit.
clamp, pile of bricks for firing, with fuel and insulating material; an earthen mound, lined with straw, for storing root crops.
cleit [G], stone storage hut, used mainly on St Kilda.
clerestory, lighting storey or range of windows in the upper part of a building, especially the nave or chancel of a church.
close [S], open area between dwelling house and steading on a farm; the passageway leading into a courtyard; the entrance passageway in a tenement.
coble [S], open flat-bottomed salmon boat with a 'shelf' at the back for the fishing net.
coffer (dam), watertight enclosure used in the construction of bridges and harbours.
coffering, sunken panels decorating a ceiling, vault or arch.
collar-beam, in roofs, a cross-beam linking rafters above wall-head level. See **couple-baulk**.
colonnette, small column or shaft.
continental impulse/Girard turbine, type of turbine moved by the impulse of a jet of water on its blades.
corbel, corbel table, projecting block of stone; range of **corbels** supporting a beam or parapet.
cortile, courtyard.
cott [S] – see **bucht**.
cottar [S], minor tenant occupying a cottage with or without land; a farm worker occupying a cottage as part of his pay.

counterscarp, defensive bank on the outer side of a ditch below the **glacis**.
couple-baulk [S], horizontal beam linking a pair of roof rafters. See **collar-beam**.
cray, cree [S], pen or small shelter for livestock, latterly used specifically to refer to a pigsty.
crosslet-loop, arrow-slit or gun-loop with cross-shaped aperture.
cruisie/crusie lamp [S], boat-shaped, double-shelled oil lamp with a rush wick.
cupola, dome; in Scotland, especially over a feature such as a stairwell.
curtain, main perimeter wall of a castle or defence work.
cutch, brown dye-stuff from India, used to preserve fishing nets.

dale, deal [S], sawn plank.
demi-bastion, bastion of two sides, instead of the usual four.
Diocletian or thermal window, semi-circular window with **mullions**.
double-pile, of a building plan, two rows of rooms deep, often expressed in double gables.
draff [S], used barley which has been malted and brewed.

Eleanor cross, crosses commemorating the twelve resting-places of the body of Eleanor (d 1290), wife of King Edward I of England, on its progress to Westminster.
embrasure, defensive opening in a wall or battlement.
emplacement, position where ordnance is installed.
encaustic tile, tile in which the pattern is formed by coloured clays, applied before firing.
enceinte, main enclosure of a castle or fortress.
entablature, upper part of a classical order above the columns, consisting of **architrave**, frieze and cornice.
Extended Defence Officers Post (XDO post), observation and control post for a submarine minefield.

fank [S], network of pens for gathering and sorting sheep.
far haaf [Shetland], distant sea.
Farnese pattern, of a window, modelled on the Farnese Palace in Rome.
fireclay refractories, heat-resistant products, made from fireclay, a type of clay found with coal measures.
fixed-jib crane, crane in which the lifting arm or jib cannot be raised or lowered.
flit [S], remove.
flitch plate, iron plate used to strengthen a composite timber beam.
form, bench, long backless seat.
fulling stocks, mechanical hammers for de-greasing and meshing the fibres of woollen cloth.

gazebo, ornamental look-out tower or summer house.
Gibbs', Gibbsian surround, door or window with a triple keystoned head beneath a cornice, and alternating blocks making up the jambs. Named after the architect James Gibbs (1682–1754).
glacis, open slope capable of being protected from the ramparts of a defence work.
glebe, farmland attached to a parish minister's manse as part of his payment.
glover [S], cupola with flight-holes, erected on the ridge or apex of a dovecot.
gof(f)er, gopher, gauffer, to make wavy, fluted, or crimped edges by a heated iron.
grieve [S], head man of the servants on a farm. Also **steward**.
grip [S], drain at the back of the animals in a byre or stable.
groin vault, tunnel vaults intersecting at right angles, the groins being the lines of the intersections.

hallan screen [S], internal partition or part wall between door and fireplace.
hammel [S], open-fronted shed (with yard adjacent) for housing cattle.
hanging/hingin lum [S], chimney set against the gable.
harling [S], form of wet-dash roughcast.
heck [S], slatted, fixed container for byre or stable feed. See also **troch**.
heckling, process of combing flax fibres in preparation for spinning.
helve or **tilt hammer**, large iron hammer which is attached to a pivoted wooden arm or helve and lifted and released for forging metal.
hind [S], skilled, male farm labourer.
hollander engine, developed in the Netherlands in the mid-seventeenth century, arriving in Scotland a century later. Also known as an engine or beater, it used rollers to produce paper pulp from linen rags.
hood-mould, projecting moulding above an opening.
horonising, horonizing, shards of whinstone left over from creating setts, set on end as an uncomfortable, lightly trafficked, footway. Often used in nineteenth-century school playgrounds.
hummeller [S], device for removing the **awns** from barley grains.

impost, horizontal moulding at the springing of an arch.
inbye [S], that part of a farm's land near to the steading.

jardinière, ornamental pot or stand for display of flowers.

keb house [S], building where orphaned lambs were confined with adoptive ewes; also used more generally to refer to a shelter for young lambs.

keiv, boiler used in cloth bleaching.
kier, vat for boiling cloth.
king-post, in roof construction, a vertical, centrally placed timber, set on a tie- or collar-beam and supporting the apex or ridge piece.
kist [S], chest, generally wooden.

lambermen [S], lambers (of sheep).
lammiehoose [Shetland], shelter for ewes and lambs.
lead [S], mill lade.
lectern, of a dovecot, rectangular structure with single-pitched roof.
level luffing crane, crane in which the inclination of the jib can be altered without changing the height of the load above the ground.
link-span, pivoted and hinged ramps for loading and unloading vehicular ferries.
loggia, a gallery, usually arcaded or colonnaded.
lucarne, small gabled opening in a roof or spire.
lunette, semi-circular window.
lustre, glass chandelier.

machair [G], coastal plain composed of blown sand.
machicolation, in castle architecture, a slot or shaft, particularly between high-level corbels.
mains farm [S], home farm of an estate, cultivated on behalf of the owner.
mansard, roof with a double or broken pitch, the lower slope being longer and steeper than the upper. Named after the French architect, François Mansart (1598–1666).
Martello Tower, a coastal gun tower developed from late eighteenth-century towers in Corsica.
monitor roof – see **clerestory**.
mortar, in military ordnance, a piece capable of firing explosive shells at high angles.
mule, machine for spinning cotton yarn, invented by Samuel Crompton (1753–1827).
mullion, vertical member between window lights.
muran [G], marram grass.

nepus gable [S], small roofed gablet rising from the front or rear wall of a building.
Newcomen engine, early type of steam engine, with a reciprocating beam, used in pumping water in mines and elsewhere.
newel, central or corner post of a staircase; a newel stair ascends around a central newel.

octostyle, of a **portico**, with eight frontal columns.

oculus, circular opening.
ogee, double or reflex 'S' curve.
ordonnance, the architectural arrangement of a building in its entirety; selection of an order and/or other classical or Renaissance details suitable for the façade of a building.

palazzo, urban palace.
parrock [S], paddock; enclosure where a ewe is confined with a lamb for fostering.
parterre, level garden laid out with formal flower beds.
peezer [S], young dove or pigeon.
pend [S], covered passage.
pentice [S], of a roof, a lean-to.
pergola, covered walk in a garden, formed with posts and joists.
perron, external forestair, usually with double curved flights.
petard, explosive device, originally bell-shaped, usually designed to destroy a defensive gate.
piece [S], snack.
piended [S], of a roof, hipped.
pilaster, flat or rectangular-section column attached to a wall in shallow relief.
pilastrade, series of pilasters, equivalent to a colonnade.
pillbox, small strongpoint, usually built of concrete and forming part of a system of defence, such as an anti-invasion **stop line** or around a military airfield.
plantation stell [S], a **stell** formed by various configurations of planted trees.
plash mill [S], water-powered mill for washing and cleaning yarn.
portico, classical porch, usually pedimented, supported by rows of columns.
postern, rear entrance to a castle or fortification.
potence [S], revolving access ladder in the centre of a dovecot.
pouring – see **bathing**.
Pratt truss, a lattice structure, consisting of upper and lower horizontal beams, invented in 1844.
proscenium arch, in theatre design, the arch and frontispiece facing the auditorium.

queen-post, in roof construction, paired vertical or near-vertical timbers, placed on a **tie-beam**.
quern [S], hand mill for grain.
quoin, dressed stone at the angle of a building.

ravelin, in defence works, an outwork placed beyond the main ditch.

reeding, the process in weaving of passing warp threads through regularly-spaced dividers, originally made from split reeds, later of iron, brass or steel.

render, protective skin of plaster or aggregate applied to the outside walls of a building.

reredos, painted and/or sculptured screen behind and above an altar.

retort, chamber in which coal is heated in order to separate gases from coke.

retting, processes whereby flax is steeped in water to loosen and separate the fibres in its stems.

revetment, revetted, stone facing or reinforcement of an earthen slope.

rick, stack of grain or hay.

rigged derrick, crane in which there is an upright post held in place by ropes or chains which are linked to fastenings in the ground.

rotative engine, engine capable of turning a shaft to drive machinery.

round, open-roofed sheep shelter, usually circular in plan.

ruskie [S], straw bee skep.

Saladin maltings, developed by M Saladin in France in the late nineteenth century, a system of accelerating malting by means of forced air and mechanical turning.

scale-and-platt stair, stair of 'dog-leg' form, with parallel flights and landings or half-landings.

sconce, here (Chapter 13), bracketed candle-holder or lantern.

scutcher [S], knife-like tool or machine for separating flax fibres in the plant's stem.

Serlian, door or window with three openings, the central one arched; also referred to as Palladian or Venetian. Named after the architect Sebastiano Serlio (1475–1554).

sheer legs, lifting device, originally consisting of two long poles linked at their upper ends and with their lower ends separated and pivoted later associated with a third (back) leg, the foot of which could be moved backwards and forwards.

shelmets, shelving or frames on a cart.

shieling, summer hill-grazing area.

sixareen, sixern [Northern Isles], six-oared open boat.

skeo [Northern Isles], stone hut for drying (eg meat).

skep [S], small, conical straw beehive.

smearer, person who smeared sheep with a tar mixture.

soiling, feeding livestock indoors on freshly-cut green fodder.

soor dook [S], whey or buttermilk.

sound detector or **mirror**, acoustic dish for detecting the sound of approaching aircraft.

spandrel, triangular-shaped space above the haunch of an arch.

spigot mortar, a type of **mortar** deployed by the Home Guard during World War II.
stanchion, upright structural member.
stathel [S], foundation for a stack of grain or hay.
stell [S], sheep pen or shelter
steward [S] – see **grieve**.
stirk, young bullock.
stook [S], sheaves of harvested corn, heaped together in the field to air-dry.
stop line, anti-invasion measure in World War II, defined by anti-tank obstacles, **pillboxes** and gun **emplacements**.
storm prow, here (Chapter 18), cinema advertising tower.
string course, horizontal or stepped moulded course projecting from a wall surface.
stugged [S], of masonry hacked or picked as a surface finish or as a key for rendering.

tailrace, artificial watercourse leading from a water wheel to a river or stream.
thirled [S], legally obliged to take grain for grinding to a particular mill.
tie-beam, in roof construction, the main horizontal timber linking the feet of principal rafters.
topiary, art of clipping shrubs and hedges into ornamental shapes.
tower-house, self-contained fortified house of three or more storeys.
trace italienne, form of artillery fortification developed in Italy in the early sixteenth century, comprising triangular or arrow-shaped bastions with gun batteries set in their flanks.
tracery, openwork decoration in the upper part of an opening.
traverser, a device with rails on a platform which itself moves on rails.
treillage, trellis, a latticed structure of interlaced bars, fastened at their intersections.
trevis(e) [S], partition separating an animal from its neighbour in a byre or stable.
troch [S], trough for animal feed. See also **heck**.
truss, braced framework, spanning between supports.
turbine, advanced form of waterwheel with rotors encased in metal to maximise the effect of the water guided to it by fixed curves or plates.
turnpike trust, body of trustees to administer toll roads.
tuyère, nozzle through which air is blown into a blast furnace or steel converter.

vortex turbine, turbine in which water flows inwards, radially, from its circumference.

warp, of a vessel, to haul a ship or boat by a rope attached to a fixed point.

Warren truss, lattice structure, consisting of angled uprights linking upper and lower horizontal beams. Patented 1848.

wort, in brewing, the unfermented liquid, derived from barley which has been malted, milled and added to hot water.

yett [S], iron grille serving as a gate or defensive door.

PART ONE
•
Introduction

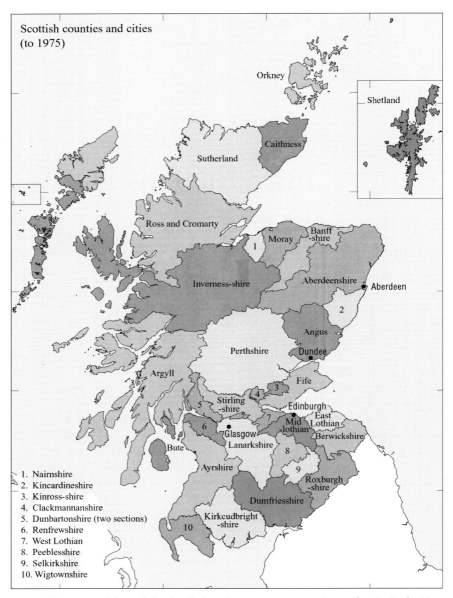

Figure 1.1 Map of Scotland showing pre-1975 counties and principal cities. Crown copyright: RCAHMS.

1 Buildings in Context

GEOFFREY STELL

The last quarter century has witnessed an intensification of interest in virtually every aspect of Scottish history and culture, a phenomenon which is reflected in – but has tended even to outpace – contemporary trends towards parliamentary devolution and political self-expression. Buildings have received their due attention in this burgeoning of knowledge and awareness, to the extent that there is now no shortage of publications relating to almost every period and type of Scottish architecture, even though such works have illuminated the country's historic built environment and its architectural 'identity' with varying degrees of clarity and penetration. So, given this massive and increasing rate of supply, what is the justification for yet another book in this field?

The answer, quite simply stated, lies in the title of this short introductory chapter. However they are shaped and designed, buildings are essentially the products of the time, space and purpose of their creation, and of the circumstances under which they have been subsequently developed, modified or abandoned. Their general characteristics and detailed features can be, and often are, viewed and studied in isolation, building by building, but there can be little doubt that our understanding of them is enhanced by an appreciation of their wider context, that is, of their historical, geographical and physical links and relationships. Scotland's buildings are capable of being grouped and interpreted in so many different ways – by period, region, style and type, for example – and each one of those buildings may occupy a place within many different contexts – aesthetic, geographical, social, economic, administrative, political and military, to name a few of the more obvious.

The following 28 essays cover a vast range of Scotland's buildings which are spread over virtually every major category and context. They have been selected and treated in a broadly thematic manner, for the most part according to function and building-type. With such coverage it is a bold, though reasonable assertion to claim that the end-result is the broadest and fullest account of Scottish building types which has ever appeared in print and which sets out, as a platform for further research, the manifold aspects of the functional tradition in Scottish building.

Many of these essays also draw attention to the human factor which underlies the creation and use of buildings, an aspect which is all too easily overlooked in more abstract and theoretical considerations of architectural forms, styles and influences. Although most of the essays are presented from the standpoint of the architectural historian, they have been arranged within a framework which seeks to give emphasis to the socio-functional aspects of the built evidence and to facilitate its inter-disciplinary use as much by ethnologists, for example, as by historians and architects. This framework consists of three main categories of building: in architectural parlance these may be classified as domestic, institutional and industrial; in broad ethnological terms the three categories are associated respectively with domestic and family life, community and public life, and working life. The approach therefore is multi-disciplinary, and fits well into an ethnological perspective.

The short sub-titles attached to these three sections – dwellings, community buildings and work places – are thus simply intended to serve as convenient and helpful signposts to these different areas of human activity. However, it is fully acknowledged that any such system of classification may also disguise or distort many of the underlying realities, complexities and inter-relationships, given that few aspects of everyday human existence lend themselves to such hard and fast distinctions. Hotels, inns and lodging houses (Chapter 8), like estate architecture (Chapter 9), for instance, are made up of units or buildings which are at once domestic and communal, while the distinctions between home and working life (Chapters 12 and 20–23, for example) are often blurred or even non-existent. For most persons engaged in farming and small-scale craft or industrial enterprises, the dwelling was also the work place, and might on occasion even serve a wider communal use.

With these reservations, however, the treatment of buildings by type remains as valid today as it did in 1976 when Nikolaus Pevsner's pioneering work on this theme was published.[1] In his words, this approach 'allows for a demonstration of development both by style and by function, style being a matter of architectural history, function of social history.'[2] Pevsner placed the emphasis firmly on nineteenth-century building types in an international setting, but felt compelled to excuse himself for what he saw as a limited and arbitrary selection of only 20 types. The possibilities are indeed almost limitless, and here, with the emphasis mainly on Scotland, any further additions would, as in Pevsner's case, 'have swelled the book to unmanageable proportions'.[3]

Partly based upon a rich and growing corpus of vernacular and industrial building studies, subject areas and disciplines which over the past half-century have grown to maturity in parallel with architectural history, the scope of this volume extends from the grandest mansions to the humblest shelters, penetrating much further down the social scale

Figure 1.2 Rural context; Ardchyle and Ben More, Perthshire, c. 1900. Crown copyright: RCAHMS, SC 507185.

Figure 1.3 Urban context; East Church Street, Fordyce, Banffshire, 1937. Crown copyright: RCAHMS, SC 710736.

Figure 1.4 High-density urban housing; 517 Lawnmarket and 10 Milne's Court, Edinburgh, 1910. Crown copyright: RCAHMS, SC 711517.

and across a far wider range of working environments and processes than most architectural scholars of Pevsner's generation would have been prepared to contemplate (figs 1.2–4). The sources from which many of these essays derive their information thus extend well beyond the conventional boundaries of Scottish architectural history, and thereby implicitly draw attention to the need for fuller works of basic reference in relation to the documentary sources,[4] to the methodology of buildings investigation, and to the geographical background.

Despite the lavish attention which Scotland's buildings have received over the past few decades, a source book and guide to Scottish building history, whether in conventional paper or electronic form, still remains a conspicuous gap, as does an atlas of Scottish building.[5] Here, so far as a national identity or sets of regional identities in Scottish building are concerned, it is left for the reader to pull together those threads which, when variously drawn out and woven, may – or may not, as the case may be – make up patterns of 'Scottishness', in both formal and informal architectural dress.[6]

NOTES

1. Pevsner, 1976 and 1979.
2. ibid, 'Foreword', np (6).
3. ibid, 9.
4. The pamphlet by Dunbar, 1969, revised as an article, idem, 1979, remains the standard reference for source material. The categories into which the sources are divided and discussed by Dunbar broadly follow those set out by Colvin, 1967: 1 Royal buildings; 2 Public buildings (a) urban buildings, (b) collegiate buildings, universities and schools, (c) prisons, (d) transport and communications; 3 Churches; 4 Domestic architecture (a) medieval castles and houses, (b) later castles and country houses, (c) town house, (d) cottages and farms; 5 Architects and craftsmen; 6 Architectural drawings and photographs, topographical views etc; 7 Standard works of reference, journals etc.
5. Stell, 1993.
6. For illustrated guides to Scottish building terms see Pride, 1975, and Pride, 1996.

BIBLIOGRAPHY

Cheape, H, ed. *Tools and Traditions: Studies in European ethnology presented to Alexander Fenton*, Edinburgh, 1993.

Colvin, H M. *A Guide to the Sources of English Architectural History*, Shalfleet Manor, Isle of Wight, 1967.

Dunbar, J G. *Source Materials for the Study of Scottish Architectural History* (Scottish Georgian Society), Edinburgh, npd (1969).

Dunbar, J G. 'Source materials for the study of Scottish architectural history', *Art Libraries Journal*, 4/3 (Autumn, 1979), 17–26.

Pevsner, N. *A History of Building Types*, London, 1976 and 1979.

Pride, G L. *Glossary of Scottish Building*, Glasgow, 1975.
Pride, G L. *Dictionary of Scottish Building*, Edinburgh, 1996.
Stell, G. Towards an Atlas of Scottish Vernacular Building. In Cheape, 1993, 159–66.

2 Continuity and Change

ALEXANDER FENTON

A decisive period for buildings in the Scottish countryside was the late eighteenth to early nineteenth century, when the farming landscape and its associated buildings were given the basic appearance which, as a result of the agricultural improvements that went on, has remained to the present day. This was a period of change that eroded much of the regional variety that had formerly been found in the fixed structures and movable features of rural buildings. Details of the new developments are to be found in Chapters 6 and 20–22 in this volume.

There was an earlier period of substantial change, and this chapter, by taking a long view in time, tries to establish its nature and approximate timing. Broadly speaking, it is the change from circularity to rectangularity of plan-form, which in terms of the divisions and uses of internal space must point to the evolution of a rather new philosophy of existence. The prehistoric Scottish evidence for housing and its structural elements, including the plan-forms and the nature of the hearth, has never been fully assessed – though it would be well worth while to do so for the available data as a whole, especially in the light of modern excavation techniques that also take environmental factors into account. Analysis of the technicalities of housing, in alliance with the accompanying artefactual and environmental evidence, should throw light on the validity, or otherwise, of such an evolution. Here, some preliminary thoughts are presented, which can then be subjected to testing.

The prehistoric remains that have received most archaeological attention are those of 'substantial' houses or structures. In a review article of 1992 about 'Society in Scotland from 700 BC to AD 200', Richard Hingley[1] looked at the social significance of a range of types of building and structure, according to a broad geographical zoning: Atlantic Scotland on the one hand, and southern, central and eastern Scotland, including the south west and much of the north east, on the other. He was critical of the social theorists who saw substantial structures as symbols of cultural dependency, as it were, the power centres, and sought to make a case for viewing them more in terms of the social organisation of households and communities.

In the Atlantic zone, the semi-subterranean 'cellular' house forms of the Neolithic and Bronze Ages, consisting of discrete clustered elements, appear to have been continued or succeeded in the Iron Age by buildings of circular form. The tower-like brochs are the most remarkable examples, and it now seems to be a widely held view that these are in essence thick-walled round-houses which have been exaggerated in height. They are typically to be found in the north and north west or west of Scotland. In the west there are also duns, which may be small, circular, fortified and roofed dwellings, or may be enclosures surrounding a dwelling or dwellings. Wheelhouses with their internal sophistication are a mark of the Western Isles and are to be found also at Jarlshof in Shetland. The Caithness name for such round or oblong, dry-stone masonry structures with stone pillars supporting the roof (or sometimes more specifically pre-broch structures related to duns and wheelhouses), is 'wag', from Gaelic uamhag, a small cave or hollow. The name may be taken to emphasise the way in which such homes were partly sunk into the ground.

It cannot be doubted that there was a great number of more humble homes in more ephemeral materials than stone, for example turf or a mixture of turf and stone, or wattle and light timbers and turf, like the incipient villages, 'nucleated settlements', that clustered around some of the brochs. Of the structure of these, relatively little is known, though later shieling huts, for example, may provide parameters for reasoned speculation. There are, however, indications of sophistication at an early date. For example, a double-skin stone building, a round-house dating to the period 1500–1000 BC, excavated at Cul a' Bhaile, Jura, had a wattle-and-post lining in the inside. It had an interior diameter of around 24 feet 6 inches (7.5 metres). There seems to have been a porch at the entrance, and the marks of cultivation were found under the house. The house was twice remodelled in the course of at least six generations of occupation, the wattle-and-post lining presumably going with the last phase before climate deterioration led to abandonment, conceivably as a response to colder conditions.[2] The concept of lining inner walls for greater comfort is far from being new.

Hingley's second area also contains some of these structures, including a broch in the Borders of Scotland, but whereas in the Atlantic zone stone and turf predominated, it seems that the substantial houses of the south and east were of timber and earth construction. The majority of those in the period being looked at by Hingley were still circular.[3] There were, however, notable exceptions, such as the rectangular timber hall at Balbridie by the wooded River Dee in north-east Scotland, measuring 85 feet 3 inches by 42 feet 8 inches (26 by 13 metres), and capable of holding thirty to fifty people. It is said to be of 'Continental' type, and dates to about 3600 BC. But apart from this example, 'houses were probably small with straight sides and round corners'.[4] The external appearance may have tended to roundness, however.

Without any deliberate intention of implying continuity, but rather of looking at construction techniques which may be comparable, reference may be made to examples of circular dwellings that have existed during the period of recorded history. Of these, the most notable are probably the shieling huts, some of which still survive (figs 2.1–2.4). The more recent ones are rectangular in shape, but the older ones were round, beehive shaped, and built of circular courses of flat stones or turf or both,

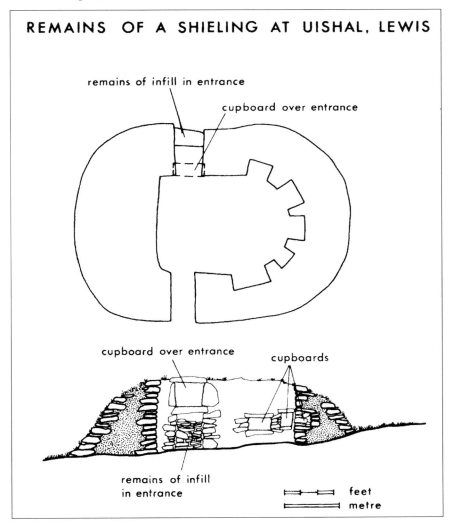

Figure 2.1 Uishal, Lewis; plan and long section of shieling hut. SLA, C4765.

Figure 2.2 Uishal, Lewis; rectangular shieling hut with gable chimney. SLA, XVIII_32_21.

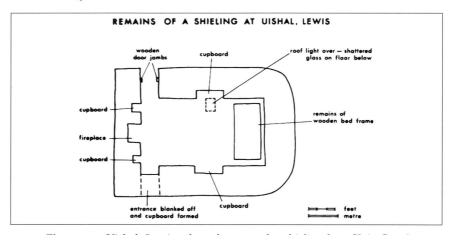

Figure 2.3 Uishal, Lewis; plan of rectangular shieling hut. SLA, C4918.

which overlapped in successively diminishing rings to make the shape of a dome. The walls could be single, or double (at least in their lower parts, before the final cantilevering), in which case there was an infill of earth between the two skins, as in an example from Uishal in Lewis. The upper part might also have been of turf only. The cleitean that are common in St Kilda are comparable structures.[5]

Shieling huts were often found in clusters, the individual units

Figure 2.4 Uishal, Lewis; general view of shieling hut. Copyright: Alexander Fenton.

being used by the tenants of the farming village from which they came to graze their stock on the summer grass of the hills (fig. 2.5). Numerous examples have been recorded by the Royal Commission on the Ancient and Historical Monuments of Scotland in Argyll. The remains of the huts that were surveyed are mostly of 'sub-rectangular' plan, though some are 'sub-circular'. It is significant in terms of time sequences that rounded forms averaging 14 feet 9 inches (4.5 metres) in diameter are often accompanied by the later rectangular examples, which average 19 feet by 12 feet 6 inches (5.8 by 3.8 metres), as at Àiridh nan Sileag in Lorn. Some huts are cellular, as in the case of one from the island of Coll, which has two sub-circular chambers with a connecting door.[6] In the 1860s, Captain Thomas made a drawing of a twin-celled shieling hut, in Gaelic, a both, at Uig in Lewis, of which the two parts were intercommunicating and each had its own door.[7] One structure in St Kilda, listed as a cleit, has an intact cell attached to it, with an internal low doorway. Each unit has its own conical, turf-topped roof.[8] Its precise function, or changing set of functions, is not entirely clear.

Shieling huts often occurred in paired units. The larger one was to live in, with a bed space (commonly filled with the fronds of young heather) whose foot was demarcated by a setting of turf or stones, opposite the fireplace which was sited near the door. The smaller one, which served as a milk house or as a temporary shelter for a new-born

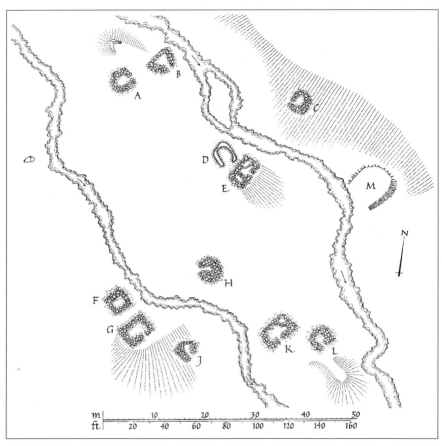

Figure 2.5 Àiridh nan Sileag, Argyll; plan of shielings, in RCAHMS, *Inventory of Argyll* 2 (Lorn), 268, fig 224. Crown copyright: RCAHMS, SC 714681.

calf, was usually backed against the bigger one, and sometimes there was an intercommunicating door. A similar duality in the use of space probably underlies the twin-cell form of shieling hut. Such cell formation, seen in the shieling huts at the simplest level of doubling, but in more permanent dwellings as an accumulation of units,[9] appears to mark a form of space demarcation or allocation which was later seen in a more specifically aligned form in rectangular buildings.

The doors of shieling huts, the use of which was confined to the summer months, were relatively insubstantial. They had no hinges and were made of wicker work, wattles, heather or bent grass. Such doors were used on other kinds of houses as well – for example in 1629 in the Borders of Scotland, near Langholm in Dumfriesshire, where three English

Figure 2.6 Jura, Argyll; shieling huts, Pennant. Tour, 1772.

travellers visited a building with walls of alternating courses of turf and stone,[10] a fire in the middle of the floor and a door of plaited wicker rods.[11]

From the archives of the Irish Folklore Commission in Dublin, it can be seen that doors in houses of the poorer type in Ireland were frequently of wicker and straw, sometimes in conjunction with a plaited mat of straw which could be hung up to inhibit the draught or which sometimes served as the door itself. A wattle door appears to be mentioned in an Irish law tract of the early years of the eighth century AD.[12]

A form of wattle walling could also be found in Scottish shielings. When Thomas Pennant was on his second tour in Scotland in 1772, his servant, Moses Griffiths, drew a number of shielings observed in the island of Jura (fig. 2.6). These were conical structures, in form like a wigwam, and one ruinous example in the foreground of his drawing suggests a structure of branches leaning against each other so as to meet at the top, and hung around with turves to give a more solid, windproof wall. There is no indication in the drawing of cross wattling, but it is scarcely conceivable that this did not exist. The doors, according to Pennant, consisted of 'a faggot of birch twigs, placed there occasionally', and the furniture was 'a bed of heath, placed on a bank of sod'.

Wattling was, in fact, quite a widespread constructional form, geographically as well as in terms of social status. Wattled buildings, known as creel-houses or basket-houses, had stake and wattle walls to which an outer skin of turf was pinned, scalewise. They could be of low or high status: there was at Kilpatrick, Dunbartonshire, a 'great house constructed of wattles' in the thirteenth century, and in the seventeenth century the MacGregor lairds of Glenstrae, Argyll, had low wattle residences, one of which was a fairly substantial building on stone foundations, surrounded by a moat.[13]

This evidence for wattle doors and for wattle and turf walling provides a complement to buildings of stone and turf, and in both cases the form could be circular, but structures including stone were more lasting and have a better chance of being picked up in the archaeological record. It is worth noting, too, that the beehive or corbelled type of building is old in origin and wide in distribution. Corbelled stone huts are to be found in Ireland at monastic settlements, in animal houses, corn-drying kilns and sweat-houses.[14] There were corbelled pig-sties in Wales.[15] In west Sweden they appear in store-, animal-, and dwelling houses and in parts of Iceland there are sheep shelters of comparable form.[16] There are plenty of examples also in the Mediterranean countries [17] and in Portugal.[18]

It should be noted that in areas where timber was readily available, shielings could be built in this material. This was the case in the central Highlands, where landowners gave permission to tenants to use timbers for shielings. The laird of Glenlyon's 'shiel' house at Finglen, Perthshire, for example, was described in the second half of the seventeenth century as 'a dale house covered with turf', and though its shape is not known, deal planks would perhaps be more suited to a rectangular than to a circular hut. The practice of transporting roofing timbers at the beginning of the season to re-roof shieling huts, or of moving old timbers to a new hut, was also known in seventeenth- and eighteenth-century Perthshire, and this again is likely to point to a rectangular or at least oblong form for the hut walls.[19]

The main period of change in the beehive hut tradition appears to coincide with the late eighteenth- early nineteenth-century lowland Agricultural Improvement period, which had its backlash on Atlantic and central Scotland. Surveys of shieling huts in Perthshire and in Sutherland,[20] for example, indicate that the old beehive huts went out of use as summer dwellings when bigger four-sided ones were erected, or were relegated to use as store houses. It is likely, however, that in the wooded central Highlands, rectangular shieling huts existed from at least the seventeenth century, and the main impact of the change from round to four-sided lay in Hingley's Atlantic zone, where stone and turf were the primary building materials. The wigwam-like Jura shielings seen by Pennant, therefore, are quite likely to have been built on the analogy of the circular stone and turf examples.

Whether round or rectangular, shieling walls could be double-skinned with a hearting of turf or earth, except for the more recently-built rectangular huts, which were mainly single-skinned. The double-skin wall also characterises the so-called blackhouses of north and west Scotland (fig. 2.7). In spite of the archaic appearance of such a wall, it could nevertheless in itself represent a form of nineteenth-century modernisation, at least for surviving blackhouses (fig. 2.8). This view is based on the fact that when the Ordnance Survey officers were surveying Lewis in 1849–52, they noted that most blackhouse walls had an inner skin of stone, but that the outside was of turf.[21]

Nevertheless, the double-skin wall is of ancient vintage. It occurs at Viking and late Norse sites in Shetland, Orkney, Caithness and the Western Isles.[22] It is found in the Faroe Islands,[23] and in buildings and farmhouses of the Middle Ages and later in Iceland.[24] At the same time, it was not the only building technique. As with the later Lewis blackhouses, it was also common for walls to be of turf with an inner lining of stone, as in the pre-Viking, Pictish, farm at the Point of Buckquoy in Orkney.[25] Sometimes the outer face – as in some of the Viking buildings at Birsay in Orkney, and at Jarlshof and Underhoull in Shetland – was of alternating courses of stone and turf.[26] Both techniques were used by the Vikings, but there is also a line of continuity from pre-Viking times until much later, for examples have survived almost to the present day, sometimes well outside areas of Viking or Norse influence.[27] Double-skin walls could be found in Norway, too: for example, in Jæren and Dalane, 'the old byres still standing [in the 1960s] are built in a kind of double wall with large stones in the outer sides and with an infill of smaller stones, sand and turf in the middle'.[28] It is clear enough that, even if the technique was known in the homeland of the Vikings, it does not follow that it was introduced to Scotland by them. Rather it is a function of the nature of the terrain and its available resources, where stone and turf abounded and where building timber was scarce.

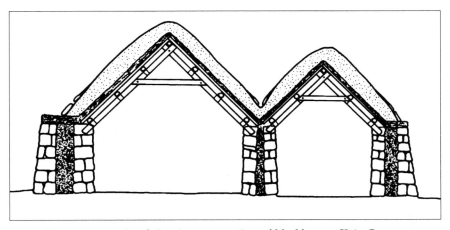

Figure 2.7 42 Arnol, Lewis; cross-section of blackhouse. SLA, C3335.

Figure 2.8 Shawbost, Lewis; ruined blackhouse wall. SLA, XIV_2_9.

The question arises as to when the emphasis on a round or oblong structure was replaced by an emphasis on a more rectangular form. It would be possible to make the story even more complicated by dividing houses with a tendency to circularity into sub-groups. For example, a recent study from Shetland[29] states that the 'oval' house is almost synonymous with Shetland prehistory. It 'was clearly a long-lived phenomenon, its durability presumably reflecting both a dependence on the same limited range of materials for building through prehistory, and the success of the type as a dwelling house. And yet, at the same time, there was incontrovertibly some development towards circular house forms ...'. But both forms could co-exist, and indeed cellular buildings were also introduced while oval and circular buildings were still in use. Different building forms could be roughly contemporary with each other.

The evidence of the beehive shieling huts shows that the older tradition could continue, especially in temporary habitations. But for more permanent dwellings, it may be that cellular structures – which could contain both rectangular and circular spaces internally, even if the outer walling presented an oblong appearance – were slowly moving towards the linearity that is the mark of rectangular buildings. The ground plans of the houses at Pictish Gurness in Orkney, Pictish Buckquoy in Orkney, and Viking Jarlshof in Shetland show the tendency to stretch and grow more rectangular,[30] and it seems that this change, at least in northern Scotland, is contemporary with the coming of settlers from the Scandinavian countries, around the late ninth and tenth centuries AD. This period, therefore, may be seen as ushering in a new phase, in which the longhouse, or byre dwelling, with man and cattle under one roof, played a prominent role.

Domesticated cattle and horses were no doubt kept in close association with their lords and masters from early times. However, it must be noted that overwintering of stock demands preparation in the way of gathering and storing winter fodder,[31] so that there would always have been a preference for letting the animals roam and forage for themselves during the cold season. The likely exception was the milk-producing cows, which were herded, tethered or housed and looked after rather better than the other cattle stock. In late eighteenth-century Islay, for example, 'None but milk cows are housed: cattle of all other kinds, except the saddle horses, run out during winter'.[32]

A German scholar, Dr W H Zimmermann, has argued that the possibilities for outwintering stock, so largely doing away with the burden of fodder gathering, was a conditioning factor in the migration of Continental people, the Anglo-Saxons, over the North Sea, to the south of Britain.[33] The milder climate, the bulk of winter biomass that resulted, and the generally better soils, were the magnets. Did they have byres in the homelands from which they came? There is a limited amount of evidence for byres in Neolithic Europe, mainly in Alpine and South German areas, including an example from Bavaria which had both a living space and, to judge by the dung accumulation, a byre. This is dated to 3496 BC. Bronze Age to Migration Period byres as parts of houses have also been identified, to judge by phosphate analysis of the floor areas, and in early medieval times, three-aisled houses with living quarters and byre under one roof remained common. Byres were also found in one-aisled houses (equivalent to the longhouse), but there appears to be no convincing archaeological proof of early byres in Anglo-Saxon England, at least in the more southerly parts.

Whether or not this theory is valid for the south of Britain, in Atlantic Scotland the climate was harsher, and byre dwellings were widespread. Examples still survive, for example at Auchindrain in Mid Argyll, where there is evidence of an original door used by both cattle and humans.[34] But how far back is it possible to find dwellings with byre and living quarters end on to each other within a roughly rectangular wall? The earliest Norse settlement at Jarlshof in Shetland, from the first half of the ninth century, had the outhouses separate from the main dwelling house. This matches the situation in Iceland, where the farmhouses had three or four rooms placed in alignment, but the stables and cow-byres were separate. The pattern was the same in the Faroe Islands, in Greenland, and at the late tenth- to early eleventh-century Norse settlement at L'Anse aux Meadows in Newfoundland.[35] It was not until the eleventh century, a period of great change in the settlement, that a byre was added to the living quarters of the parent Jarlshof dwelling house, though nearby houses built in the ninth and tenth centuries included byres from the beginning.[36] It appears from paved areas identified as the floors of byres in excavations of houses in south-west Norway that in the Iron Age people and animals could be found under one roof, so the system was at least known there.[37]

Byre dwellings do not appear to have been common or were not found in areas of Norse settlement outside the parent country, and there is no specific evidence for byres as parts of the composite longhouses (with their elements built in line) of Norway itself.[38] It seems reasonable, therefore, to conclude that the byre dwelling as adopted by Norse settlers is at least a partial reflection of a native tradition in Scotland, found also in Ireland and Wales and on to Brittany.[39] Indeed, it has been thought by some to be a 'Celtic' tradition.[40] More certainly, it is a function of climate, which was not everywhere as mild as in the south of England. But it is no simple story, for even in the Hebrides, for example in the late eighteenth century in Rhum, no hay was made, nor was a stock of winter provision laid in. According to Pennant, 'The domestic animals support themselves as well as they can on spots of grass preserved for that purpose'.[41] With the kindly support of the Gulf Stream, there was not an absolute need for winter housing and preliminary fodder gathering even there.

If the byre-dwelling form is not necessarily Norse, nevertheless there is a case for thinking of the rectangular form of building as being influenced by Norse building traditions, at least in the parts of Scotland that underwent intensive Norse settlement, since the earliest examples known through archaeological excavations are found there. But the four-sided or rectangular form was widespread in European tradition, and further research might well show an influence from the southern parts of the country. In the Atlantic zone, such is certainly demonstrable for later times, for example in the replacement of the round shieling huts by rectangular ones. Also to be considered is the influence of buildings in timber, which was clearly more widespread than had once been thought. These tend to be rectangular, though round houses in timber, for example on crannogs, are known from around the middle of the first millennium BC.[42] But on the existing evidence, the suggestion is that after a long period of comparative continuity, there was a basic change from the ninth century onwards, marked by a move to rectangular buildings, with more organised forms of living space which included animals as well as men. The next major period of change was at the time of the Agricultural Improvements.

NOTES

1. Hingley, 1992, 10, 12–14, 17–18.
2. Ashmore, 1996, 111.
3. Hingley, 1992, 28.
4. Ashmore, 1996, 12–13, 32–3, 42.
5. See Stell, Harman, 1988, 43–8.
6. For example, RCAHMS, 1971, 197–9 (Gartavaich), 200 (Talatoll); 1975, 267–8 (Àiridh nan Sileag), 271–2 (Glen Risdale); 1980, 239–40 (Càrnan Dubha, Coll), 240 (Coire Bhorradail); 1992, 469–70 (Douglas Water).
7. Thomas, 1866–68.
8. Stell, Harman, 1988, 46, Fig 18a.
9. Thomas, 1866–68.
10. The evidence for this technique is summed up in Fenton, 1968.
11. Lowther et al, 1894, 12.
12. Lucas, 1956.
13. Dunbar, 1978, 171–2.
14. Evans, 1961, 114–26.
15. Peate, 1946, 42ff, Plates 2–7.
16. Pettersson, 1954.
17. Hamm, 1971.
18. Veiga de Oliveira, Galhano, Pereira, 1969.
19. Bil, 1990, 242–3.
20. Adam, 1960; Bil, 1990; Gaffney, 1960.
21. Fenton, 1989, 11, 26.
22. Viking and later Norse settlements are usefully listed in Graham-Campbell, Batey, 1998, 155–205.
23. Dahl, 138, 140.
24. Stenberger, 1943, 176, 190; Nilsson, 292–3.
25. Ritchie, 1976–7, 174–227.
26. Cruden, 1965, 26; see also Fenton, 1968, 94–103.
27. Fenton, 1968, 94–103; Fenton, Walker, 1981, 73–76.
28. Hoffmann, 1964–5, 123.
29. Owen, Lowe, 1999, 284–6 (p 285 illustrates 19 house plans from prehistoric Shetland).
30. Ritchie, 1993, 25.
31. Zimmermann, 1999(b), 307.
32. Pennant, 1774, 1998, 217.
33. Zimmermann, 1999(a).
34. RCAHMS, 1992, 33.
35. Fenton, 1982, 231–2.

36. Hamilton, 1956, 94, 103, 158.
37. Hoffmann, 1943, 54.
38. Brekke, 1982, 51–114.
39. Meirion-Jones, 1973.
40. Campbell, 1935.
41. Pennant, 1774, 1998, 278.
42. Hingley, 1992, 27.

BIBLIOGRAPHY

Adam, R J, ed. *John Home's Survey of Assynt* (Scottish History Society, 3rd series, vol 52), Edinburgh, 1960.
Ashmore, P J. *Neolithic and Bronze Age Scotland*, London, 1996.
Bil, A. *The Shieling, 1600–1840: The case of the central Scottish Highlands*, Edinburgh, 1990.
Brekke, N G. Samanbygde Hus i Hordaland (Houses built together in Hordaland), *Årbok* (Foreningen til Norske Fortidsminnesmerkers Bevaring [Association for the Preservation of Norwegian Prehistoric Monuments]), 1982, 51–114.
Campbell, Å. Irish fields and houses: A study of rural culture, *Béaloideas*, 5 (1935), 57–74.
Cruden, S. Excavations at Birsay, Orkney. In Small, 1965, 22–31.
Dahl, S. A survey of archaeological investigations in the Faroes. In Small, 1965, 135–41.
Dunbar, J G. *The Architecture of Scotland*, London, 2nd edn, 1978.
Evans, E E. *Irish Folk Ways*, London, 1961.
Fabech, C, Ringtved, J, eds. *Settlement and Landscape: Proceedings of a conference in Århus, Denmark, May 4–7 1998* (Jutland Archaeological Society), Århus, 1999.
Fenton, A. Alternating stone and turf: An obsolete building practice, *Folk Life*, 6 (1968), 94–103.
Fenton, A, Walker, B. *The Rural Architecture of Scotland*, Edinburgh, 1981.
Fenton, A. The longhouse in northern Scotland. In Myhre, Stoklund, Gjærder, 1982, 231–40.
Fenton, A. *The Island Blackhouse*, Edinburgh, 2nd edn, 1989.
Furumark, A et al, eds. *Arctica* (Studia Ethnographica Upsaliensia XI), Uppsala, 1956.
Gaffney, V. *The Lordship of Strathavon* (Third Spalding Club), Aberdeen, 1960.
Graham-Campbell, J, Batey, C E. *Vikings in Scotland: An archaeological survey*, Edinburgh, 1998.
Hamilton, J R C. *Excavations at Jarlshof, Shetland*, London, 1956.
Hamm, F J. *Kragwölbung und Kragkuppel: Untersuchungen zu einer fortlebenden vorzeitlichen Bautechnik (Collared Arch and Collared Dome: Investigations into a surviving prehistoric building technique)*, Rome, 1971.
Hingley, R. Society in Scotland from 700 BC to AD 200, *PSAS*, 122 (1992), 7–53.
Hoffmann, M. Jærhuset (The Jæren house). *By og Bygd*, II (1943), 55–157.
Hoffmann, M. Gamle fjostyper belyst ved et materials fra Sørvest-Norge (Old byre-types in the light of material from SW Norway), *By og Bygd*, 18 (1964–65), 115–35.
Lowther, C, Fallow, Mr R, Mauson, P. *Our Journall into Scotland Anno Domini 1629, Fifth of November*, Edinburgh, 1894.
Lucas, A T. Wattle and straw mat doors in Ireland. In Furumark et al, 1956, 16–35.
Meirion-Jones, G I. The long-house in Brittany: A provisional assessment, *Post-Medieval Archaeology*, 7 (1973), 1–19.

Myhre, B, Stoklund, B, Gjærder, P, eds. *Vestnordisk Byggeskikk Gjennom to Tusen År. Tradisjon og forandring fra romertid til det 19. århundre* (*West Norse Building Customs over 2000 Years. Tradition and change from Roman times till the nineteenth century*), Stavanger, 1982.

Nilsson, A. Den sentida bebyggelsen på Islands landsbygd (Contemporary buildings in Iceland's countryside). In Stenberger, 1943, 292–3.

Owen, O, Lowe, C. *Kebister: The four-thousand-year-old story of one Shetland township* (Society of Antiquaries of Scotland Monograph, 14), Edinburgh, 1999.

Peate, I C. *The Welsh House*, Liverpool, 1946.

Pennant, T. *A Tour in Scotland and Voyage to the Hebrides, 1772*, Chester, 1774, reprinted Edinburgh, 1998.

Pettersson, J. *Kupolbyggnader med Falska Valv* (*Dome-shaped Buildings with False Vault*), Lund, 1954.

RCAHMS. *Inventory of Argyll*: Vol 1, *Kintyre*, Edinburgh, 1971; Vol 2, *Lorn*, Edinburgh, 1975; Vol 3, *Mull, Tiree, Coll and Northern Argyll*, Edinburgh, 1980; Vol 7, *Mid-Argyll and Cowal*, Edinburgh, 1992.

Ritchie, A. Excavation of Pictish and Viking-age farmsteads at Buckquoy, Orkney, *PSAS*, 108 (1976–7), 174–227.

Ritchie, A. *Viking Scotland*, London, 1993.

Sarfatij, H, Verwers, W J H, Woltering, P J, eds. *In Discussion with the Past: Archaeological studies presented to W A van Es* (The Foundation for Promoting Archaeology), Zwolle, Amersfoort, 1999.

Small, A, ed. *The Fourth Viking Congress*, Aberdeen, 1965.

Stell, G P, Harman, M, for RCAHMS. *Buildings of St Kilda*, Edinburgh, 1988.

Stenberger, M. *Forntida Gårdar í Island* (*Prehistoric Farms in Iceland*), Copenhagen, 1943.

Thomas, Captain F W L. On the primitive dwellings and hypogea of the Outer Hebrides, *PSAS*, 7 (1866–68), 153–95.

Veiga de Oliviera, E, Galhano, F, Pereira, B. *Construçones Primitivas em Portugal* (*Primitive Constructions in Portugal*), Lisboa, 1969.

Zimmermann, W H. Favourable conditions for cattle farming: One reason for the Anglo-Saxon migration over the North Sea? About the byre's evolution in the area south and east of the North Sea and England. In Sarfatij, Verwers, Woltering, 1999(a), 129–44.

Zimmermann, W H. Why was cattle-stalling introduced in prehistory? The significance of byre and stable and of outwintering. In Fabech, Ringtved, 1999(b), 295–312.

PART TWO
•
Dwellings

3 Castles, Palaces and Fortified Houses

CHARLES McKEAN

TERMINOLOGY

At some time in the early eighteenth century, countless Scottish country mansions were given the title 'castle' as a consequence of changing social aspiration and of the misreading of a European building form in Scotland within an English context. After the Romantic Movement, others adopted the title 'tower' as being redolent of medieval romance. Contemporary usage is best found in *MacFarlane's Geographical Collections*,[1] the earliest comprehensive topographical writing on Scotland, whose contents derive predominantly from the period 1630–1730.

Although there is no overall consistency among its many different authors, contemporary terminology appears to have been thus: a medieval country house would be referred to as a tower or fortalice (or both), denoting structures fundamentally vertical in plan and predominantly pre-1500 in date; a sixteenth- or seventeenth-century smaller country house would be called by its name (for example, Carnousie, Banffshire), which was also usually the way its occupier was customarily addressed; a very grand country house would be called the 'House of –' (for example, the House of Strathbogie, Aberdeenshire, later called Huntly, and now Huntly Castle); a courtyard establishment would customarily be called a 'place' or 'palace' (for example, the palaces of Auchmedden and Pitsligo, Aberdeenshire, or Culross, Fife); and a significant military establishment capable of defence would be called a 'castle', even if later converted to a residence (for example, Kinneddar and Fyvie, Aberdeenshire, and Dunbar, East Lothian).

These terminologies generally accord with the names on the 1585–95 maps of Scotland by Timothy Pont.[2] However, there are a few exceptions. 'Peel' or 'pele' was applied to lesser properties of a value under £100. 'Manor' appears occasionally in medieval charters but 'manor house' with increasing frequency in charters of the mid sixteenth century. It appears to be a legal term, for it is almost never used either as a description or as a title; Mugdock, Stirlingshire, was a manor but never called that. The term 'keep' appears to have been non-existent. Sixteenth-century documents tend to differentiate between a tower and a fortalice, the latter implying a more extensive establishment. By the sixteenth century, 'tower with barmkin' (barmkin being a form of enclosure which might accommodate livestock) is almost invariably restricted to the war-torn Borders. Many, if not most, tower-houses had become enfolded by courtyards rather than by barmkins by the sixteenth century, if not earlier, their associated buildings being very much more elaborate than cattle sheds.[3]

For the purposes of this article, the *MacFarlane* terminology will be followed. Buildings which fall into the category of what Sir William Brereton called 'laird's houses here built all castle-wise',[4] may be called châteaux,[5] for there is some slight evidence for the use of this term for mock military country mansions. They will not be referred to as castles unless this is essential for identification.

CASTLES

After the reign of David I (1124–53), Scottish defensive structures might be seen as following one of two cultures, the Anglo-Norman or the Scottish. The Anglo-Norman footprint was usually created by the principal families who arrived in Scotland in the twelfth and thirteenth centuries, families such as de Vaux, de Moravia, Brus and Comyn. Their buildings generally shared the Norman characteristics of a hierarchy of form, although the Scots variant version was arguably unique. In contrast to the normal plan characteristic in England and elsewhere of outer ward, inner ward and donjon (the lord's lodging), Anglo-Norman structures in Scotland tended to consist of a single all-encompassing courtyard, and the more vertical proportion of its buildings was possibly a consequence of Scottish climate, materials, building technology and economy. Walled outer wards are rare.

The most complete of the surviving Anglo-Norman inheritance – castles such as Bothwell, Lanarkshire, Dirleton, East Lothian, Inverlochy, Inverness-shire, and Kildrummy, Aberdeenshire – share certain common characteristics, being dry or wet moated and entered through a towered gatehouse, usually between flanking drum towers. Buildings, including lodgings, chapel and hall, lay against the curtain wall within in a precise hierarchy. One of the towers, distinguished by its greater size (except in Caerlaverock, Dumfriesshire), might be the donjon. The donjon in Bothwell, protected by its own dry moat, had its own well and chapel. It was a self-contained, vertically-planned defensive living unit, with subsidiary chambers in the wall thickness. In Yester, East Lothian, the building protected on its triangular site by ravines, was a two-tiered hall. Donjons on mottes or castle mounds on the English model such as Duffus, Moray, for example, are comparatively rare, and the horizontal proportions of Duffus more resemble a hall-house than donjon. There is a number of largely indefensible hall-houses or major houses of the period, in which, quite distinct from their English cousins, the principal, well-lit chamber is on the first floor, with a round bedroom tower adjacent. Good examples are Rait, Nairnshire, and Morton, Dumfriesshire.

Instead of presenting the narrowest front to the enemy and the world at large, later castles of enclosure moved to presenting the broadest, often sealing off a peninsula or promontory with a long curtain wall, a tower in the centre and one at each end. This plan is best exemplified by St Andrews Castle, Fife, and Tantallon Castle, East Lothian, although there were variations on the theme at Dundarg Castle, Aberdeenshire, Ravenscraig Castle, Fife, and – on a much more domestic scale – Mugdock Castle, Stirlingshire. Doune Castle, Perthshire, entered beneath a principal tower which is not unlike a tower-house, was unusual, its high walled courtyard containing a hall block linked to a subsidiary tower. This arrangement is not dissimilar to the later layout at Kilchurn Castle, Argyll, with which Doune has much in common.

A group of castles, mainly of the Gàidhealtachd, followed a different cultural pattern. They usually take the form of a stone curtain which enfolds and encloses a rocky eminence, often protected by water. Simply in the plainness of their curtain walls they appear to be natural

Figure 3.1 Mingary Castle, Argyll. Crown copyright: RCAHMS, SC 581252.

descendants of prehistoric brochs and duns, and legends about their construction rituals persist to this day. The buildings which lined the curtain wall within lacked the hierarchy of Norman buildings. In the late fifteenth and early sixteenth centuries, most of them were graced by the addition of a tower-house, representing the new living fashion of mainland Scotland. That has led to a mis-reading of them as being of 'keep and bailey' type, whereas they represent a juxtaposition of two different cultures. Examples include Castle Tioram, Inverness-shire, and Dunvegan, Skye, and in Argyll Duart, Mull, Dunollie, Dunstaffnage, Duntrune, Mingary (fig. 3.1) and perhaps also Dunaverty and Dunyvaig, Islay. It may be no accident that so many of their names begin with Dun. A Scots (rather than Gàidhealtachd) variant of the same approach are the plain castles of enclosure, with corner towers (square or round) of which Kinclaven and Moulin, Perthshire, Kincardine, and perhaps Castle Sween, Argyll, are examples. Those adapted to a particular topography include Hailes Castle, East Lothian, and Loch Doon, Ayrshire. Auchencass, Dumfriesshire, appears to have had the form of a castle of enclosure with round corner towers but with two significant differences: firstly, a substantial earth plinth; and secondly, some unusually sophisticated earthworks projecting north, added after the arrival of artillery.

FORTIFICATIONS

The primary characteristic of early fortifications was inaccessibility, with a distinct leaning towards sites that were naturally or artificially strengthened: islands, mounds, peninsulas, pendicles, promontories or cliff edges. There would be a ditch or a moat, occasionally wet, and a bridge and a drawbridge access. A sturdy battlemented curtain wall, which protected the residential buildings within, was usually flanked by projecting corner towers, and entered through the centre. Such is the general disposition in places as disparate as Ravenscraig, Fife, Tantallon, Urquhart, Inverness-shire, Fast, Berwickshire, and Dundarg. Sometimes the gatehouse became the principal defensive structure, replacing the main tower or donjon as at, inter alia, Tantallon, St Andrews, Ravenscraig, Dunstaffnage, Caerlaverock and Doune. The once splendid three-towered façade of Tantallon may have provided the chivalric inspiration for the entrance façade for the Renaissance palaces of Scotland such as Boyne, Banffshire, Fyvie, Aberdeenshire, and Dudhope, Dundee.

Gatehouses as self-contained structures varied from the substantial – such as those at the former archiepiscopal palace in Glasgow, Falkland Palace, Fife, Seton, East Lothian, and Dumbarton Castle – to the minimal, as witness the portcullis gates added to Mugdock and, in the late sixteenth century, to Edinburgh Castle. The three most distinctive gatehouses were those of Caerlaverock, the magnificent four-towered forework (1500) at Stirling Castle (the gateway design arguably an imitation of a Roman triumphal arch),[6] and the one-time Netherbow Gate, Edinburgh, probably a fifteenth-century structure, re-edified in 1546 in the manner of the Port St Honoré, Paris. Such signals of grandeur were reproduced in miniature at Thurso, Caithness, and Philorth and Tolquhon, Aberdeenshire. However, the paucity of really defensive gatehouses outside a relatively small number of royal properties, taken together with the casual attention paid to the necessity of walling the towns and cities, may be construed as evidence of the relative stability of Scotland in the later Middle Ages, remote from the dynastic wars of Europe.

The development of gunpowder and cannon, and the particular interest taken by Scots kings in the subject, led to changes far beyond the varying shapes and sizes of gunloops. The essence of the change lay in defensive structures with lower profiles, much greater mass, and cannon emplacements, which synthesised into thick, battered-base artillery walls and blockhouses containing both casemates and parapets for cannon. The entrance façade of St Andrews Castle received the addition of such blockhouses at each end, and the flamboyant sea-girt palace of Dunbar was given a huge new defensive blockhouse guarding the land entry in the sixteenth century.[7] Comparable defensive features were integrated into the passively massive entrance at Ravenscraig, Fife.

The inner ward of the three-warded Cadzow Castle, Lanarkshire, may well have taken the form of a substantial blockhouse, protected on two sides by a ravine, and on the other two by round towers. Cadzow, c. 1525, may have been the work of Sir James Hamilton of Finnart (c. 1495–1540) who possibly synthesised Cadzow's three wards into two, separating them by a dry ditch, protected not only by artillery transe, but by a caponier (an enclosed musketeer's emplacement) guarding the ditch invented by Francesco di Giorgio.[8] Great investment in outer artillery defences was restricted principally to the royal seats of Edinburgh, Stirling and Dumbarton, although in 1536 Finnart was paid for landward defences to the state prison at Blackness, West Lothian, which consisted of a new entrance protected by a virtual caponier, and casemates for substantial landward-facing cannon. In 1547, Leith became the only Scots town ever to receive integrated artillery fortifications. There, Piero di Strozzi created a new military emplacement on the most advanced European model enfolding the entire town with a star-shaped design with projecting ravelins.

By the sixteenth and seventeenth centuries, fortification had generally become more a matter of display than of defence, save in the most critical locations. There was the blockhouse at Dunbar, the great fortifications at Eyemouth, Berwickshire, the forts constructed by Cromwell in Leith, Ayr, Perth, Inverlochy (later, Fort William) and

Inverness, those erected by the Government in the late seventeenth and early eighteenth centuries – such as Ruthven, Bernera and Fort Augustus, all Inverness-shire – and the last and grandest of all – Fort George, Ardersier, Inverness-shire. Begun in 1747, it took the form of a complete military town.[9] Thereafter, defensive structures in Scotland were limited to the three Martello Towers (see Chapter 19) and the extensive reinforced concrete coastal fortifications of the two World Wars, significant remains of which surround the fleet anchorage of Scapa Flow in Orkney. Balfour Battery, South Ronaldsay, and Buchanan Battery, Flotta, are typical products of World War II.[10]

Defensiveness in Renaissance country houses, however, was – at the worst – a matter of passive defence. Few of these buildings would have been as free-standing as they appear now, having been surrounded by countless yards, walled enclosures and gardens. Not only would some of the grandest-looking gunloops not have had any field of fire to command, but many were clearly unusable. Moreover, what was later described as a wall-walk was often little more threatening than a balcony. The symbolism of gunloop and turret, crenellation and bartizan had remained as important to the mid seventeenth-century Scots aristocrat as armorial quartering did to an eighteenth-century Prussian count. Both were signs and symbols of nobility and status.

TOWER-HOUSES

The predominant Scottish medieval country house was the tower-house. At its plainest, it consisted of a chamber (or chambers) on each floor, the principal room or hall being on the first floor, and stairs and subsidiary chambers (small bed-chambers, the occasional chapel and garderobe) being formed within the thickness of the wall. Invariably, the ground floor would consist of cellars and kitchen (with a rare pit or prison).

Equivalent buildings in England had principal apartments on the ground floor, and cultural and technical imperatives may be offered as explanation of this difference. Since Scotland had little long-span structural timber, the normal form of construction remained until the mid-eighteenth century that of a stone vault, with a maximum width of about 20 feet (6.1 metres). The consequence of a mass structure was the transference of damp, which logically meant that the principal rooms should be on the floor above the vault, the piano nobile. First-floor living was, however, also the cultural norm in Europe of which, after all, Scotland formed a western part.

Thus, a tower-house would normally be three-and-a-half storeys above a vaulted ground floor of cellars and kitchens. The stronger and better built might have a second stone vault, either over the principal hall (as, for example, at Pitsligo) or over a two-storeyed space reaching to the attic storey, and split within by a timber intermediate floor. The location of these buildings was rarely influenced by serious defensive considerations. They are generally set within a fold of land, probably for shelter, usually visible from afar, and frequently near water and a quarry.[11]

Variations in house plans, increasingly frequent in the sixteenth century and misleadingly classified by geometric form – L-plan, U-plan, Z-plan, T-plan, etc – were a consequence of buildings being adapted and added to according to variations in the wealth and standing of the occupier. If one were to extend a tower of vertically stacked rooms, given the technology available, the only option would be the construction of a separately vaulted and structured wing or jamb. Thus it was that in the sixteenth century, many plain, vertical towers were elaborated; Burgie, Blervie and Brodie, all Moray, and Delgatie, Aberdeenshire, are examples.

Tower-houses were, therefore, substantial, vertically-planned country houses ideally adapted for passive defence. They generally possessed a single ground-floor door, protected by a yett (an interwoven wrought-iron gate) and a gunloop, perhaps with a projecting bartizan, turret or machicolation overhead; the doorway gave onto a stair, and led into ground-floor kitchens and cellars, illuminated and ventilated by slit windows. The main stair up to the principal floor or hall was initially contained within the thickness of the wall – and thereby tight; later the stair was housed more spaciously, in a projecting turnpike stair or wing. The same stair might lead to the top of the building and to all floors, or might terminate at the first (that is, the public) floor, access to the upper storeys being obtained via a narrower private stair in another corner, an increasingly frequent pattern in the sophisticated châteaux and villas of the Scottish Renaissance. The floors immediately above the hall belonged to the family; and those at the very top (sometimes in the roof space) to retainers and servants. Some tower-houses had spaces opening from the private floors for use as chapels, as at Niddry, West Lothian, and Affleck, Angus.

Most towers may have been initially surrounded by a barmkin or enclosure, which came to contain some of the public rooms expelled from the tower as the latter became more private. The hall at Smailholm, Roxburghshire, is thought to have been moved out to the barmkin by the fifteenth century, leaving the principal chamber in the tower to the family.[12] As at Threave, Kirkcudbrightshire, with its corner towers, battered walls and gunloops, some surrounding enclosures aspired to military status, but it was more common for the enclosure to contain household functions. The one excavated at Niddry, a tight enclosure following the line of a rocky outcrop round the foot of the huge L-plan tower, contained stables, offices, workshops, a blacksmith, a hall and perhaps something even grander in the small round towers.[13] These buildings were separated from the tower by a narrow, cobbled path.

Figure 3.2 Coxton Tower, Moray. Crown copyright: RCAHMS, SC 710990.

Tower-houses varied in scale from the diminutive (Coxton, Moray, fig. 3.2), to the large (Borthwick, Midlothian), although few reached the tall extravagance of the earl of Crawford's seven storeys and more at Finavon, Angus.[14] Being the fashionable living form, towers were added to existing curtain-wall castles in the West Highlands, to virtually all the royal palaces, and to abbeys and priories; they also emerged in the suburbs of towns as noblemen's lodgings, the high street of Maybole, Ayrshire, having had a tower-house at each end. An enormous tower was the main focus of the archbishop's palace establishment in Glasgow; a diminutive one served as the earl of Selkirk's town house in the Canongate, Edinburgh; the abbot of Crossraguel's lodging within his own Ayrshire abbey was of tower-house form, as was the captain of Dunstaffnage's new quarters in Dunstaffnage Castle, and even of the governor's lodging in the state prison at Blackness, West Lothian. Indeed, the ultimate form of tower-house was probably reached in the central tower at Blackness, which came to consist of four large stacked prison apartments, joined by a turnpike stair swelling from one corner.

Towers also varied from the plain to the elaborate, but elaboration was comparatively rare or confined to details, generally comprising a moulded doorway, a heavily corbelled parapet and armorial panels highlighted in dressed stone against a harled wall. The colour of the harling itself may have been determined by that of the local sand or by pigments imported from the Continent.

BUILDINGS OF THE RENAISSANCE

The living patterns of Scottish aristocrats and gentry changed markedly in the sixteenth century, moving from vertical to horizontal living. It coincided with a fondness for elaboration in building and decoration (as in dress) and a greater stateliness or formality in the conduct of business. Structures became much thinner, and the rooms and stairs that used to be contained within the tight geometric thickness of a rectangular tower wall now effloresced on the exterior, supported on elaborate corbels and contributing to a remarkable skyline. The influence of the substantial Scottish trading community in Europe and the Baltic may be inferred from close parallels in planning, detail and decoration between Scots and, successively, Italian, French, Danish, Polish and Dutch buildings. The horizontal mode of living was now be expressed in the 'palace' or 'place', consisting of apartments of linked chambers at the same level.

Pitsligo (fig. 3.3) offers a good instance of how such additions were made. The tower was first extended by a linking block with entrance and stair; then a palace block, or laird's lodging, consisting of a suite of rooms or chambers ending in a projecting round tower (almost invariably the bedroom stack); a long gallery – which generally faced the view –

Figure 3.3 Pitsligo, Aberdeenshire. Crown copyright: RCAHMS, SC 710721.

overlooking the garden and sea; and a suite of offices and chapel overlooking the western pleasance. Once the square was complete, the original tower was relegated to service and servants' use, if not abandoned altogether. Occasionally, as at Dudhope, the tower was demolished. That was also attempted at Pitsligo; one and a half storeys and the eastern gable were removed, but the remainder proved too obdurate.

The essence of the Renaissance plan was convenience and privacy. The pattern of the palace block is best exemplified in the majestic House of Strathbogie (Huntly Castle), a three-roomed block stacked six storeys high. The imposing public staircase rises from one corner, and in this case ascended to the top of the building. The rooms were increasingly arranged in a sequence of degrees of privacy – ante-chamber, chamber and bedchamber, although rarely using that terminology. The mid sixteenth-century projecting round bedroom tower frequently had its own private staircase, as was the case at Huntly and, among others, Balvenie, Banffshire, Kilcoy, Ross and Cromarty, and Earlshall, Fife. The laird may have occupied one floor, kitchen and cellars below, and his lady – hence that stair – the one above. It is improbable that such laird's lodgings provided all the functions of the former hall. Other buildings did that.

Some towers were extended upwards with new rooms within which to enjoy the splendours of modern living. It appears that to some families – particularly in the north east – height was one of the signals of nobility. The tower would become the harled plinth for a now regularly-fenestrated superstructure, sometimes of ashlar, sometimes aediculated. Fine lowland examples are Hoddom, Dumfriesshire, Niddry, and Preston, East Lothian. A more integrated approach of the early seventeenth century was to re-format the various accretions of earlier periods into a coherent design, a glorious superstructure sitting on a harled plinth and the transformation from one to the other sometimes signalled by an elaborate string-course. The finest examples of such reformatting are the châteaux of Crathes, Kincardineshire, Craigievar, Craigston, and Castle Fraser (fig. 3.4), Aberdeenshire (all associated with the Bell family of architects and masons), Glamis, Angus, and, in 1639–44, Innes, Moray.

In the case of Preston, East Lothian, a door was also slapped through the old tower at first-floor level in order to give access to a projecting timber balcony or gallery. Timber galleries were far from unusual, being also found in places as geographically diverse as Noltland, Westray, Orkney, Girnigo, Caithness, Craigmillar, Edinburgh, Drochil, Peeblesshire, and Carsluith, Kirkcudbrightshire. Galleries customarily faced the best view over the sea, the ravine, or the river. Anyone who may still believe that Renaissance Scots were oblivious to the splendours of the view is put right by a 1584 proposal for works to the Palace of Stirling planned specifically for the sole purpose of allowing the king to enjoy a view of his domain.[15] Projecting timber galleries were also normal in town houses. Purpose-built galleries as indoor recreation rooms became customary in country houses, sometimes doubling in use as a corridor between the main house and the guest tower, as seems to have been likely at Tolquhon, Gordon, Moray, and Craig, Angus. The great gallery planned for the west façade of the Palace of Stirling, overlooking the King's Knot and the hunting beyond, comparable to that at Blois in France, was exceptional and remained unbuilt. In some cases, as at Dunfermline Palace, Fife, galleries were distinguished by oriel windows, and in others, as at Crathes, Craigievar, Claypotts, Dundee, the Argyll lodging in Stirling, and at Duff House, Banff, the gallery was contained within the structure at roof level. This may also have been a motivation for the later high-level libraries to be found at Traquair, Peeblesshire, and the House of Dun, Angus.

Taking the air and enjoying the view at roof level was important. Claypotts, dating from c. 1560, had balconies at each end of its roof-level gallery. The bedroom towers in both Boyne and Huntly had corbelled

Figure 3.4 Castle Fraser, Aberdeenshire. Crown copyright: RCAHMS, SC 713511.

balconies, and by the early seventeenth century, a balustraded flat or viewing platform had become customary. They obviously formed part of the view of the house itself, but they also allowed the laird to show off the planted yards surrounding his house on the ground below. Save for those houses immediately commanded by a tall hill, like Balloch, Perthshire, or Melville, Fife, these viewing platforms would have provided the only opportunity to obtain an overall view of such achievements. Sometimes, elaborate square staircase and study towers terminated in a viewing platform, as, for example, in the cases of the 'Wrychtishousis', Edinburgh,[16] Innes and Pitsligo,[17] whereas sometimes they also appeared above the centre of the house, as at Hatton, Angus, Craigievar, Craigston, Aberdeenshire, and Calder, Midlothian.

These buildings were designed to be seen from a distance in a countryside which may have been largely devoid of trees. The comparison with the contemporary, low-lying, moated English manor house, invisible in its trees, its principal rooms on the ground floor, could hardly be more striking. The effulgent façades of Strathbogie, Aberdeenshire, and Balvenie, face the countryside – Strathbogie even had an outward-facing loggia – but, by contrast, their courtyard façades were the plainest imaginable, enlivened in the case of Strathbogie by a heraldic stair-tower. George Gordon's other great palace, Bog of Gight, later Gordon Castle, was even less defensible but of a rather more sophisticated design, with arcaded galleries or loggias at first-floor level.

Between 1540 and 1570, Scots houses, or châteaux, wore a distinctly French appearance, comparable to romantic, asymmetrical, and turreted mock-military country houses like Azay-le-Rideau, Touraine, France – an unsurprising flavour given the 30-year influence of French-educated queens in this country. The maps of Timothy Pont contain building elevations showing, for example, as great a concentration of such properties along the Clyde as along the Loire, an indication that Scotland was far less poor and primitive in this period than has been imagined. Perhaps the apogee of these great houses was the enormous 1596 château created by the Maitland family at Thirlestane, Berwickshire (fig. 3.5), within the grounds of an earlier English fort. Its enormous swelling sides and pepper-pot turrets, stripped even as they are now of their harl, is reminiscent of the old Louvre, Paris, or of Loches, Touraine. Of similar aspiration, but more sophisticated in form, was the Palace of Seton, East Lothian, whose plan, a domesticated version of shield-shaped Caerlaverock, was that of a Scottish Chaumont-sur-Loire. In the very late sixteenth century, however, Scottish architecture passed beyond French influence and became more rational, symmetrical, and less wilful, probably reflecting first Danish and then Dutch influences.

Figure 3.5 Thirlestane, Berwickshire. Crown copyright: RCAHMS, SC 710739.

The approach to the property was often a matter of artifice. A free-standing château occupying one corner of its enclosure and abutting walled gardens or yards, is not unusual; Amisfield and Hoddom, Dumfriesshire, and Harthill, Aberdeenshire, are examples. Gardens could be elaborate or sumptuous, witness Inverugie, Aberdeenshire, with its statues,[18] or Edzell, Angus, with its banqueting and bath houses and Dürer-inspired engravings. Several châteaux like Midmar, Fraser and Grant, Aberdeenshire, were approached between projecting wings of lower buildings. One group, of which the most distinctive examples were Braemar, Aberdeenshire, Minto, Roxburghshire, Calder and – above all – Glamis, appears to have been designed so that the approach was on the diagonal through a circular entrance tower in the angle between two wings, a feature that was later to have a profound influence on the architect, Robert Adam.

Circular, rectangular or square staircases were a dominant component of the architecture. It was customary for a rectangular stair-tower to be added to an earlier tower, and, occasionally, capped by a study, its apex sliced off by a moulded cornice to form a quasi-pediment; Eden by Banff is one such example. Renaissance villas such as Kilbaberton, Edinburgh, frequently contained the entrance in a projecting jamb or wing to the north west, offering protection from the prevailing wind. As at Greenknowe, Berwickshire, Kellie, Fife, Elcho and Menzies, Perthshire, and Hatton, Angus, and elsewhere, it became customary for the main staircase to rise only to the principal floor, further access being obtained from one or more private turnpikes beyond. This later became the pattern even for the grander houses such as Thirlestane, Brucefield, Clackmannanshire, and Kinneil, West Lothian.

The building elevations on Pont's maps imply that there was a hierarchy to Scots country houses. Those of the highest rank are shown with a full heraldic skyline, such as the now-vanished principal seat of the Grahams at Kincardine, near Auchterarder, Perthshire; by contrast, nearby Methven and Invermay are both shown as old-fashioned towers, without an heraldic skyline, as may have befitted families in decline. Two lesser ranks – but still of country house status – are shown not only with a less elaborate skyline, but also of fewer storeys and possibly without the surrounding park, such as Monorgan or Fingask, Perthshire. Pont thereafter shows four further ranks of smaller country buildings. It is clear, however, that substantial families with more than one seat did not maintain them all to the full standard. There was no need. In Angus, for instance, the Ogilvies had their fortress at Airlie, but lived in their newly-acquired château of Cortachy and hunted from the new-built hunting lodge of Forter, Angus, up the glen. The Renaissance emblematic display was for Cortachy.

Decoration
The interiors of the châteaux were elaborate. In addition to the furnishings and hangings, which could be of great richness (and in north-east Scotland often from northern Europe), sixteenth-century ceilings and walls were generally painted in vivid colours, with motifs ranging from simple armorial bearings to birds, beasts, fruit and Renaissance mythology,[19] occasionally, as at Kinneil, assuming a geometric pattern. Seventeenth-century ceiling decoration went from painted surfaces towards heavily articulated plaster ceilings. Motifs were also carved on fireplace lintels and window embrasures. The intention was not only to decorate but also to reflect and enhance light.

Likewise, where the external walls were of rubble, they would invariably have been coated, in harl, limewash or plaster, to create a monolithic structure of vertical proportion, highlighting the dressed stone

Figure 3.6 Craignethan Castle, Lanarkshire; aerial view. Crown copyright: RCAHMS, SC 710988.

details and – where there was one – an effulgent superstructure. Detail was all, and in the corbelling, turrets, oriels, cornices, parapets, fretwork, chequer-board carvings, balustrades, oculi and elaborate dormer-head designs, Scottish architecture achieved what Charles Rennie Mackintosh was later to term 'a treasury of indescribable loveliness.'[20]

Setting aside two outstanding buildings by Sir James Hamilton of Finnart, Craignethan, Lanarkshire (from 1532; fig. 3.6), and the Palace of Stirling (from 1538), a first sign of change from the Scottish pattern of having principal rooms on the first floor or piano nobile to the English ground-floor pattern appeared at Culross Abbey House in 1608, and Auchterhouse, Angus, *c.* 1633. Ground-floor cellars were gradually replaced by subordinate ground-floor rooms or 'laigh halls', of 1622 at The Binns, West Lothian, and about the same date at Liberton, Edinburgh. Clearing out former vaults to create reception rooms, such as occurred at Brucefield, Clackmannanshire, became much more frequent as the century progressed, though in grander houses the principal rooms remained on the piano nobile until the later eighteenth century.

By 1660, superstructural brilliance was being tamed, and increasingly restricted to the capping of the stairtower (more often than not with a viewing platform), corner towers, swept roofs and great ashlar chimney stacks, often set on the angle, as at Luffness, East Lothian, Kellie, Leslie, Aberdeenshire, and Innes and Brodie. Swept roofscapes and tall chimney stacks were to remain salient characteristics of Scots eighteenth-century architecture, apparent even in Robert Adam's last villa, Balbardie, West Lothian, 1792. The contrast between the use of dressed stone and coated rubble – whether harled or plastered – had evolved during the Renaissance and remained a predominant feature of most Scottish architecture outside the towns until the onset of 'rubblemania' in the 1790s.

PALACES

Scottish royal palaces, whatever their origins, all tended to assume, in the end, the approximate form of a courtyard. Two – Stirling and Edinburgh – were created from military fortresses, two – Holyrood and Dunfermline – from abbeys, and two – Falkland and Linlithgow – built ab initio. Boghouse of Crawfordjohn, Lanarkshire, was listed in 1542 as one of the principal royal seats, but is totally demolished. Almost certainly designed by Sir James Hamilton of Finnart, 1526–36, it appears to have consisted of at least two towers and a courtyard.

The castles of Stirling and Edinburgh had much in common.[21] They each comprised, at the apex of their rocks, royal lodgings, a great hall, and a chapel and offices. Stirling's great hall, of majestic scale and magnificent detail, lines the entire eastern wing of the rock. Sitting on his dais at the south end, the monarch would have enjoyed spectacular views east and west through two enormous full-height oriel windows. A 1594 chapel closes off the citadel to the north; the oldest lodgings, the King's Old Buildings, line the west flank; and Sir James Hamilton of Finnart's 1538 royal lodgings fill the southern quarter, itself a courtyard palace of noble proportion, innovative plan and with façade tableaux of outstanding humanist iconography.[22] In notable contrast to the King's Old Buildings, the principal rooms of this palace are entered from ground level.

Dunfermline Palace occupied the west wing of the monastic cloister, and this sumptuous early sixteenth-century two-storeyed block was joined across the abbey gatehouse to the guest wing of the abbey, which overlooked the adjacent river and parkland. A new lodging for Queen Anne of Denmark, the Belvedere, was added to the north to complete the court. Holyrood, the principal royal palace after the mid sixteenth century, occupied two courtyards adjacent to the abbey, entered through the twin towers of the northernmost, all rebuilt or recast in the late seventeenth century by Sir William Bruce. One entered Holyrood beneath the long gallery, a characteristic imitated in the palaces of Hamilton, Lanarkshire, and Scone, Perthshire, and – in a precise if miniature way – at Tolquhon.

Non-royal palaces earned their name 'palatium' likewise by virtue of enclosing a courtyard. Most of them – such as Crichton, Midlothian, Pitsligo or Culross – extended from an original tower to form that courtyard, and those designed in courtyard form a priori like Boyne are rarities. The components of a palace courtyard might well include the original tower, a lodging, gallery (facing the view), gatehouse, hall, loggia, chapel, and beyond, perhaps garden, tennis courts, pleasance, well, outer gatehouse, offices and guest lodgings.

The greater the symbolism, the greater the owner's status. The bishop of Moray's magnificent seat at Spynie, Moray, consists of the largest tower in Scotland added to a walled enclosure which eventually contained a splendid hall, chapel, guests' lodgings and even a tennis court. Dudhope appears to be of one period, but was, in reality, an overwhelming extension to a tower (now vanished) to the north. Fyvie may well embrace an ancient curtain-wall and earlier fifteenth-century towers, but its majestic south façade is undisputably a formal architectural composition of the late sixteenth century. With its drum-towered entrance and round tower on each corner, the plan of the palace of Boyne was probably the grandest exposition of the palace form. Its palace block, lining the south flank, had two bedroom stacks (like Inverugie), a gallery facing the river to the west, and the kitchen/hall block lining the north. All this was uncluttered by the presence of an earlier tower, for the site of the original castle, which lay closer to the sea, had been abandoned.

Castle courtyards were occasionally transformed into Renaissance palaces, most notably in Caerlaverock, and Crichton, Midlothian. Although the courtyard at Crichton had one of the most sophisticated frontages in Renaissance Scotland, comprising a splendid scale-and-platt staircase, a loggia, and a diamante façade above, the face which was presented to the outside world remained well capable of defence, albeit capped by an extraordinary skyline of square and octagonal towers, turrets, dormer windows, an unusually elaborate string-course and a two-tier stair window.

VILLAS

A villa of the Scottish Renaissance may be defined as a free-standing noble house lacking the paraphernalia of a great estate. Villas tend to be suburban and cluster near towns. Since even passive defence was given scant consideration, dispensing with even token gunloops, they took the form of miniature châteaux and symbolised the end of the transition from defensive to chivalric building. One such, Craigcrook, Edinburgh, had in all essentials an identical plan-form to that of Harthill, Aberdeenshire. The later, more sophisticated villas of the early seventeenth century tended to be of U-plan form, rather similar to the noblemen's hotels within the towns, of which the 'Wrychtishousis', and

Figure 3.7 Earl's Palace, Kirkwall, Orkney; view by R W Billings. *The Baronial and Ecclesiastical Antiquities of Scotland*, 1845–52, 1901 edn.

Kilbaberton, Edinburgh, and Pitreavie, Dunfermline, were perhaps the best examples. They differed from châteaux only in details of the court style, such as buckle quoins, ornate pediments to the windows, groups of square windows set in the angle and viewing platforms.[23]

The most elaborate non-royal palace was in reality a sub-royal nobleman's villa, namely, the home of a royal bastard, Earl Patrick Stewart of Orkney. The Earl's Palace in Kirkwall (fig. 3.7) was intended to stand at the rear of a much larger complex, like the laird's wing at Tolquhon.[24] Entered from a projecting stair wing, the great hall and the earl's bedroom on the same floor are sumptuous in scale and detail. Externally, they are identified by ashlar corbelled turrets and oriel windows in a rich composition of harl and stone. The earl's bedroom itself is an aggrandised version of the laird's tower, separate, a storey higher, and distinguished by soaring two-storeyed turrets.

DESIGNERS

The re-working of a fine group of Aberdeenshire châteaux can be attributed to the Bell family; and there is a north-eastern group with notable peculiarities in plan and groin-vaulted floors, including Towie, Gight and Craig, Aberdeenshire, and Balbegno, Kincardineshire, which have been attributed to Alexander Con of Auchry on the grounds of their vaulting and plan. Balbegno, at least, is a reworking, the vault having been inserted into an older tower. Important masons of central Scotland included Thomas and John French, Thomas Kedder (or Cadder) and Thomas Mylne (and his son John).[25] The first Master of Works Principal, as Royal Architect, was Sir James Hamilton of Finnart, who was certainly responsible for Craignethan, the royal palaces of Linlithgow and Stirling, and the hunting lodge at Boghouse of Crawfordjohn. He also appears to have been responsible in part at least for Dean Castle, Kilmarnock, Ayrshire, Cadzow, Hamilton Palace, Kinneil and many places elsewhere. Finnart was followed by his trainee, John Hamilton of Craigie, and later holders of the post included Sir William Makdowall, William Shaw of Sauchie and Sir James Murray of Kilbaberton.

NOTES

1. Mitchell, Clark, 1906–8.
2. The Timothy Pont manuscript maps are held in the Map Library, NLS.
3. Good, Tabraham, 1988.
4. Sir William Brereton in 1636. In Brown, 1973, 148.
5. See particularly McKean, 1999–2000, 3–21.
6. Glendinning, MacInnes, MacKechnie, 1996, 14–15.
7. MacIvor, 1981.
8. McKean, 1995(b).

9. Most of these buildings are well described in Gifford, 1992.
10. Burgher, 1991, 65–7.
11. Stell, 1983.
12. Good, Tabraham, 1998.
13. Proudfoot, Kelly, 1997.
14. Pont's manuscript maps.
15. Paton, 1957, 310.
16. McKean, 1994.
17. McKean, 1991.
18. Mitchell, Clark, 1906.
19. Bath. In Gent, 1995.
20. Mackintosh, 1990, 52.
21. McKean, 1997.
22. McKean, 1999.
23. MacKechnie, 1988.
24. Cox, Owen, Pringle, 1998, particularly the drawing on 578.
25. Mylne, 1893.

BIBLIOGRAPHY

Bath, M. Alexander Seton's painted gallery. In Gent, 1995, 79–108.
Brown, P H. *Early Travellers in Scotland*, Edinburgh, 1891, reprinted 1973.
Burgher, L. *Orkney: An Illustrated Architectural Guide*, Edinburgh, 1991.
Caldwell, D H, ed. *Scottish Weapons and Fortifications, 1100–1800*, Edinburgh, 1981.
Campbell, I. Linlithgow's princely palace and its influence in Europe, *Architectural Heritage*, 5 (1995), 1–20.
Campbell, I. James IV and Edinburgh's first triumphal arches. In Mays, 1997, 26–34.
Cavers, K. *A Vision of Scotland: The nation observed by John Slezer*, Edinburgh, 1994.
Cox, E M, Owen, O, Pringle, D. The discovery of mediaeval deposits beneath the Earl's Palace, Kirkwall, Orkney, *PSAS*, 128 (1998), 567–80.
Cruden, S. *The Scottish Castle*, Edinburgh, 1962.
Dunbar, J G. *The Royal Palaces of Scotland*, East Linton, 1999.
Fawcett, R. *Scottish Architecture from the Accession of the Stewarts to the Reformation, 1371–1560*, Edinburgh, 1994.
Gent, L, ed. *Albion's Classicism: The Visual Arts in Britain, 1500–1660*, New Haven and London, 1995.
Gifford, J. *The Buildings of Scotland: Highlands and Islands*, Harmondsworth, 1992.
Glendinning, M, MacInnes, R, MacKechnie, A. *A History of Scottish Architecture from the Renaissance to the Present Day*, Edinburgh, 1996.
Good, G L, Tabraham, C J. Excavations at Smailholm Tower, *PSAS*, 118 (1988), 231–66.
Gow, I, Rowan, A, eds. *Scottish Country Houses*, Edinburgh, 1995.
Hadley Williams, J, ed. *Stewart Style, 1513–1542: Essays on the court of James V*, East Linton, 1996.
Howard, D. *Scottish Architecture from the Reformation to the Restoration, 1560–1660*, Edinburgh, 1995.
Howard, D. The Protestant Renaissance, *Architectural Heritage*, 9 (1998), 1–16.
Mason, R. *Kingship and the Common Weal*, East Linton, 1998.
MacGibbon, D, Ross, T. *The Castellated and Domestic Architecture of Scotland from the Twelfth to the Eighteenth Century*, Edinburgh, 5 vols, 1887–92.
MacInnes, R, MacKechnie, A, Glendinning, M. *Building Scotland*, Edinburgh, 1999.

MacIvor, I. Artillery and major places of strength in the Lothians and the east Border, 1513–1542. In Caldwell, 1981, 94–152.
McKean, C A. The House of Pitsligo, *PSAS*, 121 (1991), 369–90.
McKean, C A. The Wrychtishousis: 'A Very Curious Edifice', *Book of the Old Edinburgh Club*, 3 (1994), 113–22.
McKean, C A. A plethora of palaces. In Gow, Rowan, 1995, 1–15.
McKean, C A. Craignethan: The castle of the Bastard of Arran, *PSAS*, 125 (1995(b)), 1069–90.
McKean, C A. The palace at Edinburgh Castle, *Book of the Old Edinburgh Club*, 4 (1997), 89–102.
McKean, C A. Sir James Hamilton of Finnart: A Renaissance courtier-architect, *Architectural History*, 42 (1999), 141–72.
McKean, C A. The Scottish château, *ROSC*, 12 (1999–2000), 3–20.
McKean, C A, Bath, M. *The Scottish Château*, Stroud, 2001.
MacKechnie, A. Evidence of a post-1603 court architecture in Scotland, *Architectural History*, 31 (1988), 107–119.
MacKechnie, A. Design approaches in early post-Reformation Scots houses. In Gow, Rowan, 1995, 15–35.
Mackintosh, C R. Scottish Baronial architecture (Introduction by F A Walker). In Robertson, 1990, 29–63.
Mays, D, ed. *The Architecture of Scottish Cities*, East Linton, 1997.
Mitchell, Sir A, Clark, J T, eds. *Geographical Collections relating to Scotland, Collected by Walter Macfarlane* (Scottish History Society, 1st series, vols 51–3), Edinburgh, 1906–8.
Mylne, R S. *Master Masons to the Crown of Scotland*, Edinburgh, 1893.
Paton. H, ed. *Accounts of the Masters of Works, 1529–1615*, Edinburgh, 1957.
Proudfoot, E, Kelly, C A. Excavations at Niddrie Castle, West Lothian, 1986–90, *PSAS*, 127 (1997), 784–842.
Robertson, P, ed. *Charles Rennie Mackintosh: The architectural papers*, Wendlebury, 1990.
Stell, G P. The Scottish medieval castle: Form, function and 'evolution', *RIAS Prospect*, 14 (1982), 14–15, and in Stringer, 1985, 195–209.
Stringer, K J, ed. *Essays on the Nobility of Medieval Scotland*, Edinburgh, 1985.

4 Country Houses, Mansions and Large Villas

KITTY CRUFT

The country house in Scotland developed from the fortified castles and tower-houses which were the centres of a way of life followed by nearly all the people of Scotland, whether as part of the clan system in the Highlands or in the political role of the lowland gentry, dependent on the Crown for power and finance. The accession of William III and Queen Mary in 1688, and the Union of Parliaments in 1707, brought about relative order, particularly in the Borders and central belt. Freedom from war and the linking of families by marriage and patronage created a complex structure of national and local life. In particular, a new interest in mining and especially in improvements on the land by the aristocracy and lairds brought in new wealth which enabled the explosion of intellectual and artistic activity in the eighteenth century.[1]

The introduction of professional architecture into Scotland proceeded slowly. As John Harris has shown, the gradual appreciation of order and rationality that began to affect country house design was concurrent with the introduction of Italianate models into England, which prepared the way for the arrival in Scotland of the double-pile plan house.[2] The influence of Andrea Palladio has been established by Howard Colvin.[3] Sir William Bruce (1630–1710) and James Smith (c. 1645–1731) played their parts in introducing Palladian architecture into Scotland by making their patrons aware of the pleasures of living in a new house in the latest style.[4] Some regularising of existing houses took place from the mid seventeenth century, principally by Sir William Bruce, politician, landowner and architect, who was 'responsible for making the practice of architecture respectable'.[5] Not only was the classical house a new departure, but the integration of house, garden and landscape contributed to a complete design.

Bruce's social and political background was just right to attract patrons whose lives were spent on the political stage in London. Leslie House, Fife, built for the 7th Earl of Rothes between 1667 and 1672, was designed by John Mylne (1611–67) and his nephew Robert Mylne (1633–1710), with advice from Sir William Bruce on all aspects of the design. The exterior of the courtyard-planned palace hid a rich and costly interior in the latest style. Thirlestane Castle, Berwickshire, built for the Earl (later 1st Duke) of Lauderdale in 1670–6, was a remodelling of his Scottish seat, where Bruce formalised the late medieval tower-house and introduced a symmetrical forecourt, pavilions and a classical doorpiece, leading to a sumptuous interior.

In 1665, at the same time as the above works were proceeding for his patrons, Bruce purchased Balcaskie, Fife. He regularised the existing house with the addition of a forecourt elevation and offices placed on the north side of the existing L-plan house, but still used crowstepped gables and ornamental gunloops to reflect the seventeenth-century character of the original house. Unfolding classicism in Scotland at this time is reflected in the south entrance doorway with its panelled and banded pilasters, and a carved basket of fruit on its segmental pediment. Balcaskie was the first of Bruce's houses to include a terraced garden and terminal vistas.[6] The first houses designed by Bruce on clear sites seem to have been Dunkeld House, Perthshire, c. 1676–84, for the Duke of Atholl, followed by Moncreiffe House, Perthshire, 1679, built for Sir Thomas Moncreiffe, and described by John Macky in 1737 as 'a neat little seat ... built of Free Stone after the Manner of the Country-Seats in the Villages about London, with a Glass Cupola or Lanthorn at Top, and very neatly wainscoted and furnish'd within'.[7] Its interest lies in the square plan divided into three parts by partition walls containing the chimney flues, a timber scale-and-platt stair placed to the right of the front hall or saloon, with a service stair positioned diagonally in the opposite corner of the house. A hipped roof with a platform covers a suppressed attic floor.

The source of the design for Moncreiffe House appears to have been Chevening, Kent, built before 1630, the plan of which was based on Rubens' Palazzo di Genova.[8] The double-pile plan of Coleshill, Berkshire, in which all the elements are perfectly proportioned and rational, foreshadows the new house Bruce designed for himself at Kinross from 1686.[9] Kinross House (fig. 4.1) was built out of the proceeds of successful political offices, as the house of a country gentleman. Bruce poured vast resources into the building of Kinross and in improving his estate, leaving it encumbered with debts. Details of Kinross such as the suppressed attic derived from Chevening, the double-pile plan from Coleshill, the giant pilasters which strengthen and accentuate the corners of the building and support the entablature, the string course and the rusticated basement, were a mixture of Italian, French, English and possibly some Dutch themes, carried out to perfection. Such textbook classicism was never repeated in Scotland with such distinction. Craigiehall House, West Lothian, completed 1699, built for the 1st Marquess of Annandale, and Hopetoun House, West Lothian, 1699–1702,

Figure 4.1 Kinross House, Kinross-shire. Crown copyright: RCAHMS, SC 710770.

Figure 4.2 Dalkeith House, Midlothian; late nineteenth-century view. Crown copyright: RCAHMS, SC 570916.

for Charles Hope (later 1st Earl) of Hopetoun, both exhibit more complex tripartite plans.

James Smith, Scotland's first professional architect, carried on the pioneering work of Sir William Bruce. His rise to the profession of architect came with his appointment as overseer of the Royal Works in Scotland in 1683, at a time when little royal building was taking place, a state of affairs which gave him time to pursue a private architectural practice. Among his major country house projects were Melville House, Fife, built between 1697 and 1700, for the Earl of Melville, Dalkeith House, Midlothian, 1702–10 (fig. 4.2), for the Duchess of Monmouth and Buccleuch, and Hamilton Palace, Lanarkshire, 1693–1701, for the Duke and Duchess of Hamilton. Dalkeith and Hamilton were remodelled, with deep forecourts flanked by pavilions and central blocks with dominant pediments and giant pilasters, to make them into large fashionable houses fitting for the newly exalted status of their owners. Melville was a completely new house built on an H-plan. The central block is of double-room width, with a large scale-and-platt staircase, and there is a great room or saloon at first-floor level.[10]

The progress of country house building proceeded swiftly between 1715 and 1745, the dominant figure being William Adam (1689–1748). His importance as an entrepreneur, businessman and architect was to dominate Scottish country house building. His first appearance was at Floors Castle, Roxburghshire, 1721–6, for the 1st Earl of Roxburghe, but more successful was the construction of Mavisbank, Midlothian, 1723, where he assisted in the design of the house developed on Palladian principles by Sir John Clerk of Penicuik. The plan is compact though spacious, with a large hall and staircase projection at the back, giving space internally for a large suite of rooms. Two-storeyed pavilions are linked to the house by arcades. Hopetoun House, started in 1723 for the Earl of Hopetoun, was a remodelling of Bruce's original house. Adam's brief was to provide a splendid suite of state rooms which he did by attaching a Palladian front to Bruce's house with the typical plan of a central block and concave screen walls to large pavilions housing the stables and ballroom. Duff House, Banffshire, designed in 1730, for William Duff (later Lord Braco and Earl of Fife), although Baroque in feeling with bold square corner towers, was planned on Palladian principles. Craigdarroch House, Dumfriesshire, 1726, for Alexander Ferguson, and The Drum, Midlothian, 1726–30, for Lord Somerville, were both to William Adam's preferred design of central block linked to two-storeyed pavilions.[11]

Variants of Bruce's tripartite plan and its development by William Adam were to produce the accepted layout for large and small houses in the country for nearly one hundred years. Brucefield House, Fife, c. 1724, for Alexander Bruce of Kennet, was one of the first post-Bruce houses to use the tripartite plan, with a scale-and-platt stair around a stone newel, at the front of one of the divisions. Houses were designed with projecting bows on the rear and side elevations, as seen at Cally House, Kirkcudbrightshire, 1763–5, for James Murray, designed by Robert Mylne (1733–1811). Wings or small pavilions were joined to compact central blocks by short walls punched through with a window or two, often a dummy window disguising a passage to the wings. This particular development was the cheap answer to the arcades and colonnades which were such expensive components of grand Palladian designs. The Palladian repertoire of Diocletian and tripartite windows, niches and oculi were joined by the Gibbs' Baroque motifs including the Gibbs' surround.[12]

Changes were already taking place in the gap between those who held land heritably, and those who were acquiring it for the first time. A rise in prosperity, assisted by the Scottish Enlightenment, led to a great increase in building activity. Large additions were made to ancestral houses, and law lords and the new rich burgesses and merchants succeeded to, or bought, estates in the country.

Architectural source-books became a requisite for success. James Gibbs' *Book of Architecture*, London, 1728, dedicated to the Duke of Argyll, consisted entirely of his own designs, and was expressly intended as a pattern book 'of use to such gentlemen, especially in the remoter areas of the Country, where little or no assistance for Designs can be procured', and his *Rules for Drawing the Several Parts of Architecture*, London, 1732, enabled correct copying of the Palladian orders. Later manuals plundered his ideas and became enormously popular.[13] One of the earliest Scottish architectural publications was George Jameson's *Thirty Three Designs with the Orders of Architecture*, Edinburgh, 1765. Perhaps more informative to the builder was a later version, *The Rudiments of Architecture or The Young Workman's Instructor*, Edinburgh, 1772, which as well as being a construction manual, contained twenty-three of the designs for small dwellings found in Jameson's original publication, 'the most of which have been actually executed in North Britain' (fig. 5.3). The success of this small volume was guaranteed, as no other such work existed in Scotland to help in designing large buildings and small country houses. The contents were updated five times, as gradually much of the Baroque ornament shown on the houses in Jameson's original version was erased, presumably to keep building costs down, but perhaps also to appeal to changing taste, leaving a simpler neo-classical style.[14]

The Indian summer of Palladianism in Scotland produced a number of important country houses. This entailed pulling down old ones unsuitable for remodelling and building new grand models. Marchmont House, Berwickshire, 1750s, for the 3rd Earl of Marchmont, by Thomas Gibson (fl 1750s), Amisfield, East Lothian, 1756–9, for Francis Charteris (later 6th Earl of Wemyss) designed by Isaac Ware (d 1766) with John Baxter as executant architect, and Penicuik House, Midlothian, designed by Sir James Clerk for himself, again with John Baxter, 1760s, all displayed the conceits of their builders.[15]

The mainstream of Scottish architecture followed the English or British pattern of national classical architecture, producing such imposing examples as Dumfries House, Ayrshire, 1754–9, for the 4th Earl of Dumfries, by Robert Adam (1728–92) and John Adam (1721–92), Auchinleck House, Ayrshire, 1760, for Adam Boswell, by an unknown architect, Paxton House, Berwickshire, 1757, for Patrick Home of Billie, attributed to Robert and John Adam, and Duddingston House, Edinburgh, 1763–8, for the 8th Earl of Abercorn by Sir William Chambers (1723–96).

Many small country houses were built in the last years of the eighteenth century and early years of the nineteenth century. A popular pastime was to purchase landed estates and build a villa, and to use it as a place for pleasure, mostly in the summer months. Glendoick, Perthshire (fig. 4.3), for Robert Craigie, Lord President of the Court of

Figure 4.3 Glendoick House, Perthshire. Crown copyright: RCAHMS, SC 710837.

COUNTRY HOUSES, MANSIONS AND LARGE VILLAS • 53

Session, by an unidentified designer, has a compact tripartite plan, and tall, hipped slightly bell-cast roof, with the principal staircase in the centre division. Stewart Hall, Bute, 1760, for James Stewart of Kingarth, has five bays with an advanced centre pediment, and Strachur House, Argyll, for General John Campbell of Strachur, built in the 1780s, is a three-storeyed oblong block with tripartite plan and two-storeyed flanking wings, with a bow to the garden. An advanced central bay with a Palladian window and a three-light window on the upper floor reflects the pattern of the original front door with its flanking side lights. Arbigland, Kirkcudbrightshire, probably designed by William Craik for himself and built c. 1755, is an oblong block with straight links to canted wings and an advanced three-bay pedimented front with giant Ionic pilasters and a perron. Wardhouse, Aberdeenshire, dated 1757, had quadrant links to two-storeyed pavilions with the entrance on the ground floor, there being no basement. Cromarty House, Ross and Cromarty, c. 1772, for George Ross of Pitkerrie, has a five-bay front with a slightly advanced central bay decorated with a Venetian window over a tripartite front door. The wings are attached directly to the main block. Culloden House, Inverness-shire, 1780, for Arthur Forbes of Culloden, is very similar in style to Cromarty House, but the two-storeyed pavilions are linked to the house by straight screen walls.

There had to be a rebellion against the orthodox nature of the Palladians. The Gothic Revival in particular was taken to the heart of Scottish proprietors and their architects. The remote ideal of the Romantic movement was pertinent to the Scottish landscape, the Rococo was absorbed from the Continent, and the Picturesque, with its Gothic elements, mixed to resemble a medieval picture.

The classic picturesque building type was the country house in the castellated Gothic style. It could express the pride of ancient lineage, and it could turn a simple residence into an imaginary palace. Inveraray Castle, Argyll, begun in 1745, for the 3rd Duke of Argyll and designed by Roger Morris (1695–1749), with the building work supervised by William Adam and his sons Robert and John Adam, was symbolic of the ancient Campbell lineage. Inveraray was the prototype of many later castles. It looked forward to Douglas Castle, Lanarkshire, 1757–61, for the Duke of Douglas, by Robert and John Adam, and was used by James Playfair (1755–94) at Melville Castle, Midlothian, for the 1st Viscount Melville, 1786, and again at Kinnaird Castle, Angus, 1790, a remodelling for Sir David Carnegie. It flourished triumphantly at Taymouth Castle, Perthshire (fig. 4.4), for the 4th Earl (later 1st Marquess) of Breadalbane, 1802, by Archibald Elliot (1760–1823) and James Elliot (1770–1810). All these houses were symmetrically planned along Palladian lines.

The Adam castle style, a unique achievement by the Adam brothers, is considered by Professor Rowan as 'amongst the most original

Figure 4.4 Taymouth Castle, Perthshire. Crown copyright: RCAHMS, SC 710989.

creations of eighteenth-century European architecture'.[16] A large number of clients was prepared to commission castles in this blend of continental, English and Scottish influences, in which Gothic pointed arches, traceried windows, buttresses and pinnacles disappeared, their places being taken by corbelled battlements, angle-turrets, hood-moulds, crosslet-hoops, and Scottish crowstepped gables; these together produced the Romantic element so important in the presentation of the style. The form was used at Wedderburn Castle, Berwickshire, 1770–8, for Patrick Home of Wedderburn and Caldwell Castle, Ayrshire, 1771–3, for Baron Mure of Caldwell, and was developed with great accomplishment at Culzean Castle, Ayrshire, 1770–90, for David Kennedy, Earl of Cassillis, and at Oxenford Castle, Midlothian, 1780–2, for Sir John Dalrymple of Cousland. It achieved complete fulfilment at Seton Castle, East Lothian, 1789–91, for Alexander Mackenzie. The compositional effect must have pleased Adam, as he could call on his work at Diocletian's Palace, Split (Croatia), for the massing and detailing at Seton.[17]

Picturesque theory had a profound and unqualified influence on later house designs. The ecclesiastical and secular buildings used as examples had functions different from those of an efficient country house. Irregularity of planning was the goal which contributed to the pictorial effect, but the arrangement of rooms produced endless discomfort and inconvenience for proprietors and servants.

Figure 4.5 Millearne, Perthshire. Crown copyright: RCAHMS, SC 710836.

The introduction of the first country house into Scotland without a balanced façade was by the English architect, Robert Lugar (c. 1773–1855), who designed Tullichewan Castle, Dunbartonshire, 1800, for John Stirling, merchant, and went on to produce two more additions to country houses in the same area – at Balloch Castle, 1809, for John Buchanan of Arden, and Boturich Castle, 1830, for John Buchanan of Ardoch. Picturesque effect was the objective of Abercairney Abbey, Perthshire, 1803, by Richard Crichton (c. 1771–1817) for Colonel Moray, but the most Gothic of these houses, with ecclesiastical elements, was Millearne, Perthshire (fig. 4.5), 1825–35, for J G Home Drummond, attributed to Richard Dickson (1792–1857) and Robert Dickson (c. 1794–1865). The miniature Inchrye Abbey, Fife, 1827, for Sir George Ramsay, also attributed to R and R Dickson, was a small Gothic gem with many pinnacles and open parapets. William Atkinson (c. 1773–1839), designed Scone Palace, Perthshire, 1803–12, for the 3rd Earl of Mansfield, producing a dignified and restrained house, which followed the monastic layout of the original palace.

James Gillespie Graham (1776–1855) was considered by his contemporaries as one of the most competent Gothic country house architects in Scotland. His access to wealthy patrons was strengthened by his marriage in 1815 to Margaret Graham, heiress of the estate of

Orchil in Perthshire. His designs for country houses in the castellated style tended towards asymmetrical planning and a picturesque aspect was often provided by a large round tower at one side or corner. At Dunninald Castle, Angus, 1819–23, for Peter Arkley, a well-to-do farmer, he plagiarised Lugar's design for Tullichewan Castle, and at Armadale Castle, Skye, 1815–19, for the 2nd Lord MacDonald, Graham incorporated two towers and some rooms of the previous house, providing a suite of state rooms, hall and staircase in a series of variously shaped towers and connecting walls, castellated at roof level. Duns Castle, Berwickshire, 1818–23, remodelled for William Hay, is set on a terrace in a landscaped park, looking like the feudal pile William Hay desired. As an artist himself, Hay probably acted as his own supervisor. At Lee Castle, Lanarkshire, 1822, for Sir Charles Lockhart, the plan reverts to the Inveraray model of central tower with corner towers, though, as Dr Macaulay has pointed out, the design for Lee was romanticised by the addition of straggling office buildings, defeating the Palladian idea of the isolated block.[18]

Tory patronage played a large part in the commissions of the architect Sir Robert Smirke (1780–1867). Kinfauns Castle, Perthshire, 1820, for the 15th Lord Gray, is strikingly situated on a small plateau in Glencarse and is of a very stark symmetrical design, to which the pale pink sandstone adds a certain degree of softness. Kinmount, Dumfriesshire, 1812, for the 5th Marquess of Queensberry, is 'in the New Square Style of Mr Smirke',[19] and Whittingehame, East Lothian, for James Balfour 1817–18, is in a simplified classical style, illustrating Smirke's admiration of the rational simplicity of Greek architecture.[20]

The early work of William Burn (1789–1870) included domestic Gothic for large additions to existing sixteenth- and seventeenth-century houses such as Saltoun Hall, East Lothian, 1818–26, for Andrew Fletcher, and Dundas Castle, West Lothian, 1818, for James Dundas. His series of Tudor Gothic mansions, included Carstairs, Lanarkshire, 1822–4, for Henry Monteith, and Garscube, Dunbartonshire, 1826–7, for Sir Archibald Campbell. Blairquhan, Ayrshire, 1820–4, for Sir David Hunter Blair, developed a style introduced to Scotland by William Wilkins (1778–1839) at Dalmeny House, West Lothian, 1814–17, for the 4th Earl of Rosebery, and Dunmore Park, Stirlingshire, 1820–2, for the 2nd Earl of Dunmore, both with English Tudor Gothic details. William Burn experimented with Tudor Jacobean elements in the 1820s, resulting in St Fort, Fife, 1829, for Captain Robert Stewart, and later Falkland House, Fife, 1839–44, for O Tyndall Bruce.[21] At Dunlop House, Ayrshire, 1831–4, for Sir James Dunlop, David Hamilton (1768–1843) revived a seventeenth-century Scots Jacobean house of considerable aesthetic merit. Experimenting with Tudor Gothic styles irrevocably led to the revival of early Scottish domestic motifs.

The term Scots Baronial was used to define the large group of buildings from the mid nineteenth century which produced the specific characteristics of Scottish fifteenth- to seventeenth-century architecture. Abbotsford, Roxburghshire, 1816–23, for Sir Walter Scott, principally designed by William Atkinson (c. 1773–1839), with suggestions by Edward Blore (1787–1879), reproduced Scottish details in a haphazard way mixed with a stock of Gothic features. The importance of Abbotsford

Figure 4.6 Castlemilk House, Dumfriesshire; main entrance. Crown copyright: RCAHMS, SC 372898.

as the Romantic prototype of the Scots Baronial style which emerged as the dominant fashion for a new country house, came about principally through the writings of Scott.[22] It took the place of Georgian Gothic which became unpopular as a style for a convenient dwelling.

On a visit to London, Burn had sought out the antiquarian architect Robert William Billings (1813–74), a pupil of John Britton (1771–1857), and had advanced the considerable sum of £1,000 to finance Billings' *Baronial and Ecclesiastical Antiquities of Scotland* (4 vols), 1845–56, containing 240 illustrations. This publication had a strong influence on the work of David Bryce (1803–76).[23] Burn introduced the Baronial style for the first time at Milton Lockhart, Lanarkshire, 1829–36, for William Lockhart, and remodelled Tyninghame, East Lothian, 1829–30, for the 9th Earl of Haddington.

The other architect in this triumvirate, William Playfair (1790–1857), designed a small number of country houses, but his early work at Grange House, Edinburgh, 1830–31, for Sir Thomas Dick Lauder, and Prestongrange, East Lothian, 1830, for E Grant Suttie, showed a far better understanding of the massing and detailing of old Scottish houses than his contemporaries, due notably to his attitude to his commissions. He undertook only one or two large commissions at once, giving him time to get to the essential details of his design.[24] Billings' publication had a 'new and specifically architectural purpose as a source book and it was soon established as the bible of the Baronial Revival'.[25] This four-volume work became the main source of authentic detail for country house design in Scotland for patrons and architects, and was used extensively by David Bryce. Billings' plates can be shown to have been used by Bryce in a number of his large house commissions. Details from Fyvie Castle, Aberdeenshire, occur at Craigends, Renfrewshire, 1857, and his additions to Blair Castle, Perthshire, 1870–2, and details from Castle Fraser, Aberdeenshire, appear at Castlemilk, Dumfriesshire (fig. 4.6), 1863–4. A fairly repetitive repertoire can be found to have been used, as for the turrets at Glamis Castle, Angus, and Pinkie House, Midlothian, the entrance tower and oriel window at Maybole Castle, Ayrshire, and the round towers at Midmar Castle and Huntly Castle, Aberdeenshire.

Country house building represented a very lucrative market in the mid nineteenth century, the majority of patrons representing the aristocracy and minor landed classes, as well as very wealthy merchants wishing to change their status in the social scale.

The country house almost resembled a village where different social units lived under one roof and the family's privacy was respected and safeguarded from servants and guests. The dining room was normally on the north or east side of the house out of the heat of the morning sun, and the hall or saloon came to reflect the medieval great hall. The children were confined in a nursery or schoolroom wing, and the

Figure 4.7 Cringletie House, Peeblesshire. Crown copyright: RCAHMS, SC 710835.

servants' accommodation was in a wing not visible from the main rooms of the house, where luggage and supplies could be delivered out of sight. David Bryce was the archetypal member of the architectural profession in Scotland, providing his clients with large, lavish country houses. They, mostly the aristocracy, passed on their satisfaction to their wealthy and influential neighbours.[26] The massing of Bryce's houses interacted with the landscape, producing a silhouette which enhanced the view from all directions. Cringletie, Peeblesshire (fig. 4.7), 1861, for James Wolfe Murray, is a perfect example of this visual impact, where from its large windows and corner bays on the principal floor, magnificent views are obtained of the parkland and distant hills. A private way to the garden from the family apartments, by a balustrade and steps, or by a conservatory, became striking features of many Bryce houses. Bryce's plan for the family's comfortable living was produced over and over again, and was carried on by the firms of Peddie & Kinnear and James Maitland Wardrop of Wardrop & Reid, and culminated in the work of Sydney Mitchell & Wilson.

The last decade of the nineteenth century and the first years of the twentieth century saw a revival of the Scots craftsmanship of the sixteenth and seventeenth centuries, not only in the continued use of Baronial styling but in the provision of different coloured stones and slates. James Gillespie prefaced his publication on Scottish domestic architecture with the words 'The problem of how an architect may follow and make use of these early traditions is not so simple, but it is hoped that the spirit of the old work may be caught, and only such features used as can be developed and adapted to modern conditions.'[27] In his publication, *Das Englische Haus*, Berlin, 1904, Hermann Muthesius set out to characterise the exemplary qualities of the English house, and importantly, pointed out that the down-to-earth qualities he so admired were not based on a style but on simple vernacular buildings. This form of architecture became the preoccupation of architects working from about 1880. Their style of architecture came to be called 'Arts and Crafts' after the name given to the Arts and Crafts Exhibition Society in 1888. Plans and elevations became the expression of utility, building materials were taken from their localities so that they did not jar, being in harmony with their surroundings, and details were based on vernacular originals and not taken from classical pattern books. Proprietors and architects became more interested in the crafts, employing plasterers, painters, carvers and sculptors to enrich their buildings with ornamental details based on nature.[28]

The architectural press did much to encourage the vernacular trend. The new magazines, *Academy Architecture*, 1889–1931, London, volumes 1–62, which published designs exhibited in the annual Academy show, *The Studio*, 1893–1963, London, volumes 1–166, *The Architectural Review*, 1896–, London, and *Country Life*, 1897–, London, all contributed to the trend.

Scottish architects slipped easily into the English Arts and Crafts movement, but maintained their own personal style in combining Scottish vernacular motifs with the theories of the movement. Muthesius considered Sir Robert Lorimer (1864–1929), as 'the first to recognise the charm of the unpretentious old Scottish buildings, with their honest plainness and simple, almost rugged massiveness'.[29]

Lorimer's first restoration was at Earlshall, Fife, for R W Mackenzie, 1892, with the conservation of the sixteenth-century tower-house and layout of a new garden, garden stores and gate lodge (see fig. 10.9). Virtually all his country house practice was the mending of old fortalices to provide country houses as the centres of substantial landed estates with lodges, outbuildings and gardens. Balmanno, Perthshire, 1916, for W S Miller, a Glasgow shipping magnate, was remodelled with a new garden belvedere, gardener's storehouse, and gate lodge. A simple restoration of Bavelaw Castle, Midlothian, 1900, improved circulation on the attic floor, added dormers, reconstructed the outhouses, and joined them to the main block with a balustraded scheme. He carried out a similar idea on a much larger scale at Dunderave Castle, Argyll, 1911, for Sir Andrew Noble. His first commission for a house on a new site was at Rowallan, Ayrshire, 1901, for A Cameron Corbett, where he produced a large Scots Baronial mansion which was never finished nor used for the lavish house parties for which it was intended. Ardkinglas, Argyll, 1906–8, also for Sir Andrew Noble, is another massive Scots Baronial pile on the shores of Loch Fyne. Rhu-na-Haven, Aberdeenshire, 1907, for J Herbert Taylor, was designed with curvilinear gables and built of silvery grey granite. For John A Holmes, stockbroker and art collector, Lorimer built Formakin, Renfrewshire, 1908, to house Mr Holmes' art treasures, but again the money ran out before this spacious Scots mansion, with a lodge and landscaped grounds, was finished.

By the late nineteenth century the Romantic outlook had changed and landed property was being acquired increasingly for sporting activities and smart country house entertaining.[30] Melsetter House, Orkney, 1898, for Thomas Middlemore, a manufacturer of bicycle seats from Birmingham, was designed by W R Lethaby (1857–1931) and built around an existing farmhouse according to SPAB (Society for the Protection of Ancient Buildings) principles.[31] Kincardine House, Aberdeenshire, 1895–6, for Mrs Pickering by D B Niven (1864–1942) and H H Wigglesworth (1866–1949), Kinloch Castle, Rhum, 1897/1906, for

Figure 4.8 Mar Lodge, Aberdeenshire, 1880s. Crown copyright: RCAHMS, SC 711076.

Figure 4.9 Gribloch House, Stirlingshire. Crown copyright: RCAHMS, SC 710849.

Sir George Bullough, by Leeming & Leeming of London (1872–1931), Mar Lodge, Aberdeenshire (fig. 4.8), 1896, for the Duke of Fife, a shooting lodge by A Marshall Mackenzie (1847–1933), and Kildonan, Ayrshire, 1914–18, for Captain Ewan Walker by James Miller (1860–1947), an immense house in the English Manorial style (which was never finished, being on too vast a scale for the financial output of the owner), all evoke a ritzy way of life. Manderston, Berwickshire, 1895–1905, designed by John Kinross (1855–1931), for Sir James Miller, is a classical house incorporating an eighteenth-century core, and is considered to be the epitome of the social country house at its best. Aultmore House, Inverness-shire, 1913, for H Miller, the owner of a Moscow department store, was designed by C H B Quennell (1872–1935) and is in the Palladian revival style, with a Scots seventeenth-century forecourt. It incorporated much new technology, including a centralised vacuum cleaner.[32]

After the 1914–18 war there was an absence of wealthy patrons necessary for an architectural renaissance, and only a handful of country houses with offices and lodges were built. Cour, Argyll, 1920, for J B Bray by Oliver Hill (1887–1962), is a fortalice type with slit windows and battered walls. Gribloch, Stirlingshire (fig. 4.9), 1937–41, for Sir John Colville, the steel magnate, by Sir Basil Spence (1907–76), is very important as an example of a house in the 1930s mode, a style rare in Scotland. With Spence's design of a traditional L-plan seventeenth-century tower house at Broughton, Peeblesshire, in 1938, for Professor and Mrs Elliot, the great days of the substantial country house were over.

NOTES

1. Smith, 1970.
2. Harris, 1985, 15–27.
3. Andrea Palladio (1508–87). An Italian, he was the first important professional architect. He revived Roman symmetrical planning and evolved a formula for the ideal villa, of a symmetrically planned central block, decorated with a portico, lengthened horizontally or curved forward in quadrants to wings containing farm buildings, thus linking the house to the landscape. Translations of his publication *I Quattro Libri dell'Architettura*, Venice, 1570, were acquired by many intellectual and travelled Scots.
4. Colvin, 1986, 169–82.
5. Dunbar, 1970, 2.
6. Gifford, 1988, 84.
7. Macky, 1723, 159.
8. Harris, 1985, 15–27.
9. ibid.
10. Colvin, 1978, 755–58.
11. ibid, 56–9.
12. A window or door with a triple keystoned head beneath a cornice, and blocks of stone punctuating the jambs. Named after the architect James Gibbs (1682–1754).
13. Harris, 1990, 208–13.
14. ibid, 245–47.
15. Macaulay, 1987, 165–73.
16. Rowan, 1985, 17.
17. Brown, 1992.
18. Macaulay, 1975, 229–52.
19. Colvin, 1978, 742.
20. ibid, 741.
21. Walker, 1976.
22. Colvin, 1978, 115.
23. Walker, 1976, 30.
24. Colvin, 1978, 645.
25. Gow, 1992, 59.
26. Rowan, 1976.
27. Gillespie, 1922, VIII.
28. Richardson, 1983, 7–9.
29. Muthesius, 1904, 1979, 621.
30. Aslet, 1982, 79.
31. Society for the Protection of Ancient Buildings, founded by William Morris in 1877.
32. Aslet, 1982, 310.

BIBLIOGRAPHY

Adam, W. *Vitruvius Scoticus*, Edinburgh, c. 1812, facsimile edn, 1980.
Aslet, C. *The Last Country Houses*, London, 1982.
Brown, I G. *Monumental Reputation: Robert Adam and the Emperor's Palace*, Edinburgh, 1992.

Colvin, H M. *Biographical Dictionary of British Architects, 1600–1840*, London, 2nd edn, 1978.
Colvin, H M. The beginnings of the architectural practice in Scotland, *Architectural History*, 29 (1986), 169–82.
Devine, T M, ed. *Lairds and Improvement in the Scotland of the Enlightenment: The proceedings of the Ninth Scottish Historical Conference, University of Edinburgh, 1978*, Glasgow, 1979.
Dunbar, J G. *Sir William Bruce, 1630–1710* (Exhibition Catalogue), Edinburgh, 1970.
Dunbar, J G. *The Historic Architecture of Scotland*, London, 1966 (2nd edn, 1978).
Ferguson, W. *Scotland: 1689 to the Present*, Edinburgh and London, 1968.
Fiddes, V, Rowan, A, eds. *Mr David Bryce, 1803–1876*, Edinburgh, 1976.
Gifford, J. *Buildings of Scotland: Fife*, Harmondsworth, 1988.
Gifford, J. *Buildings of Scotland: Highlands and Islands*, Harmondsworth, 1992.
Gifford, J. *William Adam, 1689–1748*, Edinburgh, 1989.
Gillespie, J. *Details of Scottish Domestic Architecture*, Edinburgh, 1922.
Gow, I. *The Scottish Interior*, Edinburgh, 1992.
Harris, E. *British Architectural Books and Writers, 1556–1785*, Cambridge, 1990.
Harris, J. *The Design of the English Country House, 1620–1920*, London, 1985.
Harvie, C. *No Gods and Precious Few Heroes: Scotland, 1914–1980*, London, 1981.
King, D. *The Complete Works of Robert and James Adam*, Oxford, 1991.
Lenman, B. *Integration, Enlightenment and Industrialisation in Scotland: 1746–1832*, London, 1981.
Macaulay, J. *The Gothic Revival, 1745–1845*, Glasgow and London, 1975.
Macaulay, J. *The Classical Country House in Scotland, 1660–1800*, London, 1987.
Macky, J. *A Journey in Scotland*, 3 vols, London, 1723.
McKean, C. *The Scottish Thirties*, Edinburgh, 1987.
McWilliam, C. *Buildings of Scotland: Lothian*, Harmondsworth, 1978.
Mitchison, R. *Lordship and Patronage in Scotland, 1603–1745*, London, 1983.
Muthesius, H. *Das Englische Haus* (*The English House*), Berlin, 1904 (English edn, London, 1979).
Palladio, Andrea. *I Quattro Libri dell'Architettura*, Venice, 1570, London, 1738 edn, reprinted as *The Four Books of Architecture*, New York and London, 1965.
Philipson, N T, Mitchison, R, eds. *Scotland in the Age of Improvement*, Edinburgh, 1970.
Richardson, M. *Architects of the Arts and Crafts Movement*, London, 1983.
Rowan, A. David Bryce. In Fiddes, Rowan, 1976, 11–22.
Rowan, A. *Designs for Castles and Country Villas by Robert and James Adam*, Oxford, 1985.
Savage, P. *Lorimer and the Edinburgh Craft Designers*, Edinburgh, 1980.
Smith, J A. Some eighteenth-century ideas of Scotland. In Philipson, Mitchison, 1970, 107–24.
Walker, D. Burn and Bryce. In Fiddes, Rowan, 1976, 23–30.

Figure 5.1 Haddington House, Sidegate, Haddington, East Lothian; a seventeenth-century urban L-plan house with stair in re-entrant angle, typical of domestic urban and rural work of this scale and period. Copyright: Deborah Mays.

5 Middle-Sized Detached Houses

DEBORAH MAYS

Domestic architecture in modern Scotland has enjoyed two boom periods, from 1770 to 1840 and from 1870 to 1905. The proportion of moderate-sized houses within the total number in each decade reflects changing social, industrial and regional patterns, changes which were mirrored in the form and function of designs. John G Dunbar's pioneering monograph, *The Historic Architecture of Scotland*, provides an invaluable overview of this class of building but it is intended here to develop and extend its appraisal.[1]

The first appearance of houses which can usefully be grouped as middle-sized detached came in the aftermath of the Reformation in the late sixteenth century. The numbers increased very gradually as the polarisation of society became less extreme in the wake of two principal changes: the secularisation of ecclesiastical holdings which saw a greater division of property; and the creation of feu-farm holdings by the greater landlords. A new residential building type appeared in the landscape, the homes of the gentry. In the first phase, the form of these houses was semi-defensive, as at Pilmuir in East Lothian, 1624, and evidenced the evolution from the tower-house; some were vaulted at ground floor, with gunloops, and the principal residential areas were on the upper floors.

Later in the seventeenth century, the form evolved a still more horizontal emphasis, and any tendency to embellish with pepperpot angle turrets disappeared. In these years the house often followed an L-plan, as at Ford House, Midlothian, 1680, and Auchenbowie, Stirlingshire, with the stair tower set conveniently and economically in the re-entrant angle with the entrance at the foot (fig. 5.1). It was more usual at this date for the main stair to be separated, only diverting at the first floor to a newel form enclosed in a tower. A marriage lintel, comprising the initials of the laird and his wife, occasionally appeared above the doorway as a decorative detail and was frequently used in the following century. Carved ornament was common on the seventeenth-century house. Roof pitches were steep, the gables often crowstepped with small attic windows or apex stacks, as can be seen at the Old Manse, Cromarty, c. 1690, by John Laing. Lingering references to defensive details, found in the early period, had petered out by the Restoration in 1660. Internally,

Figure 5.2 Leaston House, Humbie, East Lothian; an early eighteenth-century two-storeyed, five-bay house (with early nineteenth-century wings) which, typical of contemporary practice, mixes simple classical symmetry with traditional Scottish elements. Copyright: Historic Scotland.

the laird's house often boasted fine decoration, with painted ceilings and lively plasterwork, as at Sparrow Castle, Cockburnspath, Berwickshire, for example, and Fountainhall, Pencaitland, East Lothian.

Towards the end of the seventeenth century, the pattern emerging was for gabled, rectangular-plan, two-storeyed, symmetrical houses (fig. 5.2), a formula which would prevail. These were single-pile (one room deep) in the early years, with kitchen and parlour at ground level flanking a central stair sited directly by the entrance, or slightly to one side, and evolved to double-pile later in the century. Blairhall, by Culross, Fife, probably by Robert Bruce, is one of the earliest to demonstrate these classical proportions, displaying a five-bay form with moulded doorpiece. John Reid, author of the influential *The Scots Gard'ner*, 1683, commended this orderly form as 'but little, yet very commodious & cheap'.[2] The bays were generally grouped towards the centre, with the lintels of the first-floor windows at the level of the wallhead. Economy of space and building costs prompted the arrival of projecting staircases in the eighteenth century, in conjunction with a single-pile plan as at Whitelaw, Whittingehame, East Lothian, 1728. Many of the wealthier houses possessed a flagstone basement, and most included an attic lit by small gablehead windows. Glen, Kirkcudbrightshire, 1734, was one of the first to display rusticated quoins and a moulded eaves course. The double-pile house first appeared from about this date, still in three sections with stair at centre, predominantly with stacks at gablehead rather than at wallhead, or above central flue walls with a piended roof. Roofing materials varied according to the locality: pantiles were rarely used for domestic housing of any size until the late nineteenth century when they became widespread on the east coast, being employed earlier only in exceptional instances.[3] Orkney was one of the first to adopt slate on any scale owing to its easy local availability.

The urban terraced house developed from the sixteenth century when traditional linear town plans began to evolve, taking the form of irregular terraces fronting broad principal streets with cobbled closes leading to the back 'tenements' behind. The pattern with rigs running off can still be seen clearly in Dunbar, East Lothian, Stirling, Edinburgh and St Andrews, Fife.[4] Stone construction began on a wider scale, replacing lesser timber residences after the Reformation. It was commonly rubble in form, harled and limewashed, with exposed ashlar for ornamental detail. In Elgin and Forres, Moray, a distinctive pattern emerged with arcades to the street, but more common was the inclusion of a timber balcony at first floor, accessed from a turnpike stair and sheltering shops at ground level. These galleried balconies provided the occupant with an uncluttered forecourt, free of the peat stacks and dunghills which apparently covered much of the unpaved streets below. Glass was still an expensive luxury and the openings were part-shuttered, a form reinstated in The National Trust for Scotland properties at Culross. The semi-circular stair, projecting either onto the street or more commonly to the rear, with conical or swept pentice roof, was standard. Closes were often within a tenement, a door or gate giving onto the street. The most luxurious urban properties were on the courtyard plan, paralleling those of French 'hôtels', Argyll's Lodging in Stirling, predominantly 1632–40 and 1674, being the best known of these. Less prominent properties often presented a gable, or gables, to the street, almost always crowstepped, or featuring a tympany gable, as at Gladstone's Land, Edinburgh, and Hamilton's Land, Linlithgow, West Lothian. Nisbet of Dirleton's House, Canongate, Edinburgh, 1624 (replicated 1953), boasts a square angle projecting at the second floor, harking back to the bartizan form of fortified dwellings and indicating how gradual was the evolution even in the urban context. In time, balconies became enclosed to make provision for additional accommodation within the burgeoning burghs, and later their timber form gave way gradually to stone construction, as at 'John Knox's House' in Edinburgh's Royal Mile. Allan Ramsay's House (before alteration) was a fine example of a form common in Edinburgh, with timber street elevations distinctive of its type. The L-plan house seen in rural areas, particularly during the seventeenth and early eighteenth centuries, also found a place in the towns, as at Hamilton House, Prestonpans, East Lothian, 1626, Plewlands House, South Queensferry, West Lothian, 1641, and in the area of Edinburgh's Canongate. The interior decoration of urban houses echoed that of their rural counterparts, often displaying finely panelled interiors with painted ceilings and plasterwork ornamentation. A modest example of such a mid seventeenth-century ceiling can be seen in the Merchant's House, 49 South Street, St Andrews, another at 341 High Street, by the harbour in Kirkcaldy, Fife, formerly disguised by later ornate plasterwork.

Lowland lairds and Highland tacksmen were not the only people to commission middle-sized houses outside the burghs. The great landlords required dower houses and homes for their factors, and each parish required to house its minister. After 1663, for example, the heritors of each Highland parish were formally required to provide a 'competent' manse, from within a certain budget.[5] From the middle of the eighteenth century the Agricultural Revolution opened up the social structure still further and introduced a new generation of landholders, traditional and conservative in their taste. Mill houses and farmhouses followed the fashionable symmetrical two-storeyed pattern but with regional variations, particularly in building materials and scale: heather thatch, for example, was an indicator of the wealthy status of the owner.[6] The *Statistical Account* reported in 1798 that 'within the last ten or twelve years most of the farmhouses have been rebuilt with considerable improvement both in point of size and accommodation'.[7] The farmhouses of the north east remained single-storeyed, often with a deep attic breaking into the wallhead. In the agriculturally wealthy south east, farmhouses of three storeys had appeared by the 1790s, with between four and eight rooms. In the early years of Improved farming, the farmhouse tended to face south onto the steading, with a barn in a flanking range running at right angles. Fenton and Walker in *The Rural Architecture of Scotland* explain that, by 1802, the best farmhouses were similar to those of the clergy, but less lavishly appointed and on a smaller scale. Box beds were a common means of dividing internal space and floors were earthen, or of timber or flagstone. The island areas were less advanced with unsophisticated, two-part houses, divided by a hallan screen, surviving into the nineteenth century. The Grieve's House, East Barns, Dunbar, *c.* 1800, is a good example of the standard pattern for its area. The plan comprised a small lobby entered from a central door leading directly to a stair, and with kitchen parlour to one side (probably subdivided to accommodate a servant's bed and scullery), and living room to the other side. The first floor contained four bedrooms. Often attics were divided into two, lit by small gablehead windows, one half to store firewood, the other for extra servants, and known aptly as the barracks. Many farmhouses contained their kitchen offices in single-storeyed wings to the rear, occasionally two-storeyed with a nursery and woman servant's room above.

The first boom period for residential building in modern Scotland, from the late eighteenth century, was well supplied with pattern books to guide and advise developer and craftsman alike. None was more influential than the *Rudiments of Architecture*, issued in several reprints and three editions from 1772, comprising an expanded version of George Jameson's *Thirty-five Designs with the Orders of Architecture*.[8] The introduction to the 1992 reprint explains that by 1770, urban renewal and Agricultural Improvement were advancing at such a rate that the new mechanics, gentlemen operatives and land surveyors needed more constructive and more extensive assistance than Jameson's pioneering publication could provide. The universal popularity and dependence on the *Rudiments* during its seventy years in circulation has been obscured by its almost total disappearance. However, study of the text and an examination of the traditional building of the period provides evidence of its widespread influence. Just as Reid delighted in the rectangular box a century earlier, the *Rudiments* popularised its adaptable form, illustrating its versatility in a large number of dignified, often classically-inspired designs for middle-sized houses, predominantly with single-storeyed service wings. Designs X, XI, XIX, XX, XXI (fig. 5.3) and XVI were perhaps the most widely adopted and adapted of all those featured. Designs X and XIX served as the standard pattern for manses and farmhouses and early suburban villas, many town houses and village terraces, at least until the middle of the nineteenth century.[9] Design XX possibly served as the source for such houses as Beech House, Ormiston, East Lothian, minus pavilions, and St Ann's, Cromarty, at a slightly later date.

Enlightenment ideas spread across the country, disseminated in the more remote areas by the various agricultural societies and their journals, as well as by visits to the cities, and gave rise to such gentleman-architect designs as Charles Ritchie's Bielgrange, Stenton, East Lothian, 1803. Craftsmen's manuals and builders' companions were published in response to the demand: Peter Nicholson's numerous publications, for example, were leading texts in the field, aimed at assisting architect and craftsman, and mostly date to the early nineteenth century.

Representative examples of good eighteenth-century burgh architecture can be seen at Cromarty, Whithorn, Wigtownshire, Banff, Falkland, Fife, and Haddington, East Lothian. In Inveraray, Argyll, Robert Mylne's designs for the west side of the principal street, 1774–5, provided an impressive example for contemporaries in the stout, plain dignity of the five terraced, three-storeyed, three-bay houses.[10] William and John Adam introduced a new classical language which remained the norm

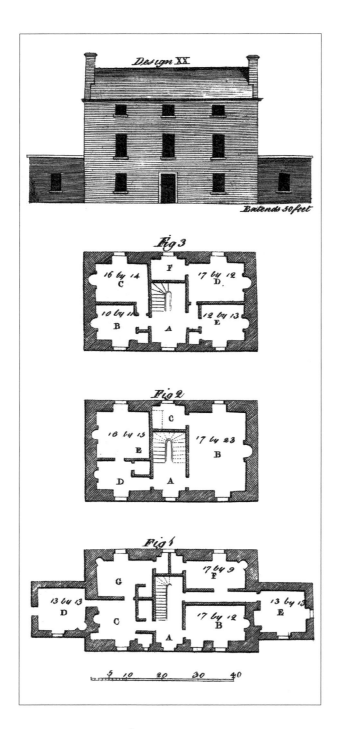

Figure 5.3
Rudiments of Architecture, 1778; designs XX and XXI.

through several generations of Scottish building, until the close of the eighteenth century. Town houses were perhaps particularly susceptible to its superior pretensions, such as Lord Milton's House in the Canongate, Edinburgh, 1755–8, an unassuming classical box designed by John Adam, now subsumed in later work. In Glasgow, the houses in Charlotte Street followed the classical form, though only No 52 remains to give evidence. Gillies's House, Inveraray, 1759–60, also believed to have been designed by John Adam, was typical of the restrained provincial architecture of the period, distinguished from its neighbours in being raised over a cellar, with keystoned round-arched door and long and short quoins.[11] The refined and exceptional eighteenth-century suburb of Inveresk, Midlothian, presented a sampler of classical designs for middle-sized houses. Shepherd House illustrated a modest variant, with the traditional form enlivened by curvilinear gableheads. Catherine Lodge offered a fine example of the restrained use of classicism and, in contrast, Manor House displayed a rich pride in the repertoire of classical details. By the early nineteenth century, provincial architects had mastered the style, informed by pattern books and education, and many fine, classical villas date from this period, more correct in their proportions and detail. The building trade experienced a revolution as the local mason styled himself 'architect-mason' and later 'architect', with a grasp of classical composition over-riding knowledge of local traditions and skills. Robert Balfour's work in St Andrews gives testimony to this, such as the elegant house at 91 North Street, *c.* 1812. James Burn of Haddington was another to produce a distinctive style early in the century: his trademark, the use of broad angle pilasters, suggests that he may have designed Yetholm Hall, Town Yetholm, Roxburghshire.

The Edinburgh New Town, in its various stages of development and following the example of work by the Woods in Bath, introduced a form of urban living of extensive and consistent grandeur new to Scotland.[12] The standard classical pattern adopted at first comprised three-storeyed, three-bay terraced properties, notably in Queen Street, grouped in pairs, sharing a flue-bearing party wall, with rusticated basements, the road built up to include cellarage. Formality became the norm for terraces on level sites after Robert Adam's design for Charlotte Square in 1791, unified palatial façades distinguishing the terraces, occasionally with flatted accommodation in the taller centre and end pavilions with classical attics, or else entirely in terraced tenemental form, as in Castle Street. The palatial formula was employed elegantly in the Grecian design for Ann Street, though here a stepped frontage and urban front gardens set the terrace apart, following a pattern favoured in the second decade of the nineteenth century. William Playfair's designs for terraces on Calton Hill and Hillside broke away from the palatial composition, removing the focal emphases. The arrangement of interiors

Figure 5.4 Royal Circus, Edinburgh. Copyright: Deborah Mays.

generally comprised a top-lit stair hall with dining room, parlour and study on the ground floor, cantilevered stone steps leading to the first-floor drawing room, which usually spanned the width of the property, and the second floor containing the bedrooms. Fine railings lined the basement recesses and oversailing steps. Heriot Row was one of the grandest terraces, overlooking Queen Street Gardens, with four variations evident in the design. The later developments moved away from the grid plan of the first New Town. The magnificent layout from Randolph Cliff to Moray Place, for example (fig. 5.4), containing the longest, unbroken terrace in Europe, with crescent and circus in its course, was designed by several architects, notably James Gillespie Graham.

Glasgow's expansion after 1750 equalled that of Edinburgh in scale, new residential suburbs springing up in response to the population boom. These were mostly the result of private, sporadic speculative development, one of the first being the residential western suburb of the tobacco barons, the Merchant City.[13] Only after 1800 did Glasgow follow the lead of Edinburgh and begin to develop its own classical residential developments, the first taste of which was the palatial terraces at Carlton Place by Peter Nicholson, 1802–18. The development of Blythswood Hill followed, to the west of the Merchant City. Blythswood New Town comprised terraces of individual three-storeyed and basement houses, distinct from their east-coast counterparts, not least by the rarity of pediments and pilasters and the heaviness of their cornices. The later developments westward, as in Edinburgh, became less formal and more varied, such as Charles Wilson's Italianate Kirklee Terrace, begun 1845, and John Bryce's Queen's Crescent, 1840. One of the most lavish residential developments was that of Charles Wilson's Park area, with views across Kelvingrove Park. In the latter half of the nineteenth century, the Glasgow terraces reached new heights of grandeur, embodied in masterpieces such as J T Rochead's Grosvenor Terrace, 1855, Alexander Thomson's Great Western Terrace, 1869, and John James Burnet's University Gardens, 1882–4.

Many existing terraced houses in provincial burghs were given a face-lift in the nineteenth century by the addition of a classical doorpiece, sometimes endearingly out of scale with the composition of the rest of the façade, the ashlar contrasting with the rubble walls.[14] Advances in the provision of glazing allowed more generous fenestration, and many windows were enlarged. The feeling against thatch as a roofing material had been growing from the late seventeenth century (thatch presented too great a fire risk and was a harbour for vermin) and it was increasing replaced with slate or pantiles depending on the wealth of the owner or the locality. In the country, existing farmhouses in wealthier areas were often enlarged with the addition of 'drawing room wings' and their windows opened up, frequently to bipartite or tripartite form. New farmhouses from the early to mid nineteenth century were set at a distance from the steading, and the old were occasionally incorporated into the farm buildings for storage, as at The Brunt, Innerwick, East Lothian.

Thomas Telford, the engineer responsible for numerous masterpieces in bridge design across the country, was the architect of Parliamentary church manses which have become distinctive components of the Highland landscape. The manses, built over three decades from 1825, followed a choice of two patterns: the two-storeyed, T-plan, and the single-storeyed H-plan, and in some of the more remote parishes these were second only in size to the laird's house.[15]

Influenced by the formal layout of the Edinburgh and Glasgow New Towns, several provincial burghs tailored their residential expansion along similar, classical lines. Sir John Sinclair of Ulbster, from 1802, provided Thurso, Caithness, with a fashionable new town, where the homes of the wealthy merchants were sited peripherally, overlooking the river, in the smart, classically detailed villas of Janet Street. Barossa Place in Perth provided a development of equal elegance, and in the same city Rose Terrace boasted a fine development with the Academy as its centrepiece. In St Andrews, the classical terrace found expression in George Rae's Playfair Terrace, 1846–52, and John Chesser's Abbotsford Crescent, 1865, the latter making direct reference to Edinburgh's New Town.

The evolution of the cottage mansion in the late eighteenth century, part of the search for the ideal landscape with rustic residences, saw the creation of such fine designs as Lasswade Cottage, Lasswade, Midlothian, and Annat House, Rait, Perthshire (demolished), both with generous bowed windows and on a scale quite at odds with the humble thatch, or Connachan Lodge, Crieff, Perthshire, endowed with elaborate timberwork. They tended to be built on the outskirts of polite burghs, or large estates as ornate dower houses.

Suburbs came into existence gradually from the late eighteenth century. John Dunbar has noted that the earliest villas (more appropriately termed villa-mansions, such as those at Inveresk, or Gayfield House and Hawkhill, both in Edinburgh, and Ednam House, Kelso, Roxburghshire) were indistinguishable from urban mansions of the same date.[16] From the 1820s, however, suburbs with a more unified layout and house type began to appear, conforming to a street pattern embroidered with garden walls. Among the first suburbs to emerge on the new pattern was Allan Park, Stirling, where Alexander Bowie and Thomas Traquair's 1–9 Wellington Place set the scene for classically detailed, dignified design.

In Edinburgh, the Newington, Grange, Bruntsfield, Inverleith and Trinity suburbs were developed from the 1820s, and in Glasgow areas such as Partickhill, Dowanhill and most of Pollokshields were also developed at the same time. The jute barons of Dundee commissioned palatial villa-mansions, such as those by David Neave in the Magdalen Yard area around 1820, and George Angus designed the Dudhope development. There were major jute baron houses in the east end around Eden Street and Springhill but these have been swept away.[17] The standard design in suburbs such as Edinburgh's Grange was for the late Georgian box, later accompanied by French-inspired and Italianate designs. The latter was perfected for houses in the fashionable resorts which grew up along the River Clyde, such as Thomson's Craig Ailey, Kilcreggan, Dunbartonshire, 1850, or John Gordon's Oakleigh Villa, Loch Long, Dunbartonshire. Playfair had mastered the style for suburban villas such as Dumphail, Moray, 1828–9, and Belmont, Corstorphine, Edinburgh, 1828–39. Castellated designs which were in fashion after 1820, and Tudor and Jacobean which flourished from mid century, were more suited to more rural suburbs or uniform estates (as demonstrated by William Burn at Biel, Stenton, East Lothian). A brand of Jacobean cottage houses appeared, again the work of Burn, the successors of the late eighteenth-century cottage ornée in seeking a smaller and more comfortable scale of living. John White, writing in 1845, commented wisely that 'Grecian villas are well adapted for suburban situations where the surrounding scenery is generally in unison with the style of architecture'.[18] Canted windows became a favoured component in designs from the mid century, occasionally with some chip-carved detail, and with good guttering and brattishing in the more exclusive areas.

Railway dormitories sprang up at short distances from the nation's burghs and cities, generally weekend settlements made possible by rail travel from the 1860s, for example, Birnam, Perthshire, where the villas at Torr Hill constitute a suburban village, or Pitlochry, Perthshire, Bridge of Allan, Stirlingshire, and the more suburban Broughty Ferry, Angus, where an earlier fashionable resort was expanded after the arrival of the railway station in 1848. Alexander Thomson's Greek Revival house, Holmwood, Renfrew, 1857, can be cited as a grander example of the villas which resulted, as can later, similarly innovative masterpieces such as Mackintosh's Hill House, Helensburgh, Dunbartonshire, for the Glasgow publisher, Blackie, 1902.

Hand-in-hand with improved transport, technological advances in the nineteenth century widened the realms of possibility in the design field. Study travel became a crucial component of architectural training, a source for design inspiration. Pattern books gained a stronger hold as the century marched on, and, together with the growing number of architectural publications and periodicals, enabled stylistic innovations to spread rapidly. John Claudius Loudon, a Scot born in Gogar, West Lothian, produced a highly influential, comprehensive tome, the *Encyclopaedia of Cottage, Farm and Villa Architecture* in 1833, with several subsequent editions.[19] Published when the wealth of revival styles in vogue ranged from Greek, Gothic, and Baronial, to Tudor and 'Jacobethan', the text was warmly received. Many of the examples came from leading architects of the day, such as David Cousin, who produced a villa in the style of a 'Scottish Manor House'.[20] Loudon urged that revival styles were not to replicate the form they were reviving, which he considered to be low art, as portrait painting is to historical painting. Provoked by a design for a 'Jacobethan' villa at Springfield, Fife, by Mr Cleland (probably James Cleland, 1770–1840), he declared that:

> ... in imitating any style we are not limited to copying particular forms; but are required to enter into the spirit of the subject or style to which they belong, and to form a new composition in that spirit, adapted to whatever use it may be required for ... Any builder may copy a style but it requires an architect to compose in it.[21]

Among the wide variety of designs included in the *Encyclopaedia*, one in particular must be singled out for its avant-garde form and subsequent popularity, the type known collectively as 'art gallery houses'.[22] The prototype illustrated was Walter Newall's Hannayfield, Dumfries, completed by 1831, a single-storeyed and basement rectangular-plan house built on falling ground and sited for the view. The use of classical details was restrained and the interior distinguished by a 'high style of finish'. The best-known examples of this type of house in Scotland are The Vine, Dundee, 1836 (fig. 18.5), and Arthur Lodge, Edinburgh, 1828, both in a Greek Revival style, but there are many others such as Brunton House near Cullen, Banffshire, Innerwick Manse, Innerwick, Westfield House, Haddington, and The Brae, Jedburgh, Roxburghshire.

Certain pattern books exercised a primarily local influence, such as William J Gray's *Rural Architecture*, 1832, which focused entirely on farmhouses and their offices in the Borders, such as Turtleton Farmhouse, Duns, Berwickshire, which he explained was 'in the style of rural architectural improvement now inspired after', that is, embellished with Tudor details, and costing a moderate £392.[23] Other texts appealed to a wealthier urban clientèle, notably John Starforth's *Villa Residences*, 1866, which included several Baronial designs and confirmed the high fashion for the style.[24] The increasingly eclectic form of many designs in currency in these years should not be underestimated. Blackie's *Villa and Cottage Architecture* is a compendium of diverse designs by some of the leading architects of the day: the villa at Grange, 44 Dick Place, Edinburgh, by R Thornton Shiells, was one of the more modest, displaying favoured period details, namely decorative fish-scale slates in contrasting bands, colonnette mullions, gutter brackets, decorative iron finials and brattishing.[25] John James Stevenson's two-volume *House Architecture*, 1880, served as a watershed, closing the High Victorian years and opening the boom period from 1880–1905.[26] Stevenson advocated comfort realised on a more compact scale: he suggests economies in design and practical solutions for every aspect of house construction.

Among the most lavish suburban villas of the second half of the nineteenth century were Avallon, Churchill, Edinburgh, probably by David Bryce, c. 1862, in full-blown Baronial, John Honeyman's Craigie Hall in Bellahouston, Glasgow, 1872, and Knox & Hutton's Linsandel House, Eskbank, Dalkeith, Midlothian, 1884, all Italianate with Greek details. James Boucher's palatial 998 Great Western Road, Glasgow, was perhaps the grandest of them all. The repertoire of styles was wide and the designs invariably eclectic. Romantic compositions by Frederick Pilkington, such as the 'Rogue Gothic' villas in Dick Place, Edinburgh, and by James Gowans, at Lammerburn, Napier Road, Edinburgh, can be compared with the Greek Revival works of Alexander Thomson on the west coast, among which is Tor House, Rothesay, Bute, 1855, and with J B Pirie's imaginative home at 50 Queen's Road, Aberdeen, 1886. William Leiper's free Baronial and French-inspired designs in Helensburgh, such as Dalmore, 1873, and Cairndhu, 1871, encapsulate the bold scale and powerful composition of High Victorian suburban design at its most confident. The period penchant for romantic Baronial (popularised by Sir Walter Scott and Queen Victoria) can be seen to full advantage in St Andrews, where George Rae's Edgelcliffe, 1864–6, David Bryce's Castlecliffe, 1869, and John Milne's Wardlaw Hall, 1865–8, illustrate the style's expressive powers. Villas by the Bryce-trained James Maclaren in Broughty Ferry followed the extensive form.

Changes in social structure, which led to reduced numbers of servants, necessitated increasingly compact and convenient residential design by the closing decades of the nineteenth century and account in part for the growth of such speculative developments as Fryers & Penmans' red sandstone 'villadoms' in Pollokshields, Glasgow, and the work of others in the southern acres of Morningside, Edinburgh. A reaction to the eclectic fussiness and 'dishonesty' of high Victorian revival design was underway. The Arts and Crafts Movement was born of this reaction, and from concerns that the prevalence of machine-made, mass-produced designs was stifling the crafts. The majority of patrons came from the swelling ranks of the middle classes.

The informal, cottagey style preferred by many of the rising middle classes saw the creation of further new suburbs, slightly removed from the urban centre and supplied by a rail link, assisted by the Cheap Trains Act of 1883. These were influenced directly by the work of English architects, their designers having worked in offices in the south, for example, George Washington Browne in the office of William Eden Nesfield, and William Laidlaw Carruthers with Ernest George. The Braid Estate, Edinburgh, where Wardrop, Anderson & Browne were feuing architects, illustrates this well, as here the first Queen Anne-style, Arts and Crafts artistic suburb appeared in Scotland, with sandstone terraces of two-bay houses on a more intimate scale. Robert Lorimer's rustic cottages in Colinton, Edinburgh, inspired by the works of James MacLaren at Fortingall, Perthshire, embodied the fashionable spirit of vernacular revival favoured by the Arts and Crafts protagonists, if on the smallest scale. Colinton is a remarkable Arts and Crafts suburb, but other towns acquired their own pockets in the style. William Laidlaw Carruthers provided the district of Inverness with a spattering of Arts and Crafts-inspired designs, including his own house, Lethington, 13 Annfield Road, 1892, with an octagonal stair tower in the re-entrant angle. The semi-rural situation of the turn-of-the-century suburb suited the cottage garden, the appropriate complement to the Arts and Crafts house. Croquet lawns, summerhouses and tennis courts graced the grounds of villas, serving leisure pursuits in the Indian summers of the Edwardian years. American influence was also apparent in the late nineteenth-century suburbs; the ideas of H H Richardson can be seen, for example, in John Murray Robertson's The Cottage, Lochee, Dundee, 1880, and the Bughties, Broughty Ferry, 1884. Similarly the shingle style was apparent in J J Burnet's Corrienessan, Loch Ard, Stirlingshire, 1887, with verandah-clad elevations.

Various publications served and responded to the changing fashions, such as W Shaw Sparrow's *The British Home of Today*, 1904, and *The Modern Home*, 1907. One of the most interesting publications, illustrating the variety of middle-sized detached housing in these years, was James Nicoll's *Domestic Architecture in Scotland*, 1908.[27] He explained how the range of available building materials had extended considerably, encouraging the use of colour on a wider scale, for example in red tiled roofs and tile-hanging. He bemoaned the arrival of the speculative developer and noted the introduction of casement windows in increasing numbers in place of the traditional sash and case, but he conceded that improved planning had come in response to 'the ever present servant difficulty', bringing the luxury of washbasins to each principal bedroom.[28]

The Scots Renaissance Revival pioneered by Sir Robert Rowand Anderson, John Kinross, A N Paterson (figs 5.5, 5.6) and others, married well with the Arts and Crafts Movement. Kinross's designs for four villas in Mortonhall Road, Edinburgh, 1898–9, and Paterson's Longcroft, Helensburgh, 1902, illustrate the success of this merger. Smaller in scale than the High Victorian villas, with lavishly crafted interiors, low ceilings and smaller rooms, they are perfect case studies for the design process of the period, each architect involving significant numbers of top craftsmen to realise their work. The Scottish revival, in on-going revision, continued to find a place for the middle-sized residence after World War I: the traditional vernacular villas in Drumbathie Road, Airdrie,

Figure 5.5 Lagarie, Rhu, Dunbartonshire; design by A N Paterson. *Academy Architecture*, 21 (1902, part 1), 100.

Lanarkshire, for example, designed by J M Arthur and built in the 1920s and 1930s, and James Bow Dunn's The Bield, Elgin, 1930. The Dundee area was adorned with villas by Thoms & Wilkie, often with a new traditional emphasis created in the detailing. The Art Nouveau style was

Figure 5.6 Lagarie, Rhu, Dunbartonshire. Crown copyright: RCAHMS, SC 713508.

also often married with Arts and Crafts and served a similar set of patrons, but enjoyed a place in wider continental developments. It flourished in Glasgow from the 1890s as the Glasgow Style, spearheaded by Charles Rennie Mackintosh, Salmon & Gillespie and John Ednie. It was too avant garde a style to be comfortable in Edinburgh (though Pentland Terrace, Fairmilehead, might be quoted as exception) or the more provincial areas, but Art Nouveau stained-glass work and chimneypiece designs appeared all over the Scottish mainland.

English styles flooded in with the greater availability of materials in the late nineteenth century. The half-timbered villa with red tiles was popular, inspired by such designs as Sir Robert Rowand Anderson's Beesknowe, Stenton, East Lothian, 1886, George Washington Browne's house for Professor Ewart, Bog Road, Penicuik, Midlothian, 1885, and William Leiper's Piersland Lodge, Troon, Ayrshire, 1898, each echoing the Old English style popularised by Richard Norman Shaw. Like Stevenson, Nicoll appreciated that modern standards of living required fresh thought in detail and planning, and advocated a more compact house form. He also issued a caveat on Englishness, warning that some overdid the removal of 'the shams, inanities and affectations of the Victorian era', so that 'We might have too much of the "simple" dwelling and be in danger of Anglicising our buildings till we have forgotten our national style'.[29]

Hippolyte Jean Blanc's half-timbered design for Warrender, Murrayfield, Edinburgh, illustrates Nicoll's points perfectly and provides a good example of the common plan form. The servants' wing was invariably lower and to the rear, possibly including a cycle house, a feature evident in several Edwardian designs. The standard provision of public rooms consisted of drawing and dining room, with one or two additional rooms, either a study or a library, and either a billiard room or smoking room. The drawing room, which tended more often to be on the first floor in Victorian houses, returned to the ground in the Edwardian years. The average middle-sized house would have three to four bedrooms, with servant accommodation or a nursery above the service wing.

The particular interest which Hermann Muthesius took in the modern free-style work of the architects Mackintosh, James A Morris of Ayr (Hinton House) and Henry E Clifford of Glasgow (Stoneleigh, Kelvinside) in gathering material for *Das Englische Haus* prior to 1904, demonstrates the impact their designs made on Britain's continental neighbours, who were struck by the forward-looking intentions of their work. The villas of Mills and Shepherd around St Andrews, Thomas Graham Abercrombie in the Paisley area, and William Kerr around Alloa, Clackmannanshire, combined to establish the popularity of a new traditional style which flourished until 1939.

Lawrence Weaver's various publications with *Country Life*, entitled *Smaller Country Houses*, featured a new genre of middle-class, residential designs: the diminutive country house. Leading manufacturers sought compact, comfortable grandeur, often with policies, to which cursory reference may be made here. One style which particularly suited the requirements of the wealthy Edwardian merchant commissioning such a house, providing a sense of antiquity, was that of the Cotswold Manor. Several appeared by the Links of East Lothian, such as John Kinross's Carlekemp, North Berwick and James Bow Dunn's Whatton Lodge, Gullane, 1910, and on those of Ayrshire, in designs such as J K Hunter's High Greenan, 1910, all expressively described as 'golfing boxes'.[30] The Scottish seventeenth-century revival styles were similarly popular for the grandest of the smaller country houses, for example, John James Burnet's Fairnilee, Selkirk, 1904–6, Lorimer's Ardkinglas, Loch Fyne, Argyll, 1906, and Kinross's The Peel, Selkirk, 1904–7.

It must be stressed that the English and Scottish revival styles did not necessarily operate in isolation, and indeed were frequently merged, particularly by those seeking a modern interpretation; H E Clifford's eclectic 'Free Style' designs, such as Crosbie House, Southwood, Ayr, 1908, are examples.[31]

Between 1904 and 1914 the market for houses over £41 in annual value fell sharply, and more noticeably in Scotland than in England or Wales.[32] The years preceding the war were bleak for many. James Steel Maitland, for example, emigrated to Canada in 1908 immediately after qualifying as an architect in order to find work. Others diversified in order to remain solvent: John Kinross established a small-scale antiques trade to supplement his minimal income, and Alfred Lochhead returned from Canada a decorator and dealer in high quality French Art Deco objets d'art.

After World War I, the emphasis fell on public housing, on a smaller, mass-produced scale, while the commissions from middle-class patrons were invariably reduced to attic additions and extensions. However, the architectural response to the changes in the social scene did not exclude the middle bracket. Although the bungalow architecture, which appeared by the acre in the 1920s and 1930s, derived from colonial prototypes, was largely smaller in scale and primarily for retirement homes, the *Daily Mail Bungalow Book*, 1922, indicated that it was still the province of the wealthier patron, designs making provision for female servants.[33] In the introduction, the scale of the demand for such 'labour-saving one-floor houses' bringing 'the emancipation of home-lovers from the home's more irksome tasks' is revealed. Mactaggart & Mickel, working in and around Edinburgh to designs by architects such as Stewart Kaye in Hillpark, were the most prolific developers of 'bungalurbia'. Charles McKean's revealing monograph, *The Scottish Thirties*, examines Scotland's acceptance of the bungalow.[34]

The demand for highly utilitarian accommodation which had given rise to bungalow architecture in the 1920s and 1930s found similar response in the Modern Movement 'New Architecture' designs for two-storeyed residences, of continental inspiration. Obertal, Largo Road, Leven, Fife, illustrates this clearly. It is a flat-roofed house with roof terrace costing about £1,000 to build (in order to benefit from the Government's £100 subsidy for every house constructed under this sum). It followed the form of the winning design by a teacher, Mrs Reid, in the 1931 Ideal Home competition.[35] The picture windows and glazing pattern of Obertal were synonymous with the period and style, but only the 'deck-style' railings of the roof paid homage to the nautical, steamboat theme favoured in many other contemporary designs, parading such elements as prow-shaped bow windows with canopies, port-hole windows, white rendered walls and curved corner fenestration, dazzling contemporaries with glittering steel, glass and concrete. The cubic form of Frank Taylor's designs for Mactaggart & Mickel on the Broom Estate,

Figure 5.7 46a Dick Place, Edinburgh. Courtesy of the RIAS.

Whitecraigs, Renfrewshire, in the 1930s, with interiors redolent of Art Deco detail, were typical of the short-lived generation of Modern Movement houses. George Lawrence's design for 4 Glenlockhart Bank, Edinburgh, 1938, was a fine east-coast equivalent, but Sir William Kininmonth's house in Dick Place, Edinburgh (fig. 5.7), 1933, with its Swedish purity, was perhaps the style's most prestigious flagship. The stunningly modern designs of Carruthers, Ballantyne, Cox & Taylor in the genre around Inverness, illustrate that the style reached the four corners of the country without diminishing the resulting impact during its limited life. McKean gives a lively overview of the breadth and extent of the Modernist designs for the middle-sized market.[36]

The architectural press continued to play a seminal role, reporting on competition designs and Ideal Home Exhibitions. Publications such as F R S Yorke's *The Modern House*, 1937, and Patrick Abercrombie's *The Book of the Modern House*, 1939, anchored the fashion, but more traditional designs continued to make their appearance. James Steel Maitland's Littlecroft (now Savoy Croft), Ayr, 1924–36, was essentially an Arts and Crafts survival house, but in terms of planning and construction it was very much more durable and economically designed than its predecessors in the style. It showed how simple the transition from traditional to modern values could be, without loss of quality.[37]

The neo-Georgian style which found favour in the Edwardian years, fared successfully before and after World War II in the hands of architects such as John Jerdan. Ian G Lindsay's Eventyr, Longniddry, East Lothian, 1936, was a Scottish neo-Georgian masterpiece. East Port, Lennoxlove, Haddington, produced in the mid 1980s, echoed the work of Clough Williams-Ellis at Portmeirion, North Wales, and revealed the style's timeless value. These houses sit comfortably in semi-rural situations, disregarding Mackay Hugh Baillie Scott's generalisation that the style was essentially 'urban and sophisticated'.[38]

Considering that the previous boom periods for residential design had bridged centuries, it would not have been unreasonable to expect some similar boom to occur towards the year 2000. However contemporary housing provision is geared towards a different market, the majority of new middle-sized designs being found in speculative theme-park estates. The few architect-designed private commissions which have arisen since 1945 have tended to be of strong character, for example, Peter Womersley's Port Murray at Turnberry, Ayrshire, 1963, or James Morris's Scadlaw House, Humbie, East Lothian, 1967, and more recently, Burnet, Bell & Partners' Atrium, Kilmacolm, Renfrewshire, 1980, and Tony Vogt's Bowhouse, Helensburgh, 1988 – all remarkable and highly original designs in the field.

NOTES

1. Dunbar, 1966. The foundation material from this text is drawn upon repeatedly in this article.
2. Reid, 1683/1988, Chapter 1, paragraph 2: 'The Model House', and Figure One.
3. Fenton, Walker, 1981, 65.
4. MacGibbon, Ross, 1887–1892, Vol 4, Chapter 9: Houses in towns, 407–10. See also Cant, Lindsay, 1946; 1947; 1948; Cant, 1975(a); Cant, 1975(b); and Cant, 1976. The Scottish Burgh Surveys by the University of Glasgow, Department of Archaeology, and more recently commissioned by Historic Scotland, provide further information and evidence on this subject.
5. Gifford, 1992, Introduction, 71.
6. Fenton, Walker, 1981, 102–21, 136–42.
7. *OSA*, xx, 1798, 148.
8. Walker, D M. Introduction to 1992 reprint of Dickson, Elliot, 1778.
9. ibid.
10. RCAHMS, 1992, 440–1.
11. ibid.
12. For further information, see Youngson, 1988; and the excellent Introduction to Gifford et al, 1984, from which some of the following information is drawn.
13. The Introduction to Williamson et al, 1990, gives an outline of Glasgow's residential development, and Gomme, Walker, 1987, contains much of value in Chapters 3 and 4, 69–102.
14. Cant, 1975(b), 198.
15. See Maclean, 1989, for further details for Telford's manses.
16. Dunbar, 1966, 198.
17. See Walker, 1958.
18. White, 1845, text to Plate XIII.
19. Loudon, 1833.
20. ibid, Figures 1569–70, 888–9: Cousin's design for a villa in guise of a Scottish manor house.
21. ibid, Design XIV, 879–880: Villa at Spitalfields, near Glasgow.
22. ibid, Design IX, 850–3: Newall's villa, classed subsequently as an art gallery house.
23. Gray, 1832, Plate VII, text 45: Turtleton Farmhouse.
24. Starforth, 1866, plates XV and XVI: Design for a Baronial villa. See also Starforth, 1853.
25. Blackie and Son, 1878, plate XXIX: Thornton Shiells, Villa at Grange.
26. Stevenson, 1880.
27. Nicoll, 1908.
28. ibid, Introduction, xi.
29. ibid, Introduction, xviii.
30. See Davis, 1992.
31. 'Free Style' is a useful umbrella term, coined by Alastair Service, 1975, to define the stylised eclectic work of certain architects from the end of the nineteenth century, part of the search for a traditional yet modern style.
32. Walters, 1918.
33. The *Daily Mail Bungalow Book*, 1922, reproduced the best designs from the newspaper's Architects' Competition for Labour Saving Bungalows, 1922.
34. McKean, 1987, a popular and informative introduction to the period archi-

tecture of the 1930s. See Chapter 10, How they lived, 154–63, on bungalow architecture.
35. In an effort to improve the standard and extent of economical designs after the War, the Housing and Town Planning (Scotland) Act, 1919, caused the Government to give local authorities funding with which to promote such a scheme of subsidy.
36. McKean, 1987, 169–82.
37. Hamilton, 1983.
38. Scott, Beresford, 1933, Chapter 1, 7.

BIBLIOGRAPHY

Beaton, E. *Ross and Cromarty*, Edinburgh, 1992.
Brogden, W A. *Aberdeen*, Edinburgh, 1986, 1998.
Burgher, L. *Orkney*, Edinburgh, 1991.
Blackie and Son. *Villa and Cottage Architecture: Select examples of country and suburban residences*, London, 1868.
Cant, R G, Lindsay, I G. *Old Elgin*, Elgin, 1946.
Cant, R G, Lindsay, I G. *Old Glasgow*, Edinburgh, 1947.
Cant, R G, Lindsay, I G. *Old Stirling*, Edinburgh, 1948.
Cant, R G. *Historic Buildings of Angus*, Forfar, 1975(a).
Cant, R G. *St Andrews: The handbook of the St Andrews Preservation Trust*, St Andrews, 1975(b).
Cant, R G. *Historic Crail*, Crail, 1976.
The Daily Mail Bungalow Book, London, 1922.
Davis, M C. *Castles and Mansions of Ayrshire*, Paisley, 1992.
Dickson, J, Elliot, C. *Rudiments of Architecture: The young workman's instructor*, Edinburgh, 2nd edn, 1778, reprinted Whittingehame, 1992.
Dunbar, J G. *The Historic Architecture of Scotland*, London, 1966 (2nd edn, London, 1978).
Fenton, A, Walker, B. *The Rural Architecture of Scotland*, Edinburgh, 1981.
Finnie, M. *Shetland*, Edinburgh, 1990.
Gifford, J, McWilliam, C, Walker, D, Wilson, C. The *Buildings of Scotland: Edinburgh*, Harmondsworth, 1984.
Gifford, J. The *Buildings of Scotland: Fife*, Harmondsworth, 1988.
Gifford, J. The *Buildings of Scotland: Highlands and Islands*, Harmondsworth, 1992.
Gifford, J. The *Buildings of Scotland: Dumfries and Galloway*, Harmondsworth, 1996.
Gomme, A, Walker, D M. *The Architecture of Glasgow*, London, 1968, 2nd edn, 1987.
Gray, W J. *A Treatise on Rural Architecture*, Edinburgh, 1832.
Hamilton, L. A Paisley architect: James Steel Maitland (1887–1982), *Scottish Georgian Society Bulletin*, 10 (1983), not paginated.
Loudon, J C. *Encyclopedia of Cottage Farm and Villa Architecture*, London, 1833, later edns 1839, 1842, 1853.
MacGibbon, D, Ross, T. *Castellated and Domesticated Architecture of Scotland*, Vol 4, Edinburgh, 1887–92.
McKean, C, Walker, D. *Dundee*, Edinburgh, 1984, 1993.
McKean, C. *Stirling and the Trossachs*, Edinburgh, 1985, 1994.
McKean, C. *The Scottish Thirties*, Edinburgh, 1987(a).
McKean, C. *The District of Moray*, Edinburgh, 1987(b).
McKean, C, Walker, D, Walker, F. *Central Glasgow*, Edinburgh, 1989, 1993, 1999.

McKean, C. *Banff and Buchan*, Edinburgh, 1990.
McKean, C. *Edinburgh*, Edinburgh, 1992.
Maclean, A. *Telford's Highland Churches*, Inverness, 1989.
McWilliam, C. The *Buildings of Scotland: Lothian*, Harmondsworth, 1978.
Mays, D, ed. *The Architecture of Scottish Cities: Essays in honour of David Walker*, East Linton, 1997.
Mays, D. 'A taste of heaven': Some picture books for the developing Victorian suburb. In Mays, 1997, 136–45.
Nicoll, J. *Domestic Architecture in Scotland*, Aberdeen, 1908.
Peden, A. *The Monklands*, Edinburgh, 1992.
Pride, G L. *Fife*, Edinburgh, 1990 and 1999.
RCAHMS. *Inventory of Argyll*, Vol 7, *Mid-Argyll and Cowal*, Edinburgh, 1992.
Reid, J. *The Scots Gard'ner*, Edinburgh, 1683, reprint Edinburgh, 1988.
Scott, M H B, Beresford, A E. *Houses and Gardens*, London, 1933.
Service, A. *Edwardian Architecture and its Origins*, London, 1975.
Shepherd, I. *Gordon*, Edinburgh, 1994.
Starforth, J. *The Architecture of the Farm: A series of designs for farmhouse and farm steadings*, Edinburgh, 1853.
Starforth, J. *Designs for Villa Residences with Descriptions*, Edinburgh, 1866.
Stevenson, J J. *House Architecture*, 2 vols, London, 1880.
Strang, C A. *Borders and Berwick*, Edinburgh, 1994.
Swan, A. *Clackmannan and the Ochils*, Edinburgh, 1987 and 2001.
Thomas, J. *Midlothian*, Edinburgh, 1995.
Walker, D M. *Nineteenth-Century Mansions in the Dundee Area*, Dundee, 1958.
Walker, F A. *The South Clyde Estuary*, Edinburgh, 1986.
Walker, F A, Sinclair, F. *The North Clyde Estuary*, Edinburgh, 1992.
Walker, F A. *The Buildings of Scotland: Argyll and Bute*, Harmondsworth, 2000.
Walters, Tudor. *Report of the Committee Appointed to Consider the Questions of Building Construction in Connection with Dwellings for the Working Classes*, London, 1918.
White, J. *Rural Architecture: Illustrated in a new series of designs for ornamental cottages, villas, exemplified by plans*, Glasgow, 1845.
Williamson, E, Riches, A, Higgs, M. *The Buildings of Scotland: Glasgow*, Harmondsworth, 1990.
Youngson, A. *The Making of Classical Edinburgh*, Edinburgh, 1966, 2nd edn, 1988.

6 Small Houses and Cottages

ANNETTE CARRUTHERS *and*
JOHN FREW

Impermanence is one of the notable features of small houses, since their construction requires a lesser investment in materials and labour than grander dwellings. The history of the small house in Scotland is one of frequent change and repair, decay and demolition, improvement and modernisation; a clear historical sequence is difficult to discern because numerous regional variations confound the chronology.[1]

Surviving structures and ruined remains provide evidence, but much has been destroyed and relatively little archaeological investigation of medieval and later Scottish domestic sites has been undertaken.[2] Vivid contemporary descriptions in travellers' tales and even government reports must be regarded with some caution since they often tend to comment on the unusual or to exaggerate from a sense of shock at unfamiliar conditions.

In general, the small house or cottage is a type of dwelling found predominantly in the country, in small burghs, or the suburbs of larger towns and cities. When the first official census was taken in 1801, only about one-fifth of the Scottish population lived in urban areas, and most of the remainder lived in the country, often in small settlements in one- or two-roomed houses.

Evidence from field survey, augmented by descriptions and the survival of stone buildings into the twentieth century, shows that the longhouse was the main form of rural dwelling over virtually the whole country until the eighteenth century, and that in some places it survived even later.[3] It had distant origins and was intimately tied to a way of life based on working the land. Tenure of land was short-term, there was no expectation that estate owners would house their tenants, and little incentive for the tenant to build anything beyond what was necessary.

Local factors are of crucial importance for the form of the small house, since the availability of building materials, the climatic conditions of the area, and the occupations of the inhabitants were the major determining elements until at least the mid nineteenth century, when improved forms of transport began to make it economically viable to take factory-made materials and goods to the most remote parts.

Before this, houses were built of readily available materials, though shortages of timber in the Lowlands necessitated imports from Norway in the seventeenth century. In many areas stone was abundant (for example, granite in the north east, granite and red sandstone in the south west), but in the far north and the western Highlands and Islands timber was scarce, especially in windy coastal areas. Here driftwood was frequently used for roofs with coverings of turf and heather, fern, rushes or straw. Turf also provided material for walls in combination with stone, clay, or timber.

Walls were usually of unshaped stone without lime mortar, often constructed with rounded corners, and since they were thick, any openings for doors and windows were deep-set. Houses were usually between 12 and 16 feet (3.6 and 4.9 metres) wide, and varied considerably in length, from 30 to 60 feet (9.1 to 18.3 metres).[4] The common thatching materials were not hard-wearing and quickly became disordered, giving a dilapidated appearance to the house.

According to the needs and means of the inhabitants, houses had one or more rooms and were built singly, in small groups ('fermtouns' in Scots-speaking areas), scattered along a trackway by the coast, or at the edges of larger settlements (fig. 6.1). If properly planned, the structure was sited to take account of prevailing winds and the need for drainage

Figure 6.1 Glen Dochart, Killin, Perthshire; long house, photograph probably late nineteenth century. SLA, W 220317.

downhill. This was important because animals were usually kept under the same roof, with the byre normally lower than the house (though sometimes animal quarters were built separately). People and animals entered by a door in the long side, turning to the byre in one direction and the living quarters in the other. The main room was only dimly lit by small windows and usually had a central fire of faintly glowing peat on a shallow platform of clay or stone. Thick peat reek hung in the roof, coating the timbers with dripping brown tar and dissipating through the thatch, which was removed every year and spread on the land to add richness.

From at least the eighteenth century, a smoke hole of timber was sometimes incorporated in the roof, showing on the outside as a round or square tapering box. Inside, this might be extended to form a canopy of lath and plaster or spars and mats above the fire to funnel away the smoke.[5] The fire itself frequently came to have the addition of a stone fire back or a wall built behind it, and this gradually led to the adoption of gable-end chimneys, which became the norm in 'improved' cottages from the late eighteenth century, though they had been known in the previous century. The low-level peat hearth with a 'hangin lum' above remained common in the kitchen into the twentieth century, but in the second room, used for storage and sleeping and sometimes for entertaining visitors, it became a status symbol to have a modern grate, and in some areas to burn coal. Often, for example in cottar houses, this room was simply a space separated from the kitchen by large items of furniture such as box-beds or a dresser.

Such houses were generally temporary in nature because of short-term employment patterns and shifting populations, but the type survived in the more remote parts of the country because it was not economic to import the additional materials needed for a modern house. A distinctive variant which survived in occupation in the Western Isles until the 1960s was the 'blackhouse', with double walls built round a core of earth to a thickness of about 5 or 6 feet (1.8 metres) and the thatch resting on top of the inner wall ledge.[6] These blackhouses often consisted of a long dwelling room with barn and byre lying to either side, the spaces all interconnecting.

In addition to dwelling houses there were ancillary buildings for other purposes. Huts, known as cleitean on St Kilda, were used to store vegetables, dried birds or fish, and shielings and bothies of a variety of forms provided accommodation for seasonal work, such as fishing or tending cattle at the summer grazing. These were simpler structures than the houses, using the same materials but creating one-roomed shelters, rectangular or circular in plan.

Rugged in appearance and lacking in services, blackhouses were regarded as curious or shocking by most outside visitors. They were, however, developed over a long period to answer a specific set of local conditions, and were suited to a life of labour on the land in areas of long winters, very high winds, and limited natural resources.[7] When they were replaced by modern structures on St Kilda from the 1860s, the residents complained of the inadequacies of the new roofing materials.[8]

'Scots housing is full of ... "survivals" of past stages of evolution, which have had their day', wrote the members of the Royal Commission on Housing in Scotland in their report, published in 1917.[9] These blackhouses of the Western Isles were among the most noted of the survivors, though there were other examples of regional variation on the old long house form in, for instance, the Strathclyde area and the islands of Islay and Jura, where the gable construction and roof shape differed from the Lewis type. The abandonment of the old houses in these areas in the twentieth century has been the end of a long process of change and innovation which started in the burghs and wealthy farming areas about 400 years ago. Pressure of space in many burghs resulted in the replacement of one-storeyed cottages in the late sixteenth and seventeenth centuries by two-storeyed houses with a forestair to the upper living quarters and either storage or the byre on the ground floor. Before this, many burgh houses were of wood and little evidence survives of their form.[10] After great fires in the late seventeenth century, regulations began to prohibit the use within burghs of flammable building materials, so houses were increasingly constructed of stone with tiled or slated roofs instead of thatch. The wealthy sea-trading towns of Fife provide many examples, but they were also found in the closes and pends of the bigger towns, such as Edinburgh, though small houses here were gradually replaced by small dwelling units within much larger tenement buildings.

The eighteenth-century impetus to intensify farming methods promoted ideas of improving houses in order to increase income from rents and to foster better workers. With this came the idea of creating villages to provide a focus for community life and commerce, including fishing, a movement which started in the areas of greatest wealth in the south east in the early eighteenth century and continued up to the 1840s.[11] It was connected with the widespread building of new roads, and concentrations of planned towns and villages appeared in the east, especially the north-east coast, and in the south west. These planned villages were inevitably influenced by the work of architects employed by the landowners, and the spread of ideas was furthered by the publication of pattern-books of houses and cottages deemed suitable for particular classes.[12]

As a result of these moves, a basic cottage type (fig. 6.2) became almost universal in Scotland as a replacement for the longhouse. The new cottage was built of lime-mortared stone, often, in the eighteenth

Figure 6.2 Shandwick, Nigg, Easter Ross, mid nineteenth century (photographed 1975). SLA, 23_9_11.

century, thatched and later re-roofed with slate or tile, or from the 1880s with corrugated galvanised steel. It had larger windows than before, usually sash windows, and often the walls were plastered or timber-lined inside, the roof space concealed by a ceiling, and the floor was boarded. The plan was generally symmetrical, with a small lobby in the centre flanked by a room on each side and a small space between. The major structural features were the gable-ends which incorporated the chimneys. The front and back walls could be built to two storeys or more, though one or one-and-a-half were more common.

In the country such houses were often built on their own, but on estates and in villages, for economy, they were put up in pairs or in rows, often in repetitive ranks of rows. They were starkly rectangular and were usually built directly onto the street to allow no space for the accumulation of rubbish, giving a neat but bald appearance. Greater variety appears in the many fishertowns developed in the eighteenth and early nineteenth centuries, especially in the north east, where the houses present their gable-ends to the sea and small gardens and meandering pathways fill the gaps in between.

Regional variations appear in the use of local stone and slate (and occasionally brick) which demand different treatments, in the employment of pantiles in the east, and in the development of techniques to combat local conditions, such as the greater use of harling and colour wash in the west as protection against the rain. The basic architecture, however, was remarkably consistent over the whole country.

These improved cottages were built for farm servants, industrial workers such as weavers and nailmakers, for miners, fishing people, and other groups. Their symmetrical plans and elevations were based upon the style of merchants' houses in the burghs and on manses, farmhouses, and small lairds' houses as developed in the late seventeenth and early eighteenth centuries and later illustrated in pattern books.[13] The small house in Scotland was thus part of a continuum with the medium-sized house, exhibiting differences only of scale and quality of construction and finish. For the poorest people, houses were often single rooms with inadequate ventilation, without floorboards, and with shared midden privies which horrified the Royal Commissioners.[14] In the country they were commonly without running water or sanitary arrangements of any kind. Miners' dwellings continued the Scottish tradition of impermanence, being abandoned when the mines were closed, though in some of the wealthier mining villages such as Newtongrange in Midlothian, houses of four or five rooms with conveniences and gardens received approval in the 1917 Report. Today, when two small houses are often knocked into one and improvements are made, the rooms seem very spacious, since they can be 20–30 per cent larger in floor area than their equivalents in England. In the nineteenth century however, most were greatly overcrowded, ill-maintained, insanitary, and depressingly damp.[15]

When improved cottages were built in remote country areas and the islands, they were usually placed next to the older buildings they replaced and these were used in turn as stores, though some remained in occupation.[16] Extra accommodation was also frequently provided in the twentieth century by redundant railway carriages, and it seems that in the north east almost every farm had one of these.[17]

Small houses and cottages in Scotland developed a remarkably consistent architectural form which gives a strong character to both country and town, but there were, of course, variations, sometimes partly caused by functional considerations and often by the involvement of architects. Toll houses, for instance, began to appear after Turnpike Acts – the first in 1713, others from 1750 onwards. These needed windows facing in two or more directions, as did canal-side houses for lock-keepers. Gate lodges (fig. 6.3) and estate buildings were frequently

Figure 6.3 Campbeltown, Argyll; lodge house, *c.* 1900. Courtesy of Argyll and Bute Libraries.

designed to suit the style of the main house, and the visitor to a large estate might be greeted by a tiny Greek temple in the 1830s or a Scots Baronial mini-castle twenty years later.[18] By the end of the century an even wider range of influences had filtered in, producing, for instance, the synthesis of Scottish and Devon vernacular in James Maclaren's cottages at Fortingall, the estate village of Glenlyon House in Perthshire.

Steadily throughout the Victorian period, however, the people of Scotland moved away from the country and into the towns and cities. The 1891 census was the first to show that more than half the population lived in areas with concentrations of more than 5,000 inhabitants, but even by 1861 it has been estimated that 57.7 per cent lived in settlements of over 1,000 people.[19] In the smaller towns the traditional two-storeyed flatted dwelling with external stair at front or back was still in use by 1917, when the Royal Commissioners reported on its many defects, some being simply problems of age and lack of maintenance. Mention was also made of recently constructed four-to-a-block, two-storeyed tenements in Kirkcaldy, Fife.[20]

Most of the movement of population was to the major Lowland cities, where pressure on land gradually squeezed out the small house in favour of the more densely-packed tenement. Even in the largest cities, however, there was still an interest in building small houses, usually with the motive of improving the lot of the skilled worker. Some were the achievement of philanthropic groups, such as the Pilrig Model Buildings Association formed in 1849, which built 44 houses off Leith Walk in Edinburgh in 1850–1. These were in terraces of flatted dwellings with access to the ground floor from the front and to the upper storey via an internal staircase reached from the back; and they had architectural pretensions, with pediments on the end houses and corniced door heads. Others were the work of self-help groups of skilled artisans. The Colonies in Stockbridge (fig. 6.4) were developed by the Edinburgh Cooperative Building Company Limited between 1861 and 1911, presumably based on the form of Rosebank Cottages, Edinburgh, designed by Alexander McGregor and built by James Gowans for 'the better class of mechanics' in 1854–5.[21] These two-storeyed houses gave entry to the ground floor on one side of the terrace and to the upper door by an outside stair on the other side. The upper flat generally occupied two floors and was therefore larger than the lower one, and this gave flexibility of accommodation since occupants could move into larger or smaller houses as finances permitted or family circumstances demanded. The same company built similar houses in several other parts of Edinburgh.

According to the 1917 Report there was much discussion about whether this form of flatted housing was preferred to cottages of the English type, self-contained on two floors with an internal stair. Such houses had been built in Falkirk and Dumbarton, sometimes as semi-detached instead of terraced houses, with front and back garden plots, and some with bathrooms. The Falkirk and Dumbarton Building Societies claimed that the occupants preferred these, while the manager of the Edinburgh Cooperative Building Company asserted that the housewife liked rooms all on one floor.[22]

The two-storeyed, four-apartment houses in Dumbarton were built in 1868 by William Denny & Brothers, apparently as part of a scheme to improve housing for their workers and to encourage skilled artisans to remain in the area. They were sold at a low rate of interest and this policy seems to have succeeded.[23] It is not clear why Scottish employers favoured the English cottage type, though it may be simply that published views on the matter were predominantly English ones. Later on, a practical example was set by Lord Leverhulme, one of the pioneers of company housing, who had firm ideas about the importance of going upstairs to bed. He paid for a pair of half-timbered red brick 'Sunlight Cottages' which were erected for the 1901 Glasgow Exhibition and remained in Kelvingrove Park. In the same year, a village in similar style was built at Rosebank on the Clyde.

It was, however, becoming clear that an acceptable standard of working-class accommodation was not attainable under existing conditions. In fishing communities it was fairly common for people to

Figure 6.4 Reid Terrace, Stockbridge, Edinburgh, 1861–2.

own their houses, and elsewhere co-operative schemes enabled a skilled élite to buy property, but ownership was rare and the vast majority of Scottish workers were paid too little to afford rents which would give the financial return required by property investors. Few employers accepted any responsibility for their workforce.[24]

Concerns about public health led to increasing legislation on housing in the late nineteenth century, and pressure on government began to build up, led by miners' groups, medical officers, and the Local Government Board for Scotland. As a result, the Royal Commission on the Housing of the Industrial Population of Scotland was set up in 1912 with the aim of prescribing 'How to provide a healthy, comfortable dwelling for every family in the land'.[25] Its massive report of 1917, detailing the inadequacies of a very large part of the housing stock, soon led to legislation and dramatic change.

The interventionist policies adopted by the post-World War I Coalition Government held profound consequences for housing in Scotland, through the creation of a public-sector programme that would dominate house production throughout the next sixty years, accounting for 253,000 out of 365,000 new dwellings completed between 1919 and 1939. The need for government involvement had been recognised well before the passing of the Housing and Town Planning (Scotland) Act in August 1919, which gave legislative support to the notion of a nationally co-ordinated programme of house building, aimed exclusively at the 'working classes' and administered through the local authorities and Scottish Board (later Department) of Health. Launched under the emotive slogan 'Homes Fit for Heroes', this aimed at a short-term target of 115,000 new houses (a figure that approximated to the shortage identified in the 1917 Royal Commission report), financed by rents, rates and government subsidies. Confirming the revolutionary significance of the 1919 Act, the new houses were expected to adhere to standards recommended by the Tudor Walters *Report on the Provision of Dwellings for the Working Classes* (1918), which had set a minimum of three apartments per dwelling unit, supplemented by a bathroom, scullery, and internal water closet.

Tenement building was permitted under the Act, but few at this stage were prepared to defend a format that was associated with overcrowding. No such stigma attached to the double cottage which, in contrast, had come to be identified with reformist ideology, with the result that it was now promoted as providing the preferred solution to problems posed by Scottish working-class housing requirements (fig. 6.5). Although architectural guidelines were issued for local authority use, these were permitted to be interpreted with some freedom, reinforcing the Garden City ethos that permeated virtually every aspect of the programme, and which extended to encourage a low density of houses to the acre (12 was the ideal) as well as the provision, in individual

Figure 6.5 34–6 Bayview Crescent, Methil, Fife (1919). Courtesy of Peter Adamson.

schemes, of a relatively broad mixture of accommodation, in the form of three-, four- and five-apartment cottage and flatted units. With few exceptions, the new houses were set back from the road to accommodate front gardens, an arrangement that contributed to the frequently commented upon 'open' quality of the estates and broke decisively with the street-side alignments favoured by eighteenth- and nineteenth-century urban cottage planning.

Criticisms were at first muted, but eventually found focus in house costs (which spiralled), rents (which were high, restricting tenancy to comparatively well-to-do families), and the fact that the great majority of houses were built of brick, which was held to be at odds with indigenous traditions of stone construction, as well as affording insufficient protection from the dampness of the Scottish climate. In the event, problems with brick construction hindered the progress of house building, which ground to a halt in the winter of 1922–3, by which time approximately 27,000 houses had been completed, less than a quarter of the total originally envisaged.

Subsequent legislation perpetuated the principle of subsidised housing but placed a ceiling on central government support, set, for example, at £9 per house per annum for a period of 75 years under the 1924 Housing Act. The obvious, if unstated, intention of this was to transfer responsibility for 'uneconomic' housing (in which the costs involved were not met by rents charged) to the local authorities. This represented a hugely significant shift of emphasis that goes a long way towards explaining the extraordinary vigour with which economies were pursued throughout the 1920s, contributing to drastic reductions in unit costs, from an average of over £1,000 per house in 1922, to under £400 in 1928, a process that was reflected in the abandonment of the four- or five-apartment double cottage in favour of a two-storeyed flatted arrangement, with four two- or three-apartment flats to a block. The extent of this realignment was revealed by the Scottish Department of Health survey of post-1919 house-building activity, *Housing of the Working Classes – Scotland* (1933). Of the 23 'model' estates surveyed, 18 incorporated exclusively tenemented or flatted accommodation, while at least three of the five that included cottages had been completed during the earliest phase of house building.

It was only, indeed, in the aftermath of the 1935 Housing Act that municipal cottage building resumed on a significant scale. Aimed primarily at the problem of overcrowding, the Act also provided a means through which to address the serious imbalance that had developed between standards of accommodation provided by local authorities in Scotland (where 73.7 per cent of new houses were of three apartments or fewer) and England and Wales (where 80.3 per cent were of four apartments or more).[26] The resultant upturn in accommodation levels – 76 per cent of houses completed in Scotland in 1938 were of four or five apartments – accompanied a conscious reversion to Garden City values, producing a succession of large-scale, low-density 'cottage' estates (as for example at Kincorth, Aberdeen, and Hayfield, Kirkcaldy), a significant number of which (such as Magdalene's Kirkton, Dundee) were distinguished by a systematic exploitation of the terrace.

Inter-war private house building, always dwarfed by the sheer scale of the public-sector programme, revived significantly in the mid-1930s to produce a mini 'boom' of over 47,000 completions between 1933 and 1938. The ready availability of relatively cheap mortgages underpinned a great deal of this activity, which found expression in a proliferation of semi-detached houses (usually described as 'villas', but which invariably represented minor variations on the public-sector double cottage) and, in particular, a phenomenal rise in demand in bungalow accommodation. Although certainly experimented with in the late Victorian and Edwardian periods, it was only in the early 1920s that the bungalow's viability as a low-cost house type was established, largely through the promotion of a simple three- or four-apartment, single-storeyed, square-plan arrangement, adaptable to rural as well as suburban needs.[27] High Architectural hostility, which repeatedly emphasised its identification with 'jobbing' builders and unplanned ribbon development[28] thus did little to undermine the popularity of a format that, although expensive to heat, was comparatively cheap to build, and afforded modest scope for individuality, usually achieved by minor variations in plan and architectural detail.[29] The verandah, adopted widely south of the border, and which provided a direct link with the bungalow's Indian origins, was employed relatively infrequently in Scotland. Like its high-rise counterpart, the sun balcony, it seems to have been viewed as a dispensable luxury by inter-war house builders, and fundamentally ill-suited to a northern climate.

The near complete cessation of house building that followed the outbreak of hostilities with Germany in 1939 placed enormous pressures on the Scottish housing stock, the extent of which was revealed by the Scottish Housing Advisory Board (1944) declaration that 470,000 new houses were needed immediately to replace 'unfit', overcrowded or destroyed dwellings, and that a further 405,000 houses had 'no independent water-closets or sanitary conveniences of any kind'.[30] As in 1918, building plans were held in check by shortages of materials and skilled labour. 'Non-traditional' construction (notably in steel, timber, poured and pre-cast concrete), already experimented with on a limited scale in the mid-1920s and late 1930s, was identified as providing a potential solution to both problems, and accounted for approximately 100,000 out of 204,000 public-sector houses (including 32,000 temporary bungalows, or 'prefabs') completed between 1945 and 1954, many under the auspices of the Scottish Special Housing Association.[31] The undiluted utilitarianism of these arrangements accorded perfectly with the economics of post-war austerity and affords a valuable insight into the pressures also placed on 'traditional' cottage construction as the Scottish housing programme geared itself up to address the challenge posed by the New Towns and the creation of large-scale peripheral estates.

Guidelines in this respect were provided by the Department of Health for Scotland's *Approved House Designs* (1945), in the form of 38 arrangements, prepared by the Royal Incorporation of Architects in Scotland and supplied to local authorities on condition that the plans would be used 'without modification'. Although markedly plainer than their immediate pre-war counterparts, these perpetuated (and in terms of floor-space, slightly surpassed) levels of accommodation adhered to in the aftermath of the 1935 Act. Further developments, encouraged by the recommendations of the Baillie Committee *Report* (1953), were, however, all in the direction of economy, achieved through a diminution of room numbers (with the result that 59.8 per cent of public-sector houses completed in 1953 were of three apartments or fewer) and range of plan-types, as well as by a process of design simplification, achieved through the progressive elimination of such standard pre-war features as the hipped roof and dormer window.

The 1957 Housing Act, by reducing subsidies for 'standard' public-sector housing, added further impetus to what had become an irresistible process, which would become inextricably associated in the following decade with the drift to high rise and the triumph of 'systems build'. The enormous amount of publicity that surrounded 1960s tower-block construction has, however, obscured the fact that the majority of public-sector houses completed in Scotland in the 1960s and 1970s exploited what were by now well-established methods of construction (notably brick and poured concrete), that were better suited to the economies of scale associated with the dominant long-term product of 1950s cost-cutting, the two/three-storeyed terrace.

The case for the terrace as a solution to municipal housing needs had been argued from the mid-1930s onwards, and hinged on the contention that it was cheap in terms of construction, susceptible to broad architectural treatment, and flexible enough to accommodate a comparatively wide range of house types, represented in the post-World War II period by cottages, flats and maisonettes. The last of these, in the form of two-storeyed units either superimposed one above the other or above a single-storeyed flat, was of comparatively recent evolution, having been brought to local authority attention by the Central Housing Advisory Committee booklet *Design for Dwelling* (1944), a source that derived additional significance from the fact it encouraged modest experiments in 'open' planning, achieved by the creation of an extended sitting room/dining area that would become a commonplace interior arrangement from the 1960s onwards.

The limited response these initiatives prompted within a steadily (after 1950) reviving private sector testified to an ingrained conservatism that Niven (1979) has attributed to near monopoly conditions produced by the dominance of the owner-occupier market by a 'handful' of building firms.[32] Changes were nevertheless accommodated, particularly when, as in the case of the gradual phasing out of the open fire and chimney (also reflected in municipal house design, and the product of the move towards all-electric interiors), these permitted considerable savings in terms of building costs. The relative importance of window openings, more or less unchanged since the early century, was subjected to scarcely less radical modification. Although the inter-war debate over the respective merits of the casement and sash-and-case window had been resolved largely in favour of the former by 1950, it was only in the following decade that its major advantage – an ability to extend horizontally as well as vertically – was exploited fully, to produce, in extreme cases, full wall-to-wall and floor-to-ceiling windows and, more commonly, the so-called 'French window', affording direct access from sitting-room to garden. Timber boarding, popularised as an antidote to post-war austerity by the Festival of Britain (1951), added a degree of informality to rendered brick arrangements that were frequently also enlivened by, amongst other features, stone facings and asymmetrical gables.

The extension of the same principle, to infiltrate conventions of layout, already anticipated by a trickle of individually commissioned 'small' (but expensive) post-war houses,[33] followed in the late 1960s and 1970s, finding fullest expression in the 'ranch house', based on an adaptation of the bungalow format, which proved to be sufficiently flexible to accommodate ideas of free planning inspired by much earlier experiments in the USA.[34] These, however, remained minority experiments that belied a general trend towards smaller and simpler houses, sustained by a renewed phase of suburban expansion, and achieved in the main by a ruthless rationalisation of the semi-detached villa (fig. 6.6) as well as by increasingly frequent experiments in terrace design, to produce what Gibbs (1989) has described as signs of 'convergence' with the scale (if not with the standard of internal fittings) of local authority housing.[35]

Appropriately, therefore, subsequent private-sector developments have involved an increasing reliance on 'alternative' building methods, evidenced, for example, by a steady rise in demand for timber-framed or 'kit' units, based on the example of low-cost, 'energy efficient'

Figure 6.6 Winram Place, St Andrews, Fife, 1974. Courtesy of Peter Adamson.

experiments in Scandinavia and North America. The criticisms that have attached to these houses have for the most part stemmed from their alleged insensitivity to local (and, in particular, rural) building traditions, and are indicative of a general shift of attitude, also generated by the perceived blandness of 1960s and 1970s estate architecture, that has found additional focus in the search for a recognisably 'Scottish' small house style. Although this latter process has resulted in a degree of tokenism, it has also inspired imaginative interpretations (anticipated by a narrower base of experiments in the 1930s and 1950s) of an increasingly thoroughly researched vernacular past, at least some of which can be associated with a now largely marginalised local-authority programme that constituted less than one third of the total Scottish output for 1979 to 1985.[36]

NOTES

1. We have defined the small house as one of up to five apartments because this is the number specified as desirable in the *Manual on the Preparation of State-Aided Housing Schemes* published by the Local Government Board in 1919. The 1911 Census showed that 53.2 per cent of the Scottish housing stock at that time was of one- or two-roomed dwellings (Rodger, 1989, 29). Figures from this Census, which provided many of the statistics for the 1917 Royal Commission Report, showed that over 88 per cent of the population lived in houses of up to five apartments. This chapter looks at houses and cottages of this size but, like the 1917 Report, focuses initially on the one- and two-roomed houses in which 53.2 per cent (about 2.3 million) of the people in 1911 made their homes. In the early eighteenth century a small laird's house could be of only four rooms (for example, Old Auchentroig, Stirlingshire, illustrated in Dunbar, 1966, 82).
2. See for example Simpson, Stevenson, 1980.
3. The longhouse survived particularly in the Highlands. See Loudon, 1831, 1193–4.
4. The width of the house was limited by the span possible with wooden couples (crucks) to support the roof.
5. Whyte, 1975, 64, mentions a house with a chimney inventoried in 1656; Fenton, 1981, describes the difficulty of dating developments because of regional variation and the lack of material evidence.
6. Fenton, 1978/1989, 11.
7. ibid, 5.
8. Stell, Harman, 1988, 19.
9. Royal Commission, 1917, 41 (this actually refers to tenements and reflects the Commission's opinion of them).
10. Mair, 1988, 25–7.
11. See Smout, 1970.
12. Early pattern books of cottages were published by Nathaniel Kent in 1775 and John Wood in 1781; see Robinson, 1983, 109.
13. See Dunbar, 1966, 82, 87–8, 190–1.
14. Royal Commission, 1917, 133–6, 139.
15. ibid, 139–40.
16. Stell says by 'single persons of lowly status', Stell, Harman, 1988, 21.

17. Hareshowe Farmhouse at Aden Country Park, Mintlaw, Aberdeenshire, has one which provided storage and sleeping accommodation for guests when needed.
18. For example, Doric temple as east lodge at Aberlour House by William Robertson, 1838; Baronial example at Brucklay Lodge, Aberdeenshire, by James Matthews 1849–54.
19. Flinn et al, 1977, 313.
20. Royal Commission, 1917, 47, 49.
21. Gifford, 1992, 265; See also McAra, 1975, 10–11.
22. Royal Commission, 1917, 49; and Osborne, 1990, 6–9.
23. Osborne, 9.
24. Royal Commission, 1917, 4, 182.
25. ibid, 4.
26. Scottish Housing Advisory Committee, 1944, 16–20.
27. See for example Phillips, 1922, 83–5, where the arrangement is costed at £700.
28. See especially Morton, 1936.
29. Described in McKean, 1987, 154–62.
30. Scottish Housing Advisory Committee, 1944, 9–11.
31. Christie, 1987, 5–6.
32. Niven, 1979, 96–7.
33. Fladmark Mulvagh, Evans, 1991, 141–4, 161; See also Penn, 1954, for contemporary 'free' plans by Basil Spence (at Longniddry), Stuart Matthew (Edinburgh) and Sinclair Gauldie (Dundee).
34. King, 1984, 249–50, 259.
35. Gibb, 1989.
36. Begg, 1987, 238.

BIBLIOGRAPHY

Barley, M W. *The House and Home*, London, 1963.
Beaton, E. *Scotland's Traditional Houses: From cottage to tower-houses*, Edinburgh, 1997.
Begg, T. *Fifty Special Years: A study in Scottish housing*, London, 1987.
Carruthers, A, ed. *The Scottish Home*, Edinburgh, 1996.
Christie, A. *A Guide to Non-Traditional Housing in Scotland*, Edinburgh, 1987.
Dunbar, J G. *The Historic Architecture of Scotland*, London, 1966.
Fenton, A. *Scottish Country Life*, Edinburgh, 1976, 2nd edn, East Linton, 1999.
Fenton, A. *The Island Blackhouse*, Edinburgh, 1978, 2nd edn, 1989.
Fenton, A. *The Hearth in Scotland*, Dundee and Edinburgh, 1981.
Fenton, A. *Country Life in Scotland*, Edinburgh, 1987.
Fenton, A, Walker, B. *The Rural Architecture of Scotland*, Edinburgh, 1981.
Fladmark, J M, Mulvagh, G Y, Evans, B M. *Tomorrow's Architectural Heritage: Landscape and buildings in the countryside*, Edinburgh, 1991.
Flinn, M et al. *Scottish Population History*, Cambridge, 1977.
Gibb, A. Policy and politics in Scottish housing since 1945. In Rodger, 1989, 155–83.
Gifford, J. *The Buildings of Scotland: Highlands and Islands*, Harmondsworth, 1992.
Gifford, J, McWilliam, C, Walker D. *The Buildings of Scotland: Edinburgh*, Harmondsworth, 1984.
Glendinning, M, ed. *Rebuilding Scotland: The postwar vision, 1945–75*, East Linton, 1997.
Glendinning, M, Page, D. *Clone City: Crisis and renewal in contemporary Scottish architecture*, Edinburgh, 1999.

Glendinning, M, Watters, D. *Home Builders: Mactaggart and Mickel and the Scottish housebuilding industry*, Edinburgh 1999.
King, A D. *The Bungalow*, London, 1984.
Local Government Board. *Manual on the Preparation of State-Aided Housing Schemes*, London, 1919.
Loudon, J C. *An Encyclopaedia of Agriculture*, London, 2nd edn, 1831.
McAra, D. *Sir James Gowans: Romantic rationalist*, Edinburgh, 1975.
McKean, C. *The Scottish Thirties: An architectural introduction*, Edinburgh, 1987.
McWilliam, C. *Scottish Townscape*, London, 1975.
Mair, C. *Mercat Cross and Tolbooth*, Edinburgh, 1988.
Morton, R S. Edinburgh town planning or laissez-faire?, *Outlook*, 1/1 (1936), 76–82.
Naismith, R J. *Buildings of the Scottish Countryside*, London, 1979.
Niven, D. *The Development of Housing in Scotland*, London, 1979.
Osborne, B D. Dumbarton shipbuilding and workers' housing, 1850–1900, *Scottish Industrial History*, 3/1 (1990), 2–11.
Penn, C. *Houses of Today*, London, 1954.
Philipson, N T, Mitchison, R, eds. *Scotland in the Age of Improvement*, Edinburgh, 1970.
Phillips, R R. *The Book of Bungalows*, London, 1922.
Pipes, R J. *The Colonies of Stockbridge*, Edinburgh, 1984.
Reid, R. *The Shell Book of Cottages*, London, 1977.
Robinson, J M. *Georgian Model Farms: A study of decorative and model farm buildings in the Age of Improvement, 1700–1846*, Oxford, 1983.
Rodger, R, ed. *Scottish Housing in the Twentieth Century*, Leicester, 1989.
Rodger, R. *Housing the People: The Colonies of Edinburgh*, Edinburgh, 1999.
Royal Commission. *Report of the Royal Commission on the Housing of the Industrial Population of Scotland, Rural and Urban*, Edinburgh, 1917.
Scottish Housing Advisory Committee. *Planning our New Homes*, Edinburgh, 1944.
Simpson, A T, Stevenson, S, eds. *Town Houses and Structures in Medieval Scotland: A seminar*, Glasgow, 1980.
Smout, T C. The landowner and the planned village in Scotland, 1730–1830. In Philipson, Mitchison, 1970, 73–106.
Stell, G P, Harman, M, for RCAHMS. *Buildings of St Kilda*, Edinburgh, 1988.
Whyte, I D. Rural housing in lowland Scotland in the seventeenth century: The evidence of estate papers, *Scottish Studies*, 19 (1975), 54–68.
Whyte, I D. *Agriculture and Society in Seventeenth-Century Scotland*, Edinburgh, 1979.

7 Tenements and Flats

MILES GLENDINNING

Across the modern developed world, most countries have created a strongly polarised pattern of housing in and around their main urban centres. In a process culminating in the nineteenth century, inner areas became dominated by dense, high blocks of vertically-stacked dwellings, and the outer edges by separate single-family dwellings. During the years after 1945, multi-storey blocks began to appear in the outer areas too.

This article is concerned with the urban element in this equation: the tenement or apartment block – a dwelling pattern which originally came into being in the late Middle Ages, in various locations across Europe. Among these, Scotland occupies a special status, having been not only a cradle of the urban tenement-building (and, for that matter, the ex-urban 'villa'), but also the place where decisive steps were first taken to adapt that pattern for massed reproduction in the industrial age. The present article presents a chronological summary of this process, with the intention of demonstrating its underlying continuities.

The towns which gave birth to Europe's urban norm of vertical dwelling agglomeration had, as a rule, one thing in common: separation from their immediate hinterland, whether by legal and trade barriers, by fortifications, or by natural obstacles, leading to constriction of space. In such centres – ranging from Paris and Venice to Edinburgh – the sixteenth century seemed to be a general watershed, a time at which vertical subdivision and multi-occupation began to shade into vertically-segregated new building. Edinburgh's membership of this pioneering category stemmed from the coincidence of its compact topography and constrained boundaries (unchanged for 450 years, until the 1750s) with a status as governmental seat of one of the earliest and most unified of Europe's emergent nation-states.

As a result of these influences, the town's population tripled between 1550 and 1650, reaching as many as 40,000 or 50,000 by the end of the seventeenth century. During that period, the abundance of good building stone (rare in Europe north of the Loire) made it possible for Edinburgh to gradually phase out its combustible medieval building pattern, which combined stone internal walls and timber projections. This process was directed by the capital's precocious official building-regulation system, the Dean of Guild Court. At first the existing, subdivided 'lands' were simply refaced in masonry, with ground-floor arcading, as in the case of Gladstone's Land (1620). Increasingly, however, massive, all-stone blocks of dwellings were substituted. In 1618 a visitor (Taylor) recorded that the High Street frontages were generally of five to seven storeys and 'all of squared stone'.[1] This official promotion of a co-ordinated monumentalism in dwelling design conformed to the architectural mainstream of the time, the court architecture, which tended to reinforce the new ideology of the national state through grandiose, martially-symbolic stone buildings.

This first chapter in the story of Edinburgh's tenement housing reached a spectacular climax immediately following the outlawing of all building in materials other than stone (in 1675). The capital was now rapidly growing not only in population, but also – despite the political uncertainties of the period – in wealth and cosmopolitan sophistication. In response, a range of unglamorous but vital seventeenth-century innovations, ranging from the evolution of a body of tenement law (for instance, concerning common gables) to dramatic advances in fuel technology (including the world's first undersea coal mine), made possible a quantum leap in tenement-building methods.

The first advance was in sheer scale. The rebuilding of individual blocks, on their narrow medieval sites, was replaced by concerted area redevelopment; the Netherlandish tradition of jostling gabled frontages was abandoned in favour of a gigantic, cubic classicism, broad as well as high. And the social meaning of the tenement changed. Makeshift expediency was swept aside by the demands of the wealthy: convenience, comfort and prestige. Milne's Square and Milne's Court (1684 and 1690; fig. 7.1), with their dignified panelled or stuccoed interiors, established a resounding precedent for fashionable living in monumental blocks which, in Steel Maitland's words, 'towered to 12 and 14 storeys at a time when Manhattan Island boasted only the wigwams of the Iroquois Indian'.[2] This pattern, which fused Edinburgh's dense urbanism with the sophistication of the Parisian hôtel, was continued early next century at the adjacent James Court (1723).

At the same time, other markers for the future were being laid down outside Edinburgh. In subordinate centres such as Glasgow (after fires of 1652 and 1677), Dundee, Perth, and Stirling, a more modest process of stone-refacing and tenement-building was proceeding. Near the capital, at places like Royston (Caroline Park), a new type of stone-built 'rural' dwelling began to appear, reviving the classical ideal of the villa-retreat from the congested city. Here, as in Paris, an inter-relationship of 'apartments' and 'villas' for the new wealthy urbanites was being established.

During the eighteenth century, the existing pattern of tenement-

Figure 7.1 Milne's Court, Lawnmarket, Edinburgh. Crown copyright: RCAHMS, SC 711518.

building was steadily developed into a mature national tradition of monumental housing and city design. Yet, since the parliamentary union with England at the beginning of the century, there no longer existed a national government to sponsor such a programme – a striking contrast to equivalent European experience, such as that of France. The gap in patronage was partly filled by burgh initiative: the co-ordinating efforts of the Convention of Royal Burghs were already helping overcome the economic consequences of the events of 1707, which had inhibited building activity for half a century.

The key decision was taken by Edinburgh Town Council in the 1760s to create a completely new development for wealthier citizens, by at least extending the boundaries of the ancient royalty. This idea, originally suggested in 1680, with royal support, was taken up in 'Proposals' published in 1752 by the Convention of Royal Burghs: financed by revenue from mining and commerce, the project was viewed as a prestige scheme to rival overseas capitals and counteract the 'pull' of London.

What was novel about the Edinburgh New Town was its extension of the principle of monumentality in house design to cover a whole, self-contained district. Paradoxically, it, and immediate predecessors such as Argyle Square (1742), had at first seemed instead to presage a different pattern of urban living: dwellings ranged in horizontal rows 'after the fashion of London'.[3] In 1791, for instance, Robert Adam was told by the Town Council to design part of Charlotte Square as 'all Lodgings, not houses to set in flates'[4] (the latter word seen here, incidentally, in transition from its old meaning of 'storey' to its present usage). But, as the New Town spread, the national mainstream reasserted itself – now in the form not of deep courtyard blocks like Milne's Court (a plan type which, just then, was becoming established as a norm for the Paris apartment block) but of the shallow 'terrace' lining a street-grid.

Although Adam's 'megastructure' projects provided a spectacular overture to this fusion of neo-classical sublime with national monumentalism, it was the austere tenements of the later New Town (some containing as many as four floors and two sunk storeys), and the vast ranges laid out in Glasgow between 1790 and 1830, in new areas south of the Clyde such as Laurieston and Hutchesontown, which decisively set the pattern for the prodigious tenement-building of the nineteenth century. At first, elements of the 'English house' image were retained (for instance, individual porticoes to main-door flats in Burn's Henderson Row, Edinburgh, 1825–6); but these were gradually whittled away. This quarter-century head start over the commencement of large-scale tenement-building in other northern European countries ensured that neo-classical homogeneity, rather than variegated 'Renaissance', would be this country's governing stylistic framework.

Such preferences were given legal backing by a key judgement in 1818,[5] which caused the rapid spread of 'real burdens' (conditions attached to disposition of property such as development land): this made possible co-ordinated tenement façade design in towns across the country, to a degree unmatched in Europe.

The tenement dwellings of the Glasgow and Edinburgh 'New Towns' were chiefly for the middle classes; they carried further the process of limited household self-containment (for instance, WCs started to appear around 1790 in Edinburgh), while providing larger rooms. But simultaneously, purpose-built tenement provision was extended to the new industrial proletariat. Smaller urban centres had already further developed the traditional two-storeyed flatted pattern (for instance, by tidying away its forestair to the back), and this new lesser-urban type had been used for late eighteenth-century planned developments in a rural setting, such as Catrine, Ayrshire, or Inveraray, Argyll. At New Lanark, Lanarkshire, however, there was a dramatic departure: the Edinburgh New Town formula was extended to the building of a complete planned industrial settlement, including tenement blocks of three to four storeys (built 1785–98) – a settlement which then (from 1799) became the scene of Robert Owen's Utopian community experiment.

Thus, by 1830, all the elements were in place for Europe's first comprehensive extension of the urban tenement tradition into the post-Industrial Revolution world, providing for both the new proletariat and bourgeoisie. It is no surprise that, on his visit to these islands in 1826, Karl Friedrich Schinkel – the foremost architect of Prussia, a country poised to unleash its own industrialisation drive – should have taken care to visit Edinburgh, New Lanark and Glasgow, where he hailed the new dwelling monumentalism for its 'magnificence' and 'purity'.[6]

During the nineteenth century, this monumental tradition became transformed into a great nationwide movement, matching others across Europe. In contrast to the following century, the main engine of this activity was not social concern but pursuit of profit – to be gained, eventually, through building and renting, but, more immediately, through speculation in building land.

Fuelled by industry-led demographic pressure – an explosion in population, and in the degree of urbanisation – the colossal strength of land speculation perpetuated, and multiplied on a vast scale, the existing dense urbanity of cities such as Berlin, Copenhagen, Vienna, or Glasgow. In this country, land speculation was mainly channelled through the process of sub-infeudation, a mechanism through which (although prohibited in Royal Burghs until 1874) men such as Sir James Steel made large fortunes.

Despite the chronological and international continuity of its general physical form, the pace of tenement development fluctuated dramatically, channelled by interest rates into booms and slumps which varied from country to country and city to city, influenced in turn by the wider economic climate. Thus, while still-industrialising Berlin lagged ten years behind Glasgow's 1873–7 tenement-building boom (when nearly 5,000 tenement dwellings were built each year in the city), the next spurt (around 1900) coincided in Germany and Scotland. This jagged pattern of development was emphasised by the small size of the firms generally involved; as on the Continent, large speculative builders only emerged late in the nineteenth century. Small builders of tenements were partly cushioned by the fact that lower interest rates were charged on loans than in the case of villa-building.

What were the broader trends which underlay these dramatic fluctuations in the market? Despite investigations by Daunton and others,[7] many things are still not clear: for instance, the extent of subletting, or tenant perceptions of rent levels. But some key processes may at least be indicated. The first was a steady extension of the scope of new building further and further 'down-market', to the point where, by 1900, new flats were being provided for all but the unskilled working class. In parallel, there was a gradual reduction in the size, but increase in the facilities and amenities, of the 'best' dwellings: thus, a degree of convergence. Yet, as elsewhere in Europe (for instance, in Germany or Austria), the creation of large new districts of either proletarian or bourgeois tenements, along with the widening of the scope of villadom on the outer edges of cities, resulted in some degree of social segregation. This gradual extension of the social-economic scope of new building was matched by rises in average rents in the late nineteenth century.

The free interaction of demand and supply was also regulated. Within the feudal system itself, the inherent pressure for once-and-for-all dense development was offset by an increasing orderliness, caused by the wide use of real burdens to restrict density and compel formal homogeneity: for instance, in Glasgow's West End, or on the Heriot Trust's 1,600-acre land holdings in Edinburgh. These orderly developments were a privately-led, less ambitious equivalent of the town-extension Fluchtlinien (Alignments) overseen by police authorities in Germany; but both systems had the effect, throughout the nineteenth century, of encouraging the expansion of existing centres rather than the mushrooming of unplanned industrial towns, as in England. The lack of vast grid blueprints like the Vienna Ring plan of 1857, Berlin's 1862 Hobrecht Bebauungsplan (House-building Plan), or Stockholm's 1860 Lindhagen plan, did not deter vigorous town expansion on roughly the same pattern in this country.

However, one official system of control did exert a considerable influence, that of building regulation. Here the foreign parallels were striking, especially in their effects on building form. In Germany, for instance, the police authorities tried, in the 1870s and 1880s, to curb poor construction in tenements by enacting strict building regulations. One effect of these was to add to building costs, and encourage bigger and more solidly built blocks. In this country, the 1862 Burgh Police Act, which developed the Dean of Guild tradition, likewise set exacting standards of construction and provision. It was adopted first by Glasgow and then (by 1880) 11 other burghs, and extended to the country areas in 1897. Further nationwide Acts, each triggering a flood of local Acts, were passed in 1892 and 1903. By the early 1900s in Edinburgh, for instance, a maximum of 12 dwellings per close was permitted, and bed recesses were banned. Local Acts in Glasgow and Paisley began to break up street blocks with ventilation gaps. Minimum stipulations as to dwelling size – significantly, by volume – were laid down. These were not very stringent: for instance the 1903 Act specified 1,000 cubic feet ($c.$ 33.9 cubic metres) for a one-roomed dwelling. Such measures, taken cumulatively, obviously added to construction costs, and were subsequently denounced by critics anxious to bring this country's dwellings 'into line' with England's cheaper, more lightly-built patterns: they claimed that tenements' 'excessive solidity' made rents 'too high'. Yet this basic corpus of controls would remain in force until the 1950s, through a period when the lowest rents (in real terms) in the country's history were being charged!

Further remarkable intervention in the workings of the market came in the form of Europe's first major area redevelopment programme, beginning with the Edinburgh Improvement Act 1827, slum-clearance in the Glasgow Improvement Act 1866, and follow-up measures in Edinburgh (1866), and Dundee (1871). Although these were far more modest than the vast sponsored programme of 'Haussmannisation' which had swept across urban France since the 1850s, it still produced substantial results. Edinburgh saw the start of a programme of Old Town remodelling in self-conscious revival of court architecture, while in Glasgow the more urgent matter of slum clearance was the focus: over 16,000 houses were cleared by 1914, and, from the 1880s, the City Improvement Trust became drawn, by default, into building some replacement tenements itself. This achievement bolstered Glasgow's worldwide reputation as a centre of 'municipal socialism', and laid foundations for momentous policy changes after 1918.

But – looking at nineteenth-century tenement buildings as a whole – what kind of physical forms, 'on the ground', resulted from all this prodigious energy? As noted above, the years around 1800 had established, as a norm, the shallow-plan 'terrace' built as part of a stylistically homogeneous neo-classical scheme. The nineteenth century generally kept to this pattern: there were no deep-plan courtyard blocks as in Paris or Berlin, or deep lots with tiny lightwells in the manner of the New York tenement; and, for all the Glasgow 'backlands' and ground-level 'saloon' accretions, there was no general practice of building large, full-height tenements on the inside of newly developed street blocks. What was also noticeably different from the Continent, and closer to England, was the profusion of external drainpipes which became normal after about 1850 (sometimes, as in Paisley, even on front façades).

And while overall order in continental town planning was undoubtedly promoted by extension plans, Fluchtlinien or grands boulevards, the picture within individual streets there was far less orderly, owing to the custom of building each tenement as an individual design, generally aligned with, but differing in detail from its neighbours. In this country, by contrast, blocks were generally a storey or two lower than in Germany, Denmark or France (or, for that matter, than the largest mid century London 'terraced houses'), and lacked equivalents to the French attics or (excepting Greenock) German 'souterrains'. But the monumentality lost in sheer height was more than regained through the architectural unity of these long ranges, which were one of the climactic elements in this country's final, resounding development of the European neo-classical tradition.

What were the key elements in this national style of urban dwelling groups, which was brought into being largely through the co-ordinating mechanism of post-1818 real burdens? The first was an insistence on masonry facing – at least on the front walls – and not just for prestige developments, as had been the case in the Edinburgh New Town, but for all new blocks, however humdrum. This new uniformity of stone-built quality was made possible not just by the nineteenth-century transport revolution, but also by great advances in the technology of stone preparation, in particular, the development of machine tools from the 1820s. Once this tradition of masonry monumentalism was established, it proved resilient in the face of further economic change: in the 1860s and '70s the response to a drop in the relative cost of brick – a material viewed, like stucco, as mean and inferior – was to use more of it, but in non-facing contexts (like timber already). This neo-classical perspective should be distinguished from England's 'High Victorian' movement for 'genuineness' in stone facing. Another consequence of the neo-classical affiliation of the nineteenth-century 'tenement style' was an avoidance of the bristling 'free Renaissance' applied details of, say, German Gründerzeit tenements, in favour of sparse architraves and string-courses, and an almost brutally severe (from a continental viewpoint) cornice and roofline.

The mid/late nineteenth-century call for greater façade modelling, in combination with the convergence in house sizes, was answered not only by a reversal of the previous pattern of uneven storey heights and even spacing of bays, and by an abandonment of overall 'palace' compositions in favour of homogeneous rows, but also by the dramatic device of the bay or oriel window. This, as an element in grouped housing, had originated in England (around 1825) as part of a move away from austere neo-classicism. However, when applied to tenement blocks in this country (first by Thomson in 1857), especially with more

Figure 7.2 Walmer Crescent, Glasgow. Crown copyright: RCAHMS, SC 710986.

vertical window-proportioning, it had a different effect: not to 'break up' long façades, but to accentuate their unity and monumentality, acting as a kind of giant order (fig. 7.2). A counter tendency, balancing massiveness and asymmetry, was introduced by the adaptation of the 'Baronial' court revival style to Edinburgh tenements in the 1860s. However, by c. 1890–1900, this tendency had been re-integrated with the classical mainstream, giving Glasgow turn-of-century streets, with their Baroque corner outcrops, a somewhat Haussmann-like air.

The shallow but broad (in European terms) 'terraced' building pattern had many consequences for tenement plan types. Robinson's recent investigation has defined two broad groups: a three- or four-storeyed category for bourgeois or proletarian occupancy in larger centres (the equivalent of Miethaus [tenement] and Mietkaserne [block of flats]), derived from the 'New Town' type, and a country-town two-storeyed type (equivalent of the Bürgerhaus [middle-class house]).[8] The first type was confined to larger centres: its normal height was four floors and sunk storey in Greenock and Leith, four floors in Glasgow, Edinburgh, Paisley and post-1880 Dundee, and three or four storeys in Clydebank. It had an internal staircase and, because of its shallow plan, no more than 20 dwellings per stair (Leith): the Glasgow working-class norm was 12 (two two-apartment and one one-apartment on each floor), but very large tenement dwellings were still built for the rich. The second type, with its external stair or (sometimes) segregated entrances, continued to predominate in lesser centres; with extra storey(s) added, it also, in Robinson's opinion, formed the basis of the mid-century rear-balconied Dundee type, and the Aberdeen tenement, with its prominent attic and prolific internal use of timber (for example, for staircases). This small-town pattern was also built in Edinburgh by philanthropic building companies for a time in the 1850s and 1860s.

How these blocks were used and experienced is less easy to reconstruct, eighty years after World War I turned upside down the urban society they housed. Among better-off people it is clear that, although the villa and 'terrace house' attracted some, the convenience of apartment life had lost none of its appeal. Here the process of qualified self-containment continued: an internal WC was universal in Glasgow middle-class tenements by the mid nineteenth century, and after 1880 bathrooms spread.

Among the increasing proportion of working-class people who could afford to rent a new dwelling, residential patterns were far more constrained by necessity. Yet the remarkable 1860 *Report* of Edinburgh's Committee of the Working Classes argued for the standard four-storeyed tenement not as a matter of expediency, but for its warmth and convenience, and as a symbol of collective social organisation and national tradition. The committee rejected the English-style suburban cottage dwelling, with its many small, use-differentiated rooms, as inconvenient and socially divisive, arguments endorsed by French workers of the time, and German commentators such as Hobrecht.⁹ Here, as in most other parts of Northern Europe, working-class life had no single spatial focus: with male culture sharply orientated away from the dwelling, the functions of cooking and sleeping seemed most conveniently and sociably housed in lofty but warm spaces, with purpose-built bed recesses. But while the big kitchen/living room was a European norm, the continental free-standing stove was not taken up in this country; instead, the traditional fireplace, combining cooking and space-heating functions, was retained.

It is difficult, now, to imagine life in these dwellings, with their combination of openness to the public domain, and intense, collective territoriality. Of course, the economic insecurity and harshness of working-class existence was reflected just as much in the frequent sub-letting of these purpose-built new dwellings as in the physical squalor of the 'made down' older houses of the unskilled poor. But this problem of over-occupancy should not be glibly conflated with national cultural patterns of habitation and social morality. A newspaper print worker, asked by the 1860 Edinburgh Committee to define his ideal urban 'working-man's dwelling', replied: 'A but and a ben, with a water closet and a soil pipe, is all I want'.¹⁰

It was only the gross social and economic distortions brought by World War I which allowed this popular and entrenched pattern to be substantially disrupted. The English Utopian alternative to dense European urbanism – the self-contained 'garden city' of cottage dwellings – was built on a limited scale in many countries (for example, in Germany from 1902), without disturbing the basic continuity of the high-density mainstream. Only in this country was it imported wholesale, as part of the centralising 'British' social and economic policies of the wartime and post-war years – notably the government's building of huge cottage housing schemes for munitions workers, and its suppression of the private housing market in favour of municipally-organised provision.

Emboldened by the war's calamitous effect on the economics of stone building, which cut the market for building sandstone by 95 per cent overnight, and seemed to point to an architectural miniaturisation of new housing, housing reformers in senior government positions issued voluminous official reports (above all, a Royal Commission report of 1917–18) which branded the tenement as an outmoded aberration from the 'norm' of the 'house' (used in its restricted English sense). These instructed the new municipal housing authorities to build, as far as possible, cottages and two-storeyed flats; tenements were to be limited to a three-storey maximum and limited in density.

In the chaotic aftermath of world war, this attempt to overthrow the previous norms seemed, for a time, to meet with general acceptance. However, the reformers' apparent success was in fact largely an illusion: within a few years, a range of pressures, ranging from the banal to the elevated, would undermine it.

No more than three years had passed since the 1919 'Addison Acts' when the reformers suddenly found that their success in transferring housing from the fragmented and conservative speculative builder to the co-ordinated municipality – the traditional focus of urban order and initiative – was beginning to backfire on them. The local authorities, led by Glasgow Corporation, were starting to have ideas of their own. Finding that, as cottages were dearer to build than tenements, they were having to charge much higher rents for dwellings which did little more than rearrange the same living space in more but smaller rooms, they began to take matters into their own hands.[11]

Defying the protestations of the weak government housing department (the Scottish Board of Health – later the Department of Health for Scotland), the burghs and counties began building large numbers of the officially outlawed two-roomed flats. For artisan tenants, these took the form of two-storeyed, 'four in a block' dwellings, which, in towns, preserved elements of the garden suburb image, while 'urbanising' many rural places for the first time (fig. 7.3). For those rehoused directly from slums or overcrowded houses, there was a return to the pre-1914 type of bay-windowed tenements on street-grid layouts – although with lower density, bigger air gaps, three low rather than four tall storeys, and 'nomenclature ... changed to the more grandiloquent one of "block of flats"'.[12] The effects of cost restrictions ensured that ceiling heights, now cut to English levels, would never be restored. So the two-apartment interwar flat, although better equipped than its 1900 counterpart, was in fact smaller.

What had happened was, in effect, a qualified resumption of the nineteenth century's steady progress in 'down-market' expansion of the scope of new rental tenement building, to include for the first time the unskilled poor. But the takeover by the municipalities of new middle- and working-class rental building as a whole (including this new unskilled 'sector'), coupled with the corrosive effects of rent control, inevitably restricted the private market in new housing. Although there was some building for rent (mostly with government subsidy) by developers such as Sir John Mactaggart, this field soon become dominated by building for owner occupation. Although some high-class tenements and many four-in-a-block flats were privately built, speculative buildings' main adaptation of the tenement tradition took a more unexpected form, conditioned by cheap suburban transport: the bay-windowed, stone-faced bungalow, whose horizontalising massiveness perpetuated key elements of the monumental tradition of dwelling design.

Figure 7.3 Glenboig, Lanarkshire, inter-war settlement of two-storeyed flats built by Lanark County Council; aerial view. Crown copyright: RCAHMS, SC 710970.

This process of instinctive survival, however, was powerfully reinforced at the end of the 1930s by a conscious tenement revival. An offshoot of the contemporary movement for Scottish cultural regeneration, this drew inspiration both from Scotland's own architectural heritage and from contemporary European Modern housing architecture. Here, a vital turning point was a 1935 continental visit, and published report, by a DHS-sponsored committee. One key member of the delegation, Edinburgh's City Architect, Ebenezer Macrae, then proceeded to denounce cottage and two-storeyed flats as anti-national building types, and began to design striking tenement schemes combining tradition and modernity, above all at Piershill (1937–41; fig. 7.4). The interwar years also saw a vigorous reaction against the use of brick, especially in the east and north east: some remedy was provided by

Figure 7.4 Piershill, Edinburgh; aerial view. Crown copyright: RCAHMS, SC 417243.

limited stone facing, some by harling, but concrete (usually in blockwork form) also began to come into its own as a potential way of perpetuating the masonry tradition.

World War II, with its fresh outpouring of initiatives of state co-ordination, seemed for a few years to reinvigorate the British housing reformist movement, above all through its new doctrine of regional planning, which promised to cut urban densities and municipal power simultaneously, by exporting population to self-contained 'New Towns'. But this movement turned out to be essentially rhetorical in character, and its first real success, a curb on Glasgow's building land through a 'green belt', only spurred the tenement revival to new heights, in a literal sense! Glasgow Corporation's first reaction to the green belt was simply to intensify tenement building on the huge peripheral schemes. Then, as the city's land supply became yet more scattered, it fervently embraced that new building form, the modern tower block, chiefly at the instigation of David Gibson, its firebrand socialist Housing Convener (between 1961 and 1964).[13]

Where the driving motor of nineteenth-century tenement building had been the profit motive, that of their municipal successors was output: to build as many low-rental dwellings as possible for 'their' people. The multi-storey building of the 'sixties was the climax of a municipal crusade which had revolutionised housing tenure. By 1964, 79 per cent of new Scots housing was municipally built (96 per cent in Glasgow). This was by far the highest figure in Europe. Even the USSR was achieving only 6 per cent, while most other West European countries were remaining faithful to pre-1914 patterns. The publicly-built share of West German output was only 2.5 per cent, and that in Belgium a mere 0.3 per cent.

Architecturally, however, the post-1945 period marked a re-convergence with the European and national mainstream. Indeed, like the Edinburgh New Town and New Lanark, Gibson's high blocks anticipated a trend which was to sweep Europe, in this case, in the later 1960s and the 1970s. This was firstly because of their sheer height. Schemes such as the 31-storey Red Road (from 1961; fig. 7.5), inspired by US skyscrapers, were unprecedented in the Europe of the early 1960s, and seemed to look forward to the public housing of Hong Kong in the 1980s. But more important than the height was the general physical pattern: a revival, in modern form, of the neo-classical willingness to impose a formal, unified pattern, and its liking for the massive aesthetic of the sublime. The rectilinear, hard-faced severity of slab-block schemes such as Sighthill (1962–9) closely paralleled the grands ensembles of France, and anticipated the extension of the same aesthetic to Eastern Europe. A later irony should, however, be noticed: the output yields of the urban multi-storey drives then inspired the rural authorities – the powerful county councils – to start their own crusades. This actually

Figure 7.5 Red Road development, Glasgow. Courtesy of City of Glasgow District Council, Department of Architecture.

caused an increase in the national proportion of cottage building, with tenement types squeezed out, for a time, between cottages and high flats.

Socially, the 1950s tenements and 1960s high blocks continued the broad interwar scope of provision, catering for all of the working-class, and some middle-class, tenants. Housing as they did thousands of unskilled tenants in thoroughly modern, yet collectively-arranged, dwellings, these blocks completed the characteristically national, and mainstream European, process by which self-sufficient dwellings were provided for all without, in so doing, atomising urban society.

Inevitably there came a time when, the public clamour for output having largely been satisfied and the worst slums demolished, much of the impetus behind multi-storey building dissipated. In 1974 a British town-planning historian, Roger Smith, speculated that 'perhaps the tide is beginning to turn and traditional multi-dwelling building in Scotland is on the wane. The anti-tenement feeling generated at the beginning of the twentieth century may at least have become effective.'[14] In fact, the very opposite was to be the case. The ending of the multi-storey drive saw a dramatic turn around in the public estimation of the previously reviled nineteenth-century tenement, of which the many survivors were largely post-1870 blocks built to Burgh Police Act standards. This reversal, in turn, had the effect of obscuring the considerable cultural similarities between tenements and multi-storey blocks.

The first phase in this movement, in the early 1970s, was rehabilitation, spearheaded by housing co-operatives and the architectural group ASSIST. Then, in the 1980s, following an influential study by the Scottish branch of the National Building Agency, the building of new tenements of traditional street-block type was begun. Although, encouraged by Glasgow planners, some tenement revival architects during this period made extensive and eventually, controversial, use of brick facing and neo-vernacular detailing, a potentially more productive way forward – a revival of concrete blockwork and generally classical forms – was indicated in 1984 by Ken McRae's competition-winning design for a 'Tenement for the Twenty-First Century', built at Maryhill.[15]

By the beginning of the 1990s, therefore, the stage seemed set for a reassertion of the continuity of the tenement tradition (on a scale intermediate between the terrace tenement and the twentieth-century 'multi'), a development which would take national housing patterns, in some respects, full circle, back to the classical monumentalism of the 1690s.

NOTES

1. Brown, 1973, 110.
2. Maitland, 1952, 44.
3. Youngson, 1966, 18.
4. Information supplied by Ranald MacInnes.
5. Gordon v Marjoribanks: Reid, K G C. Legal background, in 'Studying the Scottish Home', a conference held at the National Museums of Scotland, 5 November 1992.

6. Bindmann, Riemann, 1993, 150, 174, 197.
7. See for example, Daunton, 1983.
8. Robinson, 1991, 445–8.
9. Muthesius, 1982, 256.
10. *Report of a Committee of the Working Classes of Edinburgh*, Edinburgh, 1860, 39.
11. Information from Louise Christie, University of Edinburgh.
12. Maitland, 1952, 48.
13. Glendinning, Muthesius, 1994, Chapter 25; Horsey, 1990.
14. Smith, 1974, 238.
15. Glendinning, MacInnes, MacKechnie, 1996, 498; Horsey, 1990.

BIBLIOGRAPHY

Ballantine, W M. *Rebuilding a Nation*, Edinburgh, 1944.
Begg, T. *Fifty Special Years*, London, 1987.
Bindmann, D, Riemann, G, eds. *K F Schinkel: 'The English journey'*, New Haven, 1993.
Brown, P H. *Early Travellers in Scotland*, Edinburgh, 1891, reprinted 1973.
Corporation of Glasgow Housing Department. *Review of Operations, 1919–1947*, Glasgow, 1947.
Daunton, M J. *House and Home in the Victorian City*, London, 1983.
Department of Health for Scotland. *Report on Working-Class Housing on the Continent*, Edinburgh, 1935.
Doughty, M, ed. *Building the Industrial City*, Leicester, 1986.
Frew, J. Ebenezer Macrae and reformed tenement design, 1930–1940, *St Andrew Studies*, Vol 2 (1991), 80–7.
Glendinning, M, Muthesius, S. *Tower Block*, New Haven and London, 1994.
Glendinning, M, MacInnes, R, MacKechnie, A. *A History of Scottish Architecture*, Edinburgh, 1996.
Gomme, A, Walker, D M. *The Architecture of Glasgow*, London, 1968, 2nd edn, 1987.
Gourlay, C. *Elementary Building Construction*, Glasgow, 1903.
Gow, I R. The Edinburgh villa, *Book of the Old Edinburgh Club*, new series 1 (1991), 34–46.
Horsey, M G, for RCAHMS. *Tenements and Towers: Glasgow working-class housing, 1890–1990*, Edinburgh, 1990.
Lynch, M. *Scotland: A new history*, London, 1991.
Markus, T A, ed. *Order in Space and Society*, Edinburgh, 1982.
Meade, M K. Plans of the New Town of Edinburgh, *Architectural History*, 14 (1971), 40–52.
Morgan, N J. £8 cottages for Glasgow citizens: Innovations in municipal house-building in Glasgow in the inter-war years. In Rodger, 1989, 125–54.
Muthesius, S. *The English Terraced House*, London, 1982.
National Building Agency. *Residential Renewal in Scottish Cities*, Edinburgh, 1981.
National Museums of Scotland. Conference 'Studying the Scottish Home', 5 November, 1992, lectures by W Baird, J Hume (Stone-building technology), K G C Reid (legal background), and T C Smout (Coal technology and Old Town high building).
Plunz, R. *A History of Housing in New York City*, New York, 1990.
Rasmussen, S E. *Towns and Buildings*, Copenhagen, 1951.
Report of the Royal Commission on the Housing of the Industrial Population of Scotland, London, 1917.

RCAHMS. *Inventory of Edinburgh*, Edinburgh, 1951.
Robinson, P. Tenements: A pre-industrial tradition, *ROSC*, 1 (1984), 52–64.
Robinson, P. Tenements: The industrial legacy, *ROSC*, 2 (1986), 71–84.
Robinson, P. 'Aspects of a Scottish Flat Tradition', unpublished PhD thesis, University of Strathclyde, 1991.
Rodger, R. The Victorian building industry. In Doughty, 1986, 151–99.
Rodger, R, ed. *Scottish Housing in the Twentieth Century*, Leicester, 1989.
Simpson, A T, Stevenson, S, eds. *Town Houses and Structures in Medieval Scotland: A seminar*, Glasgow, 1980.
Simpson, M A, Lloyd, T H, eds. *Middle-Class Housing in Britain*, Newton Abbot, 1977.
Simpson, M A. The West End of Glasgow, 1830–1914. In Simpson, Lloyd, 1977, 44–85.
Smith, R. Multi-dwelling building in Scotland. In Sutcliffe, 1974, 207–43.
Steel Maitland, J. Scottish housing, *Quarterly Journal of the RIAS*, 89 (August 1952), 42–50.
Stell, G. Anatomy of the tenement, *The Scotsman* (24 September, 1977).
Stell, G. Scottish burgh houses, 1560–1707. In Simpson, Stevenson, 1980, 1–31.
Sutcliffe, A, ed. *Multi-Storey Living*, London, 1974.
Sutcliffe, A. *Towards the Planned City*, Oxford, 1981.
Walker, F A. The Glasgow grid. In Markus, 1982, 155–99.
Worsdall, F. *The Tenement*, Edinburgh, 1979.
Youngson, A J. *The Making of Classical Edinburgh*, Edinburgh, 1966, 2nd edn, 1988.

8 Inns, Hotels and Related Building Types[1]

DAVID WALKER

EARLY INNS

No surviving building in Scotland, proven to have been built as an inn, is older than the seventeenth century, although earlier origins are claimed for a number of sites such as that in Tarbet, Loch Lomond, Dunbartonshire, which has a recorded history extending back to the sixteenth century. Among other early foundations are, or were: the Dove Inn in Glasgow's High Street, which was certainly an inn in Stuart times and bore a dove with the date 1596; Tweedsmuir, Peeblesshire, where an inn was established in 1604; the St George's Hotel at Dunbar, East Lothian, which is proudly inscribed 'Aedificata 1625', although the present building was 'Renovata 1828'; and the Boar's Head, now the Gardenstone Arms, at Laurencekirk, Kincardineshire, which is dated 1638.

As inn sites, some hotels in the more ancient burghs probably have an even older origin, but without an examination of the titles – which do not always specify how the building was used – or any other records as may survive, it is difficult to say which town houses were adapted for the purpose, like the Cross Keys at Peebles, built in 1693 as the town house of the Williamsons of Cardrona, and which were purpose-built. The former Bluebell Inn at Haddington, East Lothian, three-storeyed with a corbelled stair-turret on its front elevation, is certainly of late sixteenth- or early seventeenth-century date, and had a long, though indeterminate, history as an inn, while the Stag Inn at Falkland, Fife, is dated 1630. The Falkland Arms in the same town has long been an inn, but its 1607 inscription refers to it as a house. Of similar vintage is the former Whitehorse Inn at the head of the close to which it gave its name in Canongate, Edinburgh, built or rebuilt in the earlier seventeenth century for Laurence Ord. Its central dormer, partly re-cut in 1930, is dated 1623 and corresponds with that date although some had read it as 1523. It certainly had a long history as an inn, particularly as a departure point for the London coach. Restored in 1889 and drastically rebuilt in 1962,

Figure 8.1 The Black Bull, 12 Grassmarket, Edinburgh; early twentieth-century view. Crown copyright: RCAHMS, SC 711086.

it was a long parallelogram 24 feet (7.32 metres) deep, single-storeyed, dormer-headed attic and garret to the court, with vaulted stables beneath. Its plan, as it existed before rebuilding, suggested that it could have been a purpose-built inn, being divided into two quite separate sets of lodgings with very thin partitions between the rooms. Access to these upper floors was provided by a common forestair to twin stair-towers cantilevered out in heavy timber with lath and plaster walls.

A more fashionable establishment, most probably a converted building, existed in Boyd's Close, St Mary's Wynd, in the same city, and was sometimes called Boyd's Inn. Demolished in the 1860s, it was the preferred arrival point for the better-off travellers from London. Some of these more up-market establishments were said to have had 'English rooms, where English travellers could eat and converse together.' In the Grassmarket, Edinburgh, the now-demolished Black Bull at No 12 (fig. 8.1) was probably purpose-built. In its final form it was of L-plan around a close with stables, the rear wing being of the same height as the front block and pantiled. The front block was very deep in plan, three storeys and attic high. It had a huge crowstepped gabled roof of slate with small gableted dormers, a lead flat instead of a ridge, and a massive central ridge-stack. Early photographs suggest that it may originally have had an 'M'-roof. In elevation it had twelve fairly regular widely-spaced windows, six each at first and second floor, probably of mid to later seventeenth-century date. Just off-centre in the fourth bay of the elevation was an arched pend to the stable court.

The Black Bull's general arrangement of frontal inn with a pend to stabling at the rear was to remain typical of the larger burgh inns for the next century and a half. They served as taverns, stage-coach arrival and departure points, and in many instances as mail offices. Not all inns, however, offered such extensive accommodation as the Black Bull. The English traveller Topham, arriving in the Pleasance in the earlier 1770s at what he was assured was the best inn in Edinburgh, found reception to be a room in which twenty drovers were washing down a meal of potatoes with whisky, and was informed that he and his companion could have no accommodation unless they were prepared to sleep in the same bed, and even that was in a room already occupied by travellers from the previous stage-coach. Like others unable to cope with tavern life and the arrival and departure of coaches and carriers, he found lodgings as quickly as possible with one of the city's numerous 'room setters', who provided accommodation ranging from single rooms or self-catering rooms-and-kitchen, to suites of apartments with silverplate, china, table linen and wines and spirits, as at Mrs Thomson's at the Cross in the 1750s.

Outwith the capital a number of fairly early purpose-built inns survive. The two-storeyed and attic double-gable front of the Hawes Inn at South Queensferry, West Lothian, is unmistakably late seventeenth century, but it was originally built as the house of the Stewarts of Newhalls, not as an inn. The Salutation at Perth is dated 1699 at a chimneypiece but was remodelled in the early nineteenth century, at which date it was probably heightened. A much less altered building is the hotel of the same name at Kinross, dated 1721, a tall three-storeyed, L-plan building like a tower-house. It is a well-built structure with good dressed margins at the angles, straight skews and skewputs and panelled ashlar stacks with moulded copes. The skews and thackstanes indicate that it was originally thatched.

A few other early inns survive dotted around the countryside. The two-storeyed building which was formerly the Rosslyn Inn at Collegehill, Roslin, Midlothian, is dated 1660. Its original nucleus appears always to have been three-windows wide, but it has been so much rebuilt in the eighteenth century that it is difficult to be sure of its original appearance and plan. The Sun Inn at Newbattle, again in Midlothian, is still externally very much as built in 1697 by James Chirnsyde and marks the first stage on the coach route from Edinburgh to Galashiels. It conforms to the two-storeyed, five-window symmetrical-fronted format of the smaller laird's house of that period, and is a very superior building with a moulded architrave doorpiece, raised quoins, a big ingleneuk at the kitchen and panelled stacks. Later but rather similar was the now much rebuilt, early eighteenth-century Manor Inn at Lanton, Roxburgh, which, although constructed with mud mortar, was dignified by an architraved doorpiece.

At Dunkeld, Perthshire, two very early inns survived unaltered until the 1950s but have been much renovated since. As first built, the former Duke's Inn on the south side of the Cross appears to have been a low, two-storeyed and attic building, four windows wide. A subsequent two-windows wide addition has an arched pend, once leading to stabling at the rear. At the north end of the town, adjacent to the barn court of Dunkeld House, was the inn variously known as Laird's Inn and the Atholl Inn, a still lower two-storeyed building, very thick walled and clearly early. Nevertheless, it was symmetrical-fronted with four windows and a central door at the ground floor and three slightly taller windows at the first floor. It had much less provision for horses than the Duke's Inn, however, one side of its court being occupied by an Independent (Congregational) Church. It was evidently intended for a different class of traveller.

In the old town of Dollar, Clackmannanshire, there survives another very early inn, which similarly had scant provision for the accommodation of horses; perhaps not much was required on the drove road through Glenquey. Whether the building is of late seventeenth- or earliest eighteenth-century origin is difficult to determine. It is still crowstepped in an area where straight skews were more usual by that date, but the two-storeyed, laigh floor and dormerless attic three-windows wide format seems unlikely to be earlier. Its two-windows wide annexe, presumably for the innkeeper's family or perhaps for a different class of traveller, appears to be of the same vintage.

Not all inns were as large as these. The early eighteenth-century, two-storeyed and (originally) dormerless attic inn at Weem, Perthshire, now an annexe to that built later in the century, was a simple rectangle only two-windows wide with end-stacks. On the Edinburgh-Lauder route at Blackshiels Inn, Fala, Midlothian, there is a rather similar building, now also an annexe to a later one. It is of two storeys and a very low attic with swept dormerheads, and differs from the Weem example in having a massive central ridge stack. The skews suggest it was thatched.

In the Panmure muniments there survives a drawing of *c.* 1700 by the architect Alexander Edward for a long-vanished inn at Monifieth, Angus, on the old route from Dundee to Arbroath. This was a similarly-planned but rather larger building, four-windows wide with a central door, and much more sophisticated in design and construction with a slated urn-finialled pavilion roof. The windows, however, adhered to the old pattern of fixed leaded lights over shutters rather than up-to-date sashes. Edward's drawing gives the clearest picture we have of the accommodation such a building offered. At ground floor on the west side of the central division wall was a hall-kitchen, 16 feet by 14 (4.88 by 4.27 metres), with a stair to the apartments above. On the east was a 'low chamber' with two box-beds, the rear part of the room being partitioned off as a store. At first floor the stair rose to a passage along the back of the building which led to the eastern and best room in the house. It had two beds, one partitioned off from the other, and enough space to sit and dine by the fire. The western apartment was smaller and more crowded with three beds, of which one was again partitioned off as a separate compartment with its own window. Such arrangements were no doubt fairly typical and probably represent what Topham found so unacceptable in the Pleasance some sixty years later.

POST-1750 INNS

From the mid eighteenth century onwards roads in Scotland gradually improved, partly as a result of the Commutation Acts which substituted payment for statute labour, and partly as a result of the Turnpike Acts which encouraged the creation of new routes. In the north, Major-General George Wade began work on the military roads in 1725, and through the labours of his troops and those of Lieutenant-General Jasper Clayton and Lieutenant-General Sir John Cope, with Major Caulfeild as their inspector in Inverness, the first properly made routes through the Highlands were gradually achieved. In the mid-1720s Daniel Defoe, while acknowledging the hospitality of local landowners, had advised travelling military-style with a tent, and even Wade himself initially operated from timber-framed 'hutts', which were gradually replaced by a series of 'King's Houses' at ten-mile intervals. These inns varied in design according to what was required at the time. That at Letterfinlay, Inverness-shire, was two-storeyed and three windows wide, with a single-storeyed outbuilding, but that at Moulinearn, Perthshire, the next inn north of Dunkeld, was five windows wide. Moulinearn proved too small for the number of travellers on the road and was soon extended to a ten-window frontage. That at Dalwhinnie, Inverness-shire, where Queen Victoria and her entourage found too little to eat in October 1861, was similarly much enlarged, achieving the U-plan form recorded in R P Leitch's water-colour. Although a slightly later rebuild, and converted to a country house many years ago, that at Dalnacardoch, the intermediate inn between Blair Atholl and Dalwhinnie, best preserves the spirit of the military road-building programme. A finely lettered tablet bears the following inscription:

HOSPITUM HOC / IN PUBLICUM COMMODIUM / GEORGIUS III

REX/CONSTRUIIUSSIT(sic) AD 1774 / REST A LITTLE WHILE / GABHAIF FOIS CAR TAMUILL BHIG

Like Moulinearn, it had extensive courts of stable offices at the back, but the inn itself differed slightly in format, having the Laird's Inn arrangement of three windows rather than five at first-floor level.

This network of new roads, together with harbour improvements, brought an increase in the number of travellers and a corresponding rise in demand for accommodation. In the burghs, large purpose-built inns became a clearly defined building type, usually three-storeyed rather than two, with extensive stabling. By far the largest of the earlier examples was the still-extant White Hart in Edinburgh's Grassmarket, which would appear to be of mid eighteenth-century date. It is four-storeyed and attic, near-symmetrical with its windows arranged 4–5–3, those at the end bays being very narrow and presumably for closets. Twin wallhead attic gables with big panelled stacks give it a memorable skyline, while an off-centre pend arch in the eastern half of the elevation gave access to a large but now vanished stable court, long famous as the departure point of the Lanark coach. Apart from sheer scale it was little different from High Street tenements of the same vintage. It is not altogether easy to see how it functioned as an inn and early photographs suggest that it may not have been wholly occupied as such.

As ever, Glasgow led the way in developing the larger inn as a building type. In 1754 the magistrates and council, conscious that the town had no suitable accommodation for visitors from the south, encouraged Robert Tennant of the Glasgow White Hart to build a new inn, the Saracen's Head, in Gallowgate. This was a plain symmetrical country house-like building, three storeys high and nine windows wide, the end pairs being advanced as broad piend-roofed pavilions with quoined angles. Tennant's own description of it at its opening in 1755 helpfully gives an indication of the arrangements in earlier Glasgow inns, and probably Edinburgh ones such as the White Hart in the Grassmarket. It was built:

> agreeable to a Plan given him [probably by Allan Dreghorn] containing thirty-six Fire Rooms, now fit to receive Lodgers. The Bedchambers are all separate, none of them entering through another, and so contrived that there is no need of going out of doors to get to them. The Beds are all very good, clean, and free from Bugs. There are very good Stables for [sixty] Horses, and a Pump Well in the Yard for Watering them with a Shade within the said Yard for Coaches, Chaises and Wheel Carriages ... There is a large room where an hundred people can be Entertained at one time.[2]

Although Tennant went bankrupt in 1757, the inn was for the next

25 years 'the Rendezvous of all the Nobility in the West and distinguished Strangers'. 'Balls, Suppers and County Meetings' were held in its assembly room, a handsome apartment with a deeply-coved but plain ceiling. The inn also had a 'billard' table, then fairly recently introduced from France.

A rival establishment, the Black Bull at the corner of Argyle Street and Virginia Street, was built by the Glasgow Highland Society in 1768–9, mainly from the proceeds of a sermon given by George Whitefield. When it closed in 1849 it was recorded as having a commercial room, a coffee room, a ballroom, nine parlours, 29 bedrooms and stabling for 30 horses. Whether the commercial room was an original feature is unclear, as Tennant's very full description of the Saracen's Head does not refer to one, but it was to be an important feature of nearly all nineteenth- and earlier twentieth-century hotels, enabling coachmen and travelling servants to dine separately. It also catered for commercial travellers, tradesmen and passing trade, the title 'Commercial Hotel' becoming a common indicator of premises specialising exclusively in that type of business at moderate charges by the mid nineteenth century.

In the same year that the Saracen's Head was opened (1755), the New Inn was built adjacent to the tolbooth at Aberdeen, seven windows wide and deep-planned with a platformed roof and a blind Serlian window making an ambitious show at its eastern gable. This was a fashionable establishment which included a Masonic Lodge, an arrangement which provided it with a sizable meeting hall and made it the town's principal place of assembly.

The provision for assemblies and balls in the design of the Saracen's Head, the Black Bull and the New Inn is of some significance in a British as well as a Scottish context. Both Macky and Defoe commented on assemblies as a new development in English social life in the 1720s, and there are earlier references to them in letters from c. 1715. In the principal English cities and towns very grand buildings were erected for them by public subscription. In Edinburgh an assembly room was more economically formed within an existing building in the West Bow at about that date, followed by a larger one in what is now Old Assembly Close in 1723. In Glasgow, the hall of the Merchants' House was used for the purpose, and from 1740 the Town Hall, supplemented by an assembly room built by public subscription in 1757–63. But as the assemblies held in these buildings tended to be literally socially exclusive, there was a considerable incentive for inn-keepers to provide similar facilities for a wider clientèle. The Glasgow and Aberdeen inn-keepers seem to have been fairly early in making such provision, the earliest English examples cited by Mark Girouard in *The English Town* dating from the 1770s.

Almost exactly parallel in date with Aberdeen's New Inn was the

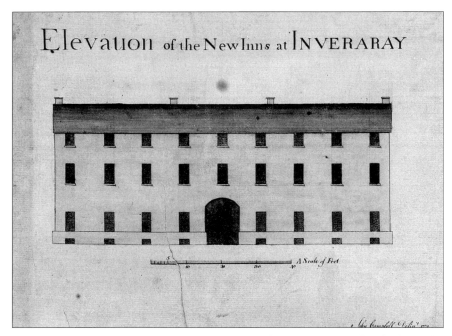

Figure 8.2 Great Inn, Inveraray, Argyll; elevation as completed, drawn by John Campbell, 1771. Reproduced by kind permission of His Grace the 13th Duke of Argyll.

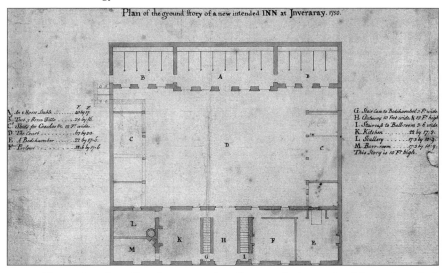

Figure 8.3 Great Inn, Inveraray, Argyll; ground plan and stables as intended, John Adam, 1750. Reproduced by kind permission of His Grace the 13th Duke of Argyll.

still larger Great Inn at Inveraray, Argyll (figs 8.2, 8.3), designed for the Duke of Argyll in 1750 by John Adam, and built between 1751 and 1756 with some amendments to the details suggested by Harry Barclay of Collairny. It consisted entirely of lodgings and did not contain any provision for assembly. Although it is now much altered internally, we have some information as to its plan and furnishing. It was three storeys high and nine windows wide, with accommodation for servants in the attics. A central pend, flanked by twin stairs to the bedchambers at first and second floors, provided access to the stable court which had generous accommodation for coaches, a measure of the degree to which the roads had improved. At ground floor, the kitchen (22 feet by 17 feet 3 inches [6.71 by 5.26 metres]) and the bar room (17 feet 3 inches by 10 feet 9 inches [5.26 by 3.28 metres]) were on the left of the pend, and a parlour (probably for dining) and a large bedchamber on the right. By October 1756 the Duke's furniture-maker George Haswell had provided four posted bedsteads, six press beds, two box-beds, 4½ dozen (54) leather-bottomed chairs, 9½ dozen (114) wooden-bottomed chairs, five chests of drawers, six square folding tables, 18 small tables, four joint stools, two presses and two kitchen dressers. Such provision was probably fairly typical of the larger establishments. Together with the relatively small parlour – which may have been for the exclusive use of the occupant in the adjoining bedchamber – it indicates that most of the guests lived and ate in their bedchambers.

Other large estates provided similar but plainer buildings, notably the Earl of Breadalbane's three-storeyed, five-windows wide inn (now Victorianised) built at Kenmore, Perthshire, shortly before 1760, and the Menzies's very similar inn at Weem, which preserves its original external appearance. The Murray Arms at Gatehouse-of-Fleet, Kirkcudbrightshire (1766), built by a David Skimming, is a better documented example and still has its original kitchen fireplace. It adhered to the older two-storeyed, five-window format, albeit on a much larger scale than at Newbattle, and like that inn it was dignified by raised quoin angles and a good doorpiece. At Laurencekirk, the Boar's Head, rebuilt by Lord Gardenstone between about 1770 and 1778, achieved a deep two-storeyed U-plan form with a small porticoed pavilion containing a library for the amusement and instruction of both locals and travellers. All of these inns were free-standing and did not require pend access to their stable yards; but a central pend arrangement very similar to that at Inveraray existed at the seven-windows wide Cross Keys, Kelso, Roxburghshire (1761), until its modernisation in 1850 and the recent demolition of the stables.

The mid-to-later eighteenth century transition to these large purpose-built inns is well illustrated at Moffat, Dumfriesshire, where the advent of regular coach services coincided with the development of the town as a spa, particularly after the discovery of a second well, the

Hartfell, in 1748. The oldest inn in the town is the Black Bull opposite the parish church. It is a low two-storeyed building, similar to the Duke's Inn at Dunkeld, and has happily survived fairly unaltered to demonstrate how radically inn design was to change in the space of half a century. By contrast, the two mid Georgian inns were tall three-storeyed buildings which stood in the High Street, re-planned as a broad central space in 1772–3. The Spur Inn, now the Balmoral Hotel, was symmetrical, five windows wide and (very unusually) built of brick with quoined angles. A still finer bigger-scaled building was the Annandale Arms (fig. 8.4) of c. 1783, also five windows wide but with architraved windows and a central doorpiece of coupled Doric columns. The pend to the extensive stables at the rear was in an adjoining two-storeyed block so that the ground floor had the unified plan with central reception that the Inveraray and Kelso buildings had lacked.

Similar but generally less ambitious buildings were to be found in towns along the main coach routes such as Coldstream, Berwickshire, where the Black Bull, later the Crown, survives, and in ports which were once significant centres for sea or river passenger traffic. Alloa in Clackmannanshire and Kinghorn in Fife were once particularly rich in eighteenth-century inns for those awaiting a passage. These have vanished, but at Annan, Dumfriesshire, once an important passenger port

Figure 8.4 Annandale Arms, Moffat, Dumfriesshire, c. 1880. Courtesy of A M T Maxwell-Irving.

for Liverpool, there is the Queensberry Arms, very similar in scale to the Annandale at Moffat, if slightly less regular in arrangement. Similar again is the relatively unaltered New Inn at Dunbar, built in 1788–91 for the Earl of Lauderdale to designs by Alexander Ponton, and clearly intended to accommodate those who had business with the earl at nearby Lauderdale House.

On the same model, but a little later in date (c. 1804), is the Duke of Buccleuch's Cross Keys Inn at Dalkeith, Midlothian, where the pend to the stabling was in the western end bay. Although primarily a coaching inn, it was similarly designed to provide for visitors of some standing and to accommodate functions connected with the estate. It was rather more architectural in treatment than its predecessors, ashlar-faced with a tripartite pilastered doorpiece, rusticated ground floor and balustraded aprons at first-floor level, like a street block in Edinburgh's New Town. Within, an elegant railed stair led to a first-floor assembly room the full width of the street frontage, with chimney pieces at each end.

Similar in style, scale and vintage to the Moffat examples was the Caledonian at Inverness (fig. 8.5). This inadequately recorded building was built in 1776 by the town's two masonic lodges with money from the forfeited estates and was originally the Masons' Hotel. It was a free-standing mansion-like building of three storeys and an attic, five windows wide with Gibbsian detail, and grandly set back from Church Street within a railed area. At the back it presented a four-storeyed frontage to the river. It had stabling for 19 horses, and after enlargement in 1822, provision for 80 beds with 'Hot and Cold Shower Baths'. Much more typical of this late Georgian three-storeyed and attic type was the Royal Hotel at Nairn, built c. 1805, six windows wide with the ground-floor openings in arched recesses and built right up to the pavement without a front area of any kind. This plain late Georgian hotel type was to persist with little change far into the nineteenth century, notably at the King's Arms, Girvan, Ayrshire, rebuilt by the Union Bank c. 1850, and at the former Royal Hotel, St Andrews, Fife, designed by George Rae in 1857 and extended, again with a pend to the stables and service court, as late as 1894.

In the smaller burghs, inn buildings tended to be the work of architect-builders rather than architects, as at the low-ceilinged Black Bull at Lauder, Berwickshire, three-storeyed with dormer heads and six windows wide. It was extended c. 1780 by a two-storeyed addition of the same height to provide a function room with a Serlian window. The Black Bull probably incorporated fabric from an earlier building, but towards the end of the eighteenth century inns generally became more spaciously proportioned with bigger windows. A particularly attractive example is the Angel at Gatehouse-of-Fleet, Kirkcudbrightshire, three storeys high with a more up-to-date, low-pitched piended roof, and three

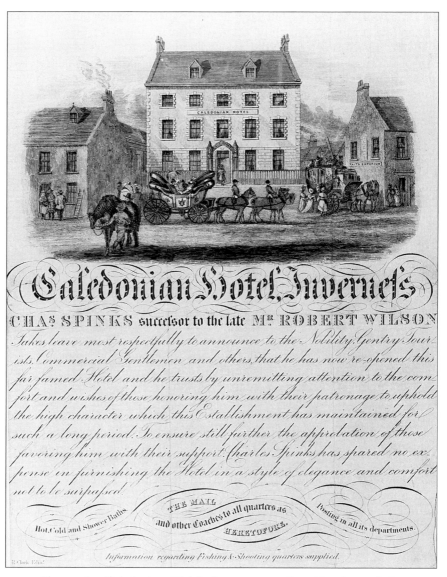

Figure 8.5 Caledonian Hotel, Inverness, c. 1830. Courtesy of Inverness Museum and Art Gallery.

windows wide with tripartite windows at the outer bays, those at first floor being Serlian. Its English-influenced style and proportions contrast instructively with the more robust northern proportions of the exactly contemporary and otherwise similar Tolbooth Hotel at Banff of 1801, which still has a steep gabled roof and old-fashioned piended dormer heads.

By the mid eighteenth century some inns were places to stay rather than just places to stop for the night. As at Moffat, the existence of a spa well was an important factor. A notable early instance was Bridge of Earn, Perthshire, where the nearby Pitkeathly wells attracted large numbers of people seeking to recuperate. A three-storeyed, three windows wide inn with lodging rooms only, built at Pitkeathly by a Mr McLaren, proved inadequate by 1818; at that date the late seventeenth-century Drummonie House was provided with a ballroom to cater for the flourishing social life there – a very early instance of a country house conversion. In Bridge of Earn itself Sir William Moncrieff had to rebuild the Moncrieffe Arms at the bridge on an enormously lengthened plan in 1788, and began offering feus for houses providing lodgings, the Cyprus Inn being built in 1790.

Substantial inn, and more particularly boarding-house, provision was made for the same purpose at Peterhead, Aberdeenshire, in the later decades of the eighteenth century. At Pannanich in the same county the discovery of a well *c.* 1760 resulted in Colonel Francis Farquharson of Monaltrie building long, two-storeyed vernacular lodgings which included an inn at the well itself, with further residential provision for those taking the waters at Tullich nearby and at Monaltrie, rebuilt by James Robertson in 1782 as what was probably Scotland's first country house hotel, long and low with castellations at the centre and ends. This too proved insufficient and from 1783 development was concentrated in Farquharson's village at Ballater where the original inn was rebuilt as The Monaltrie Hotel as late as 1860. By the mid eighteenth century what we would now describe as cultural tourism had also begun to be a factor, notably in the provision of inns at Melrose, Roxburghshire, for visitors to the ruins of the abbey, and at Penicuik, Midlothian, where the Crown Inn was built to cater for visitors to the landscaped park and buildings at Penicuik House.

Throughout the countryside great numbers of roadside inns with stable yards and steadings were built along the principal coach routes to cater for the ever-increasing volume of travellers, the inn-keeper usually being the farmer, blacksmith and postmaster as well as host. Most of these inns still exist but are now simply farms. Until the recent fire, Howgate, Midlothian, was a particularly good example, built *c.* 1743 for the Carlisle route, and comprising a rather low two-storeyed, three-windows wide, farmhouse-like building to which a large room had subsequently been added, linking it to a row of cottages. Its outbuildings included a brewhouse, probably a fairly common provision. Most rural inns followed that simple two-storeyed, three-windows wide format, often with an attic, but the larger ones tended to be four or five windows wide, very much as at the Newbattle and Lanton examples referred to earlier. Many were subsequently lengthened, as at Moulinearn, or extended at right-angles into an L-plan, the Fife Arms at the north end of Turriff, Aberdeenshire, being a notable example of such progressive enlargement to accommodate increases in trade. Of those which have remained as first built, the two-storeyed, five windows wide inn associated with the former ferry at Inver, Dunkeld, is set in a small formal square; other good examples associated with river crossings were to be found at Hyndford Bridge, Lanarkshire, and, until fairly recently, at Bridge of Dee, Aberdeen.

All of these were simple vernacular buildings but a few had greater architectural pretensions to distinguish them from farmhouses. An exceptionally good roadside inn, reminiscent of those on the A1 south of the Border, but smaller in scale, was Camphouse, south of Jedburgh, Roxburghshire, only two storeys and a sky-lit attic high, but of three very wide bays, the central one minimally advanced and pedimented. Pediments were also adopted as a distinguishing feature in eastern Perthshire, although the best examples have now been either demolished or modified. These had blind Serlian windows and urn finials, perhaps the most notable being Meikle Fardle on the military road from Coupar Angus to Dunkeld. Another example, which at some stage also housed a bank, is still to be found at Meigle, and a smaller but still very smart one, latterly the Commercial Hotel, formerly existed in Commercial Street, Alyth. Of the same general family was the considerably larger Queen's Hotel at Blairgowrie, pedimented with urns, and at Melrose, far distant, is a rather similar building, Burt's Hotel, which retains its Serlian window but has been shorn of its urns. Old photographs show that prior to its Victorianisation as a hotel by Andrew Heiton, the very large Dreadnought Inn at Callander in West Perthshire was also distinguished by urns at its skew ends. These very distinctive inns appeared to date from the 1780s and 1790s. Even more architectural than these was the still-extant inn at Ardencaple, Rhu, Dunbartonshire, a medium-sized country house by the roadside to provide an appropriate stopping point on the ducal route from Glasgow to Inveraray and Roseneath, Dunbartonshire. It has a three-storeyed, three-windows wide main block flanked by advanced single-storeyed and attic wings with arched and pedimented fronts and an equally architectural stable block at the rear.

These were fairly exceptional. More typical were the inns on the once-prosperous coach routes through West Lothian, which reflected the differing aspirations of the landowners who formed the turnpike trustees. Those at Mid Calder and Livingston, both built *c.* 1763 when the road was formed, are plain two-storeyed buildings, the former being dignified by an architraved and corniced doorpiece. Those on the northern route are rather more architectural, that at Craig being an immensely wide and deep-planned three-storeyed, three-bay block with a recessed porch, and that at Uphall, low, pedimented two-storeyed with a pilastered doorpiece, both clearly designed by architects rather than by builders. All of these had extensive stabling, that at Livingston being more impressive than the inn itself.

HOTELS AND ASSEMBLY ROOMS

The wider provision of up-to-date purpose-built hotel buildings in the cities and larger burghs came surprisingly late, much later than in the towns and villages along the coach routes. Gradually the 'room setters' had been superseded by a superior class of 'tavern'. These new taverns had begun to appear in Edinburgh and Glasgow in the 1770s. In Edinburgh the tendency had been to adapt New Town houses and tenements, such as Fortune's Tavern at 5 Princes Street opened in 1796; the Crown at No 2 and Walker's at No 3 before 1811; the Royal at No 53 sometime before 1817, when Grand Duke Nicholas and his entourage resided there; and the much larger Barry's at Nos 5–8, opened in 1821, of which Fortune's became part. At nos 35–7 was the Star (fig. 8.6), heightened shortly after 1820 and the first hotel to break the original Princes Street eaves line.

Figure 8.6 Star Hotel, Princes Street, Edinburgh, c. 1835. Crown copyright: RCAHMS, SC 466071.

Several more hotel buildings flourished in and around St Andrew Square, most notably the Douglas Hotel opened in 1830.

As in the 1750s, it was in Glasgow rather than in Edinburgh that the pace was set. In 1781 a Tontine Society – its last member died as late as 1862 – raised £5,350 in 107 shares of £50 to acquire Allan Dreghorn's handsome town hall of 1737–40 and 1758–60, and to convert it into the Tontine coffee room and hotel, the society's architect being William Hamilton who had worked for a time in London. The word 'hotel' was a novelty at that date, and it may have been its first use in Scotland. The coffee room, 74 by 32 feet (22.56 by 9.75 metres), was a handsome Adam-influenced room lit by a glazed dome, well up to metropolitan standards. The reading room formed an exchange for the merchants who congregated in its piazza, and provided newspapers and shipping news for subscribers paying £1,12s per annum. The upper floors consisted of the best residential accommodation in the city at that time. Apart from the Saracen's Head and the Black Bull, the other inns and taverns in the city were mainly converted houses or tenements in Argyle Street, Trongate and Gallowgate.

In 1782 the Earl of Abercorn commissioned the London-Scottish architect George Steuart to erect an inn and assembly room providing similar accommodation in Paisley. It was a smaller but quite remarkably smart building with an arcaded elevation unifying the different fenestration of the inn and assembly room elements, the latter having an elegant pilastered bow with a balustraded parapet as its gable. It took the Tontine name of the Glasgow establishment, although there does not appear to have been any Tontine element in the financing of it.

This combination of inn and assembly rooms became more widely adopted over the following decades, and a number of older inns were updated to match the facilities which had been provided in Glasgow and Paisley. A particularly good example is the Salutation Hotel at Perth where, c. 1805, an assembly room building with a fine Adamish classical façade was added, probably to designs by Robert Reid. Other assembly room additions of this vintage are still to be found at the Castle Hotel in Haddington and at the Royal Hotel in Coupar Angus and at Dunbar where a separate assembly room block was constructed behind the St George's Hotel. A particularly well-integrated solution was achieved at Cupar, Fife, in 1811–17 where the south side of St Catherine Street was designed by James Gillespie Graham and the local architect Robert Hutchison as a unified palace block with the County Buildings in the middle, the Tontine Hotel immediately to the east, and the Town House on the west. Accommodation and catering were thus conveniently on hand for the sheriff, the circuit judges, meetings of the Fife Hunt, the Commissioners of Supply and the Town Council.

Similar hotel and assembly room provision came late in Edinburgh, the city having assembly rooms in Buccleuch Place (1783) and George Street (1787). But in 1818 the joint-stock Waterloo Tavern & Hotel Company built a very modern hotel as part of the north-eastern block of Archibald Elliot's Waterloo Place scheme. It had a large 'travellers' room', a bar, several 'parlours' and a long coffee/assembly room for the higher class of guest, towering above Calton Road at the back. The last was a very sophisticated room, aisled on plan with Ionic colonnades. Latterly sub-divided, it regrettably disappeared, inadequately recorded, in 1971 when the long-closed hotel was rebuilt as an office block.

In Dundee at the end of the eighteenth century there were two large inns, Gordon's and Morren's, plain buildings which may have been purpose-built, and a number of taverns in older buildings in Murraygate. Merchant's Hotel, later the British, at the corner of Castle Street and High Street, a large four-storeyed block with ground-floor shops, was certainly built as such, shortly after 1800. It too was very plain. In 1826–9 the Royal Hotel was built at the Nethergate corner of Union Street, a new street planned by William Burn in 1824 and laid out by David Neave in 1825. Like Merchant's it was four-storeyed with ground-floor shops, but it was an altogether larger and more handsome astylar classic building. Following the precedent of the New Inn at Aberdeen, it operated in association with Neave's adjacent Masonic Thistle Hall in Union Street, thus providing ballroom and assembly room facilities in a similar configuration to those of the Tontine and adjoining County Buildings at Cupar.

In Perth two sizable hotels were built. The more fashionable was the George in George Street (1790), refaced with canted bays after Queen Victoria patronised it. It then assumed the 'Royal' prefix. The second was the much larger Star Hotel in Canal Street, built shortly after 1790, which survived unaltered as a tenement until its demolition some twenty years ago. It was a huge four-storeyed and attic ashlar-faced block of five widely-spaced windows with a pilastered doorpiece and a painted star emblem between the second- and third-floor windows. Within, a low hall with a semi-elliptical fanlight door gave access to a big staircase at the back. In Stirling the principal hotels were the three-storeyed Golden Lion, probably late eighteenth century in origin but early nineteenth in its present form with an asymmetrically-placed Greek Doric porch and central pedimented bays, and the much larger Royal Hotel of 1839–40, a very impressive classical building seven windows wide to Barnton Street and five to Friar Street with a Greek Doric entrance at its bowed corner. These were probably by William Stirling's practice which certainly designed the severely neo-Greek Red Lion in Falkirk's High Street in 1828. As a county town where some of the Ayrshire gentry had town-houses, Ayr was also rich in such buildings. On its High Street were the very handsome astylar neo-classical King's Arms and the Crown (*c.* 1820) which had an assembly room, but neither of these has survived.

Figure 8.7 Fife Arms Hotel, Low Street, Banff, c. 1890. Courtesy of Bodies, Banff.

In 1817 Archibald Simpson designed the Aberdeen Hotel on Aberdeen's Union Street, which seems to have been the first large purpose-built hotel in that city since the building of the New Inn sixty years earlier. Like the Waterloo Tavern in Edinburgh it was built adjacent to an urban viaduct with as many storeys beneath street level as above, and, like Merchant's and the later Royal in Dundee, it included shops. Although fairly severe in its architecture it was quite big, six windows wide to Union Street, and five to Belmont Street and the Denburn. Smaller but more stylish was the neo-Greek Union Hotel in Inverness of 1838–9 by William Robertson of Elgin, which had a pedimented façade recessed between end bays with pedimented tripartite windows. Elegant buildings such as the Union were expensive and very much for local society and travellers of the higher class. They were to remain fairly rare for several decades as the supply of relatively cheap and adaptable house blocks for the ordinary traveller proved more than adequate.

In the smaller burghs the new generation of inns was similarly well designed and detailed. Quite especially grand, like a medium-sized two-storeyed and basement country house, was the Castle Hotel at Greenlaw, Berwickshire, built to the designs of John Cunningham in

Figure 8.8 Duke of Gordon's Inn, Kingussie, Inverness-shire, *c.* 1890. Courtesy of Gargunnock Estate Trust.

1829–34 and set in a formal classical square facing the County Hall, conveniently providing accommodation and catering for county meetings, the sheriff and the circuit judges. As in the previous generation, the inns which were associated with great estates tended to be rather grander than those built by inn-keepers as they had to accommodate visitors on estate or political business, and sometimes house-party guests and relatives for whom room could not be found within the house itself. Although George Alexander's Sutherland Arms at Golspie, Sutherland (1808), was no more than a neat symmetrical two-storeyed villa with single-storeyed wings, at Cullen, Banffshire, the Earl of Seafield's architect, William Robertson, integrated town hall, post office, inn and stable court into a very grand composition in 1822–3 as the key element of his new town. At Dunkeld, the new ducal inn, the Atholl Arms (1833), similarly formed part of a new town development, here associated with Telford's bridge of 1809. A tall three-storeyed corner building, five windows wide by four, with a Greek Doric doorpiece and a pend to the stables, it offered far more spacious accommodation than the old Duke's Inn and had a fine outlook over the river.

Perhaps the grandest of these urban estate hotels was William Burn's 1843 rebuilding of the Black Bull in Banff for the Earl of Fife (fig. 8.7). Its three-storeyed, five-windows wide, ashlar-faced, Italianate façade with rusticated quoins, Roman Doric portico and long flanking two-storeyed office wings was designed as a formal approach from the town to the now-vanished gates of Duff House, and its original title was ennobled to the Fife Arms to provide a more genteel address for the earl's visitors and guests.

Such gentrification was widespread in the earlier nineteenth century, especially where the inn was in an estate village or close to the park gate, as at the Stair Arms at Oxenfoord, Midlothian, and sometimes where the inn was built solely at the charge of one of the larger estates. Particularly attractive examples of the late Georgian 'cottagey' school are to be found in Perthshire: notably at Meikleour (1820), low and broad-eaved with a Tuscan portico; Inchture (c. 1835), with a very big stable yard, probably the work of David Mackenzie and forming part of a complete neo-Tudor estate village to complement Lord Kinnaird's Rossie Priory, designed by the English architect William Atkinson in 1807; and the ducal hotel at Blair Atholl (1832), enlarged from the eighteenth-century inn by R & R Dickson as an approximately symmetrical neo-Jacobean pile with an assembly room and a large stable court with a big white horse over the entrance. Further north at Kingussie, Inverness-shire, there was once another especially stylish example, the Duke of Gordon's Inn (fig. 8.8), a low, broad-eaved Italianate box with skilfully grouped chimneys, presumably the work of the duke's architect, Archibald Simpson of Aberdeen, c. 1835. It still exists, but refaced and dwarfed by the new hotel of 1902.

With the rise of a provincial architectural profession catering for church, manse, school, farmhouse and steading, most of the rural inns along the main coach routes and the new turnpike roads were now architect-designed. They are too numerous to catalogue here, but a particularly good example, if unusual in being 'cottagey' single-storeyed, is Cairns Castle, West Lothian, on the Edinburgh-Lanark route, designed by the Edinburgh architect James Gillespie Graham in 1823 and utterly dwarfed by the huge stable court on the opposite side of the road where the coaches changed horses. Most took the form of a sizable two-storeyed villa, a fairly well-documented example, even if its architect is not known, being the Torsonce Inn at Stow, Midlothian, which may be taken as an illustration of how some of them came to be built. It was opened in 1819, one year after the completion of the new Edinburgh-Galashiels turnpike, and was built by subscription of 15 of the heritors of the counties of Edinburgh, Selkirk and Roxburgh, one of them being Sir Walter Scott who used it on his journeys into the capital. It was a very deep-planned building with a jerkin-headed roof giving good attic accommodation. It had the usual simple classic detail of its time, but it was entered at the gable from a spacious forecourt with gatepiers, rather than from the road.

As in the previous century, some of the largest early nineteenth-century hotels were to be found at river crossings and at harbours with significant passenger traffic. The new ducal hotel at Telford's Bridge in Dunkeld was just such an instance, as is the rather older (c. 1820) and larger Royal Hotel in the same town which still retains its picturesque stable-court. Other large establishments associated with Telford bridges are to be found at Helmsdale, Sutherland (1816), three-storeyed with a broad bowed gable, and Beattock, Dumfriesshire (1822), a big and quite remarkably stylish two-storeyed villa probably designed in Telford's own office. Ferries usually had sizable inns nearby, a notable example being the three-storeyed, three windows wide Balblair Inn of 1820 in Easter Ross, surprisingly old-fashioned for that date with piended dormer heads. Notable surviving examples of inns of early to mid nineteenth-century date associated with passenger harbours are General Patrick Sinclair's vernacular two-storeyed and dormered attic, seven windows wide inn at Lybster, Caithness, of 1802, the country house-like Eglinton Arms at Ardrossan, Ayrshire, designed by Peter Nicholson in 1806 as part of the Earl of Eglinton's new town, and the Duke of Buccleuch's Granton Inn, Edinburgh, designed by William Burn in 1838 as part of the duke's new harbour. The last was probably designed to greet the royal yacht, which it duly did in 1842. It was, at the time, architecturally the finest building in Scotland erected solely for hotel

Figure 8.9 The New Club, 85 Princes Street, Edinburgh, 1966. Crown copyright: RCAHMS, SC 466146.

purposes, a handsome seven-windows wide palazzo with a coupled-column Roman Doric porch, as at the slightly later Burn hotel at Banff. Its primary purpose was to serve the Burntisland ferry, and it was answered on the north shore by the Forth Hotel, reconstructed from the parish manse for the same purpose to designs by John Henderson in 1843, the project including harbour offices and houses, and a large stables block.

RESIDENTIAL CLUBS

In the mid 1830s a new building type, the residential club-house, made its first appearance in Scotland. The Edinburgh New Club had first met in Bayle's Tavern in Shakespeare Square in 1787 but quickly concluded that having its own residential premises would be a better option, particularly for country members. It first bought a house in St Andrew's Square and in 1820 considered a scheme for a club-house at 1 Princes Street, designed by William Burn on the London model with a Corinthian-columned pedimented front. This was not proceeded with, but in 1833–4 Burn visited the London club-houses and designed a very big-scaled three-storeyed and basement palazzo for a site at 85 Princes Street (fig. 8.9) overlooking the gardens. In Glasgow, the Western Club had been formed in 1824 with a similar mix of great landowners, industrialists, merchants and leading professionals. In 1839–40 it commissioned David and James Hamilton to design an equally grand club-house on Buchanan Street (figs 8.10, 8.11). This was a still taller palazzo of four storeys on a corner site, more elaborate in detail than the Edinburgh one with a skilful blend of late classic mannerist and Baroque motifs. Both these buildings had grand staircases and accommodation on a scale undreamt of in Scotland before the 1830s, with large dining rooms, coffee rooms, libraries, billiard rooms and very well-furnished bedchambers, those in the Western having sleigh beds of monumental proportions until the Club moved to smaller premises in the 1960s.

These two club-houses in some degree set the standards for the next generation of hotels, but in both Edinburgh and Glasgow only after an interval of more than a decade. The next significant building was another club-house, the Union Club in Aberdeen, a plainer palazzo of 1853 by James Matthews. But in 1855, when the Life Association constructed its great Venetian Baroque palazzo on Princes Street, Edinburgh, to designs by David Rhind and his early mentor Sir Charles Barry, the upper floors were constructed as the Bedford Hotel. It had a grand staircase like a club-house, but the scale was altogether smaller, with bedrooms cleverly packed into the mezzanine floors in its sculptured entablatures. This hotel-over-insurance office arrangement had already

Figure 8.10 The Western Club, 147 Buchanan Street, Glasgow. Courtesy of The Western Club.

Figure 8.11 The Western Club, 147 Buchanan Street, Glasgow; interior. Courtesy of The Western Club.

appeared at David Bryce's much smaller Caledonian Insurance building (1838) on George Street, greatly enlarged by David MacGibbon & Thomas Ross in 1879, and since the 1930s wholly occupied by the George Hotel.

VICTORIAN HOTELS AND CLUBS

In 1859-64 the Edinburgh High Street and Railway Access Company constructed Cockburn Street to designs by Peddie & Kinnear under Acts of Parliament of 1853 and 1860. Although this serpentine thoroughfare was very much what its promoters said it was, a direct route from the High Street to Waverley Station, it was also a major hotel speculation offering a wide range of accommodation and probably priced for those arriving in Edinburgh, comprising the upmarket Philp's Cockburn Hotel at No 1, the now vanished Middleton's City Temperance Hotel at No 14 and Inglis's City Hotel at No 69, and the still-extant but presumably more downmarket Star Commercial Hotel at No 59, the last three all having ground-floor shops. All of these were Scots Baronial as required by the 1827 Improvement Act for new developments in the Old Town.

All the Cockburn Street hotels were of moderate size, and at least one of them had originally been planned as a tenement rather than a hotel. The first and finest of the really big hotels built exclusively for hotel purposes, David Bryce Junior's magnificent five-storeyed rebuilding of the Star Hotel (fig. 8.6) on Princes Street of 1861, was clearly modelled on the New Club as extended in 1859 by his more famous uncle of the same name. Built by the Kenningtons of Kennington and Jenner department-store fame, it had the same cantilevered balcony and balustraded parapet with urns, and, although raised a storey over ground-floor shops, it was more generously provided with bay windows to enjoy the view. Within, it offered accommodation comparable to that in the great club-houses and was clearly intended for a quite different class of traveller from those using the hotels in Cockburn Street.

Although very successful, the Star was to remain unique in scale for more than a decade. The very fashionable Douglas Hotel, which numbered the Empress Eugénie among its guests, was content to add a plain bedroom block to the back of its houses at 34–5 St Andrew Square as late as 1864. The buildings which followed the Star's lead were club-houses rather than hotels. In 1865–6 Peddie & Kinnear designed the University Club on Edinburgh's Princes Street, a tall four-storeyed and basement palazzo with a peristyled bow of superimposed orders to give a 180° panoramic view of the gardens and the castle. With its colonnaded entrance hall and long dining room with screens of columns it was as grand as the New Club, but its shorter frontage meant that it had to be of greater height, a reflection of rising land values on Edinburgh's premier street.

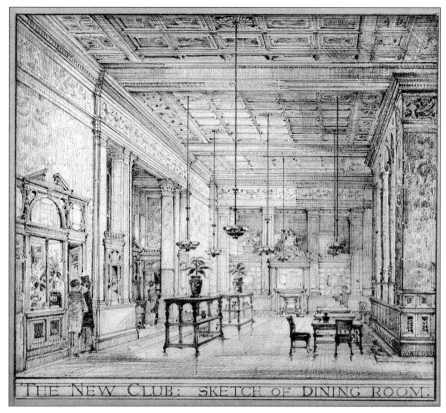

Figure 8.12 The New Club, 144-6 West George Street, Glasgow; sketch of dining room by James Sellars or assistant, c. 1877–8. Crown copyright: RCAHMS, SC 466151.

The University Club was paralleled in Greenock by Boucher & Cousland's magnificent two-storeyed palazzo in Ardgowan Square for the Greenock Club, 1867–9, in Elgin by A & W Reid's smaller but rather similar Elgin Club, 1868, and in Glasgow by the first of the purpose-built political club-houses, the Conservative on the corner of West George Street and Renfield Streets. This building had an unusual construction history. John Burnet was commissioned by the club to design it in 1868, but in the event the club allowed the project to be taken over by the Scottish Widows Fund which brought in David Bryce to redesign the elevations on the Edinburgh New Club model. The Scottish Widows Fund then occupied the ground floor and basement as its Glasgow office with the Conservative Club and the Bo'ness Iron Company on the floors above, Peddie & Kinnear being brought in as early as 1871 to make necessary adjustments to the internal arrangements. No such compromises were made at Frederick Pilkington's Eastern Club in Albert Square, Dundee, 1869–70, a mixed Romanesque and Renaissance Venetian palazzo of three tall storeys with an entrance loggia and richly decorated interiors in which the city's linen and jute magnates could imagine that they were indeed merchant princes.

None of these had much influence on subsequent hotel design, but James Sellars's tall and elegant New Club on West George Street, Glasgow (fig. 8.12; 1877–81), the result of a schism in the Western in 1866, most certainly did. At four-storey and basement with two further floors in its twin French pavilion roofs and leaded mansard it was an elevator building designed to the absolute limit of the firemaster's reach to make more generous provision for bedroom accommodation. Its subtle blend of Beaux Arts, neo-Greek and low relief Aesthetic Movement sculpture, matched by equally splendid interiors, set the pace for the best work of the 1880s and 1890s.

By the later 1870s neither the mansard attic and garret roof nor the elevator were, however, altogether new ideas. French roofs had made their first appearance as a solution to the accommodation of staff and guests' servants at Philip Hardwick's Great Western Hotel at Paddington, London, as early as 1851, but it was not until passenger elevators were introduced that they became really practicable. The first to have elevators were two London hotels, J T Knowles's Grosvenor Hotel (1860–2) at Victoria Station, where it was described as a 'lifting room', and W & A Moseley's Westminster Palace Hotel in Victoria Street of the same date, where it was an 'ascending room'. These were quickly followed by Edward M Barry's Cannon Street Station Hotel (1861) and Charing Cross Station Hotel (1864), both in London, and his Mansion, later Star & Garter, Hotel at Richmond (also 1864).

These early elevators were as much for the movement of trunks and other heavy baggage as for the convenience of guests. The earliest installation of this kind in Scotland appears to have been at the University Club in Edinburgh (1865–6), where it was located in the service stairwell, suggesting that passenger use was, originally at least, by special arrangement and limited to those with physical disabilities. Described on the plans as a lift, it was only 3 feet 6 inches (1.07 metres) square and was probably hydraulic, operated by water mains pressure, as no engine-house is shown on the drawings. It was probably also quite slow, and was electrified in 1896.

The first Scottish hotel actually built with a French attic and garret roof was the Café Royal Hotel, occupying half a street block in West Register Street, just off Edinburgh's Princes Street and conveniently close to Waverley Railway Station. It did not have a lift. Designed by Robert Paterson in 1862–3 with free Renaissance elevations of the Liverpool-

Manchester school curved into one another, its tall pavilion roof gave it a very Second Empire profile. It had a curious building history, having originally been planned as a gasfitter's showroom with speculative hotel accommodation above, as at the Life Association. In the event, the nearby Café Royal restaurant, established in 1817, acquired the whole building prior to completion. The lavish interiors with Doulton tile pictures are the work of a later hotel specialist, J McIntyre Henry, 1898–1901.

The Café Royal was to remain unusual in the height and unbroken pavilion form of its roof, which was originally richly brattished. Much more common was the four-pitch mansard variety first introduced at David MacGibbon's reconstruction of the Georgian houses which comprised the Royal Hotel on Princes Street, for the Leith magnate Don R MacGregor, in 1867. The Royal was quickly followed by the Palace Hotel at 108–10 Princes Street, built anew by John Lessels in 1869 as the upper floors of Taylor & Sons' furniture showrooms. The latter has elegant free Renaissance details and a central pavilion roof dividing its attic and garret mansard roof. Much grander than any of these would have been the railway hotels, had they been built. In 1864 a French-roofed hotel very similar to the Grosvenor, and almost certainly designed by Peddie & Kinnear, was planned for the Glasgow Union Railway's station at St Enoch Station, Glasgow, but neither the station nor the hotel was built at that time. The only record of the scheme is a poor wood engraving, but it was doubtless to have been similarly equipped. Better recorded is Peddie & Kinnear's scheme for the Caledonian Station Hotel in Edinburgh, designed in 1867–8 for the same site as the present hotel in the West End. Four main storeys and an attic and garret high, the Caledonian would have rivalled E M Barry's London and Richmond hotels with its pavilion roofs, central cupola, arcaded semi-circular oriels and fountain in its enclosed forecourt. It would also have been the first Scottish hotel to have a full-scale passenger lift or 'rising room', 8 feet (2.44 metres) square and adjacent to the main staircase. Back-to-back with this passenger elevator there was to have been a set of smaller service lifts, presumably to handle baggage, laundry, and fuel independently and/or simultaneously as on the model of those provided by Easton, Amos & Sons at the Brighton Hotel in 1864. The surviving drawings regrettably provide no details of the actual equipment proposed at the Caledonian. Still more ambitious would have been the hotel Peddie & Kinnear proposed in 1867 as part of Charles Jopp's Waverley Station reconstruction and Waverley Market proposals of 1866–73. This probably never got beyond sketch-plan stage but it was prophetic of the present building in having a great clock tower like an English town hall. In the event, a financial crisis, over-commitment on other projects and market uncertainties resulted in both these Edinburgh projects being shelved. Thirty years were to elapse before they were revived.

In the mid 1870s William Hamilton Beattie of the Edinburgh architect-builders, George Beattie & Son, a pupil of David Bryce Senior, established a commanding lead as Edinburgh's premier hotel specialist, his reputation ultimately extending as far as Lerwick in Shetland where he built the very competent Bryce Baronial-style Grand Hotel in 1887. His first Edinburgh hotel, the turreted Central on Market Street (1873), was also Baronial, designed as a westward extension of Peddie & Kinnear's Cockburn Street scheme, but thereafter his city centre hotels were free Renaissance and several floors higher than their predecessors, a development made possible by improved fire-fighting equipment. As early as 1876 Beattie rebuilt MacGregor's Royal in a taller six-storeyed attic and garret form, generously provided with canted bays like David Bryce Junior's Star. Beattie also built the four-storeyed, mansard attic and two garret levels Clarendon (1875) at Nos 104–6 Princes Street, which superseded a similar scheme by Peddie & Kinnear. It included a glass-roofed shopping arcade 105 feet (32 metres) long and 30 feet (9.14 metres) high. Both these buildings had elevators. A decade later he designed the very tall Central at Nos 121–2 Princes Street, replacing his earlier hotel of the same name on Market Street. It was fairly closely modelled on Sellars's Glasgow New Club, having leaded mansard attic and garret floors between Dutch gables.

Each of Beattie's three Princes Street hotels was really two hotels linked together, the front blocks being for the well-off and the plain tenement-like rear blocks on Rose Street Lane for commercial travellers, tradesmen and personal servants. Rather similar to Beattie's Clarendon, with their twin bay windows and French attic and garret roofs, were Robert Paterson's Windsor Hotel (1879) at No 100 and MacGibbon & Ross's short-lived five-storeyed and attic Richmond Hotel (1882) at No 30 on the corner of St Andrew Street. These were more exclusively for the better-off, with the garrets accommodating staff and personal servants. Although their hotel buildings have now mostly vanished, MacGibbon & Ross briefly rivalled Beattie as hotel specialists. They reconstructed houses at the west end of Princes Street with 'compo' Second Empire façades and an attic and garret mansard roof as the Osborne (1873); it was once a memorable Edinburgh landmark, its domed corner octagon providing splendid panoramic views of the castle and the West End. It was followed by smaller but similar reconstructions at the Great Western Hotel, 130–2 George Street (1875), and the Maitland Hotel on Shandwick Place (1876). Most of these were promoted by limited companies set up under the Limited Liability Act of 1855 and the Companies Act of 1862, in which the architects were major shareholders and often directors. They were dependent on borrowed money and several were in difficulty even before the failure of the City of Glasgow Bank in 1878. The Osborne was burned out as early as 1879. The Great

Western may never have opened as a hotel, the earliest occupancy recorded in directories being as offices.

Princes Street, St Andrew Square and Charlotte Square were so much the preferred location for hotels that it was perhaps difficult to trade successfully anywhere else. The most significant of the later hotel developments in Princes Street was the temperance Old Waverley (1883) by John Armstrong, an elevator building in which, by keeping all the floors the same height, he managed to contain eight storeys within the firemaster's reach, the upper two being in French roofs as in the Beattie and Paterson examples. It was followed in 1888 by J McIntyre Henry's reconstruction of the houses at the east corner of Princes Street with Castle Street as the gigantic makeshift of the Palace Hotel, built up to five storeys, with twin pyramid-roofed towers and immensely tall French chimney stacks. Henry was also responsible for the relatively narrow-fronted Royal British (1897), but a finer example of the fin-de-siècle genre was Cousin & Ormiston's St Andrew Hotel (1900), a tall elevator building on a postage-stamp site in St Andrew Street, just off Princes Street.

While nearly all the larger hotels had a 'commercial room' and a room for the display of goods for sale and trade meetings, arrangements for commercial travellers and sales representatives were a particular feature of the Glasgow hotels. There, the earliest really large hotels had been achieved by uniting the houses in the two terraced blocks on the north side of George Square to form two hotels, the North British Hotel on the west, and the Royal on the east. Both hotels were closely associated with the North British Railway's Queen Street Station, and were acquired and improved in 1875 in connection with the station rebuilding works, though the Royal was relinquished two years later. A description of the latter in 1894 gives a fair impression of the facilities offered by the larger mid Victorian hotels in the cities:

> reading, writing and coffee rooms all splendidly furnished ... the grand dining room at the rear is a noble saloon, eighty feet long by sixty feet wide [24.38 by 18.29 metres], and furnished in sumptuous style ... The artistic decorations here are particularly noteworthy, and the frescoed ceiling is considered one of the finest in Scotland ... The Royal [contains] many suites of private apartments, bedrooms and sitting rooms, and single bedrooms all most comfortably and tastefully appointed, and commercial gentlemen will find excellent stock-room facilities ... The commercial clientèle ... is of a very superior order, including representatives of the foremost mercantile houses at home and abroad, and special provision is made for the satisfactory accommodation of this important section of the hotel's patronage.[3]

As in Edinburgh, the great era of hotel construction in Glasgow was the 1870s, and much of it followed the same pattern of conversion rather than new building. In 1871 Murdoch & Rodger, solicitors, instructed Alexander 'Greek' Thomson to convert the very smart pilastraded tenement with shops, which he had built in 1864–6 at 172 Sauchiehall Street, into the Washington (later Waverley) Hotel. In 1874 Thomson was commissioned to reconstruct a large Georgian house block on Bath Street as the Northern Club; but in 1876, after work had begun, the project was taken over by Andrew Philp of the Edinburgh Cockburn Hotel and completed by Thomson's partner, Robert Turnbull, as the Glasgow Cockburn, with an eaves gallery and a big French roof of the Café Royal type. It included a bath club. In the same years a similar reconstruction, but without a French roof, was undertaken by the same architects at 148–50 Bath Street as the Alexandra Hotel, the new details of the ground floor and attic being executed in pre-cast synthetic stone as at E M Barry's Charing Cross.

In 1875–9 the Blythswoodholm Buildings Company constructed the vast quadrangular Blythswoodholm Hotel, occupying a complete street block on Hope Street, between Bothwell and Waterloo Streets. In size this ultimately over-ambitious project far exceeded anything hitherto built in Scotland. The site had its origin in the Caledonian Railway's parliamentary authorisation of 1873 to cross the Clyde at Broomielaw to a new station on the west side of Hope Street, opposite Gordon Street. By 1875 this route had been found to be too expensive and further parliamentary powers were obtained to move the line east of Hope Street with a station at its corner with Gordon Street. At that date the Caledonian was content for the Blythswoodholm Company to take over the original site and build what was effectively the counterpart to the Glasgow & South Western Railway's hotel at St Enoch (see below). Designed by Peddie & Kinnear, the Blythswoodholm Hotel's pilastered elevations were modelled on Thomson's successful Washington Hotel formula, but with tall French roofs encompassing a total of seven storeys at the centre of the Hope Street elevation. The ground floor consisted of shops, and, as at Beattie's Clarendon in Edinburgh, a shopping arcade filled the area between the frontal blocks and the lower grade commercial accommodation which flanked the big dining room on the lane. The Bothwell and Waterloo Street sections of the building consisted of separate stairs of business chambers into which the hotel could subsequently expand if required.

The Blythswoodholm was the largest hotel development in the United Kingdom at the time, and although the interiors seem to have been relatively simple compared with those of the great London hotels, it had an elevator. Other ambitious, new-build Glasgow hotel projects of the boom years just before the collapse of the City of Glasgow Bank were

James Thomson's giant Maclean's Windsor Hotel at 244–50 St Vincent Street (1875–8), and the still bigger but now-demolished Grand (1876–8), a £35,000 project occupying a complete street block on Sauchiehall Street, at Charing Cross. Both were elevator buildings with French-roofed attic and garret bedroom floors.

LATE VICTORIAN AND EDWARDIAN CLUBS

Except for the station hotels (see below), very little hotel building took place in either Edinburgh or Glasgow after the speculators of the 1870s got their fingers burned. But the thirty years from 1880 were to be the era of the club in both cities, and political clubs in particular. In 1880 the Edinburgh Liberals bought the burned-out wreck of the Osborne Hotel and rebuilt it as their club, that prominent Liberal architect, John Dick Peddie, by then an MP, securing the commission for his son John More Dick Peddie.

The Edinburgh Conservatives almost immediately rose to the challenge in 1882 by erecting a magnificent club-house on Princes Street as a memorial to Disraeli who had died in the previous year. Designed by Robert Rowand Anderson at a cost of £32,000, it did not have as wide a frontage as the New Club but made up for it in height, four-storeyed and basement with attic and garret floors in its big and shapely red tiled roof. Stylistically it was Aesthetic Movement influenced neo-Georgian, subtly infused with early Italian Renaissance details. The composition of its elevation was drawn from J J Stevenson's Red House in Bayswater, suggesting that Stevenson's ex-assistant, and Anderson's partner, George Washington Browne, had a hand in the design. While the third-floor balcony neatly balances the design, it was not there merely for compositional reasons, as it enabled those in the smoke room to emerge for fresh air. Within, the club had – and still has – a splendid early Italian Renaissance stair with a stained glass window by Ballantyne & Son commemorating Disraeli. The Edinburgh Liberals did not attempt to match it, but they did make a strategic move to Lessels's Palace Hotel next door in 1890.

What the Edinburgh Conservatives had done, Glasgow's had to do better, or at least more expensively. They selected Colonel Robert W Edis, evidently on the assumption that the Prince of Wales's architect must be head of the profession: perhaps some of the members also knew his Constitutional Club of 1884–6 in London and had seen his book, *Decoration and Furniture of Town Houses*, published in 1881. His now-forgotten red sandstone pile at the corner of Bothwell and Wellington Streets (1892–4) was much bigger and showier than Anderson's Edinburgh one, with five canted bays and a skyline bristling with turrets and Netherlandish gables. The original Bryce building then became the Junior Conservative Club, again on the London model.

Predictably, the first purpose-built Liberal Club was in that wealthy hot-bed of radicalism, Paisley. Designed by Alexander Thomson's pupil, the Paisley architect John Donald, in 1886 this was an ambitious building with a circled corner, Ionic columns and Sellars-inspired details. Finally, the Glasgow Liberals decided that they had to have a more than comparable building, and there were so many wealthy Liberals in Glasgow that money was no object. In 1906 a competition which seems to have been limited to Liberal architects was held, and won by the Beaux Arts-trained Alexander Nisbet Paterson who had travelled extensively in the United States. His giant five-storeyed and attic red sandstone building, completed in 1909, was the grandest of them all, with an immensely spacious interior rich in marble and plasterwork by Bankart, notably in the hall and first-floor dining room.

Military club-houses also began to appear, but the only purpose-built one was the Caledonian United Services Club at the corner of Shandwick Place and Queensferry Street in Edinburgh, designed by Sydney Mitchell in 1901. It was grand enough in its architecture, with Gibbsian windows and 'Queen Anne' gables flanking a corner cupola, but differed from the great political club-houses in having ground-floor commercial premises.

SAILORS' HOMES

Another building type with a restricted clientèle was the Sailors' Home. These were philanthropic rather than commercial buildings, built to accommodate crew between voyages and to keep them out of doubtful lodging houses.

Following the success of the very architectural one built in Liverpool in 1846, a Sailors' Home for Glasgow's dockland was first considered by the Seamen's Friend Society in 1853. Some £12,000 was subscribed, and in 1855–7 a large four-storeyed astylar Italianate corner-building with ground-floor shops was erected on Broomielaw and James Watt Street, to the designs of James Brown of Brown & Carrick with modifications by John Thomas Rochead who took over the project on Brown's retirement. The Building Chronicle of March 1857 reported that it had 130 bedrooms, a reading room, a dining room and a separate mess room for officers. The lighthouse-like tower at the corner was designed for a time-ball. It was extended in 1869, the original tower now being answered by a picturesquely-angled square one on the east. A similar but somewhat makeshift institution had existed in converted premises in Dock Street, Leith, since 1840, but after it was displaced by the Navigation School a towered Baronial edifice was built in 1883–5 to designs by C S S Johnston. It provided cabin accommodation for 56 seamen, rooms for nine officers and a barrack in the attic for up to 50 shipwrecked seamen without funds. In Dundee, the Sailors' Home was quite an elaborate five-storeyed free Renaissance building on Dock Street with a circled corner rising into a cupola. Designed by David Maclaren of Ireland & Maclaren in 1879–81, it provided cabins for 80 seamen and a chapel seating for 240.

VICTORIAN HOTELS IN DUNDEE AND ABERDEEN

There had been no new hotel development in Dundee since the building of the Royal in the late 1820s, but in 1866 the hotelier, Thomas Lamb, was persuaded to complete Reform Street by building a smart four-storeyed and attic, free Renaissance hotel to designs by the Dundee architect, James Maclaren, a pupil of David Bryce senior. It was a rather slim building because the site backed on to the ancient graveyard of the Howff. Its immediate purpose was to accommodate distinguished delegates to the British Association meeting of 1867 in the Albert Institute directly opposite. It was strictly temperance, soap being put in the hot water flagons to ensure that they were not used for toddy.

Maclaren was also responsible for the Royal Hotel reconstruction which completely absorbed the Thistle Hall, and probably also for the reconstruction and extension of Merchant's Hotel as the Royal British, c. 1880, with elegantly detailed Italianate compo façades masking as best they could the irregularities of fenestration in the buildings which had been absorbed. The history of the Queen's Hotel in Nethergate, designed by the Dundee and Perth firm of Young & Meldrum in 1878, paralleled that of the Blythswoodholm in Glasgow. Planned on the Edinburgh-Glasgow scale with tall attic and two-garret French pavilion roofs over four-storeyed French gothic elevations, and with sumptuous Gothic interiors, it was built on misinformation that Dundee's new Caledonian station was to be immediately adjacent, and was a disaster for those who promoted it. After the Caledonian Station was finally built at the foot of Union Street in 1890, Mather's Hotel was built to designs by Robert Hunter on the correct site immediately opposite, completing William Mackison's Whitehall Crescent Improvement Act development. Like the Queen's it was six-storeyed, as high as the firemaster would allow, and like Lamb's it was strictly temperance.

In Aberdeen, the first major purpose-built hotel development since the Aberdeen Hotel of 1817 was the Imperial Hotel (1869) on Stirling Street, close to the joint station and the harbour. It took its title from the Empress Eugénie, who used it en route to Abergeldie. Designed by the London-trained architect, James Souttar, it was London Gothic, but relatively low in scale at three-storeys and attic. Much higher at four-storey, basement and attic, but somewhat meagre in its architecture, was Matthews & Mackenzie's Caledonian Hotel (1880s), Renaissance with a towered and gabled asymmetrical frontage of canted oriels to Union Terrace, modelled on Beattie's Royal in Edinburgh. It was, however, much better sited than the Imperial, since it had an open outlook over the gardens in the Denburn.

RAILWAY HOTELS

Although nearly every major English railway station had been provided with a hotel as an integral part of the complex from the very beginning, most of the Scottish railway companies had at first been content to leave the provision of hotel accommodation to others; many inns and hotels which included the words 'station' or 'railway' in their titles were never owned by the company whose passengers they served.

Perhaps because of the nature of the terrain they crossed, the Highland Railway and its predecessor companies were the pioneers of hotel provision in Scotland. At Inverness, the Inverness & Nairn Railway built the hotel in Station Square, designed by the company's engineer Joseph Mitchell in 1855. It was aggrandised by the Aberdeen-Inverness firm of Matthews & Lawrie as early as 1858 as a towered three-storeyed Italianate villa, and further enlarged in 1876 when a new wing and a second tower were added by the same firm. Matthews's previous partner, Thomas Mackenzie, had already built the rather similar but tower-less station hotel in Elgin (now the Laigh Moray) in 1853 for the Morayshire Railway. The Highland Railway also included a ferry hotel at Kyleakin, Skye, in its proposals for the line to Kyle of Lochalsh, Ross and Cromarty, in 1864, and built hotels at Attadale (1868), Achnasheen (1871) and Kyle of Lochalsh (1880). The only other company to own hotels prior to 1870 was the North British Railway, which for various reasons connected with railway works had acquired eight hotels, all of which retained their original identities. None of these companies managed its hotels; all were leased to selected tenants.

Although the schemes for the Caledonian and North British station hotels in Edinburgh of 1867–8 had been shelved, that for the Glasgow & South Western Railway in St Enoch Square, Glasgow, of 1864 had merely been delayed. While the building of a hotel by a railway company was primarily a commercial decision, by the 1870s providing suitable overnight accommodation for members of royal families, British and European, who travelled on their lines, and indeed for the numerous families wealthy enough to travel in private carriages, it had assumed an importance difficult to appreciate now, and featured prominently in advertisements. However, the immediate catalyst for the highly competitive station-building programmes in Scotland in the final quarter of the century was the construction of the Settle and Carlisle line in northern England, which linked the Midland Railway to the Glasgow &

Figure 8.13 St Enoch's Hotel, Glasgow. Crown copyright: RCAHMS, SC 710767.

South Western and created a new main line between London and Glasgow (and Edinburgh via the Waverley route) in direct competition with that of the London & North Western and Caledonian Railways. The Midland and the Glasgow & South Western had the same chairman, Matthew William Thompson, and in 1867 and again in 1872 and 1873 had made determined attempts to merge, which were defeated in Parliament. But that failure did not prevent them from co-operating and running through trains from London to Glasgow; an important part of that strategy was to have very grand matching hotels at both termini.

Consequently, the design and layout of the Glasgow & South Western's offices and hotel were fairly closely modelled on Sir George Gilbert Scott's Grand Hotel at the Midland Railway's St Pancras, the ramped vehicular terrace being very similar (fig. 8.13). The commission was, rather oddly, given neither to Scott nor to any of the leading Glasgow practitioners, but to an obscure Roman Catholic church architect in Hampstead, Thomas J Willson, with a local Episcopal church architect, Miles Gibson, in charge at the site. This vast Gothic building, five-storeyed with attic and garret above its raised terrace, enclosed its train-sheds on the west and north with a frontage of 360 feet (109.73 metres) to St Enoch

Square and 500 feet (152.4 metres) towards Argyle Street, extending as far as Dunlop Street. Like St Pancras, it made extensive use of iron construction, though not in the same decorative form: at St Enoch it was enclosed in lath and plaster. When it opened on 3 July 1879, St Enoch's was the first building in Glasgow to have electric light. It did not have such a splendid entrance hall, grand stair and dining room as St Pancras, but it had 20 public rooms and over 200 bedrooms. It was even larger than the Blythswoodholm, and, more importantly, it was much more up-to-date in its fitting-out, with furnishings by Gillows, kitchens by Jeakes & Company of London (who had equipped the kitchens at St Pancras), 'ascending rooms' worked by hydraulic pressure as at St Pancras, and a central postal collection system with post-boxes on each floor.

Denied access to the Glasgow Union Railway, the Caledonian had to construct its own route across the Clyde if it were to survive the competition from the Glasgow & South Western. This it did in 1875–9, its new terminus being the Central Station at the corner of Hope Street and Gordon Street. In 1877 it had selected Rowand Anderson to design a similar building to that at St Enoch's, originally conceived only as centralised railway offices because of the building of the Blythswoodholm immediately opposite. As at St Enoch's, a Gothic design was considered, closely related to Anderson's Mount Stuart on Bute, but with an acute sensitivity to changing fashions south of the Border, a northern European Renaissance was ultimately chosen. In October 1880, when construction was far advanced, the directors made the surprise decision to complete the Central Station building as a hotel in competition with the newly-opened Blythswoodholm, having belatedly realised that the hotel at St Enoch was attracting passenger traffic from their route. The change of plan was achieved only with difficulty as the building was of very solid fireproof construction, with concrete slab floors on rolled iron beams. Everything was of the best; the awning was commissioned from Francis Skidmore of Coventry, and the principal spaces were much more architectural than those at St Enoch's with furniture designed by Anderson. A few statistics give an indication of the scale and complexity of the project. In all, the building required 7,200 square yards (6,581 square metres) of carpet, 1,200 electric bells, 5,000 feet (1,524 metres) of speaking tubes and 29 miles (46.7 kilometres) of bell-wire. Electric light was provided in the public rooms by Anderson & Munro, and gradually extended to the remainder in 1885 (by Anglo-American Brush Electric Light Company) and 1890 (by Mavor & Coulson). Stylistically, the interior work was all in a quiet François Premier manner, although the original wall coverings and paintwork probably produced a richer effect than it has now. It was even more up-to-date than St Enoch's when it finally opened in June 1885. Its superior equipment sealed the fate of the

Blythswoodholm, which closed in 1890 and was wholly converted to business chambers for rental, a reverse process to that which had originally been planned. It was not, however, until the Central was extended down Hope Street by James Miller in 1901–7 to provide almost 400 rooms and to match the facilities of the North British in Edinburgh (see below) that the interior achieved its final Edwardian opulence and surpassed St Enoch in size.

The commercial success of these hotels emboldened both the Glasgow & South Western and the Caledonian Railways to build others at their busiest stations. At Ayr, the Glasgow & South Western built a 45-bedroom free classic mansardic hotel in red sandstone to designs by their engineer Andrew Galloway in 1885–6 at a cost of £50,000, followed by a smaller 29-bedroom hotel at Dumfries (1896, replacing one bought in 1880), designed in a much simpler vein by W J Milwain. At Perth, an important transfer point between the Caledonian, the Highland and the North British Railways, there was a particular problem as Queen Victoria was in the habit of breaking her journeys there, the station's committee rooms being somewhat inconveniently made available to her as improvised dining rooms. An Act of Parliament for a jointly-owned station had been obtained in 1865 but was not implemented until 1888 when the Perth architect, Andrew Heiton, was instructed to replace his modest building of 1855 with a large Flemish Gothic structure of three storeys and an attic, with apartments suitable for the Queen's use. It opened in June 1890.

The Highland Railway had taken its Inverness hotel into direct management in 1878, its clientèle including the royal family and 'most of the nobility of Europe'. By 1886 the Great North of Scotland Railway had concluded that the volume of royal traffic on its Aberdeen to Ballater route was such that it had better have a grand hotel in line with the other Scottish companies. In Aberdeen, it found a ready-made building of sufficient architectural pretension in the Pratt & Keith drapery warehouse on Union Street at the south-west abutment of Union Bridge, adjacent to the site of the future Joint Station. This was a suitably Balmoral-like, three-storeyed attic and basement structure with a symmetrical turreted elevation to Union Street, designed by Matthews & Mackenzie as part of the Bridge Street harbour and railway access works in 1872–3. In 1886 the same architects enlarged and converted it as the Palace Hotel with a covered way from the station, and in 1891 and 1903 it was further enlarged and raised two storeys higher, increasing the number of bedrooms and private suites to 77. Although Queen Victoria does not seem to have used it, its 1899 publicity featured an impressive list of royals, which included the Prince and Princess of Wales, the Empress Eugénie, the Empress Frederick, King Carlos of Portugal and King Leopold of the Belgians. The Station Hotel itself (1900), a

Figure 8.14 North British Hotel (The Balmoral), Edinburgh, c. 1930. Crown copyright: RCAHMS, SC 466207.

four-storeyed and attic, twin-gabled design of Norman Shaw inspiration, translated into granite by R G Wilson, was added to the Great North's hotel division in 1910, apparently to remove the embarrassment of commercial and middle-class family business from the Palace.

The final round of urban station hotel building was inaugurated by the North British Railway's limited competition of 1895 for the Waverley Station Hotel, Edinburgh, soon renamed the North British Hotel, built as part of their Waverley Station reconstruction of 1892–6 (fig. 8.14). Although aesthetically finer designs had been submitted by J J Burnet, Rowand Anderson and Dunn & Findlay, the commission went to William Hamilton Beattie, his 195-feet (59.43-metre) high clock (and cistern) tower having caught the imagination of the directors and in particular that of Charles Jenner, for whom he had already built the nearby department store. Beattie and the General Manager were sent round some 16 premier hotels in Paris, Brussels, Amsterdam, Vienna and Budapest, where they particularly admired the Royal, but the basic design did not change very much as a result of these visits. Quadrangular around a 50-feet (15.24-metre) square Corinthian-columned palm court with a glazed dome, it rose five floors and an attic above the pavement with four more beneath. Its free Renaissance elevations of 180 and 190 feet

(54.86 and 57.91 metres) with bows, balconies, corner domes and Netherlandish gables are of variable quality, as are its interiors, but with its 1,600 tons of steel frame, patent fireproof floors and roofs of Stuart's granolithic, it was an impressive feat of hotel planning and engineering.

Following the precedent of the similarly located Palace in Aberdeen, it was entered from the station as well as from the pavement, access being by elevator and a winter garden corridor to a marble-columned arrival hall at lower basement level. It contained 700 rooms, more than 300 of them being bedrooms or private suites, and 52 bathrooms, a very high ratio of bathrooms to bedrooms at that date. The coffee and dining rooms were 112 feet (34.14 metres) long, the drawing room 71 feet (21.64 metres) and the writing room 42 feet (12.80 metres), and there were in addition two billiard rooms and six private drawing and dining rooms on the scale and elaboration of those of a major country house. The electrical system was designed by Professor Barr of the University of Glasgow, the heating was on the plenum system and the boilers produced 2,000 gallons of hot water per hour. For all its opulence, the commercial traveller was not forgotten, there being several stock-rooms and commercial rooms for their use and for personal servants travelling with the hotel's guests.

Beattie's brother, George Lennox Beattie, and his partner, Andrew R Scott from Burnet's office, were also responsible for the Carlton Hotel (1898), diagonally opposite on North Bridge, where the Scots Renaissance detail of Burnet's competition design for the Waverley Station Hotel was shamelessly exploited. By 1899 these developments had finally induced the Caledonian Railway to commission John More Dick Peddie and George Washington Browne to design the Caledonian Hotel around the platforms of its Edinburgh station, as at Glasgow Central, of which they had a survey made to help them determine the requirements; Sir Rowand Anderson had apparently declined to lend them the drawings. The Caledonian opened in December 1903, 14 months behind the North British. The arcaded station façades of 1890–3 were retained as its lower floors. The site, the brief, the late decision to increase its height by a further floor and the hurried circumstances of its design all made it difficult to achieve a fully unified composition. Its V-plan layout inevitably lacked the formality and spaciousness of the North British's foyer and palm court, but the décor was of greater refinement.

The other major station hotels of the 1880s and 1890s were all built by private capital. By that period the words 'station' and 'railway' had acquired a considerable cachet to which the modest privately-owned railway and station inns of the 1840s and 1850s had never aspired. It is of some interest that the proprietor of the giant three-storeyed, attic and basement hotel at the bridge in Wick, Caithness, built in 1866, thought it worthwhile to adopt the title 'Station Hotel' after the Sutherland & Caithness Railway opened in 1874. Notable later instances, built after the railway had arrived, were the neo-Jacobean Station Hotel, Oban, Argyll (now the Caledonian Hotel), by William Menzies (1880) and Alexander Shairp (1884), and the Railway (later the Royal, now the Highland) Hotel at Fort William, Inverness-shire. The latter was a very 'up-market', if stylistically old-fashioned, hotel designed by Duncan Cameron in 1895 as a larger version of the Alexandra of 1876, but it was set in spacious grounds, anticipating the essentially similar hotels built by the Great North of Scotland and the Highland Railways a few years later. The same architect was also responsible for the much more up-to-date Station (later Highland) Hotel at Nairn in 1896, northern European/early Renaissance in red sandstone, which in the lavishness of its decorative detail, some of it ceramic, certainly set out to achieve full railway company standards.

The subsequent changes of name are interesting indicators of more modern perceptions of the original titles. As architecture, perhaps the most impressive of these privately-financed station hotels was the Central at Annan, its name probably taken from the Caledonian's Glasgow hotel. Designed by the Dumfries architect, Frank Carruthers, in 1898 on a prominent corner site facing the station, its circled Ionic portico between tall canted bays rising into octagonal towers gives it a metropolitan grandeur clearly intended to answer the railway company hotels at the London, Edinburgh and Glasgow termini.

PROVINCIAL HOTELS

Although the rapid expansion of the railway network into the remoter areas of Scotland from the mid-1850s was the main factor behind most provincial hotel building in the later nineteenth century, scenic tourism had begun much earlier, particularly on the west coast. The Prussian architect, Karl Friedrich Schinkel, arriving in Glasgow on 7 July 1826, had found 60–70 steamboat notices advertising pleasure trips to Staffa and the Scottish lochs, and a coach link taking him to Loch Lomond. Although Schinkel did not himself venture so far, these steamer routes extended up to Gairloch in Wester Ross, where a half-century later the Maitlands of Tain, Ross and Cromarty, were to design an astonishingly large hotel to accommodate the ever-increasing number of tourists. At Oban, Schinkel found a comfortable inn, but there were not enough hotels on the steamer routes. At Dunoon, Argyll, it was not until 1837 and 1847 respectively that the Argyll and Caledonian Hotels were built, and not until c. 1850 that the Victoria and Royal Hotels appeared on the waterfront at Rothesay, Bute. The very busy inn at Tarbet on Loch Lomond was not rebuilt on an adequate scale until 1850. In 1838 the difficulties of the proprietor of that inn were noted by Lord Cockburn in his *Circuit Journeys*:[4]

> The inn is, for a Scotch one, very good but too small for the resort in summer and far too large for the want of resort in winter. I am always ashamed of our country for its want of hospitality, in this respect of inns, to the many strangers who now visit it.

The shortness of the season was also a problem at the Trossachs where he found the inn:

> could perhaps put up a dozen or at the very most two dozen of people; but last autumn I saw about a hundred people apply for admittance, and after horrid altercations, entreaties and efforts, about fifty or sixty were compelled to huddle together all night. They were all of the upper class, travelling mostly in private carriages and by far the greater number strangers ... I saw three or four English gentlemen spreading their own straw on the earthen floor of an outhouse, with a sparred door and no fireplace or furniture.

Sir Walter Scott's *The Lady of the Lake* had spread the fame of Loch Katrine and the Trossachs far and wide, to the point where visitors were undeterred by a journey which had to be made by road; the heroic scale and royal opening of the Glasgow waterworks undertakings of 1855–9

Figure 8.15 Trossachs Hotel ('Ardcheanochrochan Inn'); dining room, watercolour by Sarah Sherwood Clarke, 1854. Courtesy of Mr and Mrs Champion.

only served to spread that fame still wider. Lord Willoughby d'Eresby built the wonderfully picturesque inn at the Trossachs (fig. 8.15) in 1852, one year in advance of the additional influx of tourists anticipated from the construction of the Caledonian Railway to Callander. Its twin 'candle-snuffer-roofed' profile was the work of George Penrose Kennedy, an ex-assistant of Sir Charles Barry.

The Trossachs Hotel inaugurated a great era of railway-related Highland hotel building in places of natural beauty. The opening of the Perth & Dunkeld Railway in 1856 sparked off a complete new-build village for summer visitors at Birnam, Perthshire, where construction was reported as being in full swing as early as 1857. A hotel was built, much larger than that at Trossachs, but no less picturesquely composed. It contained a ballroom and a public hall, 'so large and ornate as to be one of the finest in Scotland; and has attached to it a billiard room, a bowling green and beautiful grounds'. Perhaps some of the interior work intended for the never-completed new castle of Murthly, Perthshire, had been utilised there. Sir William Drummond concurrently feued the main street for Scots Baronial terraces of genteel boarding-houses for summer visitors, the hinterland nearest the river being reserved for a private estate of sizable mansions. Development moved up the line to Pitlochry, Perthshire, when the station was opened there in 1863 and Fisher's Hotel was built, and later to Kingussie.

HYDROPATHICS

The wonder of the age were the hydropathics which tended to be found wherever a railway line coincided with natural beauty and the presence of the appropriate therapeutic waters. Although Peterhead and Pannanich had gone out of fashion at the beginning of the nineteenth century and William Burn's 1824 neo-classical township of boarding houses at Bridge of Earn had been abandoned unfinished, a few watering places had continued to flourish. Moffat remained fashionable thanks to the building of a new Bath House in 1827 and the arrival of the Caledonian Railway in 1847 at Beattock where, as at Birnam later, a particularly smart station was built for the reception of visitors to the spa. Two new hotels, the Buccleuch Arms and the Star Hotel (1860), the latter by the Edinburgh architect William Notman, were added to the mid eighteenth-century provision. Building was still more active at Bridge of Allan, Stirlingshire, where three new hotels were built: Barr's (later the Queen's), large plain classic; the Westerton Arms (1831), 'cottagey' by William Stirling I; and Philp's (1842, later the Royal; fig. 8.16), Jacobethan, probably by William Stirling II, ranking with Burn's Granton and Banff Hotels as the largest and best hotel building in Scotland of its date.

The further development of taking the waters into the alternative

Figure 8.16 Royal Hotel, Bridge of Allan, 1920s. Courtesy of Whiteholme (Publishers) Ltd, Dundee.

medicine of hydrotherapy, usually shortened to hydropathy, was first introduced by an unqualified Austrian, Vincenz Preissnitz, at his hydropathic in Grafenberg in Austrian Silesia in 1826. In Scotland, where the treatment was first introduced in the early 1840s, the promoters were often closely associated with the temperance movement or the United Presbyterian Church. Spiritual well-being was an important element of the programme and at some of them the clergy of that faith enjoyed favourable rates in return for pastoral duties. Strictly speaking, the hydropathics were medicinal establishments with a resident physician. Sometimes described as sanatoria, they were places where the patient's family could be accommodated either in single rooms or in suites and kept amused with winter gardens, croquet lawns, putting greens, racquet courts, boating, skating and curling ponds in a formal landscape of 25–100 acres (10–40 hectares), guarded by its own lodge and gates like a country house.

These hydropathics pioneered the concept of the hotel as a complete health and leisure complex. The simple virtues of rest, exercise and fresh air, as well as water, were all part of the cure, and tobacco and the discussion of business matters were firmly discouraged. Some 20 of these establishments were built in Scotland. A very few like Shandon (1876–7) on the Gare Loch, Dunbartonshire, took advantage of large mansions which were both in the right place and on the market, but most were tall three- to four-storeyed, basement and attic structures, built completely anew by limited companies of about a dozen shareholders at a cost of between £30,000 and £100,000, major undertakings which called for the services of a civil engineer and a landscape gardener as well as an architect.

The first was the Glenburn at Rothesay, a two-storeyed and basement classical villa built to the designs of David and James Hamilton in 1843 for Dr Paterson who had worked with Preissnitz. But it was not until the Companies Act of 1862 became law that hydropathic building really got under way. The much bigger Cluny Hill Hydropathic (1863) at Forres, Moray, a plain Jacobean building on the southern slope of its hill, seems to have been the first sizable purpose-built example, but the earliest of the really large establishments was the Strathearn House Hydropathic at Crieff (1868), Perthshire, designed in a mixed Jacobean and Scots Baronial style by the Aberdeen architect, Robert Ewan. It was laid out on a somewhat hospital-like corridor and pavilion plan with a central tower and a very large domed winter garden. Its total frontage was 345 feet (105.15 metres), with dining and drawing rooms 84 feet (25.60 metres) long to accommodate some 200 guests; the Turkish baths, stamping baths, rainwater baths, plunge baths, vapour baths and sitz baths had a daily requirement of 20,000 gallons of water, suitably rich in minerals, piped from a one-acre (0.4-hectare) reservoir four miles away.

Smaller, but with a remarkable bath-house block containing exotic oriental interiors, was the Bridge of Allan Hydropathic (1868), followed by the much larger Waverley Hydropathic (1869 and 1876) at Melrose, at which its architect, James Campbell Walker, pioneered the use of mass concrete in Scotland. Thereafter, more ambitious and much less hospital-like structures sprang up wherever there was a railway line and water with the appropriate minerals within piping distance. Andrew Heiton's vast Atholl Hydropathic at Pitlochry (1875) and Frederick Pilkington's Moffat (1875–7; fig. 8.17), built to accommodate 250 and 300 guests respectively, in single rooms or private suites, were both in a château manner which anticipated the Canadian railway hotels of a decade later. In the same year (1875) Pilkington also designed that at North Berwick, East Lothian, which soon became the Marine Hotel at the hands of William Hamilton Beattie.

The premier specialists in this type of building were, however, Peddie & Kinnear with the Italianate Dunblane, Perthshire (1875–6) and Craiglockhart, Edinburgh (1879), and the French Second Empire Callander (1879–80), which for a brief period of 13 years, brought a whiff of the French Riviera to the west Perthshire countryside. Other major examples were John Starforth's multi-towered Peebles (1874–81), John

Figure 8.17 Moffat Hydropathic, Hotel, Dumfriesshire, *c.* 1875. Crown copyright: RCAHMS, SC 710755.

Honeyman's Baronial Kyles of Bute (Swanstonhill Sanatorium, 1878–80) and Thomas Lennox Watson's single-towered Kilmacolm, Renfrewshire (1879). But at Oban Hills, a rambling Scots Baronial design by the little-known J Ford Mackenzie of Glasgow, hydropathic building ground to an abrupt halt when the liquidators of the City of Glasgow Bank made their final call of 2,250 per cent on the shareholders, leading to a chain of secondary bankruptcies and a prolonged economic depression. Withdrawn bond and loan capital and lack of patients and guests crippled the companies which had built them, William Hamilton Beattie's vast Morningside Hydropathic, Edinburgh, failing even before it opened.

By the early 1890s nearly all the hydropathics had been sold off to hoteliers like Philp for adaptation to orthodox hotel use. Nevertheless, the hydropathics had a profound influence on the luxury hotel market of the earlier twentieth century, even if none of them had had the foresight to include a golf course. It was only after they had been successfully operated as hotels in the 1890s that their merits began to be appreciated by both hoteliers and guests, the Ben Wyvis Hotel (1879) at Strathpeffer, Ross and Cromarty, a rambling three-storeyed and attic pile by William C Joass, being perhaps the earliest to adopt a hydropathic-type layout within spacious grounds guarded by gates.

SPORTING HOTELS

From the middle of the nineteenth century a number of quite grand estate hotels were built, mainly to accommodate those who had come to fish and shoot. These tended to be much larger buildings than those built before 1840, again a reflection of the increased numbers coming to the Highlands after the railways had made them accessible. Among the most impressive of these carriage trade buildings were two at Braemar, Aberdeenshire. The older was the immensely long Invercauld Arms with turreted central porch and log-column verandas, built piecemeal from the 1850s, probably by James Henderson of Aberdeen, and enlarged by the London architect, J T Wimperis, in 1886. More unified in design was the prettily bargeboarded, three-storeyed and attic Fife Arms, built c. 1880 and remodelled by A Marshall Mackenzie in 1898. It had broad canted bays stylishly linked together by a veranda, subsequently interrupted by the insertion of a two-storeyed Gothic porch in 1905. Both of these hotels had old inns at their cores, and had been rebuilt to at least four times their original size to accommodate Balmoral-related business; among others connected with the court, Queen Victoria's son-in-law, Crown Prince Frederick of Prussia, stayed at the Fife Arms in 1887. They stood by the roadside just as their predecessors had done, an arrangement also to be found at the huge but dourly plain Fishers' Hotel at Pitlochry as enlarged to its present three-storeyed and attic mansardic form before 1900. Pleasant formal gardens were, however, provided at the rear of these buildings.

Much the best of the mid-Victorian fishing and shooting hotels linked with great estates was the Scots Baronial Grant Arms at Grantown-on-Spey, Moray, rebuilt in 1873–5 to designs by Marshall Mackenzie, lest the queen and her entourage repeated their unexpected stay in September 1860. It too stood in the Square, just as its predecessor had done. By that date Grantown had acquired a remarkable reputation for human longevity and was being recommended to invalids and convalescents by leading London and Edinburgh physicians. The hotel had to be enlarged in 1900. Its new wing by the Glasgow hotel specialist, J M Monro, was taller, but tactfully adhered to Mackenzie's idiom.

Most later nineteenth-century hotel-building was, however, in or near towns which offered the attractions of sea air and golf. On the west coast the popularity of Oban had been boosted by a visit from Queen Victoria and Prince Albert in August 1847 when the royal yacht anchored in Oban Bay, but the impressive Italianate Great Western Hotel and its flanking crescent of 'up-market' boarding-houses, designed by the Glasgow architect, Charles Wilson, was not built until 1861–3. Enlarged in 1884 by the local firm of Ross & Mackintosh and lit by electricity from that year, the Great Western had at its rear a magnificent dining room with a canted ceiling, the décor of which was well up to London standards. The Alexandra followed in 1871, but with the arrival of the Callander & Oban Railway in 1880 the town became much more accessible, hotel development reaching the point at which it was said, almost certainly correctly, that Oban had the highest ratio of hotels to population in the United Kingdom. Almost all of these were along the waterfront and varied in height from three-storey and attic to five-storey. The Station Hotel, as described earlier, was built concurrently with the station and so was the Royal by Ross & Mackintosh. These were followed by J Fraser Sim's elegant Columba with French-roofed tower and long iron balconies and the Scots Baronial Argyll (both 1885), the Marine (Alexander Shairp, 1890) and the mildly neo-Jacobean Queen's (1891). Most of these were initially the work of locally-based architects but in 1898 J M Monro yet again took over, being brought in to rebuild and extend the Alexandra on more up-to-date metropolitan lines.

On the north-east coast a similar pattern of hotel development is to be seen, if not quite in the same concentration, at those towns which the railways had made important holiday resorts, particularly Inverness, Forres, Nairn and Tain. Several of their hotels were comparable with the best in Oban, varying in style from Scots Baronial as at Alexander Ross's Columba (1881) and Palace (1890) on the river front at Inverness, to Elizabethan and Jacobean as at the towered Marine in Nairn by James Matthews (1859), the Victoria (1864) and the Carlton (1900) in Forres by George Petrie and Peter Fulton respectively, and, very unusually, banker's Italianate at the stylish Waverley (1877) in Nairn. The best detailed of these north-eastern hotels was the Maitlands' French Gothic Royal Hotel (1872) in Tain, a striking testimony to the affluence of its summer visitors at that date.

The grandest of these late nineteenth- and earlier twentieth-century coastal hotels were nearly all directly linked to golf. In St Andrews itself, the earliest appears to have been The Golf designed by George Rae in 1863, rebuilt in enlarged form *c.* 1895. Frontage overlooking the Old Course was at a premium, resulting in the tall, elongated H-plan of Rusack's Marine Hotel, squeezed into a long narrow plot between the Links and Pilmour Links. Designed by David Henry in 1886–7 and 1891, it was old-fashioned late classic in style and, four-storeyed basement and attic, it was as high as the firemaster would allow. A much finer building, if alien to St Andrews in its red Dumfriesshire sandstone, is J M Monro's Grand Hotel (1895), occupying a big corner site at the west end of The Scores to command a grandstand view of the Royal & Ancient clubhouse and the old course. It was also four-storeyed basement and attic, strongly influenced by English south coast hotels in its concrete balconies and corner cupola, and bright and spacious within. At Lundin Links, Fife, Peter L Henderson's tall brick and half-timber hotel (1900), with jerkin-head tiled roofs and high Norman Shaw chimney, similarly reflected southern English tastes of the 1880s and 1890s, as did Marshall Mackenzie's Bay Hotel (1902) at Stonehaven, Kincardineshire. But James

Bow Dunn's Bellevue Hotel at Dunbar of the same vintage was still Scots Baronial, its white-harled walls with generous red sandstone dressings a prominent landmark. Another great coastal landmark, and by far the finest of these big late nineteenth-century seaside hotels, was the free-standing Marine at Troon, Ayrshire, tall two-storeyed and attic, symmetrical and predominantly English Arts and Crafts of 1897, with a towered four-storeyed and attic wing of 1901, all with subtle Art Nouveau nuances.

The two-storeyed and attic scale of the original block at the Marine reflected a tendency from the 1880s onwards to build lower and longer wherever there was enough site, and with still bigger windows to let in plenty of light. The change of scale was well exemplified at Callander Hydropathic after the fire of 1895, when J M Monro demolished Peddie & Kinnear's building to half its height and extended it laterally with perfunctory half-timber detail at the attic. The inspiration for the change to English Tudor half-timber revival and ultimately to a more relaxed Arts and Crafts school of design was partly the country house work of Norman Shaw, and later that of Ernest Newton, but it was also a reflection of developments in rural hotel and country club design in the United States. The introduction of this idiom to Scottish hotel design can be dated to the competition for the Clyde Yacht Club (1886) at Hunter's Quay, Argyll, for which Burnet, Son & Campbell and Thomas Lennox Watson, an ex-assistant of the English architect Alfred Waterhouse, both submitted excellent designs with half-timbered elements, the former being the more

Figure 8.18 Marine Hotel, Elie, Fife. Crown copyright: RCAHMS, SC 710759.

Figure 8.19 Panmure Arms, Edzell, Angus, early twentieth century. Courtesy of David Walker.

American in character. Watson received the commission, probably because of his connection with the yachting community through his brother, the naval architect, George Lennox Watson. It was followed by a fine series of medium-sized hotels, mostly built for golfers, though several had eighteenth- or early nineteenth-century inns at their core: Burnet Son & Campbell's original Marine Hotel at Elie, Fife (fig. 8.18), in 1889; the same firm's hotel at Lamlash, Arran; Edwin Landseer Lutyens's English-thatched additions to the inn at Roseneath for his royal patron, Princess Louise; Thomas Martin Cappon's once excellent Panmure Arms at Edzell, Angus (fig. 8.19); and Ramsay Arms at Fettercairn, Kincardineshire, all 1895 and all English Arts and Crafts; Sydney Mitchell's Ugadale Arms (1898), Machrihanish, Argyll, in the same idiom; and James Marjoribanks MacLaren's very simple harled Scots vernacular fishing and shooting hotel at Fortingall in Glenlyon, Perthshire, executed, probably with some revision, by his successors Dunn & Watson in 1891. These were all of moderate size but the concept of inexpensive harled finishes quickly spread, Alexander Nisbet Paterson's refined Lorimerian Scots vernacular Colquhoun Arms (1908–11) at Arrochar, Dunbartonshire, being the finest of the later examples of the genre.

This change of style and the ever-increasing importance of golf to the hotel market was particularly well illustrated by rival developments commissioned by the two smaller Scottish railway companies, which might not have been expected to be innovators. In 1897 the Great North

Figure 8.20 Aviemore Station Hotel, Inverness-shire, c. 1930.

of Scotland Railway built a hotel at Cruden Bay, Aberdeenshire, with electric tramway access from the main line. Designed by their in-house architect John J Smith, the hotel was a symmetrical three-storeyed and attic Scots Baronial building, solidly built of granite with a central tower in the style of 25 years earlier. It contained 55 bedrooms and was very much on the model of the hydropathics, set in spacious grounds with racquet courts and putting and bowling greens; but, crucially, it also had a championship-standard golf course laid out by Archibald Smith and Tom Morris. It closed as early as 1932 because the site was too exposed and the season too short, but the precedent was to be important. In 1899 the Highland Railway followed suit, promoting the £25,000 Station Hotel at Aviemore, Inverness-shire (fig. 8.20), a near-symmetrical three-storeyed and attic Scots Renaissance pile by Ross & Macbeth set in spacious grounds with a lochan for curling and boating. It exploited a recently laid-out existing golf course, but at Dornoch, Sutherland, the Highland astutely took advantage of the historic Royal Dornoch Course, building a lower-profile, three-storeyed and half-timbered attic hotel development designed by J Russell Burnett. Burnett was also responsible for the same company's more successful Germanically-towered Highland Hotel, again three-storeyed and attic, at Strathpeffer, in 1909–11; located close to the

Figure 8.21 Turnberry Hotel, Ayrshire. Crown copyright: RCAHMS, SC 680931.

pump room, the Highland again took advantage of an existing golf course as at Aviemore and Dornoch.

All three of these buildings were economically built in brick with simple but bright roughcast elevations, the pioneer Aviemore being the most expensively detailed. In 1904 the Glasgow & South Western Railway constructed a similar but much more expensive development at Turnberry, near Girvan, Ayrshire (fig. 8.21). It had direct light railway access as at Cruden Bay but was much bigger with two newly-created championship standard golf courses rather than one. Its very American-looking 78-bedroom hotel was the work of James Miller, himself once a Caledonian Railway staff architect. It had a spacious entrance forecourt with a winter garden corridor to the railway line and a long garden front with a portico and bowed sun rooms looking out over terraced grounds to the sea. Its affluent summery atmosphere was hugely successful, Miller being called back to extend it time and again until the mid-1920s. As at the Highland Railway hotels, the Turnberry was bright roughcast, Miller explaining to the directors that:

> Common brick with roughcast will be much less than stone and

be more effective for the situation and the purpose. Red tiles will give colour to the landscape and be valuable as non-conductors making the rooms in the attics cool in summer and warm in winter.[5]

All the bedrooms had washstands with hot and cold running water, a great convenience to both guests and staff at the time, dispensing with the need for ewers and hot water jugs and flagons, and the heating was by radiators rather than hot air.

In 1905 Miller was called in to rebuild Starforth's burned-down hydropathic at Peebles. As at Turnberry, it was built of harled brickwork, but perhaps because the winter garden had survived, he adopted a more compact three-storeyed format with a very big attic and garret red tiled roof, as against the low two-storeyed and attic profile of Turnberry. Punctuated by big stacks, this produced a more shapely and effective profile. In 1900 and again from 1904 after a fire, Sir John Burnet enlarged his 1889 hotel at Elie in the same partly American-, partly Ernest Newton-inspired idiom, but made it cleverly asymmetrical.

At Portpatrick, Wigtownshire, in 1905 the Arts and Crafts Ayr architect, J Kennedy Hunter, built another hotel and golf-course development to attract traffic to the jointly-owned Portpatrick Railway. It was a relatively inexpensive building in a very simple Scottish vernacular idiom, but despite the more distant location it too proved successful and had to be enlarged by John Dick Peddie in the following year. By far the grandest of these railway-related golfing developments was at Gleneagles in Perthshire, which was very unusually an inland location on the Strathallan Castle estate. It was very much the creation of Donald A Matheson, engineer-in-chief of the Caledonian Railway and its general manager from 1910. He built a very smart new station, promoted a subsidiary company in which the Caledonian was the major shareholder, brought in James Braid to lay out two championship courses and a small nine-hole course, and held some kind of limited competition between J M Dick Peddie & Forbes Smith and James Miller. Peddie & Forbes Smith produced a scheme for a Vanderbilt François Premier château, but Miller won with a much enlarged version of the successful three-storeyed attic and garret formula of Peebles Hydropathic, having already discovered that the two-storeyed and attic format at Turnberry was resulting in an over-extended plan. Gleneagles was much bigger than Turnberry, but it was more compact with a single large sun lounge, a central ballroom with a proscenium which could be used for theatre, and 114 bedrooms. Construction was begun in 1913 but, following the outbreak of war, was abandoned before the building was roofed. It was not until 1923 that the Caledonian obtained powers to complete it on its own; but following the grouping of the railway companies in that year, the London Midland & Scottish Railway dispensed with Miller and gave the commission for its completion to Matthew Adam, who modified the design and delegated the interior work to Charles W Swanson. His elaborate Rococo scheme was executed by Scott Morton & Company, that firm also providing the Valter patent dancing floor of the Ionic-pilastered ballroom.

Nothing quite like either Turnberry or Gleneagles ever existed south of the Border, although the Great Western Railway's Tregenna Castle Hotel, Cornwall, first leased in 1878, may have provided the original idea. What is certainly true is that the success of Gleneagles induced the Great Western to create very similar facilities on Dartmoor, Devon, by buying and extending North Bovey Manor in 1929–35; the London Midland & Scottish immediately responded by buying Welcombe House at Stratford-on-Avon, Warwickshire, where the emphasis was more on fishing.

LODGING HOUSES

At the bottom end of the social scale were the lodging houses. In the early years of the nineteenth century all Scotland's industrial towns, particularly Glasgow, Edinburgh and Dundee, had grossly over-crowded lodging houses accommodating people who had come south from the Highlands in search of work, or had immigrated from Ireland, particularly during the famine. Invariably these were fairly ancient houses which had been deserted for newer and more fashionable terrace and villa developments.

In order to provide cleaner and healthier accommodation, the Edinburgh Lodging-house Association, founded in 1841, converted houses in the West Port (1844) and Merchant Street (1849); these had accommodation for 70 and 76 men respectively. In 1847 a similar group, the Glasgow Association for Establishing Lodging-houses for the Working Classes, was founded. It adapted a four-storeyed house in Mitchell Street for 60 men, each having a private apartment, no doubt achieved by sub-dividing the original rooms. New buildings of some architectural pretension by Charles Wilson, with an entrance in McAlpine Street for men and in Carrick Street for women, just off Broomielaw, were provided in 1856.

In 1857 the Model Lodging Houses Association fitted out the newly-vacated Dundee Infirmary of 1794 as a lodging house. This Dundee association had as its president Lord Kinnaird who had supported Lord Shaftesbury's Lodging Houses Act of 1851 which encouraged the municipalities to set up model lodging houses. This became law in Scotland in 1855, but the Act proved generally ineffective until replaced by the succeeding Act of 1885. The exception to the rule was Glasgow where in 1866 the Glasgow City Improvement Trust secured Parliamentary powers

Figure 8.22 Orient House, 16 McPhater Street, Glasgow. Crown copyright: RCAHMS, SC 710763.

to run lodging houses. Between 1870 and 1884 the Trust instructed the City's Master of Work, John Carrick, to build seven further model lodging houses in Drygate, Greendyke Street, Portugal Street, Clyde (later Abercromby) Street, North Woodside Road, Hydepark Street and East Russell Street for men, together with Moncur Street for women. These varied in architectural pretension according to the mood of the trustees at the time. Drygate was severest four-storeyed, five windows wide; but Clyde Street was surprisingly handsome Scots neo-Jacobean with flanking towers, crowstepped gables and window pediments. Portugal Street consisted of two blocks, one a severe four-storeyed and basement, 11 windows wide palazzo, the other a more neo-classical block of five widely-spaced windows, tripartite at the ends. Both had ground floors of channelled masonry with arched doorpieces. Whatever they looked like outside, the City Improvement Trust lodging houses were much the same inside. The internal construction was as far as possible fireproof, right to the roof in some cases, with barrack floor dormitories divided into cubicles, and a single lavatory on each floor. As in other buildings of this class, there was a self-catering dining area with two big hot plates, a recreation room, a small shop at the entrance providing basic provisions, and security lockers. They varied considerably in size, Clyde Street housing 272 men and Portugal Street 437.

In addition, there were in Glasgow a number of lodging houses which were privately owned and run under the supervision of the city. Three of these were of considerable architectural importance. The first was the Fireproof Building Company's four-storeyed Orient Boarding House (1892; fig. 8.22) in McPhater Street, designed by the scholar-architect William James Anderson, and built of early shuttered concrete with pre-Hennebique forms of reinforcement in a kind of arcaded and crenellated bargello manner with rendered walls, the architectural detail being in red 'compo'. It accommodated 353. The Orient was followed in 1898 by Napier House in Govan Road, again by Anderson, which was also of shuttered concrete construction, the corner block being faced in red sandstone with oriels and interesting non-period details. It was built without his supervision and suffered a fatal collapse during construction, bringing about the architect's own death following a nervous breakdown. Still finer was the now-demolished Neptune Buildings at 470 Argyle Street, designed by another scholar-architect, Dr Peter Macgregor Chalmers, tall red sandstone early modern with excellent sculpture, which housed the Pitt Street Common Lodging House.

Although not of the same order of merit, another example of considerable architectural pretension was Rutland House in Govan Road, Glasgow, five-storeyed red brick with polychromatic detail at the windows, a Netherlandish gable and a circled corner with an onion dome. It was the best of several lodging houses built and owned by Thomas Paxton, and was designed by Bruce & Hay in 1896 at the very high cost of £15,000, providing accommodation for 536 men. Of the numerous buildings converted for the purpose, the most interesting was the Great Eastern in Duke Street, which was dignified by the use of the word 'Hotel'. This giant five-storeyed, 22-bay building had been built in 1842 as Alexander's Cotton Mill to designs by Charles Wilson. It had the merit of having Renaissance details and fireproof floors, and in 1907–9 Neil C Duff raised it a storey and converted it to a basic hotel for 450 men. Somewhat similar accommodation was provided on a philanthropic basis at the vast neo-Romanesque Christian Institute in Bothwell Street, the six-storeyed west wing of which was designed by R A Bryden in 1895 to provide 189 tiny bedrooms and a restaurant for the YMCA.

Nothing quite like these existed in Edinburgh where some City Improvement Trust-promoted buildings in the Grassmarket were adapted for the purpose. That in Dundee, Rose's Home at 55 Commercial Street, astonishingly occupied a domed Renaissance corner building designed in 1877 by Alexander Johnston as part of the 1871 Improvement Act programme, externally much more Grand Hotel than lodging house. West Lothian had several purpose-built examples, quite late in date and all built by groups of about half-a-dozen shopkeepers and tradesmen: Bathgate (1899), Armadale and Uphall (both 1903), and Broxburn (1904), all plainest three- or four-storeyed, divided into cubicles, which were wire-netted over for security and the free circulation of air. In other areas, however, a simple barrack arrangement with lockers was provided, some having partitioned-off areas for couples. A particularly interesting instance was to be found until the 1960s in Coupar Angus, where the Hanoverian Cumberland Barracks of 1766 had continued to be used very much as it had been during its military existence.

At a rather different and more philanthropic level were the night asylums and night refuges, the first being established in a converted granary in St Enoch's Wynd, Glasgow, in 1838. It accepted homeless people not eligible for the poorhouse and moved to plain premises in North Frederick Street built anew in 1847. Similar provision was made in Aberdeen in 1840 and in Edinburgh in 1842, but neither had their own premises. The only one of any interest as architecture was the endowed Curr Night Refuge at Dundee, a handsome three-storeyed and basement free Renaissance building in West Bell Street, designed in 1881–2 by Ireland & Maclaren to complete a late Georgian street block; matching Parish Council Chambers were added to it by the City Architect, William Alexander, in 1901.

INTER-WAR HOTELS AND ROAD-HOUSES

The inter-war years saw very little purpose-built hotel construction. In 1921 the Taymouth Castle Hotel Company acquired and adapted Taymouth Castle, Perthshire, and laid out its landscaped park as a golf course. Its aim of rivalling Gleneagles was never quite realised and it did not re-open after World War II. Nevertheless, it commenced a trend towards converting redundant country houses, which increased as motor car ownership grew, but did not become widespread until after 1950. In suburban areas and holiday resorts, large high-quality houses set in spacious grounds with plenty of room for motor cars, racquet courts and putting and croquet lawns could be acquired very cheaply. Andrew Heiton's Tom-na-Monachan in Pitlochry, which was sold at auction for £110 and converted to Pine Trees Hotel by the addition of a matching bedroom wing by Gordon & Scrymgeour in 1937, was a classic instance. In such circumstances there was little incentive to build anew.

In the city centres, particularly in Edinburgh and Glasgow, there was similarly a market in large late Georgian and Victorian terraced houses, the value of which had fallen steeply after World War I. Following the long-established precedents set in Glasgow's George Square and at the Roxburghe Hotel in Edinburgh's Charlotte Square, hoteliers acquired several in a row as the opportunity offered and linked them up.

Sometimes, as at Grosvenor Crescent in Glasgow and Learmonth Terrace in Edinburgh, they eventually acquired almost half the block. Few of them could be said to have been wholly successful conversions but they did meet the needs of a market which was now more often touring by motor car rather than travelling by train.

In general, the provision of city centre hotels in Victorian times had been more than adequate to cater for the commercial market and the diminishing number of holidaymakers still arriving by rail. But in June 1933 the London – originally Glasgow – firm of Sir John Burnet, Tait & Lorne prepared designs for the Hotel Buchanan on the site of Prince's Square in Glasgow's Buchanan Street. It was to have been a huge 11-storeyed block with horizontally banded steel-framed windows similar to their Mount Royal in London, the top two floors being stepped back with roof terraces looking out over the city. It was not built, but the concept was in some degree realised in the eight- and nine-storeyed Hotel Beresford on Sauchiehall Street, though without the Buchanan's horizontally banded fenestration and recessed top floors. Built of reinforced concrete by the architect and cinema proprietor, W Beresford Inglis of Weddel & Inglis, for the Glasgow Empire Exhibition of 1938, it was faced, cinema-style, in mustard yellow faience, its centre bays flanked by semi-circular oriels and lined out with fins in red and black. It cost £180,345, roughly half the original estimate for Burnet, Tait & Lorne's St Andrew's House in Edinburgh, which gives a measure of the ambitious scale of the project. It had some quite lavish but now vanished Art Deco interior work, notably in the cocktail bar formed as an afterthought in October after the exhibition was over. The Beresford represented the top end of the market; at the bottom was the five-storeyed Bellevue Hotel (1935–7) at 609 Gallowgate by C J McNair, with horizontally-banded steel-framed fenestration in faience, brick and cement render.

Similarly related to the Empire Exhibition was the Bay Hotel at Gourock, Renfrewshire, designed by the 'up-market' Glasgow firm of J Austen Laird & James Napier to enable a visit to the exhibition to be combined with a seaside holiday, the resort being within easy commuting distance of Bellahouston. Like the Bellevue, it had horizontally banded fenestration, once very smart in green and white. Laird & Napier were also responsible for the well-sited modernist hotel at Keil, Southend, in Kintyre, Argyll, in 1938–9, but the finest of the inter-war west coast hotels was James Taylor's Regent at Oban, built in 1936. This building anticipated the common post-World War II arrangement of a low range of public rooms with a high bedroom block rising above it. The Regent Hotel was quite stylish with rounded angles, horizontally-banded bedroom windows and a roof terrace guarded by marine-style tubular railings. The roof terrace has now been built over, two further bedroom floors having been added in recent years. The same elegant curved-angle treatment is to be seen at A MacGregor Mitchell's four-storeyed McColl's hotel (1938–9) at Dunoon, where the planning and fenestration were more conventional as a result of the incorporation of an existing building into the scheme.

In the east of Scotland the only completely new hotel building of quality was the Northern Hotel (1937) at Kittybrewster, Aberdeen, a modernist design by the much-respected architect of the Waldorf Hotel in London, A Marshall Mackenzie's son, Alexander George Robertson Mackenzie. It replaced a burned-out hotel and was of four storeys with a bowed gable and horizontally-banded steel-frame windows, all faced in granite as required by the City of Aberdeen at that time; the incorporation of a tall ballroom floor level, with cantilevered balconies on which to take the air, gave it excellent classic proportions. In the same year, Mackenzie also heightened, unified and re-styled the two buildings of the plain early Victorian Douglas Hotel in Market Street, Aberdeen, inserting an up-to-date recessed entrance bay with flagpoles.

The rapid increase in the number of motor cars on the roads from the mid-1920s onwards brought in a new building type, the road-house. As Charles McKean observed in *The Scottish Thirties*, Scotland had been somewhat deficient in buildings akin to the traditional English inn ever since the coach routes were superseded by rail travel. At those that were still trading the facilities were often fairly basic. The primary aim of the road-house was therefore to provide smart restaurant, cocktail bar and leisure facilities for the motorised classes. As most of them did not have residential accommodation they do not strictly fall within the scope of this review, but the prototype appears to have been the Bird in the Hand at Johnstone, Renfrewshire, an excellent small two-storeyed English Arts and Crafts inn, designed c.1910 by Thomas Graham Abercrombie of Paisley. The Wheatsheaf at Saughton, Edinburgh, designed by Robert (later Sir Robert) Matthew as late as 1934 still followed this low-profile Arts and Crafts format, albeit with some Art Deco details, but most were 1930s modern. Of these, Leslie Graham Macdougall's now-demolished V-plan Pantiles Hotel at West Linton, Peeblesshire, which did have residential accommodation, was perhaps the best. Its green-pantiled theme was repeated at the curvaceous ochre-brown pebble-dashed Hillburn Roadhouse, designed by the cinema architect Thomas Bowhill Gibson, which was purely a bar-restaurant with a bowling alley. Easily the best surviving example of the genre is the Maybury Roadhouse, Edinburgh, designed by Paterson & Broom in 1935, a very horizontal two-storeyed white stucco composition comprising a bowed entrance and bar block, and a long ballroom wing with the windows banded together by stripes of red brickwork. Its flat roof was originally intended as a games deck.

POST-WAR HOTELS

Of the post-World War II years still less need be said. While nearly all the older hotels had been updated by the 1930s with wash-hand basins with hot and cold water in the bedrooms, the increasing expectation of bathroom facilities en suite was not always so easily accommodated. Many of the older hotels were demolished, their demise often hastened by increasing site value for retail purposes and the difficulty of accommodating guests' cars in urban centres, a trend which had begun in Ayr in the early 1930s. Tarbolton & Ochterlony's Mount Royal (1955) on Edinburgh's Princes Street, standing on the site of Beattie's Royal with canted bays and eaves gallery, and the aggressive Caledonian Hotel in Inverness, 1965–7 by Leach, Rhodes & Walker, were the only notable instances where a new hotel was built on exactly the same site as the old. Another project with a link to earlier hotel-building initiatives was the very last of the golf-related railway hotels, Curtis & Davis's Old Course Hotel at St Andrews. This was built in 1967–8 for British Transport Hotels Ltd, the wholly-owned subsidiary of the British Railways Board which had taken over those railway hotels that had not been sold off previously. This was a surprisingly belated development given the prime sight-seeing land in railway ownership and, rather oddly, it was undertaken after the board had closed the Leuchars-St Andrews railway line. Its compact, robustly modern balconied form is now lost in shapeless additions of 1982–90.

These were, however, fairly uncommon, if not unique, developments. The new generation of hotels, encouraged by the Board of Trade to meet the expectations of transatlantic visitors and boost the tourist industry and the balance of payments, tended to be of the same basic type on both sides of the Border: a large one- or two-storeyed block containing the foyer, bar, dining and dancing facilities, and a tall slab of bedrooms with en-suite bathrooms above. Among the first were the Angus Hotel, Dundee (1963), by Ian Burke, followed by the Royal Stuart on Jamaica Street, Glasgow (1963–5), by Underwood & Partners, which had full air conditioning and a basement car park. The wall panels of its reinforced concrete frame structure were (somewhat unsuccessfully) clad in mosaic. The bleak, brick-faced Albany by James Roberts followed in 1970–3, and the Holiday Inn on Argyle Street, Glasgow, by Cobban & Lironi in 1979–82. The Edinburgh examples of the type were rather better: the Crest (1968, now the Holiday Inn) on Queensferry Road by Morris & Steedman, was skilfully sited on a landscaped knoll and the detailing of its bedroom slab was of some refinement, reflecting the American experience of its architects. Also well-sited and of above-average architecture for that time was the Post House at Corstorphine (1971) by Nelson Foley, set in the hillside with rising façades of aggregate panels.

Suburban hotels in locations with lower land values tended to be more of two to three storeys. T M Miller's Tinto Firs Hotel (1964), Kilmarnock Road, Glasgow, planned around a patio, was one of the best as well as one of the earliest of its type; most of the remainder are of depressingly poor quality.

In the early 1960s The Scottish Office promoted the grand project of Aviemore, a completely new resort alongside the old, largely within the grounds of Ross & Macbeth's Station Hotel. It was based on ski-ing in winter and other forms of leisure and recreation in summer, a theatre and an ice rink being provided. It was intended to revitalise Badenoch and Strathspey but the actuality fell short of the original concept and had unfortunate consequences for some of those involved in it. The curious eighteenth-century T-plan inn, stable and steading courts of Aviemore House, which might have given some interest to the site, were ruthlessly demolished and two new hotels built, the seven-storeyed Four Seasons and the butterfly-plan Badenoch, both 1965–6 and not much more interesting than council flats of the same vintage. These, like the remainder of the original complex, were designed by John G L Poulson. Nelson Foley, in association with Sir John Burnet, Tait & Partners, added the Post House at Coylumbridge, a rather better building, in 1970–1.

These buildings have not aged well. The Royal Stuart was converted to the Clyde Hall of the University of Strathclyde very early in its existence, the Post House at Corstorphine has been reclad as a twenty-first-century building, and the Angus Hotel at Dundee has been completely demolished.

Although the economics of hotel building remain a constraint, the last decades of the twentieth century has seen greater individuality in hotel design. In 1978, following earlier precedents with mill buildings at Saucel Mills, Paisley (Watermill Hotel, 1968) and Upper City Mills, Perth (City Mills Hotel, 1970, both by T M Miller), Crerar & Partners converted the surviving granary at Bell's Mills, Edinburgh, and built a new block as the Edinburgh Hilton. The latter hardly matched the robust quality of the granary but at least it was decently stone-faced, as was the same firm's Sheraton (1984), Festival Square, Edinburgh. Also a conservation-influenced exercise was the huge 238-bedroom Scandic Crown (1989) in Edinburgh's High Street, built on what had been a gap site since the mid 1960s. Passers-by looked on in wonder as identical imported bedroom and bathroom units were piled on top of one another to be veneered over with façades of slightly different sixteenth- and seventeenth-century character designed by Ian Begg. Although these façades would have been better if Begg had been retained to supervise them, they have, like Patrick Geddes's buildings of a century earlier, settled into being an accepted part of the Royal Mile scene, similar in scale to the buildings which had been lost.

A similar feeling for context characterises Andy Doolan's minimalist classicism of Edinburgh City Travel Inn (1997–9) on Morrison Link and the bolder Brunswick (1996) in Glasgow by Elder & Cannon. With his elegant conversion of the former Co-operative Department Store on Bread Street, Edinburgh, as the Point Hotel, Doolan has demonstrated that grand hotel spaciousness and style is still achievable if a good redundant building can be found. This exercise was repeated by David Clarke Associates in the far greater Edwardian grandeur of *The Scotsman* building on North Bridge, where long-encapsulated interiors were recovered to achieve the desired ambience. And although the new-build Holiday Inn Garden Court (1995) by Cobban & Lironi at the corner of Renfrew Street and West Nile Street, Glasgow, is architecturally perhaps a bit thin by comparison, it at least has a degree of elegance with its domed clock-tower.

Perhaps something of the style and sense of occasion to be seen in Scotland's older hotels may yet return, even in those built completely anew; the era when a big neon sign was thought to be enough would seem to have passed. In the eighteenth, nineteenth and early twentieth centuries the architecture of inns, hotels and hydropathics was the primary form of advertising, particularly before improvements in the postal service made it possible to book in advance. In burghs such as Annan, Moffat and Dunbar, the eighteenth-century inns were conspicuously better buildings than their neighbours, bigger scaled and finer detailed. Along the coach routes, the more 'up-market' inns, at least from the time of Alexander Edward's at Monifieth, tended to be subtly differentiated from farmhouses by more distinctive detailing. On the steamer routes, hotels like the towered Argyll at Dunoon and those ranged along the waterfront at Oban, were clearly designed to catch the eye on the approach to the pier. As the railways spread throughout the land and transformed coastal towns and small Highland villages into major holiday resorts, hydropathics in particular were designed to be prominent objects in the landscape as seen from the train, nearly all of them being tall towered buildings on commanding hill-top sites. Although the scale of the really big Edwardian golfing hotels tended to be lower, their bright roughcast walls and, in the case of Turnberry and Peebles Hydropathic, equally bright red roofs, were similarly designed to be seen from afar. In the cities the truly great hotels – the Star in Edinburgh, and St Enoch's, the Blythswoodholm, the Grand and the Central in Glasgow – were designed as great civic buildings, the most extreme being the North British in Edinburgh. With its outsize clock-tower, anyone unfamiliar with the city might well assume it to be the city's municipal buildings rather than a station hotel.

Finally, it is worth remembering that, as a building type, hotels and hydropathics had a historical significance in the development of building technology which has not yet been sufficiently recognised. In the 1870s and 1880s they had an important pioneering role, firstly, in the introduction of the load-bearing, rolled-iron fireproof floors necessary for the wide-span spaces at their lower floors, secondly, in building to the absolute limit of the firemaster's reach, and, thirdly, in the provision of mechanical and electrical services, all of which led directly to the achievements in commercial building and department store design in the 1890s and early 1900s.

NOTES

1. This chapter has relied heavily on the writer's field notes as Investigator/Inspector of Historic Buildings, 1961–93, on help from colleagues at Historic Scotland, particularly Dr D C Mays, on the lists of buildings of special architectural and historic interest, and on the photographic and architectural drawings archives of RCAHMS. Annette Carruthers was also very helpful. The *Buildings of Scotland* series and the *Architectural Guides* of the Rutland Press will be found to be the best published sources of information, the latter providing more illustrations but the former providing more detailed information. For other topographical and biographical works, lists of periodicals and annuals, and guidance on archival sources relevant to the theme of this chapter, see the Bibliography and Note attached to chapter 29 below.
2. Senex, 1884, Vol 2, 306.
3. Anon, 1894, 115.
4. Cockburn, 1889, 26.
5. Carter, 1990, 75.

BIBLIOGRAPHY

Anon. *Rivers of the North: their cities and their commerce*, London, 1894.
Beaton, E. *Caithness*, Edinburgh, 1996.
Beaton, E. *Ross and Cromarty*, Edinburgh, 1992.
Beaton, E. *Sutherland*, Edinburgh, 1995.
Biddle, G, Nock, O S, eds. *The Railway Heritage of Britain*, London, 1963.
Brogden, W A. *Aberdeen*, Edinburgh, 1986, 1998.
Burgher, L. *Orkney*, Edinburgh, 1991.
Carter, O. *An Illustrated History of British Railway Hotels*, St Michael's, 1990.
Christie, G. *Crieff Hydro*, Edinburgh, 1967, Crieff, 1986.
Close, R. *Ayrshire and Arran*, Edinburgh, 1992.
Cockburn, H. *Circuit Journeys*, Edinburgh, 1889.
Defoe, D. *A Tour through the Whole Island of Great Britain, 1724–26*, London, 1769, reprinted Exeter and London, 1927, New Haven and London, 1991.
Duff, D, ed. *Queen Victoria's Highland Journals*, Exeter, 1983.
Finnie, M. *Shetland*, Edinburgh, 1990.
Gardiner, L. *Stage Coach to John o' Groats*, London, 1961.
Gauldie, E. *Cruel Habitations: A history of working-class housing, 1780–1918*, London, 1974.
Geddie, J. *Souvenir of the Opening of the North British Station Hotel, 15 October 1902*, Edinburgh, 1902.

Gifford, J, McWilliam, C, Walker, D. *Edinburgh*, Harmondsworth, 1984.
Gifford, J. *Fife*, Harmondsworth, 1988.
Gifford, J. *Highlands and Islands*, Harmondsworth, 1992.
Gifford, J. *Dumfries and Galloway*, Harmondsworth, 1996.
Glendinning, M, MacInnes, R, MacKechnie, A. *A History of Scottish Architecture from the Reformation to the Present Day*, Edinburgh, 1996.
Gordon, A. *To Move with the Times*, Aberdeen, 1988, especially Chapter 12.
Groome, F H, ed. *Ordnance Gazetteer of Scotland*, 2nd edn, London, npd [c. 1894].
Haldane, A R B. *New Ways through the Glens*, Edinburgh, 1962.
Haldane, A R B. *Three Centuries of Scottish Posts*, Edinburgh, 1971.
Haldane, A R B. *The Drove Roads of Scotland*, Newton Abbot, 1973.
Haynes, N. *Perth and Kinross*, Edinburgh, 2000.
Hopkin, M. *Old Oban*, Glasgow, 2000.
Hume, J R. *Dumfries and Galloway*, Edinburgh, 2000.
Jamieson, J H. Some inns of the eighteenth century, *The Book of the Old Edinburgh Club*, 14 (1925), 121–46.
Johnston, C, Hume, J R. *Glasgow Stations*, Newton Abbot, 1979.
Jaques, R, McKean, C. *West Lothian*, Edinburgh, 1994.
Laidlaw, S I A. *Glasgow Common Lodging Houses and the People Living in Them*, Glasgow, 1956.
Lockhart, A M. *Origin and History of the Old Glasgow Club, 1900–35*, Glasgow, 1935.
McKean, C, Walker, D. *Dundee*, Edinburgh, 1984, 1993.
McKean, C. *Stirling and the Trossachs*, Edinburgh, 1985, 1994.
McKean, C. *The Scottish Thirties*, Edinburgh, 1987.
McKean, C. *The District of Moray*, Edinburgh, 1987.
McKean, C, Walker, D, Walker, F. *Central Glasgow*, Edinburgh, 1989, 1993, 1999.
McKean, C. *Banff and Buchan*, Edinburgh, 1990.
McKean, C. *Edinburgh*, Edinburgh, 1992.
Macky, J. *A Tour through Scotland*, London, 1723.
McWilliam, C. *Lothian*, Harmonsdworth, 1978.
Millar, D. *Queen Victoria's Life in the Scottish Highlands depicted by her Watercolour Artists*, London, 1985.
Nottage, J. *The Gleneagles Hotel*, London, 1999.
Peden, A. *The Monklands*, Edinburgh, 1992.
Pride, G L. *Fife*, Edinburgh, 1990, 1999.
Robertson, D. *The Princes Street Properties*, Edinburgh, 1935.
Salmond, J B. *Wade in Scotland*, Edinburgh, 1934, 2nd edn, 1938.
Senex (Reid, R). *Glasgow Past and Present*, Glasgow, 1884.
Shepherd, I. *Gordon*, Edinburgh, 1994.
Sinclair, F. *Scotstyle: 150 years of Scottish architects*, Edinburgh, 1984.
Strang, C A. *Borders and Berwick*, Edinburgh, 1994.
Swan, A. *Clackmannan and the Ochils*, Edinburgh, 1987, 2001.
Thomas, J. *Midlothian*, Edinburgh, 1995.
Topham, A. *Letters from Edinburgh*, London, 1776.
Walker, F A. *The South Clyde Estuary*, Edinburgh, 1986.
Walker, F A, Sinclair, F. *The North Clyde Estuary*, Edinburgh, 1992.
Walker, F A. *Argyll and Bute*, Harmondsworth, 2000.
Weir, M. *Ferries in Scotland*, Edinburgh, 1988.
Williamson, E, Riches, A, Higgs, M. *Glasgow*, Harmondsworth, 1990.

9 Ancillary Estate Buildings

ELIZABETH BEATON

Although the mansion or shooting lodge may be the focal point, much interest lies in the buildings designed to service the rural estate: some are large, elegant and intended as eye-catchers, others are of simpler form discreetly sited on the periphery. All mirror life and work on a Scottish country estate.

The functional estate or garden building combines two influences in form and design; the vernacular (traditional) and the architecturally conscious. The vernacular may manifest itself in materials, techniques or the function of the building, perhaps designed for a peculiar local need. Conscious design is revealed through the architectural pattern book or the individual style of an architect, reflecting both contemporary fashion and status. Of the pattern books, it was Loudon's *Encyclopaedia of Cottage, Farm and Villa Architecture and Furniture*, published in 1833 and running into further editions, that had far-reaching influence, copies apparently being held in many estate offices.

These two parallel architectural trends, the traditional and the polite, converge and overlap on the Scottish estate. Estate buildings reveal a wide variety of materials and the indigenous construction skills associated with those materials. Where the building, in particular the gate lodge and the stable, reflects polite design, selected materials may be brought in from elsewhere. Where the principal design source is the vernacular, service buildings, in particular the dovecots which span the longest historical period of any estate structure, can be read as an index of regional building resources: wood, clay, mud, stone, pebble, slate, straw and other vegetable matter.

These working buildings also reflect and record other aspects of Scottish rural life; if there is generous carriage house provision in an elegant stable range, this is an indication not only of the wealth of the family at the time of building but also of the contemporary state of roads and bridges in the neighbourhood. Generating houses, usually dating from the early 1900s, suggest the introduction of private electric power long before this energy source was available locally on the national grid.

Two books of outstanding quality cover the growth and

development of individual Scottish country estates from social and economic viewpoints, identifying associated building programmes. *Inveraray and the Dukes of Argyll* (1973), by Ian Lindsay and Mary Cosh, examines and illustrates eighteenth-century change and the growth of an ancient and aristocratic estate. Philip Gaskell's *Morvern Transformed: A highland parish in the nineteenth century* (1968), vividly describes and analyses the part played by the family of an incoming industrial magnate, both as improving landowners and sportsmen. The wide range of estate buildings, many of innovative design and material (particularly concrete), are well documented in this west coast parish history, and to a lesser extent in Angus Graham's *Skipness*.[1]

GATE LODGES

The gate lodge (fig. 9.1) is designed to shelter the man (and his family) responsible for opening and shutting the gates separating the mansion house grounds from the main thoroughfare. These dwellings were largely built in the eighteenth and nineteenth centuries, but the tradition of invigilating entrances, whether of a defensive or merely watchful nature goes further back in time. One of the earliest provisions of watchmen's accommodation is found in brochs, a building form peculiar to Scotland. These consist of circular, double-skin walled, fortified farm dwellings

Figure 9.1 Aberlour House, Banffshire; probable tea house, later gate lodge, by William Robertson, c. 1830–5, architect of Aberlour House in similar classical style, 1838–9. Copyright: Elizabeth Beaton.

which once proliferated in the Highlands and Islands in the first millennium AD. In these, the low tunnel entrances are flanked by an intra-mural guardian's cell. From there it is but a short step to the medieval castle gatehouse, becoming more grandiose and less defensive by the sixteenth century and subtly reflecting the grandeur and power within. The mural cell, the castle and monastery gatehouse are integral parts of the main edifice, but the gate lodge is isolated, as necessitated by the eighteenth-century fashion of surrounding country houses by landscaped parks.

The gate lodge is usually a cottage or pair of cottages flanking a driveway entrance, accommodating the lodge keeper and his family. The early lodges were very small, sometimes a single room, but they increased in size and accommodation towards the end of the nineteenth century. They offer a statement about, and a prelude to, the mansion, which will come into sight as the traveller proceeds through the policies.

These lodges are as varied as they are numerous. If constructed contemporaneously with their mansion, the lodge may reflect the style of that house, probably having been designed by the same architect. At Gordon Castle, Moray, the twin lodges designed by John Baxter of Edinburgh in 1791–2[2] are linked by an arched gateway, all in the finest Moray sandstone. They presage the 4th Duke of Gordon's re-modelled and greatly enlarged, symmetrically-fronted mansion, while the idiosyncratic Baronial revival arch and lodge at Thurso Castle, Caithness, gave entry to an equally martial and machicolated pile.[3] Besides reflecting the architectural essence of the principal house, lodges can be fanciful creations, expressing the peculiar whims and tastes of the patron. When additional roads and entrances were made on an estate, a new lodge might be built in one of many architectural styles. Classical, Baronial, Italianate, Scottish (and English) vernacular revival, Arts and Crafts, picturesque and even whimsical – all have their place. Of these, and amongst the last to be built in Scotland (1922), are the imposing asymmetrically paired Scots Baronial towers flanking a drive to Dunecht House, Aberdeenshire. Rising four storeys high, each with corbel-courses, wallhead bartizans, cap-house and circular stair-tower, these lodges flank magnificent decorative iron carriage and pedestrian gates set between equally imposing gate piers. The one closest to the Loch of Skene incorporates a boathouse, the lodge keeper perhaps doubling as boatman. These lodges express, in succinct architectural terms, the aspiration to Scottish lairdship by the English entrepreneur ennobled landowner.[4]

STABLES AND CARRIAGE HOUSES

Amongst estate buildings it was the stables, intended for the accommodation of riding and carriage horses, that most reflected the standing and wealth of the laird. Well-bred carriage and riding horses were expensive commodities and important status symbols, and the often elegant carriages were housed in the same range of buildings, which besides up to a couple of dozen horses, accommodated a hierarchy of grooms and stable lads, tack rooms and feed storage. The stable block, therefore, had a special architectural presence within the estate. It was intended to be seen and to impress.

Important stable ranges began to make their appearance on country estates at the end of the seventeenth century. The age of symmetry lent itself to the regularity of the hollow square, a plan form which facilitated a fine exterior yet screened the utilitarian activities associated with the care of beasts. Carriage houses would usually be expressed by round-headed arches in the outer bays of the main elevation while the horses enjoyed superior quarters. Some would be accommodated in loose boxes and some in stalls divided by shaped and finialled trevises, each fronted by an alcove fitted with a hay rack rising directly to the hay loft above, from which the feed could be dropped directly. Some of the earlier complexes were referred to as a 'square of offices', these ranges also accommodating farm and domestic functions, such as a brewhouse or bakery.

The elegant quadrangular stable block designed by Sir James Clerk with John Baxter Senior on his estate at Penicuik, Midlothian (fig. 9.2), and completed in 1766 (converted to a dwelling in 1902), included, besides stables and coach houses, a brewhouse, henhouse, doghouse, cowhouse, infirmary, hogs' sty, slaughterhouse, barn, smiths' and wrights' shops, workmen's hall and central midden; in reality combining some of the role of an improved mains farm with that of stables.[5] The porticoed front is topped by a Gibbsian steeple and the rear elevation by a domed dovecot. At Cromarty House, *c.* 1772, the U-plan stables (adapted in 1995 for use as an arts centre) were for the household only, with the carriage houses in the outer wings and the horses occupying the lofty central portion which was lit by a series of lunettes, the vaulted ceiling being supported by wooden Tuscan columns.

Where humbler draught horses shared premises with riding and carriage horses, their accommodation reflected the equine hierarchy. On some small estates, the mains steading housed both of these besides carts, the riding and carriage horses always occupying superior quarters.

Manses, which enjoyed a building boom from the late eighteenth to the mid nineteenth century, were also provided with 'offices'. Besides the minimal agricultural accommodation necessary to service the glebe lands, there were also stables for a couple of horses and a gig house. The country minister, in particular, had considerable distances to travel.

The most important stable ranges qualify as 'polite' architecture. However, the influence of the native vernacular is evident in the circular

Figure 9.2   Penicuik House, Midlothian; stables. Crown copyright: RCAHMS, SC 710950.

courts, of which the earliest is the impressive 'Round Square' at Gordonstoun, Moray, c. 1700. This fulfilled the functions associated with the 'square of offices'. The circular form was developed into elegant and practical stables by architects such as John Paterson in the turreted range, c. 1794, at Castle Fraser,[6] Aberdeenshire, and at Prestonfield, Edinburgh, by James Gillespie Graham, 1816.[7]

Great private stable courts ceased to be built in the early twentieth century. Amongst the last in the grand manner were Manderston, Berwickshire, in 1895[8] and, in the Scottish Renaissance manner, Altyre, Moray, dated 1902, both elegant complexes of impeccable workmanship designed by John Kinross for wealthy patrons.

As the motor car superseded the carriage for private transport, many stables were adapted for garage use, wide doors being slapped through and the horses' non-slip tiled floor coated with smooth surfaces. The 'motor house' took its place amongst estate buildings; an example of note is the chunky, crenellated masonry range built by the biscuit manufacturer, Sir Alexander Grant, at Logie, Moray, in 1939, though the intended chauffeur's accommodation on the upper floor was never realised.[9]

LAUNDRIES

Substantial estate households comprising family and domestic staff, who could number ten, twenty or more depending on the wealth and size of the household, obviously generated quantities of soiled clothing and household linen which required constant washing and renewal. As estate households grew, particularly in the nineteenth century, there was washing and wear and tear on clothing, domestic linen, table napery and kitchen cloths besides the aprons and uniforms of domestic staff.

Well into the twentieth century, many estates employed laundry staff and constructed specialised laundry buildings to accommodate this activity. These were usually situated near running water. Some were of a utilitarian nature while others manifested architectural pretension, particularly if sited near the main house. These buildings housed sinks for washing and a boiler or 'copper' above a small fire, which produced hot water and also served to boil 'whites' in soapy water to remove stains and bleach white linen and cotton. Skilled glazing and starching were required on dress-shirts, most cuffs and collars, table- and some other household linen. A room would be devoted to ironing, furnished with a small stove fitted with ledges on which the flat and goffering irons rested while heating. Ironing was carried out on sturdy tables covered with a blanket and sheet, the former providing a soft surface, the latter for better finishing and cleanliness. A sewing and mending room could be included in these offices though it was more likely to be in the service quarters of the mansion, under the watchful eye of the housekeeper.

In 1815–19 the 2nd Lord Macdonald of Sleat, Skye, commissioned James Gillespie Graham to greatly enlarge Armadale Castle, which he did in elegant Gothic castellated manner. This enlargement facilitated entertainment, generated work and increased the household, for which generous laundry provision was required. The Gothicised laundry (1820–2, now ruinous) was sited close by a burn, within walking distance from the rear of the mansion. Designed to enhance the built landscape of the estate, as well as discharging its utilitarian function, the Armadale laundry is a picturesque, two-storeyed structure with pointed-headed windows.[10] Expansion at Armadale was symptomatic of the aggrandizement of the Scottish country estate. In north-east Scotland, Cullen House in Banffshire grew from a simple sixteenth-century mansion house to the home of one of the largest landowners in Scotland. The laundry, almost certainly designed by William Robertson of Elgin, c. 1825, is an austerely symmetrical two-storeyed, five-bay window pattern, the fenestration equally generous on the rear (south-east) elevation, providing good light to the workers within. The building is sited on a former 'bleaching green' by the Cullen Burn, probably a

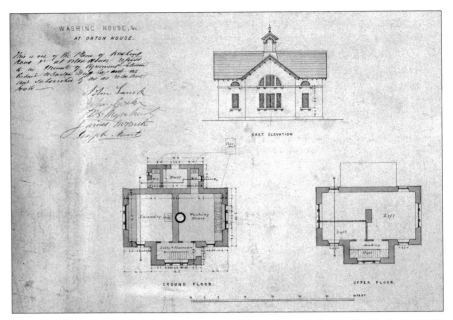

Figure 9.3 Orton House, Moray; 'washing house' design by A & W Reid, Elgin, 1855. Courtesy of Wittets, Architects, Elgin, and The Moray Council Libraries.

traditional site for household washing and possibly also used for bleaching linen woven from local flax.[11]

In Moray, the new laundry at Orton House (fig. 9.3), designed by A & W Reid, also of Elgin, 1855, was sited not far from the rear of the main house, and in due course screened from it by trees. The building is a neat gabled, barge-boarded cruciform structure, the symmetrical frontage well lit with picturesque lattice glazing and enhanced by a decorative bellcote at the gable apex. Inside, the accommodation comprised a 'washing house' with boiler and four sinks placed beneath an east-facing window and a 'laundry' (probably the ironing and finishing room) with stove for heating irons. Access to these was via a lobby from which the staircase rose to the drying attic, perhaps fitted (as at Clynelish Farm, Sutherland, 1865) with drying racks. Louvred windows on all sides of the loft gave plenty of ventilation to facilitate drying. A rear porch was assigned to 'dust', presumably for brushing muddy clothes of sportsmen (the River Spey flows through the policies and the adjacent hills provided game shooting), and separated from the laundry area in order that this activity should neither spoil nor interrupt work in progress. A WC was provided for the laundresses, while another with external access served outside workers.[12] If 'dusting' clothes was the work of male valets, then these arrangements segregated male and female staff.

Laundries were at times sited some distance from the mansion they served. The laundry for Ardtornish House, Morvern, Argyll, was at Larachbeg, a mile or so distant. The tall, single-storeyed, gabled building is constructed of concrete, an innovatory material introduced from 1871 by the estate master-of-works, Samuel Barham. This laundry is notable for its fittings, disused but still in situ in the mid 1980s. Besides the standard boiling 'coppers' there are six heated drying cabinets which draw out on runners and a spin dryer fitted in the concrete flooring, cranked by a handle. Soiled washing was transported from Ardtornish House by horse and cart, the finished work returning by the same means of transport.[13] Ardtornish was largely a sporting estate, owned by a London distiller and his family, and frequented by large house parties. Another English industrialist built Kinloch Castle, Rhum, 1897–1906. Here, the household washing was taken to Kilmory, about five miles from the castle, where on a level site near the shore and the Kilmory River a cluster of utilitarian corrugated iron huts housed the laundry facilities.

From the mid twentieth century, with ever-decreasing staff, these estate laundries were superseded first by commercial laundries serviced by mobile vans, collecting the dirty and delivering the clean washing, and increasingly from the 1950s and '60s, by the domestic washing machine powered by electricity, which improved in scope and technique as the automated 'front loader' became common to most households. These have been installed by their owners in estate mansions, as well as in their own homes by the dwindling numbers of employees. Aided by these machines, spin dryers, electric irons and fold-away ironing boards, the laundry work on most estates is now undertaken by the casual or part-time staff, 'easy-care' and synthetic fabrics reducing their task of ironing.

GIRNALS OR ESTATE STOREHOUSES

Until the end of the eighteenth century and even into the nineteenth, agricultural rents were paid partly in service and in kind. Of the latter, grain, largely in the form of barley and oatmeal, was the most usual. In turn, farm servants received meal as a recognised portion of their wages. Public servants such as school masters and ministers were also paid some of their salary or stipend in meal by their parish heritors, the very landowners who reckoned much of their income in these terms. Wealth in such a bulky form obviously required secure, dry storage, pending the issue of wages in kind or the export of the grain from the estate to realise cash. Most large estates, therefore, had their own storehouse, or

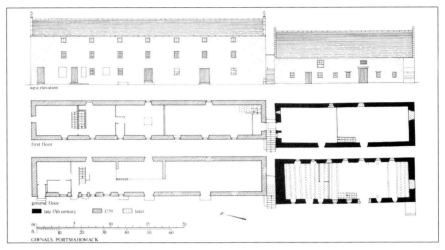

Figure 9.4 Girnals, Portmahomack, Ross and Cromarty; measured survey drawings. Crown copyright: RCAHMS, SC 714691.

girnal. Where the estate was near the sea or navigable water, these buildings were sited by the shore in order that agricultural commodities could be readily exported to urban markets.

The girnal took the form of a substantial two- or three-storeyed rectangular masonry warehouse with assembly yard at the rear where goods were unloaded. If by the coast, they fronted a sheltered beach or harbour for easy loading into boats. Girnals differ from large mains farm barns in that they have no opposing winnowing doors. They are usually provided with mural ventilation slits and forestair access to the upper floors, which are lit by regular shuttered or louvred openings. A single gable apex chimney indicates custodian accommodation. At Portmahomack, Easter Ross, 'two storehouses for the reception of rents in kind'[14] (fig. 9.4) stand beside the harbour, one of the late seventeenth century and the other dated 1779. Both were the property of the Earls of Cromartie, substantial landowners in Ross and Cromarty. The earlier girnal was constructed by George Mackenzie, 1st Earl of Cromartie (1630–1714), whose Edinburgh townhouse (Royston House, later Caroline Park) was sited conveniently near the harbour at Leith, no doubt so that he could keep an eye on the arrival of his principal source of income – his pay as a senior legal officer of the Crown, up to seven years in arrears.[15]

ICE HOUSES

Ice houses are semi-subterranean cold stores, early forms of refrigerator for food and drink. They were usually built into a natural slope conveniently sited near the kitchen offices, either egg-shaped (fig. 9.5) or a vaulted cube, approached through a short passage closed at each end by heavy insulating doors. Though often decorative landscape features in England, in Scotland domestic ice houses tended to be of a utilitarian nature, the only deference to architectural adornment perhaps being

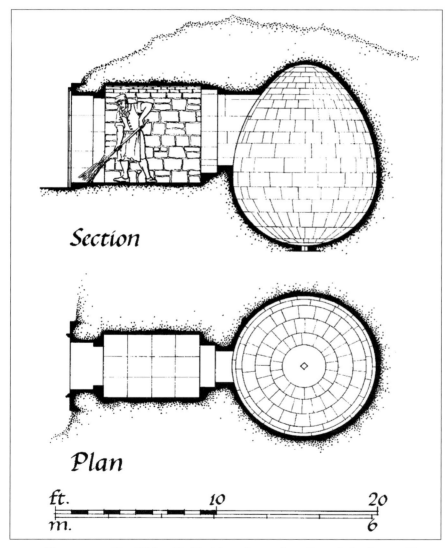

Figure 9.5 The Whim, Peebleshire; plan and section of ice house. Crown copyright: RCAHMS, SC 714584.

around the doorway. At Cullen House the round-headed entrance with blocked imposts is the sole enrichment.[16] Ice houses were normally of rubble masonry, sometimes lined with brick which possessed good insulating qualities, and filled with ice collected from nearby rivers and ponds. The compacted ice lasted for upwards of a year provided the icehouse was well drained; a sump in the base released surplus moisture. The ice surface was covered with a layer of straw and on this, or above it suspended on hooks, baskets of fish, meat and other foodstuffs were stored, while bottles were packed in chaff and set into the ice itself.

DOVECOTS

Dovecots were batteries for the breeding of pigeons. The young squabs or 'peezers' were the main ingredient of pies, stews and other dishes. They are amongst the oldest farm buildings in Scotland, established principally on estates with large households, including monastic establishments. Though encouraged in 1503 by an Act of (the Scottish) Parliament that 'ordains ilk lord and laird to make them dowecots', such were the predations of these birds on crops that, by 1617, another Act limited their construction to proprietors of lands yielding a certain yearly rental, adjacent to the dovecot or at least lying within two miles of it, the landowner being entitled to one dovecot only.[17] As a structure, the dovecot spans the vernacular and the architecturally conscious, forming an interesting record of local building materials and their use, besides entering the field of household management and farming practices.

The basic requirements of dovecots were defined by the Roman writer Varro as:

1. shelter
2. nesting facilities
3. access (for bird and man)
4. ventilation
5. defence against vermin.

Within these unchanging parameters there is a surprising number of variants of style and form spanning five centuries, all providing the necessary shelter for the domesticated dove, a descendant of the rock dove (Columba livia), which in nature lives on sea cliffs and nests in caves. The earliest are beehive cotes, circular, tapering masonry structures so called because they resemble straw bee 'skeps' or 'ruskies'. The rectangular lectern cote with a single pitch roof was a favoured form in the seventeenth and early eighteenth centuries, found in Scotland (fig. 9.6) and the south of France, but seldom in England or northern France. This type normally faced south, the tall rear wall shielding the birds as they alighted from the cold north winds. These, together with oblong,

Figure 9.6 Cadboll, Ross and Cromarty; rectangular 'lectern' dovecot. Crown copyright: RCAHMS, SC 400153.

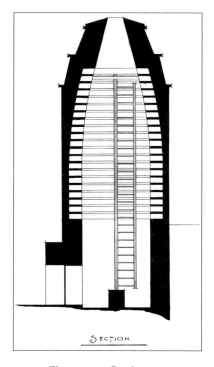

Figure 9.7 Gordonstoun, Moray; cylindrical dovecot converted from windmill. Courtesy of Wittets, Architects, Elgin, and The Moray Council Libraries.

double-pitched roof cotes, are frequently divided into two or more chambers, increasing the internal wall area for nesting boxes. Serried ranks of these small stone, wood, clay or brick cubicles are an impressive sight. Cylindrical dovecots with conical roofs (fig. 9.7) also date from the eighteenth century, and hexagonal or octagonal forms were popular by the turn of the 1800s. The free-standing dovecot lost popularity in the early nineteenth century, with pigeon accommodation provided in a loft or tower crowning the central arched entrances of the substantial farm steadings of the time, or incorporated in upper storeys of masonry hen houses.

Birds came and went through round apex holes in the beehive and cylindrical cotes, by rectangular flight-holes through the walling fronted by alighting ledges, low swept dormers or sometimes ridge or apex glovers pierced with groups of small entrances. Small openings helped to keep out hawks preying on the doves. Ventilation holes were barred for the same purposes. Ledges encircled most dovecotes, enabling birds to alight, preen and sun themselves.[18] The pigeon keeper entered by a small doorway, often fitted with a heavy lock, sometimes with an internal 'yett'. Openings were blocked when the squabs were collected from nests, their collection facilitated by a central revolving ladder or 'potence' mounted on a stone plinth. Floors were often paved to ease the collection of pigeon droppings as a rich and much-prized manure.

Some cotes also doubled as hilltop follies, particularly in Aberdeenshire. At Auchmacoy, 1638,[19] a squat beehive topped by unusual corbelled and crowstepped upper works overlooks the estate of that name. This must surely be one of the earliest follies in Scotland. The tall octagonal cote at Mounthooly, dated 1800, its wallhead parading an array of ball finials, crowns a rise above coastal Rosehearty as a landmark for those at sea.

The final phase of the dovecot was for sporting purposes. Pigeons were released from cotes or traps, their speedy flight making them challenging targets. Cotes designed for this purpose have small flight-holes with no outside alighting ledges, and the flight-holes can be closed internally by shutters. At Sandside, Caithness, there is a small dovecot from which birds were released from the upper loft.

BEE BOLES

Bees were valued as providers of honey, almost the only form of sweetness before the introduction of cane sugar from the West Indies in appreciable quantity in the eighteenth century. Bee boles are small alcoves in which the straw bee skeps or ruskies sheltered in the south-facing aspect of garden walls. Free-standing sets of shelves, with sloping roofs, also accommodated hives.[20] Bee bole provision is found in a wide spectrum of properties: abbeys, castles, lairds' houses, manses and small farms. Medieval Pluscarden Abbey, Moray, is well provided, and the late sixteenth-century pleasance of Tolquhon Castle, Aberdeenshire, has a dozen of these alcoves arranged neatly in the south-facing garden wall while at (Old) Canisbay Manse, Caithness, there are three semi-circular bee boles, cunningly shaped to accommodate the circular straw skeps. At Carmylie Manse, Angus (fig. 9.8), the bee bole is secured by a

Figure 9.8   Carmylie Manse, Angus; bee boles. SLA, C 1852.

padlocked iron grid.[21] In Strath Avon, Upper Banffshire, many smallholdings have drystone-dyked garden enclosures, to keep out both beasts and wind. Within these walls are one or two simple boles for the family bees, the bees finding nectar amongst the spring flowers of this sheltered strath or the aromatic heather-clad slopes above.

PRIVIES

Sanitation presented few problems in scattered rural communities, but as the country estate expanded, with an increased number of employees, so lavatory provision became a necessity. Before the flushed water closet, lavatories were principally the outside 'necessary house', a small building where excreta were disposed of by earth pits, buckets or running water.

In 1762 Moy House, Moray, had six different outside lavatories for use of the household hierarchy, from the 'gentlemen' to the staff.[22] In the garden of Sandside House, Caithness, a small, early nineteenth-century square rubble building with piended roof is built against a slope; it houses a two-seater bench above a dry ditch. At Mains of Eden, Aberdeenshire, a small building, modestly fronted by a screen wall, combines two bucket privies with the ash house (ashes were also retained as garden or field fertiliser) conveniently facing the house service door. This provision is in a wing of the poultry/dovecot complex built in 1852.[23]

Burn- and sea-flushing toilets provided hygienic sanitation. On the island of Shapinsay, Orkney, the Balfour family, who developed the estate in the mid nineteenth century, erected a sea-flushing two-chamber toilet against the harbour wall at Balfour Village (fig. 9.9) for the use of their tenantry. The walling, as might be expected in Orkney, is of local Rousay

Figure 9.9 Balfour Village, Shapinsay, Orkney; tide-washed toilet, in SVBWG. *Vernacular Building*, 14 (1990), 39. Courtesy of W Ashley-Bartlam.

flag, apparently a drystone construction.[24] At Samalan House, Loch Ailort, Inverness-shire, a small rectangular rubble 'little house' (roofless in 1992) is slung on two wooden beams across the burn flowing at the rear of the house. This was also the case in the 1960s at Mill of Newmill, Auchterless, Aberdeenshire, where the wooden lavatory straddled the mill-lade.

Though often utilised now for other purposes or in a ruinous state, these buildings offer an interesting insight into estate hygiene.

ESTATE ELECTRICITY-GENERATING STATIONS AND GASOMETERS

Progressive landowners availed themselves of modern energy resources before such facilities were distributed locally or nationally. Both private electricity-generating resources and gas production were used, according to taste and local circumstances. A suitably Baronial, crenellated masonry gasometer stands by the shore in Balfour village, Shapinsay (see above), providing gas for Balfour estate tenants. At Altyre, Moray, John Kinross, c. 1902, designed an electricity-generating house in a bold, Scottish Revival manner. Similar constructions can be found elsewhere, often adapted for another purpose after having been superseded by the introduction of local gas supplies and the national grid.

ANCILLARY BUILDINGS ON HIGHLAND SHOOTING ESTATES

Although deer hunting was active from earliest times, the development of the highland sporting estate as we recognise it today commenced in the late eighteenth century, reaching its apogee in the second half of the nineteenth and requiring a range of specialised ancillary buildings. By the 1850s, roads in northern Scotland were much improved and the railway was pushing northwards, facilitating the movement of tweed-garbed gentleman and their families from the south to shoot grouse and deer at their appropriate season, besides catching salmon from the river. This influx was led by Prince Albert whose sporting activities at 'dear Balmoral' were much admired by his queen. These sports were centred on either ancestral Highland homes, where large house parties were held during the sporting season or on specially constructed, and often architect-designed, shooting 'lodges' (some more aptly described as mansions) in distant glens. The building materials and workmen travelled many miles by train, horse and cart, and by boat the length of lochs, for their construction.

Lodges were owned or rented either by industrialists from the south aspiring to the status of landed gentry, or by English aristocracy

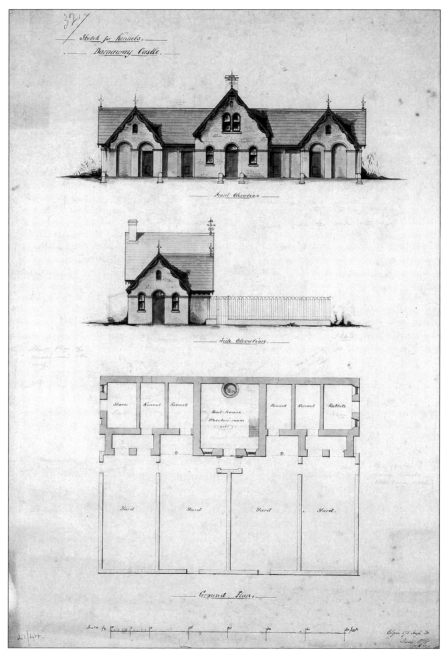

Figure 9.10 Darnaway Castle, Moray; kennels design by A & W Reid, Elgin, 1878. Courtesy of Moray Estates Development Company.

keeping up the fashion already practised by their Highland peers. Besides cottages for game keepers, stalkers and gillies, stables were required for ponies transporting the kill from mountainous moorland back to the lodge, kennels for the dogs and ventilated larders for the processing and storage of flesh, fowl and fish. Buildings used for hanging the carcases of shot deer are noted in Chapter 22. Though some of these ranges were entirely utilitarian, others were architecturally sophisticated. The Duke of Sutherland commissioned the English architect George Devey to design a game larder at Dunrobin in 1861,[25] while at Darnaway Castle, Moray, elegant kennels were matched by an equally elegant head keeper's house with integrated gunroom (fig. 9.10). The kennels had four dog cubicles, each opening onto an arcaded loggia and railed yard. Besides a 'boil-house' there was a 'watcher's-room' and two stores, one a rabbit larder (rabbits were boiled and fed to the dogs). Some kennels also had infirmaries for sick dogs. The Darnaway kennels were designed by the Elgin architects A & W Reid, 1878,[26] who like John Rhind of Inverness, received commissions for buildings on shooting estates throughout the Highlands.[27]

The isolated nature of shooting lodges necessitated accommodation for seasonal staff such as beaters, and the gillies. The range of corrugated-iron stables at Coignafearn, far up the Findhorn valley in Inverness-shire, incorporates simple living accommodation. As with so many similar buildings, the Coignafearn stables are timber-lined; this, together with the use of peat and wood as fuel, created an ever-present fire hazard, the flames difficult to quench in isolated sites, destroying lodges and the buildings that surrounded them.

BOATHOUSES

Boathouses, built for pleasure and fishing boats used on estate lochs, were largely associated with the development of the sporting estate in the late nineteenth century.[28] Most are utilitarian structures, providing minimal loch-side shelter for moored boats and storage for oars and other equipment. Some, however, were larger and more opulent; a tea room and balcony overlooking the water constituted an upper storey, where families could enjoy picnics (fig. 9.11).

Figure 9.11 John Starforth. *The Architecture of the Park*, 1890; boathouse design, plate XCIX.

NOTES

1. Graham, 1993.
2. Colvin, 1978, 100; Simpson, Simpson, 1973, 55. John Baxter (Jnr), Edinburgh, was responsible for the remodelling of Gordon Castle, 1769–82.
3. Gifford, 1992, 135. Architect, David Smith, Thurso.
4. Watson, 1985, 54; Architect, A Marshall Mackenzie, Aberdeen.

5. Buxbaum, 1989, 97; McWilliam, 1978, 387.
6. Cornforth, 1978.
7. Gifford, 1984, 640.
8. Baldwin, 1985.
9. Moray District Record Office, Forres, DCZ P8; Architects, R Neish and Forsyth, Forres, Moray.
10. Gifford, 1992, 521.
11. NAS, RHP 12879, 1822. Laundry, fronted by 'bleaching green' indicated on Cullen House estate map. The building, now 'The Old Laundry' was converted as a private house, 1985–90, by Douglas Forrest, architect. A vignette on an earlier map of the same area illustrates a building with a water wheel, probably the bleaching house (with water-powered rubbing mill) of John Christie's bleachfield. See Durie, 1977, 85.
12. Elgin Library, Moray, Wittet Collection, DAWP, 1105/1–2.
13. Pers. comm. 1983. A terrace of houses was constructed at Larachbeg in 1875, the laundry seems later; Gaskell, 1968, 90–1.
14. *NSA*, vol xiv (1840), 463.
15. Clough, 1977; Beaton, 1986.
16. Buxbaum, 1989, 109; Architect, James Adam, *c.* 1775, built at a cost of £30.00.
17. Robertson, *c.* 1958.
18. Until recently these ledges were referred to a 'rat courses' and thought to deter such vermin. This theory has now been disproved by McCann, 1991.
19. Dunbar, 1966, 91.
20. Until *c.* 1980 there was a set of free-standing beeskep shelves at Gordonstoun, Moray, accommodating ten or a dozen beeskeps. Two shelves with south facing aspect and solid rear wall, constructed of local sandstone slabs, faced south under a sloping, single pitch roof.
21. Fenton, Walker, 1981, 17.
22. NAS, RHP 9060, plan of Moy House by Colin Williamson, 1761–2.
23. Elgin Library, Moray, DAWP 629/6–8. Plans and elevations of poultry/dovecot complex, 1852; Architects A & W Reid, Elgin.
24. Bartlam, 1990.
25. Allibone, 1991, 49, 154.
26. Plans and elevations of keeper's house and kennels, Moray Estates (copies in RCAHMS).
27. Gifford, J. Architects of the Highlands in the Nineteenth Century – a sketch, *Scottish Georgian Society Bulletin*, 7 (1980), 29–48.
28. A 1751–2 boathouse on the Inveraray estate, Argyll, may probably be the earliest surviving structure in this category in Scotland; RCAHMS, 1992, no. 193.

BIBLIOGRAPHY

Allibone, J. *George Devey*, London, 1991.
Bailey, D C, Tindall, M C. Dovecots of East Lothian, *Transactions of the Ancient Monuments Society*, 11 (1963), 23–52.
Baldwin, J. *Exploring Scotland's Heritage: Lothian and the Borders*, Edinburgh, 1985.
Baldwin, J, ed. *Firthlands of Ross and Sutherland*, Edinburgh, 1986.
Bartlam, W A. Notes on a victorian sea-flushing toilet, *Vernacular Building*, 14 (1990), 39–40.
Beamon, S, Roaf, S. *The Icehouses of Britain*, London and New York, 1990.

Beaton, E. *The Doocots of Moray*, Elgin, 1978.
Beaton, E. *The Doocots of Caithness*, Dundee, 1980.
Beaton, E. Late seventeenth- and eighteenth-century estate girnals in Easter Ross and south-east Sutherland. In Baldwin, 1986, 133–55.
Buxbaum, T. *Scottish Garden Buildings: From food to folly*, Edinburgh, 1989.
Buxbaum, T. *Icehouses*, Princes Risborough, 1992.
Clough, M. Making the most of one's resources: Lord Tarbat's development of the Cromarty Firth, *Country Life*, 162 (29 September, 1977), 856–7.
Cornforth, J. Castle Fraser, Aberdeenshire, *Country Life* (17 August, 1978), 442–3.
Crane, E. *The Archaeology of Bee Keeping*, London, 1983.
Colvin, H M. *A Biographical Dictionary of British Architects, 1600–1840*, London, 2nd edn, 1978.
Douglas, R. *The Dovecots of Moray*, Elgin, 1931.
Dunbar, J G. *The Historic Architecture of Scotland*, London, 1966, 1978 (as *The Architecture of Scotland*).
Durie, A J. *The Scottish Linen Industry in the Eighteenth Century*, Edinburgh, 1977.
Fenton, A, Walker, B. *The Rural Architecture of Scotland*, Edinburgh, 1981.
Gaskell, P. *Morvern Transformed: A Highland parish in the nineteenth century*, Cambridge, 1968.
Gifford, J, McWilliam, C, Walker, D. *The Buildings of Scotland: Edinburgh*, Harmondsworth, 1984.
Gifford, J. *The Buildings of Scotland: Highlands and Islands*, Harmondsworth, 1992.
Graham, A. *Skipness: Memories of a Highland estate*, Edinburgh, 1993.
Hart-David, D. *Monarchs of the Glen: A history of deer-stalking in the Scottish Highlands*, London, 1978.
Lindsay, I G, Cosh, M. *Inveraray and the Dukes of Argyll*, Edinburgh, 1973.
Loudon, J C. *Dictionary of Cottage, Farm and Villa Architecture and Furniture*, London, 1833 (and later edns).
McCann, J. An historical enquiry into the design and use of dovecotes, *Transactions of the Ancient Monuments Society*, 35 (1991), 89–160.
McWilliam, C. *The Buildings of Scotland: Lothian*, Harmondsworth, 1978.
Mowl, T, Earnshaw, B. *Trumpet at a Distant Gate: The lodge as prelude to the country house*, London, 1985.
Naismith, R J. *Buildings of the Scottish Countryside*, London, 1985.
Peterkin, G A G. *Scottish Dovecotes*, Coupar Angus, 1890.
RCAHMS. *Inventory of Argyll*, Vol 7, *Mid Argyll and Cowal*, Edinburgh, 1992.
Ritchie, J N G. *Brochs of Scotland*, Princes Risborough, 1988.
Robertson, C N. Old Scottish Dovecots (*c.* 1958), manuscript in Library of the National Museums of Scotland, Edinburgh.
Robertson, U. Pigeons as a source of food in eighteenth-century Scotland, *ROSC*, 4 (1988), 89–103.
Simpson, A, Simpson, J. John Baxter, architect, and the patronage of the Fourth Duke of Gordon, *Scottish Georgian Society Bulletin*, 2 (1973), 47–57.
Vernacular Building, Vol 1–. Annual journal of the Scottish Vernacular Buildings Working Group, Dundee and Edinburgh (1975–).
Watson, W H. *A Marshall Mackenzie, Architect in Aberdeen*, Aberdeen, 1985.
Walker, P. Bee boles and past beekeeping in Scotland, *ROSC*, 4 (1988), 105–17.

10 Garden Layouts and Garden Buildings

TIM BUXBAUM

The most significant gardens in Scotland have been those at the centres of country estates, which were largely self-sufficient until the system fell apart in the early 1900s. Centuries earlier, deer parks, fish ponds and dovecots began to be linked with paradise gardens of orchards, clipped evergreens, mazes, sundials and fountains. Those and other elements were then developed by succeeding generations in a spiral of increasing complexity, interdependence and enrichment; maturing, falling out of favour, and being revived in a new way. The development of thousands of gardens over a period of some four hundred years has resulted in considerable diversity. A combination of shelter, political stability, fertile land and abundant labour has always been advantageous, yet gardens and garden buildings can be found all over Scotland, sometimes with one layout superimposed upon another.

In 1604 at Edzell Castle, Angus (fig. 10.1), small pavilions comprising a bathhouse and banqueting house (where only a few could gather) were constructed in the return angles of a highly decorated wall guarding a formal parterre with knots and topiary, the whole composition being overlooked by the main tower. Such gardens were formal, demonstrating man's control over nature and providing a civilised retreat from a dangerous world. They paralleled contemporary physic gardens. The fashion continued through the 1600s, as in Europe: clipped parterres of scented herbs on coloured groundworks were punctuated with topiary, rows of small trees, espaliers, planted beds, flights of steps and geometric paths (fig. 10.2). These all interacted to provide a pleasance with centre points of elaborately carved and sculpted fountains and sundials (eg Hopetoun, West Lothian, Glamis, Angus, fig. 10.3). Heraldic and armorial devices would be incorporated wherever possible. The result was best viewed from above, increasingly from artificial terraces focused on twin gazebos usually taking the form of a cube with an ogee roof and often a lower storey (eg Pitmedden, Aberdeenshire, and Aberdour, Fife). Gazebos might be internally panelled and finished with a painted

Figure 10.1 Edzell Castle, Angus; garden. Crown copyright: RCAHMS, SC 713516.

ceiling to provide a welcome retreat from the main house (eg Traquair, Peeblesshire).

Growing wealth and political stability encouraged larger gardens such as the Great Garden formerly at Alloa, Clackmannanshire (1706–14), which extended several miles in diameter and was full of terraces, statues and 'basins of water'; 32 different vistas led the eye to distant mountains and castles. In the absence of these, designers used purpose-built vista fillers – silhouettes providing no shelter but terminating a view (eg Mellerstain, Berwickshire, and Blair, Perthshire). Once a formal framework had become fashionable, avenues and double avenues of trees strode across the countryside, perhaps diverging in a patte d'oie (the shape of a goose foot), clearly demonstrating man's triumph over nature. With estate enclosure and the agricultural improvement of a neglected

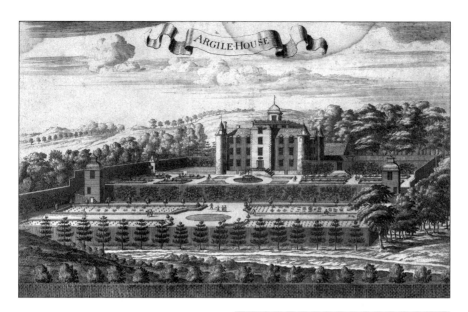

Figure 10.2 Hatton House, Midlothian; late seventeenth-century Slezer view, incorrectly labelled 'Argile House', showing high walls sheltering the terraced gardens immediately adjacent to the house.

Figure 10.3 Glamis Castle, Angus; sundial. Crown copyright: RCAHMS, SC 713501.

and sometimes barren landscape, massive tree-planting schemes began, initially in formal layouts of squares and triangles extending perhaps to hundreds of acres apiece (eg Blair Adam, Kinross-shire). Lines between the plantations reinforced vistas from the house. In a similar vein, elaborate earthworks, including artificial hills, amphitheatres and terraces, were raised to emphasize the position of the house and to provide adjacent bowling greens (eg Castle Kennedy, Wigtownshire, and Newliston, West Lothian). Garden areas were demarcated with boscage – topiary grown into great walls of architectural foliage – zoning thematic shelters or imitating buttresses. Formal water features repeated the same message: geometric pools and 'canals' were created, some featuring fountains or linked into cascades (eg Drumlanrig, Dumfriesshire, and Arniston, Midlothian). By way of relaxation from such ordering of nature, wildernesses appeared around 1715, providing a softer area of woodland with rushing streams, serpentine walks, bee skeps and statuary.

By the 1750s there was greater emphasis on agricultural improvement than on the maintenance of massive formal gardens, many of which were swept away with the Landscape Movement. The ideal became a villa in flowing and subtle parkland, with animals kept at a distance by means of a ha-ha, and the scene changing as one moved about. The estate aspired both to improved farmland and pleasure grounds, the latter seeking identity through little buildings generally dubbed follies. Yet that may be an oversimplification; many were built at points of significance in the landscape – to thrill, to beckon, to intrigue, to focus expeditions, to expedite liaisons. Some recorded events or commemorated individuals. Some provided labour during economic depression. Others were welcome stops on the estate tour. In the early 1700s that was almost a journey of fantasy, heightened by such structures as the Chinese tent with its gilded bells and dragons; after 1760 there was a growing interest in the Sublime and Picturesque matched with suitable buildings, especially decorated steadings. Later, it became an opportunity to view progress, learn from innovation and boast of success on the estate.

Classical, Gothick or freestyle temples were popular as stops on the estate tour (eg Amisfield, Dumfriesshire, and Taymouth, Perthshire). A wide range of summerhouses was built (eg Dunglass, East Lothian, and Mellerstain), together with tea pavilions (eg Auchincruive, Ayrshire), and rotundas (eg Cullen, Banffshire, and Duddingston, Edinburgh). Inevitably many designs originated from pattern books but a number resulted from genuine architectural experimentation. Decorative dairies for retirement rather than animal management combined cool practicality with pleasure (eg Taymouth, and Guisachan, Inverness-shire). Dramatic viewpoints were available from prospect and lookout towers (eg Inveraray and Lochnell, Argyll) sometimes built on a natural summit or at the very brink of an abyss yet still accessible by horse and carriage.

Commemorative towers celebrated military or agricultural success or dynastic change, provoking publicity and goodwill. Nelson and Waterloo towers belonged to the same family (eg Monteviot, Roxburghshire, and the House of the Binns, West Lothian). The walled enclosure of estates and relocation of villages marooned chapels which were adopted as family mausolea (eg Yester, East Lothian, and Balcarres, Fife), or yielded the ruins of an old family tower as a feature in the policies (eg Hopetoun, and Castle Semple, Renfrewshire). Other buildings in the policies included the aviary, pheasantry and icehouse. Some buildings were bizarre, for example, the Gate of Negapatam at Fyrish above Invergordon, Ross-shire, built to celebrate victory in India in 1781, and the Malakof Arch at Murthly, Perthshire, a hilltop Arc-de-Triomphe blown up after World War II. Also the Fort at Taymouth Castle, the Panmure Testimonial, Angus, and the Lanrick Tree, Perthshire, the latter proclaiming the return to normality of a clan forbidden to use their patronymic. Clare Hall at Penicuik, Midlothian, was never built, but would have combined bathhouse and domed library in a splendid neo-classical composition.

Sensations of awe generated by dramatic scenery were, in the later 1700s, incorporated into the estate tour to provide the visitor with a cathartic experience. One of the most dramatic examples was Hurleycove Tunnel at Penicuik House. Natural caves evoking appropriate folklore were employed, as were artificial grottoes (eg Yester and Arniston) and underground passages leading to a spectacular waterfall view (eg Acharn, Perthshire). Hermitages (eg Dunkeld, Perthshire), rarely inhabited by hermits, were lined with moss, books, mirrors and stuffed animals at an atmospheric viewpoint. Fog houses, moss houses and later root houses were of the same rustic family which extended to intricate Victorian summerhouses lined with coloured lichens, larch rods and pine cones (eg Brodick, Arran, and Drumlanrig). Harder and more brittle materials like tufa, flint and quartz were made into rustic bridges over white water and even more sober estate buildings where a sophisticated 'primitive' quality was felt to be appropriate. Formal geometric pools left from earlier times were softened to generate informal views. New lakes were dug, at least one as an excuse to build a bridge! The growing appreciation of 'natural' bodies of water and running burns engendered fishing temples (eg Duff House, Banffshire), viewing shelters (eg Colzium, Stirlingshire) and bathhouses raised to new heights as at Pitfour, Aberdeenshire, in the form of a granite-columned Temple of Theseus (fig. 10.4). Pump houses for taking medicinal waters were a transient building type (eg Inveraray). Lakes became further developed through small structures such as the curling house; sluices and improved water management made possible more elaborate boathouses with accommodation eventually running to an upstairs saloon (eg Manderston, Berwickshire, and Dougarie, Arran).

Figure 10.4 Pitfour, Aberdeenshire; bath house, 'Temple of Theseus'. Crown copyright: RCAHMS, SC 712194.

The 1800s progressed with pragmatic experimentation and growing knowledge, choice and availability of plants and building techniques. Trees from the New World were planted in a formal pinetum or arboretum (eg Scone and Blair, Perthshire) until better knowledge of them allowed a more relaxed layout. Rhododendrons were introduced and eventually spread everywhere. Brilliantly-coloured new flowers stimulated a free re-interpretation of the old parterres which had been swept away, but now related to low terraces functioning as outdoor rooms and viewing platforms (eg Manderston). The terraces themselves were transformed by the industrial production of cast stone balustrades, pavings, copings and jardinières. Foundries cast anything from garden seats to fountains to kangaroo houses if required. Garden statuary diversified from the lead figures of the late 1600s into statues of mythical beasts and domestic pets, well heads, urns and bases – sometimes of enormous size (eg Haddo, Aberdeenshire) – occupying points of importance and informality, from riverside walks to bosky groves. There were also memorial cairns (eg Balmoral, Aberdeenshire) and obelisks (eg Tyninghame, East Lothian), Eleanor crosses (eg Dunrobin, Sutherland), benches, silhouettes for target practice, and genuine antiques both brought in from abroad and rescued from Scottish demolition sites. These included mercat crosses (eg The Drum, Edinburgh, and Fingask, Perthshire), carved stones used to decorate sunken gardens (eg Arniston), or cathedral relics incorporated into garden grottoes. At Yester, the glazed

Figure 10.5 Dunmore, Stirlingshire; The Pineapple. Crown copyright: RCAHMS, SC 712284.

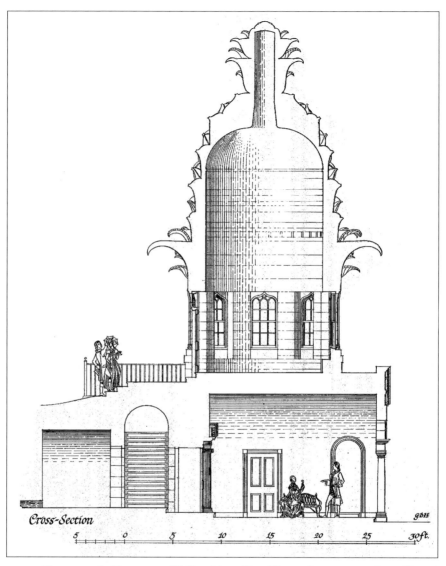

Figure 10.6 Dunmore, Stirlingshire; The Pineapple, cross-section. Crown copyright: RCAHMS, SC 712286.

clock-tower from a former Caledonian Railway station was erected in 1970 as a summerhouse.

New plants from abroad required new buildings for protection. On the estates most bedding plants were raised in great ranges of glass houses associated with the walled garden. That had existed from early times, typically as a sheltered enclosure which followers of the Landscape Movement tried to banish from sight; its position varied relative to the house. Many saw the walled garden as the focus of the estate, providing food, demonstrating control, and marking a critical point on any tour, and thus it was built in dressed stone or decorative brickwork. Some of the finest garden buildings were erected there, either as decorative pavilions (eg Preston Hall, Midlothian, Amisfield, or Dunmore Pineapple, Stirlingshire; figs 10.5, 10.6), as an apple house (eg Johnstounburn, East Lothian) or as a built-in gardener's house (eg Yester, and Gordon, Moray). The job of head gardener warranted a residence of style. Walled gardens could include water with ornamental bridges and thatched swan houses (eg Blair, Perthshire), a grotto and rustic shelter (eg Culzean, Ayrshire) and a wide variety of garden architecture. Layout and subdivision of the most elaborate examples would be carefully controlled by means of decorated gateways, boscage, statuary and subsidiary buildings (eg Guthrie, Angus, and Culzean); technical innovation introduced hot walls, flues, boilers, and an array of glass houses purpose-built for raising vines, figs, orchids, apricots, camellia and so on. There were pits for pineapple, cucumber, and asparagus, and generally a vast collection of specialist facilities served by a veritable army of staff. The most specialised buildings, such as ferneries, lily houses, orangeries and conservatories could be associated with, or quite separate from, the walled garden, and many such buildings of the mid nineteenth century were architecturally

Figure 10.7 Dalkeith House, Midlothian; conservatory. Crown copyright: RCAHMS, SC 713514.

significant (eg Dalkeith, Midlothian, fig. 10.7, and Kibble, Glasgow; see Chapter 18).

After the early 1800s many garden designers looked back to earlier formal gardens and re-interpreted them as room gardens, suggesting enclosure where privacy or ostentation was more important than defence (eg Crathes, Kincardineshire, and Earlshall, Fife). Separating gateways became self-conscious, displaying sculpture or mottoes, even fashioned into moon gates. It was not unusual for gardens of the mid 1800s to display quite different characteristics, as shown by the amazing Italianate formal layout at Drummond, Perthshire (fig. 10.8), or the quasi-medieval turreted courts at Abbotsford, Roxburghshire. At Keir, Perthshire, a rich and catholic collection of gates, pavilions, statues, urns and even a yew house throve amidst a remarkable collection of plants. Half a century

Figure 10.8 Drummond Castle, Perthshire; garden. Crown copyright: RCAHMS, SC 713515.

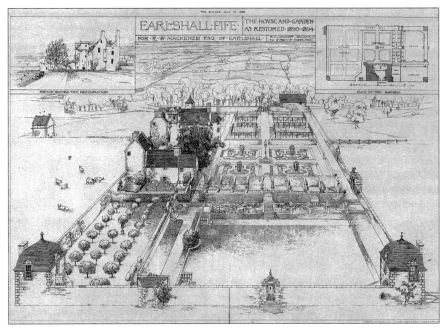

Figure 10.9 Earlshall, Fife, 1900. Courtesy of Dr Peter Savage.

later, Lorimer's designs (eg Earlshall, fig. 10.9, and Formakin, Renfrewshire) again refer to previous gardens, using enclosing walls and formal elements like fountains, sundials, gazebos, topiary, treillage, fencing, pergolas; all are re-interpreted, becoming complex and personal and enriching the experience of the visitor.

By the later 1800s, transport, communications and society were changing, and fashion percolated downwards. Merchants' houses began as a matter of course to include topiary, a flower garden and modest conservatory. Alpine and woodland gardens and the herbaceous border became more popular and more accessible to the general public, and there was a movement away from the country estate. In the half century after World War I many country houses were burnt or abandoned, their gardens becoming massively overgrown, and the buildings crumbling. Meantime a few enthusiasts looked to Japan for inspiration (eg Dalzell, Lanarkshire, and Carnell, Ayrshire), and established specialist gardens like Inverewe, Ross-shire, and outstations of the Botanic Gardens. Finally, a revival of interest beginning in the 1970s spurred much research, culminating in repair, conversion and restoration: one of the most ambitious projects was the rescue of the William Adam-designed Baroque hunting lodge at Chatelherault, Lanarkshire, with its formal gardens and menagerie. Many other rescue projects followed its success. There was even limited innovation in the creation of new gardens and garden buildings, for example, Ian Hamilton Finlay's work at Little Sparta, Dunsyre, Lanarkshire, and the moon-viewing pavilion at Kinkell, Ross-shire.

BIBLIOGRAPHY

Colvin, H M. *A Biographical Dictionary of British Architects, 1600–1840*, New Haven and London, 3rd edn, 1995.
Countryside Commission for Scotland and the Historic Buildings and Monuments Directorate. *An Inventory of Gardens and Designed Landscapes in Scotland* (report by Land Use Consultants), Glasgow, 1988.
Craven, Rev. J B, ed. *Journals of the Episcopal Visitations of the Right Rev Robert Forbes 1762 and 1770*, London, 1786.
Defoe, D. *A Tour through the Whole Island of Great Britain, 1724–26*, London, 1769, reprinted Exeter and London, 1927, New Haven and London, 1991.
Donaldson, J. *Husbandry Anatomised*, Edinburgh, 1697.
Grant of Monymusk, Sir A. Description of the present state of (Monymusk) and what hath been done to make it what it is. In *Miscellany of the Spalding Club, Vol 2* (Spalding Club 6), Aberdeen, 1842, 96–7.
Gray, J M, ed. *Clerk of Penicuik's Memoirs: Extracted by himself from his own journals, 1676–1755* (Scottish History Society, 1st series, vol 13), Edinburgh, 1892.
Hussey, C. *The Work of Sir Robert Lorimer*, London, 1931.
Kemp, D W, ed. *Richard Pococke Bishop of Meath: Tours in Scotland 1747, 1750, 1760. From the original MS and drawings in the British Museum* (Scottish History Society, 1st series, Vol 1), Edinburgh, 1887.
Langley, B. *New Principles of Gardening or the Laying Out and Planting of Parterres, Groves, Wildernesses, Labyrinths, Avenues, Parks, etc, after a More Grand and Rural Manner than has been Done Before*, London, 1728.
Lawson, W. *A New Orchard and Garden*, 1618, reprinted, London, 1927.
Lindsay, I G, Cosh, M. *Inveraray and the Dukes of Argyll*, Edinburgh, 1973.
Loudon, J C. *Observations on Laying Out Farms in the Scotch Style, Adapted to England*, London, 1812.
MacGibbon, D, Ross, T. *The Castellated and Domestic Architecture of Scotland*, Vol 5, Edinburgh, 1892.
McIntosh, C. *The Book of the Garden*, Edinburgh, 1853–55.
Mackintosh of Borlum, W. *An Essay on Ways and Means for Inclosing, Fallowing, Planting, etc, Scotland: And that in sixteen years at the farthest, by a lover of his country*, Edinburgh, 1729.
Macky, J. *A Journey in Scotland*, 3 vols, London, 1723.
Maxwell, R. *Select Transactions of the Honourable Society of Improvers in the Knowledge of Agriculture in Scotland*, Edinburgh, 1743.
Mitchell, R. *Plans and Views in Perspective with a Description of Buildings Erected in England and Scotland*, London, 1801.
Nicol, W. *The Scotch Forcing Gardener*, Edinburgh, 1797.
Pennant, T. *A Tour in Scotland and Voyage to the Hebrides, 1772*, Chester, 1774, reprinted Edinburgh, 1998.
Reid, J. *The Scots Gard'ner*, Edinburgh, 1683, reprinted 1988.

Robson, J. *A General View of Agriculture in the County of Argyll*, London, 1794.
Shairp, J C, ed [Wordsworth, Dorothy]. *Recollections of a Tour Made in Scotland, AD 1803*, Edinburgh, 1874.
Skrine, H. *Three Successive Tours in the North of England, to the Lakes, and Great Part of Scotland*, London, 2nd edn, 1795.
Slezer, J. *Theatrum Scotiae*, London, c. 1693.
Triggs, H I. *Formal Gardens in England and Scotland*, London, 1902.
Williams, C A. *Food and Drink in Britain from the Stone Age to Recent Times*, London, 1973.

11 Seasonal and Temporary Dwellings

ROGER LEITCH

Seasonal and temporary dwellings exemplify regional strands of a grass-roots building tradition and their marginality makes it vital that they be surveyed and recorded. A building can be seasonal in the sense that it is associated with a seasonal pattern of work, for example salmon fishing. Seasonal dwellings, as opposed to daytime shelters, provide accommodation at or near the place or area of work, such as deer watchers' bothies. Temporary dwellings imply impermanence, such as a tent which can be dismantled and re-erected elsewhere.

The bothy type of building epitomises seasonal dwellings. A bothy means different things to different people. One definition of a bothy given in the *Concise Scots Dictionary* is 'a rough hut used as temporary accommodation eg by shepherds, salmon-fishers, mountaineers'. James Hogg, the Ettrick Shepherd, reveals that for a time his brother lived at Entertrony, 'a wild and remote sheiling [sic] at the very sources of the Ettrick'.[1] A shepherd's bothy deep into the Fisherfield Forest in Wester Ross is mentioned briefly by Osgood Mackenzie as broken-down and thatched.[2] While remoteness was a governing factor, most shepherds occupied outlying cottages at burn-heads or the top of the glen, and could reach their flocks by walking. Shepherds visiting the formerly inhabited island of Taransay, Harris, used to stay for short periods in the main farmhouse, the only building to be maintained after the inhabitants had left. Nowadays they 'commute' by speedboat from Harris.

'Lambermen' stayed with the resident flock on sheep farms in the Borders. Their line of duty was distinctly seasonal and they also undertook clipping. In places such as the Lammermuirs old railway carriages served as temporary accommodation for the lambermen. More mobile were the sheep drovers and smearing gangs. Drovers stayed at isolated farms and cottages, or slept rough. It was known for sheep being driven along rough roads to have custom-made leather 'shoes' to protect their feet. Far less attention has been paid to the migratory gangs of smearers who moved from farm to farm before the advent of chemical dipping dispensed with their seasonal services. These men probably slept in farm outbuildings.

Figure 11.1 River North Esk, Angus/Kincardineshire; salmon netsmen in their bothy, c.1918. SLA, C 13039.

With salmon-fishers it is necessary to distinguish recreational rod fishermen from the boatsmen and netsmen who earned a seasonal livelihood from commercial salmon fishing. The latter employed a significant number of people in the nineteenth century, with ancillary activities such as coopering, net and rope making and boatbuilding. On the Tay and Earn, fishermen entered into a verbal bargains to work for the forthcoming season. Similar to those of farm servants, bargains were sealed by the payment of arles or earnest money.

The crews were housed in bothies found up and down the coast of Scotland, with higher concentrations near the mouths of the great salmon rivers, along estuaries and sometimes relatively far up-river (fig. 11.1). In the case of the Tay, there was a bothy as far up-river as Stanley, Perthshire. Bothies along the Tay were formerly called 'lodges'; on the Tweed they were referred to as 'shiels'. The city of Perth had its own lodges which tended to be of a better description than others. Certain shiels on the Tweed, which can be seen to this day, have a fording box (or look-out post for spotting runs of salmon) on a gable-end, reached by ladder.

A report of 1889 condemned the appalling state of the Tay and Earn lodges.[3] There was insufficient air space, a lack of privies, dumping of rubbish immediately outside, and a prevalence of earthen floors. Many bothies at this time were rat-infested dens. Those on the Earn were simply excavated into the riverbank and roofed with bunches of reeds. Anything of an edible nature had to be stored in glass jars and it was known for rats to encroach upon the bothy table – more likely an upturned fish-box – when the men were having their 'piece'. Time and tide dictated shiftwork, and formerly the men worked long hours in icy and damp conditions which prevailed during the early part of the fishing season from February onwards. Indeed, it was not unknown for the river ice to have to be broken in order to launch a coble.

Furnishings in the lodges were minimal: a form or bench to sit on, the ubiquitous heavy cast-iron kettle that was constantly by the fire, and double iron bedsteads on which the men slept in pairs (latterly replaced by bunk beds). Employers might supply a tyke or ticking mattress which needed replenishing with chaff at intervals in order to ensure a more comfortable rest. This the fishermen did by visiting a nearby farm, at the same time as the travelling threshing mill, where they obtained the necessary 'caff' (chaff) and straw for their tykes. The tedious pattern of work and rest was sometimes broken by a few tunes from a button-key accordion, perhaps the sole concession to light entertainment in a world without mains electricity and piped water.

A number of lodges in various stages of neglect and disrepair can still be visited near Newburgh, and extending eastwards along the Fife side of the Tay estuary to Birkhill. Generally their front elevation is perilously close to the water, which during flood tides can lap up to the door. Sometimes two different generations of bothy are side by side, the older type being an overgrown ruin. Most of those standing are rubble-built, two-roomed structures dating from the nineteenth century. Lower Taes lodge near Flisk is symmetrical in plan and elevations, and its stonework is of a particularly high order for a building of this type. There are impressive whinstone quoins at the corners and stugged sandstone rybats at the windows. As with other lodges in this vicinity, it seems probable that it was constructed by accomplished stonemasons working from a pattern book.

The surviving bothy and adjoining remains at Flisk Point infer three stages of development, available building space being limited by the physical nature of the site. The bothy is believed at one time to have been occupied by river watchers.[4]

It would be a major undertaking to locate and survey every surviving salmon bothy in Scotland. Those on the east coast that have been inspected by the author suggest that, like their river brethren, they are mostly rubble-built and date from the nineteenth century. In size, shape and form, there is a marked diversity. There are examples, as at Usan on the Angus coast, which are highly individualistic. Usan is a two-storeyed structure with bothy accommodation above. It has undergone change of use to a dwelling house and according to the householder, the external walls are nearly three feet thick.

In summary, it can be stated that salmon bothies are distinctly

Figure 11.2 Badentarbat, Achiltibuie, Ross-shire; former salmon-fishing bothy. Copyright: Roger Leitch.

vernacular in that they retain indigenous regional characteristics as well as having a specific function (fig. 11.2). When not in use they were often used as stores for nets and other gear. Redcastle, Angus, now a dwelling house, had a net loft. This bothy housed fly-net men and bag-net fishers who worked different shifts. The old bothy at Reiff, in Coigach, Ross-shire, retains a square opening at the foot of a gable through which net poles could be slotted for storage. Reiff had an open peat fire that lay bare on the floor, smoke issuing out through a hole in the roof.

Thatched roofs were very much a feature of west-coast salmon bothies in areas like Coigach and Skye. These were high-maintenance roofs and were gradually supplanted by materials such as corrugated iron and tarred felt. Corrugated iron is often perceived as being modern, when in fact it was first patented in 1828. It had early building application in relation to shooting lodges (1830s) and farm buildings (1850s).[5]

Pantiled roofs were characteristic of vernacular buildings in Fife, including salmon bothies (Leven, Pitmillie, Balmerino). Pantiles were cheaper than slates, but had poor insulation qualities and were often dislodged during gales. The salmon bothy at Pitmillie, near the village of Boarhills, was surveyed by Bruce Walker and the author in 1989, when its red pantiled roof was still intact. A feature of the roof was the incorporation of a transparent pantile which afforded light.

Mountaineers' bothies or climbing huts take us into the area of recreational use and recreation itself. Apart from one or two established mountaineering clubs who had club huts earlier in their histories, these tend to be more recent features of the hillwalking scene. Club huts can be élitist and difficult to access by non-members, while some are referred to jocularly as fronts for nocturnal shebeens. Their periodic habitation and relative isolation give rise to security problems. If a house of God in central Glasgow can be stripped of its lead, then the slates of a country cottage leased to the Eight Miles High Mountaineering Club will not deter the agile hut-breaker.

Recreational mountaineering along organised lines was formerly the preserve of the leisured Victorian middle and upper classes, as exemplified by the ranks of the Scottish Mountaineering Club and, to a lesser degree, the Cairngorm Club. Early journals of the SMC reflect a world of 'exploration' in tweed jackets, collar and tie, hobnailed boots and dubious claims of temperance. It is worth noting that the first two munroists were both clergymen – the Rev A E Robertson, also an excellent photographer (1889–1901), and the Rev A R G Burn, the first 'compleat' munroist (1914–23). At that time ministers were well remunerated, and enjoyed high status as well as long holidays. The hill activities of both Robertson and Burn are discussed in the context of the times in two fascinating books.[6] Actual mountaineering bothies were not a feature of the glens at that time and tents were heavy and unsophisticated. Our two gentlemen of the cloth availed themselves of Highland hospitality in the homes of keepers and shepherds, where they were put up for the night. Burn's diaries in particular capture a forgotten way of life in the glens.

By a strange course of events, some houses that were formerly occupied by hillmen, as well as abandoned crofthouses and deer watchers' bothies, now serve as unlocked hill bothies for today's mixed breed of walker and climber. It is very different from the 1930s. Then, the construction of massive hydro-electric dams was in its infancy and conifers had not yet blanketed many hillsides. It was the era of the Depression, when unemployment in Britain in 1933 reached 'the appalling figure of 2,400,000, plus 300,000 on poor relief'.[7] A good many working-class people from the towns and cities found an escape in the countryside, where the Scottish Youth Hostels Association introduced thousands to the pleasures of cycling and hiking. Some formed climbing clubs, which very much did their own thing, as Jock Nimlin neatly puts into context:

> We were a bit harum-scarum, I suppose, delighting in caves where we slept like tramps, taking pride in hard living, but being as comfortable as circumstances allowed. Not for us the kind of equipment people take for granted today. These were weekends of simplicity I would not have missed. Climbing was a healthy outlet.

The early [19]30s brought hard times, unemployment or dead-end jobs. Just to have a job was happiness. We climbed for adventure because we needed it.[8]

Affleck Gray, author of *The Big Grey Man of Ben Macdhui* (Aberdeen, 1970), recounted to the author that in his earlier days of winter walking in the Cairngorms he employed a wartime gas mask as protection from spindrift.

Car ownership was relatively rare until the 1960s, there being petrol rationing until May 1950. Furthermore, class exclusivity saw certain sporting estates zealously safeguard their privacy by turning back walkers. A controversial juncture in the story of hill bothies occurred in 1979 with the publication of the Butterfield survey.[9] Irvine Butterfield had set out to record any structure, intact or ruinous, which might afford emergency shelter in the Scottish Highlands. Butterfield visited some 400 such sites, listing them with short descriptions even when he deemed them unsuitable. The range was bizarre, including such gems as a ganger's hut on the Highland railway line, Redpoint fishing station south west of Gairloch, Ross-shire, and historic Garvamore, Inverness-shire (a one-time barracks for redcoats and former drovers' inn).

Butterfield effectively revealed the exact locations of many 'bothies' by providing fool-proof maps and six-figure grid references. To the more hard-nosed bothy enthusiasts, who well knew the whereabouts of their favourite bothies and howfs, this was sacrilege, and they easily foresaw an end to their previously-enjoyed seclusion. Landowners, on the other hand, were none too pleased to see their less salubrious properties touted in a guide to the dosshouses of Gaeldom, which could, for example, prove to be a boon to poachers. Some landowners' displeasure went well beyond the vehement lecture received by Iain Smart when he inadvertently disturbed an eight-hour stalk on one of the Drumochter hills (on the border between Perthshire and Inverness-shire). Some buildings were literally blown up. Others were unroofed or padlocked in a vain attempt to thwart access.

Now even isolated bothies are under pressure from 'munro-baggers'. It was very different for others, however. When the remarkable Syd Scroggie set out to discover the Clova hills, Angus, in the 1930s, he cycled there and his equipment came from ex-army stockists. Gore-Tex was unknown and an old army cape groundsheet sufficed. Portable paraffin cooking stoves were used. As Syd explained: 'I wasn't really a munro-bagger or anything like that. I just liked to disappear into the hills with some pals, preferably for a week and just float around – live in bothies and live rough.'[10] In his lively and engaging tribute to the Cairngorms, Scroggie writes that the stark environs of the former Sinclair hut and Etchachan hut were 'about as loveable as a frigid wife'.[11]

There is nothing remotely romantic about some of the meaner bothies under fireless winter conditions. Bothies used in the mountains tend to be at lower levels, down in the glens. To an outsider they may indeed conjure up words such as 'impoverished' and 'spartan'. In a sense they are, but bothy living is about a quality of life, an experience of being. It is also about being in the midst of a natural 'architecture of hill and sky' and experiencing a deeper 'harmony' as the late W H Murray described in his writings.[12]

Certain shelters have personal names. Davy's Boorach on Jock's Road, not far from upper Glen Doll, Angus, is one example, and is something of a landmark, the drystone pile with its flat corrugated-iron and turf roof being easily made out in summer.

There are still exponents of fireside storytelling to be found in bothies, howfs and bunkhouses. Tales may not always be along traditional lines but they are part of the wider, increasingly modern hill culture. In this and other respects, hill bothies are living buildings.

As with 'bothy', the word 'shieling' conjures up different meanings for different people. The popular use of 'shieling', which properly means the grazing area, relates to the actual shieling hut or bothan àiridh. It is often viewed in relation to the former movement of cattle to the upland (or moorland) pastures. The hut accommodated the herdsfolk, often young women, and could be built of turf or stone. Corbelled stone shielings with round or oval floor plans and one door are believed to be the oldest. While a good deal has been written about the shieling system, far less has been published about the actual construction of the buildings.[13]

There also existed shielings connected with fishing, kelping, and peat-flitting. In the days before decked and motorised boats, fishermen used different types of temporary accommodation in order to be nearer fishing grounds. Lobster fishermen (fig. 11.3) used huts on peninsulas at various locations and shielings in different parts of Uist and in outliers like Heisker or the Monach Isles. On Heisker the fishing shielings were roofed with marram grass or muran, which was abundant on the dunes and used as thatch on indigenous island homes. There was no peat on Heisker, so the fishermen had to flit this over by boat from North Uist. Latterly, they took to sleeping in the abandoned schoolhouse.

At the extremities of the island groups which comprise Shetland were established fishing stations in connection with the far-haaf deep-sea or great-line fishery, prosecuted from open boats known as sixareens (fig. 11.4). On return from the distant grounds the crews were accommodated in 'lodges', with bed space, living area and fire. Building materials would have been whatever was to hand. As in Uist, timber was a precious commodity, and there was an organised reliance on what the sea cast ashore in the form of driftwood. There is some evidence that the lodges were unroofed at the end of the season and the wooden laths taken

Figure 11.3 The Lag, Monreith Bay, Wigtownshire; lobster fishermen's hut. Copyright: Roger Leitch.

Figure 11.4 Stenness, Eshaness, Shetland; fishing station lodges with haaf fisherman and sixareen boats, c. 1880. By permission of Shetland Museum.

home.[14] From about the 1880s, tarred felt was overtaking thatch in Shetland and the use of this percolated through to some of the lodges.

To say that the living conditions of kelpers were a social injustice is an understatement. They were a source of concern neither to the landlords nor to their ground officers who acted as superintendents. The best weed was often found on rocky islets or remote islands and kelp huts are sometimes marked on early charts.[15] The huts in these areas were wretched hovels where the people sustained themselves on oatmeal and water and whatever shellfish they could obtain from rockpools. On the machair land near Bornish in South Uist can be seen the impressions of over 20 kelp huts. These were dug out of the blown sand, their internal walls and entrance being faced with rounded shore stones, and built up with turf. The roofs were of sods over driftwood. It is of interest that they were regarded as common property. As well as kelp-burners, they were also used by collectors of tangle and wrack, latterly migrant lobster-fishermen, and those who gathered winkles from the shore.

On Fetlar, one of the Shetland islands, peat-flitters stayed in rough temporary 'peat-hooses', ably described by Fenton and Laurenson,[16] and complemented from a wider social angle by Bruford.[17] Before starting to import household coal, the people of the island of Berneray (Harris) – an island which has no peat – used to flit peats by boat from satellite islets such as Hermetray and Stromay. The peat-flitters stayed in turf shielings which they built themselves and roofed with a lug sail stretched over driftwood or the oars from their boat. It was impossible to stand upright inside them.[18]

According to Christian Watt,[19] migrant fisher girls lived in sod-built bothies round the coast of Skye, where they acted as cooks to their menfolk who were at the herring fishing. Girls from Findochty, Banffshire, ventured as far as the island of Vatersay, Barra, where they looked after the men. These girls were accommodated in a wooden hut and were in turn watched over by an upstanding figure of maternal authority who was known as 'The Mistress'. This was in the 1860s.[20]

Purpose-built gutters' housing was a feature of the seasonal influx to Shetland for herring fishing. Single-storeyed wooden and corrugated-iron huts in rows were provided by the fish curers for the women. The last vestiges of such accommodation in Shetland have been explored by Walker and MacGregor.[21] The west coast (fig. 11.5) and Shetland stations differed from those on the east in that the curers had to build their own wharfs or landing stages, and generally also temporary accommodation for the workers. Orkney experienced a far shorter herring boom than Shetland. Isa Ritchie (b 1896) explained to the writer in 1991 how she and other girls from the village of Whitehills regularly went to Orkney to work as gutters. Here she describes living conditions in the huts at Stronsay:

Figure 11.5 Ullapool, Ross-shire; herring gutters' hut. Courtesy of Heather Campbell.

I mind the hut at Stronsay was ready for us. It was an old one and we'd to paper the walls with wallpaper we bought in a wee shop at Papa Stronsay. Again, it was just a wooden hut with wooden floor and the felt roof. We'd to scrub out the beds and floor, and clean the fireside: not everything was done for us. We had a hundredweight bag of coal a week to cook our food, but it was poor fires. We took with us a kettle, frying pan, big pot for boiling our dinner, and the curer gave us a big washing pot to wash our clothes. We also took our own dishes ... bed linen ... a frock for tidying ourselves and a Sunday frock, a skirt for putting on to work and nearly always a woollen jumper even though it was summer. We nearly always wore headscarves.[22]

Behind the self-effacement of Isa Ritchie and her generation was a depth of spirit to look ahead, not back. There was a certain reluctance to talk about the old days, bad or good. It was an attitude of mind which spoke louder than any words.

To quote the late Alan Bruford 'ethnological fieldwork can often be exciting, but though it may be pleasant to remember it is usually exhausting to do.'[23] Such fieldwork generally involves contact with people, and it can be easily forgotten that 'informants' are individuals, real people, who come from very different backgrounds and whose learning has not come from books. Some historians may derive their knowledge from collections in archives, libraries or museums, and look no further. Those studying old buildings may never see them used as they were intended.

ACKNOWLEDGEMENT

The author is grateful to Iain Smart for discussing aspects of the text, but the angle of approach rests with the author.

NOTES

1. Mack, 1995, 5, The editor glosses 'shieling' as 'a hut for shepherds on high ground'.
2. Mackenzie, 1995, 120.
3. NAS, AF56/1422, Report by Alexander Carmichael, 'Fishing Lodges on the Tay and Earn', 11 May 1889.
4. Ratcliffe, 1989, 31.
5. Verbal communication from Dr Bruce Walker, formerly School of Architecture, University of Dundee.
6. Drummond, Mitchell, 1993; Allan, 1995.
7. Johnston, 1952, 93.
8. Quoted in Weir, 1994, 103.
9. Butterfield, 1979.
10. RWL/141 (personal tape reference). Syd Scroggie recorded by R Leitch at Kirkton of Strathmartine, Angus, on 31 May 1997.
11. Scroggie, 1989, 12.
12. Murray, 1946.
13. In particular see Thomas, 1857–60; Thomas, 1866–68, 153–95; Campbell, 1943–44; Fenton, 1976, 124–46; and Bil, 1990.
14. Goodlad, 1971, 110.
15. I am grateful to Professor Emeritus J B Caird for drawing my attention to those marked on Admiralty Chart 2825 (1861), Lochs Uist and Maddy.
16. Fenton, Laurenson, 1964.
17. Bruford, 1993–98.
18. Verbal communication from D MacKillop (b 1926), Isle of Berneray (Harris), on 27 June 1996.
19. Fraser, 1988, 27.
20. School of Scottish Studies SA1983/123, Mary Murray recorded by R Leitch at Cellardyke, Fife.
21. Walker, McGregor, 1999.
22. Isa Ritchie (b 1896), former fisher girl, interviewed by R Leitch at Whitehills, BNF, on 22 May 1991.
23. Bruford, 1993–98, 50.

BIBLIOGRAPHY

Allan, E. *Burn on the Hill: The story of the first 'compleat munroist'*, Beauly, 1995.
Bil, A. *The Shieling, 1600–1840: The case of the central Scottish Highlands*, Edinburgh, 1990.

Borthwick, A. *Always A Little Further*, Stirling, 1939, new edn, 1947.
Brown, D, Mitchell, I. *Mountain Days and Bothy Nights*, Barr, 1987, reprinted 1990.
Bruford, A. Flitting peats in North Yell, *Scottish Studies*, 32 (1993–98), 50–69.
Butterfield, I. *A Survey of Shelters in Remote Areas of the Scottish Highlands*, Perth, 1979.
Campbell, Å. Kelkisk och Nordisk kultur ì møte på Hebriderna (The meeting of Celtic and Nordic culture in the Hebrides), *Folk-Liv*, 7 (1943–44), 228–52.
Drummond, P, Mitchell, I. *The First Munroist: The Reverend A E Robertson, his life, munros and photographs*, npp, 1993.
Fenton, A, Laurenson, J J. Peat in Fetlar, *Folk Life*, 2 (1964), 3–26.
Fenton, A. *Scottish Country Life* (1976), East Linton, 1999.
Fenton, A, Walker, B. *The Rural Architecture of Scotland*, Edinburgh, 1981.
Fraser, D, ed. *The Christian Watt Papers*, Aberdeen, 2nd edn, 1988.
Goodlad, C A. *Shetland Fishing Saga*, Lerwick, 1971.
Hogg, J, see Mack, D S, 1995.
Ives, E. *D Joe Scott: The woodsman-songmaker*, Illinois, 1978.
Johnston, T. *Memories*, London, 1952.
Leitch, R. The Tay and Earn salmon fisheries, *Folk Life*, 37 (1998–99), 7–32.
Mackenzie, H O. *A Hundred Years in the Highlands*, London, 1921, new edn, Edinburgh, 1995.
Mack, D S, ed [Hogg, James]. *The Shepherd's Calendar*, Edinburgh, 1995.
Murray, W H. The evidence of things not seen, *Scottish Mountaineering Club Journal*, XXIII (April 1946), 137.
Murray, W H. *The Companion Guide to the West Highlands of Scotland*, London, 4th edn, 1970.
Ratcliffe, A. 'Salmon Fishers' bothies on the Tay Estuary', BArch dissertation, Duncan of Jordanstone College of Art and Design, University of Dundee, 1989.
Scroggie, S. *The Cairngorms Scene and Unseen*, Edinburgh, 1989.
Thomas, F W L. Notice of beehive houses in Harris and Lewis, *PSAS*, 3 (1857–60), 127–44.
Thomas, F W L. On the primitive dwellings and hypogea of the Outer Hebrides, *PSAS*, 7 (1866–68), 153–95.
Thomson, I R. *Isolation Shepherd*, Inverness, 1983.
School of Scottish Studies, University of Edinburgh. *Tocher: Tales, songs and traditions from the archives of the School of Scottish Studies*, Edinburgh, 1971–.
Vernacular Building, Vol 1–. Annual journal of the Scottish Vernacular Buildings Working Group, Dundee and Edinburgh (1975–).
Walker, B, McGregor, C. Herring gutters' bothies in Shetland, *Vernacular Building*, 23 (1999), 30–46.
Weir, T. *Weir's World: An autobiography of sorts*, Edinburgh, 1994.

12 Housing for Seasonal Agricultural Workers

HEATHER HOLMES

This chapter examines the housing of seasonal agricultural workers in lowland farming communities during the nineteenth and twentieth centuries. Seasonal workers were employed on farms to provide a supply of labour at peak periods of demand in the calendar when the work could not be fully undertaken by the regular farm labour force. Generally, on the majority of farms, their employment lasted for only a short period, from a few days to a number of weeks. But in some cases it lasted for much of the period from sowing to harvesting time. Such workers were recruited from a number of sources. A large proportion came from villages and towns, and were transported daily to the farms where they undertook their work. These individuals did not require any housing, and are therefore not the subject of discussion here. However, as such workers could not fulfil all labour demands, recourse had to be made to other types of seasonal workers. These were usually migrant workers brought specially into an area. Their employment patterns determined that they had to be supplied with housing.[1] Indeed, so important was their housing that it formed a major aspect of their employment conditions and was also a perquisite within their remuneration.[2]

THE WORKERS AND THEIR HOUSING

In order to examine the housing of seasonal workers, it is necessary to look at the workers and their employment patterns.

Girls and young women from the Highlands were employed in small groups to harvest the grain crop, especially during the second half of the nineteenth century; by the first decade of the twentieth century this source had greatly declined. The exact geographical area of their employment is not known.[3]

Donegal workers were males comprising family members and neighbours from Counties Donegal, Armagh and Down in Ireland, who travelled as individuals or small groups to undertake general agricultural activities between May and November or December.[4] They were employed over a wide geographical area, which included northern England and across counties in southern Scotland, as far north as Stirlingshire, the Lothians, Fife, Perthshire and Angus.[5] During the early twentieth century they were 'estimated to approximate 3,000 workers'.[6] They had a variety of employment patterns. Some remained on one farm throughout the duration of their migration, undertaking a range of tasks; others were employed on a number of farms and did only one or a few tasks at each.[7] Their employment ceased shortly after World War II, largely as a result of the mechanisation of farm tasks.

Achill workers were women, men and teenagers, usually family members and neighbours from Counties Mayo, Donegal and Galway in Ireland, who were employed in large groups or squads of between 20 to 30 people specifically to harvest the potato crop between mid June and late October or November. A few individuals remained throughout the winter and spring to dress or grade the crop.[8] They were recruited by gaffers and employed by potato merchants located throughout central Scotland. Their employment took place over a wide area of south-western and central Scotland, from Ayrshire and Wigtownshire in the south west to East Lothian in the east and Fife, Perthshire and Angus further north. Their work was essentially peripatetic in nature: squads moved from farm to farm every few days or weeks during the course of their employment; the longest period of stay was around six weeks. At the start of the twentieth century, and indeed throughout a substantial proportion of it, between 1,500 and 2,000 workers were employed annually at this work.[9] Their employment declined in the years following World War II, but did not die out entirely until the 1980s.

Women, men and teenagers from the same recruiting districts as the Achill workers were organised into squads to undertake a range of general agricultural activities such as hay-making and harvesting the grain, berry and potato crops. Like the Donegal workers, they were employed by farmers, but similar to the Achill workers, were organised by gaffers who acted as labour contractors.[10] They moved from farm to farm, and were thus accommodated at several establishments during their employment. The extent of their migration, and its duration, is not known, although it certainly continued into the early twentieth century.

Workers employed on government schemes during periods of national crisis, such as the two World Wars, undertook harvesting work on the fruit, grain and potato crops. During World War II and the years immediately following, labourers included school children, blackcoat workers, prisoners of war, and European voluntary workers.[11]

Housing was thus supplied for a number of groups of workers, organised into differing sizes of group and social, age and sex structures. They were employed across a wide area of lowland Scotland, most groups continuing throughout the entire period under discussion. For others, the work continued for only a few years. Their accommodation had to be suited to short-term residence, in some cases for only a few days. As a result, many of the workers experienced a range of housing and housing conditions during the course of their work.

ELEMENTS OF PROVISION

The provision of housing for these groups of workers had a number of common elements. It was generally provided by the farmers on whose farms the workers were employed. Where workers were employed by gaffers and labour contractors, these supplied accommodation in partnership with the farmer. In the case of the Achill workers, the farmer provided the building and facilities, and the potato merchant supervised the workers during their stay on a farm.

Housing was located close to the workplace, usually on the farms where workers were employed, and within buildings at the farm steading, or immediately set apart from it. Whilst this supplied the greater part of the accommodation, other provision was also made. This was within buildings located beyond the farms. Where farms were near villages, a place of lodging could be found there. Harvestmen – Donegal workers – were recorded in the common lodging house at Dunbar, East Lothian, in the 1930s.[12] During the 1920s, one squad of Achill workers in the Stirling area was accommodated in a disused woollen mill, and in the following decade squads were housed in dwelling houses in villages and towns, as at Kirkintilloch, Dunbartonshire.[13] Such houses, along with disused farm servants' houses, became increasingly popular for accommodating these types of workers, especially in the eastern counties. Noted by the Caithness Committee of 1936 for casual workers in the Lothians,[14] they were widely utilised for Achill workers in that area by the 1960s.[15]

On the farms, and at their steadings, workers were accommodated in a range of buildings. The term 'bothy', 'any primitive dwelling or shelter of any kind', was generally applied to these dwelling places.[16] Although this term is widely associated with the housing of unmarried male farm servants in Perthshire, Angus and Kincardineshire [17] where the workers slept, cooked and ate as a single-sex group in a self-contained structure, not all farms had one of these specialised buildings. Even if bothies were present on farms where seasonal workers were accommodated, they might only be large enough for a small number of workers. They were, however, well suited to housing small groups such as Donegal workers or Highland women; in the 1860s, one bothy in the Carse of Gowrie accommodated six Highland women [18] and during the 1930s, at Pilmuir, Midlothian, the bothy held three Irish males.[19] When Achill workers were housed on a farm where there was a bothy, some of the squad members might stay in it. In mixed-sex squads these might comprise all the members of one sex. However, as not all could be housed, and as not all farms had such a building, alternatives often had to be found, and other buildings, all of which were specially cleaned out, were converted for use. In 1917 the Ballantyne Commission records a great diversity. For the Achill workers, they include 'lofts, barns, potato houses, byres, loose boxes, and other outhouses.'[20] Similarities are noted between these buildings and those for berry pickers.[21]

The use of these buildings generally reflected the type of farm and the broad farming regions where they were located. So distinct was the pattern of provision that it was noted from the 1890s and well through the twentieth century.[22] In Ayrshire, especially in the early potato districts and dairying areas, byres were frequently recorded,[23] and in Dunbartonshire, another mainly pastoral county, there was also a general predominance of buildings associated with livestock husbandry, with byres and stables being common. In eastern arable districts in 1907 the predominant type included 'farm houses, cottages, outhouses, granaries, barns &c.'[24] Specialised buildings, such as potato houses, used in conjunction with the sprouting of seed potatoes, were noted in early potato growing districts such as Ayrshire.[25]

While there was an emphasis on the use of existing buildings, some were specially erected for the sole purpose of housing seasonal workers. In 1917, these were of various types. For the Achill workers and berry pickers some were permanent structures; others were of a temporary nature, such as tents, removed at the end of their period of attendance.[26]

FACILITIES

The wide range of buildings, generally providing makeshift accommodation, had a number of facilities. These were basic, and centred on what was available at the farm steading or could easily be made available. Washing facilities were located at the farm pump or in the open air; fires for cooking and heating were made in fireplaces in buildings, grates on the floors of sheds or in the open.[27] Furnishings were seldom provided, although they could include tables and chairs. Beds and bedsteads were not usually supplied, and workers made makeshift arrangements.[28] Until the 1920s, Achill workers lay on straw strewn across a floor; they also had straw palliasses, and might use potato boxes to raise themselves off the floor, a provision that was noted until at least the 1960s.[29]

SUITABILITY OF THE BUILDINGS

The accommodation for the seasonal workers was not always regarded as suitable. For the early nineteenth century, Hugh Miller records that buildings were not always fit for human habitation.[30] However, similar comments could apply equally to general housing in the later nineteenth century and the first decades of the twentieth century, though by the latter date some excellent housing did exist.

The wide range of buildings had varying levels of suitability as accommodation. In 1907 the Medical Officer of Health for Renfrewshire remarked on this phenomenon: 'housing in loose-boxes should not be tolerated as they were usually dark and the floor of wooden boards, saturated with urine and foeculent matter.'[31] Housing in young beasts' byres was 'usually excellent' as they had been cleared out some months before use. Byres of 'old construction' were 'unsatisfactory'. Lofts made 'a fairly satisfactory dormitory as a rule'. Bothies, 'of course furnish suitable accommodation but from their limited size there was some tendency to overcrowding'.[32] Building materials also had an effect on the suitability of premises. Some were more satisfactory than others. In Ayrshire in 1913 Dr Elizabeth McVail notes that 'barns and potato houses, if provided with wooden floors, form fairly satisfactory sleeping places, but others, such as byres, are the very reverse. Byres have a cold and cheerless aspect and are often badly lit. They also have stone floors.'[33]

The nature of the buildings caused a number of problems for their seasonal occupants, especially during the nineteenth and early twentieth centuries. Although considered suitable for healthy persons, the buildings could exacerbate the spread of disease among the workers.[34] With mixed-sex squads, adequate provision was not always made to provide each sex with its own accommodation. There was a demand that such arrangements should be made as a necessity of decent living.[35]

CRITICISM AND IMPROVEMENT OF THE HOUSING

There exists a long tradition of criticism of the housing for seasonal workers, one also expressed in relation to the bothies of unmarried farm servants. For the seasonal workers it had its greatest emphasis from the 1890s and throughout much of the twentieth century.[36] From the 1890s, Medical Officers of Health in local authorities, especially Dunbartonshire and Stirlingshire, started to take an interest in the housing for potato workers.[37] By the following decade, that level of awareness had extended.

Figure 12.1 67 Eastside, Kirkintilloch, Dunbartonshire; interior of women's sleeping accommodation, 1937. Courtesy of the NAS, DD13/227.

In 1907 the Local Government Board for Scotland took action over housing conditions of potato workers and berry pickers; that of another group, herring gutters, received attention by the Home Office in 1906.[38] But it was not until 1912 that the housing of a number of seasonal and temporary workers had official attention drawn to it, by the Ballantyne Commission, which reported its findings in 1917.[39]

The findings of the Ballantyne Commission were important in securing better housing conditions for a number of groups of seasonal and temporary workers. These laid the basis for the 1919 Housing, Town Planning, etc (Scotland) Bill which received its Royal Assent in August that year. Under Section 45 the housing for 'navvies, harvesters, potato-workers, fruit-pickers, herring-gutters, and such other workers engaged in work of a temporary nature as the Board may from time to time prescribe' was regulated so that a legal minimum standard was provided; penalties were made for non-observance.[40] Byelaws, providing for local legislation, were adopted by local authorities between 1920 and 1925.[41] Further legislative revision followed in 1925, 1931 (as a result of a fire at Kilnford, Ayrshire, some years earlier), 1938 (owing to the death of ten workers in a fire at Kirkintilloch; fig. 12.1),[42] 1950, and 1966.[43] Although aspects of the legislation remained constant during these revisions, others changed in a number of respects, especially from 1938.

Until that time, each local authority made byelaws for each category of seasonal worker (agricultural, maritime and industrial), but from then, all groups were brought together and were covered by one set of byelaws applicable to all; their geographical application was extended to cover much of the county local authority areas of Scotland, including many burghs, even ones where there was little likelihood of workers being accommodated, and there was a wider legislative coverage.

The cumulative effect of national and local legislation was to improve the standard of accommodation; this was especially marked in 1966.[44] Not only did legislation allow for suitable and satisfactory buildings to be provided but also for a generally increased level of basic facilities. During the 1960s, facilities were brought into line with those supplied for the settled population and included piped running water, electric lighting and flush toilets.[45] The housing system also altered, especially for the Achill workers. It was not possible for each farmer to provide the new, improved standards and so potato merchants had to supply their own premises, usually cottages, which were utilised as bases from which workers were transported to their work on a daily basis.[46]

However, despite the favourable changes that were made during the twentieth century there remained a number of constant elements in the housing for seasonal workers, even until the second half of the twentieth century. Indeed, Anne O'Dowd asserts, from limited evidence of conditions during the early 1970s, that 'the accommodation given to the potato workers in some areas of Scotland was still basically an outhouse which the cattle and Irish workers used alternately'.[47] Yet that system of housing had survived during the course of the twentieth century because it proved suitable for variously sized groups of workers and, once established, could provide temporary accommodation easily; a convenient and relatively cheap measure for workers who were often employed for very short periods of time.

NOTES

1. For an examination of sources of labour see Collins, 1969 453-73; Collins, 1970, Chapters 1-3; Holmes, 2000(a), 27; Handley, 1947, 174-5; Wilson, 1921, 4.
2. For example, Bell, 1991, 78-9; Holmes, 2000(a), 185-246.
3. Devine, 1979, 349-51.
4. Department of Agriculture and Technical Instruction for Ireland, 1908, 9-10.
5. Inter-Departmental Committee on Seasonal Migration to Great Britain, 1938, 12.
6. Department of Agriculture and Technical Instruction for Ireland, 1906, 15; O'Dowd, 1995.
7. Holmes, 2000(a), 185-7.
8. Department of Agriculture and Technical Instruction for Ireland, 1908, 12-13; Holmes, 2000(a), Chapter 11; Board of Trade (Wilson Fox), 1905, 104.

9. Department of Agriculture and Technical Instruction for Ireland, 1880–1916. For later statistics see Inter-Departmental Committee on Seasonal Migration to Great Britain, 1938.
10. McLellan, 1990, 27–35.
11. Holmes, 2000(a), 27.
12. Holmes, 1995–6, 57–75.
13. ibid.
14. Committee on Farm Workers in Scotland, 1936.
15. Holmes, 2000(a), 238–9.
16. Scottish National Dictionary, 1941, 223, sv 'bothy'.
17. Fenton, 1984, 191.
18. Commission on Agricultural Labour, 1870, Mr G Culley's report, 51, para 18.
19. Oral evidence, Robert Holmes, Pilmuir, Balerno, Midlothian, January 2000.
20. Royal Commission on the Industrial Population of Scotland, Rural and Urban, 1917, 191, paragraph 1263.
21. ibid, 1917, 196, paragraph 1299.
22. For Ayrshire in the 1970s see O'Dowd, 1991, 202.
23. O'Ciarain, 1991, 78; MacGill, 1985, 75; MacGill, 1986, 139.
24. NAS. Department of Agriculture of Scotland, AF59/62, III, Housing of Potato Diggers, report for Haddingtonshire.
25. Royal Commission on the Industrial Population of Scotland, Rural and Urban, 1917, 191.
26. ibid, 196.
27. ibid, 191, paragraph 1266.
28. O'Ciarain, 1991, 79.
29. Holmes, 2000(a), Chapter 12.
30. Quoted in Royal Commission on the Industrial Population of Scotland, Rural and Urban, 1917, 187.
31. NAS. Department of Agriculture of Scotland, AF59/62, III, Housing of Potato Diggers, Report for Renfrewshire.
32. NAS. Department of Agriculture of Scotland, AF59/62, III, Housing of Potato Diggers, Report for Renfrewshire.
33. Royal Commission on the Industrial Population of Scotland, Rural and Urban, 191, paragraph 1263.
34. Holmes, 1998(b), 2–16.
35. ibid.
36. Holmes, 1998(a); Holmes, 2000(a).
37. Holmes, 1998(b); Department of Agriculture and Technical Instruction for Ireland, 1908; For berry pickers see Local Government Board for Scotland, 1911, lxxviii.
38. Royal Commission on the Industrial Population of Scotland, Rural and Urban, 1917, 199, paragraph 1326.
39. Royal Commission on the Industrial Population of Scotland, Rural and Urban, 1917.
40. Housing, Town Planning, etc (Scotland) Act, 1919, Section 45.
41. Holmes, 2000(a), Chapter 12.
42. Holmes, 1995–6.
43. See Holmes, 2000(a), Chapter 12.
44. Holmes, 1995–6.
45. Holmes, 2000(a), Chapter 12.

46. ibid.
47. O'Dowd, 1991, 202.

BIBLIOGRAPHY

Anthony, R. Herds and Hinds: Farm labour in Lowland Scotland, 1900–1939 (Scottish Historical Review Monograph Series No 3), East Linton, 1997.

Bell, J. Donegal women as migrant workers in Scotland, *ROSC*, 7 (1991), 73–80.

Board of Trade (Wilson Fox). *Second Report by Mr Wilson Fox on the Wages, Earnings and Conditions of Employment of Agricultural Labourers in the United Kingdom*, London, 1905.

Collins, E J T. Harvest technology and labour supply in Britain, 1790–1870, *Economic History Review*, 22 (1969), 453–73.

Collins, E J T. 'Harvest technology and labour supply in Britain, 1790–1870', unpublished PhD thesis, University of Nottingham, 1970.

Commission of Agricultural Labour. *Royal Commission on the Employment of Children, Young Persons and Women: Fourth report with appendix*, Parts I and II, London, 1870.

Committee on Farm Workers in Scotland. *Report of the Committee on Farm Workers in Scotland*, 1935–6, Edinburgh, 1936.

Department of Agriculture and Technical Instruction for Ireland. *Agricultural Statistics, Ireland. Reports and tables relating to migratory agricultural labourers* (Annual Reports), Dublin, 1880–1916.

Department of Agriculture and Technical Instruction for Ireland. *Agricultural Statistics, Ireland, 1905. Report and tales relating to Irish agricultural labourers*, Dublin, 1906.

Department of Agriculture and Technical Instruction for Ireland. *Agricultural Statistics, Ireland, 1907–8. Report and tables relating to Irish agricultural labourers*, Dublin, 1908.

Devine, T M. Temporary migration and the Scottish Highlands in the nineteenth century, *Economic History Review*, 32 (1979), 344–59.

Devine, T M, ed. *Farm Servants and Labour in Lowland Scotland, 1770–1914*, Edinburgh, 1984.

Fenton, A. The housing of agricultural workers in the nineteenth century. In Devine, 1984, 182–213.

Handley, T E. *The Irish in Modern Scotland*, Cork, 1947.

Holmes, H. The Kirkintilloch bothy fire tragedy of September 16, 1937: An examination of the incident and the resulting legislation, *ROSC*, 9 (1995–6), 57–75.

Holmes, H. Improving the housing conditions of the Irish migratory potato workers in Scotland: The work of the Bishops' (Gresham) Committee, 1920–1923, *Rural History, Economy, Society, Culture*, 9:1 (1998(a)), 57–74.

Holmes, H. Dr John C McVail and the improvement of the housing of the Irish migratory potato harvesters in Scotland, 1897-c.1913, *Ulster Folklife*, 44 (1998(b)), 2–16.

Holmes, H. Sanitary inspectors and the reform of housing conditions for Irish migratory potato workers in Scotland from the late 1940s to the early 1970s, *Saothar*, 24 (1999), 45–58.

Holmes, H. *'As Good as a Holiday': Potato harvesting in the Lothians, 1870 to the present*, East Linton, 2000(a).

Holmes, H. 'An abject and deplorable existence': Problems faced by Irish women migratory workers in Scotland in the early twentieth century, *Folklife*, 28 (2000(b)), 42–55.

Inter-Departmental Committee on Seasonal Migration to Great Britain. *Report of the*

Inter-Departmental Committee on Seasonal Migration to Great Britain, *1937–8*, Dublin, 1938.
Local Government Board for Scotland. *Sixteenth Annual Report of the Local Government Board for Scotland, 1910*, Edinburgh, 1911.
MacGill, P. *Children of the Dead End*, 1914, reprinted London 1985.
MacGill, P. *The Rat Pit*, 1915, reprinted London, 1986.
McLellan, R. The Donegals. In *Linmill Stories*, Canongate Classics 28, Edinburgh, 1990, 27–35.
Nolan, W, Ronayne, L, Dunlevy, M, eds. *Donegal History and Society: Interdisciplinary essays on the history of an Irish county*, Dublin, 1995.
O'Ciarain, S. *Farewell to Mayo: An emigrant's memoirs of Ireland and Scotland*, Dublin, 1991.
O'Dowd, A. *Spalpeens and Tattie Hokers: History and folklore of the Irish migratory agricultural worker in Ireland and Britain*, Blackrock, 1991.
O'Dowd, A. Seasonal migration to the Lagan and Scotland. In Nolan, Ronayne, Dunlevy, 1995, 625–48.
Royal Commission on the Housing of the Industrial Population of Scotland, Rural and Urban. *Report of The Royal Commission on the Housing of the Industrial Population of Scotland, Rural and Urban*, Edinburgh, 1917.
Scottish National Dictionary [W Grant, ed], Edinburgh, 1931–76.
Wilson, Sir J. *Report to the Board of Agriculture on Farm Workers in Scotland in 1919–20*, London, 1921.

PART THREE
•
Community Buildings

13 Buildings of Administration

CAROLINE A MacGREGOR

From the reign of King David I (1124–53) until the reorganisation of local government in 1975, the main agency of local administration in Scotland was the burgh. Burghs enjoyed special status and economic privileges, including the right to hold markets and levy tolls and customs, and their councils met regularly to discuss such matters. They also had judicial powers and responsibilities, notably the provision of secure prisons. The tolbooth or town house was the building type adopted to serve these needs. 'Head' or county burghs, such as Aberdeen and Kirkcudbright, were regularly used to hold Sheriff Courts, even though some, such as Clackmannan and Stonehaven, Kincardineshire, were not royal burghs. Edinburgh Tolbooth, meanwhile, was used regularly for Royal Courts in the sixteenth century and occasionally for the annual Convention of Royal Burghs. There are over ninety extant pre-1830 tolbooths in Scotland. Of these, only one, Crail, Fife (fig. 13.1), may predate the Reformation. Approximately twenty-five were built before the Union of the Parliaments in 1707. The vast majority, however, were constructed or rebuilt in the eighteenth and nineteenth centuries.

This chapter will focus primarily on the architectural and historical background of tolbooths, but reference will also be made to specialised buildings such as law courts, penal institutions and police stations and boxes. Generally speaking, these purpose-built structures are a product of increasing judicial specialisation in the nineteenth and twentieth centuries. Several general articles have been written on Scottish tolbooths and town houses and information on comparative municipal buildings in England and elsewhere in Europe is also available. Whilst penal institutions and Edinburgh's supreme courts have been amply covered by several articles and books, not much research has been undertaken on police stations and boxes.

The tolbooth or tolloneum was originally the place where customs or tolls were collected. However, by the late sixteenth century it had come to be the place 'quhair courtis may be halden; justice ministrat vpon thame according to thair demerites ...'.[1] The tolbooth was also the venue for Town Council meetings. During the course of the eighteenth century the term 'tolbooth' went out of fashion and was replaced by

Figure 13.1 Crail Tolbooth, Fife. Crown copyright: RCAHMS, SC 335729.

'town house'. A notable exception to this rule is Wick Tolbooth, Caithness (1828); despite its late erection, the act declaring it to be a legal jail specified that it was 'to be called the Tolbooth of Wick in all time coming'.[2] By the late nineteenth century the terms 'town hall', or simply 'public' or 'municipal buildings', were favoured.

Tolbooths are usually situated in prominent positions, often on island sites, close to the market place or cross, or at the junctions of major thoroughfares. In the past, tolbooths built on island sites were particularly prone to demolition, on the grounds that they were obstructions. In 1810, the magistrates of Cupar, Fife, campaigned for the erection of a new town house and jail partly on the grounds that the existing building 'is situated in the middle of the principal street and greatly obstructed the entry and passage'.[3] It was duly replaced by a new jail (1813) and separate town hall (1815–17). Similarly, in 1837–8 the ministers of St Andrews, Fife, complained that the town house was 'very inconveniently situated in the centre of Market Street, in the vicinity of the spot where the cross formerly stood'.[4] In this case, however, the Town Council's plan to rebuild it on a 'convenient piece of ground' had to be postponed, owing to insufficient funds.[5] More recently, the council chamber block of Newton-upon-Ayr, Ayrshire, was demolished in 1967 to permit the construction of a dual carriageway. Most tolbooths, however, have withstood the vicissitudes of time, and many have retained their original sites for centuries. Aberdeen's municipal buildings are a notable example, having reputedly stood on the north side of Castlegate since the late fourteenth century.[6] A plaque on Inverness Steeple (1789) records that it stands on the ancient site of the tolbooth, and the successive municipal buildings of Perth are also said to have occupied their traditional position since medieval times.[7]

The earliest recorded tolbooths in small burghs tended to be simply constructed and were not dissimilar to other town buildings. The majority were made from stone, and some had heather-thatched roofs (for example, Tain Tolbooth, Ross and Cromarty, 1631, demolished 1703). Gradually, however, and especially in the larger, wealthier 'head' burghs, tolbooths began to take on a more substantial and defensive appearance. Many resembled fortified tower-houses in appearance. One of the best examples of this kind is Aberdeen Wardhouse (1615–16), which has vaulted chambers, battlements and narrow window slits covered with iron stanchions. Other examples include Canongate Tolbooth, Edinburgh (1591), Musselburgh, Midlothian (c. 1590), and Crail (possibly 1517 in origin). Kirkcudbright Tolbooth (1623–9) is particularly noteworthy because of the gun-holes discovered on the third storey during recent restoration work. Pittenweem Tolbooth Steeple, Fife, also has shot-holes which may relate to fortifications ordered in 1639.[8] This fortified style of building remained popular in northern Scotland well into the eighteenth

Figure 13.2
Dingwall Tolbooth, Ross and Cromarty; reconstruction drawing. Crown copyright: RCAHMS, SC 716873.

Figure 13.3 Stirling Town House. Crown copyright: RCAHMS, SC 336990.

century, and Tain Tolbooth (1706–33) and Dingwall Tolbooth Tower, Ross and Cromarty (1732–45; fig. 13.2), are particularly good examples. Further south, the classical gradually ousted the native Scots style from the last quarter of the seventeenth century onwards. Stirling Tolbooth (1703–5; fig. 13.3) is hailed as 'one of the first Tolbooths in Scotland to be treated in a strictly Classical manner, the steeple alone conceding something to traditional Scottish taste'.[9] Dumfries Midsteeple followed soon afterwards in 1705–7. Not surprisingly, contemporaries were also aware of this change in architectural fashions. John Macky (an early eighteenth-century traveller) described Linlithgow Town House, West Lothian (c. 1668–78), as 'a very beautiful Piece of Modern Architecture'.[10] Similarly, the Rev Mr Molleson noted in 1790–1 that 'the new Town House [in Montrose, Angus], built in 1763 ... is constructed according to the modern taste'.[11] Forty-nine tolbooths are built in the classical style. Bo'ness Tolbooth, West Lothian (1775, 1857; demolished 1970), was built in a fanciful Gothic style. A few were built in a Castellated Gothic style, for example, Strichen, Aberdeenshire (1816; fig. 13.4), Kinghorn, Fife (1829–30), and Rothesay, Bute (1832–35). Sadly, a delightful rhomboid Gothic design for Huntly, Aberdeenshire (1823–32), attributed to John Buonarotti Papworth, was not built.[12] Duns Town House, Berwickshire (1816, demolished in 1966; fig. 13.5), was designed by James Gillespie Graham in a Perpendicular style. It was not until the later nineteenth century that the distinctive 'Scots Baronial' style regained favour. Examples include Huntly Town Hall (1875; 1886–7), Annan, Dumfriesshire (1875; fig. 13.6), and Hawick, Roxburghshire (1885–7).

Although style was of some importance, tolbooths were essentially functional buildings, and it was this which largely determined their form and appearance. Generally speaking, tolbooths consisted of three elements: a tower/steeple; a council chamber block; and a prison.

As already mentioned, sixteenth- and seventeenth-century tolbooth towers tended to resemble defensive tower-houses. Strength and security were important because the law held magistrates liable for the escape of prisoners. For example, when a prisoner escaped from Brechin Tolbooth, Angus, in 1612 the magistrates were themselves imprisoned. Similarly when a debtor escaped from Tain Tolbooth in 1702, the council became liable for his debts (which amounted to £1,400).[13] More commonly, town councils were fined each time a prisoner escaped. Another reason for the defensive appearance of the earliest tolbooth steeples is that Scotland was still subject to invasions by the English. Leith and Musselburgh Tolbooths were both destroyed in Hertford's invasion in 1544.[14] Linlithgow Tolbooth was also 'razed to the ground by the English at their first coming in 1650', the stones and timber thereof being 'applied towardis the workis and fortifications about the Castile of Lythgow'.[15] Given this threat, it is not surprising that town councils made strenuous

Figure 13.4 Strichen Town House, Aberdeenshire. Crown copyright: RCAHMS, SC 336453.

Figure 13.5 Duns Town Hall, Berwickshire; perspective by James Gillespie Graham, c.1816. By kind permission of Alexander Hay, Duns Castle Estate.

efforts to make their tolbooths as fortress-like and impenetrable as possible.

In addition to containing the burgh prison, tolbooth towers also housed the town's bells and clock. According to MacGibbon and Ross, 'A bell-house and bell were required by Act of Parliament to be provided in every burgh, that the council or the citizens might be summoned by its ringing before the transaction of all business of a civic character'.[16] An interesting reference to this regulation is contained in Kirkcudbright Town Council Minutes in 1642:

BUILDINGS OF ADMINISTRATION • 255

Figure 13.6 Annan Town Hall, Dumfriesshire. Crown copyright: RCAHMS, SC 710969.

and having taken into their consideration the necessity of ane steple and bellhouse to keep their knok and bell quhilk is a special ornament belonging to every burgh, and quhilk they are bound by the ancient laws of this kingdom to maintain and uphold.[17]

A new steeple was built shortly afterwards. This quotation also confirms

Figure 13.7 Falkland Town House, Fife; bell. Crown copyright: RCAHMS, SC 336942.

that steeples were a symbol of burghal status, pride and prestige. Many tolbooths still have their original bells. In addition to summoning councillors to council meetings, bells were also used to warn of invasion, as signals for workers to rise, and curfew; and sometimes also as a summons to church. The vast majority of seventeenth-century bells were cast in the Netherlands. The famous Burgerhuys foundry in Middelburg, Holland, cast bells for several Scottish tolbooths, including Haddington, East Lothian (1604), Canongate (1608), Kirkcudbright (1646), Inverkeithing, Fife (1667), and Falkland, Fife (1650; fig. 13.7). Other bells of Dutch origin include Dunfermline, Fife (which was cast by Henrick Ter Horst in 1654), Stirling (the steeple houses bells cast by C Ouderogge in 1656 and Petrus Hemony in 1669), Glasgow (Jacob Waghevens, 1554), South Queensferry, West Lothian (Adrian Dop, 1694), and Wigtown (Henrick Ter Horst, 1633). Joseph Fitzgerald believes that Dutch bells were gifted to Scottish burghs 'as a token of goodwill – hence the term Knok Hoos [clock house] borrowed from their vocabulary and used in affectionate reference to the Tolbooth'.[18] While this may sometimes be the case, there is also evidence which indicates that some burghs ordered their bells from Holland. For example, Stirling Town Council Minutes for 24 February 1668 contain the following order: 'Ordeans James Brown, theasurer to send to Holland for a new bell to the town knock as big as the tolbuith steeple will receave; the size of the said bell to be sent over'.[19]

The most plausible explanation for the large number of seventeenth-century Dutch bells in Scottish tolbooths may simply be that Scotland did not have very many good quality foundries. Eeles writes that 'There were one or two at Edinburgh, one at Stirling, one at Turriff and one at Aberdeen ...' whose bells are interesting because they are rare and Scottish, rather than for 'their intrinsic merit'. He also observes that 'foreign bells are very scarce in England, where well-nigh every town of importance possessed a foundry where bells were cast'.[20] London-based foundries, notably Thomas Mears, Mears and Stinback, G Mears, and Lester and Pack, supplied a considerable number of bells for tolbooth steeples in the eighteenth and nineteenth centuries. The number of Scottish bells increased too, most by Edinburgh founders such as John Meikle, Robert Maxwell and George Barclay. A few were cast by John Mowat of Old Aberdeen and William Coatts of Glasgow.

Most tolbooths had clocks by the middle of the fifteenth century, a situation which compares well with England and the rest of Europe. Aberdeen is believed to have had one of the first public clocks in Scotland. Aberdeen's council register contains a reference to its being sent to Flanders for repair in 1535 '... and gif it can nocht be mendit, to by thame ane new knok on the tovnis expensis'.[21] As with bell-founding, the Low Countries led the way in horological matters until the seventeenth century. Most of the surviving clocks in tolbooth steeples are nineteenth- and twentieth-century replacements. There are, however, one or two early surviving examples. Musselburgh Tolbooth clock, now on display in the foyer of the building, is traditionally said to have been given to the burgh by the Dutch States in 1496. Kirkcudbright Tolbooth clock (fig. 13.8), now in the Stewartry Museum, is also believed to be of Dutch manufacture and is first mentioned in the Council Minutes in 1576. Both clocks have only one hand. It was thought unnecessary to provide a minute hand because most clocks (if and when they worked) lost and gained a considerable amount of time each day, and in any case contemporary requirements for accuracy were low. This may explain why some tolbooths had sundials, examples of which are still to be found on the exteriors of Aberdeen Municipal Buildings (engraved UT UMBRA SIC FUGIT VITA 'Life is as fleeting as the shadow', transferred to the present building when the eighteenth-century Town Hall was demolished), Canongate Tolbooth, Edinburgh, Crail Tolbooth, Strathmiglo Steeple, Fife, and Dunbar Town House, East Lothian. Hamilton Tolbooth, Lanarkshire, and Glasgow Town House (both now demolished) are also recorded as having sundials.[22] Old Aberdeen Tolbooth, meanwhile, had a globe for telling the moon's age.[23]

The main block of the tolbooth normally contained the accommodation required by the Town Council and magistrates, namely a court room and/or council chamber, a record room, and prison. A militia store, tap-room, wine cellar, weigh/meal-house, market area and school might also be provided. The ground floor 'booths' were sometimes let as shops whose rentals contributed to the Common Good, which was

Figure 13.8 Kirkcudbright Tolbooth; clock (now in Stewartry Musuem). Crown copyright: RCAHMS, SC 710972.

used to maintain the tolbooth. During the late eighteenth and early nineteenth centuries, assembly/ballrooms were sometimes added to existing tolbooths or built into new ones. At Montrose in 1761 a group of gentlemen approached the Town Council with a proposal to build an assembly room in the town.[24] Plans for the new Town House, which included 'a handsome public room with proper waiting and retiring room adjacent', were duly prepared and the building erected between 1762 and 1763.[25] The building is known locally as 'The ba' hoose'. Similarly, Haddington Town Council collaborated with the gentlemen of the county to build an assembly room (with sheriff court room and two record rooms for the town and country below) adjoining the Town House in 1788.[26] At nearby Musselburgh the long room was widened 'to conform to plan made out by James Hay' so as to make a ballroom in 1811.[27] This reflects the fact that Scotland was becoming a more peaceful country. Town-dwellers and local landowners sought to enjoy themselves by holding dance classes, balls and suppers in their local Town House. These additions were usually funded primarily by 'the gentlemen of the county', with assistance from the public funds.

The main block was usually two to three storeys in height, Glasgow Tolbooth being unique in having five. The court room and council chambers were almost always on the first floor and were approached by an external stone forestair and parapet which performed important functions. Because the forestair was clearly visible from the market place, it was used by burgh officials to make proclamations. Kilmarnock Tolbooth (Ayrshire) forestair, for example, is recorded as having been used by the baillies and councillors to drink the king's health on his birthday. Wigtown Council was given a large wooden punch bowl for this purpose by Queen Anne at the time of the Union. The bowl was also intended to signify its status as a royal burgh.[28] The parapet, or flat area at the top of the stairs, was sometimes used as a place of public punishment. Its prominent position made it ideally suited to this task since the aim of punishment at this time was to expose the offender to as much public ignominy and ridicule as possible. Paisley Tolbooth (1757; steeple demolished 1870) had a whipping post, stocks and jougs attached to its platform.[29] Jougs were sometimes attached to the main tolbooth block at ground-floor level. Surviving examples include Kirkcudbright Tolbooth, Kilmaurs Town House, Ayrshire (fig. 13.9), and Ceres Weigh House, Fife. Executions are recorded as having taken place on the platform of Edinburgh Tolbooth from 1785 onwards. This was a special case, not being the entrance area but the roof of an annexe. The victims' heads are known to have been impaled on the tolbooth spire as an additional deterrent.

Surviving interiors and documentary sources bear testimony to the importance and significance town councils attached to decorating and furnishing their council and court chambers. A fascinating insight into the interior furnishings of a late seventeenth-century council chamber and clerk's office can be gleaned from Perth Burgh Records. The information they contain is all the more tantalising given that the council house was demolished in 1839. The records reveal that in May 1696 'thrie dusson good rushie Leather chears for the use of the Counselhouse' and 'ane Large Table for the counselhouse and ane other for the clerks chamber' were to be provided.[30] The

Figure 13.9 Kilmaurs Town House, Ayrshire; jougs. Crown copyright: RCAHMS, SC 335732.

Treasurer's Accounts reveal that the furniture was made by James Reoch, wright. Wainscot and knapwood were used to make the two tables. The chairs were made entirely from wainscot.[31] James Reoch also made 'moullars' or frames for four maps in the council house.[32] Several months later the clerk was instructed 'to wreit to baillie Rolson who is at Edr to indevoar ... to buy ane good fashionable carpit for the counsell table and if not cannot be had at Edinburgh to wreit to Londone to Mr Joseph Anstine to buy one ...'.[33] An artist, Henry Reed, was paid £100 Scots for 'the paynes and expenss of ... painting ... the Chimney piece in the counsell chamber Landsekaip and paynting of ther Bench and chimney in the Clerks office'.[34]

Exceptionally complete information is also available for the decoration of an eighteenth-century council chamber, namely the 'Great Room' of Aberdeen Town House (1750–6; demolished 1871). The Council Registers for the period reveal that the floor was to be laid with 'dales along the length of the room which will greatly add to the beauty thereof'.[35] The Town Council later agreed to finish the room 'with the best firr boxing, and to have a marble chimney from Holland, and that the space above the chimney be finished with the Town's Arms in Stucco work and Ornament', adding that they wanted the forsaid work executed 'in as genteel but easy a manner as possible'.[36] At a subsequent meeting they decided that 'the second draught of the chimneypiece ... containing a landskip was prettiest and therefore appointed the same to be execute'. John Creswell was initially appointed 'to draw a perspective of the southside of the town and harbour with a camera obscura, in order to see how far it might be proper to putt the same in the pannel above the chimney in the Townhall'.[37] In the event, William Mossman was commissioned to paint the 'south prospect of the town and harbour', on the basis 'of a draught thereof in miniature which has been shown to several of the council and approved of as being very exactly done'.[38] Mossman was paid £252 for the painting.[39] A large 'crystall lustre, with tossles, ballances & ca.' was ordered from a Mr Udny in London in 1751.[40] The Guildry Accounts reveal that the total cost, including freight, came to a staggering £768 19s.[41] Notwithstanding this, the council proceeded to order two slightly smaller lustres.[42] Additional lighting was provided by a set of twelve sconces, for which Baillie Fordyce was empowered by the council 'to write to London'.[43] In 1755 Cosmo Alexander, 'limner', was commissioned to paint full-length portraits of the Earl and Countess of Findlater.[44] Sadly, this building was demolished in 1871 to make way for the new municipal buildings. Luckily, the then Provost marked the occasion by presenting the Town Council with a watercolour by A D Longmuir of the interior of the old Town Hall.[45] This remains the only known pictorial representation of this magnificent civic interior.

Fortunately, most of the pictures and furniture, including the lustres and sconces, were preserved.

Apart from these two examples, information regarding tolbooth interiors is fragmentary and tends to relate to particular aspects of interiors such as floor coverings, wall treatments, painted panels, ceilings and furniture.

In the late sixteenth and early seventeenth centuries council and court rooms, in common with superior domestic interiors, were frequently decorated with painted ceilings. Fragments of two open-beam ceilings with different decorative schemes survive in Culross Town House, Fife. The first-floor west ceiling is decorated with thin arabesques on a white background, a relatively common beam pattern of which examples can be seen in Culross Palace and Edinburgh. The iconography of the east ceiling, on the other hand, features cherubs, draped heads, stars and faceted rectangles. The interior of Canongate Tolbooth is embellished with 'a poorly preserved but typical open-beam ceiling with arabesque decoration'.[46] The colours of all three ceilings are now faded by age, dust and smoke; originally they would have been much brighter and more striking. In the eighteenth and early nineteenth centuries decorative plaster ceilings with enriched covering became fashionable. For example, in 1768 Rutherglen town council, Lanarkshire, agreed to 'finish the Councill Hall and other rooms in the Tolbooth with a plaster work or stucco'.[47] Similarly, Glasgow's burgh records for 1766 refer to a payment made to Thomas Clayton 'stucatorian ... for plaster work in the town's hall'.[48] Another entry in 1769 records an additional payment to Clayton, this time 'for plaister, lime, and workmanship furnished by him to the tolbooth and others from 14 June 1768 to 21 March 1769'.[49] It is conceivable that this was in fact the celebrated Thomas Clayton 'who from the 1740s was the force to be reckoned with, and who gave to Scotland some of its finest plaster decoration', or one of his sons.[50] This hypothesis is strengthened by the fact that Glasgow council is known to have employed Thomas Clayton senior to decorate the ceiling and entablature of St Andrew's Parish Church in Glasgow in 1753–4.[51] Sadly, both Glasgow Town Hall and Rutherglen Tolbooth have been demolished. Nevertheless, good examples of decorative plasterwork can still be found in Stirling Tolbooth (1703–5), Falkland Town House (1800) and Kinghorn Town House (1829–30).

Rushes or straw were commonly strewn on stone-flagged floors in domestic houses from the Middle Ages onwards, to insulate and deafen noise. Meadow-sweet, on the other hand, masked unpleasant smells by releasing aromatic oils into the atmosphere when walked upon. The Lanark Burgh Accounts for 1507 reveal that rushes and meadow-sweet were purchased for the floor of the tolbooth.[52] Floor carpets (as opposed to table carpets) became popular in the eighteenth century. For example, in 1768 the Magistrates and Council of Glasgow purchased 62¾ square yards from Robert Hannah and Company, carpet makers in Glasgow, for the 'town's hall'.[53]

Council chambers and court rooms were sometimes lined with expensive wooden panelling. The eighteenth-century 'Great Room' of Aberdeen (described above), for example, was lined with the 'best firr boxing'. The chosen design was not dissimilar to an unexecuted one which had been prepared by William Crystall in 1731 for 'the Principal Roome in the New Worke of the Town house of Aberdeen'. The drawing is now in Aberdeen City Archives. Although Aberdeen's Great Room is long gone, panelling can still be seen in Dunbar council chamber and Stirling court room (fig. 13.10). The court room in Stirling Tolbooth (1703–5) is particularly interesting because, like Aberdeen, it has a landscape above its fireplace. However, it is a classical rather than a topographical scene and is painted directly onto the wood, rather than on canvas. Edinburgh Tolbooth did, however, have a picture of the town hanging in its low council chamber. It is referred to in 1677 when the treasurer was ordered to pay James Alexander, painter £15 sterling 'for helping and painting of the great draught of the good toun presentlie hanging in the laich council hous'.[54]

Council and court room walls were more commonly hung with painted panels bearing the arms of the king and/or burgh. While these formed impressive decorations, their primary purpose was to indicate the judicial nature of the room and the legal status of the council. This is evidenced by the fact that in 1686 the Magistrates and Town Council of Dunbar specified that a 'broad' (board) bearing the arms of the burgh and the date should also have the words 'justice seat for the Magistrates and Councill' painted on it.[55] Unfortunately, this particular panel does not survive, but another bearing the arms of King James VII and II, also dated 1686, still hangs in Dunbar council chambers (fig. 13.11). Hugh Mackay has suggested that this particular panel was an expression of loyalty to the Crown, following James's coronation in 1685.[56] The Town Council Minutes reveal it was painted by Alexander MacBeth.[57] An earlier example is to be found in the vestibule of Culross Town House. It bears the Arms of Charles I and the motto VNIONVM VNIO ('A union of unions'), and is dated 1637.[58] Evidence that this practice continued well into the eighteenth century is provided at Musselburgh in 1773:

> [The Baillie and Council] ... unanimously agree that two copies of the Town's Coat of Arms shall be drawn upon canvis and put in a small gilt frame and one copy placed on each side of his Majestie's Coat of Arms, in the Councill Hall; and recommend to Baillie John Cochran to find a proper qualified person for executing the Town's arms as proposed by this Act of Council.[59]

Figure 13.10 Stirling Town House; panelled interior. Crown copyright: RCAHMS, SC 337091.

Figure 13.11 Dunbar Tolbooth, East Lothian; council chamber, painted panel with royal arms of King James VII, 1686. Crown copyright: RCAHMS, SC 710982.

An undated panel hangs in Stirling Tolbooth. It shows the Burgh's later seal, a wolf couchant on a rock, with the motto STIRLING OPPIDVM ('The Town of Stirling').[60]

Inscribed lintels were similarly used to denote the function of the building, or indeed a particular room. The inscriptions vary from the stern and moralistic to the downright grim and depressing. For example, the lintel above Glasgow Tolbooth door was inscribed:

> Haec domus odit, amat, punit, conservat, honorat, Nequitiam, pacem, crimina, jura, probus[61] (This house hates evil, loves peace, punishes crime, preserves the laws, honours the good.)

Evidence that this practice was not limited to Scotland is provided by the fact that a virtually identical inscription is to be found painted along the beam behind the magistrates' bench in the court room of Much Wenlock Town Hall, Shropshire.[62] The only difference is that the English one begins 'Hic Locus' (This place), instead of 'Haec Domus' (This house). Musselburgh Tolbooth retains above the entrance to the original prison an inscribed marble lintel which is referred to in 1773:

> Likewise the Baillie and Council unanimously agrees that a marble plate of five feet eight inches in length by nine inches and a half in breadth with the Town's Coat of Arms drawn and placed in the centre ... shall be placed above the doorhead of the entry to the prison with the following inscription engraved thereon. 'Magistrates do justice in the fear of God, he that God doth fear; In vain does not to falsehead bend his ear.' Mr Gowans to execute the same.[63]

This inscription was clearly directed at the magistrates and was intended to make them aware of their duty to dispense justice fairly by reminding them of their own judgement before God. A similarly inscribed panel once hung in the council chamber of Edinburgh Tolbooth (demolished c. 1817):

> Quisquis Senator officii causa hanc curiam ingred[i]ens ante hoc ostium omnes animi affectus iram odium amicitiam abjicito nam ut aliisaequis [aliis aequus] aut iniquus qu(a)eris ita quoque Dei judicium expectabis ac sustinebis[64] (Let any judge who enters this court in the cause of duty, cast away before this door all emotions – anger, hatred, friendship – for as you do justice to others impartially or unfairly, so you [yourself] will await and undergo the judgement of God).

It is conceivable that this panel was directed at the Senators of the Court of Session who regularly used the council chamber of Edinburgh Tolbooth in the sixteenth century. Another panel, this time directed at those incarcerated in the old tolbooth of Edinburgh, once hung above the fireplace in the hall:

> A prison is a place of care, A place where none can thrive, A touchstone true to try a friend, A grave for men alive.[65]

So far as furniture is concerned, Russia leather continued to be used for council chamber chairs well into the eighteenth century. Elgin Council, Moray, had to go to considerable lengths to procure it in 1728:

> The Counsell considering that it is absolutely necessary for fitting up the new councill chamber to provide at the publick charge two dozen of chairs of Russia leather one of each dozen to be an elbow chair and that the wood for strength and durableness in the frame of the chairs be of oake and considering that there is shortly ane opportunity for a ship going from the Murray Firth to Dantzick, doe impower the Magistrates to commission for as much hydes of Russia leather as will cover the said chairs and as much oake or knappell wood as will made the frames thereof.[66]

Further evidence of the popularity of oak furniture in this period is provided by an entry in Montrose Town Council Minutes for 1767 which reveals that William Strachen junior, wright, was paid 'four pounds ten shillings sterling at the price of four square folding oak tables lately furnished by him for the use of the Council in the Town Hall'.[67] Strachen was also commissioned to make the chairs for the new Town Hall and 'three square wainscot tables five feet in breadth and six feet in length each of them so that when joined together they will make one table of eighteen feet in length.'[68] Some of Strachen's furniture survives in situ. Rutherglen Town Council commissioned seven similar interconnecting tables from two local wrights for their council room in 1777. In this case, however, the council specified that the tops were to be made from 'good mahogany, one inch thick'.[69] The present whereabouts of these tables is unknown. Tables and chairs were sometimes covered with cloths, presumably to protect them. For example, in 1677 Edinburgh town council appointed Baillie Watson 'to furnish als much grein cloth as will cover the councill table and the session tables and furmes ...'.[70] Perth Treasurer's Accounts for 1695–6 include a payment made to 'James Blair tayleour for dressing the grein cloth to the middle counsell seatt and for green threed'.[71] Evidence that this practice continued well into the eighteenth century is provided by an entry in Musselburgh Town Council Minutes in 1777 which states that a new table cloth 'having the Towns Coat Of Arms thereon' had been agreed upon.[72]

Furniture was sometimes donated to town councils for use in their tolbooths. A set of chairs, tables and forms was gifted to Campbeltown Town House, Argyll, by the local Freemason Lodge in 1769.[73] Similarly, in 1717 Alexander Dunbar of Bishopmill complimented the burgh of Elgin 'with a fine chair which is planted on the top of the oval stair of the new Tolbooth.'[74]

In addition to tables and chairs, some council chambers would have had a box for the reception of fines imposed upon councillors for being late or absent. The one used in Dundee, Angus, known as the 'Pirlie Pig', is now in The McManus Galleries (fig. 13.12). It is made of

Figure 13.12 Pirlie Pig, 1602. McManus Galleries, Dundee City Council Leisure & Arts.

pewter and is engraved and dated 14 May 1602.[75] Edinburgh Town Council seem to have had a similar box, judging by an entry in the burgh records for 1597–8:

> Ordains the thesaurer to pay for making of ane greyne boxe of irne, haiffing twa loks and two keyis, sextein pund: quhilk box beand producit in counsall ane key wes delyverit to the thesaurer ane other to Alexr. Myller and the said boxe to be putt in the counsallhous almery and the unlaws to be putt thairin ilk Fryday.[76]

The 'greyne' (ie green) box is referred to again in 1607, when it was used to hold fines imposed on 'flescheoures' for selling 'lipper swyne' (unfit meat) at the market.[77]

Council chambers sometimes had clocks in them too. Dundee's council chamber clock, the brass face of which is now in the McManus Galleries, was used to time the length of meetings. It was made by James Ivory of Dundee in the late eighteenth century. Stirling's council chamber clock has survived intact. It is a particularly fine longcase clock made by Andrew Dunlop of London about 1710–15 with geometric and floral marquetry patterns which include birds and butterflies.[78]

The responsibility for erecting and maintaining prisons in Scotland was put specifically on the burghs by an Act of Parliament in 1597. It ordained that 'within the space of three zeires, in all Burghs within this

Realme, there be sufficient and sure jailles and warde-houses bigged, vp-halden and mainteined be the Provost, Baillies, Councel, and Communities of the saidis Burrowes, vpon their awin commoun gude, or vtherwaise vpon the charges of the Burgh'.[79] This liability remained until an Act of 1839 established County Boards to take over the local management of all prisons.

Prison cells were sometimes ranged along the top storey of the tolbooth for reasons of security, and a good example is to be found in Inverkeithing. Prisons might also be located on the ground floor, as at Cupar and Musselburgh, but this arrangement had two disadvantages. Most importantly, it was less secure. Prisoners could undermine the walls in order to escape. This problem was exacerbated in areas where the local building stone was soft. For example, in 1819 the magistrates of Tain complained that although the prison walls were 'of considerable thickness, they are easily gone through with a common chisel, the stones easily giving way, and soon reduced to sand'.[80] Secondly, passers-by increasingly found the sight of wretched prisoners offensive. Most tolbooths also had a 'black hole'. This was reserved for troublesome offenders and had minimal ventilation. They were either situated under the forestair (Lauder, Berwickshire, Kilmaurs, and Kintore, Aberdeenshire), or in the base of the steeple (Strathmiglo, Stirling), or occasionally in the basement of the main council chamber block (Ceres, Dalkeith, Midlothian). Debtors were considered to be aristocrats among prisoners and were given more salubrious accommodation than felons or thieves. Debtors' cells tended to be less damp because they had fireplaces and windows, and the occupants were also free to come and go as they wished. Indeed, it was not uncommon for town councils to undertake alterations and additions to tolbooths in order to provide separate prison accommodation for debtors. For example, in 1775 Musselburgh Town Council ordered that 'proper repair' be made to the middle apartment of the tolbooth 'for the reception of burgesses that may happen to be imprisoned for lawful debts for their comfortable accommodation when imprisoned'.[81] Similarly, it was noted in 1792 that a new building had been added to the rear of Linlithgow Tolbooth, the upper part of which was designed 'for debtors, who have hitherto had no other place but the common prison.'[82]

In addition to purpose-built tolbooths, there are a small number of converted or dual-purpose buildings which do not conform to the standard type. Examples of dual-purpose civic/ecclesiastical buildings include Pittenweem (1588) and Greenlaw Church/Tolbooth Steeples (Berwickshire, 1712; fig. 13.13), and such important burgh churches as St Giles Cathedral in Edinburgh and St John's Parish Church, Perth. Pittenweem Tolbooth Steeple has been described as 'a very fine example of the intermingling of domestic and ecclesiastical work which was so

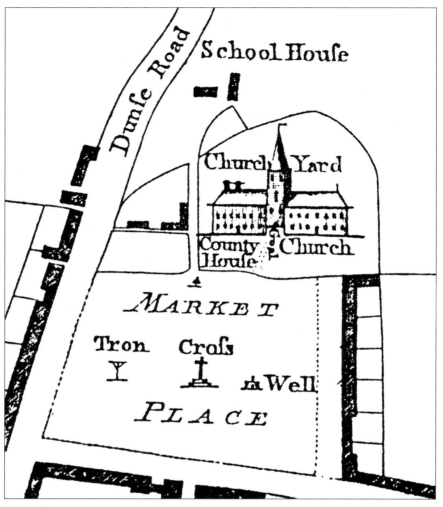

Figure 13.13 Greenlaw, Berwickshire; detail of Armstrong's map showing tolbooth steeple between county house and church, 1771.

common in the sixteenth and seventeenth centuries'.[83] The town council is believed to have convened in the steeple, which abuts the church. At Greenlaw the tolbooth tower (containing the prison cells) was centrally placed between the court house and the church. This arrangement led one writer to observe that 'Here stands the gospel and the law Wi Hell's hole atween the twa'.[84] Sadly, only the church and steeple now remain; the court house was demolished in 1830. The two western portions of the nave in St Giles Church in Edinburgh were partitioned off in 1561 to form 'the Outer Tolbooth'. The ground floor was used by the Town Council for meetings, as a storage place for 'the Maiden' (the City's instrument of execution), and on Sundays as a place of worship. This arrangement lasted until 1598 when the original layout was restored. St John's Church in Perth was similarly used by the Town Council. The Town Council considered this arrangement unsatisfactory, citing it as a reason for building a 'Counsell hous, Clerks Chamber and Pack house' in 1696–7:

> The house wher the Magistats and Councell now meets being only ane housse within the church appoynted for Ministers and Session was never found convenient it layes so levell on the street that the common inhabitants and boyes have ordinarly on the Counsel days and especialy at Michalsmas at our Elections lookt in at the windows and did hear and see all our actions which to our sad experience some few years ago occasioned a great deall of losse both to the inhabitants and patrimonie of the Burgh.[85]

This may be one reason why council chambers are usually situated on the first floor of purpose-built tolbooths. Evidence that Perth Town Council was still using the church for storage in the eighteenth century is provided by the fact that in 1767 they recommended to the magistrates 'to cause repair the High Councilhouse at the Church were the Town's Writes are kept and to make [a] proper Repository for securing them from the Dampness of the Weather, or other accidents'.[86]

Examples of converted structures include Douglas Tolbooth, Lanarkshire, St Leonard's Tower in Kinghorn (demolished), Maybole Town House, Ayrshire, the Bishop's Palace in Dornoch, Sutherland, Stonehaven Tolbooth and Fortrose Chapter House, Ross and Cromarty. Douglas Tolbooth (later known as The Sun Inn) was originally built as a domestic house in 1621 and it is not known precisely when it was converted into a tolbooth. However, the court-room on the upper floor (known as the 'Stone Room' on account of its heavy stone-slabbed floor) is traditionally said to have been in use during the Covenanting period. Maybole Town House was similarly built as a private residence. In this case, it was the 'toune house' of the lairds of Blairquhan. It was converted into a tolbooth in the late seventeenth century. The Bishop's Palace in Dornoch lay in ruins until the north-west tower was fitted up between 1813 and 1818 as a court room, record room, and jail at the expense of the Marquis of Stafford.[87] A new town jail was built nearby in 1840–50 in Scots Baronial Style. The Bishop's Palace was subsequently 'refitted and refurbished as a quaint dwelling place for English sportsmen' in 1881.[88] It is now a hotel. Stonehaven Tolbooth was built by the 5th Earl Marischal, feudal superior of the burgh in the sixteenth century, as a grain store and converted into a tolbooth c. 1600 when Stonehaven became the county town of Kincardineshire. Prior to the erection of the present Town house in Kinghorn in 1829–30 the town council met in St Leonard's Tower, which had been converted from a place of worship into a town house after the Reformation. The tower was damaged by lightning in 1822 and had to be demolished. Fortrose Chapter House was converted into a tolbooth 'either at the time of the Reformation, or ... after it had received its charter as a royal burgh in 1590. In fact, if a minute of 1688 is to be believed, the council had been in possession since the time of Bishop Leslie, that is, since about 1567'.[89] The conversion necessitated some structural alterations, including the addition of a forestair, to provide access to the first-floor council chamber.[90] The ground floor was used as a prison. The town council continued to meet in the building until 1939.[91]

NOTES

1. *The Acts of the Parliaments of Scotland*, vol 3, 1567–92, Edinburgh, 1814, 582–3, c. 83.
2. See transcript of the *Decreet and Act Declaring the New Jail of Wick to be a Legal Jail for the Imprisonment of Debtors*, dated 15 November 1828, contained in Wick TCM, Caithness District Council, 25 November 1828.
3. NAS, GD/12/32/1.
4. *NSA*, 9 (1845), 470 (Fife).
5. ibid.
6. Kennedy, 1818, 403.
7. Findlay, 1984, 61.
8. Simpson, Stevenson, 1981, 20.
9. RCAHMS. *Inventory of Stirlingshire*, 1963, vol 2, no 232.
10. Macky, 1723, vol 3, 202.
11. *OSA*, 5 (1793), 32.
12. NAS, RHP 31776/1–3.
13. Cameron, 1983, 32.
14. RCAHMS. *Inventory of Midlothian*, 1929, no 114; RCAHMS. *Inventory of Edinburgh*, 1951, no 25.
15. NAS, B48/18/115; Linlithgow TCM 12 April 1656, quoted in Mylne, 1893, 241.
16. MacGibbon, Ross, 1892, vol 5, 106.
17. Kirkcudbright TCM 1 January 1642, quoted by Smith, 1975, 214.
18. Fitzgerald, 1957, 9.
19. Clouston, 1949–50, 94.
20. Cramond, Eeles, Exhibition of bells in the Museum, Elgin, from 23 to 30 August 1902, 15–16, Reprinted from *Elgin Courant and Courier*, Tuesday 26 August 1902.
21. *Extracts from the Council Register of the Burgh of Aberdeen, 1398–1570*, Aberdeen, 1844, 150.
22. *Extracts from the Burgh Records of Glasgow, 7, 1760–80*, SBRS, Glasgow, 1912, 230; Wallace, 1981, 26.
23. Kennedy, 1818, Vol 2, 192.
24. Montrose TCM, Angus District Council Archives, 15 April 1761.
25. ibid, 20 May 1761; 12 May 1762; 16 November 1763.

26. Haddington TCM, NAS, B30/13/18, 14 December 1774; 1 November 1788; 3 December 1788.
27. Musselburgh TCM, NAS, B52/3/5, f 14, 26 January 1811.
28. McKay, 1880, 112; Brewster, nd, 39.
29. Metcalfe, 1909, 318.
30. Perth TCM, Perth and Kinross District Council Archives, B59/16/11, f 42v, f 43, 18 May 1696.
31. Perth Treasurer's Accounts 1695–6, Perth and Kinross District Council Archives, B59/19/22, 16.
32. Perth Treasurer's Accounts 1696–7, Perth and Kinross District Council Archives, B59/19/23.
33. Perth TCM, Perth and Kinross District Council Archives, f 49v, 21 December 1696.
34. ibid, f 58v, 11 October 1697.
35. Aberdeen Council Register, Volume 61, Aberdeen City Archives, f 479, 20 September 1750.
36. ibid, f 510, 15 November 1750.
37. ibid, f 570, 3 October 1751.
38. ibid, vol 62, f 136v and f 137, 1 September 1756.
39. Guildry Accounts, 1757–8.
40. ibid, vol 61, f 570, 3 October 1751.
41. Guildry Accounts, 1750–70, Aberdeen City Archives, 15.
42. ibid, Volume 62, f 261, 1 September 1759.
43. Aberdeen Council Register, 1753–1763, 26 May 1755.
44. ibid, 17 November 1755.
45. Aberdeen Council Register, volume 88, f 36.
46. Apted, 1964, 278.
47. Rutherglen TCM, Strathclyde Regional Archives, RU/3/1/12, 11 March 1768.
48. *Extracts from the Records of the Burgh of Glasgow, 1760–80,* 7, Glasgow, 1912, 230.
49. ibid, 291.
50. Beard, 1975, 85.
51. ibid, 87–8.
52. *Extracts from the Records of the Royal Burgh of Lanark, with Charters and Documents Relating to the Burgh,* AD *1150–1722,* SBRS, Glasgow, 1893, no page number.
53. *Extracts from the Burgh Records of Glasgow,* 3, Glasgow, 1912, 270.
54. Wood, M, ed. *Extracts from the Records of the Burgh of Edinburgh, 1665–80,* Edinburgh, 1950, lvii.
55. Dunbar TCM, NAS B18/13.1, 3 May 1686.
56. Mackay, 1968, 12.
57. Transcript of Dunbar TCM, NAS, B18/13/1, 19 April 1686.
58. RCAHMS. *Inventory of Fife,* 1933, no 158 (7).
59. Musselburgh TCM, NAS, B52/3/2, f 103, 16 June 1773.
60. RCAHMS. *Inventory of Stirling,* 1963, Vol 2, no 232.
61. Brown, 1973, 238.
62. See plate 12 in Tittler, 1991.
63. NAS, B52/3/2, loc cit.
64. Gray, 1925, 161.
65. Chambers, 1931, 84.
66. Cramond, 1903–8, Vol 1, 428–9.
67. Montrose TCM, Angus District Council Archives M/1/1/8, 4 March 1767.

68. ibid, loc cit, 17 April 1765, 13 November 1765.
69. Rutherglen TCM, Strathclyde Regional Archives, RU/3/1/12, 21 November 1777.
70. Wood, M, ed. *Extracts from the Records of the Burgh of Edinburgh, 1665–80*, Edinburgh, 1950, 306, 15 June 1677.
71. B59/19/23, Perth and Kinross District Council Archives.
72. Musselburgh Town Council Minutes, NAS, B52/3/3, f 154, 30 May 1777.
73. Mactaggart, 1923, 16.
74. Cramond, 1903, Vol 1, 397.
75. Maxwell, 1884, 30–3.
76. Extracts from *The Records of the Burgh of Edinburgh, AD 1589–1603*, Edinburgh, 1927, 210–11.
77. Extracts from *The Records of the Burgh of Edinburgh, AD 1604–26*, Edinburgh, 1931, 27.
78. Allan, 1990, 28.
79. *Parliaments of James I–VI, 1424–1621*, np, nd, no 273, 19 December 1597.
80. Report of the state of burgh jails of Scotland, in answer to certain queries of the committee of the House of Commons, in *Inverness Journal and Northern Advertiser*, 6 (August 1819).
81. Musselburgh TCM, NAS, B52/3/3, f 124, 17 February 1775.
82. *OSA*, Vol 14, 568.
83. MacGibbon, Ross, 1892, 149.
84. Quoted by Gibson, 1905, 156. An illustration of it appears in Armstrong's map of the county of Durham and Northumberland.
85. Perth and Kinross District Council Archives, B59/24/13/7/12.
86. Perth TCM, Perth and Kinross District Council Archives, PE/1/1/2, 2 February 1767.
87. NAS, SC9/84/20/1–12; NAS, SC9/84/34/3.
88. Groome, 1901, 362.
89. Macdowall, nd, 134.
90. Fawcett, Breeze, 1987.
91. Macdowall, nd, 134.

BIBLIOGRAPHY

Allan, C. *Old Stirling Clockmakers*, Stirling, 1990.
Apted, M R. 'Painting in Scotland from the Fourteenth to the Seventeenth Centuries', 2, unpublished PhD thesis, University of Edinburgh, 1964 (copy in NMRS).
Beard, G. *Decorative Plasterwork in Great Britain*, London, 1975.
Brewster, D. *Wigtown: The Story of a Royal and Ancient Burgh*, Wigtown, npd.
Brown, P H. *Early Travellers in Scotland*, Edinburgh, Edinburgh 1891, reprinted 1973.
Cameron, J. *Prisons and Punishment in Scotland*, Edinburgh, 1983.
Chambers, R. *Traditions of Edinburgh*, Edinburgh, 1931.
Clouston, R W M. The church and other bells of Stirlingshire, *PSAS*, 84 (1949–50), 66–112.
Cramond, W, ed. *Records of Elgin, 1234–1800*, Aberdeen, 1903–8.
Cullen, W D (Lord). *Parliament House: A short history and guide*, npp [Edinburgh], 1992.
Cunningham, C. *Victorian and Edwardian Town Halls*, London, 1981.
Fawcett, R, Breeze, D J. *Beauly Priory and Fortrose Cathedral*, Edinburgh, revised edn, 1987.

Findlay, W. *Heritage of Perth*, Perth, 1984.
Fitzgerald, J. 'Scottish Tolbooths', unpublished Rowand Anderson Studentship report, copy in the NMRS, 1957.
Gibson, R. *An Old Berwickshire Town*, Edinburgh and London, 1905.
Gray, W F. Reminiscences of a Town Clerk, *The Book of the Old Edinburgh Club*, 14 (1925), 147–81.
Groome, F H. *Ordnance Gazetteer of Scotland*, new edn, London, 1901.
Hannay, R, Watson, G P H. The building of the Parliament House, *The Book of the Old Edinburgh Club*, 13 (1924), 1–78.
Kennedy, W. *Annals of Aberdeen*, London, 1818.
Lindsay, I. The Scottish burgh. In Scott-Moncrieff, 1938, 77–102.
Macdowall, C. *The Chanonry of Ross*, Fortrose, npd.
MacGibbon, D, Ross, T. *The Castellated and Domestic Architecture of Scotland*, Vol 5, Edinburgh, 1892.
McKay, A. *The History of Kilmarnock*, Kilmarnock, 1880.
Mackay, H. The armorial panels of Dunbar Town House, *Transactions of the East Lothian Antiquarian and Field Naturalists Society*, 11 (1968), 12–16.
Mactaggart, C. Life in Campbeltown in the eighteenth century, reprinted from the *Campbeltown Courier*, Campbeltown, 1923.
Macky, J. *A Journey Through Scotland*, 3 vols, London, 1723.
Markus, T A. Buildings for the sad, the bad and the mad in urban Scotland, 1780–1830. In Markus, T A, ed, *Order in Space and Society*, Edinburgh, 1982, 25–114.
Maxwell, A. *Old Dundee*, 1884.
Metcalfe, W. *A History of Paisley*, Paisley, 1909.
Mylne, R S. *The Master Masons to the Crown of Scotland and their Works*, Edinburgh, 1893.
Pevsner, N. *A History of Building Types*, London, 1976.
RCAHMS. *Inventory of Midlothian*, Edinburgh, 1929.
RCAHMS. *Inventory of Fife*, Edinburgh, 1933.
RCAHMS. *Inventory of Edinburgh*, Edinburgh, 1951.
RCAHMS. *Inventory of Stirlingshire*, 2 vols, Edinburgh, 1963.
RCAHMS. *Tolbooths and Town-houses, Civic Architecture in Scotland to 1833*, Edinburgh, 1996.
Scott-Moncrieff, G, ed. *The Stones of Scotland*, London, 1938.
Simpson, A T, Stevenson, S. *Historic Pittenweem*, Glasgow, 1981.
Smith, J. *Old Scottish Clockmakers from 1453 to 1850*, Wakefield, 2nd edn, 1975.
Stell, G. The earliest tolbooths: A preliminary account, *PSAS*, 111 (1981), 445–53.
Tittler, R. *Architecture and Power*, Oxford, 1991.
Wallace, W. *Marking Time in Hamilton: 300 years of Lanarkshire watches and clocks by William Wallace*, npd, reprinted 1981.
Zeegers, G, Visser, I. *Kijk op Stadhuizen (A Glance at Town Houses)*, Amsterdam, 1981.

14 Worship and Commemoration[1]

NEIL CAMERON

The built manifestations of worship in Scotland, which represent over a millennium of architectural history, are amongst the most fascinating and complex aspects of Scottish culture. While the material evidence of all kinds for religious and commemorative activity extends back into prehistory, this article takes as its historical starting-point the earliest surviving stone-built churches in Scotland.

The forms and sizes of church buildings, and the sites on which they are built, can be taken to symbolise the role of the church in the community. This morphological approach tends to simplicity, however, and is best considered in a context of chronologically based analysis. This is a large subject, the approaches to which can only be outlined here.

As regards historiography, a major development is the Inventory of the Scottish Church Heritage, organised by the Council for Scottish Archaeology, to provide a comprehensive catalogue of available information on churches in Scotland. Surprisingly little new research is being done, however, despite a plethora of possible subjects and themes. In recent years the study of church architecture has not maintained the same level of popularity as domestic and, particularly, country house architecture. Indeed, the major secondary sources remain the same today as they did thirty years ago. Representing an informative combination of synthesis and inventory are the three volumes on *The Ecclesiastical Architecture of Scotland* by MacGibbon and Ross, now over a century old, and *The Architecture of Scottish Post-Reformation Churches* by Hay, published in 1957.[2] An attempt to provide a general study, *Scottish Medieval Churches* by Cruden, fails to do justice to its subject, and like the exhibition catalogue on Scottish medieval art, *Angels, Nobles and Unicorns*, may have obscured more than highlighted the potential interest of the material it discusses.[3] One article which has aerated these somewhat stagnant waters is Fernie's article 'Early church architecture in Scotland',[4] which benefits from a profound knowledge of English and continental comparative material, but which arguably contains an exaggerated sense of Scottish cultural backwardness. The authoritative study of individual buildings has been maintained by the county *Inventories* of RCAHMS and, with a greater bias towards description over analysis, by the volumes of the *Buildings of Scotland*. Scholarship on individual architects continues to provide useful information, as in the case of MacFadzean's monograph on 'Greek' Thomson.[5]

The ambiguities in dating poorly-documented buildings are illustrated in the medieval period by some of the earliest surviving ecclesiastical buildings in Scotland, which have been ascribed dates from the eighth to the twelfth centuries. They are localised in the eastern side of the country in the relatively low-lying and rich farming lands around the Forth and Tay estuaries, and include St Rule's Church at St Andrews, Fife, Abernethy Round Tower, Fife, Brechin Round Tower, Angus, and Restenneth Priory, Angus. The two round towers are free-standing structures which probably functioned as belfries (like Italian campanile) and, in turbulent times, as places of refuge for local ecclesiastics and their valuables such as books and works of sacred art. It is likely that the churches associated with these towers were small structures built of wood rather than stone and therefore relatively impermanent, probably part of a huddle of small wooden buildings including the house of the brethren. Nevertheless, both Brechin and Abernethy were important church centres, the former described as 'the great civitas of Brechin [offered] to the Lord' in the reign of Kenneth II (971–5). That the function of Brechin Round Tower was religious rather than secular is underlined by the remarkable carvings around its doorway, set at some height above ground-level in the interests of security, which show the figure of Christus triumphans and two ecclesiastical figures carved in a style more akin to Ottonian Romanesque than local Pictish art.

The question of dating is particularly important in relation to the tower of Restenneth Priory as it has been linked to a church referred to by Bede as 'a stone church in the Roman manner', built *c.* 710 at the instance of Nechtan, king of the Picts who sought advice from Abbot Ceolfrith of Jarrow, Northumberland. While the identification of the church described by Bede is not provably Restenneth, the lower masonry courses of the tower are built in a manner associated with Anglo-Saxon building traditions. Recent scholarship however has sought, on comparative grounds, to suggest that Restenneth should be dated to the early twelfth century. This is not the place to consider this matter in detail, but it bears on an important theme which runs through the history of church architecture in medieval Scotland: the extent of its debt to outside, and especially English, sources and traditions. In the dating of many Scottish churches, even those for which documentation exists, there has been a tendency to assume that they must always follow earlier English models.

St Rule's Church at St Andrews is a case in point.[6] Despite clear documentary evidence suggesting that this splendid monument was in existence in the later eleventh century and was extended in the second

Figure 14.1 St Andrews Cathedral and St Rule's Church, Fife; aerial view. Crown copyright: RCAHMS, SC 710758.

quarter of the twelfth, the latter date has generally been ascribed to the whole building. Moreover, the design of the structure, which represents one of the masterpieces of stereotomy, or stone-cutting, in early medieval European architecture, has been considered to derive from an insignificant and poorly-constructed church tower at Wharram-le-Street, Yorkshire, on the basis of dubious historical evidence. If there was any link between these two buildings, it was almost certainly from, rather than to, St Andrews.

St Andrews (fig. 14.1) was described in the eleventh century as a pilgrimage centre comparable to Santiago de Compostela, Spain, and Rome. This was because St Andrews possessed relics of an apostle, placing it at a high position in the pilgrimage league table. The importance of pilgrimage as a cultural phenomenon in the Middle Ages should not be underestimated. It was a means of raising large amounts of revenue for the Church, as pilgrims normally donated money in return for 'indulgences' and access to the area around a shrine. In addition, pilgrims required to be housed and fed, which was in major pilgrimage centres a significant economic stimulus. More nebulously, they functioned as a means of communication throughout Europe, the Latin-based culture of the Church being truly international in character. The success of St Andrews as a pilgrimage centre is underlined by the sheer scale of the new cathedral which was begun c. 1160 as a new house for the relics of St Andrew – only Norwich Cathedral has more bays – and by the extraordinary wealth it possessed by the later Middle Ages. That the importance of St Andrews has been underestimated may be in part due to the ruined state of the cathedral, which underlines the importance of considering buildings in their historical context. That the need to accommodate pilgrims could also influence the plans of churches is clearly evident at Glasgow Cathedral, where the thirteenth-century crypt was built in a form that allowed ease of circulation around the shrine of St Kentigern.

The influence of the Church in temporal affairs was also considerable. Churches were, for example, able to provide sanctuary for wrongdoers who sought protection. At St Machar's Cathedral, Aberdeen, a great stone cross stood near the Bishop's Palace to mark out the boundaries of a 'girth' or sanctuary.[7] The Church was also central to the development of mercantile culture. This was particularly the case in the twelfth century, when a number of monastic foundations, some of which possessed great estates, were formed with royal approval and assistance. The scale of monastic trading can be indicated by the fact that the Cistercian abbey at Melrose, Roxburghshire, retained permanent warehouses at Boston, Lincolnshire, to facilitate its trade in wool with Flanders. Furthermore, monastic houses held considerable resources in craft skills such as farming, mason-work, brewing and baking which assisted the development of skilled work outside the monastery, particularly through lay brethren. The sophistication of monastic life is partly indicated by the typical monastic plan of the mid twelfth century, such as that of Dundrennan Abbey, Kirkcudbrightshire, where the various buildings are laid out in a logical fashion around a central cloister giving access to all parts of the complex.

While monasteries often controlled extensive areas of land in their vicinity, cathedrals possessed a different kind of pastoral role. They were the seats of bishops, and their areas of influence were the separate dioceses into which the country was divided. The diocese of Moray, for example, covered an extensive area bounded by Dalwhinnie in the south, Glen Moriston in the west, Aberchirder in the east and the Moray coast between Garmouth and Nairn in the north. The centre of Moravian diocesan power throughout most of the Middle Ages was Elgin. In turn, the diocese was divided into deaneries, those of Moray consisting of Strathspey, Inverness, Strathbogie and Elgin. In the broadest sense like the ecclesiastical equivalent of a feudal land-owning structure, the deaneries were further divided into parishes.

This ordered ecclesiastical structure has traditionally been associated with the introduction of Anglo-Norman culture into Scotland following the Norman Conquest of England in 1066, the indigenous Celtic Church having been more cellular in its organisational structure. Queen Margaret (d 1093), wife of Malcolm III (Canmore), is credited with having begun this process. For example, she brought Benedictine monks from Canterbury to colonise her church at Dunfermline, Fife, then the capital of the Scottish kingdom. Her son David I (d 1153) was, however, crucial to the introduction of organised religious activity to Scotland. He was a particularly enthusiastic patron of monastic foundations of all orders, even introducing the Tironensians directly from France to Selkirk in 1113.

The parish church represents the ubiquitous foundation of worship in the Middle Ages. The plans of parish churches vary considerably, but always allow for two essential tasks. Firstly, they had to shelter an altar, which was essential to the performance of divine worship. Secondly, they required to accommodate the people conducting the service and those participating in worship. The area at the east end of a church which normally housed the altar, known as the chancel, is derived from the cancellos or screen which separated those conducting the service from the nave (navis, a ship) in which the congregation was housed. There was therefore a clear division between the areas occupied by the priesthood and the laity. The division between nave and chancel was often marked by a great cross or rood, and a massive stone example of early medieval date survives at Kinneil, West Lothian.

The nature of patronage was important in determining the site and relative grandeur of parish churches. The parish church dedicated to St Cuthbert at Dalmeny, West Lothian (fig. 14.2), which has the most elaborately carved doorway of any Romanesque church north of Durham, was founded by the powerful local landowner, Gospatric, who was presumably able to afford to employ the most gifted masons in the country. By contrast, the church of St Martin at Haddington, East Lothian, which is of similar date to Dalmeny and probably founded by another local landowner, Alexander de St Martin, is a plain box devoid of decoration. It is, however, notoriously difficult to draw general conclusions from evidence of this kind.

It would be incorrect to suggest that the development of church architecture in the medieval period is from simplicity to complexity, but there is some justification for seeing increasing refinement in structural form from the twelfth to the fifteenth centuries. In this sense, Scottish

Figure 14.2 Dalmeny Parish Church, West Lothian; south doorway with Romanesque sarcophagus in the foreground. Crown copyright: RCAHMS, SC 710971.

architecture should be seen as part of a north-west European tradition, where the mural structures of the Romanesque were replaced towards the mid-to-late twelfth century with the closely integrated skeletal structures of the Gothic. Nevertheless, the idea of a simple progression during the Middle Ages towards increasing structural refinement is contradicted by the evidence of late medieval churches such as St Machar's Cathedral, Aberdeen, where the nave represents a return to Romanesque simplicity.

The use of divisions and areas endowed with different levels of liturgical significance within churches – compartments used for different purposes and by persons of different status – is a constant element in buildings designed for worship. For example, the abbey church at Dunfermline had both monastic and parish functions. The choir screen allowed the monks an extensive area for their liturgy which may also have been given further emphasis by the use of spiral decoration on two nave piers. Further functional divisions were in existence at Dunfermline by the later Middle Ages when the nave aisles were divided by a wooden screen to create separate altars for different guilds.

A parish church in a prosperous late medieval burgh – for example, St Michael's, Linlithgow, West Lothian – often had a large number of chantry altars at which masses for the dead were sung according to bequests given to the church by prosperous donors. The wide plan of St Giles' in Edinburgh is partly due to the addition of chantry chapels built and endowed by prosperous individuals and guilds. The function of saving individual souls was given special focus in collegiate churches, particularly numerous in the fifteenth century. These were secular foundations established by wealthy founders to ensure that masses were sung after death for them and their families by a 'college', which normally consisted of a provost, prebendaries and a small choir. A particularly splendid example is Seton Collegiate Church, East Lothian (fig. 14.3), begun in the later fifteenth century but never completed beyond the transepts.

The influence of the great monasteries began to wane in the later Middle Ages; this is well illustrated by the limited amount of major building undertaken in Scottish monastic houses after the late fourteenth century and the small number of new foundations. The most notable ecclesiastical building works of the later Middle Ages included a series of great churches in the larger Scottish burghs such as St John's, Perth, and the church of the Holy Rude, Stirling, both largely works of the mid-to-late fifteenth century. The latter church retains its original timber roof structure and has a remarkably tall buttressed apse which makes full use of changes in ground-level from the west to the east of the building.

Scotland has not been fortunate in the survival of medieval ecclesiastical fittings and works of art, but it would be wrong to assume that bareness was always a feature of the Scottish church interior. By the twelfth century, applied decoration was already an integral feature of church buildings, as the survival of numerous brightly-painted fragments from Glasgow Cathedral demonstrates. Objects made of precious materials were also typical, and St Andrews Cathedral had a wide range of such objects, recorded in a pre-Reformation inventory. A small number of medieval church fittings still survive; for example, St John's Church,

Figure 14.3 Seton Chapel, East Lothian; interior of choir. Crown copyright: RCAHMS, SC 711085.

Figure 14.4 Dunblane Cathedral, Perthshire; choir stalls. Crown copyright: RCAHMS, SC 711071.

Perth, retains a splendid fifteenth-century brass chandelier which is probably of Flemish origin, while Dunblane Cathedral, Perthshire, still possesses late medieval choir-stalls (fig. 14.4). A European breadth of cultural awareness was possessed by some Scottish patrons, as is demonstrated by the procurement of a triptych by the Flemish painter Hugo van der Goes (active c. 1460–82) for the collegiate church of the Holy Trinity at Edinburgh.

That there are lamentably few survivals of medieval figurative art is partly explained by the zeal of the Calvinist Church, post-Reformation neglect and the destructive contribution of English invasion, most notably the depredations of the Earl of Hertford in the mid sixteenth century. The Reformation took place in Scotland some three decades later than in most of Europe, but was followed with the zeal of the late convert. The Reformers sought the destruction of church fittings such as crucifixes, altars, reredoses and statues, and in some instances their fervent work extended to the destruction of ecclesiastical buildings, which might later become quarries for local builders, as at St Andrews.

The Reformation was responsible for introducing dramatic changes to church planning. The altar situated to the east of the church, an essential feature of pre-Reformation worship, was replaced by a table at which Holy Communion was celebrated. As this table was situated amongst the congregation, rather than beyond them, a chancel was not necessary. A pulpit was required for the minister, but this was generally situated in a central position, normally against the south wall, in order that the congregation should be able to see him without difficulty. Partly because of this, the chancels of many churches fell into disuse and disrepair, such as those of Kirkliston, West Lothian, and Dunning, Perthshire. In some larger churches such as St Giles' Edinburgh, divisions were made to allow more than one congregation to use the building.

During the early improvisatory period of the Reformed Faith little new church building was undertaken. In most cases, post-Reformation churches were adapted medieval buildings, with alterations being carried out where required. Many pre-existing medieval churches were enlarged by the addition of aisles, small outshots containing family burial places, and family lofts or pews. In 1580, for example, Sir George Ogilvy of Dunlugas added a south aisle to the church at Banff to house a family monument. In function almost like a Protestant version of a chantry chapel, its medieval antecedents are underlined by its pointed window and pointed barrel-vault. Of similar date and of more obviously Gothic appearance is the Hepburn aisle at the church of Oldhamstocks, East Lothian, its traceried window an example of the survival of late Gothic forms beyond the Reformation. Dairsie Church, Fife, provides a more dramatic example of the continuing Gothic tradition. Here, the pre-Reformation symbolism of style was almost certainly conscious as it was built in 1621 at the behest of John Spottiswoode, Archbishop of St Andrews, who was a major proponent of Episcopalian belief. Its plate tracery and corbelled belfry are in the Gothic decorative idiom, and its sanctuaried plan harks back to the requirements of pre-Reformation liturgy.

Prior to the building of Dairsie, however, the end of the sixteenth century saw a small number of churches built according to Reformed planning. Most notable is the church at Burntisland, Fife (fig. 14.5), which dates from 1592–6. Almost square in plan, with a central 'nave' and an 'aisle' on each side, it has not been widely recognised that this building has a buttressed structure which maintains elements of a Gothic architectural precedent. Nevertheless, its plan manifests the requirements of the Reformed liturgy in its lack of hierarchical divisions and its emphasis is on the participation of the congregation in the conduct of the service. While the plan of Burntisland has been claimed to be of Dutch origin, a link enhanced by trade links between the coastal towns of east central Scotland and the ports of the United Provinces, it is more likely that similar solutions were found in different countries to the same problems created by the requirements of Reformed liturgical practice. What is also significant about Burntisland is its tentative use of the decorative language of the Renaissance, with its Ionic columns supporting

Figure 14.5 St Columba's Parish Church, Burntisland, Fife. Crown copyright: RCAHMS, SC 710840.

the wooden galleries which provided guild and sailors' pews. This limited use of classicising detail alongside Gothic elements is also exemplified by the Tron Kirk, Edinburgh, built in 1637 to the design of John Mylne (1611–67), the King's Master Mason, where the walls have pilasters but the window tracery is still Gothic in inspiration.

The Tron Kirk is significant in another important respect – it is the design of a named person. While it is incorrect to suggest that the relative anonymity of medieval masons and architects (perhaps best described as 'master' masons) was the result of Christian reticence rather than the more prosaic explanations of the loss of documentation and the nature of architectural patronage in the Middle Ages, it is nevertheless true that the post-Reformation period saw an unprecedented development in the idea of the individual creative personality. However, most architects remained within the practical world of mason-work and contracting, and were not exclusively designers. Mylne is a case in point, for he not only designed buildings, but was also a contractor and overseer of a quarry at Kingscavil near Linlithgow.

In using a T-shaped plan, the Tron Kirk was uncompromisingly a Reformed church. To some degree a Scottish speciality, the T-plan evolved from the addition of transeptal aisles to pre-existing medieval churches, as at Dalmeny. Usually situated on the north side of the church, the aisle often housed a laird's loft above the family burial vault. Examples survive at Careston, Angus, and Kirkmaiden, Wigtownshire, both dating from the 1630s. The T-plan remained a staple of Scottish church design throughout the seventeenth century and into the eighteenth century. The centrally planned form was also widely used, but the square layout of Burntisland was not the only solution. The Greek cross, which represented an advanced form associated with the humanism of the Italian Renaissance, was widely used in Scotland by the mid seventeenth century, as examples at Kirkintilloch, Dunbartonshire, and Lauder, Berwickshire, demonstrate. The latter, which dates from 1673 and was designed by Sir William Bruce (c. 1630–1710), who was instrumental in the development of classical country mansion design in Scotland, was also the result of the interested patronage of the 1st Duke of Lauderdale, who stipulated:

> I would have it decent and large enough, with a handsome little steeple: if any of the timber of the old church will serve, it will be so much the cheaper, but I can say now no more till I see the draught which you promise, and I would have both plan and perspective.[8]

Concern for limiting costs and a wish for a degree of control are common amongst patrons.

While during the seventeenth century all centrally-planned churches appear to have been cruciform and most of Greek-cross plan, the eighteenth and nineteenth centuries saw the use of centrally-planned churches of different types. For example, the parish church at Hamilton, Lanarkshire, built by William Adam (1689–1748) in 1732, was a circle within a Greek cross. Fully circular plans appear not to have been popular, however, with only an example at Bowmore, Islay, Argyll, surviving from the later eighteenth century. Octagonal plans were more common, and were used by both the established church as well as by dissenters. This form was used at Kelso, Roxburghshire, in 1773, a model of the building having first been prepared for approval by the heritors. It was not confined to major burghs, however, as the kirks at Dreghorn, Ayrshire, and Eaglesham, Renfrewshire, both of the 1780s, demonstrate. The plan also served the requirements of non-established congregations, being well suited to their egalitarian, communal approach to worship; Methodist chapels of this form were built at Edinburgh, Arbroath, and Aberdeen. Despite its classical antecedents, several octagonal churches were built in the Gothic style around the turn of the eighteenth century, St Paul's Church at Perth (1807) by John Paterson (1832) being a notable example.

Generally, however, eighteenth- and nineteenth-century churches maintained traditional plan-types. Rectangular plans were common both in rural and urban contexts, often being enlivened on the exterior by the use of a steeple, as at St Andrew's, Dundee, built in 1772. T-plans also continued in use, particularly fine rural examples being at Carrington and Yester, both in East Lothian. Increasingly popular in the later eighteenth century was the 'hall' church, of rectangular plan but with galleries around three walls. It has been suggested that this plan was introduced to England by Wren and then subsequently to Scotland by James Gibbs (1682–1754). This is possible, but its use on the Continent was widespread in preceding centuries as it was an obvious solution to the need to house a large congregation within a limited ground space. St Michael's at Dumfries (1745) is typical of the genre, but the apsed form of the garrison chapel at Fort George, Ardersier, Inverness-shire (1767), almost redolent of Romanesque simplicity, is a more unusual example. Despite the variety of post-Reformation church plans used in Scotland, an up-to-date and detailed study of this subject has not yet been carried out.

Also fascinating is the use of different architectural styles in post-Reformation church architecture. To some degree it is possible to see Gothic and classical forms as representing polarities possessed of a subtext of architectural symbolism. Certainly, the 'battle of the styles' became an issue bound up with attitudes to the revival of Roman Catholicism. The influential writings of A W N Pugin (1769–1832), and his adoption of the Gothic style in his own buildings, were directed towards the idea that Gothic was the appropriate style of churches and that the Classical was essentially pagan. In Scotland, the religious significance of the use of Gothic was less pronounced than in England, where by the mid nineteenth century Gothic was firmly established as part of the architectural canon of Anglican as well as RC churches. This was, of course, largely due to the fact that the established church in

Scotland was Presbyterian. In the mid and later nineteenth century, however, partly due to the influence of the ecclesiology movement which propagated interest in medieval architecture, Gothic became increasingly widespread in Scotland. For example, F D T Pilkington (1832–98), who was strongly influenced by the English High Victorian movement, worked in an imaginative Early English style, as at Barclay-Bruntsfield Church (originally a Free Church), Edinburgh (1863). The influence of Scottish medieval architecture can be seen in the design of Barony Church in Glasgow (1891), by J J Burnet (1814–1901), which is partly modelled on thirteenth-century Dunblane Cathedral, a favourite building of the writer John Ruskin.

Throughout the nineteenth century, however, the classical strain in Scottish church architecture maintained its place. As late as 1886, St George's in the Fields, Glasgow, was built by Hugh Barclay (1828–92) and David Barclay (1846–1917) in the form of an Ionic Greek temple. More idiosyncratic by far, but still within the ambit of classical form were the churches of Alexander 'Greek' Thomson (1817–75), such as the United Presbyterian Church, Caledonian Road, Glasgow (1856). This church, and Thomson's United Presbyterian Church at St Vincent Street, Glasgow (1858), were designed as part of a development which included contiguous tenements. This might be interpreted as reflecting something of the pragmatic, worldly view of religion espoused by elements of the Presbyterian movement.

The debate about the merits of classical and Gothic architecture was fuelled in part by scholarly interest in the architecture of the past. In Scotland, Billings' *Baronial Antiquities of Scotland*[9] helped to create a climate of interest in medieval Scottish architecture which was to bear fruit later in the century with the more scholarly work of MacGibbon and Ross. This level of informed interest, also subsequently exemplified by detailed articles on Scottish churches in the *PSAS*, the *Inventories* of RCAHMS and the measured drawings of the National Art Survey, helped to widen interest in the conservation of church architecture. This was in turn a stimulus to informed restoration, as in the work of P MacGregor Chalmers (1859–1922), who put his knowledge of medieval architecture to effective use in his own ecclesiastical designs.

The challenges of the redundancy, conservation and adaptation of church buildings are still with us, particularly due to the closure of churches because of falling congregations. Listing and recording have also reached a highly professional level, and awareness of the historical importance of modern churches has also been raised, as demonstrated by the outcry over the demolition of Gillespie, Kidd & Coia's striking 1960s RC church at Drumchapel, Glasgow.

The history of commemoration is closely bound up with that of worship but deserves separate consideration. Throughout the Middle

Ages graves were created inside churches as well as outside. The position of an individual's grave was largely a reflection of social status, burial inside a church being restricted and near the altar, towards the east end, being a particular privilege. The carved Romanesque sarcophagus which survives in the churchyard at Dalmeny (fig. 14.2) was probably that of the founder or one of his family, its long uncarved face indicating it was set against an internal wall. Simple grave-markers also survive from this period, however, like the examples of the simple cross-carved type at Falkirk and on Bute. During the thirteenth and fourteenth centuries recumbent ledger-slabs were used, typically decorated with incised representations of the cross and motifs such as shears and swords indicating profession or status. A particularly fine range of slabs of this type survive at the medieval cemetery at Cambusnethan, Lanarkshire.

From the relative elaboration of a funerary monument it is possible

Figure 14.6 Collegiate Church, Castle Semple, Renfrewshire; monument to John, 1st Lord Semple (d 1513) Crown copyright: RCAHMS, SC 710967.

to gauge something of the deceased person's status. Obviously, more elaborate monuments were costlier than plain ones, particularly if they incorporated precious materials. The tomb of Robert the Bruce (d 1329) in Dunfermline Abbey, sadly now represented only by fragments, was made of marble and imported from Paris. The type of carving is also significant, a figure carved in relief being a more sophisticated form than an incised design. A particularly fine example of the former is a late medieval figure of a knight at Old Kilpatrick, Renfrewshire, which is of very high technical quality. Its style is most convincingly considered within a lowland Scottish and northern English context, although figures of knights are common in the West Highlands, an area where a very distinctive style of carving arose in the later Middle Ages. The use of mural monuments was also common in the later medieval period, often for the founder of a particular church, as in the case of Lord Semple's tomb in Castle Semple Collegiate Church, Renfrewshire (fig. 14.6).

Both recumbent and mural monuments continued in use after the Reformation. The former developed a wider array of designs in the seventeenth century, armorials and emblems of mortality or resurrection being popular. The small 'weeper' figures, generally of angelic form in the Middle Ages, occasionally transformed into portraits of children in the post-Reformation period, as in the early seventeenth-century monument of Sir George Bruce of Carnock in Culross Abbey Church, Fife. With grave-slabs and gravestones there was also a gradual change from the grim representations of decay typical of the Middle Ages to the use of emblems of mortality such as hour glasses, bones and scythes. In the later seventeenth and eighteenth centuries, increasing use was made of resurrection motifs such as cherubs' heads, and garlands. These are found on monuments of all kinds, but the artistic flowering of gravestones carved for ordinary people, which often used moralising inscriptions, is a feature of the churchyards of many lowland Scottish towns.

The discouragement of burial within churches which was heralded by the Reformation was a stimulus to the construction of tomb vaults in churchyards. Greyfriars Churchyard, Edinburgh, possesses many splendid examples from the seventeenth century. While the eighteenth century saw the development of this form from mural structures into buildings, it was in the mid-nineteenth century that this type reached its apogee, with, for example, the ambitious mausolea of the Glasgow Necropolis (fig. 14.7). This was to some degree paralleled by the development of free-standing statuary as a means of commemoration, particularly fine examples surviving in Edinburgh's New Town, such as Sir John Steele's statue of Lord Melville.

The development of cemeteries separate from churches is also exemplified by the Glasgow Necropolis, but can also be seen in many small Scottish towns, as at Wick, Caithness, where the cemetery is situated

Figure 14.7 The Necropolis, Glasgow, 1870s. Crown copyright: RCAHMS, SC 711005.

on the southern town boundary. The increasing popularity of cremation in the post-war period is also a factor which represents a considerable social change from the long tradition of church burial.

NOTES

1. This text was prepared in 1996 and does not take account of later publications. Attention should be drawn in particular to the excellent book by Richard Fawcett: *Scottish Medieval Churches – Architecture and furnishings*, 2002.
2. MacGibbon, Ross, 1896–7; Hay, 1957.
3. Cruden, 1986; National Museum of Antiquities of Scotland, 1982.
4. Fernie, 1986.
5. McFadzean, 1979.
6. Cameron, 1994.
7. Cameron, 1989.
8. NAS, GD 29/1897/9.
9. Billings, 1845–52.

BIBLIOGRAPHY

Billings, R W. *The Baronial and Ecclesiastical Antiquities of Scotland*, 4 vols, Edinburgh, 1845–52.

Blair, J, Pyrah, C. *Church Archaeology: Research directions for the future*, York, 1996.

Cameron, N M. A Romanesque cross-head in St Machar's Cathedral, Aberdeen, *PSAS*, 142 (1989), 63–6.

Cameron, N M. St Rule's Church, St Andrews, and early stone-built churches in Scotland, *PSAS* (1994), 367–79.
Cowan, I B, Easson, D E. *Medieval Religious Houses: Scotland*, London, 1976.
Cruden, S. *Scottish Medieval Churches*, Edinburgh, 1986.
Dunbar, J G. *The Historic Architecture of Scotland*, London, 1966.
Fernie, E C. Early church architecture in Scotland, *PSAS*, 116 (1986), 393–411.
Hay, G. *The Architecture of Scottish Post-Reformation Churches, 1560–1843*, Oxford, 1957.
Lindsay, I G. *The Scottish Parish Kirk*, Edinburgh, 1960.
MacGibbon, D, Ross, T. *The Ecclesiastical Architecture of Scotland from the Earliest Christian Times to the Seventeenth Century*, 3 vols. Edinburgh, 1896–7.
Maclean, C et al, eds. *The Institutions of Scotland: Religious Expression* (Scottish Life and Society: A Compendium of Scottish Ethnology, Volume 12), forthcoming
McFadzean, D. *The Life and Work of Alexander Thomson*, London, 1979.
National Museum of Antiquities of Scotland. *Angels, Nobles and Unicorns*, Edinburgh, 1982.
Richardson, J S. *The Mediaeval Stone Carver in Scotland*, Edinburgh, 1964.
Proceedings of the Society of Antiquaries of Scotland, 1851–.

15 Education

ROBERT D ANDERSON

INTRODUCTION

The history of educational buildings has been determined by the constantly growing scope of education itself, and by the increasing role of the state since the mid nineteenth century. Although Scotland was notable for the early development of popular education and literacy, constant rebuilding and adaptation mean that few schools older than 1800 are in use today. There was legal provision for rural education from the seventeenth century, but satisfactory provision for the masses in the expanding towns came only when state grants to voluntary schools (mostly church-run) began in the 1830s. In 1872 education became compulsory, and elected school boards were set up to administer education in each burgh and parish. These gave way to county-wide education authorities in 1918, and after 1929 education became the responsibility of all-purpose local authorities – initially counties and cities, and from 1975 to 1996, regions. These developments led to periods of especially intensive building: the 1840s, the 1870s, the years between c. 1890 and 1914 (when secondary, technical and higher education were all expanding), and the period between 1945 and the 1970s, when demographic growth coincided with the progressive raising of the school-leaving age and the reorganisation of education on more comprehensive lines.

Most early schools were indistinguishable from the buildings around them; the first institutions with complex and specialised needs were the universities, followed later by burgh schools. Ordinary local schools followed vernacular styles, and until the early nineteenth century architects were likely to be used only by charitable endowments, which had both funds at their disposal and the desire to commemorate their founder in stone. State intervention, however, led to the imposition of central standards for constructional materials, space and hygiene, and schools came to follow uniform patterns. After 1872, the school boards normally employed leading local architects, some of whom developed large educational practices; Alexander Ross of Inverness, for example, built schools throughout the Highlands, and H and D Barclay of Glasgow were often called in for secondary or technical schools. But after 1929, it was local authorities' own architects' departments who did most of the work

RURAL SCHOOLS

A distinctive feature of Scottish education was the provision of a public school in every parish. An Act of 1696, the Charter of the Parish Schools, required the heritors to pay for a 'commodious house for a school', though it was not until 1803 that the obligation to provide a dwelling-house for the master was formally defined. In early schools, as in the many small private schools which survived well into the nineteenth century, the schoolroom was simply a large room in an ordinary house, with nothing to distinguish it externally (fig. 15.1). But the late eighteenth and early nineteenth centuries saw large-scale rebuilding of parish schools, reflecting the new wealth created by improved agriculture. The village school of this period usually consisted of a single large

Figure 15.1 Clandeboye School, Dunlop, Ayrshire; a rare example of an early schoolhouse (1641) which, as was common at the time, is located adjacent to a churchyard. Crown copyright: RCAHMS, SC 695921.

Figure 15.2 Gordon Schools, Huntly, Aberdeenshire. Crown copyright: RCAHMS, SC 710994.

schoolroom, holding up to 70–80 children, with a porch and bellcote; a ventilator on the roof was a characteristic Victorian addition. The teacher's house, whether attached or detached, was built alongside the school in the same style, but although the school came under the superintendence of the minister and presbytery, it was usually quite distinct from the church.

The work of local masons and wrights gave way to that of town architects, but architectural embellishments were common only for endowed schools, or those built by landlords as part of an estate village; the Gordon Schools at Huntly, Aberdeenshire (fig. 15.2), are an especially elaborate example of the latter. The mid nineteenth century saw a great expansion of additional schools, built not only by landowners, but also by the churches (especially the Free Church after 1843), and by employers in isolated factory or colliery settlements. One of the commonest types was the separate school for girls, under a mistress who could teach sewing as well as reading and writing, and such schools were a favourite field of action for charitable ladies.

In 1872 the school boards took over most existing schools, and their policy was to close down the smaller schools and concentrate work on the parish (now renamed 'public') school. New schools might be built to serve outlying areas, and there were constant extensions and adaptations, but most rural school boards had no need for a large building programme. The exception was the Highlands and Islands, including Orkney and Shetland, where lack of resources had always hampered the parish school system, and where there were numerous small and inadequate schools which did not meet the Education Department's demand for stone walls, slated roofs, and boarded floors. Highland school boards were thus forced to embark on much new building in the 1870s.

In industrialised parishes and small towns, schools were similar to the urban types discussed below. But the purely rural school continued to have a single teacher, teaching children of all ages. There was limited scope for pedagogic innovation, and by the early twentieth century there was further pressure, encouraged by rural depopulation and improved transport, to close down smaller schools and concentrate their work in better-equipped centres. This process was much accelerated after World War II by the reduction of agricultural employment, though limits were imposed both by the geography of Scotland and by the resistance of local communities to the loss of a valued facility. Thus, although many rural schools have been converted into private houses, many others built in the nineteenth century, both before and after 1872, are still in use.

URBAN ELEMENTARY SCHOOLS

The parish school legislation did not apply to royal burghs, and the provision of popular education there was left largely to private enterprise or charitable effort. There were many small schools run by a single teacher, but they did not need distinctive buildings and have left little historical trace. This changed at the beginning of the nineteenth century, when rapid urban growth, the problems of industrialisation, and evangelical religious concern for social problems created a new interest in elementary education. One aspect of this was the development of teaching techniques for handling large numbers, notably the 'monitorial' system which used older children as assistants. The rival Madras and Lancasterian systems each made special demands for the arrangement and furnishing of schoolrooms. In Scotland, David Stow developed his own theories, which relied less on monitors, but introduced the 'gallery', a tiered floor which became a standard feature of nineteenth-century schools, especially for teaching infants. Stow also founded the Normal Seminary at Glasgow for training teachers in the new techniques; these training colleges, with their accompanying practising schools, were prominent institutions in the cities.[1]

The promotion of new methods and the professionalisation of teaching were accompanied by the intervention of Church and State. By 1872 the voluntary school, run by a church, a charity or a philanthropic body, but receiving state grants and following a standard educational pattern, was the norm for working-class children. Episcopalians and Roman Catholics as well as Presbyterians built church schools; these were to remain outside school board control until 1918, and were usually built adjoining the church. Specialised types of urban school included 'half-time' schools for children working in factories, and 'ragged', 'industrial' or reformatory schools for destitute or delinquent children. All these developments required larger and more permanent buildings. At first, schools had only a single trained teacher, and were hall-like buildings not very different from parish schools (though without the teacher's house). Endowed schools could be more elaborate, like the 'Dr Bell's' schools (classical at Inverness, Gothic at Leith), or the free schools built by the Heriot's Trust in Edinburgh. The first stage of evolution was to add to the large schoolroom one or two classrooms which could be used for teaching smaller groups in turn, or for girls or infants. The next was to base teaching on separate classes, as more work was entrusted to assistants, either adults or teenage 'pupil-teachers' who began their training on the job. By the 1860s urban schools with several hundred pupils were becoming common.[2]

The 1872 Act had much greater impact in towns than in the countryside. Most of the existing schools were transferred to the school boards, which closed many of them down as unhygienic and obsolete. At the same time, compulsory education brought in children who had previously escaped. A very large number of schools was thus built in the 1870s and 1880s, and they became a familiar feature of the townscape. Where space permitted, schools had only one or two storeys, and the gables of the classrooms provided the main external feature. On central sites, where schools were often designed for over 1,000 children, they rose to three storeys, and their plainness matched the surrounding tenements. An asphalt playground, covered shelters, and a separate house for the janitor were standard features, as were separate entrances and staircases for boys and girls, though they sat together in the classrooms. Internally, furniture and equipment were now provided by commercial suppliers on standard patterns.

The early board schools were planned around a central hall, which allowed the head teacher to supervise the assistants in their classrooms. But after 1900 the greater availability of fully-trained staff and a new emphasis on fresh air and hygiene led to more flexible planning, with classrooms arranged along corridors or as distinct 'pavilions'. There was also a need, as primary education developed beyond the basics, for special rooms for needlework, cookery, art, or woodwork. Nevertheless, schools of the traditional kind were built in inner-city areas until World War I, and classes of 60 or more remained normal. Charles Rennie Mackintosh's Scotland Street School, Glasgow (1906; fig. 15.3), though architecturally

Figure 15.3 Scotland Street School, Glasgow. Crown copyright: RCAHMS, SC 710854.

novel, was quite orthodox in providing for 1,250 children in 21 classrooms; apart from the entrance hall, also used for drill, the only specialised space was a cookery room.[3]

Between the two World Wars, school building was mainly

associated with suburban expansion, and since space was less constricted, primary schools could be kept to one storey. Apart from some experiments with 'butterfly' plans, their internal arrangements and outward appearance remained conventional, although styles were simplified. Much greater change took place after World War II. The movement of population from city centres to council housing schemes and new towns required large building programmes, and pedagogic ideas moved away from desk-bound classroom teaching towards 'open-plan' schools with flexible interior spaces. There was much experiment with new building materials, and some use was made of standard plans and prefabricated construction. No distinctively Scottish type, however, emerged. In general, primary schools became far more cheerful and humane, and the older Victorian schools, most of which continued in use, though with smaller rolls and classes, were adapted to the new child-centred methods.

BURGH AND SECONDARY SCHOOLS

Although there were no parish schools in burghs, it was customary for the town council to maintain at least one burgh school. Traditionally this was a grammar school teaching Latin, but from the seventeenth century burghs began to appoint additional masters for more modern and commercially useful subjects. There were 'writing' and mathematical schools, and sometimes 'English' schools for elementary teaching. Each master was independent, and the schools might or might not occupy the same building. Early burgh schools did not differ externally from other town houses, and where eighteenth-century examples survive, as at Banff, Paisley and Edinburgh, they are substantial but plain. From about 1790, however, there was a movement to modernise the curriculum and house the different schools together in a new building, usually paid for by public subscription. In these 'academies', the newly fashionable name, the masters retained much independence, and the main requirement was therefore for a set of separate classrooms. The commonest pattern was a single-storeyed classical building, with a central pedimented frontispiece, low wings, and end pavilions. A hall behind the frontispiece was used for circulation and assemblies, but not for teaching. Scottish academies thus differed significantly from English grammar schools, where a large common schoolroom remained the norm. Burgh schools were also distinctly secular institutions, without chapels, and this was one reason why classical styles were preferred to Gothic. Dundee High School (George Angus, 1832–4; figs 15.4, 15.5, 15.6) is a monumental example of the classical type, and Thomas Hamilton's Edinburgh High School (1825–9), though architecturally complex, also conforms to it, as does Edinburgh Academy (William Burn, 1823–4), a 'proprietary' school

Figure 15.4 Dundee High School; view by Joseph Swan in Charles Mackie. *Historical Description of the Town of Dundee*, 1836.

Figure 15.5 Dundee High School, 1956. Crown copyright: RCAHMS, SC 710720.

EDUCATION • 301

Figure 15.6 Dundee High School, 1956. Crown copyright: RCAHMS, SC 710734.

Figure 15.7 George Heriot's Hospital, Edinburgh. Crown copyright: RCAHMS, SC 710871.

founded privately rather than a burgh school. The plan could be adapted easily to smaller schools, and a second floor could be added where there were more complex requirements, as at Dollar Academy, Clackmannanshire (William Playfair, 1818–20), and John Watson's Hospital, Edinburgh (William Burn, 1825).[4]

The residential hospitals found in certain towns were distinct from burgh schools, being founded to give free education to deserving pupils. The first was George Heriot's, Edinburgh (fig. 15.7), begun in 1628, and one of the most influential Scottish buildings of the period. It originally housed 180 boys. The other early example still in use is Robert Gordon's at Aberdeen (William Adam, begun c. 1731), but the eighteenth-century foundations at Edinburgh (the Merchant Maiden for girls, and George Watson's for boys) have lost their original buildings which, in the case of George Watson's, survived as the nucleus of the Royal Infirmary. The founding of hospitals continued into the nineteenth century, and the buildings were usually designed to impress onlookers with the founder's wealth and benevolence (for example, Donaldson's and Daniel Stewart's in Edinburgh); a few were also founded in smaller towns. But later in the century the residential principle was condemned, and the hospitals were turned into large day-schools. They had always been untypical, for most benefactors either contributed to the rebuilding of burgh schools,

EDUCATION • 303

or founded schools of a more orthodox type; Dollar Academy and Madras College at St Andrews, Fife (William Burn, 1832–4), were two lavish examples, designed to accommodate both day pupils from the town and fee-paying boarders.

Alongside the burgh schools were many private schools, especially for girls, but these were usually in converted houses, and left little trace when they vanished. Converted premises also served those boarding schools which aspired to the English public school model, with two notable exceptions: Trinity College, Glenalmond, Perthshire (John Henderson, 1847), an Episcopalian foundation in appropriate Gothic garb; and Fettes College, Edinburgh, where the boarding houses were disposed around David Bryce's central block of 1864–70.

The nineteenth century saw much reform of secondary education; it was adapted to middle-class needs, and more strictly differentiated from elementary education. The rebuilding of burgh schools was a constant process. The Scots Baronial Aberdeen Grammar School (James Matthews, 1863) is a good example of a mid Victorian rebuilding, and, as in many cases, it involved moving from a city-centre site to the residential suburbs. In 1872 the various academies, grammar schools and high schools were transferred to the school boards, who continued to develop them as modern secondary schools. Outside the large cities, burgh schools had always admitted girls as well as boys, so the new demand at this time for girls' secondary education did not generally require new schools. There was a demand, however, for more accessible and less expensive schools linked to the elementary sector, and boards responded to this in the 1880s by founding 'higher grade' schools. Their buildings followed the usual board school pattern; they were most numerous in Glasgow, where Hillhead (H and D Barclay, 1883) enjoyed especially high prestige. Other developments at the end of the nineteenth century were related to the development of science, practical subjects and recreational activities, and schools sprouted annexes to house the required laboratories, workshops, gymnasia, assembly halls, and cadet corps armouries.

The inter-war years saw a limited amount of rebuilding of existing schools, and the expansion of selected elementary schools for secondary or quasi-secondary education. Thus when education was reorganised after World War II on the principle of 'secondary education for all', the existing schools could serve either as senior or junior secondary schools (the former being selective). As with primary education, however, demographic shifts soon required large-scale building, and when comprehensive education replaced selection in the 1960s, Scottish local authorities opted for single schools taking all pupils from twelve to eighteen; combined with the co-education which was normal in Scotland, this implied very large schools. Many older schools finally abandoned

Figure 15.8 Knightswood Secondary School, Glasgow; designed 1938, begun 1954, this was one of a number of similar commissions by Gillespie, Kidd & Coia, architects. Crown copyright: RCAHMS, SC 713493.

their historic buildings for the suburbs; some continued as schools of other kinds; others found new uses as flats, council offices, or hotels. The new buildings of the 1950s and 1960s were usually curtain-walled blocks (fig. 15.8), but later designs were more flexible in composition, sometimes with neo-vernacular elements. The new schools, however, seldom had the impact on the urban scene of the old.

UNIVERSITIES

Scotland was noted for the early development of its universities. St Andrews, Glasgow, and King's College at Aberdeen were founded in the fifteenth century, Edinburgh and Marischal College, Aberdeen (fig. 15.9), at the end of the sixteenth (the two Aberdeen colleges were not combined until 1860). Of these, only St Andrews and King's College retain original buildings, notably their chapels. Alongside the chapel, teaching and lodging accommodation was provided in quadrangles or ranges, and Glasgow was rebuilt on this pattern in the 1630s. There were

Figure 15.9 University of Aberdeen, Marischal College, 1906. Crown copyright: RCAHMS, SC 696430.

no dining halls on the Oxford or Cambridge pattern, but until the early eighteenth century universities remained residential. Most students were young, were aiming at careers in the church, and were kept under strict discipline. This changed because of the growing attraction of liberal education to the upper and middle classes, the replacement of the tutorial or 'regenting' system by professorial lectures, and the successful establishment of medical schools at Edinburgh and Glasgow. As the universities expanded, students lived at home or in private lodgings, and their 'chambers' were used for other purposes. By 1789, when Robert Adam began the rebuilding of Edinburgh, the essential requirement was for a set of large lecture-rooms, along with a library, laboratories and apparatus rooms, an anatomy theatre, and several museums. All this was provided in Adam's plan, though the building was completed by William Playfair between 1817 and 1840.[5] At Glasgow, the same needs were met by ad hoc additions, until the university moved to George Gilbert Scott's building at Gilmorehill in 1870. This included, like the old college, a set of houses for the professors.

The universities enjoyed a period of expansion and prosperity from the 1870s until 1914, with University College at Dundee joining their number in 1883. The demands of science and medicine were especially pressing, and the old buildings were soon outgrown; the new medical school at Edinburgh (by Robert Rowand Anderson) opened in 1884. At Aberdeen, Marischal College was rebuilt at the turn of the century; its Mitchell Hall for graduations, like the McEwan Hall at Edinburgh (Robert Rowand Anderson, 1888–97), was symptomatic of a new taste for public ceremony. Another new feature was the growth of sport and social or 'corporate' life among students: the Students' Union at Edinburgh (Sydney Mitchell & Wilson, 1887–8) was the chief example of a new type. But the Scottish universities remained essentially non-residential, and although a few halls of residence appeared at the end of the nineteenth century, these were private ventures.

After World War I there was a brief period of expansion, and it was then that Edinburgh began its separate scientific campus at King's Buildings. But a real building boom came only in the 1960s and 1970s, in response to government policy. Student numbers increased three- or four-fold, and the provision of student residences became an essential task. At Edinburgh and Aberdeen these were concentrated in campuses, at Pollock Halls and Hillhead. Other teaching and research needs, which included a notable clutch of new libraries, were generally met by expansion on or near the existing sites. In Edinburgh and Glasgow, the resulting demolitions, high-rise building, and encroachments on the urban fabric did not pass without controversy.

Most research institutes in Scotland have been part of universities, or sponsored by the government in areas like agriculture and fisheries. The Marine Biological Research Station at Millport, Great Cumbrae, Bute (1896), is a small but distinctive example.

TECHNICAL EDUCATION

The organisation of technical and adult education can be traced to the 'Andersonian University' founded at Glasgow in 1796, and the

Mechanics' Institutes which most towns acquired from the 1820s. The latter provided rooms for meetings and evening classes, a library, and sometimes a museum, and it was only in the late nineteenth century that other institutions began to take over their functions; some of their buildings survive in other uses. Technical education became a matter of state concern in the 1880s, and in the cities large technical colleges were created either by remodelling existing institutions and endowments (for example, the Glasgow and West of Scotland Technical College based on the Andersonian University, Heriot-Watt College at Edinburgh, and Robert Gordon's Institute at Aberdeen) or from scratch (for example, Dundee Institute of Technology). Legislation of 1890 encouraged technical education in smaller towns. Most of it was in the form of evening classes, which could use ordinary schools, but the need for specialised laboratories and workshops prompted several towns to build separate technical colleges; Coatbridge, Lanarkshire, and Elgin, Moray, possess early examples. The same movement produced colleges of agriculture, domestic science, and art: the first phase of C R Mackintosh's Glasgow School of Art was completed in 1899; its Edinburgh equivalent, by J M Dick Peddie, in 1906.

A second period of expansion came only after World War II, when a largely new sector of 'further education' for adults appeared, and technical colleges developed more full-time and degree-level work. The colleges at Glasgow and Edinburgh became Strathclyde and Heriot-Watt Universities in the 1960s; Strathclyde developed a large campus in central Glasgow, but Heriot-Watt moved from its 1870s–80s buildings to a purpose-built site outside the city. Other colleges were to achieve university status in the 1990s, and massive rebuilding and the abandonment of central sites were common phenomena.

LIBRARIES

One unusual survival is the rural endowed library at Innerpeffray, Perthshire, which dates from the 1690s, although the existing building is of c. 1750. Small village libraries were not uncommon in the late eighteenth and early nineteenth centuries, and the same period saw the growth of socially-select subscription libraries in the towns. The Mechanics' Institutes, or separate 'Mechanics' Libraries', made books accessible to a wider clientele, and their book stocks were often the basis of the public libraries founded later by town councils. Free municipal libraries were authorised by an Act of 1853, but town councils were slow to take action. At Dundee, the second town to adopt the Act, the library was opened in 1869 as part of the Albert Institute (George Gilbert Scott, 1867), which was also to house a museum and art gallery. Most towns built their libraries between 1890 and 1914, usually with financial help

Figure 15.10 Gorbals District Library, Glasgow; general reading room, c. 1907. From *Glasgow Corporation Libraries*, npd (c. 1907), 35.

from the philanthropist Andrew Carnegie. Branch libraries in cities began after 1900, Glasgow being the pioneer (fig. 15.10), and these were often attractive small buildings; but a county library service serving rural areas was not established until 1918. The grandest municipal library was the Mitchell at Glasgow, originating in a legacy of the 1870s, but not built until 1906–11 (William Whitie).

The libraries of universities and learned bodies often formed fine rooms within a larger complex. The Parliament House in Edinburgh contained both the Signet Library (William Stark, 1812–18) and the Advocates' Library (William Playfair, 1830–2). The latter eventually became the National Library of Scotland, and was given its own building, designed by Reginald Fairlie in the 1930s, but not opened until the 1950s.

NOTES

1. For these developments, see Markus, 1982.
2. For the influence of teaching techniques on school planning, see especially Seaborne, 1971, and (including Scottish material) Hamilton, 1989.
3. Williamson, Riches, Higgs, 1990, 512–13. This school was opened as a museum of education by Strathclyde Regional Council in 1990.

4. The various Edinburgh schools are described in Gifford, McWilliam, Walker, 1984. John Watson's Hospital is now accessible to the public as the Scottish National Gallery of Modern Art.
5. Fraser, 1989, a comprehensive account.

BIBLIOGRAPHY

Aitken, W R. *A History of the Public Library Movement in Scotland to 1955*, Glasgow, 1971.
Anderson, R D. Secondary schools and Scottish society in the nineteenth century, *Past and Present*, 109 (1985), 176–203.
Anderson, R D. *Education and Opportunity in Victorian Scotland: Schools and Universities*, Oxford, 1983.
Anderson, R D. Scottish Universities. In Holmes, 2000.
Brown, I G. *Building for Books: The architectural evolution of the Advocates' Library, 1689–1925*, Aberdeen, 1989.
Cant, R G. *The College of St Salvator: Its foundation and development*, Edinburgh, 1950.
Fraser, A G. *The Building of Old College: Adam, Playfair and the University of Edinburgh*, Edinburgh, 1989.
Gifford, J, McWilliam, C, Walker, D. *The Buildings of Scotland: Edinburgh*, Harmondsworth, 1984.
Gilchrist, A. The parish school buildings of Upper Clydesdale, 1872–1975, *Scottish Local History*, 17 (1989), 10–13.
Glaister, R T D. Rural school buildings in the eighteenth century, *ROSC*, 14 (2001–2002), 93–104.
Hamilton, D. *Towards a Theory of Schooling*, London, 1989.
Holmes, H, ed. *Institutions of Scotland: Education* (Scottish Life and Society: A compendium of Scottish ethnology, Vol 11), East Linton, 2000.
Horn, D B. *A Short History of the University of Edinburgh, 1556–1889*, Edinburgh, 1967.
Knox, H M. *Two Hundred and Fifty Years of Scottish Education, 1696–1946*, Edinburgh, 1953.
McKean, C. *The Scottish Thirties: An architectural introduction*, Edinburgh, 1987.
Markus, T A. The school as machine: Working-class Scottish education and the Glasgow Normal Seminary. In Markus, T A, ed, *Order in Space and Society: Architectural form and its context in the Scottish Enlightenment*, Edinburgh, 1982, 201–61.
Murray, D. *Memories of the Old College of Glasgow: Some chapters in the history of the University*, Glasgow, 1927.
Seaborne, M. *A Visual History of Modern Britain: Education*, London, 1966.
Seaborne, M. *The English School: Its Architecture and Organization*, Vol 1, *1370–1870*, London, 1971.
Seaborne, M, Lowe, R. *The English School: Its Architecture and Organization*, Vol 2, *1870–1970*, London, 1977.
Simpson, I J. *Education in Aberdeenshire before 1872*, London, 1947.
Williamson, E, Riches, A, Higgs, M. *The Buildings of Scotland: Glasgow*, Harmondsworth, 1990.

16 Health and Welfare

HARRIET RICHARDSON *and*
GEOFFREY STELL

Despite enjoying an early and solid reputation for medical learning – the first Chair of Medicine in Britain having been established at Aberdeen in the late fifteenth century – by the end of the seventeenth century Scotland did not have any hospitals in the modern sense of the word. At that date hospitals were still generally seen as charitable organisations for the poor, not medical institutions, and they could range from a small poorhouse or almshouse to a great charitable school, such as Heriot's Hospital in Edinburgh.

GENERAL HOSPITALS

In the early eighteenth century moves were made to provide in-patient medical care, and the first hospital was founded in Edinburgh. The foundation of the Royal Infirmary of Edinburgh (fig. 16.1) in 1729 and, more specifically, the large new premises which were erected for it between 1738 and 1748 to the designs of William Adam, thus marked the beginnings of post-medieval hospital design in Scotland. It was also the first teaching hospital to be established in Britain outside London, enabling many of Scotland's medical practitioners to gain clinical experience at home and not to have to complete their training abroad, particularly at Leiden in the Holland.

Few other hospitals were established in Scotland during the eighteenth century, and until the end of the century Edinburgh's infirmary remained by far the largest. In Glasgow, the Town's Hospital, established as the city workhouse, made limited medical provision for the poor. It was designed and built by Allan Dreghorn and John Craig in 1732. Engravings show it to have been both smaller and simpler than the Edinburgh infirmary, although both were of U-plan layout. William Christall's design for the infirmary at Aberdeen, begun in 1740, may have been derived from the hospitals in Glasgow and Edinburgh, but no record of its appearance has been found. Early engravings of Dumfries Royal Infirmary, built in 1778, show a plain elevation in the style of a minor country house.

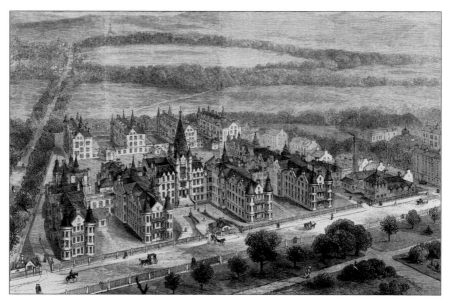

Figure 16.1 Royal Infirmary, Edinburgh; bird's-eye view of David Bryce's 1870–9 design in *The Illustrated London News*, 1 November 1879.

Dispensaries, which began to appear at the end of the eighteenth century, were often the forerunners of provincial hospitals, or supplemented existing hospitals in the major cities, offering treatment to the sick poor. In Edinburgh, for example, the Royal Public Dispensary was founded in 1776 and was followed by the New Town Dispensary in 1815. The first provincial dispensary was opened in 1777 at Kelso, Roxburghshire, which eventually, after about 130 years, was superseded by a cottage hospital.

In its original simple form, the late eighteenth-century Royal Northern Infirmary at Inverness owes much to the Glasgow Town's Hospital and shows how little hospital design had progressed over the century. By contrast, the near-contemporary Glasgow Royal Infirmary, which was built to the design of Robert Adam and opened in 1792 (the year of the architect's death), was more sophisticated and complex in both plan and elevation. A surviving Georgian hospital which shares something of the swagger of Adam's Glasgow Royal is Gray's Hospital in Elgin, Moray. It opened in 1819, having been designed by James Gillespie Graham whose name had been suggested to the building committee by the Earl of Moray, Graham having worked for the earl on his Edinburgh New Town estate.

A small group of provincial hospitals built in the second quarter of the nineteenth century generally had elegant classical elevations fronting variations of the corridor plan, closer to country house design than to other types of institutional building. Perth and Montrose, Angus, have particularly good examples of the genre dating from the 1830s. The few general hospitals built in the mid nineteenth century are richly varied. Archibald Simpson's graceful Aberdeen Royal Infirmary, built in 1840 of glittering grey granite, contrasts strongly with the fussy appearance of Dundee Royal Infirmary, created 15 years later by the English architects, Coe & Godwin. Chalmers Hospital in Banff, 1864, was a product of another English architect, William Lambie Moffatt, who made a career for himself in Scotland drawing shaped gables onto poorhouses throughout the country.

By the mid 1860s, Florence Nightingale's influence on hospital design was gathering strength, the Herbert Hospital, Woolwich, London, being one of the first hospitals to be designed on the principles which she so energetically publicised. She advocated the pavilion-plan, which comprised distinctive and long ward blocks, with windows on either side and end-towers containing sanitary facilities, with stairs and linking corridors at the other end. This plan was initially adopted widely in Poor Law infirmaries, an early example in Scotland being the Edinburgh City Poorhouse at Craiglockhart. By the closing decades of the nineteenth century it was almost unthinkable to design a general hospital on any other plan.

Small general hospitals began to be provided in lesser towns and rural areas, and in 1865 the first cottage hospitals in Scotland appeared at Crimond, Aberdeenshire, and St Andrews, Fife. A standard plan soon emerged. Male and female wards, usually in the form of single-storeyed wings, were attached to central administration sections which were generally domestic in appearance. A typical example is Leanchoil Hospital, Forres, Moray, which was established in 1888 to designs by John Rhind. The administration block at the centre is dominated by a square tower, and the main entrance gives access on either side to the linking corridors to the wards. The wards were shrunken versions of pavilion wards with sanitary facilities housed in the turret, divided from the main ward by a short, cross-ventilated lobby.

Towards the end of the nineteenth century an unusual type of circular ward enjoyed a brief period of popularity. The circular plan was evolved partly in response to a continuing obsession with ventilation, as elaborate flues could be housed in the centre, and partly as a way of taking advantage of even the most awkward urban site. A circular tower could occupy the north corner of a site and still provide for its inmates a degree of sunshine, light and air. Essentially a Nightingale ward, the beds were placed between opposing windows, and the sanitary facilities were in an annexe linked to the ward by a narrow, cross-ventilated lobby. In Scotland only two hospitals opted for this novelty: Kirkcaldy Cottage

Figure 16.2 Millbank Pavilion, Astley Ainslie Hospital, Edinburgh; a typical unit of this convalescent home associated with Edinburgh Royal Infirmary which was developed between 1925 and 1939 along similar lines to contemporary sanatoria. Courtesy of Dr Ken Liddell, Wishaw General Hospital.

Hospital, Fife, now demolished, where plans for a circular ward were first drawn up in 1895; and the Royal Alexandra Infirmary, Paisley, designed by the local architect, T G Abercrombie, and built in 1897–1900. Abercrombie's design comprised a series of rectangular pavilion ward blocks on the south side of the site and one three-storeyed circular block to the north.

Not until the early twentieth century did the basic design of patient accommodation begin to change. With improvements in the understanding of the transmission of diseases, sanitary facilities were once more allowed within hospitals and were not relegated to chilly towers. In the inter-war period, sunshine treatment, with the necessary sun-rooms and sun-balconies, was also introduced into many hospitals (fig. 16.2). It was then that the first hospitals without traditional pavilion wards emerged, with internal accommodation comprising smaller units of four or six beds placed parallel to the windows; the forerunners of the Falkirk – or racetrack – wards.

In the post-war decades, the old orthodoxy of control of infection in open pavilions was superseded by the American idea of compact, highly-serviced buildings subdivided into smaller, more private rooms. This coincided with a trend, begun by Scotland's first new Modern Movement general hospital, Vale of Leven, Alexandria, Dunbartonshire (1952–5, Keppie, Henderson & Gleave), towards more 'open-ended' hospital planning. A government-sponsored programme of architectural research in the 1960s associated this trend with a striking Modernist image: the multi-storeyed slab block of 'racetrack' wards, with services in the centre and small ward rooms around the outside in a loop. After initial experiments in the early 1950s, the first proper racetrack multi-storeyed block in Scotland was designed by Gillespie, Kidd & Coia at Bellshill Maternity Hospital, Lanarkshire (1959–62). The type was then subjected to user study in an experimental prototype block of 1963–6 at Falkirk Royal Hospital, Stirlingshire, designed by Keppie Henderson & Partners in collaboration with a Scottish Home & Health Department study team (comprising a nurse, doctor and architect).

Once it had established appropriate standards, the Department then set in train a big multi-storeyed building drive by the regional hospital authorities. This included new suburban developments, at Gartnavel General Hospital, Glasgow (1968–73, by Keppie Henderson), Aberdeen Royal Infirmary, Forresterhill (from 1964, by C C Wright, Chief Regional Architect of the North East Regional Hospital Board), and the huge brick-faced monolith of Inverclyde Royal Hospital, Greenock, Renfrewshire (1977–9, by Boswell, Mitchell & Johnston). It also included megastructural inner-city redevelopments, at Glasgow's Western Infirmary (1965–74, by Keppie, Henderson & Partners), at Glasgow Royal Infirmary (1971–82, by Spence, Glover & Ferguson), and at Edinburgh Royal Infirmary Phase 1 (1975–81, by Robert Matthew Johnson-Marshall, a partial redevelopment preceded by 20 years of discussion and abortive projects for rebuilding the entire infirmary complex).

The climax of this post-war hospital building campaign, however, represented a major change of course. Robert Matthew Johnson-Marshall was responsible for the research-led design of Ninewells Hospital in Dundee (from 1961, completed 1974), Scotland's first new teaching hospital in the twentieth century, and another extremely protracted project. At Ninewells there was a move away from the free-standing multi-storeyed slab, towards a low, dense complex disposed around service spines and courtyards, and strung out along a sloping site. The hilly contours were exploited to allow an essentially horizontal, rather than piled-up, separation of traffic and functions. The 1980s and 1990s continued this trend away from racetrack slab blocks, towards layouts which combined spreading roofs and a somewhat 'domestic' image with a continuing concentration of services. In the *RIAS Review of Scottish Architecture 1992*, SBT Keppie's low, hipped-roofed Ayr Hospital was claimed by its designers to combine 'the welcoming character and scale of a cottage hospital with the overtones of a hotel'. Following the eventual decision to abandon attempts to redevelop Edinburgh Royal Infirmary on its central site, a new complex was built on the south-eastern outskirts of the city in 1999–2002; it continued the general pattern established at Ninewells of a dense but medium-height grouping.

POORHOUSES

Some larger burghs had long possessed poorhouses or charitable hospitals. In Perth, The King James VI Hospital was founded in 1569 to provide, 'by all honest ways and means an Hospital for the poor, maimed, distressed persons, orphans and fatherless bairns within our burgh of Perth'. Destroyed by Cromwell in 1652, its replacement was begun in 1748. It remained an almshouse for the poor until 1812 when Perth adopted a system of outdoor relief only. In Glasgow, the Town's Hospital was built in 1733 as the city workhouse, and in Aberdeen a poorhouse was established in 1739 when the Royal Infirmary was founded. Edinburgh had a charity workhouse by 1742, and a poorhouse was built at Ayr in 1756.

A statute of 1579 which provided for the 'punishment of strong and idle beggars and the relief of the poor and impotent' had represented the first attempt to provide for the poor from public funds. It was administered by kirk sessions, with church collections and other revenues recognised as the Poors' Fund. The assistance provided was almost entirely outdoor relief, usually money, as opposed to accommodation. One method of raising the fund was by an assessment on the parish. The number of assessed parishes rose from only three in 1700 to eight in 1740, 145 in 1818 and 230 by 1845. The formation of the Free Church of Scotland at the Disruption of 1843 made continued administration by the Established Church impossible and an alternative system was urgently needed.

The recommendations of a Royal Commission set up in 1844 formed the substance of the Poor Law (Scotland) Amendment Act of 1845. The needs of the deserving poor were recognised, and a new type of poorhouse, akin to almshouses, where the poor would feel no shame to enter, were to be provided. The main aim of the system was to provide a home for the able-bodied, as well as for the aged and impotent poor – a strong contrast with England where, in the system of workhouses introduced under the Poor Law Amendment Act of 1834, able-bodied paupers were required to undertake menial work.

The administration of the 1845 Act was entrusted to annually-elected Parochial Boards. Although the Act allowed each parish, or adjoining parishes, with a population greater than 5,000 to erect a poorhouse, it did not make this compulsory, nor did it stipulate how the funds should be raised. Overall control was vested in a Board of Supervision in Edinburgh, which had limited powers and heard complaints of mistreatment from individual paupers. However, plans for building new poorhouses or altering existing buildings had to be submitted to and approved by the Board.

A model plan was published in 1847 by the Board of Supervision in its Second Annual Report. Its cottagey, Elizabethan style was welcoming and conveyed an impression of charity rather than punishment. The design was based on the medieval ideal presented by A W N Pugin in the 1841 edition of his *Contrasts* in which he contrasted a medieval monastic poorhouse with the characterless and bleak contemporary workhouse of the Sampson Kempthorne model; adopted for the majority of the early workhouses in England, these followed a cruciform plan, derived from Benthamite radially-planned prisons. On an H-shaped plan, the front range of the Scottish model was of two storeys instead of Pugin's three, a reflection of the generally smaller size of the average Scottish poorhouse; it also lacked Pugin's strong emphasis on the church in his monastic arrangement.

About 70 of these poorhouses were built in Scotland, most of them shortly after the introduction of the 1845 Act. In 1848 there were 14 poorhouses, including those of much earlier date. By 1859 a further 19 had been built and by 1870 there were 30 more. Although most of the designs took the model plans as their starting-point, the Board of Supervision had no power to enforce the recommended model. Even where Parochial Boards sympathised with the ideal, the cost of implementation was often prohibitive, a point made by the chairman of the Easter Ross Board in 1851:

> The Easter Ross Poorhouse is a small building of its class, and ... its cost per head of the inmates has been unusually low ... by strictly limiting everything to its lowest amount compatible with efficiency ... The only thing that can be said for it is, that it is probably just enough for its purpose, and ... it would perhaps not have been very easy to persuade the parishes in Easter Ross to consent to a more perfect and expensive building, such as that given in the second annual report of the Board of Supervision.

The Kirkcaldy and Dysart Poorhouse in Fife is a typical, small-scale version of the model, built in 1850. The plans were prepared by the Edinburgh-based architect, William Lambie Moffatt, who had been a pupil of William Burn and had worked in England with William Hurst. The building follows the basic H-shaped plan recommended by the Board of Supervision, with the front block providing living accommodation, the dining-hall and kitchen in the centre to the rear, and a single-storeyed storage range. A separate rear hospital block was built later. By 1900 this poorhouse provided accommodation for 130 residents.

A principal requirement of poorhouse planning after 1845 was the classification of the inmates. In addition to the symmetrical division between male and female sections, many poorhouses had hospital accommodation, and a separate department for children was also recommended. Once the poorhouse test was firmly established, it also became necessary in the planning of larger poorhouses to differentiate between the deserving poor and the not so deserving, and to allocate areas accordingly. One of these larger poorhouses was built in 1867 at Craiglockhart, Edinburgh. It was designed by the firm of George Beattie & Sons, whose principal architect was William Hamilton Beattie. This poorhouse has distinct Scots Baronial details, the keynote of the building being the four-storeyed octagonal entrance tower with its faceted, steep-pitched roof.

In the larger poorhouses there were generally three main sections: the asylum, the poorhouse proper, and the hospital or infirmary section. Although recognisably architect-designed, they were still quite plain. When the plans for Aberdeen's Oldmill Poorhouse, now Woodend Hospital, appeared in the *Aberdeen Daily Journal* in 1901, the report noted that:

> As the general view of the poorhouse to most people will be from the Skene Road, a few hundred yards away, it is not intended that any expense should be put upon fine masonry details, and the effect of a satisfactory composition will, therefore, be obtained by means of the grouping of the various buildings and arranging them in such a fashion as to give a suitable yet dignified appearance to the whole.

By 1900 it was clear that the Poor Law was unsatisfactory, and another Royal Commission was appointed to look into the state of destitution and distress among the poor. Its report, published in 1909, showed the failure of the poorhouse system, particularly in the provision of classified accommodation, which had resulted in an increasing number of mixed poorhouses where non-medical and medical cases occupied the same standard of accommodation. The larger poorhouses had made some provision for hospital cases but only Glasgow had attempted to solve the problem. In 1904, three new hospitals for the poor were opened at Stobhill, Oakbank and Duke Street. These institutions provided free medical care to the city's paupers, and the old poorhouse at Barnhill remained as the residence of those poor not requiring medical treatment.

Generally the conditions of the poor began to improve, directly or indirectly, following the introduction of new legislation: the Old Age Pensions Act of 1908; the National Insurance Act of 1911; and the Local Government (Scotland) Act of 1929. With the 1929 Act, Public Assistance departments were created and the old poorhouses were renamed Public Assistance Institutions. Boundary changes around Glasgow meant that Glasgow Corporation found itself responsible for a much larger area which included the former Renfrew Combination Poorhouse. In 1937 William Barrie prepared plans for a major development on this site, comprising a symmetrical layout of twelve two-storeyed blocks arranged around a horse-shoe shaped garden area. Each block contained small flats for the elderly, and some also catered for married couples. Each flat consisted of a living room, bedroom, kitchenette and bathroom. Residents had the facilities to make their own meals and, with the exception of dinner, were given rations to do so if they wished. The main block of the former poorhouse was converted to dining rooms and day-rooms. This was the first design to cater for the needs of elderly residents rather than allow for easy administration. A major step forward in housing the poor, particularly the elderly poor, Crookston Cottage Homes on the outskirts of Glasgow became a model for subsequent developments throughout Britain.

LUNATIC ASYLUMS

Progress in the understanding and treatment of mental illness was less rapid than in general medicine. The scenes of London's Bedlam depicted by Hogarth demonstrated appalling conditions for the insane, which for some observers were not acceptable. In Scotland, it was the fate of the poet Robert Fergusson, who died in the Bedlam attached to the Edinburgh Charity Workhouse in 1774, that moved Andrew Duncan to establish the Royal Edinburgh Asylum. The cells of Bedlam were cold and damp, were without fireplaces or other means of heating, and were windowless, only gratings in the door allowing any light or fresh air into them.

Between 1780 and 1840 seven Royal Asylums were built in Scotland: at Montrose, Aberdeen, Edinburgh, Glasgow, Dundee, Perth and Dumfries. Widespread public prejudice against lunatics created the greatest single obstacle for the individuals involved in these foundations. In Montrose in 1779, Susan Carnegie enlisted the help of Provost Christie to raise subscriptions to establish an asylum. At that time the insane were imprisoned in the tolbooth in the High Street, and it was Mrs Carnegie's aim to 'rid the Town of Montrose of [this] nuisance'. By providing a quiet and convenient asylum where they might receive good treatment and medical aid, she hoped that some might be able to return to society. Eventually nearly £700 was raised, and the asylum was completed in the summer of 1781. In Edinburgh, early efforts to raise funds to build an asylum met with apathy. A public appeal launched in January 1792 raised little more than £200 after 14 years. Even the last of the Royal Asylums, the Crichton Royal at Dumfries, founded in 1839, was dubbed by the local press the 'Crichton Foolery', and was bitterly attacked as a misguided and undesirable avenue for philanthropy.

Early attempts to improve conditions for lunatics through legislation were unsuccessful. A Lunacy Bill was rejected by parliament in 1817 but the reports prepared at that time revealed much about their standards of accommodation. Edinburgh was particularly bad, with overcrowding and neglect the principal causes for concern.

Of the first purpose-built asylums in Scotland, few survive today. Montrose Asylum was removed to a new site in 1866 and the original buildings were demolished, as were those of the Royal Edinburgh Asylum designed by Robert Reid in 1809. The Glasgow Royal Asylum, designed by William Stark in 1810, was probably the most influential purpose-built asylum of its time, adapting Jeremy Bentham's panopticon design to a radially-planned asylum which allowed maximum supervision of the patients by a minimal number of staff. A central tower contained the offices and keeper's apartments; in the four radiating wings single cells occupied one side and corridors the other, the latter also functioning as day-rooms. Vacated in 1843, this building was then used, until its demolition in 1908, to house the growing numbers of paupers looked after in the Town's Hospital, or City Poorhouse. The Dundee Royal Asylum (1812), also by Stark and also now demolished, was of H-plan layout, the inmates being accommodated in single cells in the single-storeyed wings on either side of a central administrative block. There were day-rooms at the end of the wings, and a matron's room was placed between the two day-rooms.

William Burn took over as architect to the Dundee Royal in 1823, and from there he went on to design the Perth Asylum (Murray Royal) in 1827 and the Crichton Royal in 1839. His designs were influenced by Watson & Pritchett's plans for the asylum at Wakefield, Yorkshire, which themselves have their roots in Stark's Glasgow Asylum. Again, a central tower contained offices whilst the patients' accommodation occupied wings radiating from it. Burn devised two different variations of the Wakefield H-shaped plan around octagonal towers, and he produced the design for the Edinburgh Royal Asylum's West House in 1839–40, the oldest part of the present hospital.

Archibald Simpson was the architect of the first Scottish asylum designed specifically for paupers, built in 1834–5 on the initiative of the managers of Gray's Hospital, Elgin. Strictly functional and a complete contrast to Gray's, this asylum was remodelled by A & W Reid in the 1860s, the accommodation provided in Simpson's building, now Bilbohall Hospital, Elgin, being difficult to determine.

During the first half of the nineteenth century, a distinctive plan type, generally termed the corridor-plan, emerged among the larger, purpose-built asylums. In these, the patients' accommodation in the wings comprised single rooms or small wards, usually on one side of the wing only, with linking corridors which generally doubled as galleries or day-rooms. The Glasgow Royal Asylum at Gartnavel, designed by Charles Wilson and built in 1841–3, best exemplifies the type and became a model for many others. There, the entrance, reception rooms and officers' accommodation were placed in the centre with long galleries on either side, running along the whole of the front between the centre and the wings. The galleries were half the width of the building and behind them were the patients' single rooms, combined with small dormitories in the end wings.

Growing numbers of pauper lunatics overcrowded the Royal Asylums and led to the setting up of a Royal Commission on Lunacy and to the Lunacy (Scotland) Act of 1857. By the terms of this Act a central Board of Lunacy was established which was empowered to inspect lunatic asylums, and to grant or remove licences, which were now necessary to run these establishments. Scotland was also divided into administrative districts, each with its own Lunacy Board which was responsible for the accommodation of pauper lunatics, either by providing a district asylum or by contracting with an existing asylum to admit pauper lunatics from their district. With the 1862 amendment to the Act, parochial boards, which formerly had possessed responsibility only for the sane poor, were empowered to cater also for lunatic paupers if they could offer suitable accommodation – usually in the form of lunatic wards attached to the poorhouse. Thus, some of the larger population centres came to build parochial asylums which were indistinguishable from the district asylums in all but administration.

The first of these new pauper asylums to the built in Scotland was the Argyll and Bute District Asylum, opened in 1863 at Lochgilphead. It was designed by the Edinburgh City Architect, David Cousin, who provided an asylum for 200 inmates. The patients' accommodation combined single rooms with ground-floor day-rooms and dormitories above. A group of other district asylums swiftly followed in the 1860s, in Ayrshire, Fife, Stirlingshire, Banff, Inverness-shire (fig. 16.3), Perthshire and Haddington, East Lothian, all built on equally modest lines; there was no scope for architectural embellishment funded from the rates. As with the poorhouses, the central authority had the power to pass or reject plans for proposed new buildings or alterations to old ones.

In the second half of the nineteenth century asylum-planning moved further away from the more institutional aspects of some of the earlier buildings. The Barony Parochial Asylum at Woodilee (Lenzie, Dunbartonshire) of 1875 marks the transition, its plan containing a greater variety of rooms for the inmates. Later, more fragmented plans allowed distinct areas of asylums to be occupied by different types of cases, a development towards domesticity that made them look even less institutional. Typical of this type is the City of Glasgow District Asylum at Gartloch, designed in 1896. Here the patients were classified and allocated to pavilions or villas, of three storeys, linked to the central service and administration buildings by single-storeyed corridors. This was also the first asylum to be designed with a separate 'hospital' in a self-contained unit, such sections being added to most existing asylums from the 1890s onwards. Providing care for patients requiring medical

Figure 16.3 Craig Dunain Hospital (formerly Northern Counties District Lunatic Asylum), Inverness; specimen ward, c. 1902. Courtesy of Highland Primary Care NHS Trust.

treatment as well as for mental illness, hospital blocks usually also had some isolation wards for cases of infectious diseases.

A further development of the linked-villa plan was the village or colony asylum, a type which ultimately derived from Gheel in Belgium. At Gheel, the shrine of the martyred St Dymphne gained the reputation of curing the insane and in the Middle Ages became a place of pilgrimage for the mentally ill. Gheel gradually developed into a mental colony and in the nineteenth century was placed under the control of a Commissioner and a Board of Governors. Its success led to the introduction of new village asylums elsewhere, particularly in Germany and the United States of America. One of the earliest was Alt Scherbitz near Leipzig, founded in the 1870s. In Britain, an influential voice in support of this system was that of Dr John Sibbald, one of the Scottish Lunacy Commissioners. In 1897 he published *The Plans of Modern Asylums for the Insane Poor*, which contained a full description and plan of Alt Scherbitz and a recommendation that it serve as a model for future asylums in Scotland. The Crichton Royal already had houses in which upper classes of patients and their servants could be accommodated, whilst many other Scottish

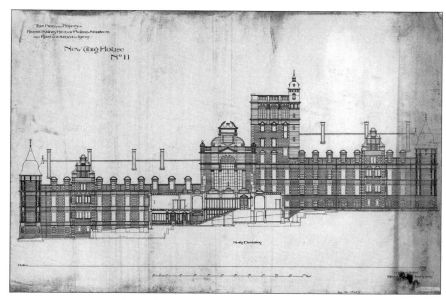

Figure 16.4 New Craig House, Thomas Clouston Clinic, Craighouse, Edinburgh; designs by Sydney Mitchell & Wilson, 1889. Crown copyright: RCAHMS, SC 712200.

asylums began to build villas on their estates. Murthly, Perth's District Asylum, was one of the first in which detached villas were provided for pauper lunatics, in about 1894.

The benefits of a stable, homely environment in the right kind of surroundings had long been advocated. W A F Browne, first Medical Superintendent of the Crichton Royal, epitomised the philosophy of asylum planners in the 1830s when he described the ideal institution as:

> A spacious building resembling the palace of a peer; airy, elegant, surrounded by extensive and swelling grounds and gardens. The interior is fitted up with galleries, and workshops, and music rooms. The sun and the air is allowed to enter at every window, the view of the shrubberies and fields, and groups of labourers, is unobstructed by shutters or bars.

This was the effect created in 1894 by Dr Thomas Clouston, the Physician Superintendent, and his architect, Sydney Mitchell, at the Craig House site of the Royal Edinburgh Hospital (fig. 16.4). Here the six new buildings constructed in the grounds of Old Craig House comprised two hospital blocks, three smaller detached villas, and the vast New Craig House, which contained central amenities such as recreation halls and dining rooms. In appearance, the buildings embodied Clouston's approach to the treatment of mental illness: variety was the keynote, hence their eclectic style, the plethora of different details, broken roof-lines and different coloured materials.

The colony system was adopted in the last of the district asylums to be built in Scotland. The Aberdeen District Asylum, Kingseat, designed by the notable Aberdeen architect A Marshall Mackenzie, opened in 1904 and was also the first asylum to successfully abandon an enclosing wall. Edinburgh's District Asylum at Bangour, West Lothian, which was designed by a leading Edinburgh architect, Hippolyte J Blanc, and opened in 1906, worked particularly well as a village asylum, once Harold Ogle Tarbolton added the asylum chapel, which has the appearance of a parish church. The colony plan also lent itself particularly well to a gently undulating site such as that at Dykebar, the site of the Renfrew District Asylum, designed by the Paisley architect, T G Abercrombie, and opened in 1909. These three institutions represent a high point of Scottish asylum design.

The needs of patients with mental disabilities took a long time to be recognised. Epileptics were admitted to asylums, but other forms of incurable mental disruption, particularly among children, met with less sympathy. As with the lunatic asylums, the first attempts to remedy the situation came from individuals. In 1852 Sir John and Lady Ogilvy

Figure 16.5 Lennox Castle Hospital, Lennoxtown, Stirlingshire. Crown copyright: RCAHMS, SC 712276.

attempted to provide education and homes for 30 imbecile children and orphans at Baldovan Institution, north of Dundee, which became solely for mentally-deficient children in 1855. In the same year a house in Gayfield Square, Edinburgh, was taken and set up as a training school for idiot children, run by Dr Brodie, and in 1859 the Society for the Education of Imbecile Youth in Scotland was formed with the intention of providing more suitable accommodation. In 1861 land was purchased at Larbert in Stirlingshire and the Royal Scottish National Institution was built. Dr Brodie was appointed as Medical Superintendent in 1862, and nine children from Edinburgh were transferred to the new building in 1863.

It was not until 1913, however, with the passing of the Mental Deficiency and Lunacy (Scotland) Act, that public funding became available for the care of mentally handicapped adults. Several institutions or colonies were then founded, notably at Birkwood, Lanarkshire in 1923, and in 1936 at Lennox Castle, Stirlingshire (fig. 16.5), which had two enormous dining-halls, each capable of seating 600 patients. At both Birkwood and Lennox Castle the institutions were centred on mansion houses in extensive grounds, in which purpose-built accommodation was provided. Gogarburn, near Edinburgh, which opened in 1931, and Hartwoodhill by Shotts, Lanarkshire, a development of the later 1930s, were newly built on the colony system.

MUNICIPAL HOSPITALS FOR INFECTIOUS DISEASES AND SANATORIA

Until the 1860s, fever patients were generally cared for within voluntary hospitals. Glasgow Royal Infirmary was typical in admitting large numbers of fever patients, particularly during epidemics; in 1818, for example, these accounted for 60 per cent of its total number of patients. For severe epidemics, like that of typhus fever in 1817, temporary premises were provided. The cholera epidemic of 1831 prompted the formation of Boards of Health, but the provision of temporary fever hospitals and small cholera hospitals, especially in coastal towns, still remained in the hands of individuals or the governors of existing hospitals. It was not until the 1860s that local authorities took responsibility for providing hospitals for infectious diseases. The first such institution to be established was in Glasgow, at Kennedy Street/Parliamentary Road, which opened in April 1865.

The Public Health (Scotland) Acts of 1867 and 1897 placed defined obligations on local authorities in the provision of isolation hospitals in Scotland. There was little real progress initially as the 1867 Act was largely permissive, simply giving local authorities the power to provide any type of hospital if they wished. In practice, it was usually easier and less expensive to enter into an agreement with a nearby voluntary hospital for the admittance of fever cases. Under the 1897 Act, local authorities found that they were legally required to provide for infectious diseases when instructed by the central authority, which by that date had become known as the Local Government Board.

Implementation of the 1897 Act was the responsibility of Town Councils, Burgh Commissioners of the Board of Police and District Committees of County Councils. They were required to appoint medical officers of health and sanitary inspectors; the Act also included measures for inspecting and dealing with general nuisances and providing sewers and public toilets. In respect of infectious diseases, the Act extended the provisions of the Infectious Diseases (Notification) Act of 1889, and gave local authorities powers to inspect premises where infectious diseases were believed to exist, to provide means for disinfecting bedding, and to regulate letting houses in which infected persons had been lodging. Most importantly, the Act stated that:

> Any Local Authority may, and if required by the Board shall, provide, furnish, and maintain for the use of the inhabitants of their district suffering from infectious diseases, hospitals, temporary or permanent, and houses of reception for convalescents from infectious disease or for persons who may have been exposed to infection.

To this end, local authorities could either build the hospital themselves, contract for the use of an existing hospital, or combine with another authority to provide a hospital. All decisions had to be approved by the Local Government Board, which in 1899 issued a memorandum relating to sites and plans of hospitals for infectious diseases. A second memorandum issued the same year recommended that:

> A hospital for infectious diseases ought to come up to the highest standard of structure prescribed for a wholesome dwelling house directed to the prevention of damp and the exclusion of ground air; while the heating, ventilation, and isolation must meet a higher standard still. In plaster work, wood work, fittings and furnishings, considerations must be given effect to, which involve an entire departure from the traditions of domestic architecture and furnishings.

The Board recommended that the administration section be larger than initially required in order to allow for future expansion, and offered general guidelines on size. An average of one bed per 1,000 of urban or 1,500 of rural population was advocated, and 2,000 cubic feet (56.6m^3) of air per bed were recommended.

Most of Scotland's infectious diseases hospitals were built or rebuilt following the 1897 Act. At this date, the most common diseases for which isolation was necessary were scarlet fever, measles, diphtheria and enteric fever (typhoid). The smaller hospitals usually made provision for the treatment of three different diseases. Smallpox cases tended to be dealt with separately, even from the main fever hospital.

Darnley Hospital, on the outskirts of Glasgow, is a typical small hospital for infectious diseases erected at this period. The decision to build such a hospital had been taken in 1894 by a joint hospitals board comprising representatives of the Upper District of Renfrew and the burghs of Pollokshaws and Barrhead, and the hospital opened in late 1897. The design, by J L Cowan, consisted of four sections: a centrally-placed administration department, two ward blocks, the laundry block and a discharge block. A corridor linked the administration to the two ward blocks on either side: one of these blocks contained one ward of 12 beds for scarlet fever cases, a smaller ward with six beds, and two single-bed convalescent wards. The other ward pavilion, principally for enteric fever, was arranged in two sections, the one communicating with the main corridor contained a ward with six beds, another with four beds and a convalescent ward. At the far end, attached but separate, was a similar section containing a ward with four beds and one with two beds, which had its own entrance and was for other diseases.

The number of infectious diseases which were identified and admitted for treatment at these hospitals gradually increased. By the mid 1920s there were 37 notifiable diseases. Special small isolation wards, where new cases could be observed and carefully diagnosed, were an important feature of hospital design, and the development of the cubicle isolation ward in the inter-war period allowed different diseases to be treated in the same ward by placing each bed in a cubicle with partly glazed walls, preventing cross-infection but allowing the nurse on duty to supervise the whole ward.

New infectious diseases hospitals of the 1930s, often built to replace a number of the small rural hospitals, were larger and more efficient than their predecessors. Hawkhead Hospital in Paisley was designed by T S Tait in the international modern style, matching the clinical ethos, and featured six detached single-storeyed pavilion ward blocks and one cubicle block. Inverurie Hospital in Aberdeenshire, which opened in 1940, and Ayrshire Central Hospital, built in 1941, were both modelled on Hawkhead. Cowglen Hospital on the south side of Glasgow was originally planned in 1933 and incorporated many new features, including cubicle wards, but construction was delayed until it was finally handed over, incomplete, to American forces during World War II.

Improved methods of treatment, the discovery of new drugs, as well as a general improvement in sanitary conditions and housing led to a decline in the mortality rate and a reduction of the length of time a patient needed to remain in hospital. As a result, the need for hospital accommodation for infectious diseases declined rapidly between 1930 and 1954, and the larger inter-war infectious diseases hospitals incorporated other specialist units. At Inverurie, for example, the former cubicle block became a maternity unit. In recent years, new infectious diseases units have been provided in general hospitals, as for example, that at Raigmore Hospital, Inverness, which was completed in 1990.

HOSPITALS FOR THE TREATMENT OF CONSUMPTION OR TUBERCULOSIS

Consumption was the scourge of the nineteenth century, and until 1882, when Robert Koch discovered the tubercle bacillus, many diagnosed as suffering from consumption did in fact have pulmonary tuberculosis, which is infectious.

In 1887, under the instigation of Dr (later Sir) Robert Philip, a dispensary for tuberculosis was opened in Edinburgh. It developed into the Royal Victoria Hospital for Consumption which opened in 1894 in Craigleith House, the first institution of its kind in Britain. From 1903, eight detached, single-storeyed ward pavilions were built in the grounds; they were designed by Sydney Mitchell on a Y-shaped plan, which created sunny and sheltered wards for the patients. The first tuberculosis hospital in western Scotland was the Bridge of Weir Sanatorium, Renfrewshire (1896, begun 1894), founded by William Quarrier and designed by Robert Bryden, the Hospital for Consumptives at Ventnor on the Isle of Wight forming the model for Bryden's plans. The movement to combat tuberculosis quickly gathered pace. Following the foundation of the National Association for the Prevention of Consumption in 1898, the Glasgow and District Branch established Bellefield Sanatorium at Lanark in 1904. Glasgow Corporation donated £500 to the Society in 1901 (the first municipal aid for work against tuberculosis in Scotland), and later granted £5,500 towards the sanatorium.

The turn of the century witnessed the introduction into Britain of open-air treatment for tuberculosis. Treatment based on exposure to fresh air and sunshine had been developed on the Continent, especially in Germany and most notably by Dr Otto Walther who, in the late 1880s, established a sanatorium at Nordrach in Baden. His Nordrach System produced a rash of similar sanatoria in Britain, including Nordrach-upon-Mendip and Nordrach-on-Dee, the first sanatorium of its kind in Scotland.

Opened in 1900, Nordrach-on-Dee was situated in an Aberdeenshire pine forest, the scent from pine trees being considered beneficial for the patients. It was designed by George Coutts of Aberdeen, and comprised a long, narrow timber-built block, local granite being used on the tall water tower. The kitchens and staff accommodation were at one end of the building, separated from the patients' area by a large dining room. The patients' rooms were on the south side with large windows allowing direct access to the balconies and verandas which surrounded the building on all but the north side.

In 1903 the Local Government Board announced that pulmonary tuberculosis was to be regarded as an infectious disease in terms of the Public Health (Scotland) Act 1897. There was little initial expansion of hospital provision, however, and in 1907 minor amendments to the legislation were enacted in order to encourage the erection of tuberculosis wards and sanatoria. Aberchalder Sanatorium, Inverness-shire, was one of the earliest county sanatoria to be built, opening in 1907, and in the same year the first municipal dispensary for tuberculosis in Scotland opened in Dundee, attached to the Royal Infirmary.

In 1912 pulmonary tuberculosis became a notifiable disease throughout Scotland, and in 1914 all forms of tuberculosis became notifiable. As a result, the provision of sanatoria became more widespread. As at the City Hospital in Edinburgh, where 50 beds had been set aside in 1906, beds for tuberculosis patients were provided at some existing infectious diseases hospitals. In Glasgow, open-air style ward pavilions, comprising small wards with verandas, were erected at Ruchill in 1914 and at Robroyston, which came into use after World War I.

Many plans for new sanatoria and extensions to existing infectious diseases hospitals were in hand in 1914, but most were postponed because of the war. The site for Mearnskirk Hospital, for example, was purchased in 1913 by Glasgow Corporation and the initial plans were approved in the following year. Halted by the war, the hospital finally opened in 1930 by which time it had become one of the last in Scotland to be built exclusively for the treatment of tuberculosis.

SPECIALIST HOSPITALS

In England, the earliest specialist hospitals were established to provide for classes of patients, such as women in childbirth, incurables and lunatics, who were excluded from most general hospitals. In Scotland, as we have seen, general hospitals such as Edinburgh Royal Infirmary were designed with a few cells for lunatics and a small maternity department. Here, the trend was for specialist hospitals to be founded by doctors who were dissatisfied with the treatment of a particular class of patient within an existing general hospital and perceived the need for additional facilities.

Specialising in distinct areas of medicine, for example diseases of the eye or ear, such hospitals often began as dispensaries before developing to take in-patients. It was also usually only in large centres of population that such institutions were established, where the numbers of special cases were sufficiently high. But from modest beginnings, specialist hospitals often became the most prestigious homes of the medical profession. With their unrivalled opportunities for clinical study of a large number of patients suffering from the same complaint, specialist hospitals came to the forefront in new treatments and research.

Any analysis of the design and planning of such hospitals, however, is hampered by the numerous different types of specialisms catered for, and by the fact that, when purpose-built, these hospitals closely resembled small general hospitals, with only minor modifications to suit particular requirements. In the out-patients' departments attached to children's hospitals, for example, the waiting rooms had to be larger than usual to accommodate the children and their parents. More fundamental differences in planning are noticeable in convalescent homes and hospitals for incurables, where the patients were not always confined to bed during the day and the average length of stay was longer than in a general hospital.

MATERNITY HOSPITALS

One of the earliest types of specialist hospital catered for lying-in women. The first maternity hospital in Scotland to be established as a separate institution rather than as a department within a larger hospital was the relatively short-lived Glasgow Lying-in Hospital which was established by James Towers c. 1790. In 1793 Dr Alexander Hamilton founded a Lying-in Hospital in Edinburgh at Park House, developed in response to the limited accommodation for lying-in women at the Royal Infirmary. Like the Glasgow Lying-in Hospital it maintained its links with the main infirmary and provided medical school students with clinical teaching in midwifery.

It was not until 1834 that the Glasgow Lying-in Hospital was re-established in make-shift premises on the upper floors of the old Grammar School in Grammar School Wynd. Later it moved to St Andrew's Square and eventually, in 1858, took up residence in a house at the corner of North Portland Street and Rottenrow. The hospital was rebuilt in 1880 to the designs of Robert Baldie, re-opening in January of the following year. It was of four storeys and a basement with the offices and kitchen on the ground floor and wards above. The wards were mostly small: on the first floor, for example, the largest ward contained seven beds; there were three small wards with two beds in each and one ward with five beds, all sharing one sanitary annexe, situated at the rear of the building.

Edinburgh's Royal Maternity Hospital was instituted in 1843 and was located in John Street. In 1878 a new hospital was built to designs by Macgibbon & Ross of Edinburgh 'in accordance with the expressed views of Sir James Young Simpson as to what such an hospital should be'. Simpson (1811–70), appointed Professor of Midwifery at Edinburgh University in 1840, had pioneered the use of ether (and later of chloroform) as an anaesthetic in childbirth, and held strong views on hospital reform and design. The Edinburgh hospital was a largely three-storeyed and basement building laid out on an L-shaped plan. The main entrance was in Lauriston Place and led into a suite of four rooms: the dispensary; two sitting rooms, one for the resident medical officer and one for the matron; and a delivery ward with three beds. In the wing running along the west side of the hospital there was a large ward of ten beds, with a small ward divided off at the south end. A similar arrangement was followed on the first floor. The hospital became known as the Royal Maternity and Simpson Memorial Hospital.

Elsewhere, maternity hospitals were slower to develop. The Aberdeen Maternity Hospital was founded in 1893, the year in which a committee was formed in Dundee to establish such a hospital there. Eventually, along with provision for infectious diseases, maternity hospitals became part of the municipal hospital system, and in some instances the two were built on adjacent sites, such as at Irvine where the Ayrshire Central Hospital comprised two sections for these very different functions.

A good example of the type of maternity hospital erected between the wars, the Elsie Inglis Hospital in Edinburgh (opened in July 1925), was designed by H O Tarbolton. His plan maximised its fine site overlooking Salisbury Crags in the provision of a suite of rooms on the south side with a balcony from which the nursing mothers could enjoy the view. A new Simpson Memorial Pavilion, with similar modernistic design features, was built in 1935–7 overlooking the Meadows on the main Edinburgh Royal Infirmary site. Increasingly, pleasant surroundings were provided within maternity hospitals to make the women feel relaxed and at home. Many later hospitals were established in converted houses, carrying the idea of homely surroundings to its logical conclusion.

CHILDREN'S HOSPITALS

The short-lived dispensary for sick children in Red Lion Square, London, which operated between 1769 and 1782, marked the first attempt to cater for the specialist needs of sick children. In Paris, Tenon had been calling for the provision of special wards for children within general hospitals in his *Mémoires sur les Hôpitaux de Paris* of 1788, and it was there that the first truly specialist hospital for children was established in 1802, *L'Hôpital des Enfants Malades*.

In Britain, the debates continued as to whether there should be separate hospitals for children or whether they should be cared for in children's wards within existing general hospitals. It was argued that the requirements of a children's hospital, in terms of construction, furnishing, equipment and organisation, were more complicated than a general hospital. Florence Nightingale, on the other hand, argued that there was no advantage in the separate clinical study of medical or surgical cases in children because they were incapable of accurately conveying their symptoms. She also considered that children were more vulnerable if they were housed together in larger numbers and pointed to the high mortality rate in specialist children's hospitals. Children's susceptibility to cross infection was certainly a major problem, and the provision of isolation wards became of paramount importance in children's hospitals.

In Scotland, the first hospital devoted to the care of sick children was the Royal Hospital for Sick Children at Edinburgh which opened in 1860 at 7 Lauriston Lane. This was followed by the opening of children's hospitals in Aberdeen in 1877 and in Glasgow in 1883.

Attempts had been made to set up a children's hospital in Glasgow as early as 1861. It was not until 1880 that funds were sufficient to purchase a house in Garnethill which was converted by Douglas & Sellars. However, proper pavilion wards could not be created in the old house, and the windows, which did not command a pleasant view, were painted with nursery rhymes, based on Caldecott's illustrations, popular

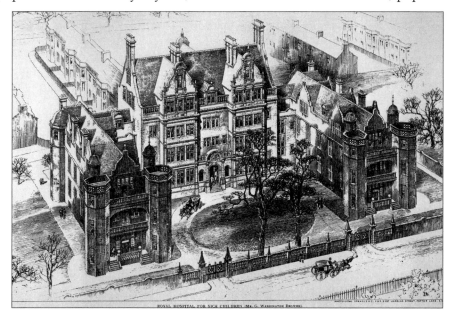

Figure 16.6 Royal Hospital for Sick Children, Sciennes Road, Edinburgh; bird's-eye view in *The Builder*, 1 January 1898.

at that time. An out-patients' department was added in 1888. The waiting hall had to be larger than in most out-patients' departments to allow for each prospective child patient to be accompanied by its parent and often by brothers and sisters as well. There also had to be space for numerous prams to be left near the entrance. It was replaced in 1914 by a large, purpose-built hospital designed by J J Burnet which no longer survives. Structural weaknesses were discovered whilst additions were being made in the 1960s. This led to the old hospital's demolition and the present eight-storeyed ward tower was erected in its place. Work began in 1968 to designs by Baxter, Clark & Paul.

In 1895 Edinburgh provided itself with a new, purpose-built hospital for children, which remains in use with various additions and alterations (fig. 16.6). It was designed by George Washington Browne and comprised a central administration section flanked by three-storeyed ward pavilions. The mortuary chapel was decorated by murals painted by Phoebe Traquair ten years earlier for the old hospital.

Plans for a new children's hospital in Aberdeen were drawn up before World War I but these were delayed until the 1920s when it was included in the Aberdeen Hospital Scheme at Foresterhill. It was the first of the buildings erected on the site, opening in 1928, and was designed by Dr William Kelly. Resembling a typical general hospital in its plan, the children's hospital comprised an administration block, four single-storeyed ward pavilions, each with 32 beds, an operating theatre and an isolation ward block with six beds. The wards were fronted by glass doors which led to a balcony.

MUNICIPAL MATERNITY AND CHILD WELFARE SCHEMES

Despite improved conditions in sanitation towards the end of the nineteenth century the infant mortality rate did not noticeably decline. The available statistics might also understate the full extent of the problem since the Registration Act permitted an interval of 21 days to elapse between birth and registration, by which time a number could have already died. The Notification of Births Act of 1907 gave local authorities the option of asking for births to be notified within 36 hours.

In Glasgow, the Committee on Health approved schemes in 1903 and 1904 to speed up the notification of births, and in 1904 a special inquiry was also begun into the general health and physical condition of children in poorer areas. Glasgow adopted the Notification of Births Act and from January 1908 this applied to dead as well as live births.

In 1915 the Notification of Births (Extension) Act made the permissive parts of the 1907 Act obligatory and empowered local authorities to prepare and operate schemes for attending to the health of expectant or nursing mothers and children under five. A grant of 50 per cent was available for establishing maternity and child welfare schemes which could include the provision of hospitals. In the same year, the Midwives (Scotland) Act was passed, under the provisions of which food and services could be made available by the local authority or by arrangement with a voluntary institution. These included day nurseries, schools for mothers, hospital and home treatment, a midwifery service and home helps, and convalescent homes.

The first large Maternity, Child Welfare and Special Treatment Centre to be built in Scotland was at Motherwell. The plan of this unit was reproduced as a model of its type by John Wilson, architect to the Scottish Board of Health, in an article on planning public health institutions, published in the *RIBA Journal* in 1922. The centre consisted of a maternity section, to which expectant mothers would go for advice and treatment, together with a child welfare centre to which they could bring their children for weighing, examination and treatment. There was also a special treatment centre for children between one and five years old, a dental clinic, skin clinic and lecture and demonstration room.

Most local authorities made arrangements with voluntary hospitals for the admission of their maternity cases. Others were quicker to provide new hospitals. Paisley and Greenock, the Middle Ward of Lanarkshire and a combination of Ayrshire authorities had all opened maternity hospitals by the mid 1920s.

EYE HOSPITALS

It was found that a high number of soldiers returning from the Napoleonic wars were suffering from trachoma, a disease of the eye, and from 1804 a number of specialist hospitals were established in Britain for the treatment of eye diseases. The first to emerge in Scotland grew out of small eye departments within the general hospitals in the 1820s and 1830s.

The Glasgow Eye Infirmary was founded in 1824, in converted premises in Inkle Factory Lane. It was largely due to two doctors, William Mackenzie and George Monteath, who were concerned at the large numbers of people suffering from eye disease and who went untreated by the Glasgow Royal Infirmary. A purpose-built hospital was not provided until 1874 when the hospital moved into a new building in Claremont Street designed by J J Burnet. An Eye Infirmary was opened in Edinburgh in 1834. In 1883 it was expanded, and changed its name accordingly to the Edinburgh Eye, Ear, and Throat Infirmary. Eye hospitals were founded in Aberdeen and Dundee in the 1830s, Aberdeen suffering a high incidence of eye injuries among the granite cutters in that area. Other eye hospitals were eventually founded in Greenock (1880) and Paisley (1889).

EAR, NOSE AND THROAT HOSPITALS

An Ear and Skin Dispensary was established in Glasgow in 1863, followed in 1872 by the Glasgow Dispensary for Diseases of the Ear. Originally situated in Buchanan Street, in 1885 this dispensary moved to a three-storeyed house at 28 Elmbank Crescent; this provided a large clinical room and a waiting room on the ground floor, a board room and lecture room on the second floor, and wards above. The adjoining house was later purchased and the hospital began to specialise in nose and throat, as well as ear, cases. By the early 1920s it had changed its name to the Glasgow Ear, Nose and Throat Hospital and had also outgrown its home in Elmbank Crescent.

In Edinburgh, ear, nose and throat cases were admitted to the Eye Infirmary from 1883 and, in recognition of its wider scope, the hospital changed its name to the Edinburgh Eye, Ear and Throat Infirmary. In Greenock, an Ear, Nose and Throat Hospital with 20 beds and an out-patients' department was built in 1937 by James Miller.

CANCER HOSPITALS

The earliest attempt in Britain to provide care for cancer patients and an opportunity for medical students to observe such cases was the cancer ward of the Middlesex Hospital in London, established in 1792. However, the first specialist cancer hospitals did not appear until the middle of the nineteenth century. The Glasgow Cancer and Skin Institution opened in 1889, being the first hospital to specialise in cancer treatment in Scotland. First established in St Vincent Street, it moved in to the former Free Church Manse in Hill Street in 1890 when it began to receive in-patients. In 1906 plans were prepared by James Munro & Sons to reconstruct two houses in Hill Street, Nos 132 and 138, and the new hospital opened in 1912.

HOSPITALS FOR INCURABLES

One class of patient generally refused admission to general hospitals was that of patients suffering from any form of incurable disease, including cancer. It was simply a practical matter of the long-term occupancy of scarce hospital beds which were needed for great numbers of short-stay patients. There was clearly a need to provide accommodation for such patients whose relatives or friends were unable to care for them. The first hospital for incurables to be founded in Scotland was in Aberdeen in 1857. Later known as Morningfield Hospital, a new purpose-built structure was created for it in 1884 to designs by William Henderson & Son. By that date, similar hospitals had been established in Edinburgh,

Perth, and Kirkintilloch, Dunbartonshire. These hospitals often bore a resemblance to contemporary houses and their domestic appearance was intended to provide pleasant surroundings for the patients who were there to end their days.

Broomhill and Lanfine Hospitals at Kirkintilloch had their roots in the Scottish National Institution for the Relief of Incurables which had been founded in 1874. The aim of Miss Beatrice Clugston, the driving-force behind the institution, was to establish a large incurables' colony between Edinburgh and Glasgow. Resources did not match her ambitions but the purchase of Broomhill House enabled beds to be provided for nearly 50 adults and 12 children in 1875.

CONVALESCENT HOMES AND AUXILIARY HOSPITALS

From the mid nineteenth century, and often in conjunction with the larger general hospitals, a number of convalescent homes were established. Not only did they give patients time in which to recuperate from illness or operations, they also enabled beds to be vacated more quickly from the parent hospitals. Later, large institutions, such as Glasgow Royal Infirmary, established auxiliary hospitals which offered accommodation for convalescent patients as well as providing treatment for certain cases.

Edinburgh Royal Infirmary was one of the first hospitals to provide a convalescent home. Situated at Corstorphine, on the outskirts of Edinburgh, it was purpose-built to designs by Peddie & Kinnear and opened in 1867. An imposing Italianate building set on high ground with a southern exposure, it was symmetrically arranged with a central administration section flanked by the patients' accommodation. The large dormitories, planned on the model of the pavilion wards at the Herbert Hospital at Woolwich, were later criticised, as it was felt that smaller dormitories or wards were better suited to convalescent patients.

The Schaw Convalescent Home at Bearsden is one of the most impressive examples of this type of hospital. Generously funded by Miss Marjory Shanks Schaw as a memorial to her brother, it was designed in 1891 by James Thomson. It operated as a convalescent home for the Glasgow Royal Infirmary in addition to the one built in 1870 at Lenzie, also designed by Thomson. The general layout of the Bearsden home received contemporary approval, although *The Hospital* considered that the architect had 'sacrificed to the artistic treatment of the exterior the practical arrangement of windows for ward purposes'. On the first floor there were two wards over the good-sized ground-floor day-rooms and a series of single bedrooms. There were two sanitary annexes at the rear, with the usual cross-ventilated lobby separating them from the main building. Staff and nurses' quarters occupied the central area on the ground floor and to the rear was the dining-hall with the kitchen and laundry adjacent.

HOSPITALS FOR WOMEN

Women's hospitals were essentially of two kinds: those which specialised in diseases peculiar to women; and those which offered treatment by female doctors.

The Royal Samaritan Hospital for Women, on the south side of Glasgow, was of the first type. It was founded in 1885 and a small hospital with just three beds opened in the following year. In 1895 a new, purpose-built hospital was designed by Ninian MacWhannel and William Rogerson. It comprised two sections, a two-storeyed block containing the wards, and an adjacent three-storeyed block accommodating the administrative offices, kitchens, and the operating theatre.

Bruntsfield Hospital was founded through the endeavours of two women doctors, Sophia Jex-Blake and Elsie Inglis, who wanted to provide medical care for women by women, and also clinical experience for young women doctors. The hospital opened in Jex-Blake's former home, Bruntsfield Lodge, in 1899. In the same year, Elsie Inglis established a small hospital on similar lines, which amalgamated with the Bruntsfield in 1910.

The Glasgow Women's Private Hospital was established to provide hospital treatment for women by women doctors. Originally situated in a converted house in West Cumberland Street, it moved to Redlands House in Lancaster Crescent in 1924. The large public rooms were converted into wards, and the attics converted into bedrooms for the nurses and domestic staff. The hospital contained an operating theatre and also had a maternity department.

OTHER SPECIALIST HOSPITALS: LOCK, ORTHOPAEDIC AND DENTAL HOSPITALS

Among the earliest types of specialist hospitals were those which dealt with cases of venereal disease. Known as Lock Hospitals, they appeared at about the same time as the early lying-in hospitals. The Glasgow Lock Hospital was founded for the treatment of 'unfortunate females' in 1805 in a house in Rottenrow which contained 11 beds. In 1846 it expanded into another building in Rottenrow where it still functioned in the late 1880s.

'Orthopaedia or the Art of Correcting and Preventing Deformities in Children' was a term introduced in the mid eighteenth century. The first hospitals to specialize in the treatment of crippled children emerged on the Continent in the early nineteenth century when experimental surgery found successful ways of operating on deformities. The English physician, W G Little, introduced these techniques to London in 1837. In the 1870s, in the new climate of antiseptic surgery, further operations were developed for the treatment of bone disease and paralysis. At the end of the century the discovery of X-rays was fundamental to the treatment of deformities caused by broken bones; the link between deformity and poverty was also made, as it was realised that children who had poor diets and dwelt in overcrowded towns required more hospital treatment. As with the treatment of tuberculosis, the importance of sunshine and fresh air became the dominant feature of hospital life.

In Scotland, a number of hospitals for crippled children were established, the largest probably being the Princess Margaret Rose Hospital, south of Edinburgh. The sanatoria-style ward blocks with verandas were designed in 1929 by Reginald Fairlie and the first two wards were opened in the summer of 1932. Two further ward blocks were in use by 1936.

The first separate dental hospital to be established in Scotland was in Edinburgh in 1860.

EMERGENCY MEDICAL SCHEME

Viewed from Whitehall, Scotland was an ideal place to send large numbers of troop or civilian casualties from England and Wales. Following the outbreak of World War II in September 1939, seven new hospitals were built in Scotland under the Emergency Medical Scheme (EMS). They were built in countryside locations, accessible from towns but in areas unlikely to be targets for bombing. They were also designed and built very rapidly with a view to their lasting only for the duration of the war and on the principle that 'buildings are much less important than personnel'. The seven new hospitals were: Ballochmyle, Ayrshire; Bridge of Earn, Perthshire; Killearn, Stirlingshire; Law, Lanarkshire; The Peel, Selkirkshire; Raigmore, Inverness-shire; and Stracathro, Angus.

The Civil Defence Act of 1939 laid the ground for the creation of the scheme under which there would be 4,800 beds available for casualties in mansion houses and hotels, a further 8,500 beds in hutted annexes to existing hospitals and 7,000 in the new EMS hospitals. In all, some 68 country houses were adapted for use as part of the EMS. These were administered by the Department of Health and used for convalescents and less serious cases. Gleneagles (Perthshire) and Turnberry (Ayrshire) Hotels were both used for convalescents during the war.

The EMS hospitals were constructed by local architectural firms to a brief from the Office of Works. Law Hospital, built by the firm of Cullen, Lochhead & Brown, formed the prototype. Work was completed by the end of 1939 and comprised four main hutted ward sections of four separate single-storeyed blocks, each having two wards. Each ward contained 40 beds, a larger complement than would have been considered suitable in peacetime. At the centre of the site were the administration block, reception, kitchen and stores. There were also separate male, female and medical staff blocks, and an isolation unit with 16 beds. Many of the EMS huts were of timber construction with asbestos sheeting on the outside which later caused extensive problems. Other huts were built of brick with pre-cast hollow slab roofs and concrete floors.

Although these new hospitals and the hutted blocks added to many existing hospitals were intended only to last for up to 20 years, the majority of them remained in use for much longer. At the time of writing, Ballochmyle, Bridge of Earn, Law and Stracathro Hospitals remain in use, albeit refurbished and upgraded to modern standards. Killearn Hospital closed in 1972, The Peel was sold in 1989 and the original brick ward blocks at Raigmore had nearly all been demolished by 1989 and a new general teaching hospital built on the site.

The allied forces also required hospital units in Scotland. The United States took over the construction of Cowglen Infectious Diseases Hospital, then nearing completion. The Norwegians had a unit at the Edinburgh Southern General Hospital, the French had units at Quothquhan, Lanarkshire, and Knockderry, Dunbartonshire, and the Canadian Navy took over the Ravenscraig Hospital at Greenock. There was also the Polish Paderewski Hospital in a former children's home in Edinburgh.

World War II also provided a stimulus to specialist surgery and medicine, particularly in areas such as brain surgery and plastic surgery which had been little developed prior to 1939. A new impetus was also given to anaesthetics, radiology, pathology and orthopaedics. Indeed, it is arguable that the success of the EMS paved the way for the National Health Service by demonstrating that a nationally co-ordinated hospital system could be effective. In Scotland, the additional hospital accommodation provided under the EMS filled the large gaps within the voluntary sector, thus increasing the attraction of a complete state system of hospital care.

ACKNOWLEDGEMENTS

The editors are grateful to Miles Glendinning and Alison Darragh for reading and commenting on a draft of this essay which was prepared by Geoffrey Stell from material supplied by Harriet Richardson.

BIBLIOGRAPHY

Catford, E F. *The Royal Infirmary of Edinburgh, 1929–1979*, Edinburgh, 1984.
Comrie, John D. *History of Scottish Medicine*, London, 2nd edn, 1932.
Aitken, R S et al (Department of Health for Scotland). *Scottish Hospitals Survey*, Edinburgh, 1945, General Introduction, Eastern Region, North-Eastern Region, Northern Region, South-Eastern Region, Western Region.
Dow, D A. *The Rottenrow, The history of the Glasgow Royal Maternity Hospital, 1834–1984*, Carnforth, 1984.
Dow, D A. *Paisley Hospitals: The Royal Alexandra Infirmary and allied institutions, 1786–1986*, Glasgow, 1988.
Dow, D A. Incessant construction beyond our reach: Two centuries of hospital construction in Scotland, *Journal AHSS*, 16 (1989), 34–58.
Easterbrook, C C. *The Chronicle of Crichton Royal, 1833–1936*, Dumfries, 1940.
Gray, J A. *The Edinburgh City Hospital*, East Linton, 1999.
Hine, G T. Asylums and asylum planning, *Journal RIBA*, 3rd series, 8 (23 February 1901), 161–84.
Levitt, I. *Poverty and Welfare in Scotland, 1890–1948*, Edinburgh, 1988.
Mackay, G A. *Management and Construction of Poorhouses and Almshouses*, Edinburgh, 1908.
Mackenzie, T C. *The Royal Northern Infirmary Inverness: The story of a voluntary hospital*, Inverness, 1946.
Mackintosh, D J. *Construction, Equipment and Management of a General Hospital*, Edinburgh and London, 1916.
McLachlan, G, ed. *Improving the Common Weal: Aspects of the Scottish health service, 1900–1984: A collation in honour of the late Sir John Brotherston*, Edinburgh, 1987.
MacQueen, L, Kerr, A B. *The Western Infirmary, 1874–1974*, Glasgow and London, 1974.
Markus, T A. Buildings for the sad, the bad and the mad in urban Scotland, 1780–1830. In Markus, T A, ed, *Order in Space and Society*, Edinburgh, 1982, 25–114.
Mitchison, R. *The Old Poor Law in Scotland*, Edinburgh, 2000.
Morrison, K, for RCHME/English Heritage. *The Workhouse: A study of Poor-Law buildings in England*, Swindon, 1999.
Richardson, H. 'Scottish Hospital Survey' (unpublished typescript in NMRS), npd.
Richardson, H. A continental solution to the planning of lunatic asylums, 1900–1940. In *Scotland and Europe: Architecture and design, 1850–1940* (St Andrews Studies in the History of Scottish Architecture and Design, 2), St Andrews, 1991, 67–79.
Richardson, H, for RCHME. *English Hospitals, 1660–1948*, Swindon, 1998.
Rilse, G B. *Hospital Life in Enlightenment Scotland*, Cambridge, 1986.
Scull, A. *The Most Solitary of Afflictions: Madness and Society in Britain, 1700–1900*, New Haven and London, 1993.
Slater, S D, Dow, D A. *The Victoria Infirmary of Glasgow, 1890–1990*, Glasgow, 1990.
Taitt, A C. *Chronicle of Crichton Royal, 1937–1971*, Carlisle, c. 1980.
Thomson, A M W. *The Glasgow Eye Infirmary, 1824–1962*, Glasgow, 1963.
Turner, A L. *Story of a Great Hospital: The Royal Infirmary of Edinburgh, 1729–1929*, Edinburgh, 1937, facsimile edn, 1979.

17 Public Services

JOHN R HUME

Study of public services in Scotland has been varied in scope and depth, and comparatively little attention has been paid to the buildings and structures associated with them. In this essay the major public utilities – water supply and sewerage, gas, electricity and telecommunications – will be discussed first, with other services, grouped by function, at the end. What the major public utilities share is the capacity to deliver at a remote point a particular service by the use of systems of pipes and wires, or today (in the case of telecommunications) fibre optics or radio-frequency waves. These systems can be local, regional or national, even international. In many cases the systems are more significant than the origination, in terms of capital invested, and in most cases the systems are only visible intermittently, if at all, between the points of production and consumption.

WATER SUPPLY

Water supply was from the beginnings of settlement a determinant of habitable sites. Even a cursory study of abandoned settlements in highland Scotland reveals that a nearby stream is a feature common to all. Larger communities might originally draw on rivers and streams for water, but these were liable to become polluted by human and animal excrement, and by industrial processes like tanning and dyeing, so recourse was frequently had to wells, which might also be geographically more convenient.[1] In the later Middle Ages some monastic communities developed more complex systems of water supply, and at Glenluce Abbey, Wigtownshire,[2] one can still see the remains of a clay pipeline, supplying the monastery. In the seventeenth century, Edinburgh[3] began taking water from Comiston springs via a timber pipeline, but it was not until the nineteenth century that large-scale provision of piped water became general. A considerable impetus to this was the development of cast-iron pipes for water mains, and of extruded lead pipes for leads into, and later, within, private houses. In Glasgow, three private companies started supplying water to the city early in the nineteenth century.[4] William Harley piped water from springs at Hundred Acre Hill to Bath Street in 1804, and the other two companies, the Glasgow and Cranstonhill waterworks companies (1807–8) used steam pumping engines to draw water from the gravels adjacent to the River Clyde (Glasgow) and from the river itself (Cranstonhill),[5] and to raise it to cisterns for gravity distribution. The Cranstonhill company supplied water for industrial as well as domestic purposes. The most remarkable industrial water-supply scheme was, however, the Shaws Water Works at Greenock, Renfrewshire (1827), which involved creating a major and several minor reservoirs, cutting a contour canal from Inverkip to Greenock,[6] and laying out two lines of falls. The 'cut' also provided water for houses in the 1880s by the construction of a parallel aqueduct and a new holding reservoir. The system is still partly in use.

A significant incentive to improving water supplies came with the growing knowledge of the mechanisms for transmission of infectious disease.[7] The public health movement of the 1830s and '40s laid great stress on purity of water supply, and began to stimulate interest in water-borne sewage disposal.[8] An early Scottish project initiated in this climate of opinion was the Gorbals Water Works, with reservoirs above Barrhead, Renfrewshire, supplying water by gravity to the southern suburbs of Glasgow, and including free water for the poor.[9] The success of this scheme led to the promotion by Glasgow Town Council of a much more ambitious notion, bringing water from the southern edge of the

Figure 17.1 Loch Katrine Waterworks, Perthshire; sluice operating mechanisms. Copyright: John R Hume.

Figure 17.2 Loch Vennachar, Perthshire; sluice house. Copyright: John R Hume.

Highlands. The source eventually chosen was Loch Katrine, Perthshire (fig 17.1).[10] The construction of the part rock-cut, part piped aqueduct to the holding reservoir at Mugdock, Stirlingshire, and of the link to the city was a major civil engineering work.

Gravity water supply became standard in Scotland, where the proximity of upland catchments to urban areas made this eminently feasible. Early schemes included Dundee[11] and Edinburgh,[12] in both cases the original facility being complemented by much larger later projects. Aberdeen alone of Scottish cities derives its water from a river, the Dee. In the last few decades pumped water has been taken from Loch Lomond to east central Scotland.[13]

The visible works associated with water supply are primarily those designed to impound water, to convey it in bulk, to purify it, and to hold it at a sufficient height to give adequate pressure in domestic mains. Water supply dams are generally of the gravity type, masses of earth with stone revetments and spillways. Good examples can be seen between Kilbirnie and Largs, Ayrshire, with smaller specimens at Girvan and Cumnock, Ayrshire, and above Strathaven, Lanarkshire. An unusual dam is the one on Loch Vennacher, Perthshire (fig. 17.2), which incorporates sluices to control the flow of water to compensate for that diverted from Loch Katrine to Glasgow.[14] This has a house running across the dam, covering the sluices. Prominent features of many reservoirs are the sluice houses perched on the top of pipes reaching to the bottom, and controlling water intakes. Some of these have architectural treatment, usually castellated.[15] The raw water from the reservoir is sometimes filtered close to intake,[16] but in other cases the filters are at a distance. The measuring houses for the Edinburgh water supply at Alnwickhill, operated in connection with the filter beds there, have an attractive rustic stone treatment.[17] There are many isolated filter houses, often brick built, which are now being superseded as EC regulations on water quality take effect.

Aqueducts are usually only visible where they cross low ground. The only significant open aqueduct for domestic water was the 'cut' at Greenock. Most others are rock-cut or in pipe. On the Loch Katrine scheme there are several aqueducts, some not easily accessible. The reinforced concrete road bridge at Killermont, Dunbarton-shire, and the cast-iron Kelvin Bridge at Hillhead, Glasgow,[18] both carry Loch Katrine mains under their decks, and many other road bridges have the secondary function of carrying water mains. The holding reservoirs at Mugdock, where the primary Loch Katrine aqueducts end, have silt-collecting basins, elaborately treated, and massive cast-iron straining wells.[19]

For the short-term storage of water at an adequate height, cisterns were first used. These might be open, as at Cranstonhill, Midlothian,[20] but were frequently closed. A large, early example is Castlehill Reservoir, by Edinburgh Castle Esplanade (now a shop).[21] A taller, classical one existed in Union Street Aberdeen (now also commercial premises),[22] and there are elaborate ones in Arbroath[23] and Montrose, Angus,[24] in castellated and Tudor styles respectively. The Montrose example is almost in the nature of a water tower, and, like an early water tower proper at Dalkeith, has now been converted to a dwelling house.[25] The majority of water towers have been built since the 1920s, as settlements have spread into upland areas. In Glasgow, one at Ruchill dates from the 1920s, and there is another early one at Easterhouse. There are several large and elaborate examples in that area, mostly dating from the 1960s, and other fine ones in Lanark and Bishopbriggs. These are all of reinforced-concrete construction.

The combined industrial water/domestic water supply systems at Cranstonhill and Greenock have already been mentioned, but it would be misleading to omit reference to the many other industrial water supply systems in Scotland. The lowland canals all functioned as suppliers of industrial water, and many streams and rivers in the Lowlands were linked to impounded catchments to regularise the flow of water for power or processing. A good example is Kinross-Loch Leven, supplying water via the River Leven for industries in west Fife,[26] and on a different scale the Levern works in Renfrewshire, fed by a series of small dams, and serving many cotton mills and printworks. Extensive lade systems also drew water for sequential use in Perth, Galashiels, Selkirkshire, and Hawick, Roxburghshire, and lades serving a number of industries were by no means rare. Water-courses designed to serve a single industry were, of course, very numerous indeed. Before the days of hydro-electric power examples included the lades serving Bonawe Ironworks, Argyll,[27] Deanston, Stanley, both Perthshire, and Catrine, Ayrshire, cotton mills and Stormontfield bleachworks, Perthshire.[28]

ELECTRICITY

The first practical generators were made in the 1860s, but it was not until the 1870s that electric lighting, using arc lights, became attractive, and then only for large spaces.[29] The first installations, not surprisingly, were for specific spaces, and were privately owned. Legislation to permit public supply was passed in the early 1880s, but it was not conducive to investment, and it was not until the 1890s, after a change in legislation, that the first central generating stations were built in Scotland.[30] By that period incandescent filament lamps were available, suitable for houses and shops, and electric motors for industrial use were being manufactured. Within a few years most urban areas had generating stations, mostly built by municipalities, and a growing number of large industrial concerns were generating their own electricity.[31] The increased demand led to the rapid development of large steam turbines which could generate electricity on a much larger scale,[32] and in 1901 the Clyde Valley Electrical Power Company was founded to generate centrally for a large geographical area outside the cities (fig. 17.3).[33] The 1900s also saw electricity being used to power tramcars.[34]

After World War I it became increasingly obvious that the many small power stations built since the 1890s and generating at many different voltages, some alternating, some direct current, were not economic, and that to develop a national market for electrical goods a standard form of electric power had to be delivered. The Weir Committee which reported on this problem, suggested the establishment of a National Grid of high voltage alternating-current power lines, supplied by a relatively small number of large power stations.[35] The larger municipal and private power stations were drawn into this, in some instances being enlarged and re-equipped to fit them for their new role.[36] Smaller power stations were gradually phased out as standardisation proceeded.[37] From the late 1920s hydro-electric generation for public consumers became fashionable,[38] but it was not until after World War II that the systematic exploitation of water power in the Highlands began.[39] The North of Scotland Hydro-Electric Board brought electricity to areas impossible to serve if commercial criteria had been applied. After the war, too, demand for electrical power from the lowlands grew apace,

Figure 17.3 Eglinton Toll and St Andrews Electricity Generating Works, Glasgow, early twentieth century.

and new coal-fired power stations were commissioned. Larger and larger stations were constructed, culminating in the building of Longannet Power Station, Fife.[40] The new large power stations had lower generating costs than their predecessors, and their completion allowed these to be closed. The first nuclear power stations were at Chapelcross, Dumfriesshire, and Hunterston, Ayrshire.[41] Hunterston A station has now been replaced by Hunterston B,[42] and a new station built at Torness, East Lothian. As these stations operate best on continuous load, pumped-storage hydro-electric power stations were built to allow nuclear power generated at night to be stored as water for daytime regeneration.[43]

The changes in electricity generation outlined above had implications for power transmission. The first National Grid (1922–35) was energised at 66 and 33 kilovolts, but from the 1940s this was supplemented by a 'supergrid' operating at 132 kilovolts, with taller pylons and more widely spaced conductors. In the days of direct current generation, substations to reduce voltage for distribution to consumers were large, accommodating rotary converters.[44] Alternating current substations were smaller,[45] and were further reduced in scale as transformer efficiency increased. Now in country areas many transformers are pole-mounted and in urban areas are frequently in the open.

PUBLIC BATHS AND LAUNDRIES

Public baths and laundries were generally products of the public health movement in the 1880s and later, but there were early examples, perhaps most notably William Harley's baths in Glasgow, supplied by spring water from Hamiltonhill.[46] Public baths took two forms: individual baths, in areas where houses did not have such facilities; and swimming baths. As fixed baths in private houses did not become common until the 1880s, and then only in middle-class homes, individual public baths were even later, and their introduction was closely linked to the developing knowledge of bacteriology, with its impetus to personal cleanliness.[47] These facilities were provided only in urban areas, most notably in Glasgow, with its well-developed 'municipal socialism'. Swimming baths appear to have been roughly contemporary in their introduction, and were sometimes linked to individual baths. Unlike individual baths, however, swimming baths have continued to be built. In the inter-war years there was a vogue for open-air pools, both heated and unheated, especially in seaside towns, but more recently covered pools have become much more popular. Though recreational swimming pools have always been popular, competitive swimming was featured in many municipal pools. Since the 1970s these two functions have been increasingly separated, with sports and leisure pools sometimes within the same complex, but often separated. The bulk of swimming pools have traditionally been owned by local authorities, but in both Glasgow and Edinburgh there are subscription swimming clubs,[48] and leisure pools are now often adjuncts to hotels.

Laundries fell into two categories. The analogues of the individual baths were the 'steamies', with individual washing booths which could be booked for a period.[49] The modern equivalent is the laundromat. As with individual baths, 'steamies' were urban phenomena, usually sited in areas of inadequate housing. Customer laundries were much more widely distributed. Some of the largest hotels and larger institutions in fact had their own laundries.[50] As with baths, the stimulus to laundry development came with recognition of the health implications of 'dirt'. Machinery was developed to wash linen on a large scale, and to iron it, and in the 1880s and '90s laundries sprang up, with astonishing rapidity, to use the new equipment. By 1914 most towns and some villages had at least one laundry,[51] and more were built in the inter-war years. The development of motor transport allowed laundries to have larger spheres of influence. After World War II the availability of domestic electric washing machines (and laundromats) ate into the domestic laundry market, especially during the 1960s. The few surviving customer laundries serve the catering and institutional trades.[52] Some hospitals and prisons retain their own laundries.

CLEANSING

Cleansing, in the sense of refuse disposal, has a very ancient history. It is closely linked to sanitary provision, another theme of this chapter. One of the principal roles of burgh councils in the medieval and early modern periods was the control of domestic and trade refuse,[53] and the management of refuse accumulation and disposal is still a major preoccupation of local authorities. In the eighteenth century and earlier, refuse could be roughly divided into two kinds: human and animal excrement; and kitchen waste (including food scraps, bones and broken utensils). The former had manurial value, and this was legally recognised, being the property of the landowner.[54] The latter, midden material, was dumped on the outskirts of communities, and its excavation and study provides us with much of the information we have on early civilisation in Scotland. Excrement, in contrast, was carefully stored in dungsteads, often in the middle of streets, until wanted to manure the 'infield', where arable crops were grown.[55] Wood and coal ashes, which contain useful minerals (especially potash), were often mixed with the excrement. This practice only ended with the identification of a connection between some infectious diseases with excrement-contaminated water. Even before this was unequivocally proved, the public health movement in the 1830s was advocating the removal of excrement from the vicinity of dwellings. One way of doing this was by adopting water carriage of sewage, and this was an important reason for developing gravitational water supply systems. The supposed economic value of human and animal manure from towns remained, however, a matter of concern, and the collection of the contents of 'ashpits' was a prime function of the cleansing departments established in the major cities from the 1860s, and remained so until the water carriage of sewage became recognised as the only proper way to dispose of excrement, during the 1880s and 1890s.[56] This was closely linked to the development of bacterial methods of sewage treatment (see below).[57]

By the time the economic components of domestic refuse were largely eliminated by water-borne systems its manurial value was already being eroded by the detritus of a society in which mass-produced goods were much commoner. The containers and packages which dominate modern refuse were becoming more significant as the nineteenth century wore on, and the town, village and township dumps of the period afford a treasure trove of bottles and ceramic containers. These were the precursors of the landfill methods of refuse disposal almost universal today. In the larger urban communities, however, in the 1880s and '90s the quantities of material to be handled were causing concern, and the first large-scale incinerators were built,[58] turning refuse into clinkers of much less bulk with some value as hard-core. Because the development coincided with the introduction of central electricity generation, in some instances the heat produced by combustion of rubbish was used to create steam to generate electricity,[59] but in general this was unsuccessful. The pattern from the 1880s until the 1970s was to build bigger and less polluting incinerators, but it is currently more economical, and felt to be environmentally less damaging, to bury refuse unburnt.

FIRE FIGHTING

Until the nineteenth century the provision of fire-fighting services was very limited, and largely in the hands of fire insurance companies, who maintained their own brigades in the larger urban centres. The increasing size and complexity of towns and cities during the Victorian period led to a greater desire for systematic fire fighting, and the city and burgh councils began to make their own provisions. At first the only fire pumps available were hand operated.[60] The smaller ones could be manhandled into position, but the larger examples were designed for horse haulage. Hoses were made of leather, riveted with copper rivets, or of canvas. By the later nineteenth century both fire-engine manufacture and hosemaking were recognised trades.[61] The houses for manual engines were small,[62] and there was no provision for firemen to live on site. Some country houses and industrial establishments had their own manual fire engines.

In the mid nineteenth century a revolution in fire fighting in the cities began. Pumped and gravitation water supplies enormously increased the quantity of water available for fire pumps. At the same time the height of buildings tended to increase. To meet the opportunities and challenges that these circumstances offered the steam fire engine was invented. Such fire engines required to be kept 'on the boil' to be available for an emergency, and so full-time firemen were recruited to serve them. There were economies in having several engines at the same base so, in place of the small houses provided earlier, stations were built with space for two or more engines, often with living accommodation above.[63] Extending ladders were also invented to allow access to the upper floors of tall buildings. When the internal combustion engine was adapted to propel road vehicles it was quickly adapted for fire engines, some of which had petrol engines for propulsion. Within a very few years, however, the pump tender had been developed, with ladder, engine-driven pump, lockers for equipment and room for men to ride in it. Descendants of this basic type still form the backbone of our fire brigade's equipment. Some steam fire stations were adapted to suit the new machines, but new ones were built. Hose-drying towers and practice towers became standard features with, in urban areas, accommodation for full-time firemen.[64] In the inter-war years the extending turntable ladder was invented which gave a stable base from which to direct hoses from a high level, as well as making rescue easier. The descendant of these is the hydraulic platform used by today's fire service. World War II, with many aviation crashes, saw the introduction on a large scale of fighting oil fires with foam, and the foam tender is now a valuable asset in both town fire-fighting and, more particularly, in airport fire services.

The increasing size, expense and complexity of fire engines has reduced the extent to which industrial concerns maintain their own fire services, but in high-risk establishments these are still used.

LIGHTING

Scotland's connection with the use of coal gas for lighting is well-known. William Murdoch is credited with having 'invented' gas lighting as a young man in Lugar, Ayrshire. Whatever the truth of the legend, it is certain that under his superintendence gas-lighting plants were made at Boulton & Watt's Birmingham works in the first decade of the nineteenth century, and some were supplied to Scottish textile mills. The use of gas for lighting streets, houses and commercial premises did not develop for another ten years or so. By 1820 both Glasgow and Edinburgh had gas light companies, and by 1840 most towns and large villages had gas lighting.[65] Experiments were also made with preparing gas from whale oil, and even with compressing it into bottles to make 'portable gas'. At first, gas companies were private concerns, but from the 1860s the larger towns took the supply of gas into their ownership, and were generally able to reduce prices, increase reliability of supply and broaden the consumer base.[66] When, during the 1880s, electric lighting began to compete with gas, the incandescent gas mantle was invented.[67] This produced a better and cleaner light, and required gas of a different quality, with greater heating power. Using higher temperatures for carbonisation, yields of gas could be increased and prices further reduced. This new style of gas was also well suited to burning in domestic appliances, such as room heaters, water heaters and cookers, and for gas engines.[68] It could most efficiently be made in vertical retorts, which supplanted horizontal retorts in larger gas works from the 1910s onwards.[69] The large works, whether using horizontal or vertical retorts, required giant gas holders, some of which remain in service.[70] They also ousted some of the smaller works. A number of medium-sized gas works were also converted to vertical retort operation.[71]

After World War II the large-scale generation of electricity put considerable pressure on the gas industry. Gas for domestic lighting was the first to go, followed by gas street lights, leaving cooking the main use (apart from industry). To economise, a gas grid was introduced in the 1950s and 1960s, allowing the closure of many small coal-carbonising

Figure 17.4 Rothesay Gas Works, Bute, 1960s. Copyright: John R Hume.

plants (fig. 17.4).[72] The production of gas by cracking by-product naphtha from the petroleum-refining industry eliminated the larger coal-carbonising plants, before the oil-gas plants were in their turn displaced by supplies of natural gas from the North Sea.[73] The cleanliness and freedom from offensive odour of oil-based and natural gas, and the relative cheapness of the latter, encouraged their use in room heating and central heating. So effective is gas in space heating that bottled propane and butane have become very popular supplements to town gas.

HYDRAULIC PLANT

Unlike gases, liquids are to all intents and purposes incompressible, so that if one applies pressure to a liquid in a pipe the same pressure will be evident at the other end. This property can be used to transmit power, and to store power in the form of liquid under pressure. Another valuable property of liquid used in this way is that pressure applied to a small surface at one end will be transmitted at the same level to a larger surface, thus effectively multiplying the pressure delivered at the far end.

These two effects, known as scientific principles in the seventeenth century, were developed for practical purposes in the nineteenth century. At first they were applied to individual pieces of plant, such as packing presses for textiles, forging presses, jacks for lifting, or jiggers for operating cranes or swing bridges.[74] In the 1850s, '60s and '70s such applications became usual. In the 1880s and '90s further applications were developed, including lifts for goods and passengers, gas-works plant and riveting machinery.[75] So useful was hydraulic plant that in docks, railway warehouses and shipyards, central power stations were installed with hydraulic mains running to the points of application.[76] In the 1890s Glasgow Corporation introduced a 'public' hydraulic system, with mains running considerable distances.[77] This remained in use until the 1960s, but was unique in Scotland.

SEWERAGE

Some reference to sanitary provision has been made in the section on cleansing. This account deals with water-borne sewage disposal, which arose from the use of the water closet. Invented in the late sixteenth century, the water closet became practical with the modifications made by Joseph Bramah (1749–1814).[78] Its use was, however, limited for a number of years by the reluctance of the sewerage authorities to authorise water closet discharge into sewer pipes, which were primarily intended for disposal of surface water. As a result, houses with water closets had to have their own cesspools, underground tanks for the collection of the output from the closets.[79] Unlike the modern septic tank, little or no purification took place in the cesspool, which had porous walls through which liquid could escape leaving solids behind – night soil – which had to be emptied regularly, by hand.

Such arrangements, though primitive by modern standards, were expensive, and the prerogative of the middle and upper classes. The lowest classes had to make do with the dry closet and chamber pot, the contents of which went onto the midden.

The more general adoption of water-borne sewage disposal began in the 1860s, by which time gravity water had reached both Glasgow and Edinburgh; by the 1880s the advantages of this system had become so clear that it became compulsory.[80] The discharge of the many water-closets into the sewers of Glasgow led in the early 1890s to the construction of intercepting sewers, running parallel to the river banks.⁸¹ These fed the raw sewage to purification plants where solids were precipitated and liquids treated before discharge into the Clyde. The sewage sludge was shipped in steam tank ships and dumped in the Clyde estuary. Glasgow was the only Scottish city with this system until recently, when Edinburgh was compelled to introduce it; both Aberdeen and Dundee send screened sewage through long pipes into the sea.

Inland, the general practice was to discharge sewage into water courses, and treatment was unknown until the 1880s and 1890s when a series of inventions made effective purification possible. The contact bed developed by Dibden⁸² was the first of these, followed by the percolating filter,⁸³ the septic tank,⁸⁴ and finally the activated sludge method.⁸⁵ Together (with the exception of the soon-superseded contact bed) these made it possible to treat sewage on any scale from the small village to the large town, and from about 1900 there was a steady introduction of sewage treatment works. A typical installation had a concrete septic tank and one or more percolating filters. Generally the latter had the rotary distributor invented by Candy & Caink,⁸⁶ which used the reaction from jets of sewage to rotate the arms. Many of these are still in use. Some larger plants used the activated sludge system.

During the last twenty or thirty years river purification boards have imposed higher and higher standards on effluent discharges, and many of the older sewage treatment works have been completely rebuilt.

TELECOMMUNICATIONS

Telecommunications as we know them today descend from the electric telegraph.⁸⁷ This was developed over a lengthy period, becoming practical in the 1830s, and receiving a great impetus from its adoption by railways as an element in train control. A critical innovation was the substitution of iron wire for copper in overhead lines,⁸⁸ which made the conductors far less subject to theft. By the 1850s railway telegraphs were commonplace, and the telegraph was being used for business and commercial purposes very extensively. Atlantic telegraph cables were laid in the mid-1850s and more successfully in the mid 1860s,⁸⁹ the subsequent extension of the overseas telegraph network being of the greatest importance to the trade of Scotland.⁹⁰

The primacy of the telegraph for overseas communication persisted until after World War II, but the invention of the telephone by Alexander Graham Bell in 1879 quickly reduced the importance of the telegraph in internal communication.⁹¹ Within a relatively short time telephones were being used in the cities, with private companies competing for custom. Glasgow was one of a number of municipalities with its own telephone system.⁹² The number of customers increased steadily after 1900, and has continued to do so. The first small manual exchanges were supplanted by bigger ones and eventually by automatic exchanges, though it was not until the 1950s and '60s that these became universal. The telecommunications revolution that has occurred since then has been less evident to users, with the exception of radio-telephones, but it has allowed telephone usage to expand markedly, especially of mobile telephones with text-messaging facilities, and has catered for the transmission of messages between computers and the creation of Internet. The use of satellites has made it possible to telephone all over the world without significant loss of clarity. The printing telegraph, familiar to earlier generations, was supplanted by the teleprinter, and now the fax machine complements it.

NOTES

1. Marwick, 1901, 1, 3, 5, 6, etc; Colston, 1890, 5, 6.
2. Stell, 1986, 150–1.
3. Colston, 1890, 9–12.
4. Marwick, 1901, 69–75, 77–80.
5. Hume, 1974, 131–2.
6. Shaw, 1984, 483–6.
7. Ferguson, 1948.
8. So long as it was believed that sewage had a real value as manure, the transition to water carriage was delayed. By the 1880s it was apparent that sewage was simply a nuisance to be got rid of.
9. Marwick, 1901, 114–15.
10. Gale, Annan, 1877, 1–3.
11. Ferguson, 1948, 150–1.
12. Colston, 1890, 5, 6; Ferguson, 1948, 151.
13. Gow, 1996, 81, Kirkintilloch.
14. Gale, Annan, 1877, plate following p 6.
15. As at Gorbals Water Works.
16. As at the Mugdock and Craigmaddie reservoirs on the Loch Katrine Scheme.
17. Colston, 1890, plate facing p210.
18. Hume, 1974, 160, 277.
19. Gale, Annan, 1877, plates following pp 22, 23 and 25.
20. Panoramic view of Glasgow, lithographed in 1863.
21. Gifford, McWilliam, Walker, 1984.
22. Groome, 1893, 12; at 474–484 Union Street, designed by John Smith in 1830 (Historic Scotland, List Description).
23. Hume, 1977, 123–4.
24. ibid, 142.
25. Hume, 1976(a), 184.
26. Bennett, 1982, 47; Shaw, 1984, 488.
27. Stell, Hay, 1984, 18, 24.
28. Hume, 1977, 268–9, 270–1, 281; Hume, 1976(b), 71.
29. Parsons, 1940, 170–83.
30. Hume, 1974, 139–140; Parsons, 1940, 154–5, plate xxii; Gifford et al, 1984, 369, 647.

31. Hume, 1976(b), 171.
32. Parsons, 1940, 170–83.
33. Hume, 1974, 140.
34. Power stations were built in Glasgow (Finkston) and Kirkcaldy specifically to generate electricity for tramways.
35. Reader, 1968, 130–6.
36. As at Yoker (Clyde Valley) and Dalmarnock (Glasgow Corporation).
37. Such as the McDonald Road and Dewar Place stations in Edinburgh, and the Pollokshaws Road and Port Dundas stations in Glasgow.
38. Hill, 1984.
39. North of Scotland Hydro-Electric Board, 1950.
40. Pride, 1990, 24.
41. Gifford, 1996, 181.
42. Close, 1992, 80.
43. There are two of these, Cruachan, Argyll, and Foyers, Inverness-shire.
44. Hume, 1974, 141.
45. Many simple sandstone-faced substations were built in Glasgow in the inter-war years, as in Lauderdale Gardens and Partickhill Road.
46. Marwick, 1901, 69–70.
47. Glasgow Corporation, 1914, 96–7 and plates between these pages.
48. Williamson et al, 1990, 280, 348, 448; Gifford, 1996, 396.
49. Glasgow Corporation, 1914, 96–9.
50. For example, the Cruden Bay Hotel; see Sangster, 1983, 34, 36, 38.
51. For instance, apart from the larger towns, there were laundries in Charlestown, Fife, Kilbarchan, Renfrewshire, Huntly, Aberdeenshire, Gourock, Renfrewshire, and Pitlochry, Perthshire.
52. For instance, the Slateford Laundry, Edinburgh, and the Initial Towels Laundry in Paisley.
53. For example, Marwick, 1901, 22; see also Ferguson, 1948, 137–43.
54. The material was known as 'fulzie'.
55. The trade in 'urban manure' was an important one for early canals and railways, including the Aberdeenshire Canal and the Dundee and Newtyle Railway.
56. The two main stimuli were the development of outfall sewers, which required a constant flow for effective operation, and the introduction of bacterial methods of sewage purification.
57. Adams, 1930, 154–265.
58. Hume, 1974, 135; Tucker, 1977.
59. Tucker, 1977, 5–27.
60. Marwick, 1901, 126, 1851 witnessed the first modern manual engine in Glasgow.
61. Hume, 1974, 191 (Pleasance Leather Works, built *c.* 1861 for John Burt, machinery belt, lace and hosepipe maker). William Herkless was described as 'fire engine maker' at 9 Shuttle Street, Glasgow, in 1872 (Post Office Directory, 1872–3).
62. One survives in Brechin, Angus.
63. Glasgow Corporation, 1914, 112–3, plates between 114 and 115; Gifford et al, 1984, 259.
64. For example, Partick Fire Station, Beith Street, Williamson et al, 1990, 372.
65. Adams, 1978, 145–7; see also Cotterill, 1976.
66. Hume, 1974, 137–8.
67. Derry, Williams, 1960, 512–13.

68. ibid, 514, 601–3.
69. Cotterill, 1980–1, 29.
70. Hume, 1974, 150 (Temple), 248 (Tradeston), 279 (Provan).
71. These included Helensburgh, Dunbartonshire, Rothesay, Bute, Dunoon and Campbeltown, Argyll.
72. Miller, Tivy, 1958, 210–2.
73. About half of Britain's natural gas comes in through the St Fergus gas terminal, Aberdeenshire, opened in 1976. McKean, 1990, 147.
74. Packing presses were in general use in Dundee in the 1860s.
75. Engineering. *Bridges, Structural Steel Work and Mechanical Engineering Productions by Sir William Arrol and Company, Ltd*, London, 1909, 210–59.
76. For instance in Queens and Princes Docks, and in the Leith Docks; also in the College Yards Station, Glasgow.
77. Hume, 1974, 134.
78. Singer, et al, 1956–9.
79. ibid, 508–9.
80. Hume, 1974, 136–7.
81. Marwick, 1901, 228–9, 233–4.
82. Adams, 1930, 147–53.
83. ibid, 182–218.
84. ibid, 163–75.
85. ibid, 219–37; Metcalf, Eddy, 1930, 636–78.
86. Adams, 1930, 194–9.
87. Singer et al, 654–61.
88. In the late 1830s.
89. Singer et al, 660–1.
90. Commercial codes were developed to allow a few words to carry much information. West of Scotland manufacturers published code words for their products in their catalogues.
91. Derry, Williams, 1960, 628.
92. Marwick, 1901, 237–8, appendix 81. The service opened in 1901, and it served a large area round Glasgow as well as the city itself.

BIBLIOGRAPHY

Adams, I H. *The Making of Urban Scotland*, London, 1978.
Adams, S H. *Modern Sewage Disposal and Hygienics*, London, 1930.
Bennett, G P. *The Past at Work: Around the Lomonds*, Markinch, 1982.
Butt, J, Ward, J T, eds. *Scottish Themes*, Edinburgh, 1976.
Close, R. *Ayrshire and Arran: An illustrated architectural guide*, Edinburgh, 1992.
Colston, J. *The Edinburgh and District Water Supply: A historical sketch*, Edinburgh, 1890.
The Corporation of Glasgow Gas Department. *The Gas Supply of Glasgow*, Glasgow, 1935.
Cotterill, M S. 'The Scottish Gas Industry up to 1914', unpublished PhD thesis, University of Strathclyde, 1976.
Cotterill, M S. The development of Scottish gas technology, 1817–1914: Inspiration and motivation, *Industrial Archaeology Review*, 5 (1980–1), 19–40.
Crompton, R H. *Early Days of the Power Station Industry*, Cambridge, 1939.
Cumming, J W. *The Greenock Cut: The story of Greenock's water supply*, c. 1873, reprinted Glasgow, 1980.

Derry, T K, Williams, T I. *A Short History of Technology*, Oxford, 1960.
Ferguson, T. *The Dawn of Scottish Social Welfare*, Edinburgh, 1948.
Fothergill, R. *Waterways of Perth*, Perth, npd.
Gale, J M, Annan, T. *Glasgow Corporation Water Works*, Glasgow, 1877.
Gifford, J. *The Buildings of Scotland: Dumfries and Galloway*, Harmondsworth, 1996.
Gifford, J, McWilliam, C, Walker, W. *The Buildings of Scotland: Edinburgh*, Harmondsworth, 1984.
Glasgow Corporation. *Municipal Glasgow: Its evolution and enterprises*, Glasgow, 1914.
Gomme, A, Walker, D. *Architecture of Glasgow*, London, 2nd edn, 1987.
Gow, B. *The Swirl of the Pipes*, Strathclyde Regional Council, Glasgow, 1996.
Groome, F. *Ordnance Gazetteer of Scotland*, London, 2nd edn, 1893.
Hill, G. *Tunnel and Dam: The story of the Galloway hydros*, Glasgow, 1984.
Hume, J R. *The Industrial Archaeology of Glasgow*, Glasgow, 1974.
Hume, J R. *The Industrial Archaeology of Scotland, Vol 1, The Lowlands and Borders*, London, 1976(a).
Hume, J R. Shipbuilding machine tools. In Butt, Ward, 1976(b), 158–80.
Hume, J R. *The Industrial Archaeology of Scotland, Vol 2, The Highlands and Islands*, London, 1977.
Marwick, Sir J D. *Glasgow: The Water Supply of the City*, Glasgow, 1901.
McKean, C. *Banff and Buchan: An illustrated architectural guide*, Edinburgh, 1990.
Metcalf, L, Eddy, H P. *Sewerage and Sewage Disposal*, New York, 2nd edn, 1930.
Miller, R, Tivy, J, eds. *The Glasgow Region: A general survey*, Glasgow, 1958.
North of Scotland Hydro-Electric Board. *Hydro-Electric Power*, Glasgow, npd [1950].
Parsons, R H. *The Early Days of the Power-Station Industry*, Cambridge, 1940.
Pride, G L. *The Kingdom of Fife: An illustrated architectural guide*, Edinburgh, 1990.
Reader, W J. *Architect of Air Power: The life of the first Viscount Weir*, London, 1968.
Sangster, A H, ed. *The Story and Tales of the Buchan Line*, Poole, 1983.
Scottish Hydro-Electric plc. *Power from the Glens*, Edinburgh, npd.
Shaw, J. *Water Power in Scotland, 1550–1870*, Edinburgh, 1984.
Singer, C, Holmyard, E J, Hall, A R, Williams, T I, eds. *A History of Technology*, 5 vols, Oxford, 1956–9.
Stell, G. *Exploring Scotland's Heritage: Dumfries and Galloway*, Edinburgh, 1986.
Stell, G P, Hay, G D. *Bonawe Iron Furnace*, Edinburgh, 1984.
Tucker, G D. Refuse destructors and their use for generating electricity, *Industrial Archaeology Review*, 2 (1977), 5–27.

18 Buildings for Recreation

CHARLES MCKEAN

For the most part, buildings for recreation are distinguished by having to cater for large congregations of the public. As a consequence, they tend to be subject to close scrutiny by the state for reasons either of public morals or of public safety. Since, to begin with, the only other large buildings of congregation were churches, new arrivals tended to be regarded as a competitive threat. In the early eighteenth century, the Church remained doubtful – at the very least – about theatres, was antagonistic to leisure activities like masques, plays, golf and football, and viewed travelling libraries as likely corrupters of morals. By a strange inversion, by the late twentieth century, places of public resort have replaced churches as the primary focus of volume activity.

Recreational buildings fall into separate categories. There are buildings for culture (theatres and concert halls), self-improvement (art galleries, museums, clubs), health (swimming pools, health centres, gymnasia), information and refreshment (public houses, coffee- and tea houses, newspaper rooms), and social interaction and entertainment (assembly rooms, card rooms, cinemas and ice-rinks). These categories are not mutually exclusive. Many buildings listed as relevant to health could be just as relevant to sport, although sport – through tennis and golf – considerably predates the preoccupation with health, fitness and swimming.

Early recreational buildings were either temporary (usually framed structures derived from fairground structures), or adaptations, such as coffee rooms and taverns occupying the lower storeys of tenements. By their very nature, they adapted well to the often inconvenient warren of small vaulted spaces that these urban structures provided. That was, indeed, sometimes their attraction. It was in a coffee-room, for example, that the East India Company set up its subscription book for what became the Darien venture.

Adaptation was also the norm for other places of entertainment. Plays, for example, were performed (when the Duke of York was viceroy) in the Holyrood tennis court, Edinburgh, presumably a covered tennis court like the surviving one at Falkland, Fife. In 1730, Allan Ramsay adapted a building in Carrubber's Close, Edinburgh, for theatrical performance, but it was in the Tailors' Hall in the Cowgate that the Edinburgh Comedians established a formal theatre in 1733, amongst whose abominations was the transformation of *Macbeth* into an operetta. Likewise, the Edinburgh Musical Society borrowed the Palladian St Mary's Chapel for its concerts until 1769.[1]

Although it seems probable that a substantial form of theatre beside Edinburgh's Calton Hill – legend has it a 'Scottish Globe' called the Playfield – was the setting of the performance of Sir David Lindsay's *Thrie Estaits* in 1554 (and for which King James VI later licensed a company of players in 1592), no trace remains.

So far as can be ascertained, the principal purpose-built recreational buildings before the late eighteenth century may well have been halls: masonic halls, like the Lodge Canongate Kilwinning in Edinburgh's Canongate (1735–6), or the Archer's Hall (begun 1776). Neither was very substantial, and both appear to have been derived from craft and guild halls, containing a large space with certain smaller ancillary spaces. Both were established in backlands, with architecture of no great pretension.

St Cecilia's Hall, established by the Edinburgh Musical Society in Niddry Wynd in 1763 to the design of Robert Mylne, was probably the first significant purpose-built recreation building (fig. 18.1). It was said by contemporaries to be designed after the great Opera Theatre in Parma.[2] Although a small, subscription-based concert hall, it was, in essence, a

Figure 18.1 St Cecilia's Hall, Cowgate, Edinburgh; interior of hall, 1960. Crown copyright: RCAHMS, SC 710756.

BUILDINGS FOR RECREATION • 359

more sophisticated version of the other halls – one single space on the principal floor, with minor and ancillary rooms supporting it. The oval hall, however, is of singular sophistication with an elliptical, domed and top-lit ceiling. St Cecilia's façade facing Niddry Wynd, although of ashlar and pedimented, is of no great purpose and is no longer used.

Between the late eighteenth century and about 1840, there arose a group of recreation rooms, athenaea and theatres. By the end of that period, self-improvement had added art galleries, museums and concert halls.

COFFEE ROOMS

Coffee rooms, reading rooms or athenaea, emerged as the foci of mercantile life, and few self-respecting large communities would be without one. A formalisation of the old coffee houses, they appear to have acted as information centres, merchants' gathering-places and, to some extent, dealing-room floors. The most grandiose was probably Glasgow's (William Hamilton, 1781), built by public subscription behind the Town Hall on the Trongate and universally allowed to be the most

Figure 18.2 Former Exchange Coffee House, 15 Shore Terrace, Dundee. Crown copyright: RCAHMS, SC 710985.

elegant of its kind in Britain if not in Europe. In addition to its 74-feet (22.56-metre) long coffee room, it contained a newsroom, billiard room and dining room, with opulent decoration and chandeliers. That Edinburgh seems to have dispensed with such a building (its merchants even rejected the Adam brothers' 1752 Exchange) may be a reflection of how little mercantile interests affected life in the capital. Instead, that function may have been performed in the clubs.

Dundee had several, beginning with the coffee room within Samuel Bell's 1776 Trades Hall. Its rival was the Baltic Exchange in Bain Square on the corner of the Cowgate. In 1828, the merchants desired to leave the Trades House for the splendid Dundee Exchange Coffee House (fig. 18.2) designed by George Smith, facing over the new docks from the bottom of Castle Street. The principal rooms (Ionic Order) were at first-floor level above squat business rooms (Doric Order) on the ground. It contained a coffee house, merchant's library and reading room, and an assembly room. Similar functions (without the assembly room) were provided in Stirling's Athenaeum (William Stirling, 1816) in a new towered building closing the vista up King Street.

As the century progressed, coffee rooms extended to workmen's coffee rooms.

ASSEMBLY ROOMS

Assembly rooms for those essential components of social life – levées, routs, balls and card parties – were quintessentially eighteenth-century buildings. Two had existed in Edinburgh's Old Town before a private one was built for George Square residents to overcome the physical shortcomings of earlier, adapted buildings. The New Town one (John Henderson, 1787), however, was splendid, consisting of saloon and ballroom on the principal floor, with subsidiary apartments beneath. Aberdeen's (Archibald Simpson, 1820), signalled by a granite Ionic portico, provided graceful support rooms and card rooms in addition to the ballroom in outstanding neo-classical form. James Adam's flamboyant Athenaeum in Glasgow's Ingram Street has been demolished. It is probably no accident that when music halls came into fashion, they were grafted onto the back of the assembly rooms – Edinburgh by David Bryce, 1843,[3] and Aberdeen by James Matthews, 1858.

The assembly rooms of smaller centres – Paisley, Inverness and Elgin, Moray, for example – comprised a large-windowed and usually handsome first-floor apartment above shops. In Dundee, David Neave's fine 1828 Masonic Hall was periodically adapted for that function.

THEATRES AND MUSIC HALLS

Theatres and music halls differed from other leisure buildings because of their scale and notorious susceptibility to fire. Indeed, no eighteenth- or early nineteenth-century theatre has survived intact, and the façade of the 1810 Theatre Royal in Dundee is one of the few left. The first purpose-built theatre may have been in Edinburgh's Canongate, but it died in a riot (a not uncustomary theatrical occurrence) to be replaced, in 1769, with the Theatre Royal, possibly the first complete building in the New Town. Glasgow's first theatre was a 1752 timber construction in Castle Street, followed by a poor construction in Dunlop Street (contemporary with a galleried brick barn for a circus in Jamaica Street), both with classical frontages. Its much more opulent successor in Queen Street (David Hamilton, 1815) had an elliptical auditorium holding 1,500 people in two tiers of 'boxes, slips and galleries'. Scenery was painted by Alexander Nasmyth.

These early theatres appear to have differed from later ones only marginally and in scale. They consisted of a galleried auditorium enclosed within a usually classical and pedimented box which contained certain front-of-house facilities. In addition to these premier civic buildings, other buildings were adapted for periodic theatrical performances. In Edinburgh, there were temporary buildings on The Mound, and in Glasgow similar buildings periodically on Glasgow Green. In Dundee, David Neave's Trinity House (or Sailors' Hall) was frequently used for the purpose.

Music halls, dominant by the later nineteenth century, derived from theatre technology. Victorian expertise in framed iron structures was developed to inflate the eighteenth-century small galleried theatre into grandiose and plush creations which reached their apogee in the Edwardian period, and in the buildings of Frank Matcham. Although the auditorium remained a galleried space facing a stage framed by a proscenium arch, these larger buildings were designed for thousands rather than hundreds, and rose from stalls to the grand circle, upper circle, gallery and, in some cases, to the upper gallery (or 'The Gods'). They included an orchestra pit, green rooms, and scenery stores behind. There was an increasingly elaborate front-of-house, with bars or entertainment rooms on each floor (fig. 18.3).

Generically, there was little difference in structure between a music hall and a theatre, although perhaps a larger orchestra pit, shallower stage and less sophisticated provision for theatre props and green rooms distinguishes the music hall. The difference lay in the externals, sophisticated versus vulgar. Matcham's 1904 His Majesty's Theatre, Aberdeen, is about as grand as a theatre could be, whereas his 1891 structure in Edinburgh's Nicholson Street, with its opening roof was,

Figure 18.3 Theatre Royal, 254–90 Hope Street, Glasgow; auditorium, c. 1930. Crown copyright: RCAHMS, SC 711072.

appropriately, an 'Empire Palace of Varieties'. The exterior had to convey a sensation of what was to be found within, perhaps most strikingly with the Alhambra (demolished) in Glasgow.

The later nineteenth and early twentieth centuries were the heyday for theatres and music halls – which were fundamentally variants of the same building type. They expanded in scale using improved methods of construction, fire escape, lighting and ventilation, and the facilities extended to coffee rooms, restaurants, bars, and, in some cases, shops. Some architects – most particularly Frank Matcham – specialised in their design. In most respects, they were to be the direct ancestors of the cinemas.

MUSEUMS AND ART GALLERIES

Museums and, later, art galleries satisfied both the wave of nationalist-inspired antiquarianism and the growing public taste for self-improvement. Patrick Blair had founded a 'Cabinet of Curiosities' in Dundee in the late seventeenth century, and the Society of Antiquaries of Scotland established its first museum off the Cowgate, Edinburgh, in 1780–1. Early collectors included Dr William Hunter, whose rooms in James Adam's Professors' Lodgings in Glasgow's High Street[4] were

BUILDINGS FOR RECREATION • 363

replaced in 1804 by William Stark's neo-classical masterpiece of the original Hunterian Museum. Museums, as can be seen from Dr William Hunter's lineage, were not originally as clearly distinguished from art galleries as was later to be the case. Dr Hunter's original exhibition was simply the display of what he had collected, which ranged from medieval manuscripts to material comparable to the display by Edinburgh's Royal College of Surgeons of anatomical exhibits.

The public museum, however, as an exhibition of material to instruct the natives in the manner of *Pickwick Papers*, was possibly first essayed in Elgin (Thomas MacKenzie, 1842). This was essentially a museum for the exhibition of material of local archaeological interest, but was later extended to include artefacts displayed by people returning home from service in the empire. Gracefully neo-Italianate (to signify learning and cosmopolitan aspiration), the building consists of a large galleried barn behind a shallow front-of-house. Exhibits were placed in display cases, or placed promiscuously where there was room.

Towns with lesser ambition, such as Banff, devoted space within their new academies for the display of objects of antiquarian curiosity (such as the skulls of Danish leaders killed at the Battle of the Bloody Pits, 1016, previously exhibited in wall-niches in St John's Church, Gamrie, Banffshire). Thereafter, a suitable ancient house was converted to museum use, similar to the conversion of Huntly House in Edinburgh and Glasgow's Provand's Lordship for just such a purpose. Indeed, by the late twentieth century, museum use had become the first use to be considered by those motivated to rescue a historic building from demolition. Such public museums were frequently matched by large private museums, assembled by landed gentry and aristocrats; and public museums were occasionally the beneficiary of large collections from such sources.

Museums in smaller communities were normally linked to art galleries, but in larger ones they acquired an independence. The largest purpose-built museum, designed by Captain Francis Fowke, was opened in 1864 as the Edinburgh Museum of Science and Art, funded with money from the Great Exhibition and containing Edinburgh University's Natural History collection which the university could no longer afford to maintain (fig. 18.4).[5] A palace of High-Victorian engineering, designed along Great Exhibition lines, it consisted of large, cast-iron, timber and glass top-lit galleries of cathedral-like grandeur, whose innovation was concealed behind the façade of an Italian palazzo signifying learning. In 1884, the national collections were given considerable assistance by the munificent donation of funds by Sir John Ritchie Findlay, editor of *The Scotsman*, sufficient to provide space for a National Portrait Gallery and a Museum of Antiquities in Queen Street.[6] The Portrait Gallery moved in in 1889, and the Museum of Antiquities two years later. It was the first time that

Figure 18.4 The Royal Museum of Scotland, Chambers Street, Edinburgh; western pavilion, 1960s. Crown copyright: RCAHMS, SC 710847.

the splendid collection begun by the Earl of Buchan and the Society of Antiquaries of Scotland in 1780 was given the privilege of display in its own purpose-built location. It consisted of large, predominantly north-lit (occasionally top-lit) galleries with much free space for the placing of cabinet and display cases.

Originally museums were designed to display objects in classification for learning purposes: but, possibly as a consequence of the link with the fine arts, museum buildings gradually moved toward the exhibition of objects as works of art in their own right. This approach reached its apotheosis in the Burrell Gallery, Glasgow (Barry Gasson, 1984), which is designed around the function of the artistic display of great works of art, thus subsuming the primacy of learning. The focus of the Museum of Scotland, by contrast (Benson and Forsyth, 1999), is upon the presentation of artefacts within their chronological context and within the powerful Scottish shell in which they are housed.

Art galleries almost invariably began with private collectors. Alexander 'Picture' Gordon's 1804 'art-gallery' house in Glasgow's Buchanan Street may have been designed by Sir John Soane, and his rivals included William Buchanan, and Archibald McLellan. McLellan's

Figure 18.5 The Vine, 43 Magdalen Yard Road, Dundee; lithograph by David Walker, 1954–5. Courtesy of David Walker.

Figure 18.6 The National Gallery of Scotland and Royal Scottish Academy, The Mound, Edinburgh, c. 1860. Crown copyright: RCAHMS, SC 466228.

own collection became the foundation of Glasgow's first public art gallery – the 1854 McLellan Galleries in Sauchiehall Street. On a smaller scale, the display of art collections led to the emergence of a particular type of suburban villa with a top-lit hall which might be christened the 'art gallery house'; of these the most distinctive were The Vine, Dundee (fig. 18.5), Harviestoun House, Clackmannan (now demolished), some of the villas in Trinity, Edinburgh, and, above all, Arthur Lodge (Thomas Hamilton, 1832) in Edinburgh.

Public art galleries emerged with an educational purpose, most notably with William Playfair's majestic (if diminutive) National Gallery of Scotland in Edinburgh, completed in 1854 (fig. 18.6).[7] A large, classical, feminine (Ionic), windowless but top-lit rectangular box, it was originally divided into two; one half the gallery, and the other half for the Royal Scottish Academy. Paintings were hung densely upon claret-coloured walls, the better to allow students, amateurs and others to enter, study, sketch and copy. Although the government had paid for the National Gallery of Scotland, it was to stimulate Glasgow into something comparable that Glaswegian architects banded together in 1854 to open the Glasgow Architectural Exhibition in 95 Bath Street, designed by

Alexander Thomson. Provisionally with a life of only three months, the exhibition lasted for three years as pressure was put upon the city council to have its own municipal art gallery to present architecture and fine and applied art.[8] The experiment proved unsuccessful, the building was turned into the Scottish Exhibition Rooms, and eventually re-cast as stables. Glasgow had to wait until 1901 for its art gallery, built in Kelvingrove, some 16 years after more northerly Aberdeen had thus added to its attractions at Schoolhill, and 36 years after private subscribers had appointed Sir George Gilbert Scott to design the Albert Institute in Dundee for similar purposes.

Museums and art galleries were sometimes extensively extended between the world wars, but in the late twentieth century there was a flowering, often National Lottery-funded, of art galleries, arts spaces and arts centres. Many were conversions of existing, and often superfluous or redundant buildings – like the church in Inverness by Sutherland Hussey. Of these, the two most notable were the 1999 Dundee City Arts Centre by Richard Murphy – a multi-function building, including art galleries, cinemas, university visual research centre and printmaking facilities – and the Lighthouse Centre for Architecture and Design, designed by Page & Park as an extension to the Glasgow Herald building in Glasgow.

PUBLIC GARDENS

Public gardens were primarily a Victorian invention. Inherited rights of public access in Scotland meant that other citizens could, should they so wish, roam freely on their burgh muir, throughout Holyrood, or elsewhere, so rendering the need for formal public parks as places of recreation less immediately necessary. Sir Anthony Weldon left a vituperative description of Edinburgh natives using Calton Hill for this purpose in 1617. The Meadows and Magdalen Green performed the same function in Dundee. Natural features often formed the principal public green space – the Denburn Valley in Aberdeen, and the celebrated walk round the Castle Rock in Stirling – so much prized by those in search of the picturesque. Unlike the others, the value of the Stirling walk was appreciated sufficiently early for it to be laid out formally with seats and vistas in the late eighteenth century, as, likewise, was the walk to Battery Point in Dundee.

Edinburgh's Royal Botanic Garden was first laid out in the neighbourhood of Trinity Hospital in the late seventeenth century, having been dislodged from Holyrood Gardens. Its purposes were demonstration, propagation and learning. Visitors, scientists, nobles and lairds went there to study plants and herbs, to enjoy the public walks of an evening, and to buy cuttings and seeds. It removed to a more spacious country setting in Leith Walk in 1767 and to its present location in Inverleith in 1824. Glasgow followed suit much later, its botanic garden having begun life in the university grounds was forced by pollution to move west, first to Sauchiehall Street, and then to Kelvinside.

When the Edinburgh New Town was first mooted by the Earl of Mar in 1728, to reappear in the 1767 design by James Craig, he suggested a civic park where Princes Street Gardens are now, around a canalised Nor' Loch. By the standards of the time, it was likely to have been a private, fenced park with entry on subscription only, possibly even entertainment gardens in the manner of Vauxhall, London. The principal open spaces that emerged in all phases of the Edinburgh New Town were private, by subscription, fenced and locked.

By the early nineteenth century, traditional, public spaces like Glasgow Green, Edinburgh's Meadows or Aberdeen's Denburn, were beginning to be formalised into some form of planned parkland, with the addition of monuments, pavilions (the Humane Society House in Glasgow's case), bandstands, fountains and avenues of trees. Walls or railings were added, where necessary, to exclude the 'riff-raff' from the nearby slums.

The great Victorian parks were different in kind. Glasgow's West End (Kelvingrove) Park emerged when the opportunity was taken to acquire the lands of Kelvingrove and adjacent areas in 1854. Part of the motivation may well have been a desire to do something about the River Kelvin, then a plague-carrying sewer, which bisected it. In due course, Kelvingrove acquired the typical accoutrements of a great Victorian park – avenues of trees, ornamental ponds, fountains, bridges and music- and

Figure 18.7 The Kibble Palace, Botanic Gardens, 730 Great Western Road, Glasgow. Crown copyright: RCAHMS, SC 710867.

Figure 18.8 The Kibble Palace, Botanic Gardens, 730 Great Western Road, Glasgow; interior. Crown copyright: RCAHMS, SC 710866.

tea pavilions. In 1860, Sir Joseph Paxton, its designer, also advised on Glasgow's Queens' Park. In 1863, Dundee's Baxter Park was opened, named after its principal benefactor and designed, like Kelvingrove, by Paxton; Aberdeen's Duthie Park followed 20 years later. In 1885, Glasgow added Victoria Park, and in Fife in 1902, Andrew Carnegie presented Dunfermline with Pittencrieff Park.

It was in the larger parks that city councils were wont to deposit their great exhibitions: the 1886 (the Meadows) and 1908 (Saughtonhall) exhibitions in Edinburgh, and the 1888, 1901 and 1911 Glasgow exhibitions in Kelvingrove. Kelvingrove House had been adapted as Glasgow's first free museum after its lands had been acquired for the West End park, but the old house was demolished on completion of the splendid new museum and art gallery, built in 1901 upon the profits of the 1888 exhibition. In 1840, Glasgow's Botanic Garden had been ready to receive visitors on its new site off Great Western Road, and in 1871 added John Kibble's great conservatory (figs 18.7, 18.8) to its attractions, on the undertaking that it be used for concerts and entertainments. The 'Kibble Palace', a glass cathedral in plan with nave, crossing, aisles and choir, remains the most extensive glass house of its kind in Scotland.

PAVILIONS, INCLUDING SPORTS PAVILIONS

Pavilions probably emerged first in Scotland as garden pavilions, prominent and necessary components of the Renaissance walled garden. These included the banqueting house in Edzell Castle, Angus (1604), and the 'apple houses' (two-storeyed corner towers, the lower floor used for utility, the upper used as a gazebo) of Craignethan, Lanarkshire, Pitmedden, Aberdeenshire, Culross Abbey House, Fife, and many late revivals, particularly in the walled gardens of Sir Robert Lorimer as at Earlshall, Fife.[9] Once gardens had broken beyond the walled confines of the Renaissance to a wilderness or estate with planned vistas beyond, new pavilions in the form of temples to Diana, Ossian's Hall, or grottoes became the fashion. They performed at least some of the functions of the pavilions added later to Victorian public gardens, namely, shelter from the weather, rest, and the taking of refreshment.

Later pavilions, particularly in public parks or by recreation grounds, would offer the taking of tea, music (also bandstands), a small zoo or aviary, kiosk, or specialist plant-growing (particularly greenhouses and orangeries). The latter were glass houses, of which the largest was the 'Kibble Palace', Glasgow, described above. They were rarely substantial structures, large music audiences almost invariably being set out in open-air amphitheatres.

Public recreation originated on common land, the burgh moors or in public parks. Golf, for example, was played on Bruntsfield Links, Edinburgh. Private sports clubs – golf clubs, tennis clubs, cricket clubs, rugby and football clubs – were predominantly a Victorian innovation, creating two new building types: the pavilion or clubhouse (sometimes the same, sometimes separate); and the stand. The former could range from a timber shack to the grandiose, according to the location, age and wealth of the club. Many such buildings, particularly in the north of

Scotland, were made from prefabricated timber components, probably transported by rail.

The earliest may be the Gothic bowling club pavilion, Stirling (1861). Typically, the early pavilion or clubhouse would be a single-storeyed, rectangular construction with a central clock tower, generically providing some changing facility, a meeting/socialising place, and a viewing platform. Quite a number, particularly in north-east Scotland, were of prefabricated timber construction. By the end of the century, wealthier ones had become lavish. Golf clubhouses expanded to include club facilities, bars, restaurants, and groundsmen's quarters. To a much lesser degree, comparable facilities were also included in the wealthier bowling, cricket and tennis pavilions, all of which tended to sport a jaunty recreational aesthetic.

Sports pavilions began to develop after the end of World War I as a consequence of the national drive towards fresh air and fitness. The provision of playing fields became more structured, and various forms of sport, particularly football and rugby, more organised at school and university level. The functional requirements of these buildings were suites of changing rooms (segregated into wings for different sexes or teams), shower rooms and lockers, and, at the apex, usually a viewing room/tea-room. Perhaps the two most distinguished are T Harold Hughes' Garscadden Pavilion for Glasgow University (1933) and A G R MacKenzie's outstanding pavilion for King's College, Aberdeen (1938).[10] Healthily white wings with changing rooms radiate from Garscadden's central tower, with a curved bow-fronted viewing/tea room, with a terrace above for viewing the game. All is dominated by the clock tower. The groundsman's flat is above. Similar white, clean-flowing lines were also adopted for golf and tennis club houses. The symbolism had moved from recreation as fun, to recreation as health.

FOOTBALL STANDS AND STADIA

Football stands and stadia began small, with open-air terracing. But as a consequence of greater investment, the volume of spectators, complications of structure, and emergency safety and public order legislation, they moved inexorably from all-standing, to part seated, eventually to full-seated. To pay for this process, clubs developed substantial megastructures which provided for media, corporate entertaining, hospitality and ancillary activities. A comparable plan was followed in the national rugby stadium at Murrayfield, whereas local rugby clubs, with neither the investment, the volume of spectators, nor the social order problem, have retained the former ways.

Figure 18.9 St Andrew's Hall, Granville Street, Glasgow, c. 1888. Crown copyright: RCAHMS, SC 710860.

CONCERT HALLS

Concert halls maintained the pattern set by St Cecilia's Hall for flat-bottomed auditorium space, adding galleries as the need became apparent. Glasgow's City Halls, built in 1841 as the principal assembly and music hall, was a large rectangular, two-storeyed clerestorey-lit space with grand staircase and front-of-house rooms to one side. Aberdeen's Music Hall (1854) was comparable in plan, whereas Edinburgh's (David Bryce, 1843) was tight and square in plan, with only one gallery; but it had a coffered ceiling and a much greater sense of architectonic space. Central Glasgow had become less salubrious by the later nineteenth century, and the richer merchants established a private concert hall in the West End – the St Andrew's Halls (James Sellars, 1871, burnt 1963; fig. 18.9). Externally, it celebrated its function with outstanding classical iconographic caryatids. Its auditorium was comparable to that in City Hall, but much larger, and its acoustics earned it a European reputation.

The Usher Hall, Edinburgh (1908) – the brewing industry's musical gift to the nation – took the form of an austere Baroque rotunda and, for the first time, introduced a gentle rake to the principal floor. There were now formal seats for ranks of massed choirs behind the orchestra platform, terminating in the organ pipes. The form appears to be a synthesis of concert hall and contemporary theatre practice, and of the Albert Hall, London. Twentieth-century concert halls in Scotland are predominantly church conversions (for example, The Queen's Hall, Edinburgh, and Henry Wood Hall, Glasgow), the exception being the Glasgow Royal Concert Hall, 1990 (Sir Leslie Martin with Robert Matthew, Johnson-Marshall & Partners). A rectangular neo-classical building strategically placed at the head of Buchanan Street, its auditorium is galleried and rectangular, more reminiscent of the City Hall than the Usher Hall, its precise configuration determined by acoustic consultants. As in football stadia, however, the ancillary space – exhibition rooms, restaurants and meeting rooms – now amount to over one-third of the entire volume. The brief required that one of the rather amorphous exhibition spaces could be used for car exhibitions!

BATHS

Baths made their splash in the early nineteenth century, often in relation to spas (for example, Peterhead, Aberdeenshire), on either a subscription basis (Seafield and Portobello Baths, Edinburgh) or a commercial basis related to leisure and afternoon tea (William Harley's baths in Bath Street, Glasgow). Harley offered three cold swimming baths (men, women, boys), five hot stretching baths each for gentlemen and ladies, and shower baths. He also provided a saloon with the latest newspapers, and shrubberies and avenues. The proposal by James Cleland to establish floating baths on the Clyde (as on the Thames and Seine) came to nothing. Public baths appeared fairly early, on the edge of Dundee's docks by 1842,[11] and, as the century progressed, they became larger and more frequent (Warrender and Glenogle, both Edinburgh, 1886 and 1897), and Whitevale, Glasgow (1899). Public baths often also provided washhouses (or 'steamies') for people who had no place to wash or clean clothes of their own (Partick 1912, Govan 1922). From the near absence of purpose-made baths outside the large cities before the 1930s, they seem to be an industrial city phenomenon. Country folk had lochs and rivers. Oddly, the experimental housing estate of Logie, Dundee (James Thomson, 1919), was designed to provide both baths and a washhouse, but the baths were dropped since each of the new houses had one of its own, and the washhouse was used only by the inhabitants of neighbouring tenements.

Baths re-appear in the later nineteenth century as private clubs (for example, the Western Club, Glasgow, 1876, and Drumsheugh Baths, Edinburgh, 1888). The Western was one of the five private swimming clubs in Glasgow 'each provided with Turkish and other baths, billiard and card rooms and a gymnasium'. Baths sometimes followed exotic, sometimes Moorish, themes, as did that of Alloa, Clackmannanshire (1895), which was presented to the town by the fabric magnate, J T Paton. Others were plain stone boxes, with an inscrutable if sometimes vaguely classical or Baroque façade to the street. They were usually top-lit, rectangular cast-iron structures, enfolding a pool surrounded by changing cubicles. Ancillary spaces have now been transformed for gymnasia purposes.

The link between public health and swimming became stronger after World War I. The baths at Prestwick, Ayrshire, were to be part of a new civic campus, whereas those in Coatbridge, Lanarkshire, were integral to what we would now call a health centre. But these were rare. Swimming pools were associated first with recreation and health, and then tourism. That was conveyed nowhere more strongly than in the development of the open-air pool in seaside towns like Burntisland, Fife, Arbroath, Angus, Stonehaven, Kincardineshire, and North Berwick, East Lothian. Perhaps above them all was Macduff's splendid open-air pool at Tarlair, Banffshire (c. 1937), on the site of a nineteenth-century spa and baths. Open-air pools were normally cut into natural seaside formations (for example, on the rock shoreline at St Andrews, Fife), although in some cases they were built up simply at ground level behind a white canopy. The viewing terraces formed a stepped amphitheatre with white edged lines, changing rooms at one end, and usually a café at the other.

The 1936 open-air pool at Portobello was different in kind. The largest pool of its type in Europe, it had seating for 6,000 visitors, and lockers for 1,284 bathers, a snack bar, a restaurant, lounge bar, open-air tea gardens and restrooms. The water of the 1.13-acre (0.45ha) pool was heated by the adjacent power station and it had Europe's first wave-making machine. It was what we would now call a 'leisure experience'. Such technological developments gradually spread to less ambitious projects. Just before World War II, both Falkirk and Kilmarnock, Ayrshire, opened new indoor pools, each with a wave-making machine.

Swimming pools in post-war years came to be a sign of civic status, as witnessed by the soaring concrete ribs of the Dollan Baths in East Kilbride, Lanarkshire; they spread rapidly to communities which heretofore had been without one. Built as increasingly large-span concrete boxes lit by clerestory glazing to avoid glare and heat gain within (for example, in Dundee and Ayr), these substantial, usually blank-walled volumes of civic obtrusiveness were found applicable to other recreation buildings like sports and squash clubs and leisure centres. The primacy of their internal function led to their architects turning their backs upon their civic settings, and local authorities short of cash were sorely tempted to buy 'leisure packages' from operators whose buildings were little more than large ugly sheds (for example, Alloa). An exception was the sophisticated and elegant Royal Commonwealth Pool, Edinburgh (1970).

With the introduction of the leisure pool (c. 1980), swimming pools changed yet again. Sometimes, the swimming pool would be segregated from the leisure pool (as in the exemplar, Perth, 1986) and sometimes serious swimming would be excluded (as in the Coatbridge Time Capsule). With its vision of water as entertainment rather than swimming, with water splashes, wave machines, flumes, palm trees, and integrated catering facilities, the leisure pool harks back to the 1930s. In Perth, the German innovation of taking heated water to the outdoor arena providing outdoor swimming was first essayed in Scotland. Smaller versions of the Perth model in Edinburgh and Shetland embraced other sports facilities (for example, squash, badminton courts, table tennis, saunas, gyms) as fashion in leisure centres continued to develop. The Inverness Aquadrome was added to an existing sports hall, whereas Greenock Waterworld (Renfrewshire) integrated an ice and curling rink.

CINEMAS

Along with all-seater football stadia, ice rinks and leisure pools, cinemas are amongst the few distinctively twentieth-century contributions to leisure building. They probably had their origins in halls, although dioramas (a form of moving pictures) had formed part of the repertoire of some music halls. Electric cinemas appeared in Scotland before World War I (Cowdenbeath, Fife, and Bo'ness, West Lothian), effectively music hall spaces, with threadbare fittings, unraked floors, and a projection room at the rear, but moving pictures became the rage during World War I. All that the operator required was reasonable projection facilities, decent fire-fighting equipment, and a large hall. Salvation Army and other halls were pressed into service, with a sheet hung at one end as a screen.

From such amateur beginnings, Scotland soon developed characteristics that made it perhaps the most movie-intensive country in Europe, if not the world. It had the two largest cinemas in Europe, the Green's Playhouse in Glasgow (4,200 seats) and in Dundee, and it also had the highest density of cinemas of any one city in Europe – Glasgow with 120, Edinburgh having over 65. Few self-respecting communities could do without at least one. What cinema proprietors aimed to do, many of them coming from a fairground background (for example, Green of Green's Playhouse and Frutin), was to attract people into their fantasy worlds and give them a cheap escape from the realities outside.

Although the cinema shared certain characteristics with music halls, there were also certain differences. Because of the highly flammable nature of the material, and the large volume of people, cinemas were soon subject to regulation in relation to fire exits, and the fire containment of the projection room. They were also generally larger than theatres.

Figure 18.10 Playhouse Cinema, Murray Street, Perth. Crown copyright: RCAHMS, SC 710992.

They were built at exceptional speed – the Perth Playhouse (fig. 18.10) was constructed in under nine months from start to finish (an unusually sophisticated deployment of steel framing) – and, because they had to compete for custom, there was great emphasis upon front-of-house imagery leading to a gaudiness, if not garishness, of the street front. The cinema very often consisted of a large brick, asbestos-roofed barn, at the rear with an opulent façade, complete with storm prow or cinema advertising tower. The latter was sometimes enhanced by neon lighting running from bottom to top (as in the County, Portobello), for the sensual and vivid use of neon in 'night architecture' was central to the design. Generally, there would be one or more shops, sometimes a café, tea room or restaurant, and, in one case in Leith, billiard rooms. On Sundays, the entire front-of house of the enormous Dundee Green's Playhouse was converted into a fashionable restaurant. Lavish cinemas were identified by one or more balconies.

Since early cinemas were normally in tenemented streets, only the façade mattered. Later sound cinemas, the 'super cinemas', had a scale and size which required free-standing sites, usually in the new suburbs. Applied decoration of Art Deco, used with a symmetrical façade and often brightly coloured tilework, gave way to a more stylish form of smooth surface upon which changing light patterns – holophane lighting – played. Generally white, the suburban cinema developed an architecture of geometric mass, and, with its attendant neon lighting, quickly became the most prominent feature of Scottish suburbia.

Similar architectural considerations applied to two other thirties' fashions: the seaside pavilion and the ice rink. Seaside pavilions were built by communities attempting to remain fashionable once holiday makers had been liberated by the car from rail destinations. They generally contained a large function/dance hall (which could also do duty as cinema), terraces and various refreshment facilities, as, for example, at Gourock, Renfrewshire, Rothesay, and Aberdeen. Ice rinks, however, exploited the latest technology to provide ice within, usually, the barest, coldest and most under-invested barns, with only a token fashionable front-of-house to redeem otherwise giant, asbestos-roofed sheds. They perpetuated the Scots habit of investing cheaply in leisure.

After World War II, the market could not sustain the number of cinemas, particularly with the arrival of specialist firms and television. Many cinema companies diversified into leisure (for example, Caledonian Associated Cinemas into CAC Leisure, and Frutin into Frutin Travel), and the buildings were either sold off for redevelopment or converted for bingo. Post-war audiences required smaller auditoria. Sometimes the 'supers' were divided into three. The individual proprietor became a rarity. 1990s cinema revival has led to the multiplex where up to 12 screens, with varying-sized auditoria, show simultaneously. Whereas it seemed in the early '70s that cinema would die out and be replaced by home videos, the cinema has proved resilient and has retained a growing market niche, particularly in the realm of small auditoria, and increasingly large front-of-house facilities. Architecturally, they are virtually indistinguishable from large suburban warehouses.

PUBLIC HOUSES

Public houses, for the most part, have always inhabited other building structures, predominantly the street-level apartment of a tenement block. Throughout history, urban bars have been howffs, with sawdust on the floor, and no women allowed. Parishioners were able to ballot on licensed premises, and such was their reputation, that until the 1970s, Bearsden and Kirkintilloch, Dunbartonshire, Dowanhill, Glasgow, and elsewhere, were 'dry'. Despite that reputation, there was a late nineteenth-century flowering of Arts and Crafts decoration in city public houses with glorious timberwork, stained and etched glass, beaten copper, gorgeous plasterwork, and occasionally decorative tilework. Thirties' pubs added chrome, bakelite and rubber to provide a more shiny and colourful, if less sensuous, interior, and some of these, for the first time, also provided for children. Suburban versions took the form of roadhouses (for example,

Maybury, Edinburgh 1937), which offered ballrooms and function suites, roof gardens and dining rooms, all aimed at the new motorist.

The most distinctive change to the Scots public house has been caused by the relaxation in drinking laws, the arrival of real ale and food, the welcoming of women and, occasionally, children. A new building-type may yet emerge – the pension (for example, Babbity Bowster, Glasgow) which combines bar, restaurant and a few bedrooms – but the majority of public houses still determinedly occupy space within other structures.

NOTES

1. Arnot, 1998, 221.
2. ibid.
3. Drawings in the Rowand Anderson Collection, Edinburgh University Library, Special Collections.
4. Denholm, 1798, 61.
5. McKean, 1987(a).
6. McKean, 2000.
7. Gow, Clifford, 1988.
8. McKean, 1994.
9. Buxbaum, 1989.
10. McKean, 1987(a).
11. Drawings in the Wellgate Library, Dundee.

BIBLIOGRAPHY

Arnot, H. *The History of Edinburgh*, Edinburgh, 1799, reprinted 1998.
Atwell, D. *Cathedrals of the Movies*, London, 1980.
Beaton, E. *Caithness*, Edinburgh, 1996.
Beaton, E. *Ross and Cromarty*, Edinburgh, 1992.
Beaton, E. *Sutherland*, Edinburgh, 1995.
Brogden, W A. *Aberdeen*, Edinburgh, 1986, 1998.
Bruce, P. *Scotland's Splendid Theatres: Architecture and Social History from the Reformation to the Present Day*, Edinburgh, 1999.
Burgher, L. *Orkney*, Edinburgh, 1991.
Buxbaum, T. *Scottish Garden Buildings*, Edinburgh, 1989.
Close, R. *Ayrshire and Arran*, Edinburgh, 1992.
Denholm J. *The History of the City of Glasgow*, Glasgow, 1798.
Fenton, A. Coffee drinking in Scotland in the seventeenth to nineteenth centuries. In Ball, D, ed, *Kaffee im Spiegel Europäischer Trinksitten*, Zürich, 1991, 93–102.
Finnie, M. *Shetland*, Edinburgh, 1990.
Gifford, J. *Dumfries and Galloway*, Harmondsworth, 1996.
Gifford, J. *Fife*, Harmondsworth, 1988.
Gifford, J. *Highlands and Islands*, Harmondsworth, 1992.
Gifford, J, McWilliam, C, Walker, D. *Edinburgh*, Harmondsworth, 1984.
Gow, I R, Clifford, T. *The National Gallery of Scotland*, Edinburgh, 1988.
Groome, F H. *Ordnance Gazetteer of Scotland*, London, 2nd edn, 1893.

Haynes, N. *Perth and Kinross*, Edinburgh, 2000.
Hume, J R. *Dumfries and Galloway*, Edinburgh, 2000.
Jaques, R, McKean, C. *West Lothian*, Edinburgh, 1994.
Kenna, R. *Art Deco Glasgow*, Glasgow, 1988.
Kinchin, P. *Tea and Taste*, Wendlebury, 1991.
McKean, C. *Stirling and the Trossachs*, Edinburgh, 1985, 1994.
McKean, C. *The Scottish Thirties*, Edinburgh, 1987(a).
McKean, C. *The District of Moray*, Edinburgh, 1987(b).
McKean, C, Walker, D, Walker, F. *Central Glasgow*, Edinburgh, 1989, 1993, 1999.
McKean, C. *Banff and Buchan*, Edinburgh, 1990.
McKean, C. *Edinburgh*, Edinburgh, 1992.
McKean, C. In search of purity: Glasgow 1849–56. In Stamp, McKinstry, 1994, 9–22.
McKean, C. *The Making of the Museum of Scotland*, Edinburgh, 2000.
McKean, C, Walker, D. *Dundee*, Edinburgh, 1984, 1993.
McWilliam, C. *Lothian*, Harmondsworth, 1978.
Mudie, R. *Dundee Delineated*, Dundee, 1822.
Peden, A. *The Monklands*, Edinburgh, 1992.
Pride, G L. *Fife*, Edinburgh, 1990 and 1999.
Shepherd, I. *Gordon*, Edinburgh, 1994.
Stamp, G, McKinstry, S, eds. *'Greek' Thomson*, Edinburgh, 1994.
Strang, C A. *Borders and Berwick*, Edinburgh, 1994.
Swan, A. *Clackmannan and the Ochils*, Edinburgh, 1987 and 2001.
Thomas, J. *Midlothian*, Edinburgh, 1995.
Walker, F A. *The South Clyde Estuary*, Edinburgh, 1986.
Walker, F A, Sinclair, F. *The North Clyde Estuary*, Edinburgh, 1992.
Walker, F A. *Argyll and Bute*, Harmondsworth, 2000.
Williamson, E, Riches, A, Higgs, M. *Glasgow*, Harmondsworth, 1990.

19 Public Defences

NIGEL A RUCKLEY

INTRODUCTION

The concept of the defence of the realm stems from the need to defend the commercial interest of the individual and ultimately that of the state. On a local scale, the community would have provided this service until the mid nineteenth century, whilst the defence of the realm on a national scale would have been either partially or wholly funded and executed by the government. National defence meant the preservation not only of a commercial base for the prosperity of the country, but also of national culture and identity against the perceived threat of a foreign power.

In Scotland, since the mid sixteenth century, these concepts can be traced in the development of private and public fortifications as the nation defended itself against the threat of invasion, primarily from England. Public defences in the mid seventeenth century were primarily to allow an English government to exercise its authority over a country considered potentially hostile to English policies. Again, in the eighteenth century, the Jacobite risings of 1715, 1719 and 1745 brought about a British imposition of order to preserve the political and religious status quo, as seen by a government based in London. With the threat of internal strife diminished, the defence of the realm against foreign invasion became of utmost importance.

In the late eighteenth century, when commerce-raiding was at its height and war with European nations gathered pace, there was a concerted effort to defend commercial ports – from the Channel coast to Orkney – either from commerce raiders or invasion, by British government schemes, as well as smaller locally-funded schemes.

A co-ordinated approach to public defence became apparent when the Victorian coastal-defence programme began in the 1860s against a sea-borne threat perceived initially from France; this produced fortifications and allied structures built to a common plan. The realisation that Germany not France threatened British interests brought about an expansion of naval bases in Scotland at the turn of the twentieth century and promoted Scotland during World War I into the forefront of naval strategy in the defence of the United Kingdom. With the naval bases of

Scapa Flow, Orkney, Invergordon, Ross and Cromarty, and Rosyth, Fife, on the east coast, and the Clyde on the west coast, Scotland's coast continued to supply important naval anchorages during World War II. Sea lochs, especially on the west coast, became merchant convoy assembly points, while Scottish airfields were used for a variety of offensive and defensive roles, including handling of transatlantic freight and the training of aircrew. Inland, parts of the western Highlands became training areas for troops and especially for commandos.

During the 'Cold War' years of the 1950s, Britain relied on an early warning system of radar that extended along the eastern side of Scotland, from Muckle Flugga off Unst, Shetland, south to the English Border and beyond. American and British Polaris-type nuclear submarines formerly used the Clyde Estuary as a base for their operations during the Cold War, but the American base at the Holy Loch, near Dunoon, Argyll, closed down in 1992. Today, only the latest type of nuclear submarines, which are armed with nuclear-tipped missiles and form the so-called 'independent' British nuclear deterrent, operate from Faslane on the Clyde. The new Trident submarines of the Vanguard class should ensure a naval presence in the Clyde for the foreseeable future.

Scotland now has its own devolved parliament able to pass laws on a variety of internal matters, but defence is not in this parliament's remit. Over the past century the defence of Scotland has benefited from a unified command structure, controlled for the good of the whole United Kingdom by the will of the British Parliament.

SIXTEENTH-CENTURY DEFENCES

During the sixteenth century, Scotland had to protect commercial and national interests from determined attacks by English forces during the Wars of the Rough Wooing in the 1540s. Private strongholds and royal castles were capable of being upgraded for defence by artillery with stone- or turf-faced Italianate bastions and their ancillary earthwork defences. Public defences were confined to the protection of towns and cities. Their defence capabilities varied greatly, but earthwork defences quickly showed their worth in artillery duels. Castles of earlier date, such as Dunbar, East Lothian, and Roxburgh, were adapted for artillery by Scottish and English engineers respectively.

The earliest known example of the wide-mouthed gun-loop in Scotland can be seen at the polygonal artillery blockhouse built at Dunbar by the Duke of Albany between 1515 and 1523. The harbour at Dunbar occupied a strategic locality as the first harbour of note between Berwick and the sheltered waters of the Firth of Forth. The blockhouse has walls up to 21 feet 4 inches (6.5 metres) thick, pierced by seven wide-mouthed gun loops, situated on an almost inaccessible sea-girt rock and linked to

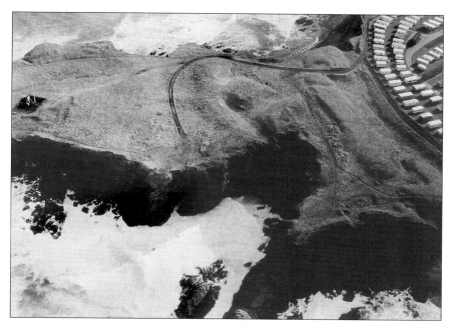

Figure 19.1 Eyemouth Fort, Berwickshire; aerial view. Courtesy of Historic Scotland.

the castle by a wing wall. Two of the surviving deep vaulted casemates retain their original smoke-vents, and all the casemates are thought to have been capable of taking large carriage-mounted guns.[1]

Between 1547 and 1560, English and Scottish, as well as French and Italian, engineers developed the then modern 'trace italienne' style of bastion for use in new forts of stone, or of stone and earth construction. This style of bastioned fortification was introduced into Scotland by the English in 1547 at Eyemouth, Berwickshire (fig. 19.1),[2] a fort that was dramatically improved by Franco-Scottish forces in 1557 after the English withdrawal.

The headland on which the fort lies is sea-girt on three sides and has two lines of defences on its landward side. The inner, English line, constructed under Sir Richard Lee, has a single, central Italianate bastion. This was an unsatisfactory arrangement as two tunnel-like casemates had to be inserted through the curtain on each side of the bastion in order to provide defensive fire along the external faces of the central bastion. The outer line of earthworks, of Franco-Scottish design, comprised a curtain-wall flanked by demi-bastions at each end.[3]

During the Wars of the Rough Wooing, a series of forts was constructed by English military forces in order to protect their lines of communication. A good example of a small fort is at Dunglass, East Lothian, built in 1548.[4] The Franco-Scottish defences of this period were the equal of, or surpassed those erected by the English invaders. Substantial remains of a fort built by the Scots in 1549 survive on the island of Inchkeith in the Firth of Forth,[5] and the prominent bastioned earthworks surrounding Luffness House, Aberlady, East Lothian, also demonstrate the quality of local work.

Two other castles illustrate the defensive improvements prompted by artillery. The reconstruction of the eastern defences of Edinburgh Castle following the partial collapse of David's Tower during the siege of the castle in 1573 included the structure known as the Half Moon Battery. It totally enclosed the remains of David's Tower and raised the eastern defences to an unprecedented height, dominating the city to this day.[6]

Figure 19.2 Blackness Castle, West Lothian; aerial view. Crown copyright: RCAHMS, SC 710838.

The defences of Blackness Castle, West Lothian (fig. 19.2), a tower house of the 1440s, were remodelled for use by artillery between 1537 and 1542. Further improvements continued intermittently until 1567, ultimately creating a thick curtain-wall pierced by wide-mouthed gun-loops.[7] The layout is in marked contrast to the Henrician fortifications along the English Channel coast, with their series of half-moon shaped towers surrounding a central keep.[8]

Town defences
In Scotland, medieval town defences would have utilised natural geomorphological features such as rivers, lochs or cliff faces, and would have used a ditch and palisade to protect the weaker side. An example of this can be seen at Stirling where the medieval burgh below the castle had strong natural defences on all but the south-eastern side. The existing fragmentary remains of the town wall, up to 25 feet (7.62 metres) high and 5 feet (1.52 metres) thick and incorporating two bastions, all date from the mid sixteenth century. Minor alterations and repairs continued until 1745. Both the wall and the bastions at Back Wall and Port Street exhibit crude gun-loops.[9]

The town defences of Edinburgh, like those of Stirling, also used natural and man-made features to secure the safety of its citizens. Documentary evidence points to the existence of gates from the late twelfth century: West Gate 1180; South Gate 1214; and the Netherbow Port 1389. Existing remains date from the sixteenth century and represent the enclosure of the suburbs of the Grassmarket and Cowgate constructed after 1514 (the Flodden Wall), while Telfer's Wall, constructed in 1620, was specifically designed to enclose ten acres (4.05 hectares) of newly acquired land. These walls were not suitable for defence in an age of artillery and were strengthened by artillery emplacements in 1650 and again in 1715. Demolition of the bastions began in 1762 and was quickly followed by the Netherbow in 1764. In the Vennel, a crenellated tower with crosslet gun-loops still forms part of the surviving Flodden Wall, while traces of the Telfer and Flodden Walls also exist on the south side of the city.[10]

Nothing now remains of the bastioned sixteenth-century trace of Haddington, East Lothian, nor of the defences of the port of Leith. The former was fortified by the English with earthworks, designed by Sir Thomas Palmer, to carry the latest artillery. Leith was begun by the French in 1548 and its 'trace italienne' system is thought to have been the earliest – by ten years – of any town in Britain.[11] Leith, held by forces loyal to the Queen Regent, Marie de Guise-Lorraine, was captured on 7 June 1560 by English forces that supported the Scottish Congregation, and a contemporary plan showing the disposition of forces at the conclusion of this eight-week siege clearly shows the rapid development of the bastioned trace.[12] It shows a fully developed bastion trace with arrow-head shaped bastions, probably master-minded by the fortress engineer, Piero di Strozzi; it also shows the final layout of the siege entrenchments and guns. The stone-faced defences were still partly extant in the early eighteenth century, but their stonework was used for other purposes, and today, two earthen mounds that formed the core of the Somerset and Pelham's batteries are all that remain above ground.[13]

Once the military need for defensive walls had receded, the town gates or ports became a restriction to developing trade and were mostly demolished. Only two survive in Scotland: the West Port, St Andrews, Fife, and the East Port or Wishart Arch, Dundee.[14] The former, constructed in 1589 on the site of an earlier gate, is thought to be a representation of the demolished Edinburgh Netherbow.[15]

In many small towns, such as Peebles, the continuous rear boundary wall of the burgess tenements served as a makeshift town wall and restricted entry to the main thoroughfares. Town walls were finally erected around Peebles some time after 1570, when a decision was taken to construct a defensive wall with towers. Repairs to the walls continued up to the second decade of the eighteenth century, and a circular tower with two wide-mouthed gun loops and a few sections of the wall still survive.[16]

FORTIFICATIONS OF THE CROMWELLIAN AND RESTORATION ERAS

The uniquely Scottish approach to artillery fortification, as illustrated by the sixteenth-century gun towers attached to the castles of Tantallon, East Lothian, and St Andrews and by the adaptation of Blackness Castle (described above), was gradually replaced by the bastioned trace that became such a dominant feature of fortress design everywhere in the period from the seventeenth to early nineteenth century.[17]

The Union of the Crowns in 1603 heralded major changes in the defence of Scotland. During the seventeenth century royal government began to take a greater responsibility for local as well as national affairs. The concept of a national regular army took root under King James VI and I, gradually rendering private strongholds ineffective, and in England, the creation of the 'New Model Army' was approved by Parliament in 1645.

Scotland's ability to secure her frontiers was directly affected by events during and after the second English Civil War. In July 1650 Cromwell used 1,600 soldiers of the New Model Army to make a pre-emptive strike into eastern Scotland in order to prevent any attempt by armies raised in Scotland, loyal to King Charles II, to invade England. These Parliamentarian forces destroyed a larger Scottish army near

Dunbar on 3 September and occupied eastern Scotland, the castle and city of Edinburgh coming under their command during December 1650. The remnants of the Scottish army loyal to the king were based at Stirling, and steadily grew in numbers during the first half of 1651. This army invaded England in late July only to be defeated, on the anniversary of the Battle of Dunbar, at Worcester, a defeat which effectively marked the end of the royalist cause. The military policies then applied by the Commonwealth and Free State over what is now the United Kingdom were determined by the English Parliament in London. They authorised the Commonwealth military engineers to embark on a co-ordinated programme of fortress construction designed to suppress an internal Scottish rebellion, rather than to defend Scotland against an external threat.[18]

The major Commonwealth fortifications at Ayr, Inverlochy, Inverness-shire, Inverness, Leith and Perth were fortifications of conquest, designed to act as bases for military government and to be linked by sea communication with England, adhering to the very same principles as the Edwardian castles built in north Wales in the later thirteenth century. The major forts were garrisoned by professional soldiers living in purpose-built military barracks that were entirely separate from civilian accommodation, a concept of a self-contained military force with its own accommodation and market place that was then entirely alien to Scotland. To complete the military pacification of Scotland, a series of twenty lesser fortifications was constructed and these were linked by a strong mobile militia and spy network.

Considered to constitute an important technical advance in fortress engineering,[19] the forts in Scotland were designed for the most part by continental military engineers – Hane, Rosworm, Van Dalem and Tessin – who had either lived in, or had extensive knowledge of, designs prevalent in the Low Countries.[20] The small fort at Perth was designed by Major-General Richard Deane.

The fortifications guarding each end of the Great Glen differed markedly in design. Oliver's Fort at Inverness,[21] situated on the eastern bank of the River Ness and close to the head of navigation for sea-going vessels, consisted of a regular pentagon, bastioned at each angle and capable of holding a garrison of 1,000 men.[22] Founded by Deane in 1652 and completed by the summer of 1657, the fort was built of stonework which in part had been robbed from local redundant churches, in part brought by ship from Aberdeen,[23] the defensive ditch having been turned into a fortified harbour for flat-bottomed boats. A 1655 description indicated the quality of the fort's construction.[24] Its slighting was authorised by Act of Parliament in 1661, but traces of the fort still survive in the form of a clock tower and vestiges of a curtain wall, surrounded by the tanks of a modern oil terminal.[25]

Inverlochy Fort, situated on the promontory at the confluence of the River Ness and Loch Linnhe, was surrounded by water on three sides, and lay near the limit of sea-borne traffic at the south-western or landward side, with demi-bastions at the remaining four angles of an irregular enceinte. Like Oliver's Fort, it was capable of allowing ships to anchor under the protection of its guns.[26]

The citadels of Ayr and Perth were founded by Deane in 1652, the former designed by Hans Ewald Tessin, chief engineer of the New Model Army, and the latter by Deane himself. Cruden described the plan of Ayr as 'a symmetrical elongated hexagon with a bastion at all six corners, an outer ditch with a deep water channel, a terraced counterscarp bank and a long glacis.'[27] Part of a bastion near the harbour survives with its salient angle crowned by a sentry box carried on eight orders of continuous corbelling. This type of construction may have been alien to Scottish practice and Cruden considered that it could have been of Dutch or German design.[28] The citadel at Perth is known to have been square in plan with a bastion at each of the corners. Its remains have been excavated prior to the development of a new flood-prevention scheme for the town and revealed the outlines of its south-western bastion.[29]

Only an entrance gateway now remains of Leith Citadel, described by John Ray in 1661 as 'one of the best fortifications that ever we beheld, passing fair and sumptuous'.[30] The strategic importance of the site on the Firth of Forth had previously been appreciated, as the citadel, a regular pentagon with bastions at each angle, was sited in 1656 on the remains of an earlier fortification. Completed in 1658 and redundant in 1660, it was demolished by the Earl of Moray in 1661.[31]

Part of a bastion of one of the lesser Cromwellian fortifications stands above the banks of the Dee at the east end of Union Street, Aberdeen, occupying the site of a royal castle of thirteenth-century date.[32]

The Restoration of King Charles II in 1660 brought major changes to Scotland's defences. He inherited a standing army in Scotland of about 2,000 soldiers who were funded by Scottish revenues and who in 1661 formally became the core of a new Scottish army. However, the cost of upkeep of the military machine was too high and reductions took place with almost indecent haste. The major citadels of Ayr, Leith, Perth and Inverness were either sold off or partially demolished, and secondary fortifications received similar treatment, leaving the Highlands denuded of first-class fortifications. Royal garrisons were kept at the fortresses of Edinburgh, Blackness, Stirling and Dumbarton.[33]

Hostilities against the Dutch were instrumental in the construction of the only fort built in Scotland during Charles's reign and one built specifically against an external threat. The site, overlooking Lerwick, Shetland, was chosen to protect the settlement and the natural harbour and anchorage of Bressay Sound from potential Dutch raids. Of roughly pentagonal shape, the fort was begun in 1665 to a design by John Mylne, Master Mason to King Charles II, and to increase its firepower over the anchorage the seaward-facing wall had two V-shaped extrusions. By the terms of the Treaty of Breda in July 1667, the slighting of the fort was ordered by the Privy Council, and in 1673 the Dutch captured and burnt the unmanned fort. It lay in a ruinous condition until the early 1780s when it was renovated and re-named Fort Charlotte (see below).[34]

The surviving garrison fortifications of Blackness, Dumbarton and Edinburgh in central Scotland were upgraded by John Slezer, the King's Chief Engineer, responsible for the fortifications and the development of the artillery train in Scotland.[35]

LATE SEVENTEENTH-CENTURY AND HANOVERIAN FORTIFICATIONS

The landing of William of Orange at Brixham on 5 November 1688 demonstrated that, given suitable conditions of weather and tide, it was possible, even in the late seventeenth century, to invade England – and by inference Scotland – by out-manoeuvring the Royal Navy. An invading army might have to undergo a longer sea voyage but the Scottish coastline with its long sea-lochs and firths studded with sheltered anchorages, together with a population potentially hostile to English Protestant rule, was attractive, and would partially help to mitigate the hazards of a such a voyage.

It took the Scottish Parliament at least a month to decide to accept William as king in Scotland. Their final acceptance of his claim in April 1689 unwittingly gave the Jacobite cause a fillip. The religious and political unrest in parts of Scotland, directed against a London-based government and Protestant royal family, meant that policing of the Highlands now had to take account of the threat of a landing by an invading army, sympathetic to the indigenous population.

Internal unrest was quick to surface. John Graham of Claverhouse, Viscount Dundee, led the Jacobites. Edinburgh Castle was held by them for three months before the Duke of Gordon surrendered on 13 June 1689, his force of approximately 120 officers and men having been reduced to about 50. The Battle of Killiecrankie, Perthshire, on 27 July 1689 resulted in a Jacobite victory but at the cost of the death of Dundee. With no leader, the Jacobite cause was lost, but the government's need to keep the Highland clans in check was clearly demonstrated.[36]

The government re-vamped the Cromwellian strategy of securing the Great Glen and the Highlands though a system of fortifications.[37] Major-General Hugh Mackay of Scourie, commander of William's Scottish army, was authorised to re-occupy and rebuild the former Cromwellian fort at Inverlochy for a garrison of 1,200 men. Within 11 days of re-occupying the site – which was not ideal – on 3 July 1689 he had constructed formidable earthworks holding tented accommodation, later replaced with wooden structures, for his troops, along with 12 guns. The enceinte was an irregular pentagon with a three-pointed bastion at the south-eastern corner, demi-bastions occupied the remaining four corners. The garrison was to police the immediate area and to supply soldiers for the minor military posts in the western Highlands. The fort was renamed Fort William, and adjacent to the fort a new town slowly became established to supply the needs of the garrison. Originally called Maryburgh, it is now Fort William.[38]

The concept of a military force policing the Highlands had already been mooted by Sir John Campbell of Glenorchy, 1st Earl of Breadalbane. His motives for the creation of a Highland militia under his command are unclear, but, perhaps to rival the emerging fortress at Fort William, in the late 1690s the earl erected the first private barracks in Scotland at his castle of Kilchurn on the shore of Loch Awe, Argyll. The completion of the barracks, which held about 200 men, around 1698 coincided with the building of Fort William with a stone façade.[39]

At Inverness, Oliver's Fort, slighted in 1661, was not rebuilt. Instead, Major General Mackay improved the barrack accommodation within the old castle of Inverness, beside the bridge over the River Ness, and strengthened its defences with demi-bastions. Work is assumed to have begun around 1690 but by 1719 the castle was out of commission, its well choked with rubbish and main buildings open to the elements.[40]

The death of the childless King William in 1702 led to another period of threatened Jacobite uprisings or invasion during the reign of his succesor, his sister-in-law, Anne. In July 1708, following the 1707 Act of Parliamentary Union between Scotland and England, the Royal Navy prevented an attempted landing at Pittenweem in Fife by supporters of King James VII.[41] To strengthen the government's hold on Scotland, now increasingly known as 'North Britain', Theodore Dury, Slezer's successor, upgraded the defences and accommodation of the government's garrisons in case of any further outbreaks of civil unrest. The outer perimeter of Edinburgh Castle was improved with extra batteries, one still known as Dury's Battery, along with the construction of the Queen Anne building for barrack accommodation. Dury left behind him a grandiose plan for upgrading the eastern defences, but that was abandoned well short of completion.[42] Dury's plans for the approach to Stirling Castle were adopted, and today the casemated Queen Anne Battery there is one of the best remaining examples of its period in Britain. At Fort William the timber accommodation blocks were replaced by stone structures, and a new governor's house and a powder magazine were built.[43]

Following Queen Anne's death and the Hanoverian succession on

1 August 1714, religious and political instability increased in Scotland, finally culminating in the Jacobite uprising of 1715, led by John Erskine, 6th Earl of Mar. However, ineffective leadership at the indecisive Battle of Sheriffmuir on 13 November 1715, and defeat of their English supporters at Preston, Lancashire, saw the effective collapse of the uprising. Although the Pretender's arrival at Peterhead, Aberdeenshire, on 22 December was followed by a token coronation at Scone near Perth, he boarded a ship at Montrose, Angus, for France on 4 February 1716 knowing that his cause was lost.[44]

The Disarming Act brought in few Jacobite arms, and a major re-organisation of the use of government forces against any future uprising in 'North Britain' was initiated. In 1716, however, probably on grounds of saving money, artillery in the major fortifications was much reduced: Edinburgh lost ten of its 50 guns; Stirling lost two; and Fort William lost 38 of its 68 artillery pieces.[45] In 1717 a new barrack was authorised to be built at Berwick on Tweed for 36 officers and 600 men,[46] and schemes for building defensible barracks throughout the Highlands were also drawn up that year. These incorporated earlier, unused plans such as the 1699 proposal to build a barrack at Ruthven, Inverness-shire, and smaller posts at Ardclach, Nairnshire, and Invermoriston, Inverness-shire. Other locations that were surveyed included Dalnacardoch, Garvamore, Lochlogan, Glenshishie and Glen Moriston, all Inverness-shire.[47]

Authorisation was given for the construction of four defensible barracks at Bernera and Kiliwhimen (Fort Augustus), Inverness-shire, Inversnaid, Stirlingshire, and Ruthven (fig. 19.3),[48] the latter occupying the mound of an earlier castle, still able to garrison troops in 1689.[49] The barracks share common design features: two large blocks occupying two sides of a square; a loop-holed and high-walled enclosure; and at diagonally opposite corners of the enclosure wall stood projecting bastion-towers, looped for muskets.[50] Work progressed slowly while the overseers of the works came and went with great rapidity. James Smith, 70 years old, was replaced in January 1719 by Andrews Jelfe who found himself replaced by Captain John Romer in 1720, 'one of the Board's finest military engineers'.[51]

Overseas attempts to destabilise the British government continued. The Spanish-derived plot by Cardinal Alberoni for an invasion of England did partially materialise in 1719. Atrocious weather off La Coruña, Spain, in March 1719 scattered the invasion fleet, but two ships unaware of this disaster reached Stornoway on Lewis, where they were joined by a handful of Jacobite supporters from Le Havre, France. With clan support and around 1,000 troops, they intended to march on Inverness. Having landed their troops on the mainland unopposed, they occupied Eilean Donan Castle at the head of Loch Alsh, Inverness-shire, but a Royal Navy

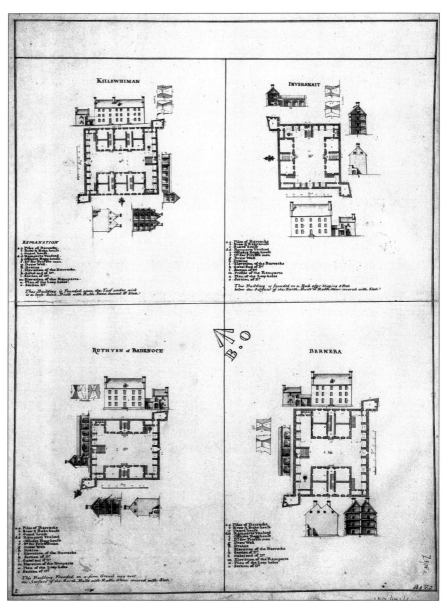

Figure 19.3 Killiwhimen, Ruthven of Badenoch, Bernera (Inverness-shire) and Inversnaid (Stirlingshire) Barracks; design drawings c. 1718. ©Crown copyright/MOD.

bombardment drove them out of the castle and eastwards. The survivors were defeated in the nearby Pass of Glenshiel, Inverness-shire, on 9 June 1719,[52] where, ironically, government mortars from Inverness were used with effect to dislodge the Spanish forces from their hill-top defences.[53]

The new fortified barracks at Inversnaid (c. 120 soldiers) and Kiliwhimen (c. 360 soldiers) were both completed, followed by Bernera (c. 240 soliders) in April 1723 and Ruthven (c. 120 soldiers) the following year.[54] The first two were topographically poorly sited and Major General Wade was fully aware of this. Under his direction a major government scheme of military road and barrack construction followed his authoritative 1724 report on military options.[55] He recommended the construction of a new fort at Inverness and, to replace Kiliwhimen barracks, another at Fort Augustus, closer to Loch Ness in order to facilitate access to Inverness by an eight-gun, 30-ton galley. Cromwell's troops had also managed to bring a gunboat on wooden rollers to Loch Ness via Inverlochy.[56]

The new Fort Augustus (fig. 19.4), designed by Romer, had an aggressive military exterior, and in plan stood four-square with bastions, of limited military capability, at each angle. Its military efficiency was reduced, however, in order to accommodate additional buildings for residential and administrative purposes, and each curtain wall was dominated by a three-storeyed range of buildings.[57] The limitations of the design were clearly demonstrated in 1746 when the fort surrendered after a two-day siege by Jacobite forces, following the destruction of a powder magazine situated within a bastion.[58]

In 1729 Wade renovated Mackay's ruinous fort at Inverness, and in 1732 Romer produced designs for the construction of an outer rampart[59] which remained incomplete in 1750.[60] During the Jacobite uprising of 1745–6, the fort was captured and destroyed. At Ruthven Barracks a new stable-block was completed in 1734.[61] Other improvements by Wade were the military roads in Scotland; he accounted for over 250 miles (400 kilometres) and about 40 bridges.[62] Outwith the Highlands the fortresses of Edinburgh, Dumbarton and Stirling had their barrack accommodation improved and their defences strengthened and brought up to date. However, although capable of sustaining a siege by a limited artillery train, none of the garrison fortresses was by European standards capable of resisting a major investment by a standing army.[63]

Nor did the forts and communications infrastructure substantially influence the course of the Jacobite rising in 1745.[64] The Hanoverian response to the landing of Prince Charles Edward Stuart on Eriskay on 23 July 1745, and the raising of the standard at Glenfinnan, Inverness-shire, on 19 August, had been to attempt to keep the uprising confined to the Highlands.[65] Initially, the Jacobites succeeded in capturing Inversnaid Barracks and laid siege – unsuccessfully – to Ruthven, but

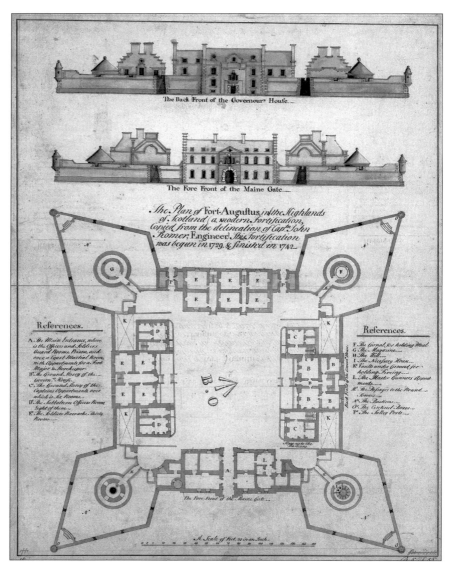

Figure 19.4 Fort Augustus, Inverness-shire; as designed by John Romer and completed 1742. ©Crown copyright/MOD.

Lieutenant-General Sir John Cope's manoeuvres with the government forces allowed the Jacobite army to head southwards, helped in part by the new military roads. The city, but not the castle, of Edinburgh was captured, and a few miles east of Edinburgh on 21 September Cope's forces were routed at the Battle of Prestonpans, leaving the Jacobites as virtual masters of Scotland. In England, however, the Jacobite cause came to naught around Derby with the army unable to swell their ranks with English support.

The Battle of Falkirk on 17 January 1746 was the last Jacobite victory in lowland Scotland, and nor, when potential freedom of action by the Jacobite army was threatened by the government forts, was there any systematic plan to subdue them. The garrison forts of Inverness, Fort Augustus and Ruthven Barracks fell to the retreating Jacobites, and only Fort William, itself poorly sited by reason of the improvements in artillery, managed to hold out, the Jacobites raising the siege in order to provide troops for an ill-planned attack on Hanoverian forces that were concentrating east of Inverness. On 16 April the forces met at Culloden Moor, where the Jacobites were routed and Jacobitism was destroyed as an effective political and military force.[66] The remnants of the defeated army were sought out by the 15,000-strong army under the Duke of Cumberland, known by many as 'Butcher Cumberland' as his policy of pillage and burning of Highland settlements left its mark long after his resignation as Commander-in-Chief in 1747.

The Hanoverian government established its authority over the remnants of the Jacobite supporters in Scotland by a mixture of new and upgraded fortifications linked by improved communications. Yet another 'Disarming Act' was also passed, and this time the wearing of tartan and use of the bagpipes were banned. The defences of Fort William and Fort Augustus were hastily repaired, but the original Fort George of 1726, in the centre of Inverness, was beyond repair, having been mined by a French artillery sergeant, L'Epice, who was 'hoist by his own petard'.[67]

The fort needed to be replaced on a strategic site in the vicinity of Inverness with sea-communications via the Moray Firth. The site of Oliver's Fort, slighted in 1661, was seriously considered, and plans for its re-fortification were drawn up by Major Lewis Marcell in 1746 and in the following year by Major General William Skinner, the recently appointed Director of Engineers to the Board of Ordnance in North Britain.[68] This site was eventually rejected because the burgh of Inverness had recently erected a harbour there, and if it were to be re-fortified, the burgh would have demanded considerable compensation.[69]

A new site for Fort George was selected later in 1747, nine miles (14.5 kilometres) east of Inverness on the promontory at Ardersier, Inverness-shire. Designed by William Skinner, it has been acknowledged as the 'finest example of British military eighteenth-century engineering and one of the outstanding artillery fortifications of Europe' (fig. 19.5).[70] Occupying an area of 42 acres (17 hectares), the fort was the largest construction project in the Highlands until the Caledonian Canal in the early years of the following century. It was designed to contain barrack accommodation for two infantry battalions (1,600 men), and an

Figure 19.5 Fort George, Inverness-shire; aerial view. Crown copyright: RCAHMS, SC 712209.

artillery unit.[71] Work began in 1748, the armament installed in 1760, and by 1769 the fort was almost finished. The original estimate of £92,673 19s 1d was greatly exceeded, and the final cost, estimated at over £200,000, represented more than Scotland's annual gross national product for the year 1750.[72]

Although overlooked by high ground one mile (1.6km) to the east, the site was out of range of contemporary howitzers. Saunders has noted that:

> The landward defences were concentrated on one front to the east, consisting of two bastions, a ravelin before the main gate in the middle of the curtain, covered-way and glacis. The two long sides of the fort were each flanked by a large bastion, and a powerful

sea battery in the centre of the curtains, and a powerful sea battery occupied the tip of the promontory. Small ravelin-like places of arms covered the sallyports on each of the long sides.[73]

Other improvements on a lesser scale were carried out elsewhere during the construction of Fort George. Rebuilt military outposts allowed foot patrols to link up within a day's march. The two ruined medieval tower-houses at Corgarff and Braemar, Aberdeenshire, for example, were converted into barrack blocks by the simple expedient of adding two wings and surrounding the building by a star-shaped wall, pierced for muskets. The former received its outer defences in 1748. Star-shaped covered ways were also added in 1749 to the barracks at Inversnaid and Bernera.[74] Strongholds such as Duart and Mingary Castles, Argyll, were considered suitable for small garrisons, along with numerous summer outposts in the West Highlands at the heads of glens or lochs such as Lochs Arkaig, Laggan, Leagh, Morar, Lag, and Rannoch, as well as in the Eastern Highlands at Tomintoul, Banffshire, and in the Braes of Angus.[75] The military road programme under Wade's successor, Major Caulfeild, also continued apace. Between 1740 and 1767 over 900 miles (1, 452 kilometres) of road and 800 bridges were constructed,[76] creating a network of improved communications which eventually helped to open the Highlands to economic growth. In the lowlands, new barrack accommodation for an infantry battalion was constructed in the last decade of the century at Edinburgh Castle.[77]

The decade after 1750 saw a reduction of military garrisons, however, and for almost the remainder of the century, the Scottish fortresses were more of recruitment centres for new regiments – the Highlands providing an unexpected source of manpower – than military outposts in hostile territory.[78] By the mid nineteenth century the role of the Highland fortified outpost had run its course. In 1864 Fort William was sold and eventually virtually dismantled, the site being acquired in 1889 by the West Highland Railway and used for a railway engine depot.[79] Traces of the fort can still be seen today but substantially more remains of Fort Augustus, sold in 1867 and then given to the Benedictines for use as a monastery.[80] The outpost at Corgarff remained in military use until sold in 1831, but Fort George still continues as a military barracks.

The rebuilding of the 1665 fort at Lerwick (see above) in the 1780s was the first Hanoverian defensive work intended to protect Scotland against a perceived external attack – in this case from the combined naval forces of Spain, France and the states of Northern Europe. Garrisoned by 270 soldiers of the Earl of Sutherland's Regiment in March 1781 and re-named Fort Charlotte it was probably still incomplete in 1783 when peace was declared at Versailles. In 1793, the newly formed Orkney and

Shetland Fencibles provided a company of about 100 men to garrison the fort, only to be disbanded in 1797 and replaced a year later by a new formation called the Shetland Fencibles who were themselves disbanded in 1814, the fortress guns being removed in 1855.[81]

During the latter half of the eighteenth century Scotland's right to defend itself was hotly debated, and finally, in 1797, the Scottish Militia Act allowed the Scottish nation to defend itself, a measure viewed in some circles simply as a ruse by the government to increase its military forces against the Napoleonic threat.[82]

FORTIFICATIONS OF THE LATE EIGHTEENTH AND EARLY NINETEENTH CENTURIES

The American War of Independence and the Napoleonic Wars dominated British military policy in the late eighteenth and early nineteenth centuries. Physical reminders of that era include smooth-bore coastal gun batteries, Martello Towers and, in southern England, the Royal Military Canal. In Scotland, there are also remains of coastal batteries constructed of stone and/or earth, as well as three Martello-type towers, two in Orkney and a third off Leith Harbour, in the Firth of Forth.

An example of a stone-built battery exists at Dunbar (fig. 19.6).[83] In 1781 an American commerce raider failed in its attempt to capture a ship in the mouth of Dunbar harbour. The episode prompted the burgh of Dunbar in the same year to erect a battery, built of local red sandstone, situated opposite the harbour mouth on Lamer Island. The seaward front of the fort was rounded and overlooked a small jetty. The gate on the landward side was flanked by small demi-bastions. Its original armament of 16 guns contained examples of nine-pounder, 12-pounder and 18-pounder pieces. In common with most batteries of that period, the work was never finished, although repair work was carried out in 1793 and again in 1808; the battery was demobilised in 1815 and the armaments removed to Edinburgh.[84] Plans of Dunbar and other batteries at Arbroath, Montrose, Aberdeen, Peterhead, Banff, Greenock and Campbeltown, Argyll, have been preserved.[85]

The Corporation of Edinburgh purchased ground in 1780 for a battery to defend Leith harbour.[86] Unfortunately, its siting did not anticipate future expansion of the dockyard and in 1807 the battery was compromised militarily by the proposed erection of a Martello Tower only 1,200 yards (1.09 kilometres) distant on its seaward side.[87] Today, only the guardhouses and part of the fort's perimeter wall stand, the interior of the fort now being occupied by high-rise flats.

All three of Scotland's Martello Towers survive virtually intact. They were built primarily to deter commerce raiders and to protect assembly points for the Baltic timber convoys which supplied the Royal

Navy with masts and spars. In 1807 the Board of Ordnance proposed that a Martello Tower be built on the Beacon Rock, commanding the approach to Leith harbour, and that it should be constructed by Edinburgh City Corporation using Board funds.[88] In 1810, the Inspector General of Fortifications, Lieutenant General Morse, revised the plan by increasing the diameter from 52 feet (16 metres) to almost 80 feet (24.6 metres), raising the height from 36 feet (11 metres) to 45 feet (13.8 metres) above water level, and altering the open gun platform to mount three heavy guns, in place of one or two.[89] The Board of Ordnance approved the new plans but issued a reprimand for not seeking their prior approval. In 1838, they took over the tower from Edinburgh Corporation; it had cost almost £17,180 and remained unfinished. Not until 1853 was it finally ready to receive its garrison of one officer and 21 soldiers along with the armament. The tower was never armed, however, and now forms part of the extension of the east pier of the docks.

A single mounting for one 24-pounder gun was reverted to in the two later Martello Towers, at Hackness and Crockness in Orkney, the construction of which was authorised in 1813 (fig. 19.7).[90] They

Figure 19.6 Fort, Lamer Island, Dunbar, East Lothian. Crown copyright: RCAHMS, SC 712207.

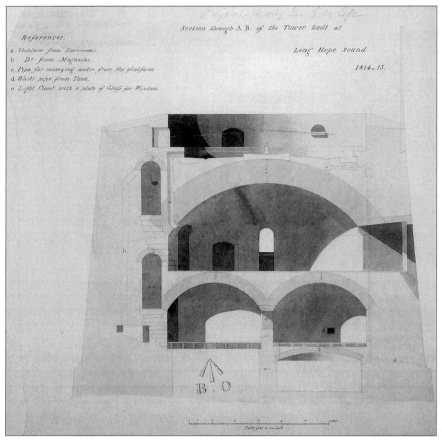

Figure 19.7 Martello Tower, Hackness, Hoy, Orkney; cross-section, 1814–15. Public Record Office, MPH1/620(5).

were built to guard the entrance to Longhope Sound which was used as an anchorage and assembly point for ships sailing in convoy to the Baltic via the Pentland Firth. Of standard English south-coast design, but of local sandstone instead of brick, each tower was built at an estimated cost of almost £5,265. They are 47 feet (14.4 metres) in diameter, 33 feet (10.1 metres) high, and on the seaward-facing side the wall thickness increases from 6 feet 3 inches (1.92 metres) to 9.5 feet (2.9 metres).

At Hackness the construction of an additional battery for eight 24-pounder guns was also begun a few months later in 1813. The towers were finished in 1815 when hostilities had ceased. Although the towers and battery were reported to have been disarmed in the 1840s, in 1866 it was decided to re-arm the towers with 68-pounder guns for which some alterations were necessary. The Hackness battery was also extensively remodelled to take four 64-pounder guns which fired through embrasures instead of over the parapet, re-armament of the battery having recommenced in 1861.[91]

FORTIFICATIONS OF THE VICTORIAN PERIOD AND EARLY TWENTIETH CENTURY TO 1918

By the mid nineteenth century, the defences of the British Isles were neglected and run down. In Scotland, for example, it was noted by the end of the Napoleonic Wars that Fort George at Ardersier could be dominated by improved artillery emplaced on the high ground one mile (1.6 km) to the east of the stronghold, but the deficiency was not remedied. The fortress's role was modified in 1860 by the addition of a coastal battery equipped with smooth-bore muzzle-loaded guns, but within another decade these were outclassed by revolutionary built-up rifled ordnance, which in turn made the bastioned fort quite indefensible, its old guns and mortar being recalled to London in 1881.[92]

After the Crimean War, when deficiencies in the ability of the Royal Navy to defend Britain from invasion were recognised and the prospect of war with France was renewed, steps were taken to defend the naval bases of southern England and South Wales from attack. Initially, Scotland did not feature in this fortification construction programme, but it did benefit from the coincidental formation in 1859 of a Volunteer Artillery Force conceived to provide manpower for new or existing batteries protecting coastal ports, towns or other strategic locations.[93] By 1880, the east coast of Scotland, including Orkney, mustered around 37 corps or battery headquarters of Artillery Volunteers compared with 23 on the western seaboard.[94] Good examples of their practice batteries can be seen at Eyemouth, Stonehaven, Kincardineshire, and Torry Point, Aberdeen.

Even as late as 1880 there was no naval base in Scotland,[95] but Scotland gradually assumed a new importance when it was recognised that the threat of invasion was moving northwards with the rise of the German Imperial Navy. The Firth of Forth came to be viewed as a valuable strategic anchorage, offering the first sheltered water north of the Humber estuary, but Scapa Flow, Orkney, was seriously used only as a naval anchorage in the first decade of the twentieth century and was virtually undefended when war broke out in 1914.[96]

The Firth of Forth was, in 1860, an anchorage leading to the capital of Scotland, itself virtually unprotected from sea-borne bombardment. To remedy this deficiency, land was purchased by the government on the island of Inchkeith as early as 1861,[97] but construction of Inchkeith's three forts, based on a design adopted by the 1860 Royal Commission on the Channel defences, was not begun until 1878.[98] As initially designed, the island's forts mounted a total of four 10-inch rifled muzzle-loading guns (RMLs). In order to defend the deep-water channel that runs close to the Fife coast, a fourth fort, equipped with two 10-inch RMLs, was constructed on the north side of the Forth, at Kinghorn, Fife. None of these forts was armed until 1881.[99] Upstream, Blackness Castle was renovated between 1870 and 1874, and converted into a new central ammunition store for Scotland. The inner courtyards were converted into covered stores, while space to the east of the castle became ammunition magazines. A new cast-iron jetty allowed conveyance of ammunition to and from ship, and on the landward side of the castle a new barrack block and other accommodation was erected.[100]

Examples of other mid Victorian coastal defences, built on previously fortified sites, can be seen at Torry Point with its purpose-built battery, and Fort George, Ardersier, which, as noted above, was modified in 1860 to receive a coastal battery of nine guns, installed in the north and west fronts of the fort.[101]

The Morley Committee of 1892 recommended improvements to the defences of the mercantile ports of Aberdeen and those within the Firths of Clyde and Tay.[102] An interesting but only partially successful earlier experiment of upgrading a fortified site can be seen at Broughty Ferry, Angus (fig. 19.8), overlooking the entrance to the Firth of Tay.[103] Broughty Castle, a ruined fifteenth-century tower-house on a cramped site on the north shore of the Tay was reconstructed between 1860 and 1861 into a self-defensible coastal battery, to a design by R R Anderson of the Royal Engineers. The rebuilt tower-house with new wing formed the barracks and administration centre, while the perimeter curtain wall and external defences were reconstructed to mount two 10-inch and four 68-pounder guns. By 1888, the site was considered to be vulnerable to attack by land or sea.[104] The addition of a controlled minefield in 1886 significantly added to the Tay's defences, and in 1905 a new battery of two 6-inch guns was emplaced outside the castle on the Castle Green. These guns were dismounted in 1932 but restored in April 1940, and later that year a second 4.7-inch gun was installed. Broughty Castle continued to serve in a coast defence role throughout World War II. Placed 'under care and maintenance' in 1945, it continued until October 1956 when all the military equipment was removed.[105]

The Firth of Forth defences on Inchkeith were strengthened by the removal, between 1891 and 1895, of the RMLs and the installation of breech-loading 4.7-inch and 6-inch guns. The first of three new 9.2-inch emplacements on Inchkeith was begun in 1892. Additional upgrading of the defences followed military reports published in 1894, 1902 and 1905. In order to meet the growing threat from Germany, the 1902 report recommended the construction of a new naval base at Rosyth.[106]

Figure 19.8 Broughty Ferry Castle and Battery, Angus; aerial view. Crown copyright: RCAHMS, SC 713517.

Changes to the pattern of anchorages for ships in the inner Forth, coupled with activity generated by the start of the new naval dockyard of Rosyth in 1909, brought further alterations to these defensive zones and prompted a seaward movement of the defences.[107] By 1914 a zone system of batteries had been introduced: an inner line of batteries at Carlingnose (Fife), Coastguard, Inchgarvie and Dalmeny (West Lothian); a middle line of heavier calibre guns at Downing Point, Braefoot (from 1915 armed with two 9.2-inch guns) and Hound Point (Fife); and the outer zone was formed by the upgraded batteries on Inchkeith (three 9.2-inch guns) and Kinghorn.[108]

Coast-defence batteries had now become highly sophisticated and included buildings specifically designed for ammunition storage, range-finding, communication by radio and underwater telegraph, searchlights, fuel, electricity generating and water storage. There were also repair shops, as well as the ever-present barrack block and guardhouse. Braefoot Battery, operational between 1915 and 1917, provides an excellent example of the range of structures that supported a two-gun 9.2-inch battery. Surviving remains include a fortified barrack block, rifle-looped blockhouses (the forerunner of the pillbox) which formed part of the perimeter defence line, gun-pits and adjacent magazines.

The upgrading of the Forth defences and the construction of the new naval base of Rosyth, regarded by the Royal Navy as the 'Scottish Portsmouth',[109] were not the only defence-related works being undertaken in Scotland prior to, and during World War I. Defences were improved on the Clyde, where in April 1915 a 9.2-inch gun was installed,[110] and at Invergordon on the Cromarty Firth, plans were afoot in 1912 to transform the anchorage into a second-class naval base.[111] In 1912 work was also started on the defences at the North and South Sutors, headlands that guarded the entrance to the new base, and this was completed by early 1914.[112] The North Sutor armaments included two 9.2- inch naval guns and four 4-inch Quick Fire (QFs) guns, while on the opposite headland one 9.2-inch naval gun and two 4-inch QFs lay in emplacements.[113]

At the fleet anchorage at Scapa Flow, no land-based guns protected the base at the outbreak of World War I. To make good this deficiency, light guns of 3- and 12-pounder size were removed from ships for shore use, and by 1915 all the main entrances to the anchorage were protected by guns of various sizes up to 6 inch. Underwater obstructions, blockships and booms were employed to protect the entrances, but through the war the defences were not adequate against determined submarine attack.[114]

In Shetland, Bressay Sound came to be protected during World War I by two 6-inch guns, one at either end of Bressay Island and two 4.7-inch guns (taken from HMS *Brilliant*) at the Knab, Lerwick. In 1917 the deep-water west Shetland anchorage of Swarbacks Minn was protected by two 6-inch guns (from HMS *Gibraltar*). Obsolete by World War II, these two guns and some of the ancillary structures still remain, and, along with the two on Bressay, are the only remnants of World War I coastal artillery in the United Kingdom still emplaced in their original mountings.[115]

Military airfields were being established during World War I. By 1913 Montrose had become Scotland's first military airfield and at Cromarty a naval seaplane base was established. In April 1914 a second seaplane base opened in the Firth of Tay at Dundee. By the end of the war over 30 stations were operational for a variety of seaplanes and land-based aeroplanes, as well as for the more cumbersome airships and kite balloons, needed for observation work.[116]

One unique aerial gunnery school, the only one of its type in the United Kingdom, was established around Loch Doon in Ayrshire, where construction work started in September 1916. The steep slopes of the hills surrounding the loch allowed rail-mounted targets to zig-zag down the hillsides to simulate the movements of enemy aircraft. The project was ill conceived, the local weather proved unsuitable for flying on too many days and the targets were unable to simulate the increasing speeds of aircraft. The School of Aerial Gunnery was closed down in 1918 after as much as £3,000,000 had been reputedly spent on the project – the sum that had passed from the Dalmellington Bank during the building period.[117]

INTER-WAR YEARS

After World War I, expenditure on defence was reduced in line with the view that Britain would not for the foreseeable future be engaged in a major continental war. This view prevailed until political necessity motivated the government on 4 March 1935 to announce a rearmament policy.[118]

After 1918, the 55 coast batteries in Scotland[119] were placed on a care and maintenance basis or kept for training. A few such as Swarbacks Minn[120] were made redundant, and a limited modernising programme was put in place for the remainder. 12-pounder Quick Fire guns were usually replaced by the more efficient twin 6-pounder Quick Fire guns that would be more capable of damaging high-speed and better armoured light craft. Where applicable, improvements were made to the 9.2-inch batteries, along with changes in gun laying and target engaging. Whenever possible, salvo firing was preferred to the previous practice of single-round firing at enemy ships. The Territorial Force that stood down in 1920 reappeared in 1922 as the Territorial Army, whose duties included the manning of coast defence batteries.[121]

In January 1937 a sub-committee of the Committee of Imperial Defence examined for the first time the threat of invasion by Germany, capable of executing either a seaborne invasion or a combination of air and seaborne attack. The committee also noted that, along with naval, air and army bases and airfields, factories producing ammunition, chemicals, aero-engines, and other military items had to be defended from air attack. The Forth, the Clyde and Scapa Flow were all mentioned in this report.[122] A report dated October 1937 on 'Home Ports, Local Naval and Seafront Defences' updated a previous report produced ten years before. It introduced a new classification of naval bases and brought Scapa Flow into 'Category A' status. The sub-committee noted however, that it would take 45 years to complete the modernising of the land defences, based on the 1937 allocation of £75,000.

In December 1937 the Cabinet recognised that security of the United Kingdom was a priority and noted the country's vulnerability to air attack.[123] Air-raid precautions had been first examined in 1924 by the government which, over the next decade, noted the growing threat that aerial bombardment posed to civilian and military personnel alike. It also recorded that it would be impossible to guarantee immunity from aerial attack.[124] The training of instructors for anti-gas precautions was introduced in 1936, and the winter of 1937 saw the establishment of the Air Raid Precaution (ARP) system. The Air Raid Precautions Act of January 1938 imposed on all local authorities the task of preparing plans for the protection of persons and property from enemy aerial attack, the cost to be shared by local authorities and central government.[125] Defence estimates of 1938-9 included £24,000,000 for civil defence, five times as much as the previous year's estimate.[126] In Scotland's major cities the old, densely-packed and overcrowded tenements represented a major fire hazard, compared with the buildings in most English cities,[127] and in Edinburgh alone the Air Raid Warden service at one time numbered over 7,000 persons. The Munich Crisis of 1938 provided an opportunity for a rehearsal of the readiness of the ARP and highlighted the strengths and weaknesses of the system.

The Civil Defence Act of 1939 required firms with more than 50 employees to provide air-raid shelters for their staff. In Scotland, government assistance for the provision of civilian shelters was limited to Dundee, Rosyth, Glasgow and Edinburgh.[128] By October 1939, Edinburgh could provide shelters for 140,000 persons, approximately one-third of its population.[129] In Dundee, records indicate that by the end of 1941, 136 trench shelters, 267 basement shelters, 2,049 surface shelters and 6,560 Anderson shelters – a total of 9,588 shelters – had been constructed for its civilian population. Elsewhere, very little shelter construction existed, in Falkirk, for example, while Stirling completed only about 300 yards (274 metres) of trench shelters.[130]

The general fear of aerial bombardment felt in the inter-war period was clearly borne out by events, as witness the effects of the wartime raids by German aircraft on Glasgow, Edinburgh, Dundee, and particularly Clydebank in the spring of 1941.[131]

WORLD WAR II

The defences of Britain between 1939 and 1945 are legion. Here, it is proposed to review only selected aspects, including defences of the coast and anchorages, airfields, anti-invasion defences, the role of radar and forms of air defence.[132]

Coast defence
A good proportion of Britain's most important naval dockyards and anchorages lay in Scotland. From Rosyth in the Firth of Forth, Invergordon in the Cromarty Firth, Scapa Flow, Orkney, Loch Ewe, Ross and Cromarty, and the naval establishments in the Firth of Clyde, Scotland was ringed with important coastal locations, all of which, together with many smaller anchorages, had to be made safe from enemy attacks. For coastal defence, a total of 66 coast batteries became operational in Scotland,[133] among which the defences of the Cromarty Firth provide a good representative example. There, on the North and South Sutors overlooking the entrance to the firth, new gun batteries were planned in 1938 on the sites of dismantled World War I batteries. At the new North Sutor battery, two 6-inch Mark VII guns became operational in February 1940, later replaced in April 1943 by two 6-inch Mark XXIV high-angle guns.[134] An additional battery at Nigg, opposite Cromarty, was completed in June 1940 and armed with two 6-inch guns.[135] The new South Sutor battery had a temporary armament from September until November 1939 when two 6-inch Mark VII guns became operational.[136]

The threat of air attack meant that no new gun positions could be constructed without adequate overhead protection, and that older gun emplacements would be either entirely reconstructed or would receive new overhead cover, examples of which can be seen over the 9.2-inch and 6-inch gun emplacements on Inchkeith.

The defences of the Firth of Forth continued their easterly progression during World War II, and by the end of the war the most easterly gun positions were located at Kincraig, Fife (two high-angle 6-inch guns), and at Dirleton, East Lothian (two 6-inch guns).[137] Radar and submarine minefields (controlled from Extended Defence Officers (XDO) Posts) completed the defences against surface ships, while anti-aircraft batteries and airfields defended the Forth from airborne attack. A submarine detection loop was run from the Isle of May,[138] and there was an XDO post at Methil, Fife, now partly buried by modern landscaping. Good surviving examples of such posts still exist on the South Sutor, Cromarty, and at Gallanachmore, Oban, Argyll.

Within the first few weeks of the war, the defences of Scapa Flow were penetrated by the enemy. At night on 14 October 1939 the German U-boat *U47* navigated past the blockships in Kirk Sound into the anchorage and sank the battleship, HMS *Royal Oak*. Of some 1,400 crew, 809 men and 24 officers died.[139] Three days later enemy aircraft twice penetrated the defences, aiming bombs at shore targets and ships, including the veteran battleship HMS *Iron Duke*, then shorn of part of her heavy armament and armour and acting as depot ship at Scapa. Damaged and beached upright, she was later made watertight and towed to Longhope, where, firmly aground, she remained in service throughout the rest of the war as headquarters for an anti-submarine drifter group patrol and a flotilla of harbour defence motor torpedo boats.[140]

In the aftermath of these raids, Scapa Flow, the premier British naval anchorage, was temporarily demoted to providing refuelling for destroyers, while the defences were reinforced by blockships, controlled minefields and booms. Before the battle fleet returned from its hurried dispersion along the west coast of Scotland, additional anti-aircraft guns, with their ancillary searchlights and barrage balloons were emplaced. Extra air cover was also provided by airfields in Orkney and Caithness. The most enduring reminders of the upgraded Orcadian defences, however, are the unique Churchill Barriers, constructed (with the aid of Italian prisoners of war) in order to block the four eastern seaborne approaches to Scapa Flow.[141]

Airfield defences
Airfields were considered prime targets for capture by an enemy invasion force and were suitably protected by varying scales of defence.[142] For defence purposes, military airfields were divided into three classes. Class 1 airfields were those which lay within 20 miles of selected ports; they were heavily defended by rings of pillboxes and barbed wire. Class 2 airfields were not strategically placed but had the capacity to repel seaborne attacks with aid of their own bombers and fighters; their close defensive capability was about 25 per cent less than a Class 1 airfield. Class 3 airfields were considered to be of only limited strategic value; they were defended by a single ring of pillboxes or in some instances only by barbed wire.[143]

Of the many varieties of pillbox associated with airfield defence, two are worthy of special mention. One type is visible at Crail Airfield, Fife, where the original pillboxes received reinforcement, possibly after the fall of Norway in 1940, with the addition of an exterior concrete skirt. The additional skirt, however, restricted the arc of fire. A second type, designed in 1940, is the Picket Hamilton retractable pillbox, 170 of which had been installed on the runways of 59 Scottish airfields by July 1941. The design failed to live up to expectations, however, and they were phased out by 1942.[144]

Anti-invasion defences
Following the evacuation of Dunkirk and the subsequent fall of France in June 1940, the Home Defence Executive instigated a major review of the defences of the United Kingdom, under General Sir Edmund Ironside, Commander-in-Chief of the Home Forces. His task was to update and instigate a method of defending the United Kingdom from invasion by enemy forces now capable of a combination of seaborne and/or airborne landings.

The vulnerable southern and eastern seaboard was destined to receive the maximum deployment of defensive structures along a zone nicknamed the 'coastal crust'. An excellent example of such beach defences, complete with anti-tank blocks, pillboxes and remains of barbed wire, can still be seen at Tentsmuir, Fife. Inland, a secondary line consisted of road blocks and pillboxes defending nodal points, and was designed to contain any breakout from a beachhead. Finally, a major series of defence lines or 'stop-lines', based on anti-tank ditches, pillboxes and natural obstacles, were established.[145] The major lines covered southern and midland England, but one line, known as the GHQ line,[146] was originally planned to run from London northwards to the Forth at Musselburgh, where it joined the eastern defences of Edinburgh. A second defence line was to have run from the upper reaches of the Firth of Forth to Stirling and inland via the rivers Forth and Teith to the west end of Loch Tay.[147]

The Chiefs of Staff reviewed Ironside's plan, a new commander, General Alan Brooke, was appointed, and on 3 August a revised plan was issued with orders to cease work on the line, except in areas required for local defence. This mainly affected the progress of the line in northern England and Scotland,[148] where limited survey work had been carried out. With the change of orders, however, the line to Edinburgh and Perth was never built,[149] though a secondary stop-line was constructed northwards through Fife from Kirkcaldy via Markinch and Ladybank to Newburgh on the south side of the Tay, then on to Perth where a blocking position was built.[150] Remains of pillboxes and anti-tank defences, including an anti-tank wall, can still be seen in and around the village of Kingskettle, Fife.

Other Scottish examples of stop-lines are the Cowie line south of Aberdeen, the blocking positions around the River Tay at Perth, and the Robertson line in Shetland that defended the landward approaches to Lerwick and Bressay Sound with anti-tank blocks, machine gun posts and spigot mortar positions.[151] Although invasion was still considered a possibility, pillbox construction continued only at selected sites from September 1941, and none was constructed for anti-invasion purposes after February 1942.[152]

Air defences
Ground-based passive air defence included the use of searchlights, barrage balloons, radar, sound detectors and bombing decoys. In Scotland, as in the rest of the United Kingdom, fragmentary remains exist of fixed anti-aircraft battery sites, rocket-launching sites known as 'Z' batteries, for non-rotating projectiles, and bombing decoy sites.[153] By their very nature, mobile gun batteries and sound-locating posts have left very little physical evidence. Good examples of anti-aircraft batteries

remain near the Braid Hills, Edinburgh, around Loch Ewe, at Mugdock, Stirlingshire, and around the Firth of Clyde. Even strategic sites of ancient ancestry were used for this purpose, as exemplified by Dumbarton Castle where a light anti-aircraft battery was deployed.[154]

Observation posts and radar

Before the invention of radar, monitoring of approaching enemy aircraft or airships in World War I was by a co-ordinated system of observation posts. The Observer Corps, a civilian organisation administered by the RAF, can trace its origin from the Metropolitan Observation Service of World War I. It was improved during the inter-war years, and on 9 April 1941 was given, in recognition of its valuable services, a new title, 'The Royal Observer Corps' (ROC), under which it operated until May 1945.[155]

The first practical demonstration in Britain of aircraft detection by 'radio location' or 'radar', as it came to be known, was given in February 1935.[156] Radar allowed surface ships to be monitored, provided the means of directing fighter aircraft, and the means of ranging anti-aircraft and coastal guns. Radio guidance systems allowed aircraft to pinpoint their positions in all weathers.

By the outbreak of war, a chain of 20 radar stations, known as 'Chain Home', covered the coast from Portsmouth to Orkney. In September 1939 only four Chain Home stations were operational in Scotland along the east coast at Netherbutton, Orkney, School Hill, Portlethen, Kincardineshire, Aberdeen, Douglas Wood, near Kirkbuddo, Angus, and Drone Hill, near Coldingham, Berwickshire. A second system, 'Chain Home Low', was installed using equipment developed from the Army's coast defence sets. The combined system was able to provide early warning of aerial attacks and was continually upgraded as new equipment became available. By July 1940, an extra four Chain Home and five Chain Home Low stations had been added to the Scottish east-coast network.[157] A review carried out towards the end of the war in March 1945 revealed that on the east coast from Berwick to Unst there were ten Chain Home, 22 Chain Home Low and two Ground Control Intercept stations (North Town/Russland, Orkney, and Dirleton). Along the west coast of Scotland by that date there had been six Chain Home stations, 13 Chain Home Low, ten Chain Home Beam (a Chain Home Low station equipped with a scanner which did not rotate) and three Ground Control Intercept stations. Remains of the buildings associated with radar installations can still be seen at numerous sites, including, for example Drone Hill and Douglas Wood, where the underground and surface structures were protected by pillboxes.[158]

Training

Scotland provided training areas for aircrews, soldiers, sailors and special forces. One spectacular example of the facilities that were created is the former Whitestone Military Range, which lies on the moors near the site of the Battle of Sheriffmuir (1715, see above). This range contains the heavily damaged remains of a purpose-built 'German Atlantic Wall', used for training troops prior to the D-Day landings in June 1944.[159]

POST-WAR DEFENCES

By mid-1945, with hostilities in Europe over, the emergency coast batteries were either dismantled or being dismantled. However, coast artillery batteries continued to be manned or put into care and maintenance until the Ministry of Defence announced in February 1956 that there were no military grounds for retaining them beyond the end of that year.[160] Ground-based anti-aircraft defences also continued to be built in Scotland into the early 1950s with five 5.25-inch, radar-controlled Heavy Anti-Aircraft Batteries put up around the Clyde at Wemyss Bay, Renfrewshire, Roseneath, Dunbartonshire, Pattiston, Renfrewshire, Limekilnburn, Lanarkshire, and Finnich, Stirlingshire.[161] All were scrapped in 1956, and in 1959 anti-aircraft guns were officially replaced by surface-to-air guided missiles.[162] Clearly, in little over ten years, the nuclear bomb had radically altered defence scenarios, rendering obsolete a number of defensive measures.

The Royal Observer Corps (ROC), stood down in May 1945, was re-formed in January 1947 for monitoring aircraft. In 1951 permission was given to re-locate 411 posts and to construct 93 new ones. In 1952 Messrs Orlit Ltd produced designs for a prefabricated concrete observation post, 10 feet (3.05 metres) by 6 feet 8 inches (2.03 metres) containing an open observation area, a roofed shelter and a store. Type 'A' design was for flat ground and type 'B' stood on 4 feet 6 inches (1.37 metres) concrete legs with access by metal ladder.[163] But the higher speeds attained by jet aircraft made it increasingly difficult for the ROC to monitor flight paths. By 1957 the majority of these posts had been decommissioned, a few continuing in use until 1965.

After 1957 the ROC assumed a new role in monitoring the effects of nuclear explosions and the distribution of radioactive fallout. A prototype underground monitoring post had been built in 1956 and, with minor modifications, became standard. A 15-feet (4.6-metre) shaft from the surface led down to a small chamber, with access to a chemical toilet. From the shaft base another door led into a rectangular 15 feet (4.6 metres) by 7 feet 6 inches (2.3 metres) monitoring room which accommodated three observers with their equipment and domestic arrangements. A total of 875 such posts were constructed between 1958 and 1964 all over the

United Kingdom, grouped in clusters to allow the triangulation of fallout plots and transmitting data to semi-sunken Protected Group Headquarters and Sector Headquarters. In November 1991 the ROC was again stood down,[164] but most of their underground posts still survive. At Letham, Angus, the fallout monitoring station lies adjacent to a type A Orlit post, while an example of a type B Orlit post can be seen at Kippen, Stirlingshire.

In the summer of 1944, plans were already afoot to reduce Britain's air defences. Radar stations were divided into two areas; a Defended Area where Great Britain faced other countries, that is, between Scarborough on the Yorkshire coast and Cornwall; and a Shadow Area, that is, the rest of Britain where the air defence framework was manned by a skeleton force but was capable of returning to full operational efficiency within two years.[165] The salient points regarding Scotland's part in the development of post-war ground radar and air defence are described below.

The Rotor Scheme, conceived in 1949 and first implemented 1950–55, was designed to track incoming fast jet aircraft across the North Sea using existing Chain Home (CH) radar for long-range detection, with Centimetric Early Warning (CEW) radar and Chain Home Extra Low (CHEL) providing additional and low-level cover respectively. The intercepting fighters used the latest version of Ground Controlled Interception (GCI) radar to lead them onto targets. In Scotland, eleven Rotor radar stations were either operational or in a state of readiness by 1955, nine along the east coast at Netherbutton (CH), Hillhead, Wick, Caithness (CH), Buchan, Aberdeenshire (GCI), Inverbervie, Kincardineshire (CEW), Douglas Wood (CH), Anstruther, Fife (GCI), Dirleton (GCI), Drone Hill (CH) and Cross Law, Berwickshire (CHEL), and two sites on the west coast at Sango, near Durness, Sutherland (CH) and Scarinish, Tiree, Argyll (GCI).[166] By early 1958, the east-coast stations included Saxa Vord, Shetland (CEW); on the west coast there was an additional site in Argyll at Kilchiaran on Islay (CHEL).[167]

On the east coast sites most equipment was buried underground to protect it from the threat of conventional and nuclear blast. On the surface, giving access to the bunker by a buried inclined shaft, was a brick-built guardhouse, usually resembling a bungalow with steep pitched roof and veranda which, in Scotland, contrasted markedly and conspicuously with surrounding structures built of local materials. On the west coast, where the threat of attack was considered to be less, structures associated with the Rotor project were not always built underground, a point well illustrated at Kilchiaran.

The Rotor operation system became obsolete in the mid 1950s with the introduction of type 80 radar,[168] while the Linesman scheme, approved in 1971, later integrated the United Kingdom's radar defence with that of the North Atlantic Treaty Organisation (NATO). Phase 1 of Linesman linked radars on Iceland with those of the Faeroes, Norway and the Netherlands, as well as those in the United Kingdom.[169] Linesman radar stations became known as master radar stations and combined the roles of control and reporting with ground control interception. In Scotland, only Buchan (closed 2000) acted as a new Master Radar Station, Saxa Vord having become a NATO Early Warning Station, operated by the RAF and having stand-by control facilities.[170]

Further developments led to the Ballistic Missile Early Warning System with the construction of the famous white mushroom radomes covering the radar equipment at Fylingdales, North Yorkshire. Today, the Improved UK Air Defence Ground Environment has one of its principal sector operations centres in Scotland at Buchan. A supplementary Airborne Warning and Control Systems adds mobility to the land-based system. Thus, in less than a decade, the defence of the United Kingdom, including Scotland, has progressed from a domestic internal system to one based on inter-continental co-operation.[171]

Finally, during the last quarter of the twentieth century, civil defence operations switched to the protection of post-nuclear authorities capable of enforcing law and order and to the rudimentary provision of welfare and food distribution from unhardened structures. A network of regional and sub-regional seats of government was created, using purpose-built bunkers or modifying existing structures from the post-war period so that they might withstand the demands of the nuclear age.[172]

Hence, at Barnton Quarry, Edinburgh, former wartime underground bunkers first became a three-level sector operations centre during the Rotor programme, and then, following the demise of Rotor, a regional seat of government; they are now defunct. A former Rotor bunker at Cross Law was also re-used as a regional seat of government. At Anstruther a former two-level bunker for ground control interception became the Scottish Northern Zone Headquarters of the Civil Defence organisation. It opened to the public in 1995 as 'Scotland's Secret Bunker'. One centre still operational as the Army Headquarters (Scotland) is the former Craigiehall bunker, on the north-western outskirts of Edinburgh, originally designed as the Rosyth/Forth anti-aircraft operations room.[173]

ACKNOWLEDGEMENT

The author and editors wish to thank John A Guy for his generous assistance in the preparation of this article.

NOTES

1. MacIvor, 1981, 107–18.
2. Saunders, 1989, 57–9. For a recent account of the 'Rough Wooing' see Merriman, 2000.
3. Caldwell, Ewart, 1997.
4. Colvin et al, 1982, 722–5 and fig 67.
5. Saunders, 1989, 58–9.
6. RCAHMS, 1951, 15–18.
7. MacIvor, 1993.
8. Saunders, 1989, 34–52.
9. RCAHMS, 1963, Vol. 2, 304–6.
10. RCAHMS, 1951, lxii-lxvi.
11. Harris, 1991, 360, note 13.
12. ibid, 359–68.
13. RCAHMS, 1951, 266.
14. Gifford, 1988, 389.
15. Walker, Ritchie, 1987, 56; 1996, 79; Stell, 1988, 62.
16. RCAHMS, Vol 2, 1967, 280.
17. Caldwell, 1984, 15–24.
18. Saunders, 1989, 102–7.
19. Tait, 1965, 9.
20. Cruden, 1981, 224–34; Tabraham, Grove, 1995, 16–20.
21. Cruden, 1981, 229–30.
22. Saunders, 1989, 104.
23. Lynch, 1991, 284.
24. Tabraham, Grove, 1995, 17–18.
25. Cruden, 1981, 231.
26. ibid, 228–9.
27. ibid, 231.
28. ibid, 231–2.
29. Roy, 2002.
30. Cruden, 1981, 234.
31. ibid, 233–4.
32. Shepherd, 1986, 28; 1996, 52–3.
33. Tabraham, Grove, 1995, 19–26.
34. ibid, 27.
35. ibid, 28–33.
36. ibid, 35–9.
37. Saunders, 1989, 107–13; Cruden, 1981, 234–45.
38. Tabraham, Grove, 1995, 39–42.
39. ibid, 42–5.
40. Cruden, 1981, 241.
41. Tabraham, Grove, 1995, 45–6.
42. ibid, 46–9.
43. ibid, 49–51.
44. ibid, 52–5.
45. ibid, 56.
46. ibid, 57–9, 61.
47. ibid, 61–5.

48. Cruden, 1981, 235.
49. Tabraham, Grove, 1995, 61.
50. Stell, 1973, 185.
51. Tabraham, Grove, 1995, 61.
52. ibid, 65–7.
53. Lynch, 1991, 331.
54. Tabraham, Grove, 1995, 62, 64.
55. ibid, 65.
56. ibid, 70, 75.
57. ibid, 78–81.
58. Cruden, 1981, 239–40.
59. Tabraham, Grove, 1995, 76–8.
60. Cruden, 1981, 241.
61. Tabraham, Grove, 1995, 81–2.
62. ibid, 70.
63. ibid, 82–6.
64. Saunders, 1981, 110.
65. ibid, 87–92.
66. Tabraham, Grove, 1995, 90–1.
67. ibid, 93.
68. ibid, 93–4.
69. ibid, 94.
70. Saunders, 1981, 110.
71. Saunders, 1989, 110–3; Tabraham, Grove, 1995, 92–100; Cruden, 1981, 242–5; MacIvor, 1988.
72. MacIvor, 1988, 7.
73. Saunders, 1989, 111.
74. Cruden, 1981, 237–8; Tabraham, Grove, 1995, 101–4.
75. Tabraham, Grove, 1995, 101.
76. ibid, 108.
77. ibid, 113.
78. ibid, 112–13.
79. ibid, 115.
80. ibid, 115–16.
81. Tabraham, 1993, 58–62. For a more recent account of Fort Charlotte see Pringle et al, 2000.
82. Tabraham, Grove, 1995, 112.
83. PRO (Public Record Office), WO 78/1150, MPH 199 (1) and (2): Plan of Lamer Island, Dunbar, AD 1811 (1) Plan and section; (2) The ground contiguous.
84. Graham, 1966–7, 188–9.
85. NLS (National Library of Scotland), MSS, Z3/56 a-h. Plans of the Napoleonic period batteries at Arbroath, Dunbar, Montrose, Aberdeen, Peterhead, Banff, Greenock and Campbeltown.
86. PRO, WO 78/1822, MPH 263: Lands between Leith and Newhaven showing redoubt built for the protection of Leith harbour in 1780. PRO, WO 78/1970, MPH 199 (1) and (3), AD 1785: Battery or redoubt at Leith, (1) Plans and section, (3) Site of battery.
87. Saunders, 1984, 471.
88. Clements, 1999, 91–3.

89. PRO, WO 78/1167, MPH 210, AD 1807 and 1810, Leith. Plans and sections of a tower proposed to be built on the Beacon Rock.
90. Clements, 1999, 93–94.
91. Ritchie, 1985, 22–3; Ritchie, 1996, 50–1. See also PRO, WO 44/540 and WO 33/10.
92. MacIvor, 1988, 9.
93. Litchfield, Westlake, 1982, 1.
94. ibid, 2.
95. Saunders, 1984, 472.
96. For the development of Scapa Flow, see Hewison, 1990.
97. Ruckley, 1984, 73.
98. Saunders, 1984, 473.
99. Smith, 1985, 93.
100. MacIvor, 1993, 23.
101. MacIvor, 1988, 17.
102. Smith, 1985, 95.
103. Mudie, Walker, MacIvor, 1970/1979.
104. ibid, 74.
105. ibid, 81–2.
106. For details of the complex re-arming of the forts and batteries in the Firth of Forth see Saunders, 1984, 469–80, Ruckley, 1984, Smith, 1985, Clark, 1986, and Heddle, Morris, 1997.
107. Clark, 1986, 50.
108. Saunders, 1984, 475–6.
109. Longmate, 1993, 425.
110. ibid, 445–6.
111. Ash, 1991, 188.
112. ibid, 192.
113. Dorman (and Guy), 1996.
114. Longmate, 1993, 445.
115. *pers com* J A Guy. The surviving gun on Hirta, St Kilda, is a beach defence gun, not a coast battery, and it was not manned by the Royal Artillery.
116. Smith, 1989(b), 9.
117. Connon, 1984, 97–108.
118. Longmate, 1993, 460.
119. *pers com* J A Guy.
120. *pers com* J A Guy.
121. Longmate, 1993, 457.
122. ibid, 463.
123. ibid, 464.
124. ibid, 438–9.
125. Jeffrey, 1992, 41.
126. Longmate, 1993, 465.
127. Jeffrey, 1991, 121.
128. Jeffrey, 1992, 52.
129. Jeffrey, 1991, 119.
130. Jeffrey, 1992, 120.
131. See Jeffrey, 1991; 1992; 1993.
132. Lowry, 1996; Wills, 1985.
133. *pers com* J A Guy.

134. PRO, WO 192/248, North Sutor Fort Record Book.
135. PRO, WO 192/246, Nigg Fort Record Book.
136. PRO, WO 192/247, South Sutor Fort Record Book.
137. Saunders, op cit, 1984, 477.
138. Clark, op cit, 1986, 52.
139. Hewison, 1990, 256–61.
140. ibid, 262–3, 293, 377.
141. ibid, 305–12.
142. For an in-depth review of the variety and roles played by the military airfields in Britain, see Smith, 1989(a), and for individual Scottish airfields, Smith, 1989(b).
143. Smith, 1989(b), 16–19.
144. ibid, 17–18. An example, complete with its pumping mechanism, has recently been recovered from Wick Airfield, Caithness.
145. Wills, 1985, 9–14.
146. Alexander, 1999
147. ibid, 86.
148. ibid, 29–31.
149. ibid, 86.
150. Wills, 1985, 71–2.
151. The southern seaward entrance to Bressay Sound was additionally defended by shore-mounted torpedo tubes, probably the only remaining example in Scotland, and a World War II coastal gun battery at Ness of Sound mounting two 6-inch guns in July 1940. For an assessment of twentieth-century defences visible in Scotland, see the bibliography under J A Guy.
152. Wills, 1985, 14.
153. Jeffrey, 1993, 88, for bombing decoy (starfish) sites around the Clyde and Glasgow. For details of numerous defence works see the unpublished works by J A Guy in the bibliography.
154. Ritchie, 1985, 68.
155. For a definitive history of the Royal Observer Corps, see Wood, 1992.
156. For a definitive history of radar see Gough, 1993.
157. ibid, 25, 28
158. ibid, 27.
159. Cowley et al, 1999.
160. Hogg, 1974, 92.
161. *pers com* J A Guy.
162. Lowry, 1996, 131.
163. Lowry, 1996, 127.
164. ibid, 127–8. The full history of all phases of the ROC role is given in Wood, 1992, including all locations of posts in Scotland.
165. Gough, 1993, 55. The complex development of post-World War II ground radar and air defence of the United Kingdom has been summarised by Gough, 1993.
166. ibid, 97.
167. ibid, 98.
168. ibid, 290–1. In January 1956 the type 80 radar designed to withstand 120 mph winds was blown off its mountings. A new mounting was developed and NATO undertook to provide the radome.
169. ibid, 239.
170. ibid, 292.

171. Lowry, 1996, 131.
172. ibid, 135–6.
173. Clark, 1986, 55.

BIBLIOGRAPHY

Alexander, C. *Ironside's Line: The definitive guide to the General Headquarters Line planned for Great Britain in response to the threat of German Invasion, 1940–1942*, Storrington, 1999.
Ash, M. *This Noble Harbour: A history of the Cromarty Firth*, Edinburgh, 1991.
Breeze, D J, ed. *Studies in Scottish Antiquity, Presented to Stewart Cruden*, Edinburgh, 1984.
Caldwell, D H. *Scottish Weapons and Fortifications, 1100–1800*, Edinburgh, 1981.
Caldwell, D H. A sixteenth-century group of gun towers in Scotland, *FORT*, 12 (1984), 15–24.
Caldwell, D H, Ewart G. Excavations at Eyemouth, Berwickshire, in a mid sixteenth century trace italienne fort, *Post Medieval Archaeology*, 31 (1997), 61–120.
Cavers, K. *A Vision of Scotland: The nation observed by John Slezer 1671 to 1717*, Edinburgh, 1993.
Clark, N H. Twentieth-century coastal defences of the Firth of Forth, *FORT*, 14 (1986), 49–54.
Clements, W H. *Towers of Strength*, Barnsley, 1999.
Colvin, H M, Summerson, J, Biddle, M, Hale, J R, Merriman, M, eds. *The History of the King's Works*, Vol 4, *1485–1660* (Part 2), London, 1982.
Connon, P. *An Aeronautical History of the Cumbria, Dumfries and Galloway Region:* Part 2, *1915 to 1930*, Kendal, 1984.
Cowley, D C, Guy, J A, Henderson, D M. The Sheriffmuir 'Atlantic Wall': An archaeological survey on part of the Whitestone Military Range, *Forth Naturalist and Historian*, 22 (1999), 107–16.
Cruden, S. *The Scottish Castle*, Edinburgh, 1960, 3rd edn, Edinburgh, 1981.
Dobinson, C. *Fields of Deception: Britain's bombing decoys of World War II*, London, 2000.
Dorman, J (and Guy, J A). *A Visit to Cromarty and Loch Ewe*, Dover, 1996.
Gifford, J. *Buildings of Scotland: Fife*, Harmondsworth, 1988.
Gough, J. *Watching the Skies: The history of ground radar in the air defence of the United Kingdom*, London, 1993.
Graham, A. The old harbours of Dunbar, *PSAS*, 99 (1966–7), 173–90.
Guy, J A. A Survey of Twentieth-Century Defences, a series of unpublished reports carried out for Historic Scotland (copies in NMRS): Orkney, 1993; Grampian, 1993; Fife, 1994; Shetland, 1995; Lothian, 1997; Dumfries and Galloway, 1998; Tayside, 1999; Borders Region and Central Region, 1999; Highland Region, 2000; Strathclyde Region, 2001; Western Isles and Inner Hebrides (2002).
Harris, S. The fortifications and siege of Leith: A further study of the map of the siege in 1560, *PSAS*, 121 (1991), 359–68.
Heddle, G, Morris, G. Charles Hill gun battery, *Tayside and Fife Archaeological Journal*, 3 (1997), 207–15.
Hewison, W S. *Scapa Flow: This great harbour*, Kirkwall, 1985, 2nd edn, 1990.
Hogg, I V. *Coast Defences of England and Wales, 1856–1956*, Newton Abbot, 1974.
Jeffrey, A. *This Present Emergency: Edinburgh, the river Forth and south-east Scotland and the Second World War*, Edinburgh, 1991.

Jeffrey, A. *This Dangerous Menace: Dundee and the River Tay at war, 1939 to 1945*, Edinburgh, 1992.
Jeffrey, A. *This Time of Crisis: Glasgow, the west of Scotland and the north western approaches in the Second World War*, Edinburgh, 1993.
Jellicoe of Scapa, Admiral Viscount. *The Grand Fleet 1914-16: Its creation, development and work*, London, 1919.
Litchfield, N, Westlake, R. *The Volunteer Artillery, 1859–1908*, Nottingham, 1982.
Longmate, N. Island Fortress: The defence of Great Britain, 1603–1945, London, 1993.
Lowry, B, ed. *Twentieth-Century Defences in Britain: An introductory guide*, 2nd edn, CBA, York, 1996.
Lynch, M. *Scotland: A New History*, London, 1991.
Lynch, M, Spearman, M, Stell, G, eds. *The Scottish Medieval Town*, Edinburgh, 1988.
MacIvor, I. Artillery and major places of strength in the Lothians and the East Border. In Caldwell, 1981, 94–152.
MacIvor, I. *Fort George*, Edinburgh, 1970, revised edn, 1988.
MacIvor, I. *Blackness Castle*, Edinburgh, 1993.
Merriman, M. *The Rough Wooings: Mary Queen of Scots, 1542–1551*, East Linton, 2000.
Mudie Sir F, Walker, D, MacIvor, I. *Broughty Castle and the Defence of the Tay*, Dundee, 1970, reprinted 1979.
Portway, D. *Military Science Today*, Oxford, 1940.
Pringle, D, Ewart, G, Ruckley, N. ' ... an old pentagonal fort built of stone': Excavation of the battery wall at Fort Charlotte, Lerwick, Shetland, *Post Medieval Archaeology*, 34 (2000), 105–43.
Redfern, N I. *Twentieth-Century Fortifications in the United Kingdom*, Vol 1: *Introduction and sources*, CBA, York, 1998.
Redfern, N I. *Twentieth-Century Fortifications in the United Kingdom*, Vol 2: *Site gazetteers, Wales*, CBA, York, 1998.
Redfern, N I. *Twentieth-Century Fortifications in the United Kingdom*, Vol 3: *Site gazetteers, Northern Ireland*, CBA, York, 1998.
Redfern, N I. *Twentieth-Century Fortifications in the United Kingdom*, Vol 4: *Site gazetteers, Scotland*, CBA, York, 1998.
Ritchie, A. *Exploring Scotland's Heritage: Orkney and Shetland*, Edinburgh, 1985.
Ritchie, A. *Exploring Scotland's Heritage: Orkney*, Edinburgh, 1996.
RCAHMS. *Inventory of Edinburgh*, Edinburgh, 1951
RCAHMS. *Inventory of Stirlingshire*, 2 vols, Edinburgh, 1963.
RCAHMS. *Inventory of Peeblesshire*, 2 vols, Edinburgh, 1967.
RCAHMS. *Scotland from the Air, 1939–1949*, Vol 1, *Catalogue of the Luftwaffe photographs in the National Monuments Record of Scotland*, Edinburgh, 1999.
RCAHMS. *Scotland from the Air, 1939–1949*, Vol 2, *Catalogue of the RAF World War II Photographs in the National Monuments Record of Scotland*, Edinburgh, 2000.
Roy, M. Excavation of the south-western bastion of Cromwell's Citadel on the South Inch, Perth, *Tayside and Fife Archaeological Journal*, 8, 2000, 145–68.
Ruckley, N A. Inchkeith: The water supply of an island fortress, *FORT*, 12 (1984), 67–82.
Saunders, A D. *Fortress Britain: Artillery fortifications in the British Isles and Ireland*, Liphook, 1989.
Saunders, A D. The defences of the Firth of Forth. In Breeze, 1984, 469–80.
Shepherd, I A G. *Exploring Scotland's Heritage: Grampian*, Edinburgh, 1986, 2nd edn (as *Aberdeen and North-East Scotland*), Edinburgh, 1996.
Smith, D J. *Britain's Military Airfields, 1939–1945*, Wellingborough, 1989(a).

Smith, D J. *Military Airfields of Scotland, the North-East, and Northern Ireland* (Action Stations 7), 2nd edn, Wellingborough, 1989(b).
Smith, V T C. Defending the Forth, 1880–1910. *FORT*, 13 (1985), 89–102.
Stell, G. Highland garrisons, 1717–23, *Post Medieval Archaeology*, 7 (1973), 20–30.
Stell, G. Urban buildings. In Lynch et al, 1988, 60–80.
Sutcliffe, S. *Martello Towers*, Newton Abbot, 1972.
Tabraham, C, ed. *The Ancient Monuments of Shetland*, Edinburgh, 1993.
Tabraham, C, Grove, D. *Fortress Scotland and the Jacobites*, London, 1995.
Tait, A A. The protectorate Citadels of Scotland, *Architectural History*, 8 (1965), 9–24.
Tully-Jackson, J, Brown, I. *East Lothian at War*, Haddington, 1996.
Walker, B, Ritchie, G. *Exploring Scotland's Heritage: Fife and Tayside*, Edinburgh, 2nd edn (as *Fife, Perthshire and Angus*), Edinburgh, 1996.
Wills, H. *Pillboxes: A study of UK defences, 1940*, Trowbridge, 1985.
Wood, D. *Attack Warning Red: The Royal Observer Corps and the defence of Britain, 1925–1992*, Portsmouth, revised edn, 1992.

PART FOUR
•
Work Places

20 Agricultural Buildings 1: Introduction; Equipment Storage and Traction

JOHN SHAW

INTRODUCTION

Agriculture has been Scotland's most widely dispersed and, arguably, its most important industry, until recent times. During the eighteenth and early nineteenth centuries Scottish agriculture became as advanced as any in the world. The buildings or steadings created to serve its needs are an essential component in the Scottish landscape and a sizeable element in its stock of historic buildings. They show considerable diversity, from the polite to the vernacular, from East Lothian mains farms to Lewis blackhouses. They are an unrivalled primary source in understanding the history of construction materials and techniques, agricultural practice and social history, over the course of several centuries (figs 20.1, 20.2).

A large proportion of nineteenth-century steadings survived substantially intact until the 1950s, largely as a result of prolonged agricultural depression, relieved only by increased demand during two world wars. Since then, grants and subsidies, as well as farm mechanisation, have encouraged new investment in buildings, through replacement, adaptation or addition. Yet many steadings still contain evidence of practices which were current one, or even two, hundred years ago.

In spite of their historical potential, little has been done to evaluate, interpret and record steadings. Since 1945 there have been only three published works of more than local scope.[1] Local studies have focused on the Lothians, east central Scotland and the north east; of these, only a few have been published.[2]

For all their variety, in regional character, in scale, sophistication or date, all but a few of the 30,000 or so Scottish steadings were built for mixed arable and livestock farming. Almost every arable farm kept some livestock; almost every livestock farm grew some root or cereal

Figure 20.1 Weem, Perthshire; glebe steading. SLA, FBS 4334.

Figure 20.2 Sidinish, South Uist, Inverness-shire; croft byre and hen house. SLA, FBS 3322.

crops. Almost all had some provision for housing the equipment and draught animals which worked the land, storing and processing harvested crops, as well as accommodation for livestock and their products. This three-fold division forms the basis of the chapters which follow. A further element, the labour required to work the farm, forms a backdrop to all and has its own components in the form of housing,[3] which is considered briefly in this chapter.

PLAN FORMS

The functional relationships between these elements have contributed to a variety of plan forms. Thus the labour involved in tending livestock is reflected in the proximity of house and byre; the siting of rick yards and hay barns relates to the land from which the crops were brought. Added to this there are constraints of topography and climate, the size of farm, the type of farming practised and the date of construction. There are many possible permutations, not all of which observe this underlying logic.

No one sequence of plan forms can be applied to all areas. There are late but simple plan forms in some districts and, elsewhere, others which are early but complex. Only the most generalised of analyses is attempted here.

The linear byre-dwelling, with or without an attached barn, is the simplest and probably the longest established plan form. It persisted in the smallest farming units until well into the nineteenth century. Where the house is of two or more rooms, the byre is associated with the but or kitchen end, with the barn either beyond the byre or adjoining the ben or best room end. The exception to this rule is Shetland, where the barn lies between house and byre, possibly on account of the direct use of grain from the barn for human or animal consumption and straw for animal feed. The addition of separate stables and cart bays, some of these latter with a granary over, can result in very long linear ranges, as at North Scotstarvit, Fife,[4] or more generally in Caithness.[5] In a common variant of the linear plan, the barn is separated and set at right angles to the byre-dwelling. This occurs not only in upland areas, but also in small eighteenth- and nineteenth-century steadings in the central lowlands.

Two areas, Orkney and Lewis, have their own distinctive layouts. In Orkney, and perhaps elsewhere in former times, byre-dwelling and barn are set parallel to each other on either side of a narrow passageway or close.[6] In the Lewis blackhouse, house and steading form an inter-connected agglomeration;[7] a similar arrangement might once have existed in Shetland.[8]

A layout frequently adopted from the eighteenth century, but of earlier origin, had barn and byre ranges set at opposite ends of the house and at right angles to it. This plan form continued to be used in the west, most notably in dairying or cattle-rearing districts, often with stable, cart bays and granary between barn and ben end, and dairy between byre and but end. Elsewhere, more elaborate plans developed.

On nineteenth-century beef livestock farms, particularly in the Highlands and the north east, the house is detached from the three-sided steading, with a byre or byres occupying the long cross-range and barn, stable, cart bay(s) and granary housed in the wings. The yard is normally occupied by a midden.

Four-sided plans were already known by the late seventeenth century in the arable south east. According to Lord Belhaven's ideal layout (1699), the house should occupy the north or south side of the square, the barn the west, the stable and byres the side opposite the house, with the entrance and any remaining structures on the east side.[9]

From the late eighteenth century onwards, the court thus formed, or a three-sided layout with enclosing walls, could be used with shelter sheds to winter cattle. Where cattle courts were used in preference to byres – mainly in the south east – the courtyard was extended to make space for walled, segregated cattle courts, with shelter sheds. Most of the remaining outer perimeter was taken up in barn ranges, stables and granary/cartshed, in proportion to the scale of the farm; one cart bay, one pair of stable stalls and one cottage seems to have approximated to 70 acres (28.3 hectares) of arable, plus or minus 20 acres (8.1 hectares). The farmhouse was usually built anew, away from the steading, and its predecessor either demolished or adapted to agricultural use.

Where both arable and beef livestock farming were important, as in the north east, there were further variations on the three-sided plan, to form H- or E-shaped layouts, courtyards or other arrangements, with both byres and grain-handling ranges occupying significant proportions of the steading, and the house close-by but detached. Such steadings were still being built in the early twentieth century. Some of the more sophisticated nineteenth-century layouts featured separate livestock and arable yards within one steading or on two detached sites.

Since the 1920s, with the exception of small-holdings[10] and a handful of home farms,[11] formally-planned layouts have given way to simple, system-built additions to earlier steadings which, with increasing difficulty, have had to be adapted to work with farming technologies for which they were never designed.

This brief outline should not be taken to imply that all steadings were planned at one date or through purposeful development. Many – perhaps the majority – have simply evolved to meet changing needs. Nor do all fit the description of one or other plan form. But without this infinite variety, without the unexpected, there would be less interest, less of a challenge, in attempting to interpret Scottish farm buildings.

WORKERS' HOUSING

As workers' housing forms a separate chapter (see Chapter 6), little need be said about it here, except in terms of its spatial relationship to the steading. In particular, there appears to be a relationship between the status of farm workers and the location of their accommodation.

Prior to the nineteenth century, the houses of married hinds – male farm servants such as ploughmen – might occupy one side of the steading square, at least in the south east where such squares existed. Otherwise, they might form a terraced row at a little distance from the steading itself. This arrangement was to become the norm, latterly with separate blocks or pairs of cottages. These might each have their own ancillary buildings: ash houses; privies; wash-houses; and pig crays. These last were most apparent in south west Scotland, the central lowlands and Caithness.

Single women, employed about the house, byre, dairy and hen-house, usually lived in, and the more substantial nineteenth-century farmhouses had servants' quarters, generally above the kitchen, accessed by their own stair and without direct access to other parts of the house. Women employed in seasonal work – singling turnips or at the harvest – were commonly drawn from hinds' households, from elsewhere in the vicinity, from the Highlands, or from Ireland (see Chapter 12). Accommodation for migrant workers was of variable quality; at best, a cottage might be set aside, at worst they had to find whatever quarters they could about the steading.

Accommodation for single men was the most basic on the farm, and might consist of no more than a loft over a stable or at the entrance to the court. Under the chaumer system, centred on north-east Scotland, the men had a sleeping apartment in the steading but took meals in the farm kitchen. Under the bothy system, centred on east-central Scotland, they prepared their own meals and occupied either a single apartment in the steading or, on the better farms, a day room with a separate sleeping apartment above.[12] In 1859, the superior accommodation at Monymusk Home Farm, Aberdeenshire, comprised a 'roomy and comfortable' kitchen, with a range of 'light and airy' bedrooms above.[13] Because of the importance of horse work in the duties of male farm workers, their housing tended to be close to the stables, even within the same building.

Where a cattleman was employed, and especially where pedigree livestock were bred, a house was provided close to or adjoining the byres or cattle courts. The cattleman's status was somewhat above that of the hind, and the standard of accommodation was correspondingly superior. Housing for inbye shepherds was generally set apart from the steading and from the hinds' houses, while the houses of hill shepherds formed the nucleus of its own, isolated group of buildings.

The larger arable farms, on which the farmer had an elevated social standing, had a house which, during the course of the nineteenth century, came to be set apart from the steading. Such farms employed a grieve or steward to take day-to-day responsibility for organising farm work. Typically, the grieve's house had a key site, commanding the entrance to the steading, with a view along the principal approach road. As might be expected, the grieve's house tended to be better appointed than those of other farm workers.

Over much of Scotland, the labour force consisted of the farmer's household, assisted by seasonal migrant workers or by neighbours. On these smaller farms, the farmhouse remained close to the steading – between byre and barn in the pastoral west, detached, but close at hand in the mixed farming districts of central and eastern Scotland.

SHIELING HUTS

In upland and island districts, where pastures far from permanent settlement were used in summer, a simple form of seasonal dwelling was used. These shieling huts were abandoned as, over the course of several centuries, the lands associated with them became permanently settled and, from the eighteenth century onwards, arable was enclosed and provision made for integrating fodder crops into rotations. The very last of these, in Lewis, were still being built and occupied within living memory.[14] In Shetland, another form of seasonal dwelling, used whilst digging peats, was still being used on the island of Fetlar in the 1930s.[15]

Equipment storage and traction

STABLES

Horse work was at the heart of improved farming: drawing implements in the field; pulling carts to or from the field or market; and powering threshing machines or butter churns. It was fitting, therefore, that accommodation for horses was generally superior to that of other beasts. Stables for riding or carriage horses were even more luxurious.

Horses had not always been so important. Oxen had been the principal draught animals in the lowlands prior to the mid-eighteenth century, and until the 1920s on small farms in the north east and Northern

Isles. Walker's study of Strathmore farms (Angus-Perthshire) found evidence of ox byres in most eighteenth-century appraising tickets: the farm of Cossans (1713), for example, had a stable and two ox byres, in addition to a cow byre;[16] it is not clear, however, which of these byres housed draught oxen and which contained oxen being fattened for market. It is quite possible that draught oxen were sometimes housed with other cattle in a common byre.

In the north west and the Western Isles, horses and ponies were used rather than oxen. On the smaller units much of the working of the land, and transport to or from it, was done by hand or back.

Late eighteenth-century references to stables indicate that they were not sub-divided, nor were there separate hecks or trochs.[17] One author, in a comparative account of East Lothian in the 1760s and 1800s, recalled with horror the conditions which had once prevailed:

> The stables – Good G** ! How often have I been alarmed and frightened out of my sleep by the fighting and screaming of a dozen horses standing loose in the same undivided stable, and the responses of the courser thundering in his trevise in the byre beyond the partition.[18]

East Lothian stables of the early 1800s were thought to be 'greatly improved' by the insertion of trevises, which separated each working pair of horses from the others. On some farms, the stable was even wide

Figure 20.3 Greendykes Farm, East Lothian; work-horse stables. SLA, FBS 7102.

enough to take two rows of stalls, tail to tail.[19] Yet neither feature was considered acceptable: only a single row of single stalls would do.[20] It was this arrangement which soon became the standard layout for stables throughout Scotland (fig. 20.3).

Stable floors had a hard surface of cobble, setts, flags, brick or, latterly, concrete, divided into the sloping floor of the stalls, a drainage channel behind, and, beyond that, a passageway. The walls, which in the best establishments were plastered, usually had a window or windows plus one or two doors onto the passageway. The windows were commonly half-glazed, the lower half having hinged, louvred or 'hit and miss' vents, for good ventilation of stables was considered to be of the utmost importance. Doors could be part-opened, horizontally, vertically or both, again with a view to controlling ventilation. Ingenious devices were developed to keep half-doors open.

Set into the passageway wall, near to the door or behind each stall, there might be alcoves where combs, brushes, medicines, a lantern or other bits and pieces could be kept, though there might also be cupboards for this purpose. The wall at the head of the stalls might have inlet vents. Against this wall, or mounted on it, were the heck for hay or straw and the troch for other food, made of wood, cast iron or, in the case of the troch, ceramics. The heck might be near head height or on a level with the troch.

Horses were tethered by means of a rope, weighted with a wooden sinker, which passed through a loop on the front of the troch. The trevises between stalls were held in place by means of a strong head-post and hind-post, generally wooden but occasionally in cast iron; the latter was more typical of house stables. The hind-post often carried harness hooks, with further pegs and saddle trees on the passage wall. The larger, more affluent farms might have a separate, heated harness room, which could also act as a day room for farm workers. A corn kist completed the stable's furnishings.

The size of a farm's stables was directly linked to the extent of its arable land, with one pair of horses to every 50–80 acres (20.2–32.4 hectares).[21] Stables tended to be situated near to the entrance to the steading, perhaps to facilitate access to and from the land.

Most nineteenth-century writers asserted that, in the interests of good ventilation, there should be no loft over the stable. In the best arable farms, notably in the south east, this was adhered to, with a row of exit vents aligned along the roof. Elsewhere, lofts continued to feature, even if this required ventilation shafts to be routed from the head of the stall, up the wall of the loft and onto the roof. In Strathmore, Walker found that stable lofts were common on farms of all sizes.[22] The loft was perhaps too useful to forego – as a store for hay, with chutes directly into the heck, and as a sleeping place for farm workers. For those stables without

Figure 20.4 Garth, near Sullom Voe, Shetland; pony pund. SLA, FBS 3213.

lofts, a separate hay house was provided, adjacent to the stable, with access to it and with a large entrance to facilitate unloading from carts.

In addition to stalls, a loose-box might be included in the stable or in a separate apartment, as a place for sick or injured horses or for mares in foal. Further removed, such an apartment might be termed an infirmary.

In other circumstances, horses might be housed in sheds with yards or paddocks (fig. 20.4). In early nineteenth-century Berwickshire there was a fashion for this, in keeping with Norfolk practice.[23] In the 1880s, on farms where draught horses were bred, they were housed in courts or boxes with yards.[24] At Dunmore Home Farm, Stirlingshire (1884), young horses occupied a detached shed, opening into a paddock which extended down to the River Forth.[25]

HOUSE STABLES

On smaller farms, the transport needs of the farmer's household could be met by using work-horses and farm carts. If riding horses were kept, they were housed alongside the other horses. The larger, more affluent farms and latterly those of lower status, kept a specialised vehicle, typically a gig or trap, which had its own, separate accommodation, a gig house or 'gig shed', in the steading but within easy reach of the farmhouse. This had an arched or lintelled opening, with lockable doors.

Figure 20.5 Allangrange Mains Farm, Ross-shire; house stables. SLA, FBS 5204.

If no additional space was available, the gig house was also used to store harness for the gig and for riding horses. Associated with the gig house were riding stables, typically with stalls for three horses: Lowe (1844) considered this to be the minimum number on a large farm: 'The farmer must himself, on any considerable farm, have a saddle horse, as necessary to the economy of time in his business' (fig. 20.5).[26] This made no provision for stabling visitors' horses. In some cases, as at Kilmux, Fife, the horses were housed in loose-boxes rather than stalls.

The fittings and furnishings were of a higher quality than those in work-horse stables: perhaps with plastered, coombed ceilings, boarded linings to walls, cast-iron rather than wooden hecks and trochs, and better surfaces to floors. Further options included a hay loft over the stable and a separate harness room, with a fireplace or stove, to help maintain a warm and not too dry atmosphere; this too might be lined with boards.

Mid nineteenth century steadings show a hierarchy of provision, linked to the extent and wealth of the farm: Morphie, Kincardineshire, had a three-stalled riding stable plus gig house; Swanston, Midlothian, had three stalls, a carriage house and a harness house; while Redden, Roxburghshire, a huge arable farm of over 1,000 acres (404.8 hectares), had a free-standing building incorporating a stable, gig house, harness room, hay loft and groom's quarters.

CART AND IMPLEMENT SHEDS

Cart sheds, typically with granaries over, were a late eighteenth-century innovation. This pattern was not invariably followed: single-storeyed cart sheds are less common but not unknown.

Suitable accommodation for carts was a growing need. As long as movements within and beyond the farm remained difficult, or where only small volumes were involved, pack-horse, sled or back transport was sufficient. But better roads, and perceptions of 'good practice' on improved farms, required something better: box carts and long harvest carts in the south east, adaptable box carts with shelmets (frames) for harvest work elsewhere. Most cart shed and granary ranges date from the middle third of the nineteenth century (fig. 20.6). For whatever reason – perhaps status – they carry a date more often than any other buildings on the farm.

The openings into the cart bays might take the form of stone columns and arches (especially in the south east), stone columns with timber or stone lintels (especially in east central Scotland) or cast-iron columns with timber lintels. Walker's study of Strathmore detected an evolutionary pattern: the earliest, from the late eighteenth century onwards, with flat or flattish stone arches; from about 1800 to 1840 (though also later), timber or occasionally sandstone lintels backed by timber or, later, iron or steel safe lintels; and from about 1850 onwards, false arches concealing timber safe lintels.[27]

Figure 20.6 Kilrie Farm, Fife; cart shed and granary. SLA, FBS 1922.

Figure 20.7 Greendykes Farm, East Lothian; implement shed and yard. SLA, FBS 7104.

The number of bays, as also the number of stalls, bore an approximate relationship to the arable acreage, with one bay to each 50–80 acres (20.2–32.4 hectares). Internally, there was seldom any sub-division between bays; unusually deep plans, such as are found on some south-eastern farms, may have been designed to store, in addition, the bodies of harvest carts or, as indicated on some contemporary plans, implements. For tools and small implements, which might easily be mislaid or stolen, one of the bays might be partitioned off and fitted with doors. Rebates for doors, cut into the stonework around the opening, help to identify implement bays, even if doors and hinges have been removed.

During the second half of the nineteenth century, the range of field machinery expanded considerably: grubbers, seeders, reapers, binders and potato diggers accumulated on the farm. Not only were they more numerous than previously, they were also mechanically more complex; additional housing was needed. Implement sheds were a response to this need. Seldom other than single-storeyed, commonly with cast-iron or timber columns, with timber, iron or steel lintels, they were characteristic post-1850 additions to the arable steading (fig. 20.7). Though they continued to be built into the twentieth century, for tractors as well as implements, the materials were now less durable or of poorer quality, common brick, corrugated sheet metal or old railway sleepers being typical. To a limited extent, cart bays could be used for tractors and trailers, though openings are too low for modern tractors and often too narrow for trailers.

SMIDDIES AND WORKSHOPS

In order to maintain equipment on a day-to-day basis, the larger nineteenth-century farms kept an apartment aside as a workshop. The workshop's essential feature was the work-bench, lit from behind by a window; there might also be a hearth or stove for heating glue and other materials. Some farms also had a smiddy, which could be used by a visiting blacksmith to carry out repairs on tools and implements plus farrier work. There was a work-bench here too, similarly lit, and a large open hearth. Home farms might also have a timber yard and sawmill. Where a steading already had water or steam power for threshing, this could also be used to drive a circular saw, as at Toxside, Midlothian.[28] The best-appointed arable steadings had more elaborate facilities: the 1,170-acre (473.7 hectares) farm at Redden, Roxburghshire (1842) had a smiddy, with a house for a resident smith, a joiner's shop with adjacent tool house and a wood yard.[29]

NOTES

1. Morton, 1976; Fenton, Walker, 1981; Maxwell, 1996.
2. East Lothian County Council. 'East Lothian Farm Buildings' (typescript), 1972; Maxwell, I. 'Functional Architecture: Hopetoun Estate, West Lothian' (typescript), 2 vols, 1974; Maxwell, I. 'Functional Architecture: Airlie Estate' (typescript), 4 vols; Walker, B. 'Farm Buildings in the Grampian Region' (typescript), 1979; Walker, B. 'The Agricultural Buildings of Greater Strathmore, 1770–1920' (PhD thesis, University of Dundee), 2 vols, 1983; White, 1991.
3. Shaw, 1994, 28–33.
4. Fieldwork, Scottish Farm Buildings Survey (SFBS).
5. Stell, Omand, 1976; Stell, 1982..
6. Fenton, 1978(a), 116–35; SFBS, Lady and Cross and Burness parishes, Orkney.
7. Fenton, 1978(b).
8. Fenton, 1978(a), 125–35; See also Roussell, A. *Norse Building Customs in the Scottish Isles*, Copenhagen and London, 1934.
9. A B C (John, Lord Belhaven). *The Countrey-Man's Rudiments*, Edinburgh, 1713.
10. Leneman, 1989.
11. SFBS. Examples located by the Survey include Laigh of Dercullich, Perthshire, Rankeillour Mains, Fife, Wellside, Moray.
12. For Angus and the Mearns see Adam, D G. *Bothy Nichts and Days*, Edinburgh, 1992.
13. Porter, J. Report on the management of a home farm, *THASS*, 3rd series 8 (1859), 226. The Society's Transactions were published under a number of titles over the years. To simplify matters, only this, the last form, is used here throughout.

14. For a full account of shielings see Fenton (1976), 1999, 130–42.
15. Fenton, 1978(a), 233–6.
16. Walker, 1983, op cit, Vol 2, 434; Walker, B. Farm buildings and archaeology: The evidence of appraising tickets, *Scottish Agricultural Gazette*, 8 (1977), 10.
17. Robertson, G. *Rural Recollections*, Irvine, 1829, 76–7.
18. Anon. View of the situation of farmers etc, *Farmers' Magazine*, 5 (1804), 400.
19. Somerville, R. *General View of the Agriculture of East Lothian*, London, 1805, 42.
20. ibid, 45.
21. Derived from estimates in later nineteenth-century county reports in *THASS*.
22. Walker, 1983, op cit, Vol 2, 438.
23. Anon. On the construction of a farm-yard, *THASS*, 6 (1805), 133; For Norfolk see Martins, S W. *Historic Farm Buildings Including a Norfolk Survey*, London, 1991, 177.
24. MacNeilage, A. System of management in breeding studs of draught horses, *THASS*, 5th series 2 (1890), 156–7.
25. *THASS*, 4th series 16 (1884), 158.
26. Low, 1844, 135.
27. Walker, 1983, op cit, Vol 2, 503.
28. Burton, W T. On the reclaiming of Toxside, *THASS*, new series 10 (1861–3), 160.
29. Grey, J. On farm buildings, *Journal of the Royal Agricultural Society of England* (hereinafter *JRASE*), 4 (1843), 13–16.

BIBLIOGRAPHY

Beaton, E. Late seventeenth- and eighteenth-century estate girnals in Easter Ross and south-east Sutherland. In Baldwin, J, ed, *Firthlands of Ross and Sutherland*, Edinburgh, 1986, 133–51.
Brunskill, R W. *Traditional Farm Buildings of Britain*, London, 1982.
Cheape, H, ed, *Tools and Traditions: Studies in European Ethnology presented to Alexander Fenton*, Edinburgh, 1993.
Fenton, A. Lexicography and historical interpretation. In Barrow, G W S, ed, *The Scottish Tradition*, Edinburgh, 1974, 243–58.
Fenton, A. *Scottish Country Life* (1976), revised edn, East Linton, 1999.
Fenton, A. *The Northern Isles: Orkney and Shetland*, Edinburgh, 1978(a), revised edn, East Linton, 1987.
Fenton, A. *The Island Blackhouse*, 2nd edn, 1978(b).
Fenton, A, Walker, B. *The Rural Architecture of Scotland*, Edinburgh, 1981.
Gray, W J. *A Treatise on Rural Architecture*, Edinburgh, 1852.
HMSO. *Farm Buildings for Scotland*, London, 1946.
Henderson, R. *The Modern Homestead*, London, 1902.
Leith, P, Spence, S. *Orkney Threshing Mills*, Dundee and Edinburgh, 1977, 4–7.
Leneman, L. *Fit for Heroes? Land settlement in Scotland after World War I*, Aberdeen, 1989.
Low, D. *On Landed Property and the Economy of Estate*, London, 1844, 135.
MacDonald, D A. Corn-drying kilns in Uist. In Cheape, 1993, 185–6.
MacDonald, J. *Stephens' Book of the Farm*, 5th edn, 3 vols, Edinburgh and London, 1908–9.
Maxwell, I. *Building Materials of the Scottish Farmstead* (Scottish Vernacular Buildings Working Group, Regional and Thematic Studies, 3), Edinburgh, 1996.

Mercer, J. Roomed and roomless grain-drying kilns: The Hebridean boundary?, *Transactions of the Ancient Monument Society*, new series 19 (1972), 27–36.
Morton, J C. *Cyclopaedia of Agriculture*, 2 vols, Edinburgh, 1855.
Morton, R S. *Traditional Farm Architecture in Scotland*, Edinburgh, 1976.
Newman, P. Kil: Variety in the design of Orkney farm kilns, *Vernacular Building*, 18 (1994), 48–66.
Omond, J. *Orkney Eighty Years Ago: With special reference to Evie*, Kirkwall, 1911.
Robinson, J M. *Georgian Model Farms*, Oxford, 1983.
Scott, Sir L. Corn drying kilns, *Antiquity*, 25 (1951), 200–2.
Scottish Vernacular Buildings Working Group. *Vernacular Building*, 1975 -.
Shaw, J P. Identifying systems within farm steadings: A Scottish case study, *Recording Historic Farm Buildings*, Reading, 1994, 28–33.
Shaw, J P. Pastures in the sky: Scottish tower silos, 1918–1939, *Journal of the Historic Farm Buildings Group*, 4 (1990), 73–4.
Shaw, J P. The girnal house at Simprim, Berwickshire. In Cheape, 1993, 197–204.
Smith, J. *Cheesemaking in Scotland: A history*, Clydebank, 1995.
Stell, G, Omand, D. *The Caithness Croft*, Laidhay, 1976.
Stell, G. Some small farms and cottages in Latheron parish, Caithness. In Baldwin, J R, ed, *Caithness: A cultural crossroads*, Edinburgh, 1982, 86–114.
Stephens, H, Burn, R S. *The Book of Farm Buildings*, Edinburgh and London, 1861.
Stephens, H. *The Book of the Farm*, 2 vols, Edinburgh, 1844 and later edns.
Transactions of the Highland and Agricultural Society of Scotland, 1799 -.
Walker, B. The influence of fixed farm machinery on farm building design in east Scotland in the late eighteenth and nineteenth centuries. In Thoms, L M, ed, *The Archaeology of Industrial Scotland* (Scottish Archaeological Forum, 8), 1977, 60–2.
Whitaker, I. Two Hebridean corn kilns, *Gwerin*, 1 (1956–7), 161–70.
White, N. *The Farmsteadings of the Bathgate Hills*, Linlithgow, 1991.
Wight, A. *Present State of Husbandry in Scotland*, 4 vols, Edinburgh, 1778–84.

21 Agricultural Buildings 2: Storing and Processing Crops

JOHN SHAW

CEREALS

Threshing and winnowing
Sheaves from the harvested grain were built into stooks in the field to dry and were then transported to the steading. It was customary in Scotland to store grain in the sheaf outdoors, in circular ricks, within a stack yard. There are references to this practice as far back as the mid fifteenth century,[1] and there is an early pictorial record of ricks at Donibristle, Fife, in a 1592 painting recording the murder of James Stewart, 2nd Earl of Moray.[2]

To keep the rick-bases dry, the ground might be ploughed into rigs,[3] though the commonest method seems to have been to make a loose bed of stones.[4] Bracken, heather, whin or coarse hay were also used. By the 1760s stone foundations or stathels, consisting of pillars and cap-stones, were being used to raise stacks off the ground, discourage vermin and improve ventilation.[5] By the early nineteenth century, cast-iron stathels were available, plus ceramic versions later in the century. There are examples of all three types in the collections of the Scottish Agricultural Museum. Ricks were thatched and roped to protect them from wind and rain, employing techniques once used in the thatching of vernacular buildings. The ricks were dismantled, one at a time, and taken to the barn for threshing.

The comparatively small size of Scottish barns arises from this practice rather than from small-scale production of cereal crops. Scottish writers could be quite scathing in their criticism of the English practice of storing crops in the barn:

> Nothing can be more absurd, than the enormous barns usually attached to all the great farms in England. Grain in the straw keeps infinitely better in the open than in close barns.[6]

As the sheaves were brought into the barn from a nearby stackyard, rather than directly from the field, there was no need for large, cart-sized openings for forking in off the cart, though there is evidence of such in the south west and in the north east. Sheaves could be brought in through a small doorway. In later machine-threshing barns, sheaves could be pitched in to the upper floor or, where topography permitted, taken in directly from higher ground. The most sophisticated nineteenth-century steadings had canopied standings, to protect the sheaves during loading into the barn.

In Orkney and Shetland, and in other areas of small-scale farming, small quantities of grain might be threshed by lashing against a rack or a stone which projected about 6 inches (15 centimetres) from the inside of the barn wall, at a height of about 3 feet 3 inches (one metre).[7] In general, however, the flail was used. Threshed grain was then winnowed, to separate the grain from the chaff. In exceptional cases, where no barn was available, both processes were conducted outdoors.

The functional requirements of flail-threshing, winnowing and temporary storage determined the form taken by barns. To prevent loss of grain, threshing floors were made of clay or timber (the latter fixed or movable), usually covering only part of the barn floor, to one side of openings in opposite side walls. An Orkney farm plan, reproduced by Omond (1911), shows a narrow clay strip, running the width of the barn, to one side of opposite doors. The strip was 3 feet 6 inches (1.07 metre) wide, with a slightly raised, bevelled surface.[8] The barn had to be of sufficient height to allow a flail to be swung; barns thus tended to be higher in the apex, if not also the side walls, than other farm buildings. The openings in opposite walls facilitated winnowing by creating a through draught; generally both openings were doors. In the barn at Newhall, Midlothian, an area 9 feet by 6 feet (2.74 by 1.83 metres), between doors, was paved, probably as a winnowing area. In exceptionally exposed districts, such as Shetland or the Outer Hebrides, a doorway plus a smaller opening opposite was the more usual arrangement.[9] Ventilation, to maintain a fresh, dry atmosphere for stored straw, was usually achieved by means of slit vents, often with differing splays on the inner side to take account of prevailing winds. In the west, triangular vents, with each side formed by a single stone, were also used. With a view to good ventilation, there was a preference for sites on the western side of the steading, the barn having a roughly north-south orientation. In major cereal-growing districts, there might be multiple barns, each compartment handling the processing of a specific crop – wheat, barley, oats and beans or pease.

Estate girnals were used by seventeenth- and eighteenth-century landowners to store grain paid as rents or the produce of lands which they cultivated directly. A group of these has been studied by Elizabeth Beaton in Wester Ross and Sutherland.[10] The so-called 'tythe barn' at Whitekirk, East Lothian, may belong to this class of building,[11] but the most remarkable is the barn at Simprim, Berwickshire, built in the 1680s by Sir Archibald Cockburn of Langton. This great barn or girnal is 210 feet 11 inches (64.3 metres) long by 24 feet 11 inches (7.6 metres) wide, and was once four storeys high. The ground floor, with five pairs of arched doorways, was used for threshing and the upper floors as granaries.[12]

Technological changes, which were to have a profound impact on barn design, first appeared in the eighteenth century, yet hand threshing persisted on small farms and crofts until the late nineteenth century. In the mid 1880s, the flail was still being used on a very few farms in Renfrewshire[13] and on small farms in the remotest districts of Lanarkshire.[14] Even in Midlothian, as late as 1878, it was said that 'the monotonous tap-tap of the barnman's flail may yet be heard'.[15] Further afield, on crofts in the far north and west, it continued to be used on at least part of the crop until well into the twentieth century.

The first technological change was the introduction of the winnowing machine, from Holland, by James Meikle in the early 1700s.[16] Though slow to be adopted, winnowing machines, or fanners, brought an end to the need for opposing doors; however, there might still be a need for access through the barn, from stackyard to barn, or from barn to straw yard – itself an eighteenth-century creation. More radical alterations came with the threshing machine, experimented with from the 1730s, and perfected by James Meikle's son Andrew in 1785.[17] Before long, fanners were being incorporated into the larger threshing machines.

Threshing machines required a power source and, for all but the smallest machines, a higher barn, having a platform or upper floor at which level the machine was fed, and a lower floor, where it discharged. Evidence of conversion, from hand- to machine-threshing and winnowing, can still be seen in some older barns. At Eaglescairnie Mains, East Lothian, the barn walls have been heightened and a horse engine built where the outer door was once situated. At Leys, Stirlingshire, the mound for an open horse walk partially obstructs one of two opposing doors.[18]

Housing for the power sources which drove threshing machines created its own distinctive structures. All of them – initially wind, water and horse or oxen[19] power – had been used before for other purposes. Wind power was expensive to install and not very reliable; except in Orkney, only large arable farms made use of it. Wind-powered threshing machines were 'becoming very common' in Berwickshire about 1810[20] and there were still seven of them in East Lothian in the late 1830s, the last of which was replaced, by a steam engine, in 1853.[21] From the Lothians northwards there was a scatter of sites up the eastern lowlands of Scotland, as far as Easter Ross. Yet the early interest in wind power was not to last; by 1841 it could be asserted that 'another windmill will never be erected in Scotland for farming purposes'.[22] Good examples survive at Dunbarrow, Angus, Hilton of Turnerhall, Aberdeenshire, and at Shortrigg and Mouswald Grange, Dumfriesshire.[23]

A far smaller wind engine was used in Orkney – possibly derived from those which were driving butter churns in the 1820s. Unlike the mainland windmills, these were simple, inexpensive structures, generally no more than a flagstone platform on two rectangular uprights. The best preserved of these is at Sanquhar, Westray, last used in 1950.[24]

Water power was more reliable and more widely used. At some sites the wheel had no housing, at others a lean-to was built over it. Those intending to use water power might go to great lengths to bring water to the threshing barn. At Crowhill, East Lothian, the wheel had to be sited some 82 feet (25 metres) below the steading, with a vertical shaft linking it to the machine.[25] If necessary, the threshing barn was built at a distance from the rest of the steading, as at Edenmouth, Roxburghshire.[26] Quite a number of water wheels are still in situ, though not always immediately apparent. A water wheel and its associated threshing machine have been recorded by Royal Commission on the Ancient and Historical Monuments of Scotland (RCAHMS) at Southwood, Lanarkshire.[27] By the late nineteenth century turbines were also being used.[28]

The buildings which housed horse engines show infinite permutations of form, roof construction and roofing materials (fig. 21.1).

Figure 21.1 Cairn Farm, Fife; horse-engine house. SLA, FBS 3517.

A study by Bruce Walker identified round, square, diamond-shaped, pentagonal, hexagonal, septagonal, octagonal and elliptical forms. Pantile or slate were the commonest roofing materials, though roofs of reed thatch were found at a few sites, such as Flatfield, Perthshire.[29] Pantiles and slate also predominated elsewhere in Scotland, but other materials, such as flagstone (Tresness, Orkney)[30] and, as original roofing, corrugated sheet metal (Leckmelm, Ross and Cromarty),[31] have also been located by the Scottish Farm Buildings Survey. In the east of Scotland, full-height openings were left between sections of wall in order to maximise ventilation; in the south west the openings are frequently confined to the upper walls, though the slenderest of roof supports – cast-iron columns – also occur in that region, in a group of sites near Castle Douglas, Kirkcudbrightshire.[32] In a few cases there are upper floors to the horse-engine house, probably once containing granaries; the example at Kersmains, Roxburghshire, has similarities to plans published by Beatson in 1798.[33] In only a very few instances has machinery survived; one such site, at Wester Gallaberry, Dumfriesshire, has been recorded by RCAHMS.[34]

Open horse-walks, consisting of a raised, circular platform, could be constructed at far less cost, and helped threshing machines to spread to smaller farms in the later nineteenth century. This was made possible by running the drive shaft underfoot rather than overhead; in late nineteenth-century sites light section malleable iron was used for draw bars in place of heavier timbers. Horse-walk platforms are still very common, particularly on small farms and crofts: there are five in a row at New Ortie, Sanday, Orkney.[35] Complete gear trains seldom survive, though two, at Conisby, Islay, and Ballyalnach, Perthshire, have been recorded by RCAHMS.[36] Occasionally more than one power source was provided for. At Stacks, West Lothian, there was both water and horse power;[37] at Courthill, Berwickshire, water and steam power.[38]

Steam power was a comparatively late introduction, the first Scottish steam-powered threshing machines having been installed in East Lothian in 1803 and in Fife in 1805.[39] Expensive to install and reliant on accessible coal supplies, steam power was slow to take on. During the 1830s it became well-established in the Merse (Berwickshire), the Lothians and Fife, but its greatest impact came in the decades around the mid century. By the 1880s, the installation of fixed steam machinery was almost a thing of the past.

Engine and boiler design became more efficient as the century progressed. In the 1850s, a typical combination comprised a vertical steam engine and a horizontal, egg-ended boiler made of iron plates, set above a separate furnace. By the late nineteenth century, where a large power source was needed, they had been replaced by flat-ended Cornish and Lancashire boilers, with internal furnaces and boiler tubes. More usually, on farms where only one to six horse-power was needed, there were compact vertical boilers, with inverted or horizontal engines.[40]

The associated buildings consisted of engine and boiler houses, usually combined, coal store and chimney. A well might be included, as a source of boiler water (as at Scar, Sanday, Orkney),[41] with the engine itself used to pump water. Engine houses were built to a high standard, often with plaster and lath linings to walls and ceilings. Chimneys were usually built in brick, on a stone base, though occasionally in stone only, as at Baillieknowe, Roxburghshire.[42] Chimneys of round, square and polygonal section have been identified. At Elie Home Farm, Fife, the chimney is concealed within a Scots Baronial tower;[43] at Bolsham, Angus, a threshing windmill tower was re-used.[44] In no case, however, is a threshing steam engine and boiler known to have survived in situ, though re-used egg-ended boilers can occasionally be seen around farms. A horizontal threshing machine engine, from Easter Ross, is in the collections of the Museum of Scottish Country Life.

Fixed steam-powered threshing belonged principally to home farms and to the larger cereal-growing farms of the east of Scotland. It became quite common in the lower Tweed basin, on both sides of the Border. It dominated Mid- and East Lothian, with a scattering of sites further up the Forth: in East Lothian, it already drove 80 threshing machines by the late 1830s, and 185 out of 373 by November 1853.[45] In 1845, Robert Ritchie noted that 'so rapid has been the extension of steam power to farms in this vicinity that, from the fine elevations round Edinburgh, more than 100 steam-engine stalks or chimneys may be observed as the landmarks of the farm, and giving a peculiar feature to the landscape'.[46] In Fife, numbers grew rapidly from the 1830s, with steam power becoming standard on most of the larger farms.[47] Further north, it was scattered through the east coast Lowlands from Perthshire to Inverness-shire. In Easter Ross, by the late 1870s, some 18 chimneys could be seen from one elevated point, some of recent origin, some established more than 25 years previously.[48] North again, through the coastal rim of Sutherland and the north-east corner of Caithness, there was a further, thin scatter. Orkney had a few more steam-powered sites, including two on Sanday and one on Stronsay.[49] The most northerly of all was on Bressay, Shetland, on a farm operated by a Berwickshire man, John Walker.[50]

The era of fixed steam-powered threshing came to an end in the late nineteenth and early twentieth centuries. In 1908, MacDonald reported that 'since about the year 1880 gas- and oil-engines have, to a considerable extent, superseded steam for driving farm machinery.'[51] Doubtless this sweeping statement excluded portable steam engines and threshing machines which had enjoyed considerable success from the 1860s onwards. The oil engine in particular enjoyed phenomenal success, and by the early twentieth century manufacturers were turning out thousands each year.[52] Little was needed in the way of housing, with either existing buildings re-used or a small brick, timber or corrugated iron hut built. From the 1920s, the availability of small, inexpensive petrol or diesel engines, such as that produced by Listers, brought the internal combustion engine to a much wider market; by this time tractors could also be used to drive fixed threshing machines.

The full-sized 'Scotch' threshing machine needed a two-storeyed building or, at least a building which was part lofted: an upper floor with space for the sheaves from one or two ricks, plus the upper part of the machine; a lower floor for the fanners and other parts of the machine. Where the slope of the ground permitted, carts could be backed up to doors in the upper floor; some barns in the south west allowed them to be driven in and through. Elsewhere, an opening was made at upper-floor level, through which sheaves could be pitched. Separate access to the ground floor allowed sacks of grain to be carried to the granary. Besides grain, the threshing machine discharged chaff, into a small compartment, and straw, into a straw barn which usually occupied the full height of the barn range. To help it on its way, in the later nineteenth century overhead conveyors were sometimes fitted, which could discharge at a number of points along the length of the straw barn.[53] Straw barns were usually windowless, but retained the slit vents which had characterised flail-threshing barns.

In addition to straw conveyors, other equipment might be run from the same power source: corn bruisers or breakers, or hummellers to remove the awns from barley grains. The arrangement at Househill Mains, Nairn, was particularly sophisticated. Here a steam engine, installed c. 1880, drove the threshing machine, a chaff cutter, root pulper, corn crusher and a 'corn blast', which blew grain through 60 feet (18.29 metres) of piping to the granary. The entire system was supplied and installed by Morton of Errol, Perthshire.[54]

Gradually, smaller machines, which could be housed in single-storeyed buildings, became available. By the 1830s, consideration was being given to hand-powered machines, without fanners or straw walkers, though it was some time later before they achieved success and widespread availability. Machines such as the Banff Foundry's cast-iron 'Tiny' could be used in existing flail-threshing barns without the need for any modification to their fabric, and with no difficulty in removing them at the end of a lease.

An alternative means of avoiding investment in fixed buildings and machinery was to use a 'portable' steam engine and machine. A portable threshing machine was the obvious accompaniment to mobile steam engines, yet there was little demand for either in Scotland where, until the 1860s, fixed machines continued to dominate, with flail threshing persisting on small farms.

The 1860s were to mark a turning point in central and south-western Scotland. An Ayrshire man, writing in 1882, reported that:

> A decided change has ... taken place during the last twenty years, through the introduction of the portable engine and mill, which in course has been superseded by the traction engine and improved mill. The use of the latter has become so general, and is so advantageous, that in many cases good horse-power, and even fixed steam threshing machines have been altogether discarded and the entire crop of the respective farms threshed by the travelling mills ...
>
> ... it may very naturally be inferred that the abolition of fixed mills is a mere question of time.[55]

There were several travelling machines in Galloway by 1875;[56] 'in some cases', in Wigtownshire, by 1885, 'no mill is put up at all, and, instead of a barn and straw house being built, merely a large straw house is put up and the whole threshing done by the travelling threshing machines. Where the large straw houses are, there is a hole in the roof through which the elevators drop the straw into the centre of the house'.[57] Even in the arable east, travelling outfits were gaining ground but demand for fixed machines in the north east and far north held up well enough for a number of north-eastern firms to continue to make these, as well as travelling mills, until after World War II.

Further developments during the twentieth century brought radical changes in the type of housing needed for grain-handling. In 1932 Lord Traprain bought a Clayton & Shuttleworth combine harvester, the first in Scotland, for use on his Whittingehame Mains and Cairndinnis farms, East Lothian. Two years later he acquired a second, Massey Harris combine.[58] Not until the late 1940s and 1950s were combines introduced in significant numbers, but their arrival made the rick yard and the threshing barn obsolete; grain was now threshed in the field. To reduce the high moisture levels – as high as 30 per cent – the bagged grain from the combine had to be dried artificially. In addition to the traditional method of kiln drying, proposals for reducing the moisture level of threshed grain by artificial means had been made as far back as the 1740s;[59] in the 1840s, one East Lothian farm used waste heat from its steam threshing engine to heat a grain drying loft via metal pipes.[60] But the general need to dry large volumes of newly-harvested grain was a problem of quite different proportions. A coke-fired vertical batch dryer was installed at Cairndinnis in 1932, to specifications from the Institute of Agricultural Engineering, University of Oxford. A second dryer, at Whittingehame Mains, was fitted up soon after.[61] By 1936 grass- and grain-drying plant was being advertised in Scotland.[62] Bagging soon gave way to bulk collection, which in turn led to further changes at the steading.

Several types of dryer had been introduced by the early 1950s, each with a heat source, a fan and a drying area. Other components then included a receiving tank for bulk grain, augers, elevators and storage bins. On some farms these have been accommodated by gutting former threshing barns, now with the characteristic roof-top housing for elevator gear. Elsewhere, wholly new structures have been erected, with steel or reinforced concrete frames clad in asbestos or metal sheets or brick. Thus, every one of the buildings which dealt with grain a century ago is now obsolete and, all too often, redundant.

Granaries
Threshed grain awaiting winnowing by hand had to be stored, as had seed corn, human and livestock food or grain for disposal at market. In post-medieval times, the simplest form of granary was a temporary structure, consisting of thick straw ropes surrounding a heap of grain and, when built outdoors, thatched in the manner of a rick. Outdoor granaries of this type – known as bykes – were noted by Bishop Pococke in Caithness in 1760, are also recorded in Orkney, and are thought to have been last used in the early twentieth century.[63] Wight observed some at Ratter, Caithness, in the 1780s:

> The method of preserving bear [barley] over year, when the price is low, is singular. Upon a spot of dry ground a foundation is made with dry straw, surrounded with ropes and straw. Ten bolls of bear are poured in, and the heap is carefully covered from the weather ... ; this is done from the want of proper houses.[64]

A similar arrangement, with grain stores built within the barn, seems to have operated in Perthshire until at least the 1750s. A contributor to the *Farmer's Magazine* of 1807, describing the agriculture of the county some 50 years previously, stated that fanners were hardly known then, so that when there was not enough wind for winnowing:

> grain in the chaff was put into what were termed 'straw sacks'. These were formed by coiling up a rope of straw, four inches thick, in some part of the barn, and throwing the chaffy grain into its area, as the rope was laid around. There it remained often for months together, till a fair wind came; during which interval some person was constantly in waiting, on the outlook for the favourable gale; and when it did spring up, every other operation was laid aside, and the opportunity seized.[65]

In eighteenth-century Galloway straw ropes or overlapping rings of sheaves were used to retain heaps of threshed grain within – and possibly outwith – the barn.[66]

Prior to the late eighteenth century, there is little to indicate that

Figure 21.2 Baillieknowe Farm, Roxburghshire; grain range. SLA, FBS 4101.

there were buildings in the steading used primarily as granaries. What little evidence there is suggests that grain was stored in lofts – in, or adjacent to the barn, around the steading or in the house – or, in small quantities, in kists. As late as 1811, fewer than one in five West Lothian farms had granaries.[67] Even in East Lothian, one of the foremost arable districts, Somerville noted in 1805 that few farmers had granaries, forcing them to send their grain, seed corn included, to market.[68]

The solution most widely adopted, from the late eighteenth century onwards, was to place the granary over a range of cart bays (fig. 21.2). This arrangement had several advantages: the grain was well ventilated, it was protected from rising damp and, to a degree, from vermin, and bagged grain, for market or for sowing, could be unloaded through trap-doors into carts.

Barclay of Urie's home farm, Kincardineshire, may have been one of the first to adopt this arrangement. Here, by 1775, one granary was situated over 'three voids for wagons, ten feet each' and a feed house, with a second granary over further sheds.[69] In practice, this combination was not always adhered to, and instead granaries were placed over stables or even cattle sheds, despite the risk of tainting stored grain. In 1778, the steading at the Pow of Drummond, Perthshire, had a large cattle shade with a loft over 'for holding grass seeds, corns or drying of lint-seed in the bowes.'[70] Stephens (1844) advocates this as a means of giving extra shelter to open cattle courts, by virtue of the extra storey.[71]

Seldom were the side walls of full height; floor space, on which grain could be spread, was of greater importance than height. To protect the grain from vermin, the junctions between walls and floors were well sealed, usually with a skirting; the walls were plastered or occasionally lined with timber and the wallheads were beam-filled. Openings in the wall, with fixed or adjustable vents, helped to maintain an equitable temperature and humidity.

Where there was no direct communication from the barn, bagged grain was carried in via an external or internal stair; a lockable door at the stair-head might incorporate a cat door. Grain left the granary by the same stair, through hatches in the floor or by a doorway over a cart bay opening. By the 1820s and 1830s, the cart shed and granary range was becoming a characteristic feature of Scottish steadings, varying in size from a single bay unit on small upland and western farms to as many as ten or twelve bays in East Lothian.

Grain-drying kilns

Though subject to the vagaries of climate, a degree of drying could be achieved by stooking crops in the field prior to stacking them at the steading. However, grain for milling had to be dried further, in a kiln.[72] An early account (1521) describes the process:

> ... a house is built in the manner of a dove-cot, and in the centre thereof, crosswise from the wall, they fix beams twelve feet in height. Upon these beams they lay straw, and upon the straw the oats. A fire is then kindled in the lower part of the building, care being taken that the straw, and all else in the house, be not burnt up.[73]

The kiln might have a secondary purpose, in drying bere which had been malted for use in brewing, or even a third, in drying flax after retting.[74] Until the eighteenth century, over most of Scotland, these kilns were associated with farm steadings rather than, as later, with mills.

The exception to this rule was south-east and east-central Scotland, where there is evidence of a much older association between mills and kilns. A plan of part of the Duddingston Burn, Edinburgh, dated 1563, depicts two mills, each with a detached, cylindrical, conical-roofed kiln.[75] Two similar kilns, both with thatched roofs, are shown beside the Heugh Mills in Slezer's view of Dunfermline Abbey, Fife (c. 1693).[76] There is further, pre-1760s evidence of mill kilns from the Borders, Lothians and Fife plus later, cartographic evidence of surviving detached, cylindrical mill kilns from lowland Angus.[77] Further investigation may reveal a wider distribution. Farm kilns might not have been totally excluded from these areas, though indications that thirlage here applied to both mill and kiln suggest that this was, indeed, the case.[78]

From elsewhere in Scotland there is ample evidence of farm kilns. In spite of regional variations in design, they had a number of features in common. Firstly, a horizontal or upward slanting air inlet, within which was set the fire, usually peat-fuelled, which heated the kiln. Secondly, an under-chamber, generally bowl-shaped but also rectangular in Shetland and Orkney, across which rested the kiln platform, much as described by Major in 1521. In addition, most kilns had a superstructure of walls and thatched roof, though a temporary covering of cloth sufficed in some areas.[79]

Within this broad uniformity, a number of regional variations are to be found in the design and positioning of the building. In Orkney, Caithness and southern Shetland kilns were appended to and entered from the threshing barn, where there might also be storage for fuel and dried grain.[80] In the Outer Hebrides, the northern Inner Hebrides and a few mainland sites, the kiln bowl was set into a platform which occupied the entire width of one end of an oval or sub-rectangular kiln-barn.[81] In Shetland, the kiln could consist of a small, rectangular structure in one corner of a barn.[82] Elsewhere, as with mill kilns, farm kilns were self-contained cylindrical structures, with or without appendages, and often sited on a slope to allow the flue to enter the kiln bowl from below. The bowl and the kiln platform were as noted previously.

The shift from farm to mill had begun in the second half of the eighteenth century. By the 1790s and early 1800s, the change was noted as having taken place in Ayrshire,[83] Stirlingshire,[84] Perthshire [85] and Angus [86] but not, as yet, in Aberdeenshire [87] or Arran.[88] By the mid nineteenth century, re-structuring through rebuilding or clearance had all but eliminated farm kilns from the Scottish mainland – Caithness only excepted. Here, and in the Northern Isles, their use in malt-drying may have prolonged their life.[89] With this change came an improvement in kiln-bed materials. Straw and timber were superseded, first by perforated ceramic tiles, then by wrought-iron sheets and cast-iron plates.[90]

Though very much the exception to the rule, a few farms retained grain-drying facilities in the form of improved, rectangular kilns similar to those which were being appended to mills. Mid-eighteenth century dates have been suggested for improved kiln-barns at Rothiemay, Banffshire, and Sandside, Caithness;[91] the rectangular kiln-barn at Ballindalloch, Banffshire, is thought to date from about 1800; a rectangular kiln formed part of the 1826 plans for the glebe steading at Ardersier Manse, Inverness-shire.[92] Erskine of Mar's innovative steading at Lornshill, Clackmannanshire, dating from the 1790s, included a 10-feet (3.05-metre) square kiln, linked to the granary by a gangway.[93] Another late eighteenth-century steading, at Primrose, Fife, also seems to have had a kiln.⁹⁴ Nor was this exclusively a landowner's idiosyncrasy: in 1811, at Wester Fintray, Aberdeenshire, the tenant, a Mr Walker, was reported to have:

> lately slated a kilnbarn, which he has built near the west wing of his court of offices. This contains a granary and a loft in one end and an iron-plated kiln in the other, which dries four bolls, or three quarters of corn at a time. It is very complete of its kind, and cost above L.150.⁹⁵

Appearances may be deceptive, however: a mid nineteenth-century kiln, at Drem, East Lothian, was not for grain-drying but for drying potato starch.⁹⁶

Continuity with past practice was most marked in the Northern Isles, where nineteenth- and twentieth-century improved, rectangular kilns have been noted in Dunrossness, Shetland, on the Orkney Mainland and on the islands of North Ronaldsay, Sanday and Westray.⁹⁷ A few farms around Wick, Caithness, had large, bottle-shaped kilns, apparently another line of development from traditional circular forms.⁹⁸

Milling on the farm
Grinding by hand, on the quern, survived in the more remote parts of the Highlands and Islands until the early twentieth century. The quern might be used in the dwelling or on a dedicated site in the barn.⁹⁹ In nineteenth-century Shetland and Lewis, small horizontal-wheeled meal mills were associated with individual settlements, though not located at the steading. In earlier times such mills seem to have operated throughout much of Highland and Island Scotland, though the usual arrangement was for all grain to be ground at a central, estate owned mill, to which particular lands were tied or thirled.¹⁰⁰

The gradual abolition of thirlages, from 1799 onwards, created scope for milling at the steading, but only rarely was this option pursued. During the nineteenth century a few steading-based grinding mills were to be found in the largest and most prestigious sites, especially where steam engines were used to drive threshing machines. There were some in Berwickshire by 1809, probably water-powered,¹⁰¹ and one or two steam-powered examples in Angus by 1813.¹⁰² Kilmux, Fife (1833), had two pairs of grinding stones for making oatmeal,¹⁰³ whilst Yester Mains, East Lothian, had, by 1845, a variety of steam-powered machinery including a pair of stones for grinding flour.¹⁰⁴ In the far north, Scar, on the Orkney island of Sanday, had its integral steam-powered meal mill and kiln.¹⁰⁵

FODDER CROPS

Hay and grass

Grasses and other wild plants, cut from river meadows, wetlands or wastes, may have been dried for hay over many centuries, but hay made from sown grasses was an eighteenth-century innovation. In general, hay was stored outdoors, in round or rectangular stacks, but there are exceptions to this rule.

Hay might also be stored in buildings close to the cattle or horses which fed on it. Lofts above byres, stables or turnip sheds were widespread but especially common in upland areas, where the loss of ventilation was made up for by exposure (fig. 21.3). Loft storage had the additional virtue of allowing direct, gravitational supply into hay and straw hecks. Where the advice of agricultural writers was heeded, as in the south east, the better nineteenth-century steadings were provided with a hay house adjacent to – but on *no* account above – the stables. These usually had wide doors for unloading from carts and, preferably, direct access into the stables.

Difficulties in securing a hay crop were especially acute in the north and west; in Argyll, in the 1790s, hay was stored in both stacks and barns.[106] Robertson, writing of Inverness-shire in the early 1800s, noted that:

> hay barns are often built open, with gable ends only as high as

Figure 21.3 Appleby Farm, Wigtownshire; bank barn. SLA, FBS 5514.

the side walls, and furnished with stakes and intertwined broom or brushwood.[107]

A group of these survive in the Kyle of Lochalsh area, one of which, in the ownership of The National Trust for Scotland, has been recently restored. These hay barns could be fitted with sloping sparred racks, on which hay or cereal crops could be spread out for drying. The same principle was applied on a grand scale at the Duke of Argyll's steading at Maam, Inveraray, Argyll, and was still being advocated as recently as the 1940s.[108] Late versions of the drying barn, with wooden louvres, were then in common use from Argyll to Ross-shire and can still be seen in Wester Ross.[109] Stripped down to its functional minimum, it formed the basis of the Richmond drying rack, invented in the 1890s by a Perthshire farmer and used to dry both hay and sheaves of grain.[110]

In parts of western Scotland, from Galloway to Argyll, there is as yet inconclusive evidence of conventional barns being used for hay storage, though there are features which suggest that some of these barns were in part, if not exclusively, used to store hay: large, cart-sized openings, unlike the small doorways of Scottish threshing barns; slit or triangular vents throughout the building; and in some cases the absence of any power source. The picture is, however, complicated by the late nineteenth-century use of travelling threshing machines, which required only straw barns rather than the more usual powered barn; this might provide a partial, alternative explanation.

Figure 21.4 Thurston Home Farm, East Lothian; hay barn. SLA, FBS 5414.

Figure 21.5 Glenochar Farm, Lanarkshire; hay barn. SLA, FBS 4704.

During the mid nineteenth century a characteristic building type began to emerge: open-sided apart from stone, brick, iron or wooden columns, sufficiently spaced and of sufficient height to admit a loaded cart or hay bogie (fig. 21.4). A particularly fine example in rubble and slate at Cally Mains, Kirkcudbrightshire, is some 156 feet (47.6 m) in length and 23 feet (7 m) wide.[111] In 1878, a report on farming in Dumfriesshire noted that:

> a great number of sheds for hay have been erected in different parts of the county ... The pillars are of timber, generally of home growth, and the roof is slated. They are usually made 12 feet [3.66 metres] high and about 18 feet [5.49 metres] wide ... The cost of the erections is about £2 for each ton of hay which they are capable of holding. Less progress has been made with these buildings in Ayrshire, but farmers are giving attention to them. They are made wider in Ayrshire than in Dumfries-shire.[112]

In Upper Clydesdale there is an interesting local variant consisting of slated roof and staved wooden sides (fig. 21.5).[113] By the 1870s, pre-fabricated metal-framed and roofed sheds were being marketed by firms such as A & J Main and P & R Fleming, both of Glasgow. Main's 1882 catalogue includes a wide selection of such buildings; at Leckmelm, Ross and Cromarty, system-made metal roofs had been used to cover an entire steading.[114] Within a few years, open-sided metal hay, straw and

Figure 21.6 Wellside Farm, Moray; tower silo. SLA, FBS 5819.

sheaf sheds had been widely adopted, especially in the dairying districts of west central Scotland.[115]

The wet summers of the 1880s, which had contributed to open-sided metal hay sheds' popularity, also encouraged experiments into alternative forms of winter feed preservation. During the mid 1880s, Britain was gripped by ensilage fever.[116] In Scotland, many experimental silos were built or adapted from other structures, but interest soon subsided. Not until the end of World War I were silos built again, this time as cylindrical towers, usually of reinforced concrete, based on North American precedents (fig. 21.6).[117] Before the outbreak of World War II, these too had fallen out of favour, as silage pits gained popularity as a cheaper alternative.[118] Significantly, the model 1930s steading at Rankeilour Mains, Fife, centred not on a tower silo but on an extensive covered pit.[119] The late 1930s also saw interest in grass-drying by artificial means. In 1936, a large Billingham (ICI) grass-drying plant was constructed at Strathallan Castle, Perthshire,[120] and by the following spring a Ransomes' plant had been fitted up at Spotts Mains, Roxburghshire.[121] The war brought development to a halt[122] and, although the idea was promoted again in the late 1940s,[123] its prohibitive cost prevented it from catching on. Since the late 1950s, however, tower silos have made a come-back, principally as concrete stave or vacuum-sealed metal silos for storing high-moisture grain, destined for cattle feed.

Roots

Turnips and other root crops were a key element in the crop rotations of improved farming. Their principal value was as a means of feeding livestock over winter and thereby returning nutrients to the soil through their manure. A proportion of the crop was consumed in the field, by sheep, the remainder at the steading, where it was cut up and fed to cattle. Between field and steading the roots might be stored in clamps before being carted to the root house, a few loads at a time.

Root houses were plain, windowless structures, generally with a large enough entrance, or entrances, to enable a loaded box cart to be backed in. Typical positions were along one side of a byre or, in the case of courts, between or at the ends of court ranges. Where topography allowed, they were situated so that there was a fall from the point of unloading into the root house, as at Boon, Berwickshire. In the 1860s, the Duchess of Atholl's farm at St Colme, Perthshire, had an ingenious system for washing turnips, involving a power-driven root slicer, supplied from root houses via inclined water chutes and an iron grating.[124]

Steaming and boiling

Steamed or boiled food is particularly associated with pig-feeding, but during the early to mid nineteenth century there was a vogue for feeding steamed food to all manner of livestock, including horses. According to one writer, in the early 1830s, 'the preparation of roots and grain for food, by steaming or boiling, is an improvement in the mode of feeding our domestic animals which cannot be too generally extended'.[125] Contemporary text books and pattern books show steaming or boiling houses with cisterns, furnaces and boilers, linked by steam pipes to rows of steaming vessels. No surviving examples of this type have been located. Later boiling houses had cast-iron boilers, either with a self-contained furnace or inset in a casing over a furnace. Occasionally, these are still found in situ. The food prepared in these later boilers was primarily for pigs and poultry, as is reflected in their locations. Poultry were sometimes housed in the apartment next to the boiler and its flue, where they could benefit from otherwise wasted heat, as at Bridgend, Linlithgow, West Lothian. Occasionally, steam boilers for threshing machines and for feed shared the same house, as at Morphie, Kincardineshire, and in Morton's published plans for Swanston, Edinburgh.[126]

Whins

The young, crushed shoots of whins (gorse) were a useful supplementary food for horses. The whins might be crushed by means of flails, mallets or on a mill. The mill, usually powered by horses, consisted of a central pivot, linked to a wooden pole which passed through a stone edge-runner or bevelled roller which ran around a stone-bedded rink. The horse was yoked to the outer end of the pole. Whin mills were in use in Perthshire by 1778,[127] Aberdeenshire by 1794,[128] and at Hillhouse, Kirknewton, Midlothian, by 1795.[129] In more recent times, a number of other sites have been noted in Aberdeenshire.[130] One of the last of these, at Quittlehead, Lumphanan, was recorded in detail by the RCAHMS.[131] A whin-mill stone from Wester Feabuie, Culloden Moor, Inverness-shire, was donated to the Scottish Agricultural Museum by Mr R H E Fraser in 1969, and one has been set up at the Rowett Institute, Aberdeen.

At least one whin mill was water-powered: machinery for crushing whins was added to a snuff mill at Woodside, Aberdeenshire, between 1764 and 1771.[132]

Other feedstuffs

Among other foods, cattle were fed chopped straw (known, confusingly, as chaff) and chopped hay. Horses were fed bruised oats and sometimes chopped hay or straw. Machines for preparing these foods were available by about 1800;[133] these might be located in the straw barn, dressing barn, hay house, granary or in the stables themselves. Already, by 1812, there are references from Dumfriesshire and Dunbartonshire to chaff-cutters and corn-bruisers powered from the same source as the threshing machine.[134]

Oil cake (the residue of pressed seeds) and cotton cake, already popular by the mid nineteenth century, required further equipment to break up the slabs in which form the cake was supplied. The best mid century farms had separate mixing rooms, where feed was prepared, often making use of power from the steam engine or water wheel which drove the threshing machine. The description given by Lord Kinnaird, in 1853, shows just how sophisticated feeding arrangements could be. A steam engine drove:

> 1st A turnip washer, from which the turnips are taken by elevators to the cutters, falling, when cut, into the tubs and wagons ready, when mixed with the chaff &c., to be conveyed away to the stock.
> 2nd Chaff-cutter.
> 3rd Corn-bruising machine.
> 4th Cake-crusher.
> 5th Pair of millstones.
>
> The prepared food from the last 3 machines falls into a store, the key of which is kept by the farmer; the steam from the engine is employed to heat a kiln for drying grain, heating water, and steaming food for horses and pigs. The only assistance the man who attends the engine has is that of a woman, who supplies the turnip-washer with turnips, and who besides feeds and has charge of a hundred sheep on boards.
>
> The cattleman, having the food thus prepared to his hand, has nothing to do but to put it into the waggon, which, by means of a railway, is easily conveyed to all parts of the building ... .[135]

New types of machine, for bruising and breaking grain, came into use from the later nineteenth century. Plate mills, roller mills and, from about 1940, hammer mills, all required power drive, which was supplied by oil or petrol engines, latterly by electricity.

POTATO HOUSES

The harvested potato crop had to be kept on the farm until sold, consumed or re-used in the following year's planting. During the first half of the nineteenth century, potatoes were stored outdoors, in pits (earthen clamps) and indoors. Potato houses, with wide doorways for cart access, were a feature of improved steadings in eastern arable districts from Berwickshire to Sutherland.

In Roxburghshire and Selkirkshire (1798) some potatoes were stored in pits, 'but most people have houses, where they can be stored in safety, by laying dry sand or saw-dust on the floor, stuffing the sides of them with straw and covering them with it or the chaff of oats'.[136] In Dumfriesshire, in 1812, it was reported that 'houses for storing potatoes may be long and narrow'; if wider, stores were subdivided with wooden partitions.[137] Opinions varied as to the relative merits of pits or houses. Alexander Fairbairn (Drum House, Edinburgh) proposed a straw-thatched house, its floor some 4 feet 6 inches (1.37 metres) below ground level, and well drained. Air vents were to be left in the walls for when needed.[138] A more detailed account of the construction of potato houses is given by Singers (1803):

> A potatoe [sic] house should be erected in a dry bank, facing southward, and high in the north end, where the potatoes are thrown at once out of the cart into the upper door, and thence descend into the body of the house. It may be as long as convenience requires, but should be only 6 feet [1.83 metres] wide at most, and not more than that in the faced wall, otherwise the increased body of potatoes take a heat about the centre. If there be suspicion of under water, as the house ought to be dug out of gravel or sandy earth, and the bulk of it to be under the level of the surface, a drain should be laid all round, at the bottom, where the wall is faced up. This faced wall should be built with stone and lime; the couples and roof should be made of oak, or very red durable wood. The gable below must be all stone and lime and also the small gable above. In this upper gable the door is wide and low, and both doors are very close.
>
> ... a house 6 feet wide, 5 deep and 20 long [1.83, 1.52 and 6.10 metres respectively], holds 60 bolls, filled to the eaves.[139]

Evidence to the Highland and Agricultural Society, gathered during the potato blight of the 1840s, indicates that potato houses were then to be found in both east and west Scotland and that it had been customary to fill them to a depth of 6 to 9 feet (1.83–2.74 metres). One of the largest, at Craiglockhart, Edinburgh, occupied a building measuring 18 by 40 feet (5.49 by 12.19 metres). Other stores, of a more improvised kind, occupied lofts, cellars and spaces partitioned off within barns, but none was able to protect the potato crop from blight. This was to be a turning point, beyond which indoor storage rapidly fell out of favour. MacDonald (1908–9) noted that deep storage in potato houses had been common before mid century, but that it had proved impossible to preserve them in such conditions thereafter.[140]

Some marginal areas retained the practice for longer. Subterranean, stone-lined stores have been noted at Migvie and Blackhills, both in Glenesk, Angus. These had low entrances, blocked openings in the opposite ends of their roofs and a covering of earth and turf.[141] Small, now roofless, stone-lined structures, excavated into hillsides, have been identified in Glen Lednock[142] and on Loch Tayside,[143] both Perthshire, and have been identified as potato houses of the Glenesk type. From Papa Stour, Shetland, there are memories of turf-built 'tattie holes', shaped like corn ricks, as well as recent (1967) evidence of thatched, part-buried stone-built 'tattie hooses'.[144]

The twentieth century brought a revival in indoor storage. By 1902, farmers in Ayrshire, Dunbartonshire, Bute and the Carse of Gowrie, Perthshire, had adopted crated storage for seed potatoes, based on practice in Cheshire.[145] New, model stores came into use in the late 1940s, but existing buildings were often adapted to serve this purpose.

NOTES

1. *The Acts of the Parliaments of Scotland*, 12 vols, Edinburgh, 1848–1875, James II, cap 37 (1452).
2. Thomson, D. *Painting in Scotland, 1570–1650*, Edinburgh, 1975, 34 and opposite 38.
3. Sinclair, Sir J. *General Report on the Agricultural State and Political Circumstances of Scotland*, 5 vols, Edinburgh, 1813, Vol 1, 27.
4. MacDonald, 1908, Vol 1, 177.
5. Cruickshank, F. *Navar and Lethnot*, Edinburgh, 1899, 13.
6. Sinclair, 1813, op cit, Vol 1, 17.
7. Fenton, A. Hand threshing in Scotland, *Acta Ethnographica Academiae Scientiarum Hungaricae*, 29 (1980), 353.
8. ibid, 380.
9. SFBS, Shetland and Uist.
10. Beaton, 1986, 133–51.
11. RCAHMS. *Inventory of East Lothian*, Edinburgh, 1924, 130–1 and fig 16.
12. Shaw, 1993, 197–204.
13. MacDonald, A. The agriculture of the county of Renfrew, *THASS*, 4th series 19 (1887), 25.
14. Tait, J. Agricultre of the county of Lanarkshire, *THASS*, 4th series 17 (1885), 42.
15. Farrall, T. On the agriculture of the counties of Edinburgh and Linlithgow, *THASS*, 4th series 10 (1877), 45.
16. Shaw, J P. *Water Power in Scotland, 1550–1870*, Edinburgh, 1984, 162.
17. ibid, 155–9.
18. Personal observation; SFBS, Denny parish, Stirlingshire.
19. Whyte, A, MacFarlane, D. *General View of the Agriculture of Dunbartonshire*, Glasgow, 1811, 73.
20. Kerr, R. *General View of the Agriculture of Berwick*, London, 1809, 161–2.
21. Stevenson, C. On farming in East Lothian, *JRASE*, 14 (1853), 291.
22. Bridges, R. On the thrashing machine, with reference to the construction of those employed in East Lothian, *THASS*, 2nd series 7 (1841), 51.
23. Douglas, G, Oglethorpe, M, Hume J R. *Scottish Windmills: A survey*, Edinburgh and Glasgow, 1984, 23, 19, 28.
24. ibid, 42; Leith, Spence, 1977, 4–7; Douglas, G, Oglethorpe, M. Orkney wind-engines: Drive for threshing, *Vernacular Building*, 7 (1988), 33–44.

25. Shaw, 1984, op cit, 161.
26. SFBS, Ednam parish, Roxburghshire.
27. National Monuments Record of Scotland, RCAHMS.
28. Carrick and Ritchie catalogue (c. 1895); MacDonald, 1887, op cit, 45, mentions turbine-powered threshing machines on Sir Michael Shaw-Stewart's Renfrewshire estate.
29. Walker, 1977, 60–2, 64.
30. SFBS, Lady parish, Orkney.
31. SFBS, Lochbroom parish, Ross and Cromarty; A & J Main catalogue, 1882, 12, copy at Scottish Life Archive, National Museums of Scotland (SLA).
32. Anderson, J. Gleanings from Galloway fieldwork, *Vernacular Building*, 12 (1988), 69.
33. SFBS, Roxburgh parish, Roxburghshire; Beatson, R. On farm buildings. In *Communications to the Board of Agriculture*, 2nd edn, 1804, 51–2 and plate 10; Other examples have been noted at Rosebank and Sandyhall, Perthshire, and Gortonlee (East Lothian); Fenton, Walker, 1981, 177.
34. Hay, G D, Stell, G P. *Monuments of Industry*, Edinburgh, 1986, 11–14.
35. SFBS, Cross and Burness parish, Orkney.
36. Hay, Stell, 1986, 11–12 (Conisby); NMRS (Ballyalnach).
37. Trotter, J. *General View of the Agriculture of West Lothian*, Edinburgh, 1811, Appendix 2 and plates.
38. SFBS, Nenthorn parish, Berwickshire.
39. Buist, G. On the agriculture of Fifeshire, *Quarterly Journal of Agriculture*, 51 (1840), 305.
40. MacDonald, 1908–9, Vol 1, 404, 414.
41. SFBS, Cross and Burness parish, Orkney.
42. SFBS, Stichill parish, Roxburghshire.
43. SFBS, Elie parish, Fife.
44. Fenton, Walker, 1981, 165.
45. *The New Statistical Account of Scotland (NSA)*, 15 vols, Edinburgh, 1845, Vol 2, 374: Stevenson, 1853, 322.
46. Ritchie, R. On the extended application of the steam–engine, or other impelling power of the thrashing machine to farm purposes, *THASS*, 3rd series 1 (1845), 142.
47. Buist, 1840, 304–5.
48. MacDonald, J. On the agriculture of the counties of Ross and Cromarty, *THASS*, 4th series 9 (1878), 171.
49. SFBS, Cross and Burness parish, Orkney; Pringle, R O. On the agriculture of the islands of Orkney, *THASS*, 4th series 6 (1874), 25, 28, 33.
50. Skirving, R S. On the agriculture of the islands of Shetland, *THASS*, 4th series 6 (1874), 256.
51. MacDonald, 1908–9, Vol 1, 424.
52. ibid, 429.
53. See Ross, J. Arrangement of machinery at a farm steading, *THASS*, 5th series 3 (1891), 183.
54. MacDonald, A. On the agriculture of the counties of Elgin and Nairn, *THASS*, 4th series 16 (1884), 79.
55. Hamilton, W S. The most economical method of threshing combined with efficiency, *THASS*, 4th series 14 (1882), 134.

56. MacLelland, T. On the agriculture of the Stewartry of Kirkcudbirght and Wigtownshire, *THASS*, 4th series 7 (1875), 27.
57. Ralston, W H. On the agriculture of Wigtownshire, *THASS*, 4th series 17 (1885), 102.
58. SLA.
59. Ritchie, 1845, op cit, 144.
60. ibid, 143.
61. Fenton, A. An early corndrier at Whittinghame Mains, East Lothian, *Industrial Archaeology*, 6 (1969), 388–1.
62. *North British Agriculturalist (NBA)*, 1936, 808, advertisement for Ransomes' grain drier.
63. Fenton, A. A note on straw-rope granaries. In *Ethnografski i Folkloristichni Izsledvaniya (Festschrift for Christo Vakarelski on his 80th birthday)*, Sofia, 1979, 144–8.
64. Wight, 1778–84, 4/1, 345.
65. Anon. On Perthshire husbandry, *Farmers Magazine*, 8 (1807), 441.
66. Fenton, 1979, 148–9.
67. Trotter, 1811, 19.
68. Somerville, 1805, 48.
69. Wight, 1778–84, Vol 2, 21–2.
70. ibid, Vol 1, 100.
71. Stephens, 1844, Vol 1, 109.
72. For grain-drying without kilns see Fenton, 1976, 94–5.
73. Major, J (trans Constable, A). *A History of Greater Britain as well as England as Scotland*, 1521 (Scottish History Society, 1892).
74. Corn kilns are known to have been used for this purpose on the island of Colonsay. Stevenson, W. Notes on the antiquities of Colonsay and Oronsay, *PSAS*, 15 (180–1), 137; At Easter Jaw, Strilingshire (NS 874746) there are slight remains of a building identified as a flax kiln on the first edition Ordnance Survey 25 inch/mile map (Stirlingshire 35.3,1862); A ruined kiln-barn adjacent to the lint mills at Linn Mill, Stirlingshire (NS 912723) may have been used for this purpose; For details of an Irish flax-drying kiln see Davies, O. Kilns for flax drying and lime burning, *Ulster Journal of Archaeology*, 3rd series 1 (1938), 79–80.
75. NAS RHP 430.
76. Slezer, J. *Theatrum Scotiae*, London, 1693.
77. Borders: Mill of Langshaw (1655), NAS GD 157/1001; Kirk Yetholm Mill (1740), NAS GD 6/1596; Ricarton Mill (1742), NAS GD 224/238/4. Lothians: Mill of Cranston (1742), NAS GD 135/94; Breech Mill (1742), NAS GD 215/216; Mill of Wester Gammelshiels (1567–8), *Registrum Secreti Signilli Regum Scotorum* (1886–1914), 20, 383. Fife: Mills of Pittenweem (1706–8), NAS GD 62/275; Shawsmill (1697), NAS GD 26/5/87; Lundin Mill (1662): Lamond Dairy, cited by Cunninghame, A S. *Upper Largo, Lower Largo, Lundin Links and Newburn*, Leven, 907, 65. Angus: Detached, round kilns are shown on plans of the following; Mill of Syde (1792), NAS RHP 83; Slaty Mill (1815), NAS RHP 505; Auldallan Mill (1790), NAS RHP 1048.
78. Mill of Netherurd, Peeblesshire (1729), NAS GD 120/258A, whole barony to maintain mill *and* kiln; Newmill of Binns, West Lothian (1745–6), NAD GD 75/536, could not get service 'aither by killn or miln'; Breech Mill, Milothian (1742), NAS GD 215/216, specification of multure payable for each kiln.
79. For a discussion of the terms used for parts of the kiln see Fenton, 1974, 243–58;

Open kilns were noted in Arran, 'These kilns are generally of very awkward construction, sometimes not covered from the rain, except by blankets supported on poles'; Headrick, J. *View of the Mineralogy, Agriculture, Manufactures and Fisheries of the Island of Arran*, Edinburgh, 1807, 314.

80. For further descriptions and illustrations see the following: Stell, 1982, 100–3; Fenton, 1978(a), 375–87; Newman, 1994, 48–66.
81. For further descriptions and illustrations see the following: Scott, 1951, 200–2; Whitaker, 1956–7, 161–70; Mercer, 1972, 27–36; MacDonald, 1993, 185–6. A kiln of this type, otherwise unknown in mid-Argyll, was found at Dounie (NR 759920); RCAHMS. *Argyll*, Vol 7, Edinburgh, 470–1.
82. Fenton, 1978(a), 376–8; Scott, 1951, 199.
83. Sinclair, J, ed. *The Statistical Account of Scotland (OSA)*, 21 vols, Edinburgh, 1791–1799, Vol 10, 491, Kirkoswald, Ayrshire.
84. Graham, P. *General View of the Agriculture of Stirlingshire*, Edinburgh, 1812, 117.
85. Robertson, J. *General View of the Agriculture of Perth*, Perth, 1799, 99.
86. Roger, Rev Mr. *General View of the Agriculture of Forfar*, Edinburgh, 1794, 20–1.
87. Anderson, J. *General View of the Agriculture and Rural Economy of the County of Aberdeen*, 1794, 80.
88. Headrick, 1807, 314.
89. Fenton (1976) 1999, 102.
90. Robertson, 1799, 99; *OSA*, 10, 491, Kirkoswald, Ayrshire; *OSA*, 15, 279, Baldernock, Stirlingshire; Headrick, J. *General View of the Agriculture of Angus*, Edinburgh, 1813, 266; Robertson, J. *General View of the Agriculture of Southern Perthshire*, London, 1794, 51; Henderson, J. *General View of the Agriculture of Caithness*, London, 1812, 107.
91. Slade, H G. Rothiemay: An eighteenth-century kiln barn, *Vernacular Building*, 4 (1978), 21–7; Beaton, E. The Sandside kiln barn, Caithness, *Caithness Field Club Bulletin* (Spring 1988), 1–4.
92. Ballindalloch; Beaton, 1988, 2; Ardersier: Cawdor estate papers. Copy of plan and specification in SLA.
93. Erskine, J F. Explanation of a plan of a thrashing mill barn, *Farmers Magazine*, 3 (1802), 320.
94. Dymock, C. 'Georgian Farmsteads of West Fife and Kinross', typescript in SLA.
95. Keith, G S. *A General View of the Agriculture of Aberdeenshire*, Aberdeen, 1811, 138.
96. Ordnance Survey Name Books: Athelstaneford parish, Haddingtonshire; Fieldwork, 1986.
97. Fenton, 1978(a), 386–7; SFBS, Cross and Burness and Lady parishes, Orkney. Sites at Nearhouse, Northskaill, Elsness, Warsetter, Tresness; Newman, 1994, 57–9.
98. See Stell, 1982, 112, ref 15 for a list of sites.
99. Fenton (1976) 1999, 102–6.
100. Shaw, 1984, 1–11; Fenton (1976) 1999, 105–6.
101. Kerr, 1809, 96.
102. Headrick, 1813, 262–3.
103. Busit, 1840, 313.
104. Ritchie, 1845, 143.
105. Pringle, R O. On the agriculture of the islands of Orkney, *THASS*, 4th series 6 (1874), 33.
106. Smith, I. *General View of the Agriculture of Argyll*, Edinburgh, 1798, 117.

107. Robertson, J. *General View of the Agriculture of Inverness*, London, 1808, 193.
108. HMSO, 1946, 65.
109. SFBS, Gairloch parish, Ross and Cromarty; communication from Elizabeth Beaton.
110. MacDonald, 1908–9, Vol 2, 202–8.
111. Anderson, 1988, op cit, 68.
112. Drennan, J. Farming in the west and south-western districts. In *Report on the Present State of Agriculture of Scotland*, Edinburgh, 1878, 85.
113. SFBS, Crawford parish, Lanarkshire.
114. A & J Main catalogue, 1882.
115. See, for example, M'Neilage, J. The agriculture of the county of Dumbarton, *THASS*, 4th series 18 (1886), 57; and MacDonald, 1887, op cit, 44.
116. Shaw, 1990, 73–4.
117. ibid, 74–83.
118. ibid, 83–5.
119. SFBS, Monimail parish, Fife.
120. *NBA*, 21 May, 1936.
121. *NBA*, 25 March 1937.
122. Cashmore, W H. Notes on farm mechanization in war time, *JRASE*, 101 (1941), 5.
123. *The Scotsman*, 3 January 1981.
124. Stephens, Burn, 1861, 489–90.
125. Liddell, D. Description of an apparatus for steaming potatoes, turnips, and grain, as food for domestic animals, *THASS*, 2nd series 2 (1831), 322.
126. Fenton, Walker, 1981, figure 94; Morton, 1855, Vol 1, plate 21.
127. Wight, 1778–84, Vol 1, 169.
128. Anderson, 1794, op cit, 116.
129. Robertson, G. *General View of the Agriculture of Midlothian*, Edinburgh, 1795, 14.
130. Ritchie, J. Whinmills of Aberdeenshire, *PSAS*, 9 (1924–5), 128–42.
131. Hay, Stell, 1986, op cit, 18.
132. Morgan, P. *Annals of Woodside and Newhills*, Aberdeen, 1886, 17, NAS RHP 814 (1763).
133. Dickson, R W. *Practical Agriculture, or a Complete System of Modern Husbandry*, London, 1805.
134. Singers, Rev W. *General View of the Agriculture of Dumfriesshire*, Edinburgh, 1812, 136; Whyte, Macfarlane, 1811, op cit, 74.
135. Lord Kinnaird. On covered farm-steadings, *JRASE*, 14 (1853), 339.
136. Douglas, R. *General View of the Agriculture of Roxburgh and Selkirk*, London, 1798, 99.
137. Singers, 1812, op cit, 93–4.
138. Anon. Digest of various papers on the preservation of the potato for domestic use over the year, *THASS*, 2nd series 7 (1845), 185.
139. Singers, W. On the introduction of sheep farming to the Highlands, *THASS*, 3 (1807), 592.
140. MacDonald, 1908–9, Vol 2, 321.
141. SLA.
142. Information from Piers Dixon, RCAHMS.
143. Personal observation, 1995.
144. Fenton, 1978(a), 418–19.
145. Spier, J. Boxing seed potatoes, *THASS*, 5th series 14 (1902), 124–50.

BIBLIOGRAPHY
See Chapter 20.

22 Agricultural Buildings 3: Livestock Housing and Products

JOHN SHAW

BYRES

The close association between dwelling and accommodation for cattle has deep historical roots. Its physical manifestation, in the byre dwelling, persisted into the twentieth century in small-scale steadings in marginal areas. In its most basic form the byre and the dwelling shared a common entrance, without internal sub-division. From here its nineteenth- and twentieth-century evolution first involved partitioning the byre from the dwelling, then providing independent entrances and finally, under the influence of legislation, eliminating direct access between the two.

The byre was principally for winter use. In summer, the cattle were herded by day on grazings away from unenclosed crops and housed in the byre overnight. In upland areas they might go further afield, to shieling grounds. In winter they were fed and housed indoors, bedded on straw, bracken, heather, turf or whatever else was available – always allowing for competing requirements for thatch or, in the case of some straws, feed. The used bedding, with the animals' dung, was removed to the midden, generally in the following spring as enrichment for the coming year's crops. In Lewis, this might entail breaking through the byre gable wall; the hole so formed was known in Gaelic as the *toll each*, horse hole, as the manure would be taken from here by cart or horse-creels.[1] Elsewhere, as in Aberdeenshire and Orkney, the muck could be forked out onto the midden through a muck hole. From the early nineteenth century there were experiments in soiling, that is, feeding cattle indoors all the year round. The byre also came to be the place where cows were milked, during summer while this was still a seasonal activity, but eventually throughout the year.

The size of dairy byres – and of the farms themselves – was controlled by the number of cows which could be milked by the women-folk of a family farm or, as was often the case in the south west, by women working for a 'bower', who in exchange for use of facilities, pasture and winter feed, paid the farmer an agreed sum per cow or part of the dairy produce.² In practice, the limit was about 20 to 30 cows. On home farms, such as Hatton, Renfrewshire, where two bowers had charge of separate herds, this number could be doubled.³ However, most byres housed beef, rather than dairy cattle.

Externally, byres were plain structures, generally of lower height – and status – than stables, and with few if any windows, though latterly some roof lighting was provided. All but the most primitive had drains discharging outside. Improved byres also had vents: small openings set high in the wall or, latterly, above the feed troughs, occasionally slit vents of the type associated with barns. These acted as inlets. Projecting or integral ridge or roof vents acted as outlets, and on slate roofs ventilation might be further enhanced by overlapping slates to a lesser extent than was normal. For the most part byres were single storeyed, though occasionally lofts or upper storeys might be used to store hay, keep poultry or even house workers.

Pedigree herds, such as those housed in the church-like 1870s byre at Floors Castle, Roxburghshire, or the extensive 1920s byre at Laigh of Dercullich, Perthshire, enjoyed far superior accommodation, in keeping with their status.

Internally, the byre had raised stalls, with a surface of cobble, flag, brick or, from the late nineteenth century, concrete. At the heads of the stalls were stone, wooden or, latterly, ceramic feed troughs, occasionally with hay 'hecks' above, though generally all feed, hay included, went into the trough. Behind the stalls ran a drain or 'grip' and beyond that an access passageway. The more sophisticated nineteenth- and twentieth-century byres had, in addition, separate feeding passages along the heads of the stalls.

Eighteenth- and early nineteenth-century practice was for stalled cattle to face the ends of the building. On most farms this was superseded by turning them to face a side wall, with an entrance and passage on the opposite side. In larger byres, including William Harley's model byre at Willowbank Dairy, Glasgow, the stalls occupied both sides, with a central passage for mucking and milking, in addition to feed passages at the beasts' heads.

Tethering systems in byres took a variety of forms. In the simplest form, cattle were tied to a perforated stone or stake, projecting from the wall a little above the trough, as in Orkney and Shetland, where examples can still be seen.⁴ In some Caithness byres, cattle were tied to a staple fixed into the rim of the trough. They might be tethered to a fixed upright stake with a movable ring, as in south-east Scotland; these were still being used on farms in Peeblesshire and Midlothian in the early 1960s and survived on a disused East Lothian farm until the mid 1980s.⁵ From western Scotland, there are late nineteenth-century references to yet another system:

It is not long since it was customary to secure the cows by the head in a sort of pillory affair. Each cow was held by the neck between two upright sticks, one of them hinged at the lower end to let the cow's head in or out. When the cow's head was in place the movable stick was secured at the top with a peg, and the animal was so fixed so far as getting backwards or forwards, or even looking over her shoulder, was concerned.[6]

A byre with this type of fitting can be seen at Auchindrain, Argyll.[7] Other examples have been located in the Inner and Outer Hebrides.[8] However, the majority of byres now have slide bars, mounted on the stall division or 'trevis', associated with chains around the necks of the animals.

Though not essential in some earlier tethering systems, most nineteenth- and twentieth-century byres were fitted with trevises, in some cases triangular but more commonly rectangular. These might be in stone, timber or latterly brick or concrete.[9] Two beasts occupied the space between trevises, occasionally with a further subdivision between troughs. On a smaller scale the same arrangement might house 'stirks' or calves.

During the nineteenth and twentieth centuries concern over hygiene, and technological change, brought further refinements in the design of dairy byres. One of many town dairies, that of William Harley, at Willowbank, Glasgow, had pointed the way in early nineteenth century:

> There were several cow-houses, of which the largest was 94 feet [28.65 metres] long and 63 feet [19.20 metres] wide. Each was constructed on the same principle, with the wide-spanned slated roofs supported on internal cast-iron columns. The windows were designed to open so as to increase the circulation of fresh air in warm weather and the walls were carefully plastered for winter insulation. An even temperature of 60–64° Fahrenheit was maintained all the year round. The floors were partly paved with stone and partly boarded, as this was thought to be best for the cows' forefeet. Channels covered with cast-iron gratings in the floor collected the valuable liquid manure which was sold to market gardeners for 5s to 8s a ton.[10]

Not until the late nineteenth and early twentieth centuries, with legislation introduced in 1885, 1899, 1925 and 1934, were measures taken to ensure that a high minimum standard was achieved: each animal was to have at least a specified minimum space; troughs and trevises, floors and walls were given a smooth finish; spaces between walls and roofs were beam-filled; and both lighting and ventilation were improved. Water was piped to individual drinking bowls, which refilled automatically. All dung and refuse was to be removed twice daily, to a distance of at least ten yards (over nine metres), and placed in a roofed store.[11]

With the introduction of machine milking, first successfully applied by William Murchland at Dykehead, Hurlford, Ayrshire, in 1889,[12] byres were fitted with vacuum pipes and sheds to house engines and pumps appended to them. Mechanisation accelerated milking, hence the doubling up of dairy byres so evident in the south west. In the arable counties of the east, a move into dairying in the 1920s and 1930s led to the use of cattle courts as housing, with other buildings converted to milking parlours, milk rooms and sculleries.[13] The most sophisticated inter-war dairy steading, built in 1939 at Fenton Barns, East Lothian, was described as 'the most modern and clearly planned dairy building that one can conceive'. Here was a huge, well-lit byre for 180 Ayrshire cows, ten washing stalls and a milking parlour with 24 stalls. Milk from the parlour flowed through glass tubes to a cooling and bottling plant. Ancillary rooms included a laboratory, laundry and drying room, a room for making yoghurt, office, staff rest room, cold store and sterilising room. An observation lounge enabled members of the public to observe milking in progress.[14]

Since the 1920s,[15] but more especially since the 1970s, the general introduction of parlour milking has rendered the milking byre obsolete.

DAIRIES

Besides its close connection with the byre, the dairy also had strong ties with the domestic economy. The premises were physically close to, if not within, the farmhouse, and much of the labour – generally female – was drawn from it.

The dairy's principal function was to store milk, in as fresh a state as possible, and to use it in butter- and cheese-making. The market for whole milk remained small. Production for urban markets was confined to the immediate vicinity until at least the mid nineteenth century, town- and city-based dairy herds remaining a common feature until the late nineteenth century. To produce a marketable commodity, therefore, more distant, rural dairy farms tended to rely on converting milk into butter and cheese. Aiton (1825) calculated that most of Glasgow's milk supplies came from within a two-mile radius of the city, with 800–1,000 cows kept within the city itself; farms between two and 12 miles distant specialised in butter-making.[16] As the century progressed, improvements in transport brought a far wider area within range.

Of whatever form, the dairy had to have a cool, clean environment, and the means to clean utensils: in architectural terms, a milk room and a scullery. Additional apartments were needed to ripen cheeses: a cheese room, and, if cheese was to be made in large quantities, a press room equipped with cheese presses.[17] At an early date, and thereafter on farms producing only small quantities, much of this activity could be accommodated within other buildings. According to Aiton (1825) 'till of late, the operations of the dairy were carried on in the sooty and dirty hovels which were then inhabited by the tenants, – and the house-wife, whilst she was sinking her arms to the elbows in the milk or curd, was alternately cooking for the family, performing the duties of the nursery, and aiding the removal of the dung from the byre'.[18]

Where there was no separate scullery, water for cleaning vessels, or stones to warm the water, might be heated on the kitchen fire. On small dairy farms of the 1820s, the milk room and scullery, facing north and south respectively, were wedged in between the kitchen and byre.[19] The cheese press was commonly sited outside the house rather than in a press room; in the absence of a cheese room, cheeses were dried on a table in the kitchen then matured in the dairy, a garret or the barn, or even, in Aberdeenshire, the gig-shed.[20]

North Ayrshire developed an early specialisation in dairying and especially in the making of Dunlop cheese, reputedly first made in the 1680s.[21] By the 1790s, 'every steading of farm houses' had 'an apartment by itself for a milk house and every convenience suited to it'.

Elsewhere in Scotland, even where dairy produce was solely for home use, dairies became part of the house and steading during the late eighteenth and nineteenth centuries: in the angle between house and byre; in a lean-to or wing behind the house; or, in grander houses, as part of a service wing.

For the milk house, a north-facing site, or a site with north- and east-facing windows, was preferred, as a means of minimising warmth from direct sunlight. In the more formally architectural steading, the milk house might also have a veranda or projecting roof to provide shade. Windows were designed to give further protection: by means of louvres (for shade and ventilation); and inner windows of canvas, later copper wire mesh or perforated zinc (to keep out insects).[22] Wherever possible, there were vents in walls and roofs to promote air circulation; a high ceiling, without an upper floor, was considered best practice. For practical reasons, many farm milk houses fell short of this ideal.

Cold, plain surfaces were preferred: flagstone, tile, later concrete floors, plastered walls and ceilings, sometimes with encaustic or glazed tiles for a few feet above the milk shelves. On home farm dairies in particular, this gave occasion for elaboration and decoration,[23] though there were also traditions of floor decoration using white pipe clay.[24] The milk-shelves, in stone, flagstone, slate or even marble, ran around the walls at a height of about two feet (0.6 metres). The space below was used for storage. Occasionally additional tiers might be provided. On the shelves the milk was left to cool and separate in 'boynes' – wide, shallow dishes of wood, ceramics or metal – until the milk was sold or used in butter- or cheese-making.

Developments in the late nineteenth century did away with the need for shelves and boynes. Instead of cooling in the milk house, the milk was poured over a milk cooler (sometimes loosely termed a refrigerator), consisting of a water-filled corrugated metal frame. This was available by the 1850s but was still said to be little used by the mid 1880s.[25] Refrigerated bulk milk tanks, into which milk could be piped directly from the milking machine, were first tried out at Auchencruive, Ayrshire, during World War II.[26] Since then, but more especially since the 1960s, these have almost entirely replaced bucket milkers and coolers.

The centrifugal cream separator, available from 1879, made boynes obsolete, but came into use only slowly; in 1886, only two commercial creameries, at Dunragit, Wigtownshire, and Sorn, Ayrshire, were making use of it. The buttermilk left over from churning whole milk was more marketable than the skimmed milk left by the separator.[27] Nether Abington Farm in upper Clydesdale, Lanarkshire, sent buttermilk by rail as far afield as Newcastle and Dundee, but this was exceptional; local markets were generally sufficient.[28]

The scullery also required clean, plain surfaces, but its environment was quite different from that of the dairy: there was no need to exclude direct sunlight, and it absorbed heat and humidity from the boiler. In a built-in or free-standing boiler, water was heated for washing dairy utensils or warming milk for use in cheese-making.[29] By the late nineteenth century there were pressurised boilers, generating steam to clean utensils in a steam chest. Where dairying was no more than a minor sideline, the scullery and its boiler might serve other domestic needs. In the most elaborate layouts, there might be an additional apartment for drying equipment and vessels.

Only the grandest of dairies had a separate churn room; in normal circumstances butter was churned in the milk house or even in the kitchen. On commercial dairy farms the hand-operated plunge- or rotary churn became too labour intensive for the volume of milk being processed. Other sources of power – wind, water, horses or oxen – had been applied to threshing in the 1780s. Soon afterwards they were extended to dairies as a means of driving the butter churn. Martin (1794) noted that, in Renfrewshire, 'where they can command water, they make use of it, the churn lying horizontally'.[30] Besides Renfrewshire, there were early nineteenth-century references to water-powered churns in Fife[31] and Aberdeenshire.[32] On John Ballantine's farm, about twenty miles from Edinburgh, there was a separate churning house in which 'two [plunge] churns are employed, driven by a walking beam with two arms, attached to the water wheel of the thrashing machine, and they work at the rate of thirty eight or forty strokes in the minute'.[33] Elsewhere, there were open horse-walks with apparatus for driving churns – plunge churns at Commonside, Renfrewshire,[34] a later, earthenware churn at Tailend, Ayrshire.[35] George Robertson recalled that in Ayrshire, by 1811, there were 'several ... mills, if they may be so called, driven by horses for churning milk'.[36] In Orkney, by 1822, there were butter churns driven by small wind engines, originally devised by George Firth of Sanday some four years previously.[37] Steam power is known to have been used in Lanarkshire by the mid 1880s.[38]

Cheese-making involved other processes: curdling the milk; pressing the cheeses; and storing them until ripe. Here too, where only small volumes were being made, no special accommodation was necessary. The vat for curdling could be used in the kitchen, with the hearth as a source of heat. The stone cheese press generally stood outdoors, occasionally built into a wall, as at Easter Eninteer, Aberdeenshire. The indoor press at The Hill, Dunlop, Ayrshire, dated 1760, was an early exception to this rule. Another indoor press, formally at Braeminzion, Cortachy and Clova, is now at the Angus Folk Museum.[39] Portable cast-iron presses were being made at Shotts Ironworks, Lanarkshire, by 1835;[40] these could be housed in the milk house. Any suitable loft, in the house or steading, could be used to store ripening cheeses; some early nineteenth-century genre paintings show what appear to be cheeses stored high up in kitchens.[41]

On commercial cheese-making farms, there could be separate vat rooms, press rooms and cheese rooms, though such elaborate provision was exceptional (fig. 22.1). In practice, the cheese room was the apartment most widely adopted. The widely used term, cheese loft, is not always appropriate in a Scottish context. On some farms the cheese room did occupy an upstairs space, but in the major cheese-making county of Ayrshire, and possibly elsewhere, a ground-floor site was not unusual.[42]

James Fulton (1861) gives an outline of the cheese room's characteristics:

> The cheese room requires an atmosphere neither very dry nor very moist, the exclusion of light and the admission of fresh air under proper control, so as to exclude it when cold and damp; but the principal condition is an equitable warm temperature ranging between 60 and 70.
>
> When the temperature requires to be maintained artificially, heat is applied by means of the stove; or if the room be conveniently situated, by hot water pipes connected with the kitchen fire.
>
> The cheeses are turned daily for about ten or fourteen days; afterwards at longer intervals, extending as the drying and ripening advance. Some cheese-rooms are furnished with a cheese-turner

Figure 22.1 Ross Farm, Kirkcudbrightshire; cheese-making dairy and bower's house. SLA, FBS 5710.

for the cheeses requiring daily turning, the oldest cheese on the frame being removed as each new one comes from the press, and such an inexpensive and useful piece of furniture no cheese-room should be without.[43]

Nether Abington in Lanarkshire was one of the best-appointed dairy farms in Scotland. In 1885 it was reported that 'the [cheese] room is fitted up in the most approved style, with turning shelves, and is lined with wood, which is not so much affected as plaster by changes in weather. The temperature is kept at 58 to 60 by means of hot water pipes'.[44]

Exemptions from Window Tax (1795–1851) provide ready clues to the location of former milk houses and cheese rooms, as the names had to be written in letters at least one inch (three centimetres) high, on their outside window lintels in order to qualify.[45]

As noted previously, dairies on home farms offered an opportunity for fanciful elaboration and rich ornamentation. Unlike other parts of the farm, which could be dirty, smelly and staffed by men, the dairy was scrupulously clean, odour-free and, as also the poultry yard, worked by women (fig. 22.2). All of which made it acceptable as a place of resort for women of the landowner's household. Scotland has, or has had, some very fine ornamental dairies, amongst them Balboughty, Perthshire, on the home farm of the Scone estate. Built in 1790, altered in 1810, it had marble work surfaces, half-tiled walls and an arched, stained-glass

Figure 22.2 Dercullich Home Farm, Perthshire; dairy. SLA, FBS 1012.

window in the east wall.[46] It was customary for such dairies to have an architectural theme: Gothic at Millearne, Perthshire; rustic, with a cladding of white quartz, at Taymouth Castle, Perthshire; castellated, in a design for Castle Forbes, Aberdeenshire;[47] and an 'attractively mannered Alpine style' at Guisachan, Inverness-shire.[48]

During the late nineteenth and twentieth centuries changes took place which all but eliminated the farm dairy. Improved transport – first railways, then road haulage – extended the range over which milk could be transported, with a corresponding outward move of butter- and cheese-making zones. By the late nineteenth century, there were farm dairies with lowered bays for loading carts with milk or buttermilk cans. Better transport also allowed milk to be delivered to a central creamery: one of the first, the Dunragit Creamery of 1882, was consciously sited beside a railway.[49] By 1891, there were at least five other creameries, including two in the cities of Edinburgh and Dundee.[50] More were to follow, leading to the virtual elimination of commercial, farm-based butter- and cheese-making. Farm bottling extended sales of milk, but of recent years this too has declined. Hygiene legislation led to a re-arrangement of traditional layouts, with the dairy no longer directly accessible from house or byre. Earlier layouts can still be seen, especially where commercial dairying is no longer practised, but the bulk milk tank, linked to automatically cleaned milking appliances, is virtually all that remains of the working dairy on the farms of the 1990s.

CATTLE COURTS

An alternative method of wintering cattle evolved during the late eighteenth and nineteenth centuries, one which could help consume straw on arable farms and make more manure. In its earliest form, the accommodation consisted of the yard formed by the buildings of the steading, bedded out with straw, with or without a shelter shed and a wall to close off any remaining openings.[51] Late eighteenth- and early nineteenth-century sheds were open-fronted, with round or rectangular pillars to support the front of the roof. Shelters of this kind (since demolished) have been noted by Walker in Strathmore;[52] they appear in Loch's early nineteenth-century plans of Sutherland steadings and Singer (1812) refers to them in Dumfriesshire, with sandstone pillars, strong lintels of red pine and slate roofs.[53] Good examples can still be seen there, as at Blackrig, Lochmaben (fig. 22.3).[54]

As the nineteenth century progressed, the cattle court took on a more purposeful form (fig. 22.4), with a yard or yards set aside, each with its shelter shed to which there was a single, wide entrance. Contemporary published plans of Sunlawshill, Roxburghshire (1851),[55] and East Barns, East Lothian (1847),[56] show this level of sophistication. In some cases, the sheds had an upper floor, principally for storing hay, though occasionally, if inappropriately, for grain storage; additional open-fronted sheds might be built against the side walls of the yards. Within any one yard the cattle were free to move at will and to feed

Figure 22.3 Blackrig Farm, Dumfriesshire; cattle shelter. SLA, FBS 5235.

Figure 22.4 Edenmouth Farm, Roxburghshire; cattle court. SLA, FBS 2931.

from troughs and racks. In a further development, feeding passages were provided between courts, with hinged or sliding doors into troughs and direct access to turnip stores. The very best nineteenth-century steadings, such as Dunmore, Stirlingshire,[57] St Colme, Perthshire,[58] Redden, Roxburghshire,[59] and Thurston, East Lothian,[60] had their own internal railway systems to facilitate feed transport.

By the mid nineteenth century, opinion had started to shift in favour of covered courts, principally as a means of preventing nutrients from being washed out by rain.[61] New courts built after about 1850, and earlier open courts, were nearly all roofed over. The exceptions were principally in areas of low rainfall, and plentiful supplies of wheat and barley straw: Moray, where partly roofed courts were still current in the 1880s;[62] and East Lothian, where many courts remained open.[63] Typical of other cereal-growing districts in the south and east, courts were also very uncommon in the west and upland districts. In Aberdeenshire, much of the grain crop was oats, the straw of which was too valuable as feed to waste as bedding; here, byres remained the customary form of beef cattle housing, though 'coorts' were found on bigger farms.

A smaller version of the court – the 'hammel' – enjoyed some popularity in the Merse during the early nineteenth century. Sir John Sinclair came across hammels on Mr Robertson's Ladykirk, Berwickshire, estate and thought well enough of them to publish plans in his *Code of Agriculture*. The same arrangement was also used for housing horses.[64]

A few may survive there in modified form. A similar design, of robust construction, for individually-housed beasts, continued to be used for bull pens. Bulls and calves might also be housed in loose boxes or even in the byre. For other classes of livestock, boxes – small compartments within a larger building – had limited appeal in nineteenth-century Scotland. Lord Kinnaird's farm, Perthshire, built about 1850, had roofed courts which could be subdivided into boxes by means of wooden spars which slotted into sockets on cast-iron columns. The courts could also be fitted with slatted floors, a feature well ahead of its time.[65] It was to be another 100 years before this form of covered court came into general use, comprising raised shed, divided into pens, slatted floor and manure pit below.[66] Mucking out and feed distribution, as with so many other tasks about the steading, can now be performed by automated systems.

POULTRY HOUSING

Two considerations influenced the location and character of poultry housing: firstly, access from the house for feeding and egg collecting; and secondly, the need for warmth, principally as a means of prolonging the laying season. However, not until the twentieth century did poultry keeping become any more than a minor domestic element in farm production.

Capons, hens and poultry in general were part of the rents paid in kind during the seventeenth and eighteenth centuries; at least part of their time was spent about the house, attracted by its comparative warmth. Genre paintings and written accounts suggest that, for small tenants and cottars at least, having poultry about the house continued into the nineteenth century.[67] The hens might also nest and roost in the byre, where the warmth of the cattle added to their comfort and egg-laying capabilities. This was still current in Shetland in the 1870s, and is recalled as having persisted in South Uist until about 1900.[68] On small farms in south-west Scotland, it was still then 'a quite common occurrence to find the henhouse stowed away on the couple-baulks, and blocking up a part of that space, both to the detriment of our feathered friend and at the risk to limb of the women-folk who have to see after them. The belief that such a position is of advantage to the poultry was widely prevalent not so very long ago. It is not altogether dispelled yet, but, thanks to the preachings of sanitarians, it is now pretty well a thing of the past'.[69] Some Orkney byres had flagstones built across a corner, about five feet (1.52 metres) above the floor, as nest sites, with roosts above the cattle stalls.[70] On the same principle hens might be housed in an upper floor over a pig sty. This arrangement was common in the north of England and two such sites have been noted in Roxburghshire and Selkirkshire.[71]

During the nineteenth century the hen house became a recognisable, if still small and domestically-orientated, component in new steadings.[72] Access might be by way of a small door within a conventional door, or by a small, separate door, at nest box height and reached by stone steps or a removable ramp. The lowest steps, in such cases, were omitted to discourage rats. In addition to a roost, the hen house might contain wooden nest boxes or, not uncommonly, stone or brick nesting alcoves, set into a wall. The need for warmth was still recognised, by siting the hen house beside a dairy scullery, boiler house or domestic premises. Andrew Gray's 1849 plan for New Mains, Berwickshire, included a boiler flue which ran around the walls of the hen house behind the nest boxes – a form of central heating at a period when the farmhouse itself had to make do with open fires.[73] Hen houses with their own hearths have been identified on the island of Stroma, Caithness, and at Gesto and Uiginish on Skye, Inverness-shire.[74]

On home farms, more elaborate arrangements were made. Eastfield, East Lothian, once the home farm for Lord Balfour's Whittingehame estate, has a two-storeyed hen house with a yard enclosed by high railings. On the upper storey flagstone has been used for both floors and skirtings.[75] Sinclair (1814) describes poultry houses of considerable extent on the farms of 'gentlemen of fortune':

> In such cases, there is a keeper's house, and a range of small apartments for the different kinds of fowls, having a projecting roof in front towards the south, supported by small pillars of iron or wood, at from 3 to 4 feet [0.91–1.22 metres] from the wall, to afford shelter to the fowls in wet weather. All these usually open into a court covered with sand and gravel in or near which is a pond or rivulet of water. The inclosure of this is generally about seven feet high, and sharp pointed, to prevent the fowls from flying over it, the larger the court the better; it being understood that all sorts of poultry thrive best when allowed to range over a variety of ground. The apartments are sometimes warmed in winter by a steam-pipe from the keeper's house, carried through the roofs.
>
> Where poultry are fed on steamed potatoes, a boiler for that purpose may be so placed, as to serve at same time for heating the houses with steam.[76]

Stephens (1871) noted an increasing interest in poultry breeding and rearing on home farms and advocated a courtyard with a poultry house in five divisions: one for hens and turkeys; one for ducks and geese; a hatching house; a laying house; and a chicken house.[77]

On the smallest farms and crofts there were no such luxuries but there was perhaps greater freedom for the birds. Hens in Lewis, Harris and the Uists spent their summers in turf or turf and stone huts near the shore; one such was photographed at East Gerinish, South Uist, in the 1970s.[78] Small, detached hen houses, rubble-built and with thatched roofs, have been noted on crofts in Shetland, Aberdeenshire, the north-west mainland and the Uists, where several survive.[79]

The declining fortunes of cereal-growing, from the late nineteenth century onwards, brought a need for diversification and increased interest in commercial poultry farming. Small, low-cost portable structures were preferred, in timber or whatever other materials came to hand. The scope for large-scale rearing was greatly increased by the invention of thermostatically controlled incubators (1883),[80] and, during the twentieth century, by new, intensive systems. The deep litter system, under which droppings were left in the hen house for a year or more, first appeared in the late 1920s.[81] Battery housing, for both broilers and laying birds, dates from the 1930s; Orkney's first battery unit, in use by 1939, housed 324 birds, a far cry from the conditions of their byre-roosting predecessors.[82]

HOUSING FOR OTHER BIRDS

The dovecot, once the status symbol of landowners (see Chapter 9), later descended the social hierarchy to feature in steadings of the eighteenth and nineteenth centuries, despite the condemnation of pigeons by protagonists of agricultural improvement. The customary siting in grander, more formal steadings was over the principal entrance, where it gave justification to an otherwise purely decorative architectural feature. Particularly fine examples, some domed, some with spires, are to be found in Fife and the Lothians.[83] In more modest establishments, the dovecot might consist of a few openings in the apex of a gable. Where only one or two openings exist, these may have been intended for barn owls, though this view has been contested. Whether for doves or owls, the probability is that these were once encased on the inside to limit access further into the building.

Ducks and geese led a freer existence than domestic poultry. At Glenhead, Denny, Stirlingshire, they were provided with a small slab-roofed shelter outside the garden wall.[84] In Orkney and Shetland, geese were turned out onto the common during summer, but were confined at night, for their own safety, in pens or small huts.[85] In some Orkney houses brood geese were allowed into the house to nest in alcoves set into kitchen walls or sited beneath settles and 'binks'.[86] As has been noted already, composite housing, for more than one species, was a feature of home farms and country houses: hens, doves and pheasants at Culzean, Ayrshire; and hens, ducks, geese, turkeys and guinea fowls at Dunkeld, Perthshire.[87] On a less ambitious scale, the steading at Scar, Sanday, Orkney, had an arrangement whereby ducks and geese could wander onto the road and hens into the rick yard.[88] Seagreens, St Cyrus, Kincardineshire, had a slated, conical-roofed poultry house, with large boxes for geese at ground level, smaller boxes above for hens and a 'doocot' in an upper floor.[89]

HOUSING FOR PIGS

For the most part, until the twentieth century, pigs had a status similar to that of poultry. They were kept in small numbers as an element in the domestic economy rather than as a major livestock enterprise.

In pre-enclosure times, pigs were herded with other animals and housed at night. We know little of these structures, though they can be assumed to have been of the most basic form. By the late eighteenth century there had emerged a standard type of pig house which was to continue in use for more than a century: a low roofed shed with an opening into an adjoining yard. The yard was enclosed by a low wall, often pierced by a food chute which served a trough in the yard; this spared the pig feeder from entering into the yard, as pigs could be boisterous feeders. One or two such sties or 'crays' were to be found on most nineteenth-century farms as well as beside farm workers' cottages. In a few instances, the low shed constituted the ground floor to a first-floor hen house.

Exceptionally, where larger volumes of feed were available, pigs might be kept in greater numbers. The association of pigs with grain

Figure 22.5 Low Gameshill Farm, Ayrshire; pigsties. SLA, FBS 0824.

processing had a long history, with pigs featuring prominently in the rents of seventeenth- and eighteenth-century grain mills.[90] The first large-scale industrial distilleries of the late eighteenth century, Kilbagie and Kennet, Clackmannanshire, kept pigs on an extensive scale as a means of consuming draff.[91] Potato-growing districts also kept pigs in larger than usual numbers, in order to consume unsaleable produce;[92] but it was on dairy farms that the closest, most widespread links developed (fig. 22.5). Whey from cheese-making and buttermilk from butter-making were both fed to pigs, though near to urban or industrial centres a ready market grew up during the later nineteenth century for buttermilk or 'soor dook' as a drink for human consumption.

By the late nineteenth century, a new type of pig house was coming into use where more than a very few pigs were kept. Writing in the late 1890s, Richard Henderson noted that 'the old-fashioned pig-house ... is now at a discount in the dairying districts. The dairy farmer, or the "bower" prefers to have the pigs housed in a place that is completely roofed over.'[93] The new type of house was already in use in Wigtownshire by the mid 1880s, though opinion differed as to whether the old or new type was preferable.[94] Covered houses were said to give pigs better protection against sunburn in summer and from cold in winter. Typically, the house consisted of one or more rows of pens plus a feeding passage.

Scotland's first large-scale commercial creamery, established in 1882 at Dunragit, was able to keep many pigs:

> The whey, with the buttermilk that has been left, is run in underground pipes to the piggeries, of which there are a great number. Three large wooden houses, the sides being of disused railway sleepers, and the roofs covered with felt, have been erected on the south side of the railway away from the creamery, thus keeping all bad odours which might be likely to arise away from the milk. Those houses have a walk up the centre, and each side of this walk is divided into pens, some large and holding as many as fifteen pigs, others smaller and holding only four or five. One of these houses has been specially prepared as a breeding place, it being intended to rear as many as possible. The houses can hold about five hundred pigs.
>
> A large wagon which runs on rails is used for feeding; the meal having been mixed with the whey, before being put into this, a flexible pipe is attached to a tap on the bottom of the wagon, the end of the pipe is inserted into the trough and the tap turned on, thus reducing the labour in this department to a minimum. The manure from the pigs is sold to Mr Broadfoot, Drochduil, on whose farm the piggeries were built.[95]

While this was an exceptionally large and sophisticated unit for its time, changes in the early twentieth century did turn pig keeping into an intensive, commercially-orientated activity, just as with poultry keeping. Disposal of dairy produce as whole milk weakened links between pig-keeping and dairy farms, but this was compensated for by cheaper grain and cheap imported concentrates. Pigs were reared and fattened in whatever accommodation could be assembled or adapted from other uses. On the best farms, an improved form of covered pig house, based on Scandinavian practice, first appeared in the 1930s:[96] Cumledge, Berwickshire, for example, had a new pig house based on the Danish system in 1934.[97] During the 1950s and 1960s larger units, with semi-automated feeding and mucking, were introduced. Current public opinion has favoured a return to more open, free-range pig farming, using simple field shelters.

MANURE STORAGE

The management of manure at the steading was a recurrent theme in improved farming. The old practice of allowing 'muck' to accumulate in the open, in front of the house, was roundly condemned as distasteful but, equally, as wasteful.

The yard continued to be the repository for muck cleared from stables and byres, but there were improvements, in the form of a lowered, hard-surfaced, well-drained base for the midden and a dry access way separating it from the buildings. Cattle courts accumulated manure (dung and trampled straw) underfoot, until removed to the fields the following spring. One of the principal motivations behind the roofing over of courts in the later nineteenth century was to protect manure from dilution by rain. Even where open courts were retained, drainage and rainwater goods were used to good effect.

From about the 1850s, where byre housing predominated, the midden might be roofed over, provided with vents or larger openings, and glorified with the title of 'dungstead'. A good example can still be seen at Crakaig, Sutherland. Steading plans prepared by James D Ferguson (1850) featured a walled yard for horse dung, with a patent felt roof to protect it from rainfall, plus a similar, roofed yard for cattle dung, with large openings in the end and sides for wheelbarrow and cart access. The yards were to be paved with small stones and closed off with iron gates.[98]

Better management of drainage allowed liquid manure to be collected too, in a central tank from which it could be pumped out for spraying onto the fields. As early as 1814, a Mr Alexander, near Peebles, had sewers laid to convey cattle urine to an earth-filled pit, 36 feet (10.97 metres) square and 4 feet (1.22 metres) deep.[99] James Kininmonth,

Inverteil, Fife, built a small liquid manure tank at his steading in 1831 and a much larger one, some 72 feet (21.95 metres) long, in 1839.[100] The most ambitious farmer in this respect was probably a Mr Brown, Libberton Mains, Lanarkshire, who in the 1860s installed a steam engine to pump liquid manure from large tanks to irrigate 6 acres (2.43 hectares) of land. The scheme was not a success.[101]

A bewildering array of new fertilisers came into use in the first half of the nineteenth century. In the 1840s a farmer in Ayton parish, Berwickshire, had a mill which ground bone-meal fertiliser for his own use and for sale.[102] Around mid century, South American guano – accumulated, dried bird droppings – was especially prized. Steadings built at that time might have an apartment, with a cart-sized entrance, to store it, an example of which can be seen at Edderston Farm, Peeblesshire.[103]

SLAUGHTERING AND PRESERVING

Killing houses

Manure was not the only animal end product to be dealt with at the steading. A small number of animals were once slaughtered there, for farm consumption, and, on the larger farms, an apartment might be set aside for this purpose. The killing house had a hard, impermeable floor, with a drain, a tethering point and some means of hoisting carcases. For obvious reasons, these were best situated away from livestock housing: at Boon, Berwickshire, for example, the killing house was built onto the outside wall of the threshing engine house.[104] On home farms in the Highlands, facilities might be used for butchering and storing game.

Skeos

Salting was the usual process for preserving meat from animals slaughtered on the farm for use there. Generally, no particular buildings were dedicated to this purpose, but in Shetland, beef and mutton were once preserved by air-drying in 'skeos'. According to Hibbert (1822) 'A skeo is a small square house formed of stones without any mortar, with holes through which the air may have a free passage; for which purpose the building was erected on a small eminence, being at the same time protected from the rain by a roof'.[105] An example of such a structure was identified by the Scottish Farm Buildings Survey at Gruna, Northmavine parish, in 1995.

BUILDINGS AND STRUCTURES FOR SHEEP HUSBANDRY

Contrary to a widely-held view that sheep farming has left few permanent remains, there is ample evidence of a variety of buildings and other structures associated with the seasonal round of wintering, lambing, washing, shearing and protecting against vermin.

Substantial medieval sheep houses, such as those identified in the Pennines and the North York Moors, almost certainly existed on the monastic granges of the Southern Uplands. Kempson [106] draws a distinction between two terms used in monastic charters: 'bercarie', signifying sheep cotes or houses; and 'falde', signifying folds or open wattle pens. The records of Melrose Abbey, Roxburghshire, include permissions to build a sheep-house (berchariam) at Lower Colmslie (1162x1165) and at Whitelee (1174x1190).[107] A further search of cartularies, followed up by field investigation, might help to locate physical evidence.

Sheep cotes

Before the native dun-faced and white-faced sheep were displaced by Blackfaced and Cheviot flocks in the later eighteenth and nineteenth centuries, most farms kept a few sheep which were housed at night, more especially in winter and spring. References to sheep housing of this kind can be found as early as the sixteenth century: the *Court Book of Carnwath*, for example, records that in 1586 one Robin Wrech was found guilty of having a 'scheip hous begit on the comewne'.[108] Late eighteenth- and early nineteenth-century writers chart the decline of the native breed and, with it, the custom of sheep housing. At that date these were remembered as things of the past in the Lothians, Lanarkshire and Stirlingshire,[109] as being under pressure from Cheviots in Caithness,[110] and still current in most of the Highlands and north east.[111] Perhaps the most vivid recollection comes from John Naismith, writing in the 1790s of earlier times in Hamilton parish, Lanarkshire:

> These sheep were constantly attended by a boy or girl during the day, whom they followed to and from the pasture, and penned at night in a house called the *bught*, which had slits in the walls to admit the air, and was shut in with a hurdle door ... .[112]

These houses, known variously as 'cotts', 'bughts' or 'crees',[113] were thickly bedded out to accumulate manure, which was valued as a powerful fertiliser. None is known to have survived, but Ryder [114] refers to comparable 'cots' in areas of lowland England such as Herefordshire. A sheep cot from Ederveen, at the Netherlands Open Air Museum, Arnhem, shows similarities of use and purpose.

The simple form and rough construction of the Shetland 'lammiehoose', in which both lambs and older sheep could be wintered, might give the impression of a survival from earlier sheep houses (fig. 22.6). However, its origins seem to be no earlier than the nineteenth century, when cheaper feed, better transport and improved demand made winter feeding and housing more viable.[115] According to the *New*

Figure 22.6 Heylor, near Ronas Voe, Shetland; lamb house. SLA, FBS 3216.

Statistical Account for Sandsting and Aithsting 'some build small houses to keep lambs in during the night, and in which they feed them, night and morning, with hay or cabbage, and occasionally with a few coarse seeds and cut potatoes. The more general practice is to keep them around the fire in the dwelling house'.[116] Surviving examples, still used in Shetland, are long, low buildings, generally entered from the gable end and fitted with racks along the side walls. Their construction is crude, occasionally with turf gablets and turf roofs, with or without thatch.[117] In terms of function, they are analogous to the hogg houses of the Lake District.[118] Lammiehooses might each accommodate 25 beasts; one large example at North Banks, Papa Stour, measured 20 feet 2 inches (6.15 metres) by 7 feet 10 inches (2.39 metres).[119]

Winter shelter and housing

On upland farms the new, improved breeds – Blackfaced and, more especially, Cheviot sheep – still required some shelter in the worst winter conditions. Circular, unroofed 'stells' or rounds were said to have been introduced to the Southern Uplands after the severe storms of 1620.[120] Yet according to Hogg[121] there had not been a single stell in Ettrick, the heartland of Border sheep farming, when the grandfather of the then Duke of Buccleuch had acquired the lands. Mid eighteenth century references to stell-building in Roxburghshire (1747)[122] relate to enclosed plantations rather than to structures. Stone stells are known to have been introduced – or re-introduced – to Eskdalemuir, Dumfriesshire in the second half of the eighteenth century.[123] Findlater, writing of Peeblesshire (1802) defines them as 'circular spaces, of area proportioned to the size of the flock, enclosed by a 5 or 6 feet [1.52 or 1.83 metres] wall of stone or sod, without any roof.'[124] During the early nineteenth century, their use was championed by Lord Napier and they were constructed extensively in the Borders.[125]

Stells remain a conspicuous feature in upland landscapes, not only in the Borders but as far north as Caithness, indeed, in all those districts colonised during the nineteenth century by Border sheep-farmers and their Blackfaced or Cheviot flocks. One such farmer, Thomas Purves, took a lease of Rhifail, Sutherland, in the late 1850s and had built 'a good many stells' by 1880.[126] Typically, surviving stells have drystone walls, with a turf cope, though one late and unusual example at Moorfoot, Midlothian, is brick-built, with a cope of concrete. They take a variety of forms, occasionally rectangular and often with projecting arms.

Another variant was built on Lord Napier's farm and illustrated in his *Treatise on Store Farming*. The stell house was similar in form to the stell but was of smaller diameter, had thicker walls and a thatched conical roof,[127] yet they seem not to have caught on. A ruined and now roofless stell house can be seen on Napier's Ettrick farm, Selkirkshire.

Where ample winter feed was available, in the form of hay and turnips, improved forms of sheep house, sometimes termed 'standing folds', came into use, on similar lines to cattle courts. These may be associated with the spread of Cheviot sheep, which were thought to be less capable of finding food in snow than were Blackfaces.[128] Andrew Wight noted an early example at The Murrays, East Lothian, in the 1780s, which he thought preferable to all others he had seen:

> It is placed in the middle of a spacious field of old grass: the form of it is a large square, shedded on every side, which forms a court within, on which the dunghill is reared. The sheep are fed with turnips every night during winter.[129]

A number of early to mid nineteenth century sites of this type have been studied in Midlothian.[130] Of the 25 sites identified, most were square or rectangular, with a few of circular or octagonal form. Typical dimensions were about 65 feet 6 inches (20 metres) square. All but three consisted of an open central yard, with sheds around; the remainder were fully roofed. Only two of the 25 were still extant; a third well-preserved site can be seen at Keith Hill, East Lothian. By the 1850s, sheep houses of this type were also to be found in Berwickshire, Peeblesshire, Dumfriesshire, Angus and Caithness.[131] Steadings incorporating sheep shelters (cotes) and yards are illustrated by Sinclair (1814) and Waddell (1831).[132]

The 'plantation stell', as described by Scott,[133] had at its centre a similar structure. For whatever reason – possibly a tendency to harbour disease or parasites – these buildings fell out of favour. One late example, at Pitcraigie, Rothes, Moray, was built c. 1880 as part of a complex which also included lambing sheds, wool store and shepherd's house.[134]

Modern sheep houses originated in one built by John Ross at Millcraig, Easter Ross, c. 1900. The Millcraig shed measured 110 by 60 feet (33.53 by 18.29 metres), with a single-span corrugated metal roof; the lower walls were built in concrete and the upper in timber, fitted with pivoted ventilating doors. The shed had cart access for mucking out, stores for hay, straw and roots, with chaff-cutters and turnip-cutters powered by a one horse-power petrol engine.[135] As in other types of farming, large, multi-purpose sheds are now widely used in sheep farming. One of the largest, built at Easter Buccleuch, Selkirkshire, in 1984, is some 328 feet (about 100 metres) in length.[136]

Lamb and lambing houses

Winter housing was not the only building type associated with sheep farming. At lambing time, temporary shelters might be erected around lambing pens. In the Southern Uplands and possibly elsewhere, permanent 'keb houses' were provided, within which were 'parrocks' (pens) for confining orphaned lambs (or one triplet lamb) with an

Figure 22.7 Shepherdscleuch, Selkirkshire; keb house. SLA, FBS 0417.

adoptive mother. Keb houses vary from roughly-built, low narrow structures to substantial, well-constructed buildings (fig. 22.7).[137]

Keb houses might be situated by the shepherd's house or connected to dispersed stells. In 1821 the farm of Bowerhope, Selkirkshire, had keb houses at all but one of its eight stells:

> most of them are 12 feet by 6 [3.66 by 1.83 metres]; the door in one end and a straight barred flake across the other end, at nearly 3 feet [0.91 metres] from the wall, which, when subdivided in the middle with a small flake 3 feet [0.91 metres] long gives two places of nearly 3 feet [0.91 metres] square for confining one ewe and lamb in each.[138]

The stell keb house advocated by Scott (1888) was somewhat more substantial:

> ... a small oblong house built of stone and lime, say 15 feet long by 12 feet wide [4.57 by 3.66 metres] within the walls. The walls may be 18 inches [0.46 metres] in thickness and 6 feet [1.83 metres] in height. A door at one end would give entrance, and all the light required. The roof, to be at once efficient and cheap, may be of corrugated-iron sheets ... .[139]

During autumn and winter the keb house might double up as hay store.[140] The alternative approach involved making permanent the centralised pens erected for lambing. Borthwick (1873) favoured a permanent yard and open-fronted lambing shed, fitted with enough parrocks, each of 3 feet square (0.91 metres), to take three days' lambing during winter storms.[141] Watson and More (1928) illustrate a similar arrangement, with the addition of a hut, where a shepherd and a lambing man could spend alternate nights;[142] a former living-van, now in the collections of the Scottish Agricultural Museum, once served this purpose.

Prior to the early nineteenth century, ewes were milked, and the milk was used in butter and cheese making. Such limited information as exists suggests that the 'ewe buchts', where milking took place, were either unroofed or roofed stone or turf structures.

Smearing and dipping sheds
A succession of means were used to protect sheep from parasites. Smearing, with combinations of butter and tar, was the customary method, into the early nineteenth century in the southern uplands and for somewhat longer in the Highlands. Smearing houses, perhaps with a boiler or a hearth to prepare the tar, are known to have existed from the Borders to Sutherland, though only one possible survivor has been identified. Napier (1822) states that the smearing house should be 'well paved and lighted, clean, airy and dry;'[143] on most farms the reality probably fell far short of this specification. Smearing was succeeded by 'pouring' or 'bathing' and, latterly, by dipping, which was eventually made compulsory in 1906.[144] Dipping pools, with their associated pens, are generally open-air structures; there is a very fine slate-roofed dipping shed, with flagstone floor, on the farm of Ribigill, Sutherland.[145]

At certain times of the year, such as at dipping or shearing, flocks had to be gathered and sorted. This was especially important in crofting areas where sheep ran on common grazings. The pens or 'fanks' where this took place consisted of stone-walled or wooden-fenced enclosures, further subdivided into compartments and passages. The fank might be situated by the steading, near to the principal shepherd's house or on sheltered ground at the head of a valley. Complex and substantially-built stone fanks can be found in upland sheep farming districts from the Borders to Caithness. Other structures, such as lambing sheds or dippers, might be found in association with them.

Wool stores
Shearing usually took place outdoors, though sheds might be used if available. The fleeces were rolled and either stored loose or packed into bags or sheets until sold at market or through itinerant wool buyers. Wool stores required certain characteristics: according to Scott (1886), 'wool should only be stored in good barns or houses. Vermin, dust and damp soon render it an inferior article.'[146] Napier (1822) states that 'a

large loft, lathed and plastered all round and over-head, will afford the most proper place.'[147] Wool stores form part of steadings illustrated by Waddell (1831) and MacDonald (1908–9).[148] There are substantial ruins of a nineteenth-century wool store on Lord Napier's Thirlestane estate, Selkirkshire.[149]

According to Waddell, 'the barn or cow-house is frequently made use of for holding it (ie the wool) till sold'.[150] MacDonald (1908–9) notes that 'on large farms a wool-room is provided, but in many cases the wool is stored in a granary or outhouse'.[151] It has been suggested that the upper floors of a group of buildings in the Lochalsh district, Inverness-shire, of bank-barn form, may have served as wool stores. Similar buildings have been identified elsewhere in the north west, from mid Argyll to Gairloch, Ross and Cromarty, yet the documentary evidence to clinch their exclusive use as wool stores has not been forthcoming. The possibility that at least some of these buildings might have served as both barns *and* as wool stores cannot be ruled out.

Finally, the shepherd's house and its related buildings might, in itself, constitute a steading in miniature. Most shepherds were allowed to keep one or more cows, hence the need for byres; they might use a horse or pony, hence the need for stabling, with further accommodation for working dogs. In the Borders, a characteristic arrangement consists of a single range, gable-ended at the house end, piend-roofed at the livestock end. In addition, there might be various sheds and pens of the kinds referred to above, plus a hay barn, typically with corrugated sheet metal walls and roofs, but often making use of old railway wagons and containers.

NOTES

1. Fenton, 1978 (a), 14.
2. For a more detailed account see Campbell, R H. *Owners and Occupiers*, Aberdeen, 1991, 90–1.
3. Macdonald, 1887, op cit, 43.
4. Fenton, 1978(a), 430–2.
5. SLA, photographs of byre tethering stakes at Mid Hartwood, West Lothian and Culzeat, Peeblesshire; Prior to its conversion for housing in the mid 1980s, the house byre at Hallhill, East Lothian had tethers of this type.
6. Henderson, R. Dairy buildings, *THASS*, 5th series 11 (1899), 45.
7. Dunbar, J G. Auchindrain: A mid-Argyll township, *Folk Life*, 3 (1965), 64 and plate 16.
8. Douglas, G J. Byre fittings: A note on open-framed stalls, *Vernacular Building*, 10 (1986), 31–4.
9. Brick stalls were noted at Mumrills, Falkirk in the late 1960s, SLA.
10. Robinson, 1983, 88.
11. MacEwan, H A. *The Public Milk Supply*, Edinburgh, 1910, 142; Chalmers, C H. *A Review of the Milk Supply of Scotland*, Edinburgh, 1952, 13.

12. *Farming News & North British Agriculturist (FN&BBA)*, (3 February 1939), 17.
13. Chalmers, 1952, op cit, 5.
14. *FN&NBA*, 30 June 1939.
15. Harvey, N. *A History of Farm Buildings in England and Wales*, Newton Abbot, 1970, 182–3.
16. Aiton, W A. *A Treatise on the Dairy Breed of Cows and Dairy Husbandry, with an Account of the Lanarkshire Breed of Horses*, Edinburgh, 1825, 65.
17. MacDonald, 1908–9, Vol 2, 476.
18. Aiton, 1825, op cit, 146.
19. ibid, 76.
20. I am grateful to Professor A Fenton, EERC, for providing this information.
21. Smith, 1995, 31–3.
22. Henderson, 1899, op cit, 88–9.
23. See, for example, Robinson, 1983, 92–100.
24. Denholm, R M. Flagstone design: Some details of the collection being made by Scottish Home & County, *Transactions of the Glasgow Archaeological Society*, new series 9 (1938), 95–101; SLA, see, for example, 37/8/11, a posed photograph of a Lanarkshire dairy maid, standing on a pipe-clay decorated doorstep.
25. Partridge, M. *Farm Tools through the Ages*, London, 1973, 221; Speir, J. Dairying in Scotland, *THASS*, 4th series 18 (1886), 318.
26. SLA, The original bulk tank is in the collections of the Museum of Scottish Country Life, Kittochside.
27. Harvey, N. *The Industrial Archaeology of Farming in England and Wales*, London, 1980, 109; Speir, 1886, op cit, 321.
28. Tait, 1885, op cit, 62.
29. Aiton, 1825, op cit, 77.
30. Martin, A. *General View of the County of Renfrewshire*, London, 1794, 14.
31. *NSA*, Vol 11, 526, Ceres, Fife.
32. Keith, 1811, 219.
33. Ballantine, J. Remarks on churning butter, *THASS*, 3rd series 1 (1845), 25.
34. The mechanical remains are now in the collections of the National Museums of Scotland.
35. SFBS, Dunlop parish, Ayrshire.
36. Robertson, G. *Rural Recollections*, Irvine, 1829, 562.
37. Firth, G. Description of a milk churn worked by the impulse of wind upon sails, *THASS*, 2nd series 6 (1824), 622–5.
38. Tait, 1885, op cit, 62.
39. SFBS, Hill, Dunlop.
40. Anon. Description of a cheese-press, *THASS*, new series 4 (1835), 52.
41. See, for example, the Geikie sketch books in the National Galleries of Scotland.
42. Henderson, 1899, op cit, 11, 94.
43. Fulton, J. On the best mode of making Dunlop and Cheddar cheese and on the comparative advantages of these two varieties, *THASS*, 3rd series 9 (1861), 59.
44. Tait, 1885, op cit, 61–2.
45. Stephens, 1844, Vol 1, 217.
46. Walker, 1983, op cit, Vol 1, 365.
47. Robinson, 1983, 92–100.
48. Hay, Stell, 1986, op cit, 16–18.
49. *NBA*, 10 October 1883.
50. ibid, 23 September 1891.

51. See, for example, the 1819 plan of Kintradwell, Kildonan, Sutherland in Loch, J. *An Account of the Improvements on the Estates of the Marquess of Stafford*, London, 1820, Plate 19.
52. Walker, 1983, op cit, 572.
53. Singer, 1812, op cit, 90.
54. SFBS, Lochmaben parish, Dumfriesshire.
55. Gray, 1852, Plate XIV.
56. Report on some features of Scottish agriculture, *JRASE*, 2nd series 7 (1871), 156.
57. Tait, J. The agriculture of the county of Stirling, *THASS*, 4th series (1884), 158.
58. Stephens, Burn, 1861, 490.
59. SFBS. The feed wagon was still on site when visited in 1993.
60. Personal fieldwork. Both track and feed wagon were still in place in 1995.
61. See, for example, Kinnaird, 1953, op cit, 336–43.
62. MacDonald, 1884, op cit, 102.
63. Personal fieldwork; Skirving, R S. On the comparative value of manure made with and without cover, *THASS*, 3rd series 11 (1865), 210–3.
64. Sinclair, Sir J. *A Code of Agriculture*, 2 vols, London, 1813, Vol 1, 24–7, and plate opposite 22.
65. Kinnaird, 1853, op cit, Vol 14, 339–42.
66. Strictly speaking this was a re-introduction via Norway, in 1955, of a practice tried out some hundred years previously by Lord Kinnaird, amongst others. See Harvey, 1970, 150–2, 229, 233–4.
67. See, for example, David Allan's illustrations to Ramsay's *The Gentle Shepherd*; also interior of a weaver's house on Islay in Pennant's *Tour*.
68. Evershed, H. On the agriculture of the islands of Shetland, *THASS*, 4th series 6 (1874), 208; School of Scottish Studies, University of Edinburgh, D J MacDonald MS, 51, 1956, 476.
69. Henderson, 1899, op cit, 57.
70. SLA, MS1978–67/52, Peter Leith.
71. For some Yorkshire examples see RCHME. Houses of the North York Moors, London, 1987, 184; This arrangement can be seen at the Home Farm, North of England Open Air Museum, Beamish, County Durham'; SFBS, Hadden, Sprouston, Roxburghshire; SFBS, Gair, Ettrick Selkirkshire.
72. England, J. On the rearing and management of domestic poultry, *THASS*, new series 4 (1835), 141–9.
73. Gray, 1852, 75 and Plate 15.
74. Beaton, E. Poultry palaces: Notes on Highland heated henhouses, *Moray Field Club Bulletin* (1982), 46.
75. Personal fieldwork.
76. Sinclair, J. *General Report on the Agricultural State and Political Circumstances of Scotland*, 5 vols, Edinburgh, 1814, Vol 1, 151.
77. Stephens, 1844 (1871), Vol 1, 260–2.
78. SLA.
79. SFBS and SLA.
80. Robinson, L. *Modern Poultry Husbandry*, London, 1951.
81. ibid, 29–31.
82. ibid, 34–7; *FN&NBA*, 20 January 1939.
83. For illustrations of dovecots over entrances in the Lothians see Morton, 1976, 35 and 41; For Fife examples see Tovey, J St J. 'Architecture and the Agricultural Revolution: Home farms and their estates in North Fife *c*. 1800 to *c*. 1850', MLitt,

St Andrews, 1989; Robinson, 1983, Plates 106, 107, has examples from Ross and Cromarty and Inverness-shire.
84. SFBS, Denny parish, Stirlingshire.
85. Fenton, 1978(a), 508.
86. Fenton, 1978(a), 147–50.
87. Robinson, 1983, 101–3.
88. SFBS, Cross and Burness parish, Orkney.
89. Walker, 1983, op cit, 359.
90. Shaw, 1984, op cit, 32.
91. Erskine, J F. *General View of the Agriculture of Clackmannan*, Edinburgh, 1795, 66.
92. Somerville, 1805, op cit, 47.
93. Henderson, 1899, 104.
94. Ralston, 1885, op cit, 125.
95. ibid, 128.
96. Harvey, 1970, op cit, 191.
97. *NBA*, 29 March 1934.
98. Ferguson, J D. Report on the best mode of saving and applying liquid manure of farmsteadings, *THASS*, 3rd series 4 (1851), 334–5.
99. Sinclair, 1814, Vol 2, 527.
100. Kininmonth, J. On the construction of tanks, *THASS*, 3rd series, 2 (1847), 292–5.
101. Tait, 1885, op cit, 18.
102. *NSA*, Vol 2, 141, Ayton parish, Berwickshire.
103. SFBS, Edderston farm, Peeblesshire.
104. Gibb, R S. *A Farmer's Fifty Years in Luderdale*, Edinburgh, 1927, 59.
105. Hibbert, S. *A Description of the Shetland Isles*, Edinburgh, 1822, 417.
106. Cited by Ryder, M L. *Sheep and Man*, London, 1983, 447.
107. Gilbert, J. The monastic records of a Border landscape, 1136 to 1236, *Scottish Geographical Magazine*, 99 (1983), 10.
108. Dickinson, W C, ed. *The Court Book of the Barony of Carnwath* (Scottish History Society, 3rd series 20), Edinburgh, 1937, 188.
109. Robertson, 1829, 180, Lothians; *OSA*, Vol 2, 184, Hamilton parish, Lanarkshire.
110. Sinclair, J. *General View of the Agriculture of the Northern Counties and Islands of Scotland*, London, 1795, 196; Henderson, 1812, 206.
111. Robertson, 1794, 14; Leslie, W. *General View of the Agriculture in the Counties of Nairn and Moray*, London, 1913, 322; Donaldson, J. *General View of the Agriculture of the County of Nairn*, London, 1794, 14; Smith, 1798, op cit, 240; Anderson, 1794, op cit, 82; Henderson, J. *General View of the Agriculture of the County of Sutherland*, London, 1812, 104.
112. *OSA*, Vol 2, 184 (1792).
113. *Cott*: Robertson, 1794, 69; Henderson, 1812, 104; Leslie, 1811, 322. *Bught*: *OSA*, Vol 2, 184 (1792); Robertson, 1829, 180. *Cree*: Bremner, S. 'North-east Caithness dialect list', SLA.
114. Ryder, 1983, op cit, 498.
115. Recollections of L G Johnson, Setter, Mid Yell, Shetland, SLA MS 1978.70.
116. Cited by Fenton, 1978(a), 447.
117. SFBS, Northmavine, Yell and Unst, Shetland.
118. Brunskill, 1982, 78–9.
119. Denyer, S. *Traditional Buildings and Life in the Lake District*, London, 1991, 80–2.
120. SLA.
121. Hogg, W. The statistics of Selkirkshire, *THASS*, 2nd series 3 (1832), 293.

122. Hardy, J. Notes concerning Oxnam parish, *History of the Berwickshire Naturalists Club*, 11 (1885–6), 110.
123. Duncan, A. A treatise on the diseases of sheep, *THASS*, 3 (1807), 533–4.
124. Findlater, C. *General View of the Agriculture of the County of Peebles*, Edinburgh, 1802, 194.
125. Napier, W J. *A Treatise on Practical Store Farming*, Edinburgh, 1822, 118–22.
126. Macdonald, J. On the agriculture of the county of Sutherland, *THASS*, 4th series 12 (1880), 72.
127. Napier, 1822, op cit, 132 and Plate 82.
128. Aiton, W. *General View of the Agriculture of the County of Ayr*, Glasgow, 1811, 480–1.
129. Wight, 1778–1784, op cit, Vol 4, 422–4.
130. Callander, R. Sheep houses in Midlothian county, *Vernacular Building*, 12 (1988), 3–13.
131. Berwickshire: Ordnance Survey 1st edn 6"/mile (NT 462528); Peeblesshire: Ordnance Survey 1st edn 6"/mile (NT 261465); Dumfriesshire, Angus: Stephens, 1852, Vol 1, 222; Caithness: Purves, J. On shelter for sheep, *THASS*, 3rd series 1 (1845), 399.
132. Sinclair, 1814, Vol 1, Plate 16; Waddell, Mr. Designs of farm buildings drawn up under the direction of a committee of the Highland Society of Scotland, *THASS*, 2nd series 2 (1831), 387 and Plate 14.
133. Scott, J, Scott, C. *Blackfaced Sheep*, Edinburgh, 1888, 186–7.
134. MacDonald, 1884, 52.
135. MacDonald, 1908–9, Vol 3, 389–99.
136. SFBS, Ettrick parish, Selkirkshire.
137. ibid; Tweedsmuir parish, Peeblesshire.
138. Napier, 1822, op cit, 123–4.
139. Scott, 1888, op cit, 188.
140. ibid, 190.
141. Borthwick, H. On lambing and the diseases incident thereto, *THASS*, 4th series 5 (1873), 75.
142. Watson, J A S, Moore, J A. *Agriculture: The science and the practice of British framing*, 2nd edn, Edinburgh and London, 1928, 508–9.
143. Napier, 1822, op cit, 194.
144. Wallace, R. *Farm Livestock of Great Britain*, 4th edn, Edinburgh, 1907, 660–1, 642.
145. SFBS, Sutherland estate; Douglas, G. Ribigill steading, Tongue, Sutherland, *Vernacular Building*, 18 (1994), 30–2.
146. Scott, C. *The Practice of Sheep-farming*, Edinburgh, 1886, 139.
147. Napier, 1822, op cit, 195.
148. Waddell, 1831, op cit, Plate 14; MacDonald, 1908–9, Vol 1, 134 and Plate 2.
149. SFBS, Ettrick parish, Selkirkshire.
150. Waddell, 1831, op cit, 387.
151. MacDonald, 1908–9, Vol 3, 385.

BIBLIOGRAPHY

See Chapter 20.

23 Workshops: Small-Scale Processing and Manufacturing Premises

JOHN SHAW

The term workshop, as used here, applies to any small-scale building, or group of buildings, in which raw materials are processed or goods manufactured. The materials or goods might be foodstuffs, other animal products, textiles, timber, metals or minerals. As this last group is dealt with elsewhere in this volume, it is given only minimal treatment here. This definition encompasses building types with other, long-established generic names, such as mills, smiddies or bakeries. Such buildings are generally under the supervision of a skilled tradesman (or occasionally woman), who may be self-employed and for whom this work is the principal occupation. Production depends more on manual skills than on powered machinery.

Workshops occupy an intermediate position in at least two senses: firstly, between the extraction of raw materials and the distribution and marketing of finished products; and, secondly, between strictly home-based and factory production. Homes and factories are dealt with elsewhere in this volume, as are premises for food processing on the farm or at the harbour-side.

Workshops have attracted little attention from architectural or other historians, for whom, all too often, they have been of only marginal interest: too vernacular for industrial historians, too industrial for historians of vernacular building, and insufficiently polite for many architectural historians. Nor do earlier, contemporary works contribute much by way of illumination. Of necessity, therefore, this section will be confined to an exploratory investigation.

HISTORICAL AND GEOGRAPHICAL DEVELOPMENT

An implicit chronology underlies, and underlines, a progression from domestic to workshop to factory production. Yet the true picture is rather more complex. Some industries were factory-based from their inception; these were comparatively late arrivals, requiring a substantial investment in plant and specialised labour. Others, such as the garment trade, were late in adopting factory production. Even within the textile industry, weaving, one of the first to mechanise and industrialise, remained a partially workshop-based sector well into the nineteenth century. In ironworking there was a striking contrast in scale and organisation between smiddies, both rural and urban, and vast iron-smelting works, both of which continued to operate into the twentieth century. Throughout industrialisation, therefore, workshop production held a continuing importance.

There is a more clear-cut chronology in the development of trades and the premises which they occupied. By late medieval times, a number of them were already well established: milling, baking and brewing; weaving, fulling, dyeing and tailoring; tanning, shoe-making and glove-making; wright-work, smith-work; and pottery. The range was extended during the sixteenth and seventeenth centuries with the introduction of, or first references to, framework knitting (New Mills, near Haddington, East Lothian, 1680s),[1] sawmills (Abernethy, Inverness-shire, 1630),[2] iron mills (Limekilns, Fife, 1630s),[3] and premises for the commercial manufacture of soap (Leith, c. 1620),[4] paper (Dalry, Edinburgh, 1590),[5] sugar (Glasgow, 1667),[6] glass (Wemyss, Fife, 1610),[7] brick and tiles (Blackness, West Lothian, pre-1709)[8] and rope (Glasgow and Newhaven, Edinburgh, c. 1690).[9] Eighteenth-century additions included snuff mills (by the 1740s),[10] lint mills (Paisley, 1726)[11] and wool-carding mills (Stoneywood, Aberdeen, 1790),[12] as well as the mechanical bleaching and printing of textiles (Ormiston, East Lothian, 1730,[13] and Bonhill, Dunbartonshire, 1795).[14] As yet, almost all production took place in premises of workshop scale.

The second half of the seventeenth century had seen the introduction of what might be considered to be the earliest factory buildings in connection with woollen manufactories. The 'great manufactory stone house' at New Mills, near Haddington, was three storeys high, 101 feet (30.8 metres) in length and 21 feet (6.4 metres) wide, a very ambitious building for its time.[15] Textile manufacturing premises on this scale were not to be seen again in Scotland until the last quarter of the eighteenth century, when cotton, later woollen and flax spinning, mills, of factory proportions, sprang up all over lowland Scotland. Prior to this time, the only other unit which might have merited the term factory was Carron Ironworks, near Falkirk, established in 1760.[16]

Nor was there a sudden move out of workshop-based production; some idea of its continuing importance during the nineteenth century can be gauged from the 1871 Factory and Workshop Returns. Of the 30,139 sites enumerated, some 28,451 came under the terms of the Workshop Regulation Act or a mixture of workshop and factory legislation. Within this total were some very sizeable sub-groups: 5,623 hand-loom weaving shops, long after hand-loom weaving had gone into decline; 3,266 boot and shoe-making premises; 3,160 making women's apparel and 2,763 occupied by tailors and clothiers; 2,786 carpenters' and joiners' shops; 1,935 bakeries or confectioners; 1,887 miscellaneous metal-working premises, most of them smithies. The list of smaller categories goes on and on.[17] Nor is this the whole picture, for buildings not covered by the Acts included grain mills (some 2,300 in the mid nineteenth century,[18] about half the total of a century or two earlier), sawmills (about 700 sites)[19] and lint mills (only a handful still at work after peaking at over 400 in the early 1800s).[20]

More is known about the location of tradesmen than about their workplaces. The basic trades of smith, wright, miller and weaver were already well established in both burgh and countryside by the early eighteenth century. Where raw materials were bulky or heavy, there were clear benefits in processing near source. Thus slate-cutting or charcoal-burning took place at the quarry or wood; given access to peat or coal fuel, lime-burning remained close to the quarries from which it was sourced. The need for water to power millstones, fulling stocks, flax scutchers or saws gave more of a rural emphasis to certain trades than might otherwise have been the case – at least until the nineteenth century when small-scale fuel-efficient steam engines came into general use.

Most trades, but especially those producing consumer goods, were represented in burghs, as reflected in the scope of trades incorporations. Here there were markets of a scale and wealth to support crafts and trades, plus the means to regulate both entry to trades and marketing of goods. Sources such as the *Register of Deeds*[21] and the *Register of Testaments*[22] indicate that already, by the 1660s, Edinburgh and its associated burghs of Leith and Canongate had specialised tradesmen making candles, glasses, combs, guns, hats, clocks, watches, locks, tobacco pipes, musical instruments and gold or silver work.

Workshops producing consumer goods, or processing animal by-products such as hides and skins, retained an urban bias. But during the eighteenth and early nineteenth centuries, a number of factors shifted the emphasis in other trades in favour of rural locations. Burgh privileges, which had artificially concentrated manufacturing in towns, were gradually eroded; agricultural improvement and village foundation offered new markets and new locations;[23] with official encouragement, the textile industries prospered, and even when spinning became mechanised, from the 1770s, increased output of yarn brought a boom in demand for as yet unmechanised trades such as heckling or weaving.[24]

From the seventeenth and early eighteenth centuries, there is also evidence of regional specialisation: small-scale ironworkers such as nailers and griddle-makers in the Upper Forth;[25] leather-workers such as saddlers, shoemakers and glovers in the upper Merse, Berwickshire;[26]

wood-workers such as squarewrights, cabinet-makers, even ship-builders beside the lower Spey.[27] At a more local level, there was a cluster of 'piggers' or potters in the Forth carselands south east of Stirling.[28]

FABRIC AND SETTING

There was considerable variety in the physical setting and architectural character of these minor industrial workplaces. Some activities involved no buildings: basket-making and tin-smithing by travellers;[29] cloth bleaching at the burn-side; sawing in saw-pits or lime or charcoal burning in clamps.[30] The simplest of structures were open-fronted shelters, or 'ludges', of a kind once used by masons. For example, the construction of Cowane's Hospital, Stirling (1637), began with the purchase of 50 deals and nails to build a workers' lodge.[31] Open-fronted sheds continued to be a feature of masons' yards and were still being used by slate cutters at Ballachulish quarries, Argyll, early in the twentieth century (fig. 23.1).[32] These provided no more than shelter from the elements but had the virtue of mobility.

Some trades, involving simple technologies and small-scale materials, could be accommodated within the home or in minor buildings appended to it: textile trades, such as heckling, carding, spinning and weaving;[33] shoe-making, tailoring, hand-knitting and other costume making; nail-making[34] – all these could be carried on in or around the home. The positioning of premises in relation to the house might be

Figure 23.1 Ballachulish quarry, Argyll; slate-trimmers' shelter. SLA, C 00689.

alongside, in the same range (as with loom shops), or in cellars or ground floors – as seems to have been the case in seventeenth- and eighteenth-century urban settings. Small detached buildings to the rear of the house were especially typical of semi-rural villages; the little loom sheds still being used on Harris and Lewis crofts are perhaps the last significant survival of this arrangement.

In a further development, in town and village buildings of the nineteenth and early twentieth centuries, the marketing function of retail trades gave rise to buildings in which the manufacturing workplace operated as an adjunct to shop premises; bakeries, tailors' shops and shoe-repair shops are still current illustrations of this practice. The spaces to the rear, or above – sometimes no more than a single room – might still be occupied as a home. More complex, or less compatible trades – brewing, soap-boiling, tanning, glass-making – occupied backlands or were squeezed into closes with entry through a pend. Where space permitted, buildings were arranged around a courtyard.

In terms of scale, materials and construction, workshops tended to adopt the same vernacular idiom as domestic and agricultural buildings: generally one, sometimes two and occasionally three storeys in height; clay-mortared, later lime-mortared rubble walls; thatched, later pantiled or slated roofs. Local variations – clay walls, turf, stone and turf – applied to small-scale industrial buildings as much as to other vernacular types. Only exceptionally, with estate workshops in villages or parks, did a suggestion of polite architecture feature in such functional structures: the Earl of Haddington's sawmill, at Tyninghame, East Lothian, with its crowstepped gables and diamond-paned windows, is a good case in point.[35]

During the later nineteenth and twentieth centuries the availability of materials, usually considered too inferior for domestic use, helped create a more distinct architecture. As estate plantations matured, timber became available as a walling material, supplemented by imports and recycled railway sleepers which were used, for example, to build a smithy at Kingussie, Inverness-shire.[36] A particularly common style used wide deals, set vertically, with narrow timbers covering their joints. Corrugated sheet metal, already available by the 1840s, and corrugated asbestos, during the twentieth century, served as cladding and as cheap replacement roofing – even on the roofs of previously thatched houses. Iron, later steel, could replace timber in frame and roof construction. Common and sand-lime brick were adequate for additions or alterations to workshops, but were not thought appropriate for the external walls of houses, unless concealed beneath a layer of render.

ANALYSIS OF BUILT FORM

Construction in low-status materials may be typical and indicative of later workshop premises, but there are other, more subtle, characteristics which set apart even those earlier premises of an otherwise domestic style.

These characteristics arise from functional requirements. The buildings were made for making things, and the processes housed might have required any of the following: the storing, moving, heating, cooling or drying of materials; manual or powered equipment for working materials and the light to work by; and, perhaps, a modicum of human comfort in terms of ventilation. It is proposed to look in turn at each of these functional elements, and especially at those types of building to which they impart a particular character.

Storage

Relatively stable materials, such as stone, required no covered storage; the scatter of material for re-working around some of the last working blacksmiths' shops suggests a similar attitude towards metals.

Indoor storage might be provided for raw materials, for partially-worked materials, or for finished products. Parts of other buildings might be set aside; lofts were one such option, being out of the way of workplaces and especially suited to materials such as those textiles which required a dry environment. Conversely, the cool, damp environment of a cellar was well suited to storing beer. The volume of materials on site might not justify separate stores, and even where it did, this might not necessarily give rise to distinctive building types. By the 1760s, lint mills were being provided with storage sheds for incoming flax [37] (in place of outdoor storage in stacks) but we know nothing of their appearance. Grain mills and saltworks had their granaries and salt girnals. Wood-working premises had open-sided or slatted stores which sheltered timber from the rain but allowed free air circulation to discourage decay and assist seasoning. These were sometimes sited at ground level, with the workshop above.

Movement

The movement of materials – onto and off site, into and between processes – also had an impact on the appearance of buildings. For many trades, doorways of domestic scale were adequate but in others the scale of the objects being worked on required higher or wider openings. Larger openings might also be needed to admit carts or other vehicles. In buildings of more than one storey, external doorways in upper floors, perhaps with hoists above, admitted materials without their having to pass through working areas. Grain mills, for example, might have external or internal hoists, the latter passing through trap doors which were opened by the ascending load. Their internal layout well exemplified the ways in which gravity could be exploited to assist the movement of materials from process to process. In ropeworks, the movement of the manufacturing process, drawing out and twisting lengths of rope, gave rise to exceptionally long, narrow buildings. Sawmills incorporate another feature related to movement, their open sides or ends allowing heavy logs to be manoeuvred onto the saw bench.

One final aspect of movement merits a passing reference. Collection of materials or distribution of products might require stabling (often with a hayloft over), plus cart or wagon bays or, latterly, garaging.

Heating

Most types of small-scale manufacturing premises used heat as part of their processing. Kilns, and the less sophisticated clamps, were built exclusively for that purpose (fig. 23.2). As an adjunct to corn mills, rather than farms, kilns first occurred in south-east and east central Scotland, and were already widespread by the sixteenth and seventeenth centuries. The kiln's role here was to dry grain in preparation for grinding, its essential components being a furnace below with a bed for the grain above. These earlier kilns were circular, and a good example survives at Preston Mill, East Lothian; improved, rectangular kilns date from the later eighteenth or nineteenth century and form part of the mill building.

Figure 23.2 Cashlie, Loch Tay, Perthshire; lime kiln. SLA, FBS 4224.

Figure 23.3 Mill of Montgarrie, Aberdeenshire. SLA, JPS 26.

An intermediate form, in which the attached kiln is round-ended, occurs in Angus and East Perthshire, as at Barry Mill, near Dundee.[38] Other regional variations can be found in roof and ventilator form (fig. 23.3). A subsidiary role for corn kilns was in drying malted grain prior to grinding; latterly maltings, breweries and distilleries had their own kilns, typically with pagoda-form roofs.[39]

Kiln drying of timber was a more rapid alternative to air seasoning. A small rectangular kiln, sited at a safe distance from other wood-working plant, can be seen at the bucket mill, Finzean, Aberdeenshire.[40] Bottle-shaped, brick-built kilns were a feature of potteries, as at Portobello, Edinburgh, or Dunmore, Stirlingshire; more slender 'cones' heated materials at glassworks, as at Alloa, Clackmannanshire.[41] George Lowrie, clay-pipe maker in Kirkcaldy, Fife, had more modest premises, with a kiln situated across the garden from his house in Sands Road.[42] Other kiln types, for lime-burning and brick- and tile-making, are dealt with in Chapter 25, and smoke-houses, for curing fish, are referred to in Chapter 26.

Direct heating of materials, within the workshop itself, was a feature of metal-working premises. Workers in precious metals – goldsmiths, silversmiths – needed only a small furnace,[43] as did small-scale working in base metals, such as nail-making. In blacksmith's shops, however, the wide, projecting raised hearth is a dominant and characteristic architectural feature: sometimes with a wide and heightened chimney stack to give extra draught – as at Kirkton Manor, Peeblesshire;[44] occasionally with an outshot to the rear of the building to give sufficient width to house the handle of the bellows, as at Cousland, Midlothian.[45]

The built-in Scotch ovens of nineteenth- and twentieth-century bake houses were, as the name implies, of Scottish origin. Starting from a separate coal-fired furnace, the flue wound its way around the outer surface of the oven before venting through a conventional chimney-head.[46]

Heating materials in liquids, or liquid materials, was a process common to many small-scale manufactories. Bleachfields had their keivs,[47] paper mills and dyers their vats,[48] breweries their mash tuns,[49] soaperies and sugar houses each had their boilers and saltworks their pans.[50] Each of these found architectural expression in fixed furnaces, boiling vats and chimneys – though the last of these might be indistinguishable from those of conventional hearths.

Finally, and leaving aside the luxury of space heating, other trades needed heat sources for minor purposes: to heat tailors' irons, cabinet-makers' glue or wax for shoemakers' thread. All this could be accommodated on conventional hearths or, latterly, on the specialised workshop stoves manufactured by Scottish foundries.

Cooling, airing and drying
Cooling, airing and drying featured in far fewer types of workshop than did heating. In architectural terms, these processes usually involved replacing part or all of the walls on one storey or more with louvred vents. Usually upper floors were used, on account of their greater exposure. Drying houses already featured amongst bleachfield buildings by the 1750s; one appears on a plan of Saltoun bleachfield, East Lothian (*c.* 1760),[51] and in a late eighteenth-century engraving of Falkirk bleachfield one of the buildings has louvred openings in its upper floor, almost certainly for this reason (fig. 23.4).[52] Similar arrangements were to be found in breweries (to ventilate the cooling vats),[53] in tanneries (to dry the tanned hides),[54] in mills making paper by hand, and, as at Newburgh, Fife, in premises for drying warp threads dressed with starch in preparation for weaving.[55] Long, single-storeyed drying sheds, with louvred sides, were a feature of tile works where they housed moulded clay in preparation for firing.[56] Reference has already been made to grain drying in kilns and to cool conditions in cellars; ice houses are also mentioned in Chapters 9 and 26.

Ventilation of workplaces was less of a concern, though some provision might be made where heat or dust levels were exceptionally high. Such windows as existed were generally fixed. Smithies might have doors with lower and upper halves, of which the upper could be opened for ventilation. In pantile districts their loose-fitting, poorly-insulating

Figure 23.4 Falkirk, Stirlingshire; bleachfield, late eighteenth century.

qualities made them well suited to smithy roofs; the better-insulated slate roofs were often fitted with ridge or side vents, or half-slated in the same manner as cattle sheds. We know little of the working environments of flax-hecklers, other than that they were notoriously dusty; in some of the last hand-heckling shops, in Dundee, each work stance had its own window opening which latterly, at least, was fitted with an extractor fan.[57]

Equipment and lighting
Central to the requirements of the workshop, as here defined, were its working equipment (fixed or movable), the space in which to use it, and the light to work by. These requirements influenced layout and built form.

Daylight was of particular importance. The light given off by cruisie lamps, rush lights and candles was of limited value except in fine, close-up work, such as home-based costume-making, silver-smithing or watch-making. Oil lamps (including paraffin from the 1850s) and coal-gas lamps (in towns, from the early nineteenth century) presented the same fire risk in some working environments as had earlier light sources.

Fenestration patterns were often closely linked to work stances. The characteristic opposing (back and front) windows of handloom shops are found in twos, fours, and occasionally sixes, each window corresponding to a loom stance. Repetitive lines of windows, a feature

Figure 23.5 Auchtermuchty, Fife; weaver's house and loom shop. SLA, JPS 47.

of larger premises for weaving or heckling, show the same relationship between light source and individual workspace (fig. 23.5).

The premises of many trades, but more especially those using hand tools, contain fixed workbenches below and beside windows, most commonly at the front of the building. The windows may be small in scale and number – perhaps only one or two – or, at the other extreme, as in some woodworking shops, may extend continuously across one or more elevations. Blacksmiths' and joiners' shops often feature a characteristic glazing pattern, in which small, overlapping panes are fitted into vertical panels; such small panes would have been inexpensive to install or, in the event of breakage, to replace (fig. 23.6).

This direct relationship between fenestration and work stance is far less apparent where powered machinery, rather than manually operated equipment, is in use. Here, the intention was to create sufficient ambient light to work by, only achieving levels which may seem intolerably low to modern eyes.

The presence of one or other power source – animal, water, wind, steam or combustion – has its own effects on form, with horse gin, wheel or wheelhouse, windmill or engine and boiler house plus stack as part of the built assemblage. Both horse engines and wind engines were used in water pumping (from the sixteenth and seventeenth centuries)[58] and (from the 1780s) on farms.[59] Yet their application in processing and

Figure 23.6 Gullane, East Lothian; smithy. SLA, S 8817.

manufacturing was very limited, with a few sites grinding corn,[60] malt[61] and tan-bark,[62] plus, in the case of wind engines, sawing timber (from 1695),[63] turning wood (nineteenth century)[64] and driving bleachfield or lint mill machinery (late eighteenth century).[65] A characteristically Scottish form of windmill, dating from the seventeenth century, consisted of a stone-built tower and an adjoining vaulted chamber.[66]

Water power was undoubtedly *the* prime mover in small-scale manufacturing until at least the 1830s, but had little impact on built form, other than in wheelhouses – which were by no means universal – and the need for well dressed stonework on walls exposed to water. Outside its original role in pumping water, the steam engine found few other applications prior to *c.* 1800. During the course of the nineteenth century the availability of increasingly compact, efficient and powerful steam engines contributed to the success of town and city workshops, and also to the rise of factory and large-scale mill production; this, however, takes us beyond the range of this chapter.

SCOPE FOR FURTHER STUDY

This necessarily brief account of workshops demonstrates that there is still much scope for research. This has been a very significant class of building, with a very long history and a relevance to both rural and urban environments, yet most sites are now threatened by obsolescence, demolition or radical alteration. Some research, documentary and field, has gone into wind- and water-powered premises;[67] industrial archaeologists, notably Butt,[68] Donnachie[69] and Hume,[70] have included these and other types of site in their surveys; the Scottish Industrial Archaeology Survey, originally under Hume at Strathclyde University, later at the Royal Commission on the Ancient and Historical Monuments of Scotland, Edinburgh, has recorded sites of this type in its threatened buildings and thematic surveys. Blacksmiths' shops have received more attention than most other workshops and, given their solid construction and domestic scale, have been partly saved through conversion. The smithy at Cousland is being preserved by a trust, with its contents, as a working smithy museum. Another trust is striving to do the same for the bucket mill at Finzean. Yet there are other classes of building about which very little has been researched or recorded, such as rural joiners' shops or the small complexes of buildings in closes and yards on the backlands of towns and inner cities – all of which are especially vulnerable. A collective effort might yet ensure that these, and the recollection of their human context as workplaces, are not completely lost.

NOTES

1. Gulvin, 1983, 62–3.
2. Shaw, 1984, 95.
3. ibid, 89.
4. Scott, 1911, 130.
5. Shaw, 1984, 54.
6. Scott, 1911, 133.
7. Fleming, 1938, 95–100.
8. Shaw, 1990, 28.
9. Scott, 1911, 173–4.
10. Mitchell, 1908, 631.
11. Shaw, 1984, 171.
12. ibid, 282.
13. ibid, 222.
14. ibid, 342–3.
15. Scott, 1905, 56.
16. See Campbell, 1961.
17. Return of Factories and Manufacturing Establishments Subject to Factories, &c. Regulation Acts and Workshop Regulation Act, *Parliamentary Papers*, 1871, 62, 165–92.
18. Shaw, 1984, 499.
19. ibid, 505.
20. ibid, 519, 184.
21. National Archives of Scotland, Edinburgh.
22. ibid.

23. For further information on village planning see Lockhart in Parry, Slater, eds, 1980; and Smout. In Phillipson, Mitchison, 1970.
24. See Durie, 1979: and Murray, 1978.
25. Shaw, 1984, 84–7. Additional information from entries in the Register of Testament and Register of Deeds.
26. Information derived from entries in the Register of Testaments and Register of Deeds.
27. ibid.
28. Caldwell, Dean, 1986, 105–12.
29. For illustrations, see material in the SLA, NMS.
30. For illustrations of former lime-burning clamps near Carlops, Peeblesshire, see SLA.
31. Harrison, 1994, 2–3.
32. Fairweather, npd, plate between pp 3 and 4.
33. On loom shops within the home see Lloyd, 1989, 66–8.
34. See Graham, 1961, 117–9 and Plates 6–7. A nailer's shop to the rear of a house at Chartershall, Stirlingshire, is noted in Hume, 1976.
35. See photograph in Hume, 1976, 125.
36. Kerr, 1986, 24–6.
37. Shaw, 1984, 175–6.
38. For illustrations of regional variation in kiln form see Shaw, 1982, 8–24.
39. Hume, 1976, 17, 94, 135; Hume, 1977, 26, 29, 162, 180, 261. Hay, Stell, 1986, 31–60.
40. Walker, 1982–3, 76–82.
41. Hay, Stell, 1986, 171–4.
42. Kirkcaldy Museums, 1979.
43. A relief carving of a goldsmith's workshop, c. 1630, is reproduced in Dalgleish, Maxwell, 1987, 8.
44. Hume, 1976, 204.
45. See Hay, Stell, 1986, 20, plan of smithy with bellows outshot at Kirkton Manor, Peeblesshire; plan and photograph of same at Nether Horsburgh, Peeblesshire.
46. Kirkland, 1927, 302.
47. For further details of bleaching processes see Shaw, 1984, 221–6.
48. For further details of paper-making processes see Thomson, 1974.
49. For further details of maltings and brewery buildings and processes see Donnachie, 1979, 100–12.
50. For a full account of salt making in Scotland See Whatley, 1987.
51. Shaw, 1984, 244.
52. SLA.
53. Hume, 1976, 77, Coldstream Brewery Berwickshire.
54. Hay, Stell, 1986, 19: Keith, Banffshire, and Ayr; Butt, Donnachie, Hume, 1968, 16, Lanark.
55. Personal fieldwork.
56. Hay, Stell, 1986, 167–8.
57. Watson, 1990, 82–5.
58. Shaw, 1984, 62, 64.
59. ibid, 158–160.
60. Douglas, Oglethorpe, Hume, 1984.
61. Shaw, 1984, 16.
62. ibid, 467.

63. ibid, 99.
64. Crieff, Perthshire: [Porteous, A. *A History of Crieff*, Edinburgh, 1912, 189.]
65. Culgroat, Wigtownshire: [NAS, NG1/1/25, 19 July 1786.]
66. Douglas, Oglethorpe, Hume, 1984, 11–18.
67. ibid; Shaw, 1984.
68. Butt, 1967.
69. Donnachie, 1971.
70. Hume, 1976, 1977.

BIBLIOGRAPHY

Butt, J Donnachie, I L, Hume, J R. *Industrial History in Pictures: Scotland*, Newton Abbot, 1968.
Butt, J. *Industrial Archaeology of Scotland*, Newton Abbot, 1967.
Caldwell, D H, Dean, V E. Post-medieval pots and potters at Throsk in Stirlingshire, *Review of Scottish Culture*, 2 (1986), 105–12.
Campbell, R H. *Carron Company*, Edinburgh, 1961.
Dalgleish, G, Maxwell, S. *The Lovable Craft, 1687–1987*, Edinburgh, 1987.
Donnachie, I L. *A History of the Brewing Industry in Scotland*, Edinburgh, 1979.
Donnachie, I L. *Industrial Archaeology of Galloway*, Newton Abbot, 1971.
Douglas, G, Oglethorpe, M, Hume, J R. *Scottish Windmills: A survey*, Edinburgh, 1984.
Durie, A J. *The Scottish Linen Industry in the Eighteenth Century*, Edinburgh, 1979.
Fairweather, B. *A Short History of the Ballachulish Slate Quarry*, Ballachulish, npd.
Fleming, J A. *Scottish and Jacobite Glass*, Glasgow, 1938.
Graham, A. A note on the making of nails by hand, *Scottish Studies*, 5 (1961), 117–19.
Gulvin, C. The origins of framework knitting in Scotland, *Textile History*, 14 (1983), 57–65.
Harrison, J G. Wooden huts and shelters in seventeenth-century Stirling, with an example of a hingin' lum, *Vernacular Building*, 18 (1994), 2–6.
Hay, G D, Stell, G P. *Monuments of Industry*, Edinburgh, 1986.
Hume, J R. *The Industrial Archaeology of Scotland*, Vol 1, London, 1976.
Hume, J R. *The Industrial Archaeology of Scotland*, Vol 2, London, 1977.
Kerr, D. *Railway Sleeper Buildings*, Dundee, 1986.
Kirkcaldy Museums. *George Lowrie, Clay Pipe Maker*, Kirkcaldy, 1979.
Kirkland, J. *The Baker's ABC*, Glasgow, 1927.
Lloyd, G. Scottish linen hand-loom weavers' houses, *Vernacular Building*, 13 (1989), 66–8.
Lockhart, D G. The planned villages. In Parry, M L, Slater, T R, eds, *The Making of the Scottish Countryside*, London, 1980, 249–70.
Mitchell, Sir A, ed. *Macfarlane's Geographical Collections* (Scottish History Society 53), 1908.
Murray, N. *The Scottish Handloom Weavers, 1790–1850*, Edinburgh, 1978.
Scott, W R, ed. *New Mills Cloth Manufactory, 1681–1703* (Scottish History Society XLVI), 1905.
Scott, W R. *Joint-Stock Companies to 1720*, Vol 3, 1911.
Shaw, J P. An introduction to the technology of meal milling in Scotland, *Scottish Industrial History*, 6 (1982), 8–24.
Shaw, J P. Dutch – and Scotch – pantiles: Some evidence from the seventeenth and early eighteenth centuries, *Vernacular Building*, 14 (1990), 26–9.
Shaw, J P. *Water Power in Scotland, 1550–1870*, Edinburgh, 1984.

Smout, T C. The landowner and the planned village in Scotland, 1730–1830. In Phillipson, N T, Mitchison, R, eds, *Scotland in the Age of Improvement*, Edinburgh, 1970, 73–106.
Thomson, A G. *The Paper Industry in Scotland*, Edinburgh, 1974.
Walker, B. The bucket meal, Finzean, Aberdeenshire, *Folk Life*, 21 (1982–3), 71–82.
Watson, M. *Jute and Flax Mills in Dundee*, Tayport, 1990.
Whatley, C A. *The Scottish Salt Industry, 1570–1850*, Aberdeen, 1987.

24 Mills and Factories

MARK WATSON

The factories described here date from the period of the Industrial Revolution onwards. This is conveniently held to have begun with the foundation in 1759 of the first integrated coke-fired ironworks at Carron, Stirlingshire. Only a few large manufacturing places of employment, such as paper mills, existed prior to this date and fewer still substantial remains have been detected above ground. No full account of Scotland's industrial prowess or pioneering place in the development of many technologies will be attempted, but an impression of the richness of the physical evidence may yet be obtained. For details of the locations of the buildings mentioned below, readers should consult John Hume's two-volume *Industrial Archaeology of Scotland* (1976 and 1977), which is nearly comprehensive and fully map-referenced.[1]

This chapter will consider, largely from personal observation, generic aspects common to all industrial building, features and facilities (fig. 24.1), and then will look at the special characteristics of various building types. Regional variation may be noted both in terms of industries associated with particular places, and more meaningfully, in industries that are found across Scotland and further afield.

BUILDING FEATURES AND FACILITIES

Walls
It is to early industrial buildings that historians of modern architecture look for the crucial transition from load-bearing walls to a full metal frame. Externally at least, the eighteenth- and nineteenth-century factory looks traditionally built with load-bearing walls, albeit on the large side. Glass curtain walls were made possible following the development of the iron frame, but their use was tempered by consideration of climatic conditions found in Scotland.

A small number of twentieth-century buildings have glass curtain walls; examples include dyeworks at Tay Works, Dundee, and at Turnbull's, Hawick, Roxburghshire, both of 1920, the Gassing Mill at Anchor Mills, Paisley, and the lamp-testing tower of the Luma Lightbulb Factory, Glasgow, of 1936.[2] Harland and Wolff's Clyde Foundry in Govan,

Figure 24.1 Kirkcaldy, Fife; aerial view of an urban industrial landscape: flax and flour mills, lineoleum - and engineering works. Crown copyright: RCAHMS, SC 711520.

Glasgow, was both the largest foundry and also probably the largest glass-clad building in Europe when completed in 1923. It operated at a loss for most of its life until closed in 1964.[3] Sir E Owen Williams completed in Albion Street, Glasgow, the third of his *Daily Express* printing works by sheeting the elevation, above reinforced concrete cantilevers, entirely in glass and vitrolite.

Local stone was the principal nineteenth-century building material, and the architectural expression of the building depended mainly on the rhythm, proportion and repetition of simple elements, principally windows. This new architecture has been described as 'the functional tradition'.[4]

Brick was used from an early date for specialised structures such as kilns. Sir Richard Arkwright built the first Stanley cotton mill, Perthshire (1786), of brick on a stone base, to resemble his Masson Mill in Derbyshire. In Glasgow, Houldsworth's and Old Rutherglen Road Mills, of 1804 and 1816, were entirely of brick, the former with a giant order of pilasters reflecting the rhythm of the bays inside. This technique helped to obtain narrow wall thicknesses at the windows and was a motif widely adopted in mid- and late Victorian western Scotland.

Polychromy, promoted by John Ruskin, was simply achieved by the manufacture of specially shaped yellow-white or buff bricks for quoins, jambs, sills, corbels and lintel arches. The earliest dated example yet found was at Greenock, Renfrewshire, in 1859. Allan & Mann of Rutherglen Pottery (active there 1861–87)[5] marketed a patent range of buff bricks, to be found on a rash of buildings erected in eastern Scotland in 1864–6, culminating in Cox's Stack, Dundee. A Greek key course is picked out on Hayford Mills near Stirling, and the type is also found at Glentana, Burnside and Ochilvale Mills in Alva, Clackmannanshire. The style was much favoured by Bruce & Hay, architects for most of the Scottish Co-operative Wholesale Society factories in Kingston and Shieldhall, Glasgow. The culmination was Templeton's Carpet Factory, Glasgow, by William Leiper, 1882–92, modelled on the Doge's Palace, Venice, and as colourful as the Axminsters it produced.

Reinforced concrete (discussed below) was more often than not masked by external brick walls. Mass and pre-cast concrete was sometimes used for exterior walls as early as *c.* 1870. Glen Nevis (1878, mass concrete), and Bruichladdich (1881, concrete block) distilleries are examples, respectively in Fort William, Inverness-shire, and Islay, Argyll, where dressed stone was relatively expensive.

Floors
When the Industrial Revolution began, the only known means of bridging between load-bearing walls was by timber beams and joists, or by stone vaults. The latter had the advantage of incombustibility but were otherwise dark and dank. Stone spine walls or thick pillars made them impractical for most industrial processes. Stone vaults were therefore reserved for use in the lower parts of warehouses, bonded stores and breweries.

Timber continued throughout the nineteenth century to be the most common means of flooring industrial buildings. Its resistance to fire could to a degree be enhanced by lath and plaster or by fitting beads at the joins between ceiling boards. Alternatively, by laminating timbers together it was possible to produce floorboards 2 to 4 inches (5–10 centimetres) thick that would char rather than burn. This 'slow burning' technique became standard in America but is uncommon in Scotland. Quayside Mills, Leith, offered a rare example.

Timber beams, as at Rosebank Distillery, Falkirk (1864), could be strengthened by sandwiching iron or steel flitch plates.[6] They could alternatively be tensioned by wrought-iron rods running either side of the beam, to meet at one or two cast-iron stirrups. This – the composite or trussed beam – was sometimes used to obtain the necessary spans over woollen mules (for example, at Shaws Water Mill, Greenock) in those weaving sheds where wide column spacing was required (such as the Dundee and area factories built by Robertson & Orchar) or to achieve exceptional floor loadings in sugar refineries. The exterior of the building may carry a regular series of cast-iron tie plates, often mistaken by surveyors as evidence of attempts to remedy defects in the building.

Spans of much more than 30 feet (9.14 metres) required intermediate supports. Square timber posts were ready to hand and continued to be used in warehouse construction for much of the nineteenth century. To spread the load and avoid crushing, it was found advisable to wedge timber pads above and below the post.

Cast-iron columns offered much greater compressive strength and therefore could be more slender. First used in Russia, and in the galleries of English churches, the oldest industrial building yet found to contain them is the Bell Mill at Stanley Mills, Perthshire, of 1786. The columns are cruciform with pads, and without interconnecting spigots. Subsequent cotton mills in Renfrewshire (for example Cartside, 1792) have cylindrical columns with separate capitals into which the columns socket. Some have a simple bell capital form. Others have flanges to spread the compressive pressure that was still taken by the timber cross beam, or longitudinal beam, or joist.

The next evolutionary stage was the interconnecting column, which transferred the load through the floors and by-passed the beam. The column might spigot through the beam, not a very satisfactory arrangement and generally adopted only where beams met anyway, or where the column was clamped between two beams. The other method is the saddle, where the load would be transferred through a base plate and down either side of the beam to the next column. Again, this seems first to have been adopted in textile mills in the 1820s, was found at Morton's Bond, Dundee (1836), and was almost universal in maltings in the mid to late nineteenth century.

The development of the iron frame is of fundamental importance to modern architecture. The system was developed in Derbyshire cotton mills but first applied to iron beams in Ditherington Flax Mill, Shrewsbury, in 1797.[7] The technique was not slow to reach Scotland. The conduits appear to have been via Manchester to Glasgow (1804), respectively the biggest and second biggest cotton spinning cities in the world, and via Leeds to Dundee (1806–7) and Aberdeen (1808), which similarly disputed the title of world flax-spinning capital.[8]

The Manchester manufacturer, Henry Houldsworth, erected a cotton mill in Glasgow in 1804–5 that in many respects resembled the first iron-framed cotton mill, the Salford Twist Mill of 1799–1804.[9] Both were engineered by Boulton & Watt and had seven floors carried on two rows of hollow cylindrical columns through which exhaust steam warmed the building. Both have been demolished, but Houldsworth's seems to have set the pattern for other large cotton warp-spinning and thread-twisting mills. One of 1816 is on Old Rutherglen Road, Hutchesontown, Glasgow, with twin rows of cast-iron columns of diminishing thickness on each brick-arched floor. Other iron-framed cotton mills dating from 1823–33 survive at Stanley and New Lanark, Lanarkshire.

There are exceptions to the use of brick jack arches in fireproof construction. A small number of mills in northern England, and some naval dockyard buildings, have stone flags laid directly onto a grid of cast-iron joists. Grandholm Mill, Aberdeen, has a similar construction, an alteration to the original mill of 1792, as does the link between Mills 3 and 4 at New Lanark, c. 1833–40.

The form of the cast-iron beam was the subject of debate. A more or less parabolic hog-back was favoured, the centre deepest in order to eliminate a weak point. Usually this was hidden within the depth of the floor. Sometimes, such as when carrying a water tank, or mash tuns in a brewery, the hog-back was inverted, becoming a fish belly, or was eliminated in favour of parallel sides. The oldest beams were of an inverted T-section, and tended to have a narrow bottom flange. A cambered underside features in the kit supplied by Fenton Murray & Wood of Leeds to Broadford Mill, Aberdeen, of 1808. Iron-framed buildings of the 1820s and 1830s favoured flat bottom flanges of around 6 inches (15 centimetres). Experiments in Manchester by Tredgold (published 1822) and Hodgkinson (1830) each put the case for an upper flange giving an I-section beam.[10] Hodgkinson's beam has a smaller top than bottom flange, the latter also having a parabolic form. William Fairbairn applied these to at least three mills in northern England in the 1830s but the earliest application in Scotland was a little later. Barbush and Broadlie flax mills in Renfrewshire have the type, built in the 1850s, and Dundee engineers Robertson & Orchar built several mills in the 1850s–60s employing the formula. Other engineers seemed content to adopt the I-section with parallel-sided rather than parabolic bottom flanges.

Exceptionally, a cast-iron beam might be required to span more than 30 feet (9.14 metres) or to carry an eccentric load. One solution was to tension the cast-iron beam in a wrought-iron truss. There are interesting examples at Seafield Works, Dundee, by Robertson & Orchar, at Port Dundas sugar refinery, Glasgow, and at Andrew Barclay's Caledonia Engineering Works, Kilmarnock, Ayrshire.

Figure 24.2 Arthur Street Works, Greenock, Renfrewshire; interior of pattern store showing timber joists (cut through) on wrought-iron beams and cast-iron columns. Crown copyright: RCAHMS, SC 712300.

William Fairbairn disapproved of this device, and promoted instead full wrought-iron beams.[11] These may be identified by their narrower section, the absence of curves and by the profusion of rivets with which the larger type of beam, or bressummer, was fabricated. The section is usually an I, with equal top and bottom flanges, and sometimes a box. They are often to be found above shops and pends but their regular use in industrial buildings is relatively rare (fig. 24.2). Fairbairn pointed to the example of a sugar refinery built with concrete arches between wrought- or malleable iron beams.[12] The beam's earliest use so far noted in Scotland seems to be in an 1855 part of Eglinton Engine Works, Glasgow, which made machinery for sugar refineries. Wrought-iron beams are found in combination with cast-iron beams and joists at sugar refineries at Bonnington, Leith, 1864, and at Port Dundas, Glasgow, 1866.

A Glasgow variant found in commercial buildings, such as Gardners', Jamaica Street, 1855, was the McConnel patent beam, which consists of two parallel wrought-iron bars separated by small iron castings, or blocks of wood, with timber joists threaded between them.

Mild steel girders were first rolled in 1879 by Redpath & Brown and came into use in composite buildings in the 1880s and 1890s. Steel is easily confused with wrought iron, which it supplanted, but it rusts much more readily. Initially, steel was used in a limited way, in short joists between substantial cast-iron beams or wrought-iron girders. Most of the distilleries built in the 1890s' boom possess timber floors, on which concrete would be laid to facilitate malting, on steel beams slotted through the saddles of cast-iron columns. The first true steel-framed building in Scotland, in which the steel columns rise independently through floors that are bolted to them, is the Scotsman Printing Works, Edinburgh, 1899–1902.[13]

Concrete was used in 1849 for upper floors in Alexander's thread mill, Glasgow, by Charles Wilson, corrugated iron serving as shuttering between the cast-iron beams. Mass concrete does not seem to have been taken up again until the 1880s in the mills built for thread magnates Coats & Clark in Paisley by architects Woodhouse & Morley. Both built in 1886–9, Ferguslie No 1 Mill and the Anchor Domestic Finishing Mill each have steel beams bearing a concrete filler joist floor.

Forms of reinforced concrete had experimentally evolved in Britain, but the creation of monolithic structures by forming continuous columns and beams and spirals was foreign.[14] François Hennebique patented and aggressively marketed his French system of hoop and stirrup reinforcement in the 1890s. The first Scottish building erected to his system, after piles and a bridge in Dundee of 1902–3, was the Sentinel Works pattern shop of 1904, Jessie Street, Glasgow (fig. 24.3), precursor to Lion Chambers. Both are exceptions to the rule that such buildings were generally clad in brick or masonry. Although part of Ladhope Mills, Galashiels, was rebuilt in 1912 with reinforced concrete piers fronting the Gala Water, externally expressed reinforced concrete was to remain rare in Britain.

Other systems include those of Coignet (from 1904) and Considère (at for example the Linthouse Shipyard office in Govan). A British system – E P Well's – was used at Portobello Chocolate Factory, Edinburgh,

Figure 24.3 Sentinel Works, Jessie Street, Glasgow; pattern shop in Hennebique Ferroconcrete, 1904. Crown copyright: RCAHMS, SC 710977.

brick clad and with a flat roof, in 1908. Reinforced concrete was particularly promoted for its strength, fire resistance and ability to be kept clean, and so was favoured in printing works, bonded warehouses and food factories. The curves and cantilevers made possible by the material saw its use in water towers, silos, loading bays and quaysides.

American influence came to bear with the 'Daylight Factory', on a more ambitious scale in Scotland than almost anywhere outside Detroit: Truscon (established in London in 1907) built to the Kahn System (patented 1903); Weirs Foundry (1912) and the Albion Motor Car Works in Glasgow, (1913–14); the Park Motor Car Body Works, Kilbirnie Street, Glasgow (1914); munitions factories; and most spectacularly, the Arrol-Johnston Motor Car Works in Dumfries, (1912–13).

Roofs

A 40-degree pitch suited Scottish slates. Shallower pitches might be clad in felt, paper or sheet metal. The roofs of early continental industrial buildings tended to be of very substantial timber trusses over a greater pitch; they had mansards in France or Belgium, while Russian and Scandinavian factories had shallower pitches clad in sheet metal. Monitor or clerestory roofs, sometimes stacked in two tiers, were favoured in New England, on the European continent and in Gloucestershire, but not in northern England or Scotland. The double-decked attic is, however, found in Borders woollen mills, where widths of 45 feet (13.72 metres) were common. The upper attic was most often for storage and the lower was a top-lit working area. Dormers are almost unknown, except over hoists and toilets.

Roof structures drew on a carpenter's repertoire of king-post, queen-post and collar-beam. Dunfermline's damask warehouses have wide-span, scissor-braced timber trusses. Laminated timber arches are rare in Scotland. A few Belfast roofs, lightweight timber lattice trusses in a segmental curve, by D Anderson & Son, Belfast, made their way across the North Channel and achieved spans of up to 100 feet (30.48 metres), generally in the period 1900–30 (fig. 24.9, left).

With an ever-increasing span as the aim, wrought-iron ties were introduced to timber trusses, vertically to substitute for king- or queen-posts, or diagonally from the centre of the collar beam to transfer stresses to the wallhead. This lightened the structure, and once the diagonal struts had been replaced by wrought-iron angles and the tie beam had likewise been replaced by tension rods, the entirely wrought-iron roof truss was created. They can be found in train sheds and over mills of the 1849–51 period; examples include Alexander's, Duke Street, Glasgow, Baltic Works, Arbroath, and Edward Street Mill and Tay Works, Dundee. Later steel roofs followed the same principle in forming triangles of compression and tension members.

A simpler variant is the Polonceau truss, developed in France. A cast-iron strut, often enriched by acanthus-type scrolls, projects from the mid point of each principal, and serves as the key for a triangulated truss of wrought-iron tension rods. This economical solution was favoured by Robertson & Orchar, Dundee engineers, in the 1860s–80s.

Cast-iron roof trusses are plentiful in the Dundee area, and are recorded in 16 spinning mills in the city, two in Kirkcaldy, Fife, and in part of Pullars Dyeworks, Perth.[15] The first of these was erected in 1850, and the last in 1885. Arches could span right across or, more often, would adopt a nave and aisle arrangement that prompted remarkable Gothic tracery. This form of roof is not found in similar building types elsewhere in the world.

Single-storeyed sheds are described below, under Engineering Works and Weaving Factories. In the twentieth century, standard light-weight steel trusses predominated. The areas they covered were flexible and multi-purpose, with a minimal number of obstructions. In the later twentieth century, designers sometimes sought to exclude natural light altogether. There have been only a few interesting variants in reinforced concrete – for example Waverley Mill, Innerleithen,

Peeblesshire, and the mercerising shed at Anchor Mills, Paisley, Renfrewshire – or shell concrete, such as in Shettleston, Glasgow, for the Gourock Rope Company and in Perth at Dewar's whisky bottling plant.

Power
Physical evidence of horse mills within industrial complexes has rarely been noted in Scotland, but they certainly were used in the eighteenth and early nineteenth centuries.[16] In Kirkcaldy, East Bridge Flour Mill has a circular building of the appropriate diameter, and at 29 St Clair Street an indigo mill (for dyestuffs) was worked by horse. Wind power was surprisingly little used, to the extent that it is possible to produce a definitive inventory of surviving Scottish windmill towers – none with sails and machinery intact – mainly for threshing and pumping, a few for grinding linseed (for example, The Shore, Leith).[17]

Water wheels are much more conspicuous, thanks to the necessary infrastructure of lades, or leads, and tailraces. They also were of extraordinary importance to the Industrial Revolution in Scotland, and many sites initially located to exploit water in the eighteenth and nineteenth centuries (up to and including those in Walkerburn, Peeblesshire, 1855) continued to develop in the age of steam.[18]

Apart from the horizontal mills that dotted the burns of the Highlands and Islands, mills with vertically-turning wheels may be classified according to the point at which the water hits the wheel: overshot; high and low breast; and undershot. The wheel pit would generally be lined with impermeable ashlar, and if low-breast, might be curved to maximise the impact of the water. The regular power required in spinning or paper mills was obtained from the width of the wheel, but diameter was not in itself of much value. Broad wheels survive at Tower Mill, Hawick, Philiphaugh Saw Mill, Selkirkshire, Perth City Mills, and Keathbank and Ashgrove Flax Mills in Blairgowrie, Perthshire. The more common narrow, but sometimes large diameter, wheels tended to be reserved for grinding, threshing and crushing mills where an intermittent action did not matter.

The biggest wheels were developed by Hewes & Wren and William Fairbairn of Manchester in the 1820s–30s, using rim gearing and suspension construction, like the spokes of a bicycle wheel. An example of 1826 from Woodside Paper Mill, Aberdeen, was until recently exhibited in the Royal Museum of Scotland. The greater scale of these wheels meant re-location of the wheelhouse from within the structure of the textile mill to detached wheelhouses parallel to the mill. At this period, lades were redirected from smaller internal to large external wheels at Stanley, Perthshire, Catrine, Ayrshire, and Fereneze, Renfrewshire, cotton mills, all founded in the eighteenth century. For example, Grandholm Flax Mill, Aberdeen, of 1793–4 had two 20-feet (6.1-metre) diameter wheels, 8 and

9 feet (2.44 and 2.74 metres) wide, in the basement, besides four other wheels in the hackling, bleaching and mechanics departments. In 1826 the two 20-feet (6.1-metre) wheels were supplanted by a 120-horsepower wheel with peripheral drive, 25 feet (7.26 metres) in diameter and 21 feet (6.1 metres) wide in a detached wheelhouse. Connected to this system were three steam engines of 40 (*c.* 1820), 60 (1826) and 70 (*c.* 1833–54) horsepower, which barely sufficed to counter the irregular water supply consequent upon disputes with neighbouring water users. The wheelhouse now contains Boving turbines installed in 1938.

Turbines were first developed in France in the 1820s and taken further in Scotland in the 1830s and 1840s. The Whitelaw or Scotch turbine was first made in 1839, and the James Thomson inward-flow Vortex turbine in 1850, a line continued to this day by Gilbert Gilkes of Kendal.[19] Continental impulse, or Girard, and American reaction, or Francis turbines, were soon also being made under licence by several Scottish engineers, such as Thomas Aimer of Galashiels. They channelled the available water efficiently and frequently replaced wheels. As they took up less space, they might be floored over and forgotten when no longer required. Several still exist in the lades of woollen mills in the Borders. Large examples by Boving of Sweden are also to be found in New Lanark, Paton's Johnstone Mill, Renfrewshire, and Ladhope Mills, Galashiels, Selkirkshire.

Steam engines transformed Scottish industry, ensuring that the momentum first given to the Industrial Revolution by water power did not falter once the best sites had been exploited. Boulton & Watt installed 24 rotative engines of the sun and planet type in Scotland between 1786 and 1802: nine cotton mills, four distilleries, three flax mills, two flour mills, one bleachworks, one glass works, one brewery, one forge, one chemical works and one paper mill.[20] Already other makers were competing, and after 1800 and the expiry of Watt's patent the number burgeoned.

Only one rotative beam engine survives in situ in Scotland, at Garlogie, Aberdeenshire.[21] The engine houses of many others are extremely informative. House-built engines had their entablature of timber or cast iron spanning wall to wall within the mill or factory. These often survive, sometimes with the columns (frequently Doric) that supported them and the platform that wrapped around the rocking beam. The masonry foundation for the cylinder and slots for the flywheel and the condenser allow a conjectural reconstruction of the engine, even if all the iron work is missing. Other engine types – the A-frame, side-lever, grasshopper, true vertical, inverted vertical and horizontal – tended to be free-standing and self-contained. Once removed, there is less archaeological information to be gained other than the groove for the flywheel and the odd bolt.

The engine house was the showpiece of the mill. Mosaic floors, tiled walls and elaborately finished ceilings, covered or with hammer-beams, are not uncommon. Most ornate were those of J & P Coats in Paisley. The exterior would be highlighted by a tall window, often arched and with removable timber framing through which parts of the engine could be swung. Dundee and district flax mills by Umpherson & Kerr of 1828–1850 adopted modish tripartite windows, possibly modelled on the Exchange Coffee Rooms, also built in 1828. Several beam engine houses in Borders woollen mills of 1860–80 carry cast-iron water tanks on their roofs for condensing, instead of using cooling ponds.

Gas engines and oil engines were attractive to light users of power, such as distilleries and malt barns, in the late nineteenth to mid twentieth centuries. They persist, particularly in the Northern and Western Isles.

Power transmission
Rotary power was transmitted from the power source to the machine by way of vertical driving shafts that were of timber in the eighteenth century, to be supplanted in the early nineteenth century by square- and then circular-section cast iron. Direct drive from below evolved in the early nineteenth century into a more flexible arrangement of a primary vertical shaft serving secondary line (from 'lying') shafts, initially of cast iron (remnants of these were found in Stanley Mills) and soon afterwards by wrought iron. Complete surviving examples are few and far between. Dangerfield Mill in Hawick (1872–3; fig. 24.4) is now almost unique.

Rope drives started to appear in the 1870s and were advantageous when power was carried some distance (for example after 1882 across the lade at New Lanark) or ropes were readily available, as at Gourock Rope Works, Port Glasgow, Renfrewshire. Rope alleys are also found in sheds, such as at Grandholm Works, Aberdeen, and Waverley Mills, Innerleithen. However the system was not as widespread in Scotland as in Lancashire.

On the factory floor rotary motion was transmitted down, and sometimes up, from the line shafts by flat belt pulleys of leather, cotton or gutta-percha, the last two being special products of Stanley Mills and Dick Brothers of Glasgow respectively.

The arrival of electric motors was marked initially by the placing of these at the ends of line shafts. The first all-electric jute spinning mill was Hillbank Linen Works, Dundee, in 1907. Individual electric motors attached to each machine are now the norm in all factories.

Chimneys
Variously, and evocatively, described in Scotland as 'lum', 'stack', or 'stalk', the chimney was at once landmark and symbol, for good and bad, of the Industrial Revolution.

Figure 24.4 Dangerfield Mills, Commercial Road, Hawick, Roxburghshire; lineshaft bearing in mule mill. Crown copyright: RCAHMS, SC 712307.

The first stacks were simply oversized domestic chimneys tacked on to buildings, rising in the case of the mills at Grandholm Works, Aberdeen, from the cores of stair towers. As the draught increased with the efficiency of engines and boilers, so the building of chimneys became a specialised task.[22] The earliest stacks were usually of stone, but might at their upper stage be of brick. By 1830, brick predominated in Scotland, and free-standing chimneys began to sprout. A square section remained the norm until c. 1870. From the 1860s to the 1880s, and as late as 1899 in the case of Anchor Mills, Paisley, octagonal brick, sometimes polychrome, thanks to the use of buff-coloured angles, offered a variant. Circular-section stalks first shot up in the 1850s and were to dominate the period from 1880 through to 1960 (Kincardine Power Station, Fife). Tallest in the world at one time was Tennant's Stalk of the St Rollox chemical works, Glasgow.

These chimneys were originally topped by cornices and oversailers to prevent smoke from drifting down their sides. In many cases these were removed when steeplejacks effected basic repairs by shortening the stalk. Three chimneys built at Dens Works, Dundee, were in the form of Egyptian obelisks. Cox's Stack, Lochee, Dundee, 1865, is an Italian campanile 282 feet (85.95 metres) tall that conveyed smoke from 39 boilers via flues tunnelled through Camperdown Works.[23]

Boilers

The earliest types of boilers were 'haystack' (circular) and then 'waggon' (a flat-bottomed tube). There followed the 'egg end' (a tube domed at each end), the Cornish (a tube containing a single flue) and Lancashire (twin flue). Multiple-tubed boilers followed in the later nineteenth century, developed first in marine engineering. They are occasionally to be found re-used as water or oil tanks.

A boiler house is often distinguishable by a series of arched openings corresponding to the number of boilers installed via those arches. The arches were also useful for the delivery of coal and the achievement of draught. Flax mills of the 1820s–30s and woollen mills of the 1850s–80s might have the rooms above these designated 'drying rooms', using in the flax mills (for example, Chapel Works, Montrose, Angus, 1828) slatted wooden floors and in the woollen mills grids of perforated iron (for example, Nether Mill, Galashiels, Selkirkshire).

An economiser might intercept smoke travelling from boiler to chimney and use it to warm water on its way to be turned into steam. They were generally supplied by the patentees, Green's of Wakefield, Yorkshire.

Ventilation

Ventilation was commonly achieved via roof and window. In some circumstances, such as where humidity was required in cotton spinning, windows would be hard to open. Fixed-pane cast-iron framed windows were universal in England, but the openable sliding sash and case window held sway in Scotland from the 1780s to the 1870s.

Louvres tend to indicate the location of specific functions in tanning, brewing, paper-making and the drying of starched warp threads prior to linen weaving. Louvred ridge ventilators sit over abattoirs at, for example, Linlithgow, West Lothian, Tain, Ross and Cromarty, and Edinburgh.

Mechanical fans for the extraction of dust either below floors or above machines tended to be a consequence of factory legislation applied to different branches of industry in the later nineteenth century.

Additional humidification was an important element in some spinning and weaving processes, the comfort of the operatives coming second to the need to avoid breakage of yarn. This was obtained from a variety of artificial heat sources, such as warm air cockles (as at Stanley and New Lanark Mills), but where steam was plentiful, hot water pipes became standard.

Artificial light
Gas lighting was pioneered by William Murdoch, and initially his plant was supplied by Boulton & Watt, Broadford Works in Aberdeen being an early recipient in 1814. As they permitted the lengthening of hours worked, gasworks were particularly valued in rural water-powered sites. These included Stanley and New Lanark Mills in the early 1820s, and other, very small sites, such as the flax mills at Auchenblae, Kincardineshire, and Hatton, Angus.

Electric lighting was applied early to water-powered sites, such as Comelybank Mill, Galashiels, where water turbines could easily be devoted to generating electricity. So, from the 1880s onwards, long before town gas gave way to electric street lighting, gas was replaced in mills by electricity, which was safer and gave stronger light. Electricity also permitted social control; all lights in New Lanark village were extinguished at 10pm from a single switch within the mill.

Sanitation
The problem of achieving satisfactory sanitation in large buildings packed with people taxed some of the more enlightened factory managers, and was graphically described in 1819 by William Brown at East Mill, Dundee.[24] A long-disused cast-iron bowl and pipe survives at the Old End, Johnstone Mill, as did timber seats in a triangular shaft projecting from Cartside Mill, both in Renfrewshire. Attic dormers in Chapel Works, Montrose, Angus, and Brothock Mill, Arbroath, Angus, lie over earth closets. Two-seater water closets are found over the lades of sawmills, at, for example, Kinghorn, Fife, originally a plash mill.

Giving relief in dusty environments, drinking fountains offered a hygienic alternative to the provision of common pitchers and to concerns about the transmission of disease, but their provision could be variable.

Walls in textile mills were generally plastered on the hard wall surfaces and received an annual lime-wash as disinfectant. A dado line about 6 feet (1.83 metres) up the wall and columns might be decorated with stencils, beneath which a blue or a green paint would be applied in order to limit glare below eye level. Floor surfaces varied. Stone flags, quarried in Angus, covered the floors of local fireproof mills. Quarry tiles were more frequently used in English and Irish flax mills, both as a consequence of the relative cost of flagstone and because of their utility in resisting water from wet spinning, the dry process being more common in Scotland. Cotton mills, such as at New Lanark, used tiles, and in the top floor cast-iron plates. Those in Paisley from the 1880s–1900s had maple floorboards fixed to the otherwise fireproof floor to give a grip to barefoot mule spinners.

Hoists
The American, E G Otis, made lifts safe for passenger use in 1854 by adding a braking system, but factory hoists and goods lifts were in use well before then, and without the benefit of such safety features. A small amount of the power employed in a mill could be directed by gearing or pulleys to haul a lift or 'teagle'. Goods in textile mills were often bumped down stairs in baskets. Sugar refineries carried materials of such weight that a substantial shaft would run down the well of the stair, but efforts were made to transfer materials there, as in breweries, in liquid and granular form by gravity. Travelling cranes in engineering works would transfer power from stationary engines by ropes, pulleys or chains until the advent of electric motors.

Hydraulic lifts raised loads from below by a ram. Hydraulics were first applied to a crane in Newcastle by William Armstrong in 1846. Most examples date from the 1860s–90s and are especially found close to docks or where there was access to a grid of pressurised water, although mains water would suffice and could be pumped to raise pressures. Presses and other machines in the metal, textile finishing and paper industries could also be hydraulically powered.

In warehouses and granaries without access to mechanical power, reliance fell on hand winches that projected from the building, sometimes enclosed by lucarnes (more common in England than Scotland), and on spiral chutes for bagged goods. Where the product could be transferred loose, such as in grain mills or maltings, Archimedean screws, band conveyors and bucket elevators might be used. The elevator at a malt barn might be crowned by a pagoda to match that over the kiln. In World War I, overhead runways and self-delivering hoists came to the aid of female employees in all kinds of factories and warehouses, and stayed thereafter.[25]

Bell, clock and water towers
The first generation of industrial workers was unfamiliar with factory discipline. Bells and clocks had to be used to summon each shift and also became invaluable to the working of the Factory Acts. Bell towers were the architectural foci of New Lanark, Lanarkshire, Catrine, Ayrshire, Stanley, Perthshire, and Grandholm Mills, Aberdeen. A domed and columned cupola was favoured, failing which a simple 'birdcage' might sit over a Palladian pediment or a gable. Dens Works, Dundee, has one of its two bell towers exactly modelled on those of S Maria della Salute, Venice. However whistles, 'bummers' or hooters employing exhaust steam sufficed at most factories.

A combined clock and bell tower dominates the 32-acre (12.95-hectare) Camperdown Works, Dundee, and there was a second substantial clock at the entrance. Perhaps unique in an industrial complex, the tower held a peal of six 'sweet and sonorous' bronze bells, all cast at Whitechapel, London, which chimed workers into and out the complex.

Clock towers dominating Wilton Mill, Hawick, Sanderson & Murray's skinworks in Galashiels, and Singers' Kilbowie Works, Clydebank, Dunbartonshire (1884, claimed to have the largest clockface in the world, at 26 feet (7.93 metres) diameter), served as municipal statements. However, most mill clocks had single faces to be viewed from the courtyard, such as within Abbotshall and Coal Wynd flax mills in Kirkcaldy. Kilncraigs Mill, Alloa, Menstrie Mill and Clock Mill, Tillicoultry, all Clackmannanshire, each have clocks as architectural foci. New Lanark's clock was a uniquely ingenious device recorded in 1795 as being regulated by the water wheel.[26] At Stanley Mills, each flat had a vertical slot in a wall which seems to have contained a timepiece regulated from a central point.

Towers also served as tanks for process water or for sprinkler systems. Early cast-iron grids carried tanks in Grandholm, Aberdeen, and Stanley Mills. Elongated water towers with balustraded parapets were added *c.* 1880 to Cartside, Crofthead and Johnstone mills in Renfrewshire.

Offices
The counting house was, in the early stages of the Industrial Revolution, relatively small and unassuming, and the main commercial activity would be carried out in market places far from the mill or factory. On site, there was need only to provide some managerial accommodation, and a pay office containing a safe. The telegraph in the 1850s and the telephone from the 1880s allowed commercial transactions to be conducted at a distance. The office at the works might therefore grow to accommodate clerks. Then, from the 1880s, the general adoption of limited liability required the addition of a directors' boardroom and architectural ostentation.

Figure 24.5 Kilncraigs Mill, Alloa, Clackmannanshire; aerial view, c. 1940. The school, office and yarn stores are in the foreground; scouring and dye-works behind, beside the tower; power station with timber cooling tower; five-storeyed woollen mills and single-storeyed worsted spinning mills; wool stores on the left. The Alloa corn mill and Youngers brewery are also in this view. Crown copyright: RCAHMS, SC 710864.

What is often presumed to be the office within a factory, owing to its possession of a greater degree of architectural ornamentation, proves to have fulfilled a number of other functions. The long Scots Baronial frontage to the Carron Ironworks, for example, also contained workshops, pattern shops and stores. Dunfermline's damask warehouses served also as showrooms, designing and embroidery rooms. Those at St Leonards, St Margaret's and Pilmuir Works are particular handsome Italianate buildings, lavishly fitted internally right down to the mosaic floors of the toilets.

Probably the most ostentatious office attached to a Scottish factory is that at Kilncraigs Mill, Alloa (fig. 24.5), seat of Paton & Baldwin's multi-national empire of hand-knitting wools. In fact, it obtains much of its presence from the fact that the Edwardian Baroque office is a thin skin added in 1904 to an existing very deep woollen yarn store that contained a massive range of stock. A directors' stair lined with marble and panelling, with a lift and a similarly lavish lavatory, is adjacent to a tiled 'wally close' stair for the workforce but separated from it by stained glass doors. A pneumatic messaging system transmitted orders from the office to the warehouse. The architect for this, and the yet more stylish modern movement wareroom of 1936 alongside, was William Kerr.

A & S Henry had a similar stair and smaller boardroom inserted in 1913 into an existing Scots Baronial-style Victoria Road Calender Works, Dundee, of 1874. A large counting house was lined with row upon row of clerks' writing desks from which jute factories in Calcutta were controlled. Verdant Works in Dundee contains a perfectly preserved smaller version of a commercial office with porter's lodge.

Half-time schools
New Lanark's School (1817) and New Institution for the Formation of Character (1809–13) are justly celebrated. The galleries installed in each by Robert Owen must have been partly in order to show off his model citizens to visitors. Although most textile mills required extensive child labour in the early nineteenth century, few other schools or apprentice houses of the period have been identified. A much smaller example with a bellcote served Valleyfield Paper Mill, Penicuik, Midlothian, from 1840.

An Act of 1844 sanctioned what was already happening, the part-time employment of children aged between eight and thirteen. The Half-Time School was a particularly Scottish educational phenomenon that arose in towns where the textile industry predominated and young workers were at a premium. It was a pragmatic solution that sought to meet the spirit of the Education Acts without denting employers' pockets. Children aged 10–14, later increased to 11 and then to 12 after 1891, spent alternate days or half a working day at school, and half of their time at menial tasks in the mill. Some schoolrooms were to be found improvised in attics of mills or warehouses. In most cases, towards the end of the century, arrangements would be made with the school board to provide tuition part time in church or board schools.

In a very few instances, the mill school was a source of pride to the paternalist mill owner, something that he would show his visitors. In rarer cases still, high academic standards were achieved. Mary Slessor, the missionary, was for example a mill girl at Baxter's of Dundee, whose school building is no longer extant but where the railings, gatepiers and horonised playground survive. Simple T- or cross-plan schools are found at Camperdown Works, Lochee, 1884, and Seafield Works, Dundee, 1890.

Architecturally more distinctive half-time schools include those for Alexander's in Duke Street, Glasgow, Panmure Works, Carnoustie, Angus (both Italianate), Kilncraigs Mills, Alloa, and St Leonards Works, Dunfermline (Gothic). Ferguslie Mill School, Paisley, is a pre-eminent example of a showpiece school. This very elaborately detailed Jacobean Renaissance building offered high-ceilinged, well-ventilated classrooms around a superbly panelled central hall.

Social facilities
The provision of communal dining, adult educational and other facilities

at New Lanark, while under the management of Robert Owen, 1800–25, was an exceptional experiment in social engineering. Its two substantial community buildings owe stylistic inspiration to the layout of Lancasterian schools, but there really is no parallel anywhere in the first half of the nineteenth century.

Budgett Meakin's 1905 study of model factories across the world alludes to Templeton and Coats as model employers, and particularly to the former for its provision of reading rooms and a concert hall for employees.[27] By the time C A Oakley was conducting his surveys into Scottish industry, 1934–53, more attention was being paid to social welfare by some of the firms singled out for approval.[28] Outstanding in this respect was the Wallace Scott Tailoring Institute, Cathcart, built in 1914 to designs by J J Burnet, and set in 16½-acre (6.68-hectare) gardens that anticipated New Delhi. It contained a concert hall, dining hall and on every wall an improving slogan, painted in soothing colour schemes. This is now the headquarters of Scottish Power.

BUILDING TYPES

Engineering works

In Scotland, famed for architectural cast-iron work and for engineering advances, the terms foundry and engineering works were often synonymous, while also retaining their specific meanings within the larger complexes.

Thus, in William McKinnon's Spring Garden Ironworks, Aberdeen (fig. 24.6), the foundry is one part of a larger complex that evolved from 1798 to 1908 and produced a variety of precision engineered products, particularly coffee plantation equipment. It was also capable of large structural castings, such as those for the Carron Bridge near Aberlour, Banffshire. At the centre were the twin cupolas in which pig iron was re-melted and then transferred by post cranes to be cast in beds of sand in the long foundry range. The post cranes were replaced in the late nineteenth century by a travelling crane, and fumes were extracted via a long clerestory ventilator. The adjacent boiler department was of three timber-framed bays. The assembling of larger items of machinery in the later nineteenth century required the construction of a taller three-bay erecting shop, the central bay served by a travelling crane flanked by galleries for the lighter work. A *c.* 1908 perspective view shows that the earlier tradition of assembling machinery outdoors in a courtyard by means of sheer legs (as, famously, at Carron Ironworks) co-existed with the newer covered methods.

Some small foundries, such as Falcon Foundry, Inverness, never developed covered shops, but remained as two-storeyed shops around open courtyards. The foundry at New Lanark made use of pulleys fixed

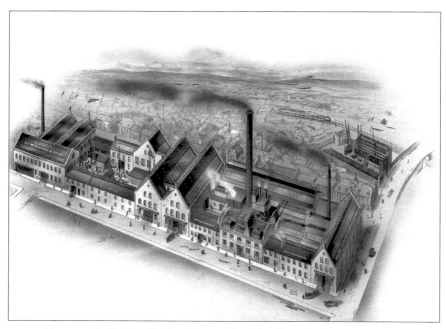

Figure 24.6 Spring Gardens Ironworks, Aberdeen; bird's-eye view, *c.* 1908–20. Crown copyright: RCAHMS, SC 710974.

to two masonry arches spanning the yard for the assembling of larger items, like waterwheels, that were cast there. Boiler works might be equipped with towers for hoisting and riveting the boilers. Larger foundries could operate more reliably under cover, and switched from post cranes and sheer legs to travelling cranes. Wide spans over long rectangular buildings served by travelling cranes required complex timber roofs, trussed with wrought iron (as at Ferryhill Foundry, Aberdeen, and Ward Foundry, Dundee) on buttressed masonry walls. Pierced walls to form arcades at Eglinton Engine Works, Glasgow (figs 24.7, 24.8), allowed transfers from one shop to another. The next step, linking shops together, saw the use of substantial I-section cast-iron stanchions to an extent unparalleled elsewhere in the world.

Marine engine works required tall iron-framed structures for the assembling of inverted vertical triple expansion engines, pioneered on the Clyde. Randolph & Elder, at Centre Street, Glasgow, 1858–60,[29] and then at Fairfield (Kvaerner Govan), 1869–71, built colossal and complex systems of cast-iron stanchions and struts to carry timber galleries between bays of 50 feet (15.24 metres) tall and 50 feet (15.24 metres) wide fitting shops. Dundee Foundry's 1870 marine engine works possesses iron-framed and brick-arched galleries beside a wide fitting shop.

Figure 24.7 Eglinton Engine Works, Cook Street, Glasgow. The four-storeyed building is the pattern shop and store over the ground-floor weighing machine shop; engine house between it and the original (1855) erecting shop. Crown copyright: RCAHMS, SC 712289.

Figure 24.8 Eglinton Engine Works, Cook Street, Glasgow; interior of heavy machine shop. Crown copyright: RCAHMS, SC 712293.

Ebenezer Kemp moved from Dundee Foundry to manage the Alexander Stephen engine works erected at Linthouse, Govan, 1871–4, designed by J F Spencer. Linthouse (now transferred to the Scottish Maritime Museum at Irvine) and Lithgow's, Port Glasgow, Renfrewshire, dispensed with galleries in favour of a single-storeyed layout with lower aisles for the lighter work. At Linthouse, alternate roof trusses are suspended from wrought-iron hangars, to enable passage of materials from nave to aisle. Crane rails at Linthouse were originally timber, but heavier demands required steel substitutes in 1882.[30] Malleable iron rails had from the start been fitted at Fairfield. The first bay at Andrew Barclay's Caledonia Works, Kilmarnock, Ayrshire, has heavy, fish-bellied cast-iron crane rails. On the other hand, a moulding shop added to Dundee Foundry in 1882 was cast-iron framed, with galvanised iron walls but a timber crane rail.[31]

Caird's (later Clark Kincaid's) Marine Engine Works, Arthur Street, Greenock, was a particularly complete example that evolved on a single site from 1835 to 1886. The main erecting shop was remarkable for the full 40 feet (12.19 metres) height of the cast-iron stanchions externally expressed to the street, and for the 52 feet (15.85 metres) span of its king-post roofs. Running off from it, and to a downward gradation in height, were the boiler and machine shops, wrapped round the original turning and finishing shop of 1835–39. The latter had a floor suspended from the roof – a similar building existed at Carron Ironworks – in order to give clear space below. Lithgow's Kingston Engine Works, Port Glasgow, is smaller but also sports I-section cast-iron stanchions.

Other engineering works, such as Rose Street Foundry, Inverness (1894), Douglas & Grant's, Kirkcaldy (fig. 24.1, middle distance), Brown Brothers, Leith, and Penman's Boiler Shop, Glasgow (The Caledonian Ironworks and 1888 International Exhibition building is to be re-erected at Bo'ness, West Lothian) possess the same characteristics. These have not been noted in England, although there were examples on Tyneside and at Chatham Royal Naval Dockyard, Kent. A workshop in Miskolc Ironworks, Hungary, and a similar form of marine engineering building in Istanbul, Turkey, offer international parallels, and may prove to be instances of technological transfer. Typical English and continental European engineering works of the period such as Garrett's Long Shop at Leiston, Suffolk, Le Creusot ironworks, France, and Budapest Ontodei Foundry Museum, Hungary, had cranes supported on timber posts.

After 1890, steel came to supplant cast iron, and the great boom in European engineering was almost entirely within steel-framed engineering shops. Blackness Foundry, Dundee, Hall Russell's Aberdeen Ironworks and Miller's London Road Foundry, Edinburgh, were amongst those in Scotland internally rebuilt with steel frames in the early twentieth century. The most interesting Scottish example is that modelled on the

Figure 24.9 Douglas Fraser Works, Orchard Street, Arbroath, Angus. Left to right: Belfast roofs, Inch Flax Mill, machine- and erecting shops, Westgate Works (small jute mill and Alpargartas shoe factory), reinforced-concrete pattern shop, foundry and jute warehouses. 1947 advertisement. Crown copyright: RCAHMS, SC 713505.

mansard form of Peter Behrens' AEG turbine factory in Berlin, Germany (1909) at Barclay Curle's diesel engine works, Whiteinch, Glasgow, 1914.

Light engineering and millwright works were often small-scale workshops in back courts. Larger ones might be arranged around a central lightwell, such as D J MacDonald's, St Roques Works, Dundee. Larger still, Douglas Fraser's Westburn Foundry in Arbroath (fig. 24.9) was built in four phases, each with two tiers of concrete and steel galleries off a central well carried on masonry piers (1890), paired cast-iron columns (1896, 1900) and reinforced concrete (1921), the latter with a Belfast roof. The wells carry travelling cranes. The 1922 pattern shop is of flat slab reinforced concrete on mushroom columns. Biggest of all, Singers built at Clydebank in 1884 the world's largest sewing machine works with multi-storeyed blocks geared towards American-style mass production and dominated by a massive clock tower. It had 8,000 employees in 1937.[32]

Two pioneering reinforced concrete engineering buildings were pattern shops for Alley & MacLellan's Sentinel Works (designed 1903 by Archibald Leitch, built 1904 to Hennebique's patent) (fig. 24.3) and J & G Weir (1912, by Truscon to Kahn's patent) in Glasgow. The beam and slab construction follows a regular grid, allows plenty of light – hence the term 'daylight factory' – and is absolutely fireproof, especially valuable where timber patterns are stored. The result is a precociously modern aesthetic.

Foundries in and around Falkirk were mostly draughty single-storeyed casting shops behind two-storeyed pattern shops. That for Carron Ironworks was the longest, incorporating boardroom and warehouse into a long Baronial pile of 1876.[33] Walter MacFarlane's Saracen Foundry, Glasgow, had, apart from foundry buildings and an ornate Gothic iron showroom, extensive sheds in a production-line arrangement. Flow-lines had evolved to minimise costs of transferring parts and materials long before Henry Ford 'invented' them.

Early locomotives were made by a surprising range of engineers in small yards, some, like Wallace Foundry, Dundee, far from railway lines. Mid Victorian locomotive works at Springburn and Cowlairs, Glasgow, tended to have timber roofs supported by paired cylindrical cast-iron columns. Steel was adopted in the later nineteenth century, enabling one locomotive to be lifted over another. Inverurie, Aberdeenshire, presented a complete early twentieth-century railway town with well-planned flow-lines within the granite-clad, steel-framed and north-lit works.

Early motor car factories were improvised in most countries. The Scottish factories by contrast showed a remarkable confidence but all were short-lived. The oldest in Britain, Madelvic in Granton, Edinburgh, 1898–1900, is galleried, having a single-storeyed assembling area between two-storeyed steel-framed blocks.[34] The Argyll Motor Works, Alexandria, Dunbartonshire, 1905–6, had a sophisticated single-storeyed layout behind an exuberant Baroque front littered with motoring insignia. Arrol Johnston began at Dumfries in 1912 in a reinforced concrete-framed, multi-storeyed daylight motorcar factory on Albert Kahn's Detroit principles, and smaller versions were built in Glasgow in 1913 at Kilbirnie Street (for coachbuilder William Park) and at Scotstoun (Albion, demolished). Yet the quality of the Scottish buildings could not compensate for the central market that was dominated by more humdrum single-storeyed factories of the English Midlands.[35] The Scottish car factories were soon adapted to other uses.

Tyre production lasted longer, and saw the construction at Inchinnan, Renfrewshire, in 1929 of the India Tyre Factory by Wallis, Gilbert & Partners. It follows a colourful Egyptian theme, and is paralleled only by the architect's Firestone (demolished) and Hoover Factories in London.[36]

The creation of industrial estates from 1937, initially at Hillington, Glasgow, saw the need for a generic standard factory design by E G Wylie, supplied with either ridge or north-light roof.[37] They were laid out to garden city principles. A small number of light engineering works built in the mid twentieth century are fronted by good examples of contemporary architecture. NCR and Timex in Dundee (1946 and later), for example, have office blocks and canteens asymmetrically placed to mask standard assembling sheds, all within manicured grounds and playing fields.[38]

Thereafter, new engineering works became architecturally anonymous. The object of designers has too often been to attempt to mask rather than to exploit the bulk of engineering shops. Creditable recent exceptions have been AI Welders, Inverness, Barr & Stroud, Thalys, Govan, and Cummins Engine Works, Shotts, Lanarkshire.

Forges
Helve hammers, whether driven by water or steam, were inefficient and were overtaken by steam hammers, and then hydraulic presses, but continued to be used for small-scale spade manufacture until relatively recently. One workshop has been transferred to the Summerlee Heritage Trust, Coatbridge, Lanarkshire.

Every engineering works, and many other sizeable industrial concerns, had a smithy equipped with one or more hearths. The buildings were often well ventilated and clad only in insubstantial timber or corrugated iron; they were also set apart as a fire hazard. The introduction of steam hammers from the mid nineteenth century provided an extra incentive for distancing them from the foundry. When J & C Carmichael erected the Dundee Steam Forge in 1861 and equipped it with a Rigby hammer, care was taken to place it a sufficient distance from their Ward Foundry in order to avoid breaking the moulds. Similarly specialised forges were established at Anderston and Parkhead in Glasgow, Dumbarton, Paisley and Kirkcaldy (the polychrome Ingleside steel foundry/Fife Forge, established 1873). The Chieftain Forge at Bathgate, West Lothian, was a particularly interesting survivor, established in 1877, and until recently crammed with an apparently haphazard layout of presses, drilling and shearing machines.

Spinning mills
Cotton mills could be as many as seven storeys tall. Scottish mills were on the large size compared to those in England, which varied considerably from the Arkwright pattern in the late eighteenth century. The earliest were water-powered and contained water frames for the production of water twist yarn. Soon, at Catrine, Ayrshire, for example, they would be complemented by mills for the production of weft yarn, which was better produced on mules. Experiments in applying power to mules took place in New Lanark in the 1790s.

Fireproof construction came to Scotland at the beginning of the nineteenth century and was the norm for flax mills after 1833 and for cotton after *c.* 1840. However, mills could still be constructed with timber floors in each of these branches of the textile industry in the 1860s, at Blairgowrie and at Anchor Mills, Paisley, respectively.

In the woollen industry, fireproofing was the exception rather than the norm. This appears to have been on the grounds of cost, availability of materials and crucially, the optimum spacing of columns around the mulegate that did not favour the use of single brick arches. Four-storeyed mills in the Borders and in Clackmannanshire built in the 1860s are mostly arranged four-bays deep, so as to accommodate two pairs of mules facing each other, and would be either one or two mules in length, with a stair tower at one end or at the centre. Their length was fixed by multiples of five or six bays, corresponding to the length of a mule. Only two examples (in Dumfries and Clackmannan) existed of the progression to mules arranged transversely. This Lancashire type of spinning mill features in cotton mills erected in the 1880s at Ferguslie Mills, Paisley, Renfrewshire, Carstairs Street, Glasgow, and at Crofthead Mill, Neilston, Renfrewshire, using either concrete on steel or multiple brick arches between cast-iron beams. The Mile End Mill at Anchor Mills, Paisley, Renfrewshire, 1899, is a ring mill and is thus also relatively narrow compared to mule mills; it was given a lively profile of turrets and balustrades.

Frame-spinning meant that relatively narrower mills were possible in the flax and jute industries. Proportionately more effort was spent in preparatory processes, in lower floors and outshots, than in spinning. The mills were generally three bays deep, one frame per window. They were therefore of a less uniform size than a mule-spinning mill and were better able to receive classical or Italianate architectural embellishment in the form of cornices, pediments and campaniles.

Single-storeyed spinning mills were a logical outcome of the application of flow processes to textile manufacture, the first being in Dundee in 1862, followed by Chapel Works, Montrose, in 1865. New integrated jute mills of 1872–4 in Dundee – and India – were of this type. A few woollen spinning mills were also single-storeyed, besides the worsted spinning mills at Kilncraigs Mill, Alloa (fig. 24.5), there is just one cotton-spinning mill of this type within Anchor Mills, Paisley.

Weaving factories [39]
Hand-loom weaving factories co-existed with domestic production, and benefited from economies of scale and the division of labour, even if virtually all of it was by hand and foot. The first linen and hemp factories established in the late eighteenth century for improvement purposes in the Highlands, at Invermoriston, Inverness-shire, and Cromarty, for example, were of this kind, long two-storeyed ranges arranged around a quadrangle. Some factories for the production of carpets or sailcloth would continue to be worked by handloom until the 1870s. Surviving buildings seen in Montrose, Arbroath and Dundee, and Newburgh, Cupar, Kirkcaldy, Fife, have in common with breweries, tanneries and paper mills a louvred drying loft, in this case to dry the starched warp threads prior to weaving. At Pilmuir Works, Dunfermline, the warp loft has a kiln floor. Ideally the looms would be placed on an earthen floor, to maximise humidity, and so access to the upper warping lofts would be by external stair. Handlooms would co-exist with power looms still longer within the woollen industry. Large buildings that contained

them survive at Forest and Dunsdale Mills, Selkirk, and Gala Mill, Galashiels.

First to mechanise was the cotton industry. Early power loom factories might be in multi-storeyed buildings, such as the six-storeyed Graham Square, Glasgow, 1825. Later in the century a few multi-storeyed weaving factories might still be built to carry 'slow' power looms – Botany Mill and Comelybank Mill, Galashiels, and Ettrick Mill, Selkirk (1874), are examples – before the speed of new looms would translate into vibrations that shook buildings. Fireproof construction permitted some power loom linen weaving factories to have multi-storeyed sections for preparatory processes (for example Edward Street Mill, Dundee, Alma and Baltic Works, Arbroath) and in rare cases actually for weaving.

Aside from a unique cotton-weaving shed by James Smith at Deanston, Perthshire – a series of domed groin vaults, to be imitated only at Marshall's Temple Works in Leeds, Yorkshire – the majority of early cotton-weaving sheds had short spans. Banner Mill, Aberdeen, c. 1828, had spans of only 12 feet (3.66 metres). Shed spans varied considerably according to whether power was to be transmitted from above, so requiring closely-spaced columns to carry shafting, or from below, thus allowing a much lighter space. Bays in sheds spanned from 12 to nearly 40 feet (3.66 to 12.19 metres); the former had a steeper pitch of skylight and was apparently more common in northern England and western Scotland than in eastern Scotland, examples being found at Newmilns and Darvel, Ayrshire, Dalmarnock, Glasgow, and Auchterarder, Perthshire. Shallower pitches or sometimes equal-sided roofs were preferred in the linen, jute and lace industries. East lights were favoured in Dundee, in order to catch the morning sun, rather than the north lights that were preferred further south. In Kirkcaldy there may be found broad-span weaving sheds by Dundee engineers (N Lockhart's Linktown Works by Charles Parker, 1858) and short-span sheds by Glaswegian engineers (Victoria Linen Works by Anderston Foundry, 1871), an illustration in built form of the junction of two cultural spheres of influence. In Dunfermline the broad-span shed, mostly by Robertson & Orchar of Dundee, held sway owing to the location of the jacquards over the loom, as was also the case in Forfar and Kirriemuir, Angus. Arbroath's indigenous weaving sheds were lower.

Regional characteristics may likewise be found within the tweed industry. Roxburghshire weaving sheds of the 1870s and 1880s have broad spans and high roofs, affording the opportunity for overhead walkways and places for reeding and other processes preparatory to weaving. Selkirkshire, Peeblesshire, and Clackmannanshire weaving sheds are more conventional, with shorter spans. Hayford Mills in Stirlingshire has remarkable cast-iron trusses by Davie of Stirling, which were used also at Templeton's in Glasgow. In the Western Isles weaving is carried out in tiny sheet-metal clad sheds built around treadle looms.

Hosiery factories
Hand- and foot-operated stocking shops could be of sufficiently large size to be considered factories. Buccleugh Mills in Hawick is the largest of this kind, three-storeyed with a turnpike stair at the rear and rows of small square windows, behind each of which was a stocking framework knitter using a bowl of water to magnify his vision. A further three smaller workshops are identifiable in Hawick and Denholm, Roxburghshire. Machine-powered stocking frames were tried first in a large building at Wilton Mills, Hawick, in the 1850s, but were only widely adopted at the end of the nineteenth century. Some of the larger mills now occupied by the industry have evolved piecemeal behind fairly large windows, for example, Peter Scott, Hawick, or were purpose built as tweed weaving sheds behind corbelled, two-storeyed fronts, such as Glebe and Eastfield Mills, Hawick.[40]

Floorcloth and linoleum works
The first in Kirkcaldy was built by Michael Nairn in 1847. Linen canvases 25 yards (22.86 metres) long were stretched, painted or trowelled, then sent up to be printed in a loft and seasoned in a drying chamber.[41] Drying was by means of fresh air until the 1860s when hot air was added. The factories were therefore exceptionally tall. In 1875 Nairn started to produce linoleum, applying linseed and cork to a jute backing. His South Factory on Victoria Road (fig. 24.1, top right), is distinguished by very high arched windows. The still-functioning Scottish Linoleum Works, across the railway line, has a mobile calender that inlays the pattern into the linoleum, which is then taken to very high stoves in a building of *c.* 1890. A high reinforced-concrete tower of the 1920s is used to mix ingredients.

Dyeworks
Dyeworks are marked out by their ridge ventilators (as at New Lanark), and half slating (as at Bridgehaugh, Selkirk), essential to extract moisture. At Grandholm Works, Aberdeen, the dyeing kiers were direct-heated externally, their locations indicated by low arched openings. Relatively large areas of glazing were required. Reinforced-concrete framed dyeworks were introduced in the 1920s at Tay Works, Dundee, and Victoria Dyeworks, Hawick, as processes were centralised and technology became more complex. Pullars of Perth developed a big mail-order and dry-cleaning business that required substantial blocks fronting the streets.

Bleachworks
These works needed fresh water and open space for bleaching greens, so tended to be in rural locations, many of them north of Perth and Dundee. Some buildings dating from the early nineteenth century have bellcotes to mark them out as the focal point of little communities, and some structures might be louvred. Beetling mills and plash mills would be water powered. Those bleachworks that continued to function into the twentieth century experienced radical change as new and larger plant was installed.

Tenement factories
The Merchant City area of Glasgow, bounded by Trongate, Queen, Ingram and High Streets, is replete with evidence of a peculiar building type let to a variety of textile manufacturers and wholesalers from the early and mid nineteenth century. An example was Queens Court, 62 Queen Street, built in 1833 and occupied up to the 1860s by eleven manufacturers of sewn muslins, ginghams and cravats. At the end of the century the area was reinvigorated by the arrival of tailors from Eastern Europe, many being Jews escaping pogroms.

Typically a four- or five- storeyed building fronting the main grid of the city, a tenement factory was pierced by a central pend which opened onto a courtyard in which two projecting wings faced each other. Solidly built pilastered doorpieces gave on to tenement stairs and a profusion of hand-painted signs. In some cases panelled electric lifts and internal partitions were supplied. On floors above, usually timber carried on iron columns, and occasionally McConnel patent beams, rooms would be let to tailors, furriers and similar craftsmen. Natural light was maximised as far as possible, and services shared. Sprinkler systems and fireproof doors were installed to cut down the ever-present hazard of fire spreading from one lessee to the rest. There are some parallels with buildings found in Paisley – primarily shawl factories – and urban lace factories in Nottingham and Lyons, France, but not those in Darvel and Newmilns, Ayrshire, where the factories could be larger.

Paper mills
Paper mills have a long history in Scotland. A mill was set up at Dalry, Edinburgh, in 1590. Valleyfield Mills, Penicuik, Midlothian, founded in 1708, was to remain one of the largest in the country. Apart from an insatiable demand for paper from a literate population, another factor in the significance of the paper industry in Scotland was the plentiful availability of process water and linen, and then cotton, rags as raw material. Alongside small vat mills using hollander engines to boil the rags, there came investment in Fourdinier paper machines for the continuous production of machine-made paper. The first of these was at

Peterculter, Aberdeenshire, in 1811. The number of machines installed in Scotland had grown to 32 by 1832 and 76 by 1860.[42]

A consequence of ever-longer paper machines was the provision of still longer buildings to contain them. They had ridge ventilators and relatively open sides. Most of the paper mills still in operation appear to have had alterations made over the years in order to access and replace these machines. Some buildings at Fettykill Mill, Fife, are fireproof. Other characteristics of paper mills (for example, Markinch, Fife) include tall chimneys, owing to the amount of steam and heat required by the processes. There might also be tall water towers, as at Kilbagie, Clackmannanshire, and Bullionfield, Angus, near Invergowrie, Perth- shire. At Thomas Tait's works, Inverurie, Aberdeenshire, a tower nearly 100 feet (30.48 metres) tall produced acid for use in the quick-cook sulphite method of producing paper from wood pulp, a process patented in the 1880s.[43] The raw material largely switched from rags to esparto grass in the 1860s and in some cases to wood pulp in the early twentieth century and to recycled paper in the late twentieth century.

Recognisable architectural style was limited at most of these paper mills, except those that had extensive finishing requirements. The special products of the Chirnside Paper Mill (1842 and 1857), Berwickshire, and the long three-storeyed ranges of the Culter Paper Mills, Aberdeenshire (envelopes, c. 1870), resulted in each case in shaped Jacobean gables.

Tanneries

Tanneries were, from medieval times, amongst the more noxious of neighbours. In the first stage of the tanning process the hides would be separated from the epidermis and flesh in pits of lime and guano. Then they would be soaked in tanning pits, generally lined in stone, for between nine and 12 months in a tannin-rich liquor. Tannin was obtained from oak bark ground in bark mills, frequently water powered, although horse mills or small steam engines might also be utilised. Almost all the rest of the labour was manual. In the twentieth century this vegetable tannage process was in most cases supplanted by faster chrome tannage in automated drums.

The hides would then go for further treatment by curriers and to be enamelled. The hides were cured in buildings having upper floor rows of louvres randomly punctuated by windows. The largest recently-remaining example was at Tullibody, Clackmannanshire, but there are others in Mill Street, Ayr, Croft Street, Dalkeith, Midlothian, and Hyde Park, Keith, Banffshire. A later variant might rely on horizontally pivoting windows, as at Silvermills and Currie, Edinburgh.

Shoe factories are found in Kilmarnock and Maybole in Ayrshire, and also in Arbroath (Grant's and Abbey Street Works). Leather was also worked into other articles, such as belting to drive machinery at Progress Works, Dundee.

Apart from cattle hides, other animal skins that might be treated in tanneries include sealskins, a by-product of whaling (for example Arctic Tannery, Dundee, where stone-lined tanning pits exist), and sheep pelts. Sanderson & Murray of Galashiels, established in 1844, became one of the largest fellmongers in the world, controlling in one year half the Australian wool clip. Their Buckholmside Works suffered fires in 1873, 1882 and 1923 but extended to more than four acres (1.48 hectares) of floor space.

Grain and flour mills
Grain mills are found all over rural Scotland but are now rather rare in urban situations. Perth City Mills provides the best remaining example, combining grain and flour milling in the Upper and Lower City Mills. British millstones belong to the group that were driven from below, whereas Dutch, Danish and Northern German millstones were generally powered from a spur wheel over the stones. John Smeaton made many improvements, introducing iron into wheel and gear construction in the late eighteenth century, which were taken up to such an extent that no all-timber wheel survives in Scotland.[44] Most of the machinery found in grain mills of undoubted antiquity dates to the late nineteenth and early twentieth centuries. Mills tended to have two or three pairs of stones, rising to five at the largest, such as Montgarrie Mill, Alford, Aberdeenshire.

Wheels at threshing mills tend to be set against the long side of the barn, corresponding with the shape and rotary power needed by the threshing mill. Grain mills, on the other hand, usually present their gables to the wheel. The strength of the gable gave a firm bracing for gears and stones. Flour mills are fewer in number and tend to have the wheel in the centre of the building (eg Hawick, Roxburghshire).

Roller milling was invented in Hungary and revolutionised flour milling in Britain in the 1860s. Each city, and each major port, acquired its roller mill. An Italian castellated style was favoured for these vertically-laid out buildings, so arranged as to minimise handling. Scotstoun, Regent and Victoria Mills in Glasgow are good examples of the type, the first two on old water-powered sites, the latter in the Tradeston grid. At Washington Street, Glasgow, there is a brightly polychrome rice mill.

Kilns
These are used in grain mills to dry barley prior to milling, and in a maltings to arrest germination. The flooring might by of perforated tile (rather rare in Scotland) or cast-iron plate. Towards the end of the

nineteenth century wire-mesh floors, first devised in Germany, tended to be substituted for the iron plates and might be placed on more than one level. These would overlie splayed vaults of stone or brick springing from a central firebox. Kilns trace their ancestry to the kiln barn and are characteristic of Scotland, north-east England and Northern Ireland. They are not found in grain mills in central and southern England.

The external form of the kiln varied according to the region. The earliest forms might be detached from the mill or malt barn, as for example, Preston, East Lothian, and Glendale, Skye. In the south of Scotland the kiln would be rather understated, identifiable by a small ridge ventilator and skews sitting on top of a fire-barrier wall. In Angus the kiln might be semi-circular, as for example at Barry. Further north the kiln would have greater prominence, with tall ventilators topped by weather vanes. Elgin architect C C Doig is credited with devising in the 1890s the pagoda form that became a trademark for distilleries but is also found on some grain mills, such as Milton, Ross and Cromarty. Pagodas were sometimes also employed to picturesque effect on top of grain elevators adjacent to the kiln.

Maltings

The key ingredients of beer and whisky are malt and water. Malt is artificially germinated barley, the germination arrested in kilns. This is

Figure 24.10 Bonthrone Maltings and Brewery, Newton of Falkland, Fife. Crown copyright: RCAHMS, SC 710984.

Figure 24.11 Slateford Maltings, Edinburgh; floor malting in progress. Crown copyright: RCAHMS, SC 712295.

performed in a floor maltings, or in a distillery in the 'maltbarn', which may be in a completely different location from the brewery or distillery.

The plan form was generally rectangular, although some brewery maltings were remarkably deep. At one end would be the kiln (or kilns, sometimes an additional one dried barley prior to malting; fig. 24.10). The malt floors proper are indicated by a regular series of windows, often fully or partly louvred. Inside (fig. 24.11) regular iron columns carried concrete laid over timber floors, after 1890 on steel beams threaded through the saddles. The attic, and sometimes the top floor, would contain grain bins, the grain or barley distributed by belt conveyors. At the opposite end to the kiln were the steeps, stone or cast-iron tanks in which the germination process began.

Pneumatic drum maltings were introduced in 1897 at Speyburn Distillery, Moray, hence a relatively short building, but still with a pagoda kiln. They were followed later in some distilleries by Saladin maltings, generally inserted into existing buildings.

Breweries
Scotland was a major exporter of beer. The largest eighteenth-century

Figure 24.12 Craigmillar Brewery, Edinburgh; timber-clad cooling unit. Crown copyright: RCAHMS, SC 710855.

buildings in Banff and Thurso, Caithness, are breweries, comprising in each case long, three-storeyed maltings forming one side of an inward-facing square, to offer some security. At Cromarty the brewery, provided to offer an alternative to whisky for the natives, comprises two closely-spaced, three-storeyed ranges, one the brewhouse with central chimney stacks rising through the building, the other a floor maltings. Belhaven, Dunbar, East Lothian, 1719, rebuilt 1887, has a picturesque profile all its own. A relatively horizontal layout remained the norm for the smaller nineteenth-century breweries, such as Craigie Brewery, Dundee, or Davidson's in Coldstream, Berwickshire.[45]

The nineteenth-century innovation in brewing techniques was the tower brewery which exploited gravity in transferring materials from one department to the next. This gave the opportunity to dress the building in the national castellated Baronial style, as at Boroughloch, Craigend, Holyrood and Abbey breweries in Edinburgh and Lochside Brewery, Montrose (1889), with the louvred French-pavilion roof embracing the chimney. A simple stone tower served W B Thomson's Brewery, continued until recently as the Gleneagles Maltings, Blackford, Perthshire.

Brick was often utilised in an Italianate style, which again suited the asymmetrical layout of a brewery: polychromatic at Ballingall's Park Brewery, Dundee, and Maclay's Thistle Brewery, Alloa; monochrome at Caledonian Brewery and the McEwan maltings at Slateford, Edinburgh.

Standard features of breweries before the advent of refrigeration were large and elevated areas of wooden-sided louvres to facilitate rapid cooling of the wort, still found at Thistle and Craigmillar Breweries (fig. 24.12), Edinburgh, and often in even the smallest brewhouses, such as that located in a court off South Street, Duns, Berwickshire. Argyll Brewery, Cowgate, Edinburgh, had both vaulted cellars and an elevated louvred cooler.

Lager brewing paved the way for Bohemian and German influence on world brewery design. Vast cooling cellars under breweries of substantial size saw the use of German national romantic styles from Milwaukee to Tsingtao. Scotland already had its own distinctive brewing style but Tennents in Glasgow began lager brewing in 1885 and built a new lager brewery in 1890. The Arrol Brewhouse of the 1960s was a striking piece of modernism in Alloa.

Historic brewing plant is to be found at Traquair House, Peeblesshire – recently enlarged to supply more than the laird's own requirements – and direct-fired riveted coppers are at Maclay's and Caledonian breweries, Alloa (fig. 24.13) and Edinburgh respectively. The fermenting vessels of Lochside Brewery, Montrose, have survived by dint of their use as washbacks when converted to a distillery.

Malt-whisky distilleries

Malt-whisky distilleries are well known emblems of Scotland.[46] Their early form, as found in Islay and Campbeltown, Argyll, was generally within a square courtyard to resist theft and tax evasion. Some of the earliest, as at Kilbagie and Kennetpans, Clackmannanshire, were of strikingly industrial scales. Late nineteenth-century distilleries in Speyside, Moray (a remarkable number of them built in 1897), tended towards a more linear layout, aligned with their railway sidings and current thinking on flow processes. Gravity and vertical-flow processes were not significant in distillery architecture. Power requirements were light, and small waterwheels usually sufficed. Single-cylinder steam engines survive at Ardmore, Islay, Longmorn, Moray, and Auchentoshan, Dunbartonshire. Double distillation in pot stills distinguishes Scotch whisky from the triple distillation of Irish whiskey. A few distilleries retain worm tubs as opposed to condensers (Dallas Dhu, Moray, Rosebank, Falkirk, and the newly installed one at Dalwhinnie, Inverness-shire). Floor maltings and their trademark pagoda kilns are described above.

MILLS AND FACTORIES • 545

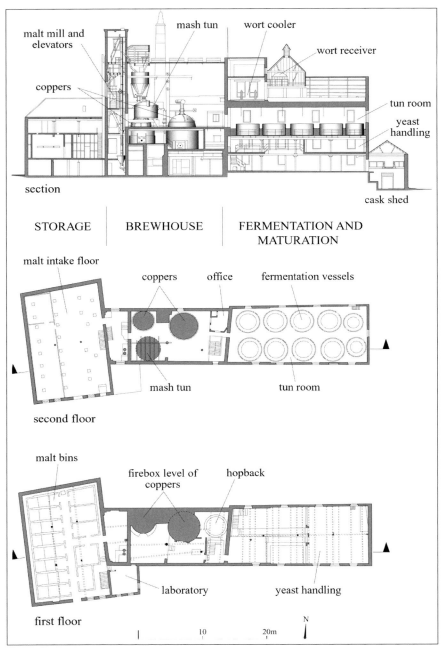

Figure 24.13 Maclay's Brewery, Alloa, Clackmannanshire; plans and section through brewhouse. Crown copyright: RCAHMS.

Grain-whisky distilleries
Relaxation of the Corn Laws in 1827 led to the development of continuous distillation, first with the Stein still at Kilbagie, Clackmannanshire, and Cameronbridge, Fife,[47] and then the Coffey still, patented in 1832. The latter comprises two columns, analyser and rectifier, 40 feet (12.19m) high. The patent still-houses, accordingly tall and almost square in section, dominated grain distilleries at for example Cambus, Clackmannanshire, Kirkliston, West Lothian, and the Caledonian, Edinburgh (1855). Their productive powers meant very large maltings and bonded stores, often with crowsteps continuing the Scottish theme.

Sugar refineries
Sugar refineries are few in number but of special importance for their scale and for the world leadership given by Greenock. The special needs of sugar refineries drove forward advances in structural engineering.

Eighteenth-century sugar houses were distinguished by their immense height relative to other dimensions. As the floors were of timber, and loads were heavy, low ceilings were a consequence of the need to maximise the ability of the building to carry weights. Verticality resulted from the need to exploit gravity to assist the transfer of liquids from one section to another. In 1812 Howard's patent vacuum pan allowed the crystallisation of sugar at lower temperatures. From this era is an eight-storeyed refinery at Regent Quay, Aberdeen, now converted to offices.

Four refineries survive from the mid Victorian period, besides a polychrome flat-iron shaped part of Glebe refinery, Greenock. They are characterised by a precocious use, in 1865–6, at Bonnington, Leith, and Port Dundas, Glasgow, of wrought-iron beams in place of the cast-iron that would be expected in contemporary textile mills, by cast-iron roof tanks (extant at Westburn/Berryards Refinery, Greenock, 1851–2, and Bonnington, Edinburgh) and in all cases by an immense height of seven or eight storeys. Brick-panel and pilaster-pier construction was required to support the height and weight of the refineries at Bonnington and Westburn, which still partly collapsed during construction. Form followed function but could not always be guaranteed to perform as Scottish engineers explored the potential of new building materials.

NOTES

1. Hume, 1976, 1977; for updates see Trinder, 1992, McDonald, 1996.
2. The Luma Factory was a joint SCWS-Swedish venture, following the example of the similar factory in Stockholm. See Watson, 2000.
3. Moss, Hume, 1986, 224, 236–7.
4. Richards, 1958.

5. Douglas, Hume, Moir, Oglethorpe, 1985.
6. Barnard, 1969. Cross Arthurlie Mill in Barrhead was also so treated, see Clark, 1982.
7. Skempton, Johnson, 1962.
8. Watson, 1992.
9. Hay, Stell, 1986, 86–9.
10. Fitzgerald, 1988.
11. Fairbairn, 1857.
12. ibid, 166–9.
13. Research by Andrew Jackson, York.
14. Stratton, 1999.
15. Watson, 1990, 54–69, 182–3.
16. Major, 1985.
17. Hume, Oglethorpe, 1984. See also, Douglas, Oglethorpe, 1986.
18. 1984; Shaw, 1982.
19. Crocker, 2000.
20. Boulton and Watt Collection, Birmingham Reference Library
21. Hay, Stell, 1986, 131–5.
22. Douet, 1990.
23. Watson, 1988, 127–45.
24. Hume, 1980, 65–6.
25. Woodfield, 1921, 2,
26. Chapelle, 1990.
27. Meakin, 1905.
28. Oakley, 1937; Oakley, 1953.
29. Hay, Stell, 1986, 115–30.
30. Research by James Grant and Lance Smith for the Scottish Maritime Museum.
31. RCAHMS. Drawing AND/170.
32. Oakley, 1937, 242–3.
33. Watters, 1998, 121–3.
34. Collins, Stratton, 1993, 248–57.
35. Dodds, 1996.
36. Skinner, 1997, 128–34.
37. Earnshaw, 1999.
38. NCR, 1996; and Stratton, Trinder, 2000, 91.
39. In Dundee parlance, the factory is for weaving and the mill for spinning. Notwithstanding, see Watson, 1990. This distinction does not apply in wool mills.
40. Research by David Roemmele (dissertation for the Ironbridge Institute), 1997.
41. Grant, Mechan, Seymour, 1992.
42. Thomson, 1974.
43. Reid, 1990, 44–5; and Hay, Stell, 1986, 181.
44. Smith, 1981.
45. The main source of information regarding historic breweries has been the Scottish Brewing Archive (SBA) Newsletters, edited by Charles H McMaster, and latterly Alma Topen, renamed the *SBA Annual Journal* from 1998.
46. See Hay, Stell, 1986, 31–62.
47. Moss, Hume, 1981, 79.

BIBLIOGRAPHY

Barnard, A. *The Whisky Distilleries of the United Kingdom*, London, 1887, reprinted 1969.
Clark, S. Cross Arthurlie Mill, *Scottish Industrial History*, 5/1 (1982), 58–62.
Chapelle, K. The New Lanark water clock, *Scottish Industrial Heritage Society Newsletter*, 21 (Summer 1990), 17–24.
Collins, P, Stratton, M. *British Car Factories from 1896*, Godmanstone, 1993.
Crocker, A. The Rolt Memorial Lecture 1999: Early water turbines in the British Isles, *Industrial Archaeology Review*, 22/2 (2000), 83–102.
Dodds, A. *Making Cars*, Edinburgh, 1996.
Douet, J. *Going Up in Smoke: The history of the industrial chimney*, London, 1990.
Douglas, G J, Hume, J R, Oglethorpe, M K. *Scottish Windmills: A survey*, Edinburgh and Glasgow, 1984.
Douglas, G J, Hume, J R, Moir, L, Oglethorpe, M K. *A Survey of Scottish Brickmarks*, Glasgow, 1985.
Douglas, G J, Oglethorpe, M K. Windpumps and Windmills: New Information, *Industrial Archaeology Review*, 9/1 (1986), 82–6.
Earnshaw, N. The establishment of Scottish industrial estates: Panacea for unemployment?, *Scottish Industrial History*, 19 (1999), 5–20.
Fairbairn, W. *On the Application of Cast- and Wrought-Iron to Building Purposes*, London, 1857.
Fitzgerald, R S. The development of the cast-iron frame in textile mills to 1850, *Industrial Archaeology Review*, 10/2 (1988), 127–45.
Grant, G, Mechan, D, Seymour, V. *That Queer-Like Smell: The Kirkcaldy linoleum industry*, Kirkcaldy, 1992.
Hay, G D, Stell, G P, for RCAHMS. *Monuments of Industry*, Edinburgh, 1986.
Hume, J R. *The Industrial Archaeology of Scotland*, Vol 1, *Lowlands and Borders*, London, 1976.
Hume, J R. *The Industrial Archaeology of Scotland*, Vol 2, *Highlands and Islands*, London, 1977.
Hume, J R, ed. *Early Days in a Dundee Mill*, Dundee, 1980.
McDonald, M R, ed. *A Guide to Scottish Industrial Heritage*, Glasgow, 1996.
Major, J K. *Animal-Powered Machines*, Princes Risborough, 1985.
Mays, D, Moss, M, Oglethorpe, M K, eds. *Visions of Scotland's Past: Looking to the future, essays in honour of John Hume*, East Linton, 2000.
Meakin, B. *Model Factories and Villages: Ideal conditions of labour and housing*, London, 1905.
Moss, M, Hume, J R. *The Making of Scotch Whisky*, Ashburton, 1981.
Moss, M. Hume, J. *Shipbuilders to the World: 125 years of Harland and Wolff, Belfast, 1861–1986*, Belfast, 1986.
NCR (Scotland). *Cash Advance: The story of the NCR in Scotland, 1946–1996*, npp, 1996.
Oakley, C A. *Scottish Industry Today*, Edinburgh and London, 1937.
Oakley, C A, ed. *Scottish Industry*, Glasgow, 1953.
Reid, J. *Mechanical Aberdeen*, Aberdeen, 1990.
Richards, J M. *The Functional Tradition in Early Industrial Buildings*, London, 1958.
Shaw, J. An introduction to the technology of meal milling in Scotland, *Scottish Industrial History*, 5/1 (1982), 8–24.
Shaw, J. *Water Power in Scotland, 1550–1870*, Edinburgh, 1984.
Skempton, A W, Johnson, H R. The first iron frames, *Architectural Review*, 131 (March 1962), 175–86.

Skempton, A W, ed. *John Smeaton, FRS*, London, 1981.
Skinner, J. *Form and Fancy: Factories and factory buildings by Wallis, Gilbert and Partners, 1916–1939*, Liverpool, 1997.
Smith, D. Mills and millwork. In Skempton, 1981, 59–82.
Stratton, M. The Rolt Memorial Lecture 1997: New materials for a New Age: Steel and concrete construction in then north of England, 1860–1939, *Industrial Archaeology Review*, 21 (1999), 5–24.
Stratton, M, Trinder, B. *Twentieth-Century Industrial Archaeology*, London, 2000.
Thomson, A G. *The Paper Industry in Scotland*, Edinburgh, 1974.
Trinder, B. *The Blackwell Encyclopedia of Industrial Archaeology*, Oxford, 1992.
Watson, M. Jute manufacturing: A study of Camperdown Works, Dundee, *Industrial Archaeology Review*, 10/2 (1988), 127–45.
Watson, M. *Jute and Flax Mills in Dundee*, Tayport, 1990.
Watson, M. Matthew Murray and Broadford Works, Aberdeen: Evidence for the earliest iron-framed flax mills, *Textile History*, 23 (2), 1992, 225–42.
Watson, M. Change for the better: Luma lamp factories, Glass-clad modernism and reworked textile mills. In Mays et al, 2000, 156–72.
Watters, B. *Where Iron Runs Like Water! A new history of Carron Ironworks, 1759–1982*, Edinburgh, 1998.
Woodfield, C H. *The Mechanical Handling of Goods*, London, 1921.

25 Mines, Quarries and Mineral Works

MILES K OGLETHORPE

Scotland's extraordinary mineral wealth has, more than any other factor, been responsible for the extent and sophistication of her economic development during the last 300 years. A complex geological legacy has provided a selection of sedimentary, metamorphic and igneous rocks yielding fossil fuels, metallic ores, building stones, and even sand suitable for glass manufacture.[1] This chapter is devoted to Scottish extractive industries which have attempted to exploit this natural wealth, and to an extent to some of the extraction processes associated with the minerals in question.

Of the minerals present in Scotland, coal is undoubtedly the most significant, and was itself the fuelling force behind the formidable industrial base that developed during the nineteenth and twentieth centuries.[2] Closely associated with coal have been the iron and steel industries, but other important though less visible links can be made with the oil-refining, town-gas, chemicals, heavy-ceramics and construction industries. Further afield, a variety of non-ferrous metallic minerals have been exploited successfully, of which lead has been perhaps the most important. Equally significant has been stone quarrying, which has yielded some fine building stones, slates and flagstones, and which has also provided raw materials for road building, lime burning, and a number of chemical and metallurgical industries.

One of the most important common features of the buildings associated with extractive industries is that they have usually been relatively temporary, being built with only a limited envisaged lifespan. Indeed, in many cases, they are of less interest than the people, machines and equipment that they housed. Conversely, many of these industries were responsible for providing large quantities of materials used in the construction of buildings beyond the scope of this chapter, but which are covered elsewhere in this book.

For this reason, more attention will be paid here to the industries and processes themselves, although reference will be made to buildings where possible or appropriate. Similarly, although all the industries mentioned are significant, the dominant position of coal justifies special attention.

SALT

Extractive industries date back to the beginning of human settlement in Scotland, but it was only in the early seventeenth century that larger-scale industrial activity began to develop, based primarily on coal extraction. One of the most important associated industries of the time was the boiling of salt,[3] which was highly valued not only as a condiment, but also as a preservative. Salt-making sites, such as that partly surviving at Preston Island near Culross, Fife (fig. 25.1),[4] usually comprised several salt pans located at the sea shore, water being driven off by coal-fired fireboxes. Scottish salt works were located along the Firth of Forth,[5] and on the west coast at places such as Saltcoats, Ayrshire, and Cock of Arran, and even as far north as Orkney (from the 1630s).[6] One of the most famous sites at Prestonpans, East Lothian, continued production until 1959.[7]

Figure 25.1 Preston Island, Culross, Fife; aerial view. Crown copyright: RCAHMS, SC 710852.

LIME

An early industry also associated with coal extraction, but not exclusively so, was the burning of lime. Lime kilns can be found throughout Scotland wherever limestone is available, and were usually associated with fertiliser for agricultural improvement, which often required the neutralising of the high acidity in Scottish soils. Also important was the growing number of private and public building ventures, which significantly increased the demand for mortar, of which lime was a crucial ingredient.

Although lime was sometimes produced in 'U'-shaped rubble, earth or brick enclosures referred to as clamps (examples of which can still be seen near Lassodie, Fife, and Carlops, Peeblesshire), the bulk of lime production was accounted for by shaft kilns. These usually comprised brick- or stone-lined shafts which were loaded from the top with limestone and fuel (mostly coal, coke or peat), the burned lime being withdrawn from one or more draw holes in the base of the kiln. The draw holes were recessed into sometimes corbelled draw arches, and were often fitted with iron grates, arches and lintels.

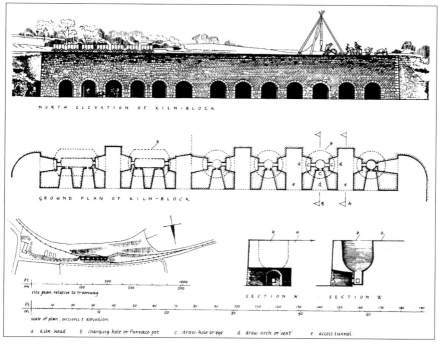

Figure 25.2 Murrayshall Limeworks, Cambusbarron, Stirlingshire; measured survey drawings. Crown copyright: RCAHMS, SC 357861.

Most kilns were small, and served farms or estates. The simplest examples were circular in plan, and were built from drystone rubble. Their appearance depended on the availability of local stone. Cruder examples, such as those found in parts of Aberdeenshire, Angus and Perthshire, were often built from river boulders, whilst others in Caithness were built from local flagstone.[8] Despite falling into disuse, many examples can still be seen throughout the Scottish countryside.

Larger banks of kilns were built by landowners, and were developed into integrated industrial operations, a fine example of which was that of Closeburn, Dumfriesshire.[9] Others, such as those at Dunure, Ayrshire, and on the north Sutherland coast, were located where raw materials and the lime itself could be shipped by sea. Many of the biggest limekilns were, however, established where coal and limestone were found close together, and were frequently associated with ironworks (as at Wilsontown, Lanarkshire), where lime was required as a flux in the iron-smelting process. These larger-scale operations usually incorporated a large free-standing kiln, or impressive banks of kilns such as those of Murrayshall, Stirlingshire (fig. 25.2),[10] and Esperston, Midlothian. The largest surviving bank of limekilns in Scotland is at Charlestown near Dunfermline, Fife. Dating from the 1760s, the complex was developed by the Earl of Elgin's estate, and comprises a bank of 14 limekilns, each with several draw holes, and tunnels to the rear provided access to more draw holes at the back of the kilns. Also associated with the kilns are large quarries and limestone mine workings, a purpose-built harbour, and the remains of a horse-drawn tramway (the Elgin Railway).[11]

METALS

With widespread improvements in transport in the mid eighteenth century came the expansion of metalliferous mining activity in Scotland.[12] One of the greatest zones of activity to develop was the old-established metal-producing area of the Southern Uplands in south Lanarkshire and nearby north Dumfriesshire, centred on the villages of Leadhills and Wanlockhead. The two villages accounted for most of Scotland's lead production, and also yielded a variety of other metals, including silver. Production continued until the early twentieth century in Leadhills, but although complex ruins and spoil heaps remain, no complete structures relating to the mines or the processing buildings and smelters survive. However, as at Wanlockhead, many of the miners' houses survive intact, and the library is said to be the oldest subscription library in the world.

Neighbouring Wanlockhead itself is now home to the Museum of Lead Mining (run by the Wanlockhead Museum Trust), and in addition to impressive but ruinous remains, includes a preserved water-powered beam engine,[13] which was used to pump water from the mines. Further

Figure 25.3 Bonawe Ironworks, Argyll; furnace. Crown copyright: RCAHMS, SC 710856.

to the south west at Woodhead near Carsphairn, Kirkcudbrightshire, the ruinous remains date from 1838,[14] and include a smelter with inclined flues leading to chimneys on the adjacent hillside, indicating another substantial lead-mining and processing operation.

Elsewhere in Scotland, a variety of metals was mined with varying degrees of success. However, by far the most significant to be exploited successfully was iron ore, which was found in clayband and blackband form in many parts of the Scottish coalfields. Prior to the exploitation of these ores, the Scottish iron industry had been confined mostly to the western Highlands of Scotland, although small-scale smelting operations based on bloomery furnaces had existed further afield.[15]

The attraction of western Highland Scotland for the early iron industry was the availability of deciduous woodland suitable for the making of charcoal. The most famous of the charcoal iron-smelting furnaces, at Bonawe, Argyll (fig. 25.3), has been taken into State care by Historic Scotland. The site has retained its rubble-built furnace buildings, charcoal and ore sheds, barkhouse, and a range of workers' dwellings. At this and the other similar furnaces at Glen Kinglass and Craleckan, Argyll, iron ore (haematite) and flux were brought in from England by the English and Irish owners. The furnaces, which began production in the mid eighteenth century, had relatively short lives; Glen Kinglass closed in about 1738, Craleckan in about 1813, and Bonawe in 1874.[16] In the case of Craleckan, charcoal capacity was diverted after closure to serve the gunpowder industry, for which Argyll became famous.[17] Gunpowder manufacture was itself an important industry, providing explosives not only for road, canal and railway building works, but also for extensive use in mining and quarrying ventures. The remnants of the Scottish gunpowder industry were eventually to be taken over by Nobel's Explosives, which itself evolved into Imperial Chemical Industries Limited, Britain's entire gunpowder capacity subsequently being concentrated at Ardeer on the coast of Ayrshire in the 1930s.

The use of charcoal in smelting iron ore was rendered unnecessary by the Darbys' discovery in 1709 that coal could be used as a fuel in its place.[18] In Scotland, the Carron Ironworks was first to adopt this technology, heralding, from the 1760s, the beginning of a new industrial era.[19] It was, however, the discovery by David Mushet in 1801 of blackband iron ore, and the subsequent introduction of the hot-blast process by James Beaumont Neilson (patented in 1828), that transformed the Scottish iron industry.[20] Thereafter, the extraordinary coincidence of easily extracted shallow coals and ironstones led to rapid growth of pig-iron production in central Scotland. The greatest concentrations of the industry were to be found in the district of Monklands near Glasgow, further to the east around Shotts, Lanarkshire, and to the west in Ayrshire.

A number of large integrated companies developed, examples including The Coltness Iron Company [21] and The Shotts Iron Company.[22] In general, these companies owned several coal and ironstone mines, limestone quarries and mines (and the associated limekilns), coke ovens, and ironworks containing blast furnaces, blowing-engine houses, and foundries. During the latter half of the nineteenth century, a substantial malleable-iron industry grew from the pig-iron industry,[23] but with the gradual introduction first of the Bessemer process (1856), and then the Siemens Martin open-hearth process (1867), so mild steel took over from wrought-iron production. This trend accelerated towards the end of the nineteenth century, as did the exhaustion of cheap, easily mined Scottish coal and ironstone. With the depletion of local iron ore, the focus of iron and steel production began to shift westwards away from Monklands, although specialised foundries survived elsewhere, particularly in the Falkirk area.

During the twentieth century, steel became one of the major strategic industries of Scotland, and the material itself grew to become a dominant building medium, with structural steel companies such as Sir William Arrol & Co rising to prominence.[24] As was the case previously with iron, steel was also a crucial component in a huge armaments industry.

The ownership of the steel industry eventually became concentrated within the Colville Group of companies, who had begun business at the Dalzell Works in 1870, and whose many works in Glasgow, Glengarnock, Ayrshire, and Motherwell, Coatbridge and Gartcosh, Lanarkshire, employed over 18,000 people in 1956.[25] A specialised steel-tube industry had also concentrated under the ownership of Stewart & Lloyds Ltd.[26] Several phases of nationalisation and privatisation have since overseen the rapid rationalisation and demise of the Scottish steel industry, which, after the closure of the integrated works at Ravenscraig, Renfrewshire, in 1992, had been reduced to Colville's first factory at Dalzell in Motherwell.

Comparatively little in the way of coherent remains now survives of the Scottish iron industry. The foundations of the Summerlee Ironworks at Coatbridge have been incorporated within the Summerlee Heritage Trust's industrial museum, and similarly, the Italianate blowing-engine house at the Dalmellington Ironworks, Ayrshire, is being developed as part of the Dalmellington and District Conservation Trust's industrial museum at Waterside. Elsewhere, perhaps the best surviving monument to the iron industry is the long L-plan rubble-built furnace bank and associated buildings of the Shotts Ironworks, which dates from 1802.[27]

The survival of buildings relating to the later steel industries is even more precarious, primarily because they are prone to a form of auto-cannibalism that is rarely witnessed in other industries. Modern steelworks were usually built predominantly from steel, and once closed, tend to be stripped out, the machinery being transferred or sold to other steelworks where it can be re-used. The buildings themselves are merely recycled, first being cut down and sold for scrap, and then being melted down in the furnaces of other steelworks in Britain and overseas.

CLAY

The iron and steel and other modern industries could not have developed as they did during the eighteenth and nineteenth centuries without agricultural improvement, which simultaneously greatly improved agricultural production and stimulated human migration to towns and cities where new industries required labour. As has already been discussed, the increased availability of coal and limestone encouraged the burning of lime for use as a fertiliser. Similarly, brick and tile works multiplied as a result of the availability of cheap coal, producing pantiles and field drainage tiles, which played a vital role in draining and improving agricultural land throughout Scotland.[28] These works were to operate on a larger scale in the rapidly industrialising areas of the central coalfields, producing bricks for an expanding construction industry, and, most significantly, fireclay refractories for the iron and steel industries, and other heat-intensive industries such as glassworks, and gasworks.[29] Fireclays were also used to produce a wide range of sanitary-ware products, ranging from salt-glazed fireclay sewer pipes to sophisticated toilet furniture and architectural fireclay ware.

Brick, tile and sanitary-ware manufacture required kilns in which clayware could be fired. Until the nineteenth century, kilns tended to be primitive clamps or updraught intermittent structures (single kilns). Firing became more sophisticated with the development of downdraught rectangular and round kilns, and with the horizontal-draught Newcastle kiln. Output was further improved by the introduction of continuous kilns, the most prolific of which was the barrel-arched Hoffmann kiln, variations of which included the Belgian kiln. Examples of more sophisticated continuous kilns were the transverse-arch Dunnachie and Staffordshire kilns, the former being the first known example of a kiln using producer gas. Most kilns were built from brick, the hotter parts being constructed from fireclay brick. Most originally were fired by coal, but later used other fuels such as gas or oil.[30]

BUILDING MATERIALS

During the nineteenth century especially, rapid urbanisation and rising consumer demand led directly to the growth in demand for high-quality building materials, and to the development of stone quarries and mines throughout Scotland. Glasgow and Edinburgh, for example, developed sandstone quarries within or close to city boundaries, but as demand expanded,[31] quarries were opened further afield in Dumfriesshire and Ayrshire. Quarries in Aberdeenshire,[32] Argyll and Kirkcudbrightshire supplied high-quality granite, whilst slate quarries in Argyll[33] were also significant, as were the flagstone quarries of Caithness[34] and Angus.

Although stone, and later brick, comprised the main raw materials of the construction industry, the availability of cast and wrought iron greatly enhanced building possibilities. Factories, warehouse and commercial buildings could all be larger and stronger with the application of internal iron frames, beams and columns. Their exteriors could also be adorned with ornamental ironwork, such as that produced by the Shotts and Carron Iron companies.[35] With the introduction of steel from the 1860s onwards, iron was gradually replaced as the preferred material for the construction industry, and even larger spaces could be created within buildings, especially for the heavier industries such as engineering and shipbuilding, and, of course, the steel industry itself.

ALUMINIUM

The arrival of the twentieth century brought new industries and materials to Scotland, one of the most important being aluminium.[36] Scotland's first aluminium smelter was established at Foyers near Inverness in 1886, and was followed by smelters in the western Highlands at Kinlochleven, Argyll (1909), and Fort William, Inverness-shire (1924), electricity in each case being supplied by hydro-electric schemes of increasing scale. One of the more interesting features of the Kinlochleven plant, which in 1993 was thought to be the oldest operating aluminium smelter in the world, was that it included a predominantly rubble-built (now disused) carbon plant for the manufacture of electrodes essential to the smelting process. In addition, much of the extensive concrete work of the Blackwater Dam and conduit was associated with Robert McAlpine, a Scottish pioneer in the use of concrete.

Other Scottish aluminium factories were to include a rolling mill at Falkirk, built in 1944, an alumina and red oxide works at Burntisland, Fife, which was completed in 1917. The last major Scottish development was that of the Invergordon Smelter, Ross and Cromarty, in 1972, which was not bestowed with its own hydro-electric power supply, and which was closed in 1981 because of the prohibitive cost of electricity bought from the Hydro Board.

PETROCHEMICAL PRODUCTS

Amongst the new extractive and extracting industries to develop during the twentieth century was Scottish Dyes at Grangemouth, Stirlingshire, later to be incorporated into Imperial Chemical Industries Limited. Similarly, British Petroleum grew from roots developed by Young's Paraffin Company, also establishing a strong base in Grangemouth, where important innovations included the cracking for the first time of petroleum to produce ethanol during the 1950s.[37]

One hundred years earlier, James Young had patented the process of destructive distillation of coals and shales, and in 1851 established a paraffin-oil works in partnership with Messrs Meldrum & Binney in Bathgate, West Lothian. During the ensuing decades, a substantial oil-shale industry grew in the Lothians. The process involved baking oil shale in retorts, the design of which became more sophisticated as the century progressed.[38] A wide variety of petrochemical products and by-products was produced, but the most obvious and lasting visual evidence of the industry has been the huge quantities of spent shale produced by the process, much of which was piled high on the distinctive pink waste heaps of West Lothian. The oil-shale industry was eventually forced out of business by competing crude-oil products from the modern oil industry, production finally ceasing in 1962.[39] As with most extractive industries, little survives of the buildings, and even the huge spoil heaps are gradually being landscaped and reduced in size.

COAL

Despite its greater size and geographical spread, the physical remains of the Scottish coal industry are also disappearing rapidly. The coals found in Scotland are from the Carboniferous period (with the exception of the small coalfield at Brora, Sutherland, which is Jurassic), and are found in two distinct sequences. These are the Upper-Series Coals of the Productive Coal Measures, and the Lower-Series Coals of the Carboniferous Limestone Series. Early larger-scale mining activity concentrated on the shallow and exposed seams mostly from the Upper Series, especially where blackband ironstone was also found in abundance. The coalfields extended from Ayrshire and North Dumfriesshire in the west, through Lanarkshire, Clackmannanshire and Stirlingshire, to Fife and the Lothians in the east. There were also outcrops further afield, such as those on Arran and at Machrihanish, Argyll. The types of coal mined ranged from high quality anthracite and coking coals, through to gas, steam and house coal, and the relatively poor quality, high-ash Upper-Hirst coal currently used by the electricity generation industry. The coal was put to a variety of uses, some of the biggest markets being the town gas[40] and coke industries, whose by-products spawned an extraordinarily diverse chemical industry.

Coal extraction has occurred for many hundreds of years, but the first references are in monastic charters. Early mining tended to involve extracting from exposed strata at outcrops or in valleys or on coasts. Shallow pits were also dug, and subsequently belled out along seams of coal, forming 'bell pits'. Areas of bell-pit activity have survived in places such as Thornton, Fife, and Muirkirk, Ayrshire. It was common for coal to be raised to the surface with the assistance of a horse engine, and history books go to great lengths to describe the appalling working conditions of these and later mines, and the indiscriminate use of women and child labour. Horse engines were also used successfully to drain shallow mineworkings, as at Sir George Bruce's famous undersea Moat Pit near Culross, where water was lifted by a chain of buckets as early as the 1590s.[41]

As for buildings associated with coal mines, nothing substantial survives from early periods of mining. The transient and small-scale nature of activity meant that there was no need to build permanent structures. It was only with deeper mines that haulage, drainage and ventilation dictated the need for strong stone buildings to house engines, and for significant capital investment in longer-term mining facilities at

Figure 25.4 Lady Victoria Colliery, Newtongrange, Midlothian. Crown copyright: RCAHMS, SC 710830.

the surface. Newcomen engines were installed with increasing frequency to drain mines in the early eighteenth century, early examples being those at Stevenston, Ayrshire.[42] By 1784, after improvements to steam power by Boulton & Watt and others,[43] the double-acting steam engine had been perfected, and enhanced drainage and winding permitted even deeper shafts.

The development of the steam engine not only aided the extraction of coal, but massively enhanced the market for coal as steam power was applied to a wide range of industries. Demand was further assisted and enhanced by the expansion of canals and railways, and by the growth of the iron industry. During the nineteenth century, collieries grew in number, size, and sophistication, especially those that were owned by some of the larger companies. However, it was only towards the end of the century that, with shallow coals becoming exhausted, larger more capital-intensive collieries were required to exploit the Lower-Series coals, which were often over 2,000 feet (610 metres) below. At this time, a number of the larger companies began to plan model collieries, the finest examples of which included Michael Colliery at East Wemyss, Fife (Fife Coal Company), Lady Victoria Colliery at Newtongrange, Midlothian (Lothian Coal Company; fig. 25.4), and Barony Colliery at Auchinleck, Ayrshire (Bairds and Dalmellington; fig. 25.5). These and other projects were also accompanied by housing developments and villages for miners.[44]

The major components of these more permanent mines included mine headframes (usually fabricated from rolled-steel sections),[45] steam

Figure 25.5 Barony Colliery, Auchinleck, Ayrshire. Crown copyright: RCAHMS, SC 710863.

or electric winding-engine houses, electricity power stations, air-compressor houses (for tools powered by compressed air), fan houses for ventilation, pumping engine houses for drainage, boiler houses, and offices. One particular feature of the more modern mines was circular or oval shafts, which replaced earlier rectangular patterns. The new shafts were an improvement because they allowed more space for service pipes and cables, were more stable, and provided more space between the shaft sides and the cages, aiding ventilation by permitting a freer flow of air. Also of importance was the introduction of wire ropes, which greatly enhanced the safety and efficiency of colliery winders.[46]

The coal-mining industry was at the forefront of social change in Britain, miners in particular playing a crucial part in the evolution of the British trade-union movement. In addition to the major issue of safety, appalling working and living conditions promoted concerted action, one of the results of which was the formation of the Miners' Welfare Fund under Section 20 of the Mining Industry Act (1920). The Fund was derived from a contribution of one penny per ton of output from every coal mine. The remit of the Fund was wide, covering improvements in the provision of miners' facilities within mines, and the building of a wide range of amenities within communities associated with the mines.

During the decades that followed, a large number of projects supported by the Fund were completed in Scotland. Within the mines, impressive pit-head baths, canteens and medical centres were built, particularly during the 1930s. Some displayed fine period architecture, as was the case at Michael Colliery where the baths contained 122 shower units for 2,552 men and 10 for the 96 women employed at the mine.[47]

Outside the mines, properties were purchased to be converted into convalescent homes for miners, and a wide range of miners' schools, institutes, sports facilities and children's playgrounds were built throughout the Scottish coalfields. Given the policy of demolition and clearance of the surface buildings of closed coal mines that has persisted in the later decades of the twentieth century, the buildings of the Miners' Welfare fund have, as a rule, easily outlived those of the mines with which they were once associated.

In general, the surface buildings of the coal mines tended to be brick-built, especially during the twentieth century. Many coal companies owned brickworks which utilised blaes (blase, shale waste) from the coal mines. The use of cheap, rapidly erected steel-framed, brick clad buildings, such as those at Frances Colliery near Dysart, Fife, was common, especially where screening and washing coal was required. As the century progressed, so the scale of mines increased, as did the use of reinforced concrete. By 1939, several large coal companies had ambitious plans for 'superpits', such as that completed by the Fife Coal Company at Comrie in 1940, using mostly German equipment.[48] After the end of World War II the industry was nationalised, the National Coal Board (NCB) assuming responsibility for 275 mines, of which the 196 largest mines were taken into public ownership.[49]

Under the care of the NCB Scotland's coal industry was transformed. A huge reconstruction programme, begun in 1955, involved the closure of 144 mines, many of which were in the once-productive Lanarkshire coalfield. Development was to occur at 430 existing pits, with 60 new or reconstructed short-term surface drift mines being sunk to offset the shortfall whilst 14 new sinkings were completed. Many of the planned superpits had been formulated by the original coal companies, such as the disastrous Rothes Colliery at Thornton, Fife. Others, such as the equally unsuccessful Glenochil Colliery near Alloa, Clackmannanshire, were entirely NCB inspired.

A number of the large new projects were successful, growing to produce an annual output of over one million tons. These included Killoch, Ayrshire, Bilston Glen and Monktonhall, Midlothian, and Seafield, Fife. Several reconstructions were successful, such as Barony and Frances. Others were not so lucky, such as the potentially lucrative Michael Colliery, which was the biggest pit in Scotland in 1947. Michael was eventually closed in 1967 after a fire killed nine miners and destroyed

Figure 25.6 Kinneil Colliery, Bo'ness, West Lothian. Crown copyright: RCAHMS, SC 711519.

many million tons' worth of development work. Similarly, Kinneil near Bo'ness, West Lothian (fig. 25.6), never achieved its potential, and Polmaise and Manor Powis, Stirlingshire, were comparative failures after risky bids to exploit the anthracite that had been anticipated because of the presence of the Stirlingshire Dolorite sill. In contrast, the coking coal output of collieries at Cardowan, Glasgow, and Polkemmet, West Lothian, was sustained for longer than might have been expected.[50]

The reconstruction of the Scottish coal industry was a bold strategy which was to be overtaken by events beyond the industry's control. The market for coal, which at one time seemed insatiable, contracted rapidly in the post-war years with the expansion of the use of electricity and the national grid, the introduction of alternative fuels such as oil and gas, and the nuclear energy programme, and also because of falling demand from a contracting manufacturing base in Scotland. Despite these setbacks, the new generation of superpits represented the Scottish coal industry at its peak. Extremely powerful electrically-powered winding engines allowed much larger quantities of coal (usually in skips instead of mine cars) to be hoisted from the deep workings thousands of feet underground. The winding engines were housed in impressive reinforced concrete towers, sometimes over 200 feet (61 metres) high. Covered concrete

Figure 25.7 Prestongrange Colliery, East Lothian; Cornish-beam pumping engine and engine house. Crown copyright: RCAHMS, SC 711082.

walkways allowed miners to walk directly from the baths, and the token and lamp rooms to the pithead. The baths and changing rooms were salubrious compared with those of older mines, providing copious quantities of hot water, and heated lockers to dry damp clothes. High-quality canteens were provided, as were medical facilities and rescue stations.

The National Coal Board's vision of the modern Scottish coal industry is well illustrated in the following paragraph extracted from one of its publications of the period:

> The new collieries, with their tall winding towers, present a picture of dignity and elegance conspicuously absent from the rather dingy, exposed and untidy-looking pits of the past. In grouping the various buildings, the Board's architects have produced an arrangement in keeping with contemporary standards of individual design. The modern colliery has become what at one time seemed impossible – a landmark of arresting interest, in close harmony with its surroundings, and flanked with shrubs and lawns any landscape gardener might be proud of. [51]

In the years after these words were written, the Scottish coal industry was reduced by 1993 to the single mining complex serving Longannet Power Station in West Fife, and the superpit at Monktonhall, which had passed into private ownership. Elsewhere, almost all the surface buildings of the superpits and older mines had been systematically cleared away, despite, on some occasions, their protected status. More significant still had been the growth of open-cast coal mining, and, to a certain extent, afforestation and land reclamation. In many areas, these activities have totally eradicated the remains of old surface buildings and entire mining landscapes.

A number of important coal mining buildings have survived, however, and appear to have a secure future. Of these, the rubble-built beam-engine houses at Devon Colliery, near Alloa, and Prestongrange Colliery, East Lothian (fig. 25.7), are perhaps the most well-known individual buildings. The most important and complete complex remains that of Lady Victoria Colliery at Newtongrange. Built by the Lothian Coal Company in 1890, it retains many of the components of a model coal mine. More important, perhaps, is the fact that it now contains the Scottish Mining Museum. In addition to fine collections of mining artefacts and exhibitions, the museum has also amassed the finest and most important library and archive of the coal industry in Britain. Its value will become all the more apparent as the remains of the coal industry in Scotland, and Britain as a whole, are erased from the contemporary landscape.

NOTES

1. For the economic minerals of Scotland, see the British Geological Survey publications for Scottish regions and Craig, 1991.
2. For general sources on the industrial development of Scotland, see the works by John R Hume listed in the bibliography. Bremner's book (1868) gives a

valuable insight into the workings of industry in the mid nineteenth century. Other useful general sources include Hamilton, 1932 and Oakley, 1953.
3. For an account of the Scottish salt industry, see Whatley, 1987.
4. The Scottish Industrial Archaeology Survey's record of Preston Island is deposited in the NMRS, reference no MS/500/2/1.
5. See also Lewis, 1989, 362–70.
6. Whatley, 1982(a), 89–101; Fenton, 1997, 211.
7. Hume, 1976, 123–4.
8. Hume, 1977, 54–6; Butt, 1967, 99–100.
9. Clarke, 1987, 5–22.
10. Hay, Stell, 1986, 164–6.
11. Hume, 1976, 132.
12. For metalliferous mining in Scotland, see Burt, Waite, Atkinson, 1982, Part I, 4–19 and Part II, 140–57.
13. Hay, Stell, 1986, 136–8.
14. Donnachie, 1971, 127–9.
15. For the early iron industry in Scotland, see Macadam, 1887, 89–131.
16. Hay, Stell, 1986, 105–15.
17. RCAHMS, 1992, 482–3 (Clachaig), 486–93 (Furnace), 494–5 (Kames), 495–6 (Lochfyne).
18. Gale, 1969.
19. Hume, 1961.
20. Thompson, 1982, 27–41; Corrins, 1970, 233–63.
21. Carvel, 1948.
22. Muir, nd.
23. Thompson, 1983(a), 10–29; 1983(b), 31–49.
24. Arrol, 1909.
25. For individual steel works in Scotland, see Colville Group of Companies, c. 1956; Payne, 1979.
26. Stewarts & Lloyds Limited, 1953.
27. Hume, 1976, op cit, 179–80; and Muir, nd.
28. Fenton, 1999, 19–23.
29. Sanderson, 1990.
30. Douglas, Oglethorpe, 1993.
31. Bunyan, Fairhurst, Mackie, McMillan, 1981.
32. Dick, 1949; Donnelly, 1974, 225–38; Donnelly, 1979, 228–38.
33. Richey, Anderson, 1944; Viner, 1976, 18–27.
34. Omand, Porter, 1981.
35. Muir, nd; Campbell, 1961.
36. British Aluminium Company Limited, 1955.
37. Porteus, 1970, 123–205.
38. Carruthers, Cardwell, Bailey, Conacher, 1927, 240–64.
39. Cameron, McAdam, 1978.
40. Burgh of Hamilton, 1931; and National Coal Board, 1958, 72–3.
41. Bowman, 1970, 353–72.
42. Whatley, 1977, 69–77.
43. For the development of the steam engine, see Cossons, 1975, 79–118.
44. Hume, 1990, 19.
45. For details of a colliery steel headframe, see MacAdam, 1946, 308–14.
46. National Coal Board, 1958, 77–9.

47. Extensive reports on these projects including plans and photographs, can be found in the *Annual Reports* of the Miners' Welfare Fund, starting in 1921.
48. Reid, Crawford, McNeill, 1939; and Muir, nd.
49. National Coal Board, 1958.
50. Halliday, 1989.
51. National Coal Board, 1958, 88. The person responsible for designing most of the new collieries was Austrian architect, Egor Riss.

BIBLIOGRAPHY

Arrol, Sir William, and Company Ltd. *Bridges, Structural Steel Works, and Mechanical Engineering Productions*, London, 1909.
Bowman, A I. Culross Colliery: A sixteenth-century mine, *Industrial Archaeology*, 7/4 (1970), 353–72.
Bremner, D. *The Industries of Scotland: Their rise, progress and present condition*, Edinburgh, 1868, reprinted Newton Abbot, 1969.
British Aluminium Company Limited. *The History of the British Aluminium Company Limited, 1894–1955*, London, 1955.
British Geological Survey. *British Regional Geology: The tertiary volcanic districts of Scotland*, 3rd edn, Edinburgh, 1961, Chapter 8, Mineral Deposits, 117–19.
British Geological Survey. *British Regional Geology: The south of Scotland*, 3rd edn, Edinburgh, 1971, Chapter 11, Economic Geology, 103–6.
British Geological Survey. *British Regional Geology: Orkney and Shetland*, Edinburgh, 1976, Chapter 12, Economic Geology, 117–24.
British Geological Survey. *British Regional Geology: The Grampian Highlands*, 3rd edn, Edinburgh, 1978, Chapter 15, Economic Minerals, 88–92.
British Geological Survey. *British Regional Geology: The Midland Valley of Scotland*, 3rd edn, Edinburgh, 1985, Chapter 16, Economic Geology, 145–53.
British Geological Survey. *British Regional Geology: The northern Highlands of Scotland*, 4th edn, Edinburgh, 1989, Chapter 16, Economic Minerals, 177–81.
Bunyan, I T, Fairhurst, J A, Mackie, A, McMillan, A A. *Building Stones in Edinburgh*, Edinburgh, 1987.
Burgh of Hamilton. *A Century of Gas Supply, 1831–1931*, Hamilton, 1931.
Burt, R, Waite P, Atkinson, M. Scottish metalliferous mining, 1845–1913: Detailed returns from the mineral statistics, *Industrial Archaeology*, 16 (1982), Part 1, 4–19, Part II, 140–57.
Butt, J. *The Industrial Archaeology of Scotland*, Newton Abbot, 1967.
Campbell, R H. *Carron Company*, Edinburgh, 1961.
Cameron, I B, McAdam, A D. *The Oil-Shales of the Lothians, Scotland: Present resources and former workings*, London, 1978.
Carruthers, R G, Cardwell, W, Bailey, E M, Conacher, H R J. *The Oil-Shales of the Lothians*, Memoirs of the Geological Survey of Scotland, 1927, 240–64.
Carvel, J L. *One Hundred Years in Coal: The history of the Alloa Coal Company*, Edinburgh, 1944.
Carvel, J L. *The New Cumnock Coalfield: A record of its development and activities*, Edinburgh, 1946.
Carvel, J L. *The Coltness Iron Company: A study in private enterprise*, Edinburgh, 1948.
Clarke, R J. The Closeburn limeworks scheme: A Dumfriesshire waterpower complex, *Industrial Archaeology Review*, 10/1 (1987), 5–22.

Colville Group of Companies. *A Technical Survey of the Colville Groups of Companies*, London, c. 1956.
Corrins, R D. The great hot-blast affair, *Industrial Archaeology*, 7/34 (1970), 233–63.
Cossons, N. *The BP Book of Industrial Archaeology*, Newton Abbot, 1975.
Craig, G Y. *Geology of Scotland*, 3rd edn, London, 1991.
Diack, W. *Rise and Progress of the Granite Industry in Aberdeen*, Aberdeen, 1949.
Donnachie, I. *The Industrial Archaeology of Galloway*, Newton Abbot, 1971.
Donnelly, T. The Rubislaw granite quarries, 1750–1939. *Industrial Archaeology*, 11/3 (1974), 225–38.
Donnelly, T. Structural and technical change in the Aberdeen granite quarry industry, 1830–1880, *Industrial Archaeology Review*, 3/3 (1979), 228–38.
Douglas, G J, Oglethorpe, M K. *The Brick, Tile and Fireclay Industries of Scotland*, Edinburgh, 1993.
Fenton, A. *Scottish Country Life*, Edinburgh, 1976, new edn, East Linton, 1999.
Fenton, A. *The Northern Isles: Orkney and Shetland*, Edinburgh, 1978, new edn, East Linton, 1997.
Gale, W K V. *Iron and Steel*, London, 1969.
Glenboig Union Fireclay Co Ltd. *Glenboig*, 6th edn of catalogue, Glasgow, nd.
Halliday, R. *The Disappearing Scottish Colliery: A personal view of some aspects of Scotland's coal industry since nationalisation*, Edinburgh, 1989.
Hamilton, H. *The Industrial Revolution in Scotland*, London, 1932, new impression, 1966.
Harvey, W S. Lead mining in 1768: Old records of a Scottish mining company, *Industrial Archaeology*, 7/3 (1970), 310–72.
Hay G D, Stell, G P, for RCAHMS. *Monuments of Industry: An illustrated guide*, Edinburgh, 1986.
Hume, J R. *Scotland's Industrial Past: An introduction to Scotland's industrial history, with a catalogue of preserved material*, Edinburgh, 1990.
Hume, J R. *The Industrial Archaeology of Scotland*: vol 1, *Lowlands and Borders*, London, 1976.
Hume, J R. *The Industrial Archaeology of Scotland*: vol 2, *Highlands and Islands*, London, 1977.
Lawson, J. *Building Stones of Glasgow*, Glasgow, 1981.
Lewis, J. The excavation of an eighteenth-century salt pan at St Monance, Fife, *PSAS*, 119 (1989), 362–70.
McAdam, R. Frances Colliery reconstruction schemes, part one, *Colliery Engineering*, 23 (1946), 308–14.
McAdam, R. Frances Colliery reconstruction schemes, part two, *Colliery Engineering*, 24 (1947), 10–16.
Macadam, Ivison W. Notes on the ancient iron industry of Scotland, *PSAS*, 9 (1887), 89–131.
Miners' Welfare Fund. *Annual Reports*, 1921–.
Muir, A. *The Fife Coal Company Limited: A short history*, Leven, Fife, nd [c. 1947].
Muir, A. *The Story of Shotts: A short history of the Shotts Iron Company Limited*, Edinburgh, nd [c. 1952].
National Coal Board. *A Short History of the Scottish Coal-Mining Industry*, National Coal Board: Scottish Division, Edinburgh, 1958.
Oakley, C A ed. *Scottish Industry*, Glasgow, 1953.
Omand, D, Porter, J. *The Flagstone Industry of Caithness*, Aberdeen, 1981.
Payne, P. *Colvilles and the Scottish Steel Industry*, Oxford, 1979.
Porteus, R. *Grangemouth's Modern History, 1768–1968*, Grangemouth, 1970.

Reid, W, Crawford R, McNeill, K H. *The Layout and Equipment of Comrie Colliery, Fifeshire*, Cardiff, 1939.

Richey, J E, Anderson, J G C. *Scottish Slates* (Wartime Pamphlet 40), London, 1944.

RCAHMS. *An Inventory of Argyll*, Vol 7, *Mid Argyll and Cowal*, Edinburgh, 1992.

Sanderson, K W. *The Scottish Refractory Industry, 1830–1980*, Edinburgh, 1990.

Smith, D L. *The Dalmellington Iron Company: Its engines and men*, Newton Abbot, 1967.

Stewarts and Lloyds Limited. *Stewarts and Lloyds Limited 1903–1953*, London, 1953.

Thompson, G. The iron industry of the Monklands: An introduction, *Scottish Industrial History*, 5/2 (1982), 27–41.

Thompson, G. The iron industry of the Monklands (continued): The individual ironworks, *Scottish Industrial History*, 6/1 (1983a), 10–29.

Thompson, G. The iron industry of the Monklands (continued): The individual ironworks II, *Scottish Industrial History*, 6/2 (1983b), 31–49.

Viner, D J. The marble quarry, Iona, Inner Hebrides, *Industrial Archaeology Review*, 1/1 (1976), 18–27.

Whatley, C A. The introduction of the Newcomen engine to Ayrshire, *Industrial Archaeology Review*, 2/1 (1977), 69–77.

Whatley, C A. An early eighteenth-century Scottish saltwork: Arran, c. 1710–1735, *Industrial Archaeology Review*, 6/ 2 (1982), 89–101.

Whatley, C A. Scottish salt making in the eighteenth century: A regional survey, *Scottish Industrial History*, 5/2 (1982), 2–26.

Whatley, C A. *The Scottish Salt Industry 1570–1850: An economic and social history*, Aberdeen, 1987.

26 Harbours, Docks and Fisheries

JOHN R HUME

The maritime buildings and structures of Scotland have been reduced substantially in numbers during the past thirty years owing to changing patterns of trade, methods of cargo handling, fishing techniques and modes of inland transport. Enough survives, however, both of the structures themselves and of photographs and drawings, to make sensible analysis of the development of maritime technology possible. Scotland's very varied topography made very different design solutions appropriate to different locations.

The earliest marine works could only be located in less exposed positions, and most of the early Scottish ports were at river mouths on estuaries,[1] or at the heads of navigation of river estuaries.[2] Surviving harbours of seventeenth-century configuration are the old harbours at Portsoy, Banffshire, Dunbar, East Lothian, Crail and St Andrews, Fife.[3] All have or had two piers protecting a basin. Parts of other harbours may incorporate masonry of similar date, generally cubic stones being employed. Where the ground could not take masonry construction, wooden piles were used, as in the original Broomielaw Quay in Glasgow.[4] The nearest existing parallel is the harbour at Palnackie on the Urr Navigation, Kirkcudbrightshire.[5] Early eighteenth-century harbours, such as Cockenzie, East Lothian, followed their predecessors in layout, and simple two-pier basins continued to be built until well into the nineteenth century;[6] improved techniques of construction led to their adoption in more exposed positions. Some simple basins were equipped with slots in the pier ends to allow planks to be inserted for protection in storm conditions: these 'booms' may still be seen at Crail and North Berwick, East Lothian.[7]

HARBOURS

The tempo of harbour construction increased in the late eighteenth and early nineteenth centuries, with improved internal transport by road, canal and wooden railway, and increased trade in agricultural produce, industrial raw materials and manufactured goods. Government encouragement of fishing also helped. Improved standards of engineering

Figure 26.1 Banff Harbour, c. 1880. Crown copyright: RCAHMS, SC 712196.

design enabled more durable works to be constructed, with Thomas Telford,[8] James Bremner,[9] John Rennie [10] and Robert Stevenson [11] among the pioneers. Where they could be afforded, larger stones, more closely jointed, with improved underwater profiles provided resistance to heavy seas in all but the most exposed situations, as at Wick, Caithness,[12] and Portpatrick, Wigtownshire.[13] Bremner devised the system of vertically-jointed masonry which allowed smaller pieces of undressed stone to be used, even in exposed locations.[14] Of equal importance was a major change in the layout of harbours, with an outer basin terminating on the landward side in a sloping beach (sometimes revetted), and an inner basin entered from one side of the outer basin (fig. 26.1).[15] The waves entering the harbour gave up most of their energy in breaking on the slope, so that the water in the inner basin remained relatively undisturbed. Particularly fine examples of this layout are Keiss, Caithness (fig. 26.2), and Burghead, Moray.

PORTS

Though small harbour construction continued throughout the mid nineteenth century, the development of railway transport and of steamships tended to focus trade on a smaller number of larger ports. The first railway ports were ferry terminals at the ends of railways: Granton, Edinburgh, Burntisland and Tayport, Fife, and Broughty Ferry, Angus.[16] Later in the century, rail-served ports were developed to deal

Figure 26.2 Keiss Harbour, Caithness; fish-curing house and stilling basin with vertically-set masonry wall. Copyright: John R Hume.

particularly with the export of coal[17] and iron,[18] the import of timber[19] and iron ore,[20] and the consignment of fish to London and other English markets.[21] The scale of the works at these harbours could be very large indeed, and mass concrete was extensively used from the 1870s onwards. Early large-scale concrete harbour works may be seen at Buckie, Banffshire, Fraserburgh, Aberdeenshire and (abandoned) at Sandhaven, Aberdeenshire, and later ones at Portknockie, Findochty, and Macduff, all Banffshire, Anstruther, Fife, and Port of Ness, Lewis.[22]

DOCKS

At the same time as concrete was revolutionising harbour design it was also being adopted for the construction of docks. A few docks had been built in Scotland before the 1860s, including Rennie's East and West Old Docks at Leith,[23] three docks at Dundee,[24] and smaller ones at Grangemouth, Stirlingshire, Dysart, Fife, Ayr, West Wemyss, Fife, and Montrose, Angus.[25] There had been plans for others, but it was not until the 1870s that expanding world trade, linked to more economical steamships and sailing vessels and better rail communication, made it imperative to build bigger docks. From then until World War I large

Figure 26.3 Princes Dock, Glasgow; containers being loaded onto cargo ship from floating crane, c. 1970. Copyright: John R Hume.

new dock projects followed each other in a steady stream: Glasgow's Queens, Princes (fig. 26.3) and Rothesay docks;[26] Dundee's Victoria and Camperdown; the James Watt Dock and partly abortive Great Harbour at Greenock, Renfrewshire; and the competing Grangemouth, Stirlingshire,[27] Methil, Fife,[28] and Leith[29] docks, designed for exporting coal. Smaller, older docks and harbours were also modernised, as in Troon, Ayrshire, Burntisland, Arbroath, Angus and Bo'ness, West Lothian.[30] The moving forces in this massive construction boom were partly civic-inspired harbour trusts, but largely railway companies. Most were docks with gated entrances, but some, as in Glasgow, were more properly termed basins.

The construction of dry docks, which had begun in Scotland at Port Glasgow, Renfrewshire, in 1762,[31] paralleled the building of the larger, and some smaller, ports. Glasgow, Leith, Greenock and Dundee all have large dry docks,[32] and there are smaller docks in Peterhead, Aberdeenshire, and Grangemouth, and even tiny ones at Maryhill, Glasgow, on the Forth and Clyde Canal,[33] and on the island of Raasay, Inverness-shire.[34] The Grangemouth dry docks are converted from dock-entrance locks.[35] In many other localities, and especially for smaller craft, slipways were and are used, notably in Aberdeen, Dundee,

Inverness and most of the larger fishing ports on the east coast.[36] Aberdeen is unique in having two side-loading floating docks for repairing fishing boats.

After World War I dock construction became more sporadic. On the Clyde the only new dock built since 1918 is the King George V dock at Shieldhall, planned as the first of a series along the south bank of the river.[37] The Clyde graving dock at Greenock is the only dry-dock of this period.[38] On the east coast major extensions took place at Leith to create Scotland's largest dock proper.[39] A new entrance lock at Grangemouth allows larger vessels to use this busy oil port. At Peterhead, Torry, Aberdeen and Montrose new oil bases have been created. In contrast to dock building, post-World War II construction of harbours has been considerable, especially in the Highlands and Islands, where the desire to encourage fishing has resulted in many new harbours being built and old ones extended.[40]

PIERS

Piers may be components of harbour works, but frequently have an independent existence. In exposed locations they may function partly as breakwaters, but in sheltered waters are generally the means of access to relatively deep water. Leaving aside jetties, as a specialised form of pier, free-standing piers seem to have first become popular in the late eighteenth and early nineteenth centuries, though their antecedents still survive in Stromness, Orkney (fig. 26.4), and Lerwick, Shetland, in the form of little private piers attached to the houses and stores of merchants.[41] In a few instances these stores rise sheer from the quay edge. Though this phenomenon is now confined to these northern ports, it was once more widespread. Slezer's view of Ayr shows similar features, and Kirkcudbright still has traces of this layout.[42] Stone piers of the late eighteenth and early nineteenth centuries survive in the Cromarty Firth at Balintraid and Belleport,[43] on the Clyde at Dunoon, Argyll, Millport, Great Cumbrae, Bute and Largs, Ayrshire, on the Forth at Limekilns, Fife,[44] and at various places in the western Highlands. An interesting and distinctive group of piers are those with inset ramps, designed for the shipment of cattle, as at Grass Point, Mull, and Otter Ferry, Loch Fyne, Argyll.[45] Simple masonry piers continued to be built for much of the nineteenth century: good late examples are at Craignure, Mull, and Ballantrae, Ayrshire.[46]

A considerable impetus was given to pier construction by the introduction of steamboats. Their mode of operation, with relatively brief calls to transfer passengers and light goods, allowed for a different approach to pier building. The sheltering function became irrelevant, and most steamer piers became simply modes of linking deep-enough water

Figure 26.4 Stromness, Orkney; private quay with house and warehouse. Copyright: John R Hume.

with the shore. Some of the earliest steamboat piers were of masonry construction, as at Largs, and Kilmun and Campbeltown, both Argyll,[47] but after 1840 most were either of open-timber or wrought-iron construction. The remains of these structures can be seen throughout western Scotland (including Loch Lomond) and in the Forth estuary.[48] A few remain in commission, most notably Dunoon and Kilcreggan, Dunbartonshire, and Tighnabruaich, Argyll.[49] Revenue from these piers was collected from individual users, so due-collecting booths were provided, and these often survive.[50] In some instances piers were built with masonry at the shore end, as in the gunpowder factory pier at Kames, Argyll, and the pier at Ellanbeich on Seil, Argyll.[51] Buildings on piers were rare, apart from small waiting rooms and bothies for staff, but there were elaborate buildings at Rothesay, Bute, and Dunoon.[52]

In a class of their own were the railway piers, designed to handle very large numbers of people. On the Clyde, the competition between three railway companies, all with their own steamer fleets, led to the erection of some very elaborate interchange facilities. The simplest among those surviving latterly was at Fairlie, Ayrshire,[53] but within the harbour at Ardrossan, Ayrshire, and at Wemyss Bay, Renfrewshire, Craigendoran, Dunbartonshire, Princes Pier, Greenock, and Gourock, Renfrewshire,[54] there were massive facilities, capable of accommodating numbers of steamers. Wemyss Bay is the only one surviving reasonably intact, though the structure below deck level has been renewed. In the West Highlands the railway piers and their facilities at Oban, Argyll, Mallaig, Inverness-shire, and Kyle of Lochalsh, Ross and Cromarty, were more modest, though the adjacent stations were fairly elaborate.[55]

In modern terminology railway piers would be described as ferry piers, but the nature of the steamer services operated by the railway companies was rather different from the point-to-point routes of today. Ferries were, however, important features of the Scottish road system from a very early date, the Queen's Ferry over the Forth being the most famous.[56] Ferryboats were generally small vessels, and so as to accommodate different stages of the tide, or flood and drought conditions, ferry piers were generally ramped. They might be incorporated in larger harbour works, as at Burntisland and Invergordon, Ross and Cromarty,[57] but were frequently free-standing. Thomas Telford designed some as part of his Highland road improvements.[58] Notable early nineteenth-century ferry piers include the Town Pier at North Queensferry, Fife,[59] and the Telford piers at Inverbreackie, Meikle Ferry and Ardelve, Ross and Cromarty, and Little Ferry, Sutherland.[60] A few ramped ferry piers of basically nineteenth-century construction survive in use, as at the Corran (Inverness-shire)-Ardgour and Luing-Seil, Argyll, ferries.[61] More recent ferry piers for roll-on roll-off ferries are similar in conception but generally of concrete construction.

On the Clyde specialised ferries operated.[62] There were chain ferries at three locations, and the ramps for two of them survive, at the Erskine and Renfrew ferries. There were also elevating-deck ferries, of which the shore facilities still exist at Govan and Finnieston, Glasgow. The little double-ended foot ferries operated from wooden stairways, and the remains of those for the Kelvinhaugh, Glasgow, ferry can still be seen. Some of the larger roll-on roll-off ferries operating in open-sea conditions gave link-span connections with land, and these may be seen at most of the ferries operated by Caledonian MacBrayne and P&O in the Clyde estuary, the western Highlands and to Orkney and Shetland.[63]

CARGO HANDLING

So far this account has been concerned with the construction of the basic dock, harbour or pier: its quay walls, sea defences (where necessary) and quay surfaces. In the case of many smaller harbours there was never any other fixed provision for handling cargoes, but in the larger docks and harbours buildings and equipment were generally installed.

The oldest surviving buildings are small warehouses on or beside harbours, as at Dunbar, Elie and Pittenweem, Fife, Portmahomack, Ross and Cromarty and Kirkwall, Orkney.[64] These were probably built as granaries, some possibly as rent houses. There are later examples of buildings designed for grain storage at Burghead, Moray, and Avoch, Ross and Cromarty.[65] At the inland port of Port Dundas, Glasgow, on the Forth and Clyde Canal, there were multi-storeyed warehouses from the 1790s (see fig. 27.4),[66] and plans of docks proposed for Glasgow a little later show warehouses lining docks as in the major English ports.[67] In the event, only Rennie's docks at Leith had this type and layout of warehouse,[68] and then only on one side. At Victoria Dock, Dundee, there is a small multi-storeyed block,[69] and there are – or were – isolated large warehouses at the Greenock docks.[70] The overwhelming majority of dockside warehouses were designed as one- or two-storeyed transit sheds, intended as short-term secure storage for goods en route to or from vessels.[71] Most of these have been demolished to free quaysides for modern use, but there are good examples in Glasgow, Leith, Grangemouth and Dundee.

One specialist type of dockside warehouse which had a late flowering was the granary. Grain warehouses in Scotland were mainly constructed in streets close to port facilities,[72] apart from those already mentioned, but the introduction of bulk handling of grain using pneumatic elevators made it attractive to store grain in silos rather than in heaps on floors. The first granary on the Clyde at Meadowside had both a multi-storeyed warehouse section and a silo section: later additions on this site were all of the silo type.[73] Leith was the other port with large grain silos.[74]

FISHING INDUSTRY

Other harbour buildings worthy of note include those provided for the fishing industry. Of these the largest are the curing sheds, which were used for the storage of barrels for salting herring, and sometimes also as cooperages. These are widely distributed, with good examples in the north at Portormin, Keiss and Scrabster, Caithness,[75] and in Orkney at Gill Pier, Westray, on Burray and at St Margaret's Hope.[76] There are off-quay curing houses in numbers at Wick,[77] and various ruined examples at, for instance, Helmsdale, Sutherland, Whaligoe and Lybster, Caithness, and at Burghead.[78] Ice houses are the other numerically-important group of buildings used by the fishing trade. Some of these are close to, if not part of, harbours, but many are linked to beach or estuarial fisheries, especially for salmon. They have been extensively studied by Bruce Walker.[79] Good harbour-side examples exist at Keiss, Portormin, and Chanonry Point, Ross and Cromarty, and Brora, Sutherland,[80] and examples close to harbours at Phillips' Harbour and Castletown, Caithness, Findhorn and Hopeman, Moray, Portree, Skye, and Helmsdale.[81]

Apart from ice houses and curing sheds, buildings used for fish

Figure 26.5 Tugnet, Speymouth, Moray; salmon-fishing station with gear store and boiling house to left and fishermen's two-storeyed accommodation block to right. Copyright: John R Hume.

processing included boiling houses, smoke houses and canneries. Boiling houses were used for parboiling salmon as a means of preserving them for shipment to market. The fish were boiled in iron tubs, which may still survive in a fishing station at Invershin, Sutherland. The buildings that housed boiling tubs form part of fishing stations at Findhorn and Tugnet, Moray (fig 26.5), Beauly, Inverness-shire, and Phillips' Harbour,[82] characterised externally by large chimneys. Smokehouses were used primarily for kippering herring and cold-smoking haddock. They were formerly very common, and not only in fishing ports. Until the 1960s there were two in Glasgow.[83] The greatest concentration surviving is in Aberdeen and Torry, where they are still used for making Finnan haddock.[84] Elsewhere, there are isolated examples. In the 1970s there were single houses or small groups in Fraserburgh and Peterhead, Aberdeenshire, Wick, Girvan, Ayrshire, and Portnancon, Sutherland.[85] The size varied widely, as did the design of the vents. Early photographs of the Aberdeen ones show vents like miniature distillery malt kilns,[86] but the recent survivors had fixed louvred vents or swivelling cowls like ships' ventilators. In Arbroath there are a number of smokeries making 'smokies' by hot-smoking haddock. These are pits with oak-chip fires sunk into the ground. [87] Canning of fish was most prominently carried on by Maconochie Foods of Fraserburgh in some rather nondescript buildings overlooking the harbour.[88]

CARGO-HANDLING DEVICES

In Scotland mechanical devices for cargo handling seem to have been a relatively late phenomenon. So far as evidence survives, there were few parallels to the wooden cranes of early modern harbours in England and on the continent, though a carved panel dated 1678, formerly in Tolbooth Wynd, Leith, shows such a crane, worked by a treadmill.[89] Fixed hand cranes with cast-iron frames and wooden jibs, capable of lifting one or two tons, were in use by the 1820s,[90] and were manufactured in numbers until the 1880s and '90s.[91] Most have disappeared, but examples survived until recently at Ellanabeich, Burntisland, and Lossiemouth, Moray, and Rodel, Harris.[92] An unusual wooden post version is still to be seen at Ollaberry, Shetland.[93] Mobile variants were to be found at Fortrose, Ross and Cromarty, and Hopeman, and the latter is now preserved.[94] The hand crane was developed in the 1850s and '60s to lift very heavy weights, but all those built, mainly if not solely on the Clyde,[95] have long gone. A late, smaller example survived until recently at Lossiemouth.

From the 1860s steam cranes came into fairly general use. Fixed heavy steam cranes were built in Dundee, Troon (fig. 26.7), Glasgow and Leith,[96] and mobile steam cranes were used extensively in Aberdeen and Leith, and less extensively in Glasgow and Paisley.[97] There was an unusually large mobile steam crane at Govan Graving Docks, Glasgow.[98] At the same time as the large hand and steam cranes were being installed, an alternative was to use sheer legs, a tripod of iron, one of which could be moved to tilt the other two. Sheer legs were used mainly in shipbuilding and repair yards, an early large example being at Napier's Lancefield Engine Works in Glasgow.[99] The only survivor is a relatively small one at Inverness.[100]

In the 1860s hydraulic power, in which water was pumped at high pressure from a central power station to the point of use, was adapted for cranes. One survives, at Leith,[101] but there were large installations there and in Glasgow. Hydraulic power was also much used for operating lock gates and swing bridges. Using armoured hoses, hydraulic power could even be used for mobile cranes, as at Prince's Dock, Glasgow.[102] The pumping stations for these installations survive in Glasgow, Leith and (until recently) in Greenock.[103] Another application of hydraulic power was the actuation of coal hoists, large steel structures for lifting loaded coal wagons, discharging their contents down chutes into ships, and lowering the empties. These were to be found at all the major coal shipping ports, notably at Methil, Burntisland, Grangemouth and Rothesay Dock, Glasgow, and individually elsewhere. The last of them did not disappear until the early 1970s.[104]

Hydraulic power was, however, rather cumbersome, and the development of reliable weatherproof electric motors during the 1890s led quickly to the adoption of electric cranes. Much ingenuity was

Figure 26.6 Troon Harbour, Ayrshire; steam crane and puffer. Copyright: John R Hume.

exercised in developing new designs of mobile cranes of various types including 'level luffing' cranes, which allowed loads to be transferred more expeditiously in stowing or unloading ships.[105] For heavy lifts earlier fixed-jib cranes and sheer legs were supplanted by giant cantilever cranes capable of positioning loads of up to 200 tons within a ship without the business of warping the vessel up and down the quay.[106] Most of these cranes were installed on the Clyde, but there were three at Rosyth Naval Dockyard, Fife. Derrick cranes, extensively used in quarrying and building construction, were less often used in ports, but there was a heavy-lift crane of this type in Clydebank shipyard, Dunbartonshire,[107] and the type was more recently used in handling containers at Leith and Ardrossan. At Greenock, Grangemouth and Leith, specialised container-handling cranes were installed in the 1960s and '70s,[108] and at General Terminus and Hunterston, Ayrshire, ore-handling gantry cranes were built.[109] A very specialised type of electrically-driven cargo-handling device was the grain elevator, to be found at Meadowside Granary, Glasgow, Leith, and at Kirkcaldy.[110]

The electric crane has persisted as a phenomenon, and rail-mounted electric cranes of some age could until recently be seen at Windmillcroft Quay, Glasgow, Ardrossan and Leith.[111] At Leith, Grangemouth, King George V Dock, Glasgow, and James Watt Dock, Greenock, there are more modern cranes. Elsewhere, mobile cranes have taken the place of these once very distinctive harbour features.[112]

NAVIGATION AIDS

Lights to aid navigation have ancient antecedents, and as a group form a link between harbour works and free-standing lighthouses. Harbour lights may well have existed in primitive form prior to the nineteenth century, but they did not become common until techniques of generating and focusing light were developed.[113] Such lights frequently functioned solely to mark the entrances to harbours, but there are many instances of pairs of 'leading' lights positioned to guide vessels along a preferred route into the harbour. The back light could often be some distance away from the harbour, as at Buckie, Crail and Gourdon, Kincardineshire, and the lights themselves might be relatively small, as at Portpatrick. Harbour entrance lights are very varied in size, form and construction. The earliest survivors are stone, such as Port Logan, Wigtownshire, Arbroath, North Queensferry and South Queensferry, West Lothian. Cast iron was popular in the mid and late nineteenth century, as at Burntisland, Buckie, Ayr, Newhaven and Wick. With the vogue for mass concrete harbours, lighthouses too were built of this material, the most noteworthy being at Aberdeen, Fraserburgh, Buckie, Macduff, Burghead and Wick. Twentieth-century lights include reinforced concrete examples at Granton and Leith. Recently plastic-lensed electric lights on poles have been introduced, often as replacements of older and more individual structures.

Surviving lighthouses detached from harbours are older.[114] Originally coal fires were used to provide illumination, and there are remains of coal-fired lighthouses on the Isle of May, Fife, and Little Cumbrae, Bute. The merchants of Dumfries erected a daymark at Southerness, Kirkcudbrightshire, in 1748–9 (converted to a lighthouse c. 1811) and those of the Clyde at the Cloch, Renfrewshire, in 1797. By far the largest providers of lighthouses in Scotland were and are the Commissioners for Northern Lights, formed in 1786, whose first lights were at Kinnaird Head, Aberdeenshire, Mull of Kintyre, Argyll, and Barra Head, Inverness-shire. Of these, the first-named was built in a tower-house, and a castellated style was used in new lighthouses on Inchkeith, Fife, and the Isle of May.

Gradually, a lighthouse style developed, with tapering circular-section towers and corbelled railed galleries for cleaning the lanterns. The first major lights had oil lamps, and systems of mirrors to focus the light into narrow beams. These were superseded first by combinations of mirrors and lenses, then by lenses alone. Flashing beams, to distinguish lighthouses from each other, were achieved by clockwork mechanisms. These changes resulted in alterations to the design of lanterns. The first lighthouses were all on land. Bell Rock, Angus, was the earliest in the world to be built on a half-tide rock, and subsequently rock towers were constructed at Skerryvore and Dubh Artach, Argyll. The support shore stations for these at Arbroath, Hynish, Tiree, and Erraid, Mull, survive, Hynish being the most elaborate. The engineers to the Commissioners were, for well over a century, members of the Stevenson family, and each individual or partnership developed his or their own style of building. Robert Stevenson had bellied cast-iron diagonal lattice railings on his galleries, and Alan favoured Egyptian-style keepers' houses,[115] plain railings and simple corbelled galleries. Later members of the family built towers that were simpler in design, and keepers' houses with steel and concrete flat roofs. Twentieth-century manned lighthouses reverted to square towers – the very last being Strathy Point, Sutherland. In recent years a programme of automation has been in progress, and all the Board's lights are now unmanned. In earlier cases the lens systems were removed; more recent practice has been to retain and electrify them.

In addition to the major towers, mostly built in stone, and a few in brick, there are many minor lights. Some of these are stone-built, like Port Askaig and the William Black Memorial[116] on Mull, but more are of iron or steel construction. Recent minor lights, both new and replacement, are made of plastic.

SHIPBUILDING

The buildings of shipbuilding yards are now much fewer in number than they were even twenty years ago; and technical change has swept away many of the older buildings even in the surviving yards. From photographs and drawings, however, it is possible to piece together something of a sequence of development.[117]

The early wooden shipbuilding yards had little in the way of buildings, but when iron shipbuilding began, the working of the material called into being sheds and workshops of various kinds. The plate-bending rolls and punching and shearing machines, and the steam engines to drive them, were best under cover, and the hot working of frames and blacksmith goods required shelter, and relative darkness, so that the temperature for working hot iron could be judged.[118] The sheds were often open-sided, iron-framed, with brick or timber walls, and frequently with Belfast roofs. The shops for such trades as copper-smithing, woodwork and plumbing were usually enclosed. A & J Inglis at Pointhouse, Glasgow, had until closure an iron-framed timber-clad joiners' shop of considerable size and sophistication.[119] The most elaborate buildings in a shipyard were, however, the offices, where the design drawings were produced and the complex co-ordination of component ordering and cost supervision was undertaken. The offices were also the places where clients were entertained and launch parties held.[120] Early photographs of the Meadowside shipyard of Tod & McGregor in Glasgow show an elaborate three-storeyed red and white brick office block.[121]

With ships becoming larger and more complex and steel replacing iron as a constructional material during the 1880s, shipyard buildings developed quite rapidly. Steel-framed sheds with corrugated iron cladding and roof glazing became standard for new construction from about 1890, as larger and more sophisticated tools were introduced, often electrically or hydraulically driven.[122] The power stations built to supply these tools became yard features. Trades which had been accommodated in simple sheds generated specialist buildings, some of which, built in brick, could until recently be seen at Govan shipyard. Offices, too, became grander, the finest undoubtedly being the Govan offices of the Fairfield Shipbuilding and Engineering Company.[123] Though most shipyards grew from relatively small beginnings, Yarrows moved to Scotstoun and a brand new yard in 1907, and some of the steel-framed buildings erected then are still in use, though complemented, and in some instances supplanted, by later structures.[124]

The capital equipment of the yards built before and during World War I lasted with comparatively little change until the 1950s, when welding began to replace riveting as a hull-building technique. It became attractive to put together large pieces of hull under cover before assembling them on the building berth, and very large steel-framed sheds were constructed for this 'prefabrication'. In Lithgow's shipyard at Port Glasgow a 'panel line' was also set up for making repetitive sections of hull prior to 'block assembly'. The only important new building type to emerge subsequently was the covered berth, a revival of an idea first tried in the 1850s at Meadowside. At Yarrow's Scotstoun yard frigates can be built, fitted out and dry-docked all under cover.[125]

The handling of materials in shipyards underwent a transition similar to that of cargo handling in docks. Block and tackle sufficed in wooden shipbuilding yards, and these were supplemented by rigged derricks and sheer legs.[126] Steel-framed derricks, introduced in the 1880s, remained in use until after World War II. Stephens at Linthouse had a wood-framed gantry over a berth, with a steam crane, in the 1870s, and both Beardmore's Dalmuir, Dunbartonshire, yard and Fairfield's yard had steel gantries. These proved expensive in first cost and maintenance, and were quickly supplanted by fixed tower cranes capable of lifting 20 tons or so. Tower cranes were adequate until sub-assembly fabrication came in, when high-capacity rail-mounted cranes were introduced. In one instance, Lithgow's Glen Yard, a Goliath crane capable of lifting 225 tons was installed. The lifting devices described above were used for handling hull parts: for fitting out ships with engines and boilers heavy cranes and sheer legs, as described in the harbour crane section, were used, either 'public' cranes in harbours, or 'private' cranes belonging to individual firms.

BOAT BUILDING

Boat building was certainly practised in prehistoric Scotland, as the discovery of many dugout canoes demonstrates. The apparatus for boat building is relatively simple and inexpensive, so small boatyards must once have been very common. The recent survivors, however, have for the most part been built or rebuilt to construct the larger classes of fishing boats, or yachts, or lifeboats.[127]

As with shipyards, most of the larger boat-building yards constructed vessels in the open, on slipways, and covered accommodation was reserved for the woodworking tools. A few fishing-boat building yards have, or had, covered slips, including one of the Buckie yards, and the Macduff yard, a yard at the Old Harbour, Dunbar, and the Girvan yard. Most of the other yard buildings tended to be fairly nondescript. A characteristic feature, which might be under cover, was the steam box, used for steaming hull planks to soften them. Yards designed for yacht building and repairing (including motor yacht and launch building) were more elaborate, with covered building berths normal. Millers' yard at St Monance, Fife, had two covered slipways with weatherboarded covers, and Morris & Lorimer's and Robertson's yards at Sandbank, Argyll, had extensive sheds. Morris & Lorimer's yard had a system of traversers to move yachts around the yard, as well as a slipway. Silvers' motor yacht yard at Roseneath, Dunbartonshire, had a very large range of corrugated-iron sheds. The largest boat building yards were those designed for lifeboat construction; Mechans in South Street, Scotstoun, had extensive single-storeyed sheds,[128] and McAlister's yard at Dumbarton has a large three-storeyed red and white brick shed. The smallest boat builders operated in sheds, using simple tools, and building boats capable of being manhandled. One in Eaglesham Street, Glasgow, survived into the early 1970s,[129] and a tiny yard in Grimsay, in the Outer Hebrides, was still working in 1991.

CUSTOM HOUSES AND HARBOUR OFFICES

Custom houses and harbour offices, sometimes combined in one building, are, apart from a few shipyard offices, the most elaborate buildings considered in this chapter. The custom houses at Leith, Greenock, Dundee and, more modestly, Glasgow, have all the solidity and strength of the Scottish classical revival, a style most appropriate for these great public buildings.[130] The finest is indubitably William Burn's Greenock Custom House recently refurbished by the Property Services Agency. Its grandeur expressed Greenock's position in the early nineteenth century as the principal port on the Clyde and focus for trade with the West Indies and United State in dutiable goods such as rum, sugar and tobacco. The remaining Scottish custom houses are more modest. Aberdeen's was originally an eighteenth-century town house, with vigorous detailing in the style of James Gibbs. There are neat villa-style buildings in Stornoway, Lewis, Bowling, Dunbartonshire, and Dunbar.[131] In Dundee, the custom house was situated in the same building as the harbour offices; elsewhere they were separate. Leith's is surprisingly modest, but the Aberdeen harbour office is both larger than the custom house and more elaborate, with a cupola that is a dominant feature of the inner harbour. By far the grandest, and one of the most opulently detailed buildings in Glasgow is the Clyde Navigation Trust's offices in Robertson Street, designed by John Burnet and his son, with a corner cupola and sculpture of the most exuberant kind.[132] Harbour offices for the minor ports are generally modest, often single-storeyed buildings basically for the collection of dues, as at Kirkwall, Stromness, Arbroath and Ayr.[133]

SALMON FISHERIES

River- and beach-based coastal salmon fisheries have produced icehouses and boiling houses of the types dealt with above, but have also generated their own distinctive buildings and groups of buildings. Of these, the most numerous are the bothies where fishermen waited for the night state of the tide, or weather conditions, to allow a catch to be attempted. Some of these are tiny, ramshackle structures, like that at Ardgay on the River Shin, Ross and Cromarty. One with crowstepped gables at Buchanhaven, Aberdeenshire, is apparently sixteenth century in date.[134] The best group is on the North Sea coast in Angus and Kincardineshire. At Montrose there is a two-storeyed, bow-fronted building, and further up the coast are several one- and two-storeyed buildings, the latter with forestairs.[135] Near Cardoness House, Gatehouse of Fleet, and at Balcary, Kirkcudbrightshire,[136] there are small single-storeyed bothies, the former with diamond-paned windows in the cottage-ornée style. On the Tay there are small two-storeyed bothies at Stanley, Perthshire, now used by fly fishers, but presumably originally for commercial fishing, and lower down between Perth and Errol, there are several fishing stations with comparatively modern single-storeyed buildings, some with railways with hand-propelled trolleys for moving fish boxes.[137] At Cairnie Point is an older two-storeyed fishing station. On the Moray Firth there is a large fishing station at Macduff.

Other structures are characteristic of different modes of fishing. The most generally distributed are drying poles for nets. These are sometimes fixed, and sometimes rigged for the season only. Their size and layout vary from station to station, depending on the type and dimensions of nets used. On some coastal stations 'fixed engines' are used to catch fish. These are arrangements of poles with nets attached. Most of these 'stake net' systems exist to guide the fish into traps. The surviving examples of this technique are mostly on the Solway Firth, some of the net systems, as at Balcary, Kirkcudbrightshire, being of considerable size.[138] To preserve the natural fibre nets used formerly they were commonly 'barked', that is, soaked in hot solutions of bark extract (usually cutch, imported from India). The cast-iron tubs generally used for barking were often merely perched on stones with a fire-grate beneath, but sometimes a permanent setting was used. Recently, survivors were at Port Errol, Aberdeenshire (fig. 26.7), and Maidens, Ayrshire.

In a majority of cases salmon fisheries do not have directly associated housing. There are, however, some instances both of housing integrated with other fishery-related buildings, and of detached houses as part of a fishing complex. Good examples of the former are the stations at Invershin, Badcall, Sutherland, and Tugnet (see fig. 26.5),[139] where the boiling houses adjoin dwelling houses (at Invershin there is also a pair of ice houses in the steading). At Findhorn the boiling house is beside a single house, but there are two other ranges of houses. The fishing station at Berriedale, Caithness, has a terrace of houses, a bothy, and an ice

Figure 26.7 Port Errol, Aberdeenshire; salmon-fishery net-boiling tub. Copyright: John R Hume.

house in a row. At Broom of Moy, near Forres, Moray, the village houses were formerly a fishing station.

The active fishing stations in Scotland are declining in number as fishing rights are bought up by the Atlantic Salmon Trust, and as fish farms erode the financial viability of fishing for wild salmon. Such fish farms have generated little in the way of permanent structures. The cages in which the fish are penned are essentially ephemeral, and for the most part fish farms make use of existing shore facilities. Where accommodation is required, it is anonymous in design and erected as cheaply as possible.

In conclusion, it should be said that the diversity of the subject matter of this chapter is such that few firm generalisations may be drawn. The innate durability of harbour works means that they tend to survive, except where pressure on land dictates otherwise. Even then, in many instances the structures are buried rather than destroyed. Thus, a broad spectrum of such works survives. In the case of other maritime buildings and structures, the forces are the same, but have different effects. Metal structures, having scrap value, tend to go quickly after abandonment, while stone or brick buildings may well survive long after their original function has gone, unless pressures dictate otherwise. This affects what evidence one can adduce to support general conclusions. Because the activities concerned were product- or geographically specific, the scope for standardisation was limited, and one of the delights of studying maritime structures is indeed their variety. The main linking features are in fact the styles of individual engineers, builders or architects, as in lighthouse design, steel-shed and harbour construction, which can be quite strongly expressed.

NOTES

1. For example, Aberdeen, Ayr, Dumbarton, Dundee, Greenock, Inverness, Leith and Kirkcudbright.
2. For example, Dumfries, Perth and Stirling.
3. Graham, 1968–9, 233–5, 226–8, 267–70.
4. McUre, 1736, 285–6.
5. Hume, 1976(a), 149.
6. For example, Portsoy New and Sandside (both c. 1830) and Seatown, Buckie (1855–7).
7. Graham, 1968–9, 226–8, 257–9.
8. Telford, 1838, plate 16 illustrates plans of works at Highland harbours.
9. Mowat, 1980, 10–11.
10. Boucher, 1963, 122–37 is a list of all Rennie's works, including harbour works. It includes schemes which were not built, as well as realised projects.
11. Leslie, Paxton, 1999, 40–1.
12. Stevenson, 1874, 45–8, plates xi, xii.
13. Donnachie, 1971, 180–4.
14. For example, Keiss, Ham and Castlehill in Caithness.
15. Stevenson, 1874, 7.
16. Bennett, 1983, plates between 31 and 32, 57–8.
17. For example, Burntisland, Girvan, Grangemouth, Methil and Troon.
18. For example, Ardrossan and Ayr.
19. For example, Bo'ness and Grangemouth.
20. Most notably Rothesay Dock, Clydebank.
21. Especially Aberdeen, Eyemouth, Fraserburgh, and Lossiemouth.
22. Hume, 1976(a), 46, 139, 145, 252; Hume, 1977, 142.
23. Lenman, 1975, 64–6.
24. ibid, 64–6.
25. Hume, 1976(a), 46, 139, 145, 252; Hume, 1977, 142.
26. Riddell, 1979, 205–15 (Queen's Dock), 219–29 (Prince's Dock), 232–8 (Rothesay Dock).
27. Lenman, 1975, 149–50.
28. Russell, 1982.
29. Lenman, 1975, 116–20, 122–31.
30. Hume, 1976(a), 55, 128, 260; Hume, 1977, 121.
31. Monteith, 1981.
32. Hume, 1974, 262; Hume, 1976(a), 187–8; Hume, 1977, 132.
33. Hume, 1974, 149.
34. Hume, 1977, 214.
35. The Garvel dock was built to serve a dock that was never completed; Hume, 1976(a), 212.

36. Hume, 1976(a), 224; Hume, 1977, 176, 184, 201, 207.
37. Riddell, 1979, 252–7.
38. ibid, 334–5.
39. Lenman, 1975, 220–2.
40. For example, new harbours at Kinlochbervie and on the island of Berneray, and extensions to harbours at Ullapool, Scalloway and Lerwick.
41. Hume, 1977, 248; Allardyce, Wilson, 1992, 6, 8 et seq.
42. Immediately to the west of the present quay.
43. Hume, 1977, 289, 294.
44. Hume, 1976(a), 65, 133.
45. Hume, 1977, 153.
46. Hume, 1976(a), 48.
47. McCrorie, Monteith, 1982, 19, 68.
48. For example, at Otter Ferry, Loch Fyne; Croggan and Salen, Mull; Kames, Kyles of Bute; Balmaha and Rowardennan, Loch Lomond; and Aberdour, Fife.
49. McCrorie, Monteith, 1982, 35, 36, 49–50.
50. For instance, at Kames, by Tighnabruaich; Craigmore, Bute; and Strone.
51. Hume, 1977, 60, 167.
52. McCrorie, 1997, 37–50.
53. McCrorie, Monteith, 1982, 13, 15, 18, 20, 28.
54. ibid, 23.
55. Hume, 1977, 155, 204, 290.
56. Brodie, 1976(a).
57. Hume, 1976(a), 128: Hume, 1977, 294.
58. Telford, 1838.
59. Hume, 1976(a), 137.
60. Hume, 1977, 290, 292, 294, 317.
61. Weir, c. 1988, 20, inside front cover.
62. Duckworth, Longmuir, 1972, 138–47.
63. McCrorie, Monteith, 1982, 15, 18, 23, 50, 71, 79.
64. Hume, 1976(a), 117, 134; Hume, 1977, 246, 297.
65. Hume, 1977, 230, 284.
66. Hume, 1974, 167–8.
67. Plan by Alex Farmer, c. 1815, in Glasgow City Archives T-CN14/266.
68. Hume, 1976(a), 189.
69. Hume, 1977, 132.
70. Hume, 1976(a), 210–11.
71. Riddell, 1979, 95, 266; Moss, Hume, 1976, 22–4.
72. Hume, 1974, 231–2.
73. French, 1947, 41, 44; Riddell, 1979, 242–3, 270–1.
74. Lenman, 1975, 129, plate facing 128.
75. Hume, 1977, 192, 195, 199.
76. ibid, 231, 241, 250.
77. ibid, 201 (32).
78. ibid, 193, 319.
79. Walker, 1976, 563–72.
80. Hume, 1977, 192, 199 (35), 292, 310.
81. ibid, 189, 193, 319, 320–1.
82. ibid, 188, 228 (37).
83. Behind West Street Underground Station, and in Falfield Street.

84. Hume, 1977, 87–8.
85. ibid, 99(34), 114, 314; Hay, Stell, 1986, 26–30.
86. Donnachie, Hume, Moss, 1977, 63.
87. Walker, Hay, 1985, 16.
88. Hume, 1977, 98; Oakley, 1953, 189.
89. Jackson, 1983, 4, 98; RCAHMS, 1951, fig. 29, plate facing LVII.
90. Hume, 1976(a), 218.
91. Loudon Brothers, *c.* 1900, 140, 141, 144.
92. Hume, 1976(a), 128; Hume, 1977, 167, (60), 213–14, 229.
93. Hume, 1977, 305–6.
94. Hume, Storer, 1997, 63–4.
95. Glasgow City Archives T-CN14/302A and B.
96. Lenman, 1975, plate facing 64; Hume, 1974, 263: Hay, Stell, 1986, 191; Riddell, 1976, 262.
97. Malden, 1991, 25.
98. Hume, 1974, 262, plate 86.
99. Hume, Moss, 1975, 27.
100. Hume, 1977, 207.
101. Hume, 1976(a), 187–8.
102. Moss, Hume, 1975, 59.
103. Hume, 1974, 237–8, 262.
104. Lenman, 1975, plate facing 144; Riddell, 1979, 235.
105. For example Riddell, 1979, 267.
106. As was necessary with fixed-radius steam cranes; RCAHMS, 1998, 7, 23, 34.
107. *Engineering*, 1909.
108. Lenman, 1975, plate facing 128.
109. Riddell, 1975, 269, 358–9.
110. ibid, 271; Lenman, 1975, plate facing 81, plate facing 128.
111. The Windmillcroft Crane is the last survivor of the many rail-mounted cranes in the upper Glasgow harbour.
112. As at Ayr and Irvine.
113. Hume, 1997; Hague, 1977, 81–5.
114. Munro, 1979, 29–38, 44–50; Hague, 1977, 75–8, 80.
115. Dunbar, 1966, 220 (Cromarty).
116. Hague, 1977, 86–7.
117. For instance, in Hume, Moss, 1975; Walker, 1984.
118. Patrizio, Little, 1994, plate 12. This is one of Stanley Spencer's paintings of shipbuilding in Port Glasgow during World War II.
119. Butt, 1966, 120.
120. Hume, Moss, 1975, plates 38, 39, 74, 75.
121. This was the predecessor of the surviving office building at the site of the shipyard, Hume, 1974, 271.
122. Hume, 1976(b), 158–80, passim.
123. Hume, 1974, 261, plate 55.
124. Moss, Hume, 1975, 98–9.
125. Walker, 1984, 144–5.
126. Hume, Moss, 1975, plate 79.
127. Wilson, 1968, 139–43, 146–54.
128. Hume, 1974, 94.
129. ibid, 159.

130. Hume, 1976(a), 185, 210.
131. Hume, 1977, 299: Hume, 1976(a), 107.
132. Gomme et al, 1987, 193; Teggin, Samuel, Stewart, Leslie, 1988, 31–3, 40.
133. Hume, 1977, 246; Hume, 1976(a), 46.
134. Howard, 1995, 95–6.
135. Hume, 1977, 226 (Nether Warburton), 277 (Nether Woodston).
136. Hume, 2000, 133.
137. Brotchie, 1965, 97: Hume, 1977, 272, 280.
138. Hume, 2000, 133.
139. Hume, 1977, 228 (13), 312, 315.

BIBLIOGRAPHY

Allardyce, K, Hood, E M. *Scotland's Edge*, Glasgow and London, 1986.
Allardyce, K, Wilson, B. *Sea Haven*, Kirkwall, 1992.
Anson, P F. *Fishing Boats and Fisher Folk on the East Coast of Scotland*, London, 2nd edn, 1974.
Bennett, G P. *The Great Road Between Forth and Tay*, Markinch, 1983.
Boucher, C T G. *John Rennie, 1761–1821*, Manchester, 1963.
Brodie, I. *Queensferry Passage*, Linlithgow, 1976(a).
Brodie, I. *Steamers of the Forth*, Newton Abbot, 1976(b).
Brotchie, A W. *Tramways of the Tay Valley*, Dundee, 1965.
Buchan, A R. *The Port of Peterhead*, Peterhead, nd (c. 1973).
Buchan, A R. The engineers of a minor port, Peterhead, Scotland, 1772–1872, *Industrial Archaeology Review*, 3 (1979), 243–57.
Campbell, J. *The Aerofilms Book of Scotland from the Air*, London, 1984.
Donnachie, I. *The Industrial Archaeology of Galloway*, Newton Abbot, 1971.
Donnachie, I, Hume, J, Moss, M. *Historic Industrial Scenes: Scotland*, Buxton, 1977.
Duckworth, C L D, Longmuir, G E. *Clyde River and Other Steamers*, Glasgow, 3rd edn, 1972.
Engineering. *Bridges, Structural Steel Work and Mechanical Engineering Productions by Sir William Arrol and Company, Ltd*, London, 1909.
French, W. *The Port of Glasgow*, Clyde Navigation Trust, Glasgow, 1947.
Gomme, A, and Walker, D. *The Architecture of Glasgow*, London, 1968, 2nd edn, 1987.
Graham, A. Archaeological notes on some harbours in eastern Scotland, *PSAS*, 101 (1968–9), 200–83.
Graham, A. Old harbours and landing places on the east coast of Scotland, *PSAS*, 108 (1976–7), 322–66.
Graham, A, Gordon, J. Old harbours in northern and western Scotland, *PSAS*, 117 (1987), 265–352.
Graham, A, Truckell, A E. Old harbours of the Solway Firth, *TDGNHAS*, 3rd series 52 (1976–7), 109–42.
Graham, A. Some old harbours in Wigtownshire, *TDGNHAS*, 3rd series 54 (1979), 39–74.
Hague, D B. Scottish lights. In Thoms, 1977, 75–90.
Hay, G D, Stell, G P, for RCAHMS. *Monuments of Industry*, Edinburgh, 1986.
Hendy, J, Cousill, M. *Ferries of Scotland*, Kilgetty, 2nd edn, 1993.
Howard, D. *The Architectural History of Scotland: Scottish Architecture from the Reformation to the Restoration, 1560–1660*, Edinburgh, 1995.
Hume, J R. *The Industrial Archaeology of Glasgow*, Glasgow, 1974.

Hume, J R. *The Industrial Archaeology of Scotland*, Vol 1, *The Lowlands and Borders*, London, 1976(a).
Hume, J R. Shipbuilding machine tools. In Butt, J, Ward, J T, eds, *Scottish Themes*, Edinburgh, 1976(b), 158–80.
Hume, J R. *The Industrial Archaeology of Scotland*, Vol 2, *The Highlands and Islands*, London, 1977.
Hume, J R. *Harbour Lights*, Edinburgh, 1997.
Hume, J R, Moss, M. *Clyde Shipbuilding from Old Photographs*, London, 1975.
Hume, J R, Storer, J D. *Industry and Transport in Scottish Museums*, Edinburgh, 1997.
Jackson, G. *The History and Archaeology of Ports*, Tadworth, 1983.
Lenman, B. *From Esk to Tweed: Harbours, ships and men of the east coast of Scotland*, Glasgow and London, 1975.
Leslie, J, Paxton, R A. *Bright Lights: The Stevenson engineers, 1752–1971*, Edinburgh, 1999.
Louden Brothers. *Illustrated Catalogue*, Glasgow, c. 1900.
McCrorie, I. *Dunoon Pier: A celebration*, Glendaruel, 1997.
McCrorie, I, Monteith, J. *Clyde Piers: A pictorial record*, Greenock, 1982.
McUre, J. *A View of the City of Glasgow*, Glasgow, 1736.
Malden, J. *Let Paisley Flourish*, Paisley, 1991.
Monteith, J, McPherson, R. *Port Glasgow and Kilmalcolm from Old Photographs*, Greenock, 1981.
Moss, M S, Hume, J R. *Beardmore: History of a Scottish industrial giant*, London, 1979.
Moss, M, Hume J R. *Glasgow as it Was, Vol 3: Glasgow at work*, Nelson, 1976.
Moss, M, Hume, J R. *The Workshop of the British Empire*, London, 1977.
Mowat, J. *James Bremner, Wreck-Raiser*, Thurso, 1980.
Munro, R W. *Scottish Lighthouses*, Stornoway, 1979.
Oakley, C A, ed. *Scottish Industry*, Glasgow, 1953.
Patrizio, A, Little, F. *Canvassing the Clyde: Stanley Spencer and the shipyards*, Glasgow, 1994.
RCAHMS. *The City of Edinburgh*, Edinburgh, 1951.
RCAHMS. *The Sir William Arrol Collection: A guide to the Scottish material held in the National Monuments Record of Scotland*, Edinburgh, 1998.
Riddell, J F. *Clyde Navigation: A history of the development and deepening of the river Clyde*, Edinburgh, 1979.
Russell, I. Randolph Wemyss and the development of Methil as a coal port, *Scottish Industrial History*, 5/2 (1982), 42–51.
Stevenson, T. *The Design and Construction of Harbours*, Edinburgh, 2nd edn, 1874.
Teggin, H, Samuel, I, Stewart, A, Leslie, D. *Glasgow Revealed: A celebration of Glasgow's architectural carving and sculpture*, Glasgow, 1988.
Telford, T. *Atlas to the life of Thomas Telford, Civil Engineer*, London, 1838.
Thoms, L M, ed. *The Archaeology of Industrial Scotland* (Scottish Archaeological Forum 8), Edinburgh, 1977.
Walker, B. Keeping it cool, *The Scots Magazine* (September 1976), 563–72.
Walker, B, Hay, E R. *Focus on Fishing: Arbroath and Gourdon*, Dundee, 1985.
Walker, F M. *Song of the Clyde: A history of Clyde shipbuilding*, Cambridge, 1984.
Weir, M. *Ferries in Scotland*, Edinburgh, c. 1988.
Wilson, G. *Scottish Fishing Craft*, London, 1963.
Wilson, G. *More Scottish Fishing Craft*, London, 1968.
Wilson, G. *Scottish Fishing Boats*, Beverley, 1995.

27 Roads, Canals and Railways
JOHN R HUME

Roads, canals and railways are all primarily concerned with moving people and goods within a land mass. This transport function is now so self-evidently dominant that it may blind us to the significance of roads and railways, and to a much lesser extent canals, which were vital modes of communication. Before telegraphs and telephones were invented, road and – from the 1830s – rail were in most circumstances the fastest way to convey information, and the movement of letters and people had an importance now difficult to realise.

ROADS

After the Romans and until the eighteenth century, when increasing economic activity made road improvement essential, 'made' roads seem to have been rare outside towns. There were, however, some bridges in the medieval and early modern period, at focal points on major rivers,[1] and in some other locations.[2] These survivors may be only a small proportion of those built, but it seems likely that such structures were never numerous. Characteristics of bridges built up to the early eighteenth century include massive piers with triangular cutwaters, often supporting refuges, semi-circular or pointed arches,[3] sometimes ribbed,[4] narrow carriageways, and steeply-inclined road profiles.[5]

The first eighteenth-century scheme of road building followed some time after the abortive Jacobite rising of 1715, and was supervised by General George Wade.[6] His bridges were slightly more sophisticated than their predecessors, and the roads linking them were durable, if crude. The military road system was revised and extended after the Jacobite rising of 1745, with construction overseen by Major Caulfeild. Though some of Caulfeild's bridges differed little from Wade's bridges, a few were more sophisticated, though the semi-circular or near semi-circular arch was still preferred, and the sophistication lay primarily in the architectural treatment. From the 1760s road and bridge building developed rapidly, with stimulus from the demands of long-distance cartage and passenger stagecoach services.[7] By the 1790s the lowland counties were well served with made roads and their associated bridges, constructed partly by turnpike trusts, and partly by statute labour, either direct or commuted to money payments.[8] The larger bridges increasingly reflected in their design the influence of French engineering.[9] Some roads in Argyll and Perthshire were part-funded by grants from central government.[10]

After 1800 the road structures designed by Thomas Telford and John Loudon McAdam were generally adopted.[11] Telford planned an enormous scheme of road and bridge building in Scotland, handsomely supported by government. The first of these roads were in the Highlands, complementing, and in some instances supplanting, the Wade and Caulfeild system.[12] The central trunk linked Perthshire with Wick and Thurso, Caithness, and there were numerous subsidiary roads, including some on the islands and some strategic bridges filling gaps in the existing system. Telford's culminating achievement was a detached road system from Glasgow to Carlisle and Edinburgh, with a Lanarkshire cross road, all engineered to a high standard,[13] and with the intention of expediting troop movements to quell any uprising stemming from the post-Napoleonic War depression. For the most part, Telford's Highland bridges were plain, but solidly, even elegantly designed. His major strategic bridges, and many of those on his lowland system, were of refined design and construction.[14]

Although Wade, Caulfeild and Telford are the best known of the Scottish road and bridge builders, other engineers made notable contributions to the formation of an effective road network. These included John Smeaton (fig. 27.1),[15] Robert and William Mylne,[16] Alexander Stevens,[17] John Rennie[18] and Robert Stevenson,[19] whose bridges are among the finest in the country. The railways slowed down, and ultimately halted, the development of long-distance road transport, but they stimulated short-haul services, and by focusing traffic in towns and cities provided a powerful stimulus to road and bridge building there. Many of these bridges after 1870 were designed to carry horse (later electric) tramways, as well as increasing volumes of conventional road vehicles.[20] Most of these new bridges, and others in the countryside where railways did not penetrate, were of conventional masonry construction (aided by stone-sawing machinery), but some were of cast-iron arch, suspension or other wrought-iron and steel types.[21]

The use of turnpike trusts to construct and improve roads has already been alluded to. Even where roads themselves were not turnpiked, bridges were frequently financed on the toll principle. Tollhouses were numerous, and a surprising number still survive, dating from the mid eighteenth century onwards. They vary in size and shape. Some are plain rectangular cottages, in Ayrshire often with eaves, unlike local vernacular buildings.[22] A few were circular,[23] and many had polygonal or rounded bays or ends.[24] The great majority were

Figure 27.1 Bridge over the River Tay, Perth; by John Smeaton (1766–72 with later cantilevered footpath). Copyright: John R Hume.

single-storeyed, but there are a few two-storeyed houses, mostly Telford built.[25] A handful have classical or Gothic features.[26] Other fixed facilities included coaching inns, with associated stabling,[27] mileposts and direction indicators.[28] In towns and cities, coaching inns usually functioned as lodgings on a more general basis, but in some places it was necessary to create single-purpose inns.[29] Many were built by landowners, and were named accordingly. A few inns retain coach houses.[30]

The railway-supported urban growth of the later Victorian period resulted in the extensive use of horse-drawn vehicles in Scottish cities. Wealthy individuals could have their own stabling, coach house and coachman's flat with a mews.[31] Tradesmen, too, could have their own equipages. A characteristic feature of the period was, however, the growth of large stabling blocks associated either with cartage or with carriages for hire. As horses can walk up ramps, many of these were multi-storeyed, with vehicles on the ground floor, horses on one or more upper floors, and feed on the topmost floors.[32] Similar facilities were provided for horse tramways, with a tendency towards linear arrangement in order to simplify track layout in large establishments.[33]

The advent of motor vehicles and electric trams altered the style of fixed provision. A few garages were built early in the twentieth century with ramped access to upper floors,[34] but, generally, where central garaging for buses, lorries or cars was provided, the buildings were single-storeyed.[35] Similarly, electric-tram depots were single-storeyed. In both garage and tram depot provision there was much re-use of premises designed for horse haulage. Private garaging was analogous to private stabling, but blocks of 'lock-ups' for rent were also provided. Since the 1950s, the dominance of private cars has created a massive demand for car parks, with large areas of ground paved and arranged as formal parking. In the cities, multi-storeyed car parks, analogues of the ramped stabling of the Victorian era, have been built, generally as commercial ventures. Modern offices and commercial developments in city centres usually incorporate one or more floors of covered car park.

When bus services developed beyond the purely local, in the 1920s and '30s, central bus stations, like railway stations, were provided in larger communities.[36] Some of these had overall roofs; others had individual platform awnings. Post-World War II bus stations were of the latter type, with a marked tendency towards the segregation of passengers and vehicles. Modern bus operations tend to avoid central facilities so bus stations are less likely to be built.

After World War I, the combined availability of war-surplus commercial vehicles, of expertise in motor-vehicle maintenance, and of ex-munitions plant suitable for vehicle manufacture led to a rapid escalation in the volume of road motor traffic.[37] From the early 1920s road improvement became a priority, and new roads were constructed, as well as old ones being widened and re-aligned. The new bridges were commonly of reinforced concrete, sometimes to profiles resembling those of masonry bridges, but often to novel designs: bowstring (fig. 27.2 shows

Figure 27.2 Bonar Bridge, Sutherland; bowstring steel arch bridge, the third on this site, by Crouch & Hogg (1973). Copyright: John R Hume.

a steel version of this type), truss and cantilever.[38] The roads themselves were generally wider and straighter than their predecessors, and included three- and four-lane single carriageways,[39] and the first dual carriageway roads in the country.[40] Significantly, many of the new roads were designed to cater for peak holiday and day-trip traffic.

CANALS

Canals in Scotland differed significantly from their English counterparts. Three of the Scottish canals – the Forth and Clyde (Grangemouth, Stirlingshire, to Bowling, Dunbartonshire), the Crinan, Argyll, and the Caledonian, Inverness-shire – were primarily ship canals, with relatively large locks, deep and wide channels, and opening bridges. Even the canals that were not designed for ships were wide canals with substantial engineering works: narrow canals did not exist in Scotland. Four canals – the Forth and Clyde, the Union (Edinburgh-Falkirk, Stirlingshire), the Caledonian and the Crinan – still survive wholly or largely complete, and of the others, two (the Aberdeenshire Canal, and the Glasgow, Paisley and Ardrossan Canal) are now railways, and the third, the Monkland Canal, Lanarkshire, is in part a motorway.[41]

Canals had much in common with roads and railways in that the earthworks were similar, and fixed overbridges, where they existed, were also comparable. Locks were, however, specific to canals, and aqueducts had distinctive characteristics. The locks on the lowland canals had parallel-sided chambers and wooden gates very much in the English manner, as did the Crinan. On the Caledonian, however, the lock chambers have walls curved in section, some of the gates were originally cast-iron framed, and the lock gates were all operated by windlass rather than by the balance beams of the smaller canals. The aqueducts on the deep Forth and Clyde and Caledonian Canals are very massive, the latter appearing as road tunnels.[42] Robert Whitworth's Forth and Clyde aqueducts are noted for their curved spandrels and wing walls;[43] later Forth and Clyde aqueducts, built to accommodate tramways,[44] had the ascititious arches introduced by Telford. In these structures the main channel/carriageways were carried on deep segmental arches and the footways/towpaths on much shallower semi-circular arches.[45] The largest Scottish aqueducts, on the Union Canal, resemble the railway viaducts they foreshadowed by twenty years, with cast-iron troughs instead of railway tracks.[46]

Bridges on the Union and Paisley Canals were similar to

Figure 27.3 Forth and Clyde Canal between Dalmuir and Old Kilpatrick, Dunbartonshire; typical two-leaf bascule bridge and bridge-keeper's house. Copyright: John R Hume.

Figure 27.4 Port Dundas, Forth and Clyde Canal, Glasgow; view by Joseph Swan showing warehouses and masts of sailing ships, c. 1830.

contemporary road bridges. The Monkland Canal bridges were wooden, on stone abutments. On the Forth and Clyde two-leaf bascule bridges (fig. 27.3) became standard, and on the Caledonian two-leaf swing bridges. Details of the first Crinan bridges are not known. From the 1870s moving bridges on the ship canals were renewed, at first on the Crinan, then on the Forth and Clyde and Caledonian. Except on the Forth and Clyde, all the replacements were wrought-iron or steel swing bridges; the exceptions were single-leaf electric bascule bridges at Blairdardie, and Temple, Glasgow, and Torrance, Stirlingshire. New railway crossings of the canals demanded heavy-duty swing bridges, of which the most impressive are the two crossing the Caledonian at Clachnaharry and Banavie, Inverness-shire.

Other canal structures include: lock- and bridge-keepers' cottages, mostly small and plain, but on the Caledonian sometimes rather grand; stables,[47] of which the finest are on the Forth and Clyde;[48] offices of which the grandest is at Port Dundas, Glasgow;[49] warehouses and transit sheds (mostly privately owned (fig. 27.4)); cranes, mainly hand-operated;[50] and even lighthouses on the Crinan and Caledonian Canals.[51]

RAILWAYS

Of all modes of transport in Scotland, railways created the largest number of structures. Many of these had parallels in those built for roads and canals, especially most over-and-under bridges and masonry viaducts, anticipated by aqueducts and some high-level road bridges of the 1820s. Crossing-keepers' cottages, of which there were many on the early railways, had parallels with toll cottages. New building types did emerge, however, most notably stations, engine and carriage sheds and signal boxes. As railways developed, so too did bridge and viaduct construction, with first wood and cast iron then wrought iron and later steel becoming increasingly important. In the last phase of new railway building, mass concrete was introduced.

Early railway under-and-over bridges were usually of masonry construction, though both materials and detailed design were variable. Most bridges were straightforward in construction; segmental and elliptical arches were favoured for the majority of crossings, but semi-circular arches were used where there was sufficient headroom, as, for example, in Winchburgh cutting, West Lothian, on the Edinburgh and Glasgow Railway. Skew bridges became common where existing routes had to be crossed at an angle; examples include Underwood Road, Paisley, and at various points along the Edinburgh-Berwick line of the North British Railway. A few bridges, particularly on the Aberdeen Railway, were built with ribbed arches, and some had ornamental treatment to suit the requirements of influential landowners, most notably at Guthrie Castle, Angus (fig. 27.5), and Castle Grant, Moray.

From the 1840s, cast iron came into favour for over-and-under bridges, both in arch and girder form, and though the material became discredited for girder construction for railway traffic, cast-iron overbridges continued to be built until the 1880s by the Glasgow and South Western and the Great North of Scotland Railways. By that period, however, wrought-iron and steel plate and truss bridges were more usual, both having been introduced in the 1860s. With wrought-iron construction, lattice and plate girders were commonest, and some good early lattice-girder bridges can be seen on the Stobcross branch of the North British Railway, now part of the north-side electrified suburban railway in Glasgow. The availability of heavy wrought-iron and steel sections led to the introduction of trusses with fewer members, such as the Warren (single and double) and Pratt types, examples of which can be seen in the bridges on the Glasgow-Dumbarton section of the Lanarkshire and Dunbartonshire Railway. In the 1890s and early 1900s, both brick and concrete came into favour, the latter in mass form, such mass-concrete bridges being built on the Lanarkshire and Ayrshire Railway and on the West Highland Extension Railway. Brick had been used for arch rings in bridges from about 1840 on, for instance, the northern section of the Aberdeen Railway.

Viaduct design paralleled that of bridges to some extent. Most masonry viaducts were in fact multiple small-to-medium sized arch

Figure 27.5 Guthrie Castle estate, Angus; castellated railway bridge and gate lodge, 1839. Copyright: John R Hume.

bridges. A few transcended that formula, most notably Ballochmyle, Ayrshire, and Dunglass, East Lothian.[52] Brick viaducts were built in small numbers in a style similar to multi-span masonry structures. The alternatives to masonry in the 1840s and '50s were cast-iron or timber structures. The former were arched in form, and replaced timber

Figure 27.6 Connel Ferry Viaduct, Loch Etive, Argyll (1903). Copyright: John R Hume.

structures in some instances on the Scottish Midland Junction and Aberdeen Railways. The latter were mostly of two generic types, laminated timber arches and trestles,[53] but there was one very striking cantilever timber bridge at Jamestown, Dunbartonshire,[54] and on the Waverley route a few short low-level viaducts consisted simply of baulks of timber on timber piles.[55] All the timber viaducts, bar a late example at Moy, Inverness-shire,[56] and all the cast-iron viaducts except two,[57] were replaced by wrought-iron or steel structures from the 1880s. The wrought-iron and steel viaducts were generally similar in construction to contemporary bridges. As with single-span bridges, bowed trusses were a feature of several 1860s bridges, as can be seen in the replacements of 1840s bridges at Aberdalgie, Perthshire, and near Carstairs Junction, Lanarkshire. The second Union Bridge in Glasgow is an example of a viaduct with arched trusses, used for the sake of appearance. Two large steel cantilever bridges – the Forth Bridge, West Lothian-Fife, and Connel Bridge, Argyll (fig. 27.6) – are in a class of their own. Contemporary with the use of mass concrete for bridges was the adoption of this material for viaducts, especially where economy in construction costs was important. The first concrete viaduct in Scotland was at Killin, Perthshire, and others were built on the Lanarkshire and Ayrshire and West Highland Extension Railways.

Stations took many and varied forms. For the purpose of analysis, they can be divided into two groups: principal stations in cities or large towns; and what may be loosely called wayside stations. In the early railways station provision was minimal – a house, presumably with a space for booking and waiting,[58] or in the case of the Dundee and Newtyle, platforms with awnings.[59] By the mid 1830s combined houses and awnings had been introduced, and in the railways opened in 1840–2 large classical terminals with cast-iron train sheds, on split-level sites, were the standard; examples are Bridge Street, Glasgow, Greenock and Haymarket, Edinburgh. At first, few of the railways of the 1840s mania could afford anything better than wooden sheds as terminals, as was the case with South Side and Buchanan Street, Glasgow, and Lothian Road, Edinburgh. It was not until the 1870s and '80s that remodelling of the system created large city-centre stations and major through stations such as Central, Queen Street and St Enoch Stations in Glasgow, Dundee West and Tay Bridge Stations, and Perth General. Further re-working between 1890 and 1914 produced other city centre stations, like Edinburgh Waverley and a major extension to Glasgow Central, which still serve the railway network.

So far as 'wayside' stations are concerned (fig. 27.7), a pattern was set in the later 1830s that persisted for most new buildings until the 1880s, that is, of combined houses and station accommodation, usually stone, but occasionally brick-built. These structures differed widely in style and complexity; some through stations had large and complex buildings, but most had simple, usually small structures. In the 1860s some wooden stations, with separate agents' houses, were built,[60] and this style became common in the 1880s and '90s.[61] Some elaborate wooden stations were constructed, good examples of which included Banchory and Ballater, Aberdeenshire, Callander, Perthshire, Carr Bridge, Inverness-shire, and Dunrobin Castle, Sutherland. The period 1880–1905 also saw the use of steel for the construction of frameworks and glazed awnings, the Caledonian Railway and its associated lines using steel extensively in, for instance, Wemyss Bay, Renfrewshire, Stirling, Gleneagles, Perthshire, and stations on the Lanarkshire and Ayrshire Railway. On the West Highland Extension Railway reinforced concrete was, uniquely, used in station building construction, as can be seen at, for instance, Glenfinnan and Arisaig, Inverness-shire.

Goods and mineral stations were usually associated with passenger stations, but in larger towns and cities were sometimes separate. Some early railways had facilities for discharging minerals into bins immediately below the track,[62] but direct discharge into horse (later steam and motor) lorries became standard. Buildings associated with the

Figure 27.7 Dunkeld and Birnam Railway Station, Perthshire (1856 and 1863). Copyright: John R Hume.

handling of goods took a variety of forms. Commonest was the goods shed, which varied in size and was simply covered accommodation for exchanging goods between road and rail. The very large goods shed at Montrose, Angus, still survives, but many have been demolished; conversions include those at Busby, Lanarkshire-Renfrewshire, and Carluke, Lanarkshire. In population centres, goods sheds could also serve to house rail-to-rail interchange, as bulk consignments were re-distributed for regional delivery. The large goods stations in Glasgow, Edinburgh, Dundee, Aberdeen and Perth all performed this role, and in Falkirk, the North British Railway built a large shed for rail-to-rail trans-shipment of castings from the many local foundries. In some instances specific provision was also made for bulk traffic – grain stores, spirit and beer stores and potato sheds being the commonest.[63] Cranes for handling the goods were often provided in sheds, and generally provided in goods yards.[64]

Signal boxes are a feature of railway operation almost without parallel – the berthing signals on Clyde piers, dock control cabins and airport control towers being the only comparisons. The earliest signal boxes were shelters for signalmen who operated signals and points on the spot, and one such structure which survived into the 1970s was at

Hilton Junction, just south of Perth. Remote control of signals and points by wire and rod enabled route control to be centralised.[65]

Workshops required for maintaining and constructing railway locomotives, carriages and wagons were of the same overall character as general engineering works, those for locomotive building and repair having heavy shops with overhead cranes for handling heavy components, such as boilers, or even whole locomotives and tenders.[66] Sheds for cleaning and storing carriages were also provided, notably on the early railways, though most carriage stock was kept in the open or, overnight, in covered railway stations. Early carriage sheds have survived latterly at Alford, Aberdeenshire, and Ladybank, Fife; larger, more modern carriage sheds were at Gushetfaulds, Glasgow, and in Dundee.

The most elaborate provision of buildings for railway vehicles was for locomotives. Sheds, sometimes of great size, and incorporating workshops for repairs, were distributed throughout the system, and could be distinguished by their roof-ridge ventilators to allow smoke to disperse. In a few instances, as at Burntisland, Fife, Kittybrewster, Aberdeen, and Inverness, such locomotive sheds were circular, with rails radiating from a turntable. Many locomotive sheds also had coaling stages, with high-level sidings to take coal to a level from which it could easily be used to fill tenders or bunkers.[67] In the late inter-war and post-war years some of the larger sheds were equipped with tall, reinforced-concrete overhead bunkers from which locomotives could be coaled quickly and mechanically.[68] Some, too, as at Polmadie and Corkerhill, Glasgow, had mechanical ash-disposal equipment. Sand, used to counteract slippery rails, was an important consumable required in small quantities, and sand driers were features of the larger locomotive depots such as St Rollox shed, Glasgow.

Water supply to locomotives was a major requirement throughout the system. Sometimes local mains supplies could be used, but generally, even where they were used, holding tanks were installed so that large volumes of water could be discharged very quickly into locomotive tanks. The stone or brick bases of these tanks still survive in many locations. Large water tanks were a feature of the major locomotive depots and stations, and there is a fine survivor on the south side of Waverley Station, Edinburgh. However, smaller ones were found at many wayside stations. The tanks were usually made of cast-iron sections bolted together. Some tanks had water distributing 'cranes' attached to them but most water was distributed through free-standing cast-iron water columns, with leather pipes, 'bags', to channel the water into the locomotive tanks. Many columns had integral cast-iron stoves to prevent freezing, as was the case, for instance, at the south end of Hawick Station, Roxburghshire. At two locations – at Strawfrank, south of Carstairs, and at Carronbridge, Dumfriesshire – water-troughs were provided, from which water could be picked up by locomotives while travelling at speed. The troughs were long shallow metal tanks between the rails, and both sets of Scottish troughs were installed by the London, Midland and Scottish Railway. The tenders of many locomotives were fitted with scoops which could be lowered to collect water from the troughs.

When diesel and electric locomotives and railcars replaced steam locomotives the requirement for motive power replenishment and maintenance facilities declined sharply, and the nature of the residual need changed, so that the dirtiness of steam locomotive preparation and repair was replaced by something much closer to motorcar maintenance. Since the initial diesel and electric schemes of the late 1950s and early 1960s there have been several generations of revised or new maintenance depots, as, for instance, at Polmadie, Glasgow, and at Craigentinny and Haymarket, Edinburgh.

NOTES

1. For instance, Aberdeen, Ayr, Glasgow, Haddington, Jedburgh, Musselburgh, Partick and Stirling. Ruddock, 1984.
2. At Pencaitland and near Haddington (East Lothian).
3. The most notable pointed-arched bridges are Brig o' Balgownie, Aberdeeen, Pencaitland, and Abbey Bridge, near Haddington (East Lothian).
4. For instance, Bridge of Dye, Kincardineshire, Upper North Water Bridge, Angus and Kincardineshire, and Bridge of Dee, Aberdeen.
5. For example, Glasgow, Ayr and Stirling.
6. Taylor, 1976, 17–23.
7. Ruddock, 1979, 80–104.
8. ibid, 233–243; Dunlop, 1959, 155–8.
9. Singer, Holmyard, Hall, Williams, 1958, 451–3.
10. For instance, Bridge of Awe (Argyll) and Forteviot Bridge (Perthshire).
11. Singer, Holmyard, Hall, Williams, 1958, 527–8.
12. Haldane, 1968, passim.
13. Gibb, 1935, 318–19.
14. Telford, 1838, for example, bridges on the Glasgow-Carlisle and Lanarkshire roads on plates 53–7, of Highland bridges on plate 51, the Dean Bridge, Edinburgh, plate 62, and Lothian Bridge, Pathhead, Midlothian, plate 64.
15. Bridges at Banff, Coldstream, Berwickshire, and Kirkconnel, Dumfriesshire.
16. Bridges in Edinburgh (Old North Bridge), Glasgow (Broomielaw Bridge) and Inveraray, Argyll (Aray Bridge).
17. For example, Bridge of Dun, Angus, Drygrange Bridge, near Melrose, Roxburghshire, and Hyndfordbank, near Lanark.
18. For example, New Bridge, Musselburgh, Midlothian, and Kelso Bridge, Roxburghshire.
19. Leslie, Paxton, 1999, 42–5; his bridges included the New Bridge, Stirling, Annan Bridge, Dumfriesshire, and a bridge over the North Water at Marykirk, Kincardineshire.
20. For instance, the present Partick, Kelvin, Hillhead, Gibson Street and Pollokshaws bridges in Glasgow and some of the Aberdeen urban viaducts.

21. For instance, Hume, 1976, 32; Hume, 1977, 69–73.
22. Hume, 1988, front cover.
23. Hume, 1974, plate 73; Hume 1997(a), 147.
24. Hume, 1976, 151; Hume, 1977, 76, 96; Stephen, 1967(a).
25. Hume, 1977, 102, 262, 287.
26. For instance, the tollhouse at Barnhill, Perth (perhaps designed by Sir Robert Smirke), Hume, 1977, 278, and the bridge toll house at Boat o' Brig, Banffshire-Moray, both classical, Hume, 1977, 75; and a Gothic tollhouse at Kelton, Donnachie, 1971, 108.
27. For example, hotels in the centre of Moffat, Dumfriesshire, Banff and Kelso. For inns and hotels generally, see below, Chapter 29.
28. Stephen, 1967–8(b).
29. For example at Auldgirth, Dumfriesshire, and at Taynuilt, Argyll.
30. One of the best of these is the Old Brig Inn, Beattock, Dumfriesshire, built in connection with Telford's Glasgow-Carlisle road. Whitfeld, 1984, passim.
31. Hume, 1974, 111–4: Hume, 1997(b), 184–94.
32. Hume, 1974, 109–10.
33. Hume, 1974, 158 (Botanic Gardens Garage).
34. There were notable examples in Edinburgh (at Haymarket Terrace and Dundas Street) and Dundee.
35. Early examples were at Killermont, Glasgow, Kilmarnock, Ayrshire, Ayr (open) and St Andrews, Fife. More recent stations are St Andrew's Square, Edinburgh, and Anderston and Buchanan, Glasgow.
36. For instance, Moss, 1979, 160–1.
37. Good varied examples at Advie, Moray, Tomatin, Inverness-shire, Dinnet, Aboyne, Hatton of Fintray, Aberdeenshire, Dunglass, East Lothian, Dunfermline, Fife, near Blairgowrie, Perthshire, and at Grantown-on-Spey, Moray.
38. The old A8 was a good example of a three-lane highway; the A77 of a four-lane non-dual carriageway.
39. The Glasgow-Dumbarton road was an example of a pre-1939 dual carriageway road, as was the southern approach road to the Kincardine Bridge, Stirlingshire (1936).
40. Notably, the A77 (for Clyde coast traffic) and the A82 (for traffic to Loch Lomond and to the Highlands).
41. Lindsay, 1968.
42. Cameron, 1972, 54–64, 67.
43. Hume, 1974, plates 77, 78.
44. ibid, 116–7.
45. Dean Bridge, Edinburgh, and Lothian Bridge, Pathhead, Midlothian, see Telford, 1838, plates 62, 64.
46. Purves, 2000.
47. Cameron, 1972, 68.
48. Hume, 1974, 277: Hume, 1976, 106–7, 156.
49. Hume, 1974, 167; Butt, 1967, 206.
50. One survives in situ at Muirtown Wharf, at the top of the Muirtown Locks on the Caledonian Canal at Inverness.
51. Hume, 1997(b), 10, 12, 14, 17, 18.
52. Hume, 1976, 65, 76–7. Both with large semi-circular main spans.
53. Booth, 1971–72, 2, 5, 13–14: Thomas, 1967, 75 (Perth).
54. Blyth & Blyth, Edinburgh, photograph album.

55. NAS, RHP 17322.
56. Hume, 1977, 211.
57. Carron Bridge over the Spey, Banffshire, and Uddingston Viaduct, Lanarkshire, where the original bridge was retained to support a footpath.
58. For instance, at Lochwinnoch, Renfrewshire, and Kilwinning, Ayrshire, on the Glasgow, Paisley, Kilmarnock and Ayr Railway, and on the Dundee and Arbroath.
59. Davey, 1975, 41.
60. For example, on the Inverness and Perth Junction Railway and at Ballater, Aberdeenshire.
61. For instance, on the Moray Coast line of the Great North of Scotland Railway, and on the Cathcart Circle Railway.
62. For example, at the Glebe Street terminus of the Garnkirk and Glasgow Railway, and at stations on the Edinburgh, Perth and Dundee Railway.
63. There were large grain stores at Buchanan Street and Queen Street, Glasgow, and smaller ones at, for example, Duns, Berwickshire, Blairgowrie, Perthshire, Eassie, Angus, and Cupar, Fife.
64. Examples surviving in the 1970s included Dingwall, Tain, Ross and Cromarty, and Montrose. The crane in Arbroath, Angus, was still there in 2000.
65. For instance, Thomas, 1967, 91–3, 95.
66. As at Cowlairs and St Rollox Works, Glasgow, and at Perth, Inverurie, Aberdeenshire, and Inverness.
67. For instance, at Hurlford, Kilmarnock, Ayrshire, St Rollox and Dawsholm, both Glasgow.
68. As at Haymarket, Edinburgh, Tay Bridge, Dundee, Kipps, Coatbridge, Lanarkshire, and Polmadie and Eastfield, both Glasgow.

BIBLIOGRAPHY

Baxter, B. *Stone Blocks and Iron Rails (Tramroads)*, Newton Abbot, 1966.
Bennett, G P. *The Great Road between Forth and Tay*, Markinch, 1983.
Biddle, G, Lock, O S. *The Railway Heritage of Britain*, London, 1983.
Booth, L G. Laminated timber arch railway bridges in England and Scotland, *Transactions of the Newcomen Society*, 44 (1971–72), 1–22.
Breeze, D J, ed. *Studies in Scottish Antiquity*, Edinburgh, 1984.
Bruce, W S. *The Railways of Fife*, Perth, 1980.
Butt, J. *The Industrial Archaeology of Scotland*, Newton Abbot, 1967.
Butt, J, Donnachie, I L, Hume, J R. *Industrial History in Pictures: Scotland*, Newton Abbot, 1968.
Cameron, A D. *The Caledonian Canal*, Lavenham, 1972.
Cameron, A D. *Getting to Know the Crinan Canal*, Edinburgh, 1978.
Davey, N. *Dundee by Gaslight*, Dundee, 1975.
Day, T. Samuel Brown in north-east Scotland, *Industrial Archaeology Review*, 7 (1985), 154–70.
Donnachie, I. *The Industrial Archaeology of Galloway*, Newton Abbot, 1971.
Dow, G. *The Story of the West Highland*, London, 2nd edn, 1947.
Dunlop, A I et al, eds. *Ayrshire at the Time of Burns*, Ayr, 1959.
Fenton, A, Stell, G, eds. *Loads and Roads in Scotland and Beyond: Road transport over 6,000 years*, Edinburgh, 1984,
Fleming, G, ed. *The Millennium Link*, London, 2000.

Fothergill, R. *Bridges of the Tay*, Perth, npd.
Gibb, Sir Alexander. *The Story of Telford*, London, 1935.
Gordon, G, Dicks, T R B, eds. *Essays in Scottish Urban History*, Aberdeen, 1983.
Haldane, A R B. *New Ways through the Glens*, Newton Abbot, 1973.
Haldane, A R B. *The Drove Roads of Scotland*, Edinburgh, 1968.
Hay, G D, Stell, G P for RCAHMS. *Monuments of Industry: An illustrated historical record*, Edinburgh, 1986.
Hume, J R. *The Industrial Archaeology of Glasgow*, Glasgow, 1974.
Hume, J R. *The Industrial Archaeology of Scotland*, Vol 1, *The Lowlands and Borders*, London, 1976.
Hume, J R. *The Industrial Archaeology of Scotland*, Vol 2, *The Highlands and Islands*, London, 1977.
Hume, J R. Cast iron and bridge-building in Scotland, *Industrial Archaeology Review*, 2 (1978), 290, 299.
Hume, J R. Telford's highland bridges. In Penfold, 1980, 151–81.
Hume, J R. Transport and towns in Victorian Scotland. In Gordon, Dicks, 1983, 197–232.
Hume, J R. *Vernacular Building in Ayrshire: An introduction*, Ayr, 1988.
Hume, J R. Building for Transport in Urban Scotland. In Mays, 1997(a), 146–60.
Hume, J R. *Harbour Lights*, Edinburgh, 1997(b).
Hutton, G. *Monkland: The canal that made money*, Ochiltree, 1993.
Johnston, C, Hume, J R. *Glasgow Stations*, Newton Abbot, 1979.
Kernahan, J. *The Cathcart Circle*, Falkirk, 1980.
Leslie, J, Paxton, R A. *Bright Lights: The Stevenson engineers, 1752–1971*, Edinburgh, 1999.
Lindsay, J. *The Canals of Scotland*, Newton Abbot, 1968.
Martin, D. *The Monkland and Kirkintilloch Railway*, Kirkintilloch, 1976.
Mays, D C, ed. *The Architecture of Scottish Cities*, East Linton, 1997.
Mays, D C, Moss, M S, Oglethorpe, M K, eds. *Visions of Scotland's Past: Looking to the future*, East Linton, 2000.
Moss, M S, Hume J R. *Beardmore: History of a Scottish industrial giant*, London, 1979.
Nelson, G. *Highland Bridges*, Aberdeen, 1990.
North of Scotland Hydro-Electric Board. *Loch Sloy Hydro-Electric Scheme*, Glasgow, 1950.
Penfold, A, ed. *Thomas Telford: Engineer*, London, 1980.
Public Works Roads and Transport Congress. *British Bridges*, London, 1933.
Purves, S. Aqueduct construction in the Canal Age: A case study. In Fleming, 2000, 23–31.
Robertson, C J A. *The Origins of the Scottish Railway System, 1722–1844*, Edinburgh, 1993.
Robertson, S J T. *The Glasgow Horse Tramways*, Glasgow, 2000.
Ruddock, T. *Arch Bridges and their Builders, 1735–1835*, Cambridge, 1979.
Ruddock, T. Bridges and roads in Scotland, 1400–1750. In Fenton, Stell, 1984, 67–91.
Ruddock, T. Crossings of the Clyde and Kelvin. In Williamson, Riches, Higgs, 1990, 617–30.
Ruddock, T. Telford, Nasmyth and Picturesque Bridges. In Mays, Moss, Oglethorpe, 2000, 134–44.
Sanderson, M H B. *The Scottish Railway Story*, Edinburgh, 1992.
Singer, C, Holmyard, E J, Hall, A R, Williams, T I, eds. *A History of Technology*, 5 Vols, Oxford, 1956, Vol 4, *The Industrial Revolution, c.1750–c.1850*, 1958.

Stephen, W M. Toll-houses of the Greater Fife area, *Industrial Archaeology*, 4 (1967), 248–54.
Stephen, W M. Milestones and wayside markers in Fife, *PSAS*, Vol 100 (1967–8), 179–84.
Taylor, W. *The Military Roads in Scotland*, Newton Abbot, 1976.
Telford, T. *Atlas to the Life of Thomas Telford, Civil Engineer*, London, 1838.
Thomas, J. *Scottish Railway History in Pictures*, Newton Abbot, 1967.
Thomas, J. *The Railway History of Britain*, Vol 6, *Scotland: The Lowlands and the Borders*, Newton Abbot, 1971.
Thomas, J, Turnock, D. *A Regional History of the Railways of Great Britain*, Vol 15, *North of Scotland*, Newton Abbot, 1989.
Thomson, D L, Sinclair, D E. *The Glasgow Subway*, Scottish Tramway Museum Society, Glasgow, 1964.
Whitfeld, E. Victorian mews in Edinburgh. In Breeze, 1984, 360–90.
Williams, W. *Scotland's Stations: A travellers' guide*, Gartocharn, 1988.
Williamson, E, Riches, A, Higgs, M, eds. *The Buildings of Scotland: Glasgow*, Harmondsworth, 1990.
Worling, M J. *Early Railways of the Lothians*, Dalkeith, 1991.

28 Structures Associated with the Retail Trade

OWEN F HAND

Structures associated with the exchange of goods in Scotland have varied in size and purpose throughout time, from the temporary trestle tables of market traders (fig. 28.1) dealing with a single commodity to modern hypermarkets retailing a wide range of foodstuffs and products. Between these extremes lies the history of the retail trade, along with that of the property conversions and purpose-built premises necessary for such trade. This aspect has been largely neglected by historians and has been left instead to economists and social commentators.

An understanding of developments in the retail trade requires knowledge of prevailing social and economic conditions. In the late sixteenth century Scotland had an estimated population of one million,[1] the vast majority of whom were rural dwellers. Their needs

Figure 28.1 The Barras and Paddy's Market, Glasgow, SLA, S 13744.

were supplied through the local market town and its craftsmen, making the area self-providing, if not quite self-sufficient. Surplus produce from the fermtouns of the area was brought to markets, official or unofficial, to be converted into cash or bartered for other requirements. Importation of goods from outside the kingdom was the right of royal burgh merchants, and such goods were sold in markets or at travelling fairs around the country. Packmen and chapmen carried goods from the towns to outlying regions, returning with country-made products to be sold in town. This system of trade, which had been in operation since the twelfth century,[2] only changed when demographic pressure and increased wealth created an accelerated urban expansion.

THE MERCAT CROSS AND TRON

Mercat crosses can be regarded as the first structures associated with trade, although in reality they were more than that. A cross became the centre of the community and was the focal point of many aspects of civil administration. Where there were two crosses, as in Glasgow, Dumfries and Aberdeen, one was generally known as the fish cross.[3] The fish cross in Aberdeen was regarded as being of lesser importance and this is likely to have been the case in other burghs.

It is most probable that the earliest mercat crosses were manufactured from wood, although the only existing example is at Kilwinning, Ayrshire, where there is a wooden cross-head, all other market crosses being in stone. Through time, wooden crosses were replaced with examples in locally-quarried sandstone. The simpler kind of Scottish market cross consisted of a shaft of stone, standing on a flight of circular or octagonal steps. Towns such as Banff and Kinross favoured cruciform-shaped crosses. Grander varieties had tall stone shafts standing on imposing circular, hexagonal or octagonal sub-structures 10 to 16 feet (3.05 to 4.88 metres) high.[4] The top of these structures formed a platform with an ornamental stone parapet, reached by an internal stair, from where proclamations were made. Larger crosses are usually associated with the more important burghs but the example of Prestonpans, East Lothian, confounds this as a general rule. Decoration on crosses took the form of carving, gilding or painting. Heraldic beasts, especially the unicorn, were favoured as ornaments by royal burghs, and other headings included globes, urns, sundials, square tops, celtic crosses and weather vanes.

Also to be found in or close to a burgh market was the town tron or public weigh beam. King David II decreed in 1364 that a tron, under the supervision of a tronar, be set up in each burgh for weighing all commodities brought into town, and for determining customs fees due on exports. From the records of two sets of repairs carried out on

Dundee's tron by 1420, it can be ascertained that the balance of the beam was made of iron supported by a wooden log;[5] the scale pans were timber boards supported by ropes. Town weights, made of lead, were inspected regularly, as were those used privately. These latter weights were obliged to carry the town seal and punishment for false weighing generally included banishment from the burgh. Public proclamations were also made from the town tron.

The creation of burghs from the twelfth century made trade an urban activity, conducted as close as possible to the mercat cross. Vendors used a variety of makeshift trestle boards to display their goods, and there can have been little to distinguish stalls used for different trades. From an early date attempts were made to position individual trades in specific areas, and by 1477 various portions of Edinburgh's High Street and Cowgate were allocated to trades within which they alone could conduct business.[6] Burgh magistrates set the days and hours of the markets, established prices for goods, and, through the offices of town bailies, supervised general conduct.

BOOTHS

It is impossible to know when 'shops', as we understand the term, came into existence. The word originally meant a booth, but all structures from which trade was conducted were known as shops. Cellars and converted stables were used as workshops by craftsmen, and warehouses by merchants, from at least the sixteenth century. Booths erected under the galleries of houses acted as shops, as an example from Edinburgh in 1570–1 illustrates. The house in question was built of clay and consisted of 'ane eird hall, ane chalmer, ane stabill, ane fore buith'.[7] The 'fore buith' in this case will have been used as a shop, and it was common to find many such booths under the galleries of houses.

In spite of the existence of fixed retail outlets, until well into the nineteenth century the main trading areas were market places. Three early terms associated with trading structures have survived: 'crame' from the fifteenth century; 'laigh' from the seventeenth; and 'luckenbooth' from the eighteenth.

'Crame' refers to a booth, stall, or tent where goods are exposed for sale.[8] These were generally to be found around the edges of a market place or clustered around the walls of a church. The majority of crames were temporary structures erected for market days or fairs and could be rented at a charge of around ½d a day. It is likely that they were no more than simple stalls built around trestle display boards. Items sold from crames were as varied as the traders operating them.

'Laigh', meaning 'low lying', forms part of compound nouns such as laigh shop, laigh hall and laigh land. When applied to a building it is the part 'below the rest of the house, used as a store, cellar, shop or frequently, especially in Edinburgh, as a restaurant'.[9] A Glasgow reference of 1780 stated that 'most of the shops had underground premises, called laigh shops which were let separately'. Laigh shops are known to have existed in Edinburgh, Glasgow, Aberdeen, Dundee and Berwick.

A 'luckenbooth' was a trading booth which could be locked when not in use, indicating that it was a more permanent structure than a crame. Edinburgh's luckenbooths continued to stand until 1817 and were well recorded. The building that contained them was erected in 1440 as a two-storeyed tenement with the name of 'Buith Raw'.[10] When described at the beginning of the nineteenth century it was 'a four-storeyed tenement, timber-fronted for the most part'. Although the early name suggests a row of booths, the building was a permanent structure of small shops with living accommodation attached. The name 'luckenbooths' was applied to the eastern end of Dundee's Overgate where it joined the High Street, and also exists as a place-name for part of the North Row, Peebles.

Other related structures include merchant's booths, some of which nestled under the galleries of houses as already mentioned. They were hut-like affairs, most probably similar in construction to the original luckenbooths. Wooden shutters in place of windows were divided horizontally, the lower halves hinged downwards to form counters and the tops folding upwards to act as covers and display places. In time, many of these booths became permanent extensions to main buildings, acquiring a conventional front door and window. Where additional space was required they extended into the neighbouring front room. Attempts by magistrates to control such building practices failed in the face of opposition from determined burgesses, as the following Edinburgh example of early 1583 illustrates:

> Decernis and ordanis Jhonne Richertsoun, saidler, to remove and tak down his treyne chop laitlie biggett under the stairis of his land on the west syde of Nudries wynd, becaus the sam is contrair to guid nychtbourheid and the Kingis hie street is narrowet thairby; and the said Jhonne beand personally present ansuerit and declairit that he wad nocht do the sam, thay mycht do as thai pleisit.[11]

Hanging signs in front of shops identified the trade practised within and were beneficial before houses were numbered. By 1751 Edinburgh magistrates deemed that these too were obstructive and where they extended beyond the building, they were ordered to be taken down. Once shops moved away from constricted town centres, and after pavements were laid, signs ceased to be a problem. Even into the 1950s it was common to see a model fish hanging outside a fishmonger's shop, a mortar and pestle outside a pharmacy, a red and white candy-striped pole outside a barber's and three brass balls outside a pawnbroker's.

MARKET PLACES

As trade increased, market places in some towns were relocated to areas where there was space to expand. It was, therefore, not unusual for market crosses to be re-sited. By the eighteenth century the population of many towns had increased dramatically and where the central locations of market places were obstructive to the expanding flow of traffic, they were re-sited. Market crosses were also replaced or in some cases demolished. Dundee's early cross, for which no description survives, was removed from Seagait to Marketgait before being replaced by a grand stone cross in 1586. The new cross, set on an octagonal structure, survived in its position until being removed in 1777. All that now remains of the sixteenth-century cross, in its new location south of the town churches in the High Street, is the shaft which has been set on a new base and has been provided with a reproduction capital. The original carvings on the shaft are still visible, showing the date 1586, the town motto *Dei Donum* and an emblematic pot of lilies.[12]

In the eighteenth century Scottish burghs developed an increased sense of civic pride and industry. Influences from Europe, often arriving via England, created a desire for grand buildings and greater cleanliness. New buildings were erected within old town centres but the most significant advances were in urban expansion. As sections of the population moved to the suburbs of towns, shopkeepers moved with them to supply their needs.

Edinburgh's planned New Town of 1767 involved the erection of magnificent dwelling houses to the north of the city with accommodation being provided on back streets for the better types of artisans. The area was designed as residential, but by 1797 many of the properties were serving as commercial premises, with shops and warehouses at ground and basement levels, similar in practice to the Old Town.[13]

Towns continued to promote markets and to try to control their use, since much of their revenue was derived from them. Meat and fish trades were moved from residential areas but this encouraged the establishment of shops selling those items. Fruit and vegetable markets, being less offensive, continued to be located close to town centres. New markets were opened to supply expanding areas, mainly with foodstuffs. A division emerged in retail trade, with markets trading in perishable goods and shops generally carrying more durable items. Manufactory-produced goods allowed shopkeepers to become involved in trades previously limited to craftsmen, and specialist shops, such as shoemakers, emerged with ready-made stock.

After many relocations, Edinburgh's main fruit and vegetable market was housed in purpose-built premises between Princes Street and Waverley Station. Prior to this however, there was much debate between the two railway companies that shared Waverley Station and Edinburgh Town Council over the siting of the market. In an attempt to satisfy all three parties, plans were drawn up which allowed for the site to be covered with a cast-iron platform at a height of 31 feet (9.45 metres) on which the market was to be held, with the railway station situated underneath.[14] An envisaged lack of light and ventilation in the station caused this plan to be rejected. Waverley Market was eventually built in 1866 and was paved and sectioned into stances, making use of decorative cast iron. Above the stalls ran a gallery also made of iron. In 1874 the entire building was covered over with an ornamental terraced roof. Competition from shops eventually made the Waverley Market redundant as a market place and for many years it was used only occasionally as a venue for exhibitions, sporting contests and entertainments. In the 1980s the building was given a new lease of life when it was converted into a shopping precinct on three floors, serviced by elevators and escalators. The retail units, of varied floor area, are uniform in height and are almost entirely glass-fronted for display purposes. Headboards are used to give the name and/or nature of the business but central management exercise control over possible changes, insisting on uniformity of design.

Princes Square in Glasgow, which was originally a galleried courtyard of merchants' offices, has been given treatment similar to Waverley Market and converted into a shopping precinct. In Princes Square the decorative iron galleries have been kept as access to the shops, and the courtyard has been covered with a glass roof. Both Princes Square and Waverley Market are occupied by shops dealing in high-profile luxury goods, which gives an indication of the manner in which city-centre shopping is developing.

There has been a revival of interest in market buying in recent decades, but primarily as a leisure activity for bargain hunters and sightseers. Large markets, where traders hire open stalls and sell from trestle tables as in earlier days, have become established on rural sites in close proximity to cities and areas of high population. Trading is generally restricted to Sundays and is heavily dependent on both the car-owning population and public transport to provide potential buyers. As with earlier markets, the goods sold vary as much as the vendors. The popularity of such markets is established by the fact that groups travel long distances on organised bus trips to attend them.

SHOPS

Shop design and retailing, being attached to supply and demand, could not advance unless supported by increased spending. The eighteenth century brought greater wealth to sections of society, and inspired new forms of conspicuous consumption among the upper classes. Fashion industries prospered and trades such as haberdashers, perfumers and umbrella makers appeared. A traveller to Edinburgh in 1792 praised the 'splendid shops that lined its streets', and referred to the city as a seat of fashion. The real upsurge, however, in retail trade did not come until spending power reached a larger proportion of the population.

Single shop units, operating from converted house properties and cottages, were the centres of retail trade in the early nineteenth century. The outward appearance of the buildings remained much the same until well into the century, with the probable exceptions of fascia boards stating the nature of the business, and trade signs. In the absence of a fascia board, the stonework itself was often painted. Inside, the only specialist furniture required was a counter and some shelves on which to store paper-wrapped stock. After the 1820s, architects began including commercial properties in their plans and continued the practice of having shop properties below residential units. The new purpose-built shops, with living accommodation attached, continued to be in single units but were generally larger than the converted variety. Moulded strings or cornices across the tops of shop units became a feature of their design, dividing the shops from the rest of the building. Another innovation of the latter period was the emergence of double-windowed fronts with a window on each side of a central door.

Window tax and duties on the manufacture of glass inhibited advances in shop design. Until the 1830s the main type of glass available was 'crown', spun into discs of no more than five feet (1.52 metres) in diameter with a dimple in the centre and much loss in cutting.[15] Rolled glass was produced only in small panes. Both types of glass were used in window manufacture but had to be set in astragal frames. Even after the introduction of gas lighting from the early 1800s, shop interiors were darkish places and goods were displayed either outside or near the door. Some of the more successful shops had gas lamps placed outside their windows to cast greater light. When excise duties were removed in 1845, and after glass became available in large sheets, shop fronts on the more fashionable streets began to change. Three-dimensional façades in carved wood, stone, or plaster mouldings were erected in front of many properties, and decorative window frames gave character and distinction to individual shops. Large windows not only provided extra light but became display areas. In converted basements, the windows were often built over to extend storage areas.

An increased volume of trade during the nineteenth century brought significant variations to retailing and shop design. The re-investment of profits and speculative investment allowed businesses to develop in a variety of ways. Independent traders continued to function, but with additional capital many expanded by acquiring neighbouring properties to enlarge their premises.

FROM CO-OPERATIVES TO SUPERSTORES

Multiple retailers emerged, who took over properties in different areas, creating chains of single shop units retailing the same goods and operating under central management. The co-operative movement, where shoppers bought shares in the company and enjoyed dividends from profits, was a system imported from Rochdale, Lancashire, in the 1850s.[16] It was introduced partly to realise the economic benefits of 'combination' purchasing and partly to combat the adulteration of foods by independent grocers. Co-operative organisations, of which there were many, were generally under-financed and as a result operated from whatever premises they could afford to rent. Gradually, by amalgamation of separate organisations, they began to benefit from increased membership accompanied by economy of scale. Financial success provided development capital which was used to build their own premises, often with tenements above. In the new buildings, food provision continued from single unit shops, while in trades such as drapery and furniture, the policies of department stores were followed. Co-operative organisations adopted uniform paint colours and shop fronts for their premises.

The example set by the co-operatives was quickly followed by individual retailers. Thomas Lipton, the most successful of the chain store grocers, opened his first shop in Stobcross Street, Glasgow, in 1871. His second shop, also in Glasgow, came about three years later, and by 1878 he announced that he would be opening branches in Aberdeen, Edinburgh, Greenock and Paisley. Lipton's retail policy was to create small internal markets specialising in groceries, cooked meats and dairy produce. The Lipton empire expanded through the United Kingdom before becoming international, and his high-street shops could be identified by their uniform fronts, in glass and tiles with the Lipton name in gilt lettering.

Lipton's expansionist policies were copied in the market, though to a lesser extent, by a succession of grocers from Glasgow and surrounding areas. One such company was Templetons, who originally adopted the publican practice of siting their premises on corner sites. The strategy behind the policy was to draw customers from both streets, but corner sites proved dusty and unsuitable for grocers' premises. When shopping habits began to favour main-street premises, corner sites were bad for display by showing only half their front to passing trade. Corner shops continue to be favoured by small general merchants in urban and industrial working class areas, where by the nature of their business they were traditionally called 'Johnny (or Jenny) Aathings'.

Climatic conditions have had their effect on retailing, with traders attempting to benefit from supplying as many goods as possible under one roof. The nineteenth century produced a variety of such premises in the forms of bazaars, arcades and department stores.

Bazaars operated in the manner of outdoor markets for non-foodstuffs by having a selection of single unit stalls within the same building. The name, which is from the Persian 'bazar', meaning 'a market', was chosen to alert customers to the variety of goods on sale. The principal items on offer were dresses, accessories, millinery and dress materials.

Arcades or covered streets, often with glass roofs, were built with uniform-fronted shops on each side. Glasgow had eight arcades during the nineteenth century, the first, Argyle Arcade, having been built in 1827. The external aspects of arcades were generally grand affairs and Glasgow's Wellington Arcade, which connected Sauchiehall Street with Renfrew Street, was said to have Doric entrances. Edinburgh's only arcade of the period was built in 1876, and connected Princes Street with Rose Street.[17] It had a frontage in Italian style carved in 'freestone', a glass roof 'supported on perforated girders of lacework pattern', and a floor of 'Austrian marble, laid in alternate squares of black and white'. The ground floor consisted of seven shops on each side of the passage; access to the galleries was by a central staircase. Stirling's Crawford Arcade, built in 1882, was unique in Britain in that it included a theatre, the Alhambra, at the centre of its development. Later arcades have been incorporated into large developments such as office or hotel buildings where shops are secondary to the main function of the building.

Department stores serve the purpose of indoor markets, with several autonomous departments dealing with different branches of trade within the same building. Business is controlled by departmental managers under direction from central management. The volume of business conducted by department stores requires them to be housed in large buildings which have either been custom-built or created by conversion of many properties into one. Scotland adopted the concept of such trading early, with the partnership of Charles Jenner and Charles Kennington opening its store on Princes Street, Edinburgh, in 1838. The business subsequently came under the control of the Jenner family and has been known by that name since then. The business was conducted from a converted house until 1885 when major renovations were undertaken. The façade of the present building is in carved sandstone and was inspired by the Bodleian Library, Oxford. Caryatids are used as columns and are meant to represent the importance of women as customers.[18]

From the middle of the twentieth century retail business has undergone a revolution in both trading practices and shop design. The growth of corporate retail businesses has created a situation where the number of outlets is diminishing, although not in volume. There has been a general decline in the number of small trading units, the only growth in this section being among chain-store outlets and others operating under franchise, such as Body Shop, the Sock Shop and Knickerbox. These shops may be housed in traditional high-street premises or units in newly-created shopping precincts or arcades. In all cases they have adopted the standardised fronts of the main company's identity.

Outlets for food supermarket chains are to be found in most towns, trading from large stores in purpose-built premises on one level and with car parking facilities available. There are also retail warehouses, trading in furniture, do-it-yourself accessories and electrical goods, built on much the same design as food supermarkets. Retail warehouses tend to group together in out-of-centre sites. Building design concentrates on functionality, with cost limitation dominating the choice of materials, generally brick or concrete. Windows are to be found only in food stores which are close to a main thoroughfare and where internal light attracts passers-by. In other cases, internal wall space is too valuable as shelving area to be used for windows. The lack of windows also gives added security.

A step up from the retail warehouse park is the megacentre or regional shopping centre. The high population density of central Scotland makes the area attractive to such developments which provide facilities for supermarkets and independent traders within one building. Such a building may cover in excess of 200,000 square feet (4.6 acres or 1.86 hectares), and there are proposals for centres ten times this size.[19] All aspects of retail trade are provided for in these megacentres, including cafeteria facilities and convenience food outlets with one or two supermarkets and perhaps a department store. As with warehouse stores, the external shells of megastores are considered unimportant and are built for cost effectiveness. Inside the buildings, which are constructed like large arcades, decoration can take many forms and involve features such as tiled walkways, water fountains, potted plants or even sculptures. Much of the floor space is taken up by the stores whose needs are consulted at the planning stage. Units are of standard height and are rented by floor area, although the building may contain more than one floor.

Rural shopping has more or less followed the pattern portrayed in early urban shopping with the conversion of existing buildings to form shops. Newly-built premises have also followed the urban pattern of functionality over design with the truly isolated shop requiring no more than a basic shell in which to act as a general trader.

Owen Hand, musician, singer, song-writer and ethnologist, died 6 February 2003, aged 64. Owen worked as Research Assistant with the European Ethnological Research Centre from 28 November 1993 until the end of 1995, when he had to give up owing to illness. He wrote this chapter and acted as co-editor of the *Compendium* volume, *Oral Literature and Performance Culture* (Volume 10).

NOTES

1. Smout, 1987, 111.
2. Torrie, 1990, 22.
3. Adams, 1978, 36.
4. Torrie, 1990, 42.
5. *Chamber's Encyclopaedia*, 1898, sv, 'market cross'.
6. Edinburgh, 1929 178.
7. RCAHMS, 1951, lxvii.
8. *Scottish National Dictionary*, sv 'laich'.
9. ibid, sv 'crame'.
10. RCAHMS, 1951, 127.
11. *Extracts from the Records of the Burgh of Edinburgh, AD 1573–1589*, SBRS, Edinburgh, 1882, 263 (30 January, 1582/3).
12. Torrie, 1990, 42.
13. Macrae, 1947, 18; Aitchison, 1798.
14. Edinburgh, 1929, 183.
15. MacKeith, 1986, 28.
16. Maxwell, 1909, 21.
17. Geist, 1983, 250.
18. Grierson, 1938; NAS, 1/769 contains details of construction work undertaken at Jenners between 1899 and 1950.
19. Aitken, Sparks, 1988, 8.

BIBLIOGRAPHY

Adams, I H. *The Making of Urban Scotland*, London, 1978.
Aitchison, A. *The Modern Gazetteer*, 2 vols, Perth, 1798.
Aitken, P, Sparks, L. *Retail Change in Scotland*, Stirling, 1988.
Chambers, R. *Edinburgh Merchants and Merchandise in Old Times*, Edinburgh, 1859.
Chamber's Encyclopaedia, Edinburgh, 1898.
Devine T M, Mitchison, R, eds. *People and Society in Scotland*, Vol 1, Edinburgh, 1988.
The Edinburgh and Leith Annual Post Office Directory, Edinburgh, 1819– .
Edinburgh, Lord Provost and Council. *Edinburgh, 1329–1929*, Edinburgh, 1929.
Ewan, E. *Townlife in Fourteenth-Century Scotland*, Edinburgh, 1990.

Fenton, A, ed. *Scottish Life and Society: A compendium of Scottish ethnology*, Vol 5, *The Food of the Scots*, forthcoming.
Geist, J F. *The History of a Building Type*, Massachusetts, 1983.
Grierson, M G, for Jenners Ltd. *Hundred years in Princes Street, 1838–1938*, Edinburgh, 1938.
Hartwich, V C. Patterns in small shop frontages in Dundee, *Vernacular Building*, 7 (1981/2), 10–25.
Levy, Herman. *The Shops of Britain*, London, 1947.
Lynch, M, Spearman, M, Stell, G, eds. *The Scottish Medieval Town*, Edinburgh, 1988.
MacKeith, M. *The History and Conservation of Shopping Arcades*, London and New York, 1986.
MacNaughton, E G. A Highland pharmacy, 1882–1972, *ROSC*, 3 (1987), 67–76.
Macrae, E J. *The Heritage of Greater Edinburgh: A report for the City of Edinburgh*, Edinburgh, 1947.
Marwick, W H. Shops in eighteenth- and nineteenth-century Edinburgh, *The Book of the Old Edinburgh Club*, 30 (1959), 119–41.
Mathias, P. *Retailing Revolution*, London, 1967.
Maxwell, W, ed. *First Fifty Years of St Cuthbert's Co-operative Association Limited, 1850–1900*, Edinburgh, 1909.
RCAHMS. *Inventory of Edinburgh*, Edinburgh, 1951.
Scottish National Dictionary [Grant, W, ed], Edinburgh, 1931–76.
Smout, T C. *A History of the Scottish People, 1560–1830*, London, 1987.
Stevenson, D, ed. *The Diary of a Canny Man, 1818–1828*, Aberdeen, 1991.
Torrance, D. *Wee Shops*, Glasgow, 1988.
Torrie, E P D. *Medieval Dundee: A town and its people*, Dundee, 1990.

29 Business and Commercial Buildings[1]

DAVID WALKER

EARLY MERCHANTS' EXCHANGES

Until at least Georgian times trading was generally conducted in the open air, either at the mercat cross as the symbol of a burgh's trading privileges or, as in Leith, at the quayside. The earliest buildings erected for the purpose appear to have been at Leith where Bernard Lindsay, then owner of the King's Wark, constructed an arcaded 'piazza', or merchants' shelter, some time between 1606 and 1612.

At a subsequent date, still in the seventeenth century, a 'bourse' or 'burss' was built to the south of the King's Wark on Queen Street (fig. 29.1) where, in the words of Maitland's *History of Edinburgh*, 'the People used to resort to treat of their affairs'. This was a three-storeyed building with a pend to its court, gablet dormers and a very high roof. Its elevation to Queen Street had a regular first-floor pattern of large windows and small closet windows which must have related to the compartmentation of the interior on that side of the building. Maitland also describes it as having three piazzas (that is, an arcade of three arches) on the 'South Side', presumably the south side of the court. These probably represented the Lindsay structure. The Queen Street section of this building survived until 1929 but its interior does not appear to have been adequately recorded.

A further bourse, consisting mainly of a large yard, existed for the import of timber just off the Shore at Timber Bush (that is, Bourse). In 1644 and 1650 this was 'fortified with a strong stonern Wall at the Expence of Eleven thousand seven hundred Marks Scottish'. Of this security wall the northern section on Tower Street survived into the mid-twentieth century to be noted, though not recorded, by RCAHMS. It had large shuttered openings for drawing in the deals from the quays, built up at the time of the RCAHMS visit.

The only fixed equipment that market place or quayside trading required was a public weigh-beam or tron (see also Chapter 13). In the sixteenth and seventeenth centuries these were normally simple open-air post-and-beam structures but one of the two Edinburgh trons was under cover, as on the European continent, the upper or Butter Tron, built in

Figure 29.1 Leith Bourse, 7–9 Tolbooth Wynd, Leith; view of (Queen Street) rear of building, c.1929. Reproduced by courtesy of Edinburgh City Libraries.

1612–14 and reduced in height after damage in the Civil War. It was a plain rectangle but it had a fine Renaissance doorpiece. Since Leith had a Weigh-house Wynd from an early date it would appear that its public weigh-beam at the Shore was similarly under cover.

EARLY MERCHANTS' OFFICES AND WAREHOUSES

For most commodities the merchant's house was also his warehouse and counting house. Andrew Lamb's House in Water's Close in Leith is now the best illustration of how such a house functioned, with wide doors and undivided warehouse floor space at the ground floor and further provision for storage in a capacious attic with a loading door. The finest example to survive long enough to be recorded was Robert Gourlay's House in Lawnmarket, Edinburgh, built in 1569 and demolished for the formation of George IV Bridge. At this very tall house the domestic accommodation was raised up over two warehouse floors. Both of these floors had six very large, close-spaced windows, evidently for the inspection of high-quality goods. As has been observed by Deborah Howard, Gourlay's house appears to have been modelled on those in the Dutch and Baltic ports. At the still surviving Baillie McMorran's House, also in Lawnmarket, the warehouse accommodation was in the attic and probably also in the cellarage below courtyard level. In several of the closes in the Old Town of Edinburgh the ground-floor cellars have wide doorways of the same type as at Lamb's House, finely moulded with lettered lintels, indicative of the quality of the goods once held within them.

Bulkier commodities required warehouses. In Dundee, Daniel Defoe commented on the tree-lined avenue between the town and the harbour with 'very good warehouses for merchandises, especially for heavy goods' along one side. These are not identifiable in any early views, but from Crawford's map of 1776 Skirling's Wynd, demolished in the 1790s for the building of Castle Street, may be identified as corresponding with the description. Similar warehouses must have existed in Leith and the other older Scottish ports but few survived into the nineteenth and twentieth centuries to illustrate what they looked like. The only survivor from the sixteenth century is that at Limekilns, Fife, an oblong structure of two superimposed vaults with a handsome doorpiece dated 1581. The building corresponds with that date although the pediment appears to have been reset. The Great Custom House at the Earl of Winton's harbour at Cockenzie, East Lothian, now unfortunately burnt out, was a much larger building. It was clearly fairly ancient, low two-storeyed and relatively wide with a huge loft under a very big roof. Its floors rested on very heavy timber joists given intermediate support by a central row of timber stanchions.

At St Giles Street in Leith is a rather later-looking survivor, the large 'Blak-Volts' of the Logans of Coatfield, built in 1682 over two 70-feet (21.34-metres) long vaults excavated in the sand, which appear to have been in existence in 1587. The superstructure is of three storeys, raised to four in 1785, and uniquely retains a fine early eighteenth-century ground-floor room where the Logans received their clients. A large walled courtyard contains a draw-well and a bottling shed dated 1689. Until the 1960s the area around Timber Bush and south of Bernard Street was densely packed with tall four- and five-storeyed warehouses of similar character, although only Robert Mylne's great tenement of 1677-8 at 8-14 Shore appeared to be older than the mid eighteenth century. Of Timber Bush one range of relatively late date survives. None of these warehouses, so far as is known, contained an office comparable with that at the Black Vaults. All the older examples in the Timber Bush area were of similar construction to each other with timber beams and stanchions, as at the earlier warehouses in London's dockland and the great naval dockyards in England.

Other warehouses of similar height and vintage survive, or survived until recently, at such harbours as Annan, Dumfriesshire, Ayr, Bo'ness, West Lothian, North Berwick and Dunbar, East Lothian, Banff and Portsoy, Banffshire, Burghead, Moray, and Peterhead, Aberdeenshire, where a rebuilt, but still quite old, example bears the date 1616. The finest example of the genre appears to have been the Packhouses or Public Warehouse at Dundee, an excellent three-storeyed symmetrical building of 1755 with warehousemen's offices flanking the central entrance. Designed by the Leven architect and mill-wright William Robertson, it had bold provincial Baroque details and a piended roof, and was flanked by single-storeyed pavilions, the eastern containing the public weigh-beam which also served the adjoining Greenmarket.

In the far north, early warehouses survive in considerable numbers, although most of them were girnals, estate storehouses for rent paid in grain rather than in cash, and not really within the scope of this review (see Chapters 9 and 21). Nevertheless, there can have been little difference between warehouse and storehouse as a building type and they serve to illustrate what the earlier warehouses looked like. The oldest survivors at Burray (1645) and St Mary's (1649) in Orkney, and at Portmahomack (Alexander Stronach, 1699) in Ross and Cromarty, are all two-storeyed with a forestair to the upper floor. The later examples are more usually of three storeys with a small office partitioned off for the storeman. That at Stromness, Orkney, built for James Gordon of Cairston, was very unusually for rice.

Unique to Orkney and Shetland is the arrangement of merchants' house at right-angles to the shore, forming a unit with a small warehouse and its own pier, known in Shetland as lodberries; all of these are, or were, on quite a small scale. German merchants' booths are also to be found in Shetland, built from the sixteenth century onwards. The best example at Symbister on Whalsay has been restored in recent years, a simple two-storeyed building, its flank elevation slotted as a loading bay with a double-wheeled pulley hoist corbelled out below the wallhead.

TRADE, GUILD AND MERCHANT HALLS

In the absence of most of their other buildings, by far the clearest indicator of the prosperity of the trade and merchant classes in the seventeenth and eighteenth centuries were the halls of their incorporations. These buildings were for the regulation of the trade and membership of the incorporation, and for social and charitable purposes, rather than for the actual transaction of business. The oldest of which we have pictorial knowledge appears to have been the Trades Hall in the Shiprow in Aberdeen, which only just survived into the era of photography. This was a two-storeyed structure of *c.* 1600 and had a square entrance jamb with a corbelled stair turret in its re-entrant angle, like a towerhouse.

In Edinburgh, each craft or trade had its own hall and of these several survive. The oldest is that of the Hammermen who fell heir in

1553 to Magdalen Chapel which had been built as recently as 1537. It was reconstructed for their purposes in 1613–17 when the arched ceiling and arcaded panelling were installed, the latter for recording benefactions. A steeple with a spirelet was added by the mason John Tailefer in 1620–5. Next is the Tailors' Hall, built in 1621, a large L-plan building like the Aberdeen Trades Hall, but three-storeyed rather than two, and raised to four storeys in 1757. The 1621 inscription recorded that it was 'for meeting of thair craft ... in trust in Gods goodness to be blist and protected'. Although the jamb contained a stair, the first-floor convening room had its own central forestair. Of the interior work only the architraved door of the hall itself survives. At the Skinners' Hall (1643), a T-plan structure with an octagonal stair tower in the re-entrant angle like a patrician mansion, the interior work of the convening room has also disappeared, but at the tall and slim four-storeyed Candlemakers' Hall of 1722 it survives at the top floor, a simple room lined with fielded panelling recording the mortifications of its members. A very similar convening room survived until the 1960s at the still-extant hall of the Wrights' Incorporation in Perth (1725).

Predictably finer than the trades or guild halls were those of the merchants. In Glasgow, the merchants hived themselves off from the trades in 1605, building the Merchants' House in Bridgegate in 1651–9, much the finest of the earlier halls. The relevant act book has unfortunately been missing since at least the 1840s, making it impossible to determine who designed it, but a drawing made prior to its destruction in 1817 shows that it was a symmetrical two-storeyed, nine-bay building with pedimented dormerheads and a central pend. The pend arch was flanked by columns and surmounted by a bold aedicule containing sculptures of a ship, decayed merchants and an inscription, all rather more classical in character than John Clerk's exactly contemporary work at the city's Old College. Behind the hall rose the still-extant Gothic survival steeple, completed with some difficulty by 1665 when the clock was installed by John Brodbreidge. Within, the building comprised four shops and offices at the ground floor and a fine convening room, 30 feet by 80 (9.14 by 24.38 metres), with a model of a ship suspended from the ceiling. In addition to the benefaction boards, in the panelling there was a painted board inscribed with quotations from the Scriptures, with 'directions expressive of the principles on which the trader might buy or sell with a safe conscience'.

The Edinburgh merchants obtained their royal charter setting them up as a separate company from the trades in 1681. They did not build a new house, acquiring a decade later the large quadrangular mansion of Viscount Oxenfoord in the Cowgate. Their convening room was, however, more richly appointed than that of the Glasgow merchants, hung with embossed Spanish leather provided by Baillie Alexander Brand

Figure 29.2 Royal Exchange, City Chambers, High Street, Edinburgh, c. 1800; from Thomas Shepherd. *Modern Athens*, 1829.

whose Prestonfield hangings survive to give an impression of what it must have looked like.

In 1680–1 Thomas Robertson built Edinburgh's original Exchange in Parliament Close, immediately opposite the Cross (see also Chapters 13 and 18). This was erected to the designs of Sir William Bruce and had a range of 'piazzas' as at Leith for the convenience of the merchants. However, it is recorded that the merchants preferred to bargain in the open air at the Cross, except in the most inclement weather. Shortly after a fire in 1700 the Exchange was rebuilt to a height of eleven storeys and attic, five storeys being below the level of Parliament Square. As reconstructed, it had a Doric pilastered open arcade to the square along its eastern elevation, probably the piazzas of the Bruce Exchange retained and repaired. This building seems not to have been much used as intended either. Its piazza was, however, found to be a useful shelter in wet weather and the same facility was provided at Dundee Town House, designed by William Adam in 1732–4, and at Allan Dreghorn's Glasgow Town Hall of 1736–8. Both these buildings had the same close relationship with the town cross, that at Dundee replacing the former tolbooth on the same site and that at Glasgow immediately adjacent to the tolbooth of 1625–7 (see Chapter 13).

Despite the failure of the merchants to use Robertson's building, a new exchange, also opposite the Cross, featured in Lord Provost

Drummond's *Proposals for Carrying on Certain Public Works in the City of Edinburgh,* 1752. Plans were obtained from John Adam for a massive Gibbsian three-sided square of buildings, four storeys high above street level around a court 83 by 89 feet (25.30 by 27.13 metres), screened from the street by a single-storeyed arcade of shops. Another piazza for the accommodation of the merchants ran along the north side. It was to have been a multi-purpose building. The pedimented and Corinthian pilastered section at the centre of the north elevation formed the Custom House (fig. 29.2). When completed in 1760, the remainder was yet again not used as originally planned. It was eventually sold off as shops, coffee houses, offices and flats, and remained so until re-acquisition as the City Chambers, the north range being bought in 1811 and the remainder between 1849 and 1893.

Although the Edinburgh Wright Incorporation had built a fine hall with a Serlian window at Mary's Chapel in Niddry Street in 1737 (the name commemorated their chapel in St Giles), it was not until the last quarter of the eighteenth century that there was a further campaign of building by the incorporations. In 1776–8 a surprisingly large hall was built for the Nine Trades in Dundee by the local architect, Samuel Bell. This was a tall two-storeyed, rectangular building on an island site. An attached tetrastyle portico of Ionic columns formed its pedimented gable facing High Street, and a big cupola sat astride its roof. The ground floor consisted of shops and offices, with a very large convening room and committee rooms above. In Glasgow, the Guildry or Convenery of the Fourteen Incorporated Trades, which had variously used the Parsonage of Morebattle, the Tron Church, the laigh session-house of the cathedral and various inns for their proceedings, commissioned plans for a hall of their own, first from James Jaffrey and then from Robert Adam, adopting the plans of the latter. Jaffrey's scheme would have comprised a Corinthian-columned hall with ground-floor shops and a tall spire, clearly intended as a challenge to the Merchants' House. Adam's scheme was more practical with a timber and lead dome rather than a spire, ground-floor shops, a first-floor great hall with a domed ceiling, and flanking committee rooms at both first-floor and attic levels.

In 1788, the Merchant Company of Edinburgh built a new hall in the recently formed Hunter Square, designed by the younger John Baxter who had studied at the Academy of St Luke in Rome. While other institutions such as the Physicians had begun to move across the Nor' Loch into the New Town, at that date the merchants clearly still thought it important to be near the Tron and the Cross, even though the Cross itself had vanished in 1756. Although a simpler building of three bays with fluted Doric pilasters, Baxter's hall was as sophisticated in design as Adam's Glasgow hall and has happily been returned to its original design. As in the other halls of this vintage, the ground floor consisted of shops. Above was, and still is, a fine convening room. It had chimney pieces at each end and was lit by three tall arched windows. The top floor consisted of a large flat.

BANKS

Trades halls were built at several other burghs, most notably at Dumfries, a handsome free-standing pedimented building (1804), still with a piazza, by Thomas Boyd, and at Arbroath (1814) by David Hill, but structures purpose-built for the transaction of business indoors were to remain few until the 1820s. The earliest appears to have been Sir William Forbes's Bank, a tall plain five-storeyed block just to the south of Parliament Square, Edinburgh, built in 1779. The other banks all occupied houses converted for the purpose, such as the Royal Bank in Old Bank Close, Edinburgh, for which William Adam prepared designs in 1744, and the British Linen Bank's still-extant premises in Tweeddale Court, Edinburgh.

It was not until 1800, 21 years after Forbes's bank, that the next purpose-built bank buildings were undertaken. In that year, the Bank of Scotland and the Aberdeen Bank commissioned new premises. Both were much more architectural than the Forbes building. At considerable cost the Bank of Scotland, designed by Richard Crichton and Robert Reid, was set on the axis of the newly formed Bank Street at the head of The Mound, Edinburgh, a strategically-chosen location close to the Parliament House and serving the New Town as conveniently as the Old. Raised on a massive under-building containing the strong rooms, it was a free-standing, Corinthian columned and pilastered three-storeyed building of late Adam country-house like character. The only feature to indicate its more public role was a shallow dome. Within the building a square hall with screens of columns led through an apse to a spacious telling room, rising through two storeys with Serlian windows overlooking the New Town. The telling room's elongated octagonal plan may have resulted in a somewhat impractical counter layout since it was subsequently found necessary to rebuild it square.

The Aberdeen Bank's office (1801) on Castlegate, Aberdeen, designed by the Haddington architect and bridge-builder, James Burn, was a Roman Doric version of the well-tried model of John Webb's building at Old Somerset House, London. It was simpler on plan with a rectangular public office. Smaller but no less architectural was John Paterson's Leith Bank, 1805–6, a miniaturised version of his country house at Coilsfield, Ayrshire, with an Ionic-columned bow and the telling room at the back, the first example of such an arrangement in Scotland.

These buildings remained unique in scale and pretension for the next 15 years. The Commercial Bank was content to reconstruct and extend the former Assembly Rooms in Edinburgh's New Assembly Close as a Roman Doric, villa-like structure designed by James Gillespie Graham in 1813, while in Dundee, the Dundee Banking Company's premises were limited to a well fitted-out, Ionic-columned interior which formed the ground floor of St Paul's Episcopal Chapel (1812). But in 1820 an English company, the Norwich Union Insurance Society, built a handsome business house at 32 Princes Street, Edinburgh, for its agent in the city, James Bridges. Designed by Thomas Hamilton, it was only three bays wide but had an attached portico of Corinthian columns which formed the model for several early insurance offices. It was followed by the Ship Bank's premises at 9 Glassford Street, Glasgow (1825, by David Hamilton). These were rather more ambitious, having a three-storeyed, five-bay frontage with a giant order of Ionic pilasters which formed the central pavilion of a 19-bay block of shops, houses and business chambers. Still grander was the Ayrshire Bank's head office (c. 1830) on New Bridge Street, Ayr, by Thomas Hamilton, again three-storeyed but with an Ionic colonnade at the upper floors.

From the mid 1820s all the major banking houses, with the single exception of the Royal Bank, undertook a programme of branch bank building. Typically, a branch office consisted of a plain classical two- or three-storeyed and basement building, three to five bays wide, with a ground-floor telling room, waiting room and agent's office, a strong room, and, on the upper floors, a house for the agent – a type which, particularly in Aberdeenshire, persisted into the 1870s and even beyond. Since the agent was usually also a local solicitor, accountant or land agent, many bank buildings included a separate office as his own professional premises. Two branches were, however, exceptional buildings and set the pattern of future head-office design in having giant entrance porticoes rather than single-storeyed porches: the former Commercial Bank (1825), Stirling, Greek Doric and designed by James Gillespie Graham; and the much more sophisticated Greek Ionic Royal Bank (1827) in Glasgow by Archibald Elliot II (fig. 29.3). The latter formed the centrepiece of a much larger development promoted by the bank, Royal Exchange Square. This comprised a formal quadrangle of business chambers planned around the Royal Exchange itself (see below).

In the five years after 1838 all the Glasgow banks either built or bought new head-office buildings. Those of the Clydesdale Bank and the Western Bank, both by David and James Hamilton (1840–1), on Queen Street and Miller Street respectively, were sizeable three-storeyed and basement, box-like late classical structures which provided the model for the larger branch banks from the 1840s onwards. Both these buildings had astylar elevations, although the Clydesdale had a Doric-columned porch. Much more ambitious was the Union Bank headquarters of 1838 on Virginia Street, designed by Robert Black, which had a pedimented

Figure 29.3 Royal Bank of Scotland, Royal Exchange Square, Glasgow. Crown copyright: RCAHMS, SC 710760.

Roman Doric portico at first-floor level, but in 1841 the Glasgow & Ship Bank commissioned a still grander building from David and James Hamilton. This incorporated the Virginia Mansion on Ingram Street which was refaced with a five-bay, Roman Doric colonnade like that of an American plantation house, with symbolic figures at its balustraded parapet. A top-lit, single-storeyed telling room was constructed at the back, with a tetrastyle portico on the axis of Virginia Street. In 1843 the Glasgow & Ship Bank was taken over by the Union Bank which then moved to its grander Ingram Street premises, the Union's original head office on Virginia Street being sold as a ready-made headquarters for the City of Glasgow Bank.

In Edinburgh, a similar rivalry in head-office buildings was inaugurated by the building of the Edinburgh & Leith Bank on the corner of George and Hanover Streets, designed by David Bryce in 1841. This had a simple ground-floor telling room at the front of the building, but at its first and second floors it was grandly Corinthian-pilastered with distyle porticoes at the end bays of the George Street frontage and cantilevered and balustraded balconies at the first-floor windows. It soon proved too small and was extended on the Hanover Street frontage in 1847.

In 1843 the Commercial Bank commissioned David Rhind to design

Figure 29.4 Royal Bank of Scotland (formerly Head Office, Commercial Bank of Scotland), 14 George Street, Edinburgh; perspective by David Rhind, c. 1845. Courtesy of the Royal Bank of Scotland.

an Edinburgh head office (fig. 29.4) which would eclipse anything hitherto built in Edinburgh or Glasgow, the old Physicians' Hall on George Street being acquired to provide a sufficiently generous site for a free-standing building. The design of its front elevation was in some degree modelled on William Henry Playfair's original Corinthian scheme for the Surgeons' Hall in the same city, having a magnificent hexastyle portico with pediment sculpture by James Wyatt, and flanking foot gates. In its internal arrangement it was modelled on Hamilton's Glasgow & Ship Bank in having a top-lit telling room at the back, but it was altogether grander, the approach being through a magnificent double-height atrium with superimposed orders of columns. The telling room had a glazed dome on pendentives providing excellent natural lighting, an important consideration at the time. It attracted a great deal of publicity and served as the model for several head offices in the dominions.

In sheer magnificence, inside and out, the Commercial Bank was rivalled only by David Bryce's British Linen Bank head office (1845–51) on St Andrew Square, Edinburgh (figs 29.5, 29.6), a project somewhat coyly referred to in the directors' minute book only as a new telling room. This had Corinthian columns with salient entablatures bearing statuary, like a gigantic fragment of Imperial Rome. As at the Commercial Bank, the telling room was at the back, approached through an immensely grand stairhall. It was even richer than Rhind's at the Commercial Bank, with Peterhead granite monoliths costing £1,000 each, the price of a sizable villa at the time; those of the Commercial Bank had been of marbled wood, an economy corrected in 1885. Bryce adopted the same arrangement of telling room at the smaller Exchange Bank of Scotland (1846) at 23 St Andrew Square; of this building, which incorporated the carcass of a house built in 1770, there now survives only the superbly detailed palazzo frontage with Ionic porch and cantilevered first-floor balcony.

In 1857 John Dick Peddie reconstructed the Royal Bank's

Figure 29.5 British Linen Bank, 37 St Andrew Square, Edinburgh, 1890s. Crown copyright: RCAHMS, SC 465220.

Figure 29.6 British Linen Bank, 37 St Andrew Square, Edinburgh; interior of banking hall, 1890s. Crown copyright: RCAHMS, SC 465222.

headquarters to match these developments. The Royal Bank had bought Sir Laurence Dundas's house in 1825 and reconstructed the interior for bank purposes to designs by William Burn in 1836. This was now further reconstructed to produce a double-height atrium of superimposed orders similar to that of the Commercial Bank, but the telling room was an altogether more modern concept without columns. Its dome had a clear span of 62 feet 4 inches (19 metres) achieved by the use of wrought-iron ribs. Star-shaped glazed coffers produced an absolutely even light over the entire floor space.

Large city sub-head offices tended to follow a similar format of front office building and top-lit, single-storeyed telling room to the rear. Architecturally, these were no less ambitious than the head offices, but since board rooms and apartments for head office staff were not required, the upper floors tended to consist of a house for the agent, and in a very few of the larger buildings business chambers for rental with one or more separate entrances. The most notable buildings of this type in Edinburgh were the branches of the Western (1846) and Union (1874) Banks by David Bryce on St Andrew Square and George Street

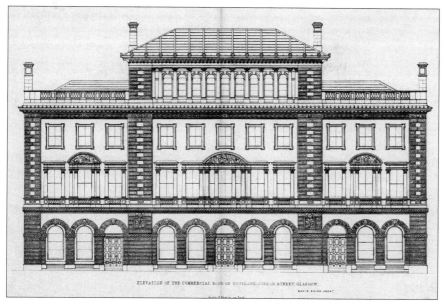

Figure 29.7 Royal Bank of Scotland (formerly Commercial Bank of Scotland), 8 Gordon Street, Glasgow; elevation as designed, 1854. Courtesy of Aonghus McKechnie.

respectively. In Glasgow, they comprised the British Linen Bank (1840) on Ingram Street by D & J Hamilton, the National Bank (1847) on Queen Street won in competition by the London architect, John Gibson, the Commercial Bank (1855) on Gordon Street (fig. 29.7) by David Rhind, and the Bank of Scotland (1869; fig. 29.8) on St Vincent Place by J T Rochead, and in Dundee the British Linen Bank (1854–8) by David Cousin. All of these were three-storeyed palazzi, five to nine bays wide, and generally similar in format except for three of the Glasgow buildings: the British Linen Bank, which had a boldly circled corner; the National Bank, which was two-storeyed with a rusticated façade of superimposed orders and set in a quadrangle of business chambers; and the Commercial Bank, the finest of them all, which had Farnese-pattern tripartite windows at first-floor level and a Corinthian-colonnaded attic floor.

In the north, a few exceptional buildings departed from the Edinburgh and Glasgow models. Archibald Simpson's original Town & County Bank (1826) on Union Street, Aberdeen, still followed the model of Burn's Aberdeen Bank, but his Graeco-Roman North of Scotland Bank (1840–2) on Castlegate, Aberdeen, was more imaginative. Its elevations were derived from Cockerell's London & Westminster Bank in London, but the obtuse-angled corner was skilfully turned by a quadrant portico of giant Corinthian columns. Internally, it had a frontal telling room with

Figure 29.8 Bank of Scotland, 2 St Vincent Place, Glasgow, c. 1890. Courtesy of Department of Architecture, University of Strathclyde.

superb Greek detail. In Inverness, Thomas Mackenzie's Caledonian Bank (1847) was Graeco-Roman Corinthian of the Rhind variety, but with the portico elevated to first-floor level as a decorative feature rather than as an entrance, as at Black's original Union Bank headquarters in Glasgow. More conventional, but certainly magnificent, was David Rhind's palazzo headquarters on St John Street in Perth for the Central Bank (1846), the bank house above being particularly palatial with a staircase as grand as a London club house.

Provincial branch bank offices of the 1840s–70s tended to follow the palazzo format on a smaller scale. William Burn's branch banks for the Bank of Scotland became subdued Italianate rather than neo-Greek at Stirling and Kirkcaldy, Fife (both 1833) and Dundee (1842). His successor, the contractor-architect James Smith of Darnick, last of the Master Masons to the Crown, followed similar classical-Italianate models at Perth (subsequently the Clydesdale Bank), but adopted a simple Georgian survival manner in the more villa-like branches at Airdrie, Lanarkshire, and Blairgowrie, Perthshire.

The Bank of Scotland's branch-building programme was paralleled by the British Linen Bank and the Commercial Bank. The British Linen

Figure 29.9 High Street, Irvine, Ayrshire, showing Royal Bank of Scotland, *c.* 1890. Crown copyright: RCAHMS, SC 714297.

employed George Angus at the large three-storeyed branch at Kirriemuir, Angus (1839), and the two-storeyed classical villa branch at Tain, Ross and Cromarty (1845), and after his death David Cousin, who designed the fine palazzi at Arbroath, Angus, and Moffat, Dumfriesshire. The Commercial Bank employed first Gillespie Graham and from about 1838 onwards David Rhind, architect of many provincial palazzi in a quattrocento or cinquecento manner, notably those at Ayr, Perth, Stirling, Peterhead and Kirkwall, Orkney. From 1855 onwards, and particularly after the collapse of the Western Bank in 1857 created a market, the Royal Bank employed Rhind's pupil, John Dick Peddie, to design a great number of very stylish palazzi in a style close to Rhind's, notably at Hawick, Roxburghshire, Duns, Berwickshire, Irvine (fig. 29.9), Ayr and Maybole, Ayrshire, Leith, Portobello, Edinburgh, and Montrose, Angus; and in 1858 Peddie's partner, Charles George Hood Kinnear, secured the patronage of the Bank of Scotland which commissioned similar, but rather plainer buildings.

The National Bank, with David MacGibbon as its architect, followed a consciously nationalist line with Scots Baronial branches at Alloa, Clackmannanshire, Falkirk, Stirlingshire, Forfar, Angus, and Montrose, the first three all 1862, the fourth 1864. These set a trend which was adopted by Peddie & Kinnear at a considerable number of branches for both the Royal Bank and the Bank of Scotland, especially on difficult corner sites.

In its early years the Union Bank employed Robert Black, architect of the fine branches built at Stirling and Moffat in the 1840s, and subsequently William Railton of Kilmarnock, Ayrshire, who designed the handsome Renaissance buildings at Dundee (1869) and Dumfries (1875). Railton also acted in a similar capacity for the rival Clydesdale Bank. His smaller branch offices, invariably Italianate for the Union Bank and more often Jacobean for the Clydesdale, are much less distinguished.

While the Caledonian Bank built a number of very smart branches, notably A & W Reid's neo-classical one at Elgin, Moray, and Mackenzie's palazzo at Forres, Moray (1854), bank design in the north of Scotland tended to remain very conservative. William Henderson built a great many plain, two-storeyed Georgian survival branches for the Union Bank (which had absorbed the Aberdeen Bank) and James Matthews a similar number of equally plain, manse-like structures for the North of Scotland Bank. This simple villa type was not unknown elsewhere, all of the banks occasionally providing villa-like structures for their smaller country-town and village branches, Peddie & Kinnear's Italian Gothic Bank of Scotland in the Square at Stonehaven, Kincardineshire (1862), being a particularly good example. Some were built in an even more informal L-plan, 'cottagey' villa form. Half of the ground floor was simply given over to bank purposes, the entrance vestibule being shared with the bank house.

INSURANCE OFFICES

Insurance head offices tended to follow the same format as those of the banks. Initially, the Edinburgh examples were on a smaller scale with three windows wide frontages adapted from Georgian houses, the prototype being Thomas Hamilton's pioneer Norwich Union Building (1821) in Princes Street as described above.

Eighteen years later, David Bryce's Corinthian Standard Life and Caledonian Insurance Buildings (1838–40) on George Street still followed the same model. His exactly contemporary and similarly-sized Insurance Company of Scotland Building on the same thoroughfare was, however, more economically astylar and set the style for the appreciably more ambitious five-bay palazzi of the North British & Mercantile on Princes Street and Edinburgh Life (now the Royal Society) on George Street (both 1843) in partnership with William Burn. Both of these followed the branch-bank formula of a frontal business hall between twin entrances, one for the company's office and one for the premises above. The finest buildings of the genre were David Rhind's Venetian Baroque Life Association Building (1855; fig. 29.10) on Princes Street, where the upper floors were rented as a hotel, and John Dick Peddie's Reform Club-like Scottish Provident Institute (1860) on St Andrew Square, both in Edinburgh.

The corresponding financial institutions in Glasgow of the same

Figure 29.10 Life Association of Scotland, 81–3 Princes Street, Edinburgh, 1966. Crown copyright: RCAHMS, SC 466092.

date tended to be housed in converted Blythswood town houses or in purpose-built blocks of less marked architectural character until the 1860s and 1870s, and did not reach their full flowering until the 1890s.

EXCHANGES

Although the Edinburgh exchanges had not been successful, there was a greater demand for such facilities in Glasgow. An exchange and public news-room was opened in 1770 and was quickly superseded in 1781 by a coffee room with similar provision in the Tontine Building in Trongate – the former Town Hall – the arcaded piazza of which had formed the merchants' shelter since 1740 (see above).

By 1827 the Tontine had been found inadequate and a New Exchange committee of merchants and ship-owners chaired by James Ewing of Strathleven bought the Cunninghame Mansion on Queen Street from the Royal Bank, then developing its garden as Royal Exchange Square. This committee commissioned David Hamilton to completely reconstruct and extend the mansion as the Royal Exchange at a cost of £60,000. In completely re-cased form, the original mansion provided merchants' and underwriters' offices, a deep octostyle Corinthian portico being added to provide the traditional open-air merchants' shelter. A large aisled, Corinthian-colonnaded news room, 130 feet (39.62 metres) long, 60 feet (18.29 metres) wide and 30 feet (9.14 metres) high, was added to the rear. At the time it was the grandest exchange building in the United Kingdom, London's not being rebuilt for more than a decade. As if that were not enough, the merchants built their giant Merchants' House on Hutcheson Street as a further symbol of mercantile power. Designed by Clarke & Bell as a related architectural unit to their immediately adjacent City & County Buildings, its nine-bay façade had a high-level Greek Corinthian colonnade with a richly sculptured frieze at the entablature, far grander than any bank portico. The interior contained a large hall and dining and committee rooms.

Prior to the building of the Royal Exchange, a Tobacco Exchange had been constructed in Virginia Street in 1819. Subsequently remodelled to an 1840-ish form as the Sugar Exchange, it consisted of a narrow, two-tiered hall of small offices, Ionic-colonnaded at the upper level with an auctioneer's box over the entrance. In 1875–7 a very large Stock Exchange was provided at the corner of Buchanan Street and St George's Place, with Early French Gothic elevations adapted by John Burnet senior from a portion of William Burges's London Law Courts competition design. It comprised two trading floors, a great hall 60 feet (18.29 metres) by 50 feet (15.24 metres) long and a clearing house on the upper floor, 80 feet (24.38 metres) by 50 feet (15.24 metres) with a glazed dome. As at the Royal Exchange, news room and telegraph facilities were provided.

Edinburgh and Aberdeen lacked comparable facilities, although a new trades' hall, Trinity Hall, was built in Aberdeen's Union Street to neo-Tudor designs by William Smith in 1846. When Edinburgh's Merchant Company decided to move from the Old Town to the New in 1879, it was content to buy David Bryce junior's City of Glasgow Bank in Hanover Street, completing his intended scheme as late as 1901. A large free Renaissance Stock Exchange designed by John McLachlan was

Figure 29.11 Royal Exchange, Dundee, c. 1880. Dundee University Archives.

eventually built in Edinburgh on North St David Street in 1890, replacing adapted buildings on the same site.

Similar provision to Glasgow's, if on a smaller scale, was, however, made in Dundee. In 1828 the Exchange Coffee Room was built to designs by George Smith. It provided a large hall, news room and library facilities over ground-floor shops, finely detailed with superimposed Greek Doric and Ionic orders. By 1855 this had been found inadequate and a much larger exchange was built in Albert Square (fig. 29.11), closer to the linen and jute manufacturers' and merchants' offices in Cowgate, Panmure Street and Meadowside. Designed with a tower in the manner of a late Flemish Gothic cloth hall by David Bryce, it comprised a large hammer-beam roofed news room hall over ground-floor offices. A matching terrace of merchant and stockbroking offices formed part of the project, fronting what was originally a lawn of un-enclosed market stances marked out with stakes. These were roofed over when the Merchants' Shelter was constructed in the same style in 1881, to designs by John Murray Robertson.

In the 1840s and 1850s a considerable number of corn exchanges were built on the model of those in England, some of them being financed by a combination of local agricultural interests. Among the earliest was Stirling (1838) but the first really big one was that in Edinburgh's Grassmarket, a magnificent Mannerist palazzo designed by the City Architect, David Cousin, in 1848. Behind the façade of offices was a spacious galleried hall with a well-detailed open timber roof providing market stances. It established his reputation for such buildings, excellent English

Figure 29.12 Edinburgh Corn Exchange, 35 Constitution Street, Leith. Crown copyright: RCAHMS, SC 617623.

neo-Tudor versions of the same formula being designed by him at Dalkeith, Midlothian (1853) and Kelso, Roxburghshire (1855), and, later, a Scots Jacobean version at Melrose (1863). That at Haddington, East Lothian (1853), by the local architect Francis Farquharson, was plainer, but with a similarly planned and constructed market hall. In Glasgow, Hugh Barclay built a magnificent corn exchange with a Corinthian portico on Hope Street in 1858, but as it was replaced as early as the 1890s, record of its roof structure is lacking. Built in the same year was the corn exchange at Dundee, designed by the local architect Charles Edward, who was also responsible for that at Arbroath (c. 1858). Both of these were in an Italianised Georgian survival manner: Arbroath was finer as architecture, but Dundee was remarkable for its wide-span arched iron roof like a railway train shed. Only slightly later were Leith (1860, figs 29.12–15), domed and arcaded classical by Peddie & Kinnear, Cupar, Fife (1861), with a fine tower by Campbell Douglas & Stevenson, and the rogue Baronial Hawick (1864), by J T Rochead. Most of these buildings played an important role in local public life, being regularly used for political rallies, public meetings and shows of various kinds; Dundee's, known as the Kinnaird Hall, was the city's only public hall until the 1920s.

Figure 29.13–15 Edinburgh Corn Exchange, 35 Constitution Street, Leith; sections of frieze. Crown copyright: RCAHMS, SC 617626, SC 617628, SC 617629.

BUSINESS CHAMBERS AND WAREHOUSES

Although Glasgow's Royal Exchange had been a tribute to the immense wealth of the city's merchants by 1830, the members of that community were at first relatively slow to spend as much on their offices and warehouses as on their residences, despite the ambitious proposals of James Adam for mercantile houses planned around a corn exchange in the early 1790s. A few purpose-built commercial blocks of that vintage are, or were, to be found in the eastern area of the centre, now known as the Merchant City. Many of these were terraced house or tenement-like structures in courts, as at Virginia Court (1817), South Exchange Court (1830) by Robert Black, and Princes Square (c. 1845) by John Baird. But by the time Royal Exchange Square had been planned in 1827-30 (originally by Archibald Elliot II but with subsequent involvement by Robert Black, John Thomson, David Hamilton and his son-in-law, James Smith), a three-storeyed format with pilastered ground-floor shops had been achieved, with spacious common stairs leading to suites of rooms at the upper levels. Hamilton's fine St George's Chambers at 151-7 Queen Street followed the same model, as did George Angus's ambitious Reform Street development (1834) in Dundee, where distyle in antis Ionic porticoes (two columns fronting a portico) distinguished the office entrances from those to the shops. Shop premises now had top-lit saloons extending to the rear, an arrangement then becoming a common feature of Edinburgh New Town houses where the ground floors had been converted to shops and offices. Still similar in general arrangement, despite their Parisian inspiration, were Bothwell Chambers (1849), 4-28 Bothwell Street, Glasgow, by Alexander Kirkland and John Bryce, where the ground floor became a Corinthian-pilastered arcade.

In general, the earliest blocks of business chambers and high-class warehouses in Glasgow, Paisley, Edinburgh and Dundee tended to be astylar classical, but in a few instances Baronial was adopted. The Baronial trend originated in Edinburgh, primarily as a result of the Edinburgh Improvement Act of 1827. This Act had specified old Scots or Flemish for the new streets in the Old Town, notably at Melbourne Place by George Smith, 1840 (partly domestic), and the earlier blocks on George IV Bridge, namely, the Highland Society Building by John Henderson, 1836, and Nos 21-5 of c. 1845. In Glasgow, J T Rochead designed the much more consciously Baronial City of Glasgow Bank Buildings (1855), 74-92 Trongate, a terrace of business chambers over shops, while J A Bell was responsible for Victoria Buildings (1858) on West Regent Street, built for the great merchant house of Orr Ewing. Neither had any remarkable features internally. By far the most original of all the early office blocks was David Cousin's 'Old Flemish' India Buildings (1864) on Victoria Street, Edinburgh, constructed on a difficult curved frontage site around a splendid domed atrium, circular on plan with gallery access to the suites of offices. Peddie & Kinnear's earlier Cockburn Street, a serpentine development of shops, hotels, business chambers, newspaper offices and domestic flats built for the Railway Access Company in 1855–64, falls into a rather more mixed category. Except for *The Scotsman* building (see below), none had anything very unusual in the internal arrangements, but the street was remarkable for the introduction of Baronial country-house features to the urban scene – most particularly since, in 1855, the practice had not yet begun the long series of houses which it was to build in the Baronial style.

Closely allied in architectural character to the early blocks of business chambers were the Glasgow textile warehouses for the supply of fine cloths, mainly on a wholesale basis. Because of the loss or subsequent reconstruction of the earliest examples, the origins of this type of building are now somewhat difficult to establish. But from the obituary of James Carswell in *The Glasgow Herald* of 25 February 1856 we learn that he and his brother William had an important role, the writer making a careful distinction between the sites which they built on their own account and those where they acted as contractor. They arrived in Glasgow from Kilmarnock in 1790, and had probably learned their business on the construction of cotton mills. They were described as the first to introduce cast-iron columned construction into Glasgow warehouse buildings, 'as well as cast-iron fronts and facings', although nothing answering the latter description survived long enough to be recorded. The account in *The Glasgow Herald* specifies almost the whole of Candleriggs as their work, including Commercial Court with its surrounding warehouses, which 'was the model on which so many courts have been built, by which business is conducted apart from the public street'. Other courts described as having been built by them included Albion Court, Ingram Court and Queen Court, together with much of Cochrane Street, Miller Street and George Street east of Montrose Street.

All of these were plain ashlar-fronted, four-storeyed buildings, but at Forbes Place in Paisley the shawl manufacturers built a street of really fine astylar classic terraces of offices and warehouses in 1835–8. By the mid 1840s, if not earlier, the Paisley example was being followed in Glasgow. As architecture, several of these buildings were as rich as the best bank buildings, expressing the prestige of the firms and the quality of their goods, and setting both the style and the standard for the city's subsequent commercial architecture.

The first of the really large and elaborate blocks appears to have been James Wylson's Canada Court (1848) on Queen Street. It was quickly followed by James Salmon's much more sophisticated warehouse at 81 Miller Street (1850) for Archibald McLellan, the great art collector, the adjacent Clapperton's at 61–3 (1854) by John Burnet senior, and just to

Figure 29.16 37–51 Miller Street, Glasgow. Crown copyright: RCAHMS, SC 710766.

the south at 37–51 the most remarkable of them all, Alexander Kirkland and James Russell's Tillie & Henderson's (1855; fig. 29.16), a Venetian Renaissance building with its court open to the street in order to secure the best possible light for the inspection of cloth. At Clapperton's, Burnet had adopted a different solution to the same problem by planning his building around a central cast-iron columned atrium with inspection shelving projecting into the well from the gallery balustrades. Much larger than any of these were MacDonald's Sewed Muslin warehouses on Hanover Street which, as built by the elder John Baird in 1854, originally extended into George Square and right across the street block into Queen Street. These had good Italian Renaissance elevations with a repetitive bay design, partly because further expansion was planned. Very different in style – but not in internal construction – from any of these were the Campbell Warehouses (1854) at 115–37 Ingram Street where, at the request of the clients, the English architect-historian, R W Billings, was brought in to redesign John Baird's Renaissance elevations in a toughly detailed Scots Baronial form.

In parallel with these large masonry structures there were a number of important experiments in which iron construction was extended from the internal structural frame to the façade itself, the aim being to get as much natural light as possible into deep-planned buildings. The first of these appears to have been the still-extant Kemp's at 37 Buchanan Street, Glasgow (1853), by James Boucher and James Cousland, architects closely associated with Walter Macfarlane's Saracen Foundry. It was immediately followed by 72 Jamaica Street, Glasgow (1854), probably the building designed by William Spence in that year. Both of these followed the same formula of superimposed colonnades of slim iron columns bearing girder entablatures with a balustraded parapet at the top, but the Jamaica Street building had a much longer frontage. In 1855 William Lochhead, the architect partner in the great department store firm of Wylie & Lochhead, applied the principle to department store design, dividing the façade of his Buchanan Street store into three bays of iron colonnades between masonry pilasters; and in the same year he built a long, four-storeyed warehouse block at 50–76 Union Street, Glasgow, with cast-iron colonnades between masonry piers of superimposed columns. Finally, and still in that same year, the elder John Baird and the ironfounder Robert McConnel designed the iron building at 36 Jamaica Street, Glasgow, in which the principles of the façades of the Crystal Palace of 1851 were applied to warehouse design.

Surprisingly, the very successful formula of the Baird building was not adopted elsewhere, although cast-iron façade experiments were to continue for more than a decade. At his iron-façaded warehouse of 1857 at 60–6 Jamaica Street, Glasgow, Hugh Barclay adopted a scheme of three giant semi-elliptical arches, while Baird's successor, James Thomson, made his iron façade at 217–21 Argyle Street (1862) seem like a very slim arcaded masonry structure. As at 36 Jamaica Street, McConnel's patent beams were adapted: these were made up of wrought-iron flanges and a cast-iron web which could be cut to size and assembled to the required length on site. Similar in concept to James Thomson's building, but with a masonry arcaded lower façade, was John Honeyman's F & J Smith warehouse (1872) on Gordon and Union Streets, which had Venetian Gothic-inspired tracery. By that date McConnel's patent was less in favour; Honeyman adopted a timber web between the flanges, while Peddie & Kinnear had begun using 'French beams' (or rolled-iron inverted U-beams with bottom flanges) at their City of Glasgow Assurance building on Renfield Street designed in the same year.

The iron-framed construction of these buildings had a considerable effect on the design of commercial buildings in Glasgow generally, and a number of large blocks with masonry façades were constructed on the same principle, their undivided floor space being used either as wholesale or retail warehouses or partitioned according to the needs of tenants. The most important of these were designed by Alexander Thomson who made extensive use of McConnel's patent beams. His first city block, 99–107 West Nile Street (1857), was a relatively small three-storeyed office building with ground-floor shops and details akin to his tenements, but 72–80 Gordon Street (1859), 1–11 Dunlop Street and 59–61 Argyle Street (1863), the Grecian building at 336–56 Sauchiehall Street (1867), the Egyptian Halls at 84–100 Union Street and the posthumously executed 17–23 Watson Street, 118–26 Bell Street and 39 Watson Street (designed 1873 or 1874, built 1876) were all iron-framed structures with ground-floor shops and pilastraded or tabernacle-framed elevations with eaves galleries. Like the West Nile Street building, Thomson's Cairney Building at 40–2 Bath Street (1860) appears to have been of a more cellular construction with heavy timber floors. Its façade followed the banded masonry and aedicule pattern of his best tenement blocks, but with a columnar eaves gallery and a glazed attic lighting Cairney's stained glass studios. The lower floors consisted of 'counting houses'.

OFFICE BUILDINGS

The great majority of the buildings specifically erected in Glasgow for office use were, as at the Cairney Building, still cellular in construction so that the principals could have coal fires. Most were speculatively-built blocks comprising suites of chambers for rental, accessed by common stairs, over ground-floor shops. Two of the earliest of such buildings were Charles Wilson's 144–52 Buchanan Street (1847), Italianate with rhythmically grouped round-arched windows, and perhaps the first really large such building, James Brown's severely astylar four-storeyed and attic block with quattrocento first-floor windows at 101–15 Union Street (1850).

Speculative office development did not, however, really take off until the 1870s when many of the original houses and tenements in Glasgow's Blythswood area were redeveloped. Among the earlier of these new buildings erected for rental were two blocks designed by Peddie &

Kinnear. The first was in St George's Place returning into West Nile Street, built in 1872 for the Scottish Lands & Building Company, one of the new property companies which had come into being as a result of the Limited Liability Act 1855 and the Companies Act 1862. It was a tall, simple but distinguished building of four storeys with a ground-floor restaurant for the adjoining Stock Exchange, Italianate in character with arched fenestration and a bracket cornice. Peddie & Kinnear's next block at 20–40 Gordon Street (1873) was altogether more ambitious with pilastraded elevations over ground-floor shop and bank premises. It was strongly influenced by the work of Alexander Thomson and was built for rental. A third block by the same firm at 13–15 Drury Street (1874) was an interesting early Italian design with arched recesses.

But the prime specialist in such buildings was James Thomson of Baird & Thomson, who was responsible for several blocks in Glasgow for the china and glass merchants D & J McDougall, built between 1873 and 1880; these were on St Vincent Street (101–03, 130–36), Buchanan Street (71–79), Union Street (72–82) and Argyle Street (26–34). Thomson also designed the even more extensive ranges of offices enclosing the Royalty Theatre on Sauchiehall Street. All of these were rather similar in design with Corinthian pilastered elevations over ground-floor shops.

More monumental examples without shops were James Sellars's fine Bank of Scotland Chambers at 24 George Square, Glasgow (1874), continuing the elevation of Rochead's bank referred to earlier, and 87–97 Bath Street, Glasgow, designed by Alexander Thomson just before his death in 1875. Such buildings now tended to be much deeper on L-, U- and H-plan arrangements, often with a block of similar height on the rear lane according to the opportunity presented by the site, the lightwells in the later examples being of white glazed brick. The accommodation within consisted typically of suites of rooms comprising a large public office with a counter, one or more partners' rooms and a safe, usually with shared water-closet facilities. Prior to the introduction of electric light in the 1890s, securing sufficient light in such buildings was a problem, alleviated by borrowed light from glazed partitions and etched glass doors.

In Edinburgh, the need for such accommodation had been largely provided by converting New Town flats and houses. The only substantial office developments subsequent to India Buildings were Thornton Shiells's late classic block at 9–16 George IV Bridge (1868) and Robert Raeburn's York Buildings in Queen Street (1878), both a comfortable three-storey and attics above street level as against Glasgow's four and five storeys; the former was built to provide accommodation adjacent to the courts and the latter to provide stockbrokers' chambers for the exchange immediately opposite on St David Street.

The situation was similar in Aberdeen where no new office building of any distinction was constructed for rental. But important developments took place in Dundee, where, under the Dundee Improvement Act of 1871, unified four-storeyed and attic street blocks of business chambers over tall cast-iron columned shop fronts capable of accommodating mezzanines were constructed on the model of Baron Haussman's Paris. These were mainly to designs by the Burgh (after 1888 City) Engineer, William Mackison, and the Edinburgh City Improvement Trust's architect, John Lessels. Three further important blocks on the new streets (Commercial Street, fig. 29.17, Whitehall Street, Whitehall Crescent and Victoria Road) were William Alexander's astylar Renaissance Victoria Chambers on Victoria Road, John Murray Robertson's Thomsonesque Greek India Buildings, Bell Street (both 1874 for linen and jute merchants) and the early French Renaissance Calcutta Buildings on lower Commercial Street (1877) by Maclaren & Aitken for ship owners.

These Dundee buildings differed from the Glasgow examples in being simple street blocks with top-lit shop or office saloons at the rear, and no lightwells. On plan they were little different from superior tenements in having closes leading to common stairs at the back, the higher levels being in some instances intended for domestic occupation from the beginning. Several of the Whitehall Street blocks were originally constructed wholly as mansion flats rather than as offices, again a reflection of lower demand and lower land values.

Figure 29.17 73–97 Commercial Street, Dundee, 1897. Dundee Central Library Photographic Collection.

HEAD OFFICES

Although the pure High Renaissance and neo-Roman formulae of the 1840s and 1850s continued to be employed for major bank offices down to the mid 1870s without any loss of quality or vitality, notably at Bryce's Union Bank (1874) in George Street, Edinburgh, and William Spence's Clydesdale Bank (1876) at Dundee, by the 1860s and 1870s Baroque and freestyle elements were being introduced into the design of some of the larger bank and company head office buildings, even if the overall character at first tended to remain either Italianate or Franco-Italianate. The first to reflect the change was James Matthews's severely rectangular Aberdeen Town & County Bank (1862) on Union Street, Aberdeen, which had non-academic freestyle elements and even touches of polychromy in its granite Corinthian pilastered elevations. In 1865–70 the Bank of Scotland recast its 1801 Edinburgh head office to Franco-Italian Baroque designs by David Bryce, its profile being enlivened with a raised attic, a broken pediment, a taller dome and airy cupolas, and the interior being similarly aggrandised with a double-height atrium and telling room to challenge its rivals in the New Town.

In Glasgow, John Burnet senior was responsible for two bank head-office developments which were stylistically innovative. The first of these was the Clydesdale Bank (1870–3) on St Vincent Place, won in competition. Although not free-standing and shorter in frontage, it echoed Bryce's remodelling of the Bank of Scotland head office in its preference for a more picturesque treatment, in contrast to the disciplined High Renaissance of Rochead's just-completed branch for the same bank a few doors to the east. Its telling room was, however, unusually spacious, a great domed cortile of arched openings and superimposed orders. The second was the reconstruction of the Union Bank (1875–7) on Ingram Street, a design of great subtlety and distinction in which Burnet returned to a more severe High Renaissance profile but with a richer ordonnance of superimposed orders and sculpture in a manner close to that of the London architect, Charles Robert Cockerell. As at the Town & County Bank, there were touches of polychromy in the red granite shafts of its columns and pilasters.

Had it been completed, the City of Glasgow Bank headquarters (1877–8) on Glassford Street by Campbell Douglas and James Sellars would have been the grandest of all the Scottish banks. It was not free-standing but its Corinthian colonnaded elevation was paralleled only at Clarke & Bell's Merchants' House on Hutcheson Street. As at Burnet's Clydesdale Bank, there were to have been picturesque pedimental gables of London 'Queen Anne' derivation on the skyline, but these were never constructed.

In the same period a number of important company headquarters buildings were commissioned. In Glasgow, the great shipping firm of J & C Burns built the Venetian Renaissance 30–4 Jamaica Street to designs by John Honeyman in 1864, and in 1880 a related company, Burns-Aitken, built a large club-house like Renaissance office on West George Street and Wellington Street to designs of John McLeod. A further exceptional building was the tall free Renaissance palazzo, in the London-Manchester rather than the Glasgow manner, built on St Enoch Square for the distiller William Teacher by James Boucher in 1875. In Dundee, two of the linen manufacturers, Don Brothers and Cox Brothers, erected impressive headquarters buildings, both to the designs of James Maclaren. Don's (1873) was on Barrack Street, and Cox's (1886) on Meadowside, both cleverly planned on awkwardly angled sites.

In Inverness, the Highland Railway erected a particularly handsome astylar Renaissance headquarters building on Academy Street, by Matthews & Lawrie (1873–5). It was the first Scottish railway company to make adequate centralised provision for its headquarters staff, closely followed in 1877 by the Glasgow & South Western Railway as part of its St Pancras-like St Enoch Hotel development (1876–9), by the London architect, T J Willson. The Great North of Scotland Railway's offices on Guild Street, Aberdeen, by Ellis & Wilson, followed in 1881. The Caledonian Railway's headquarters in Glasgow, planned in 1877 as part of its Central Station development, would have been the most ambitious, but of that more will be described below in a different context.

PUBLISHING AND PRINTING PREMISES

From the middle of the nineteenth century onwards, the leading Scottish newspapers began to be significant patrons of architecture. In 1859 Sir John Leng of *The Dundee Advertiser* built a neat two-storeyed astylar Renaissance editorial office and printing works on Bank Street to the designs of Charles Edward, progressively enlarged by Charles and Leslie Ower to a vast palazzo of four storeys and twenty-five bays to accommodate Leng's other journals. In similar vein was the original three-storeyed office (1874) of *The Dundee Courier* on Lindsay Street, designed for the proprietor Charles Alexander by his son William.

In Edinburgh, *The Scotsman* built new early Flemish Gothic offices as part of Peddie & Kinnear's Cockburn Street scheme in 1860. It had a tall twin-gabled front of five storeys proudly bearing its masthead as a sculptured frieze. Larger than either of the Dundee examples, it had its own telegraphic link to London, and included a large library, a type foundry, a composing room 150 feet (45.72 metres) long and a machine shop 80 by 40 feet (24.38 by 12.19 metres), containing three Walter presses which printed and folded 36,000 copies per hour. *The Scotsman* building was quickly followed by two further large newspaper offices built adjacent to one another in St Giles Street, that of *The Edinburgh Courant* at 12–18, and that of *The Daily Review* at 20–4, both by David Bryce and both four-storeyed and attic with machine shops in the basements beneath. In style they were Scots neo-Jacobean, with twin pepper-pot angle turrets at the *Review* gable towering high above Market Street.

In 1870–2 *The Glasgow Herald* followed with equally substantial new buildings, their Glasgow offices and machine shops being designed by John Baird II in 1870, and their Edinburgh office (1872), which formed the lower part of a tenement by Thomas Gibson adjacent to the *Courant* office in St Giles Street. Its fine galleried interior is still extant. By 1879 the Baird offices in Glasgow had been found inadequate and a very handsome Renaissance structure at 65 Buchanan Street was constructed to designs by Campbell Douglas & Sellars, drawing the style of its superstructure from their City of Glasgow Bank (see above). Equally prominently sited, although not of the same order of merit, was *The Aberdeen Daily Free Press* building (1887) on Union Street, Aberdeen, by Ellis & Wilson, in their somewhat odd free style.

These were, of course, all exceptional buildings for city newspapers which provided a national and international as well as a local news service, and had large print-runs. Beyond these, nearly every provincial town, or group of towns, however small, had one, if not two newspapers primarily concerned with local news and advertising which were published once or perhaps twice a week. Several had buildings of some architectural pretension on prominent sites, such as the four-storeyed Govan Press building (1888) at 577 Govan Road, Glasgow, by Frank Stirrat, and the *Oban Times* building (1883) on the Esplanade at Oban, Argyll, by J Fraser Sim. Most, however, were located in closes or courts as at Montrose and Ayr, or were ordinary shop premises with printing shops at the back. Robert Smail's at Innerleithen, Peeblesshire, which published *The St Ronan's Standard* with a print-run of 800 copies, survives almost intact to illustrate the latter type, having a white brick single-storeyed machine shop with an attic case room to the rear, built astride a mill lade to power a flatbed letterpress machine. In that respect it was exceptional; most later nineteenth- and early twentieth-century printing shops were equipped with small steam, oil or gas engines, but in the smaller burghs such as Kirriemuir hand typesetting and even hand presses survived until just after World War II.

Newspapers apart, publishing and printing tended to be in lower-rise buildings, with frontal offices of two to three storeys and single-storeyed machine shops to the rear. In Edinburgh, much the grandest of these was John Lessels's Parkside Works (1871) for Thomas Nelson & Sons, free early Renaissance with a shapely tower. Also of some pretension was R & R Clark's (1883) on Brandon Street by John Chesser, neo-Jacobean and more like a genteel endowed school than a printing works. W & A K Johnston's Edina Works (1878) was also architecturally treated, but more functionally so, in pilastered red and yellow brick and terracotta. In Glasgow, the most architectural publishing and printing premises were those of Blackie at Villafield Works, Stanhope Street, begun c. 1830, where the peripheral buildings, mainly the work of Alexander Thomson (1869) and his chief draughtsman Alexander Skirving (1883, 1896), were Greek, and rose as high as five storeys. Still larger but less architectural was Collins's Herriot Hill works, begun c. 1861, with four-storeyed and attic peripheral buildings.

The basic arrangement of frontal offices and studios with machine shops and warehousing to the rear continued into the twentieth century, notably at Gordon & Dobson's Art Nouveau publishing offices and printing works (1901) at Darnley Street, Glasgow, for Miller & Lang, and Harry Ramsay Taylor's Geographical Institute (1909) in Duncan Street, Edinburgh, for Bartholomew's, neo-classical with a two-tier portico re-used from Thomas Hamilton's Falcon Hall nearby.

TALL OFFICES

From the mid 1870s, buildings higher than the prevailing three- to four-storey and attic scale had been permitted in Glasgow as a result of improved fire-fighting equipment. A total height of 100 feet (30.48 metres) was now possible, James Boucher's five-storeyed and attic Arthur Warehouse (1875) at the corner of Ingram and Miller Streets being among the first, along with the St Enoch Station and the *Glasgow Herald* buildings referred to above. Similar changes were made in Edinburgh's regulations, but at first the higher scale tended to be adopted only for hotels on prime city centre sites because of the practical limitations of elevators at that date.

Apart from the Glasgow & South Western Railway's combined hotel and office buildings at St Enoch's, which were five-storeyed and attic elevated on cellarage, the only really large office building in Glasgow to adopt the new scale in the 1870s was the Caledonian Railway's Central Station buildings, somewhat surprisingly designed by the Edinburgh architect, Robert Rowand Anderson, in 1878–80 in a refined amalgam of early Italian and Northern European Renaissance styles, five-storeyed and attics high with spinal corridors. It stood an attic storey higher than St Enoch's and had fireproof floors of concrete on malleable iron joists throughout, whereas at St Enoch's only the lower floors were fireproof. Stylistically, it was the first Renaissance building in Glasgow to break with the prevailing Italianate and Franco-Italianate idioms, a break which became even more marked when, in 1880, it was decided to complete it as a hotel rather than company offices as originally planned. In scale, the completed building immediately exceeded that of Peddie & Kinnear's recently completed Blythswoodholm Hotel (1875–9), just opposite at 91–115 Hope Street. Occupying a complete street block, the Blythswoodholm followed the London hotel pattern of capacious French roofs to add attic floors to the accommodation, and it included stairs of business chambers on its Bothwell and Waterloo Street frontages. The Blythswoodholm subsequently became the first really large office block in the city when the hotel was converted to Central Chambers by J M Dick Peddie in 1890.

The response of the banks, insurance companies and other institutions to these developments varied according to location, particularly in respect of land values. In 1874 the Merchants' House, expropriated from their Hutcheson Street building by the requirements of the Sheriff Court, built a new house at the corner of George Square and West George Street to the designs of John Burnet senior as a continuation of the Rochead-Sellars Bank of Scotland development, making its principal accent the fine domed corner tower. The most impressive institutional building of the 1880s was again from the Burnet firm, his son John James's giant French Beaux Arts Clyde Trust Building (1883) on Glasgow's Robertson Street, designed like a bank head office with magnificent board and committee rooms at the upper floors and a top-lit telling room at the rear. Had it been completed as planned, its façade would have rivalled that of the City Chambers in extent as well as in grandeur. Also French, but of a very different school with Thomsonesque details, were the amazingly rich façades of the Scottish Legal Life's building (1884) at the corner of Wilson and Virginia Streets in the same city, by Alexander Skirving. Although in an obscure location, it was one of the earliest city centre blocks to be built of red sandstone and its tall domed and circled corner (the original dome was pointed in profile) was an early instance of a treatment widely adopted a few years later.

A similar concern to achieve the highest standards of detailing was to be seen at A Marshall Mackenzie's granite Northern Assurance Building (1885) at the corner of Union Street and Union Terrace in Aberdeen, again with a boldly circled corner, and at J M Dick Peddie's much more conservative Bank of Scotland palazzo (1884) at 101–3 George Street, Edinburgh, both of which departed from the old formula in having splendid business halls at the front of the building rather than at the rear. Both the Northern and the Bank of Scotland adhered to the three-storeyed and basement scale traditional for such buildings. In Edinburgh, where land values had soared only on Princes Street, that scale remained the norm for bank and insurance offices until well into the next century, as at J M Dick Peddie and George Washington Browne's three-storeyed Aesthetic Movement neo-Georgian Standard Life (1897), George Street, the three-storeyed basement and attic neo-Jacobean

Scottish Equitable (1899), St Andrew Square, and the four-storeyed attic and basement neo-Baroque North British & Mercantile (1903), Princes Street. As late as 1908 the domed Roman Ionic Edinburgh Life, George Street and Hanover Street, also by J M Dick Peddie but now working with his future partner J Forbes Smith, was of only three very tall storeys with a fourth concealed behind the parapet. Browne's four-storeyed François Premier British Linen Bank branch (1905) on the George Street-Frederick Street corner was also of moderate height but, as originally built, included a shop and business chambers for rental.

In central Glasgow land values were very much higher, although nearly a decade was to elapse before the advances made in the design and construction of tall buildings in the late 1870s could be fully exploited. The City of Glasgow Bank crash in 1878, the full effects of which were not felt until 1881, resulted in a general drying up of capital which lasted for nearly a decade and brought the property investment companies into particular difficulty. Greater attention was, however, given to fire-proofing, particularly after the disastrous Wylie & Lochhead and Fraser department store fires, and hollow tile and early patent concrete floors were widely adopted for the larger and more expensive buildings. Elevators remained a serious practical limitation as hydraulic lifts which ran on water mains were slow, and only the largest buildings could justify the running costs of the small steam or gas engines and the high-pressure hydraulic gear required to achieve a quicker ascent.

But in 1889–90 the five- to six-storeyed scale of the Caledonian Railway's Central Hotel became the prevailing height for new commercial buildings and, as in London, early Renaissance became the predominant style. The first and largest of this new generation of office blocks was William Leiper's richly sculpted François Premier Sun Life building at 117–21 West George Street, Glasgow, designed in 1889 but not built until 1892–4. It was six storeys high and had an Otis electric elevator, probably the first in Glasgow. Its splayed angle rose into a corbelled and domed octagonal angle tower, which together with that of Sydney Mitchell's Commercial Bank extension (1888) at the corner of Gordon Street and Buchanan Street, inaugurated a fashion in Glasgow for towered corner treatments which lasted into the second decade of the twentieth century.

The Sun Life building was paralleled by two further insurance blocks, the Prudential buildings (1890) at the corner of West Nile and West Regent Streets in Glasgow, and in the corner of St Andrew Square and St Andrew Street, Edinburgh (1892), both by the London master of the new early Renaissance vogue, Alfred Waterhouse, and, also, by the *Evening Citizen* newspaper building (1889) on St Vincent Place in Glasgow, sophisticated early Italian at the lower floors and Netherlandish above, designed by Waterhouse's former assistant, the Glasgow architect Thomas Lennox Watson. All three of these were of the red sandstone from

Dumfriesshire rather than the local buff sandstones, and all were, in various roof forms, five storeys and attics high with elevators, the *Evening Citizen* having hydraulic passenger and freight lifts installed by Otis.

Still higher was yet another newspaper building, the Mitchell Street section of the *Glasgow Herald* complex built in 1893–5 to designs by Charles Rennie Mackintosh, then an assistant in the offices of Honeyman & Keppie. At six full storeys it rose slightly higher than its immediate predecessors and, although gabled Scots Renaissance at the top, its second and third floors – originally warehouses for rent – showed marked early modern tendencies in having oblong windows. These, although still timber-sashed, were the predecessors of a type which in steel casement form was to become the norm for lane elevations. The *Glasgow Herald* building was also unique in being equipped with a system which could spray the façade with water in the event of a fire in the buildings opposite, Mitchell Street being unusually narrow at this point, no wider than many of Glasgow's lanes.

Tall buildings on confined sites such as that of the *Glasgow Herald* had become more practicable since the introduction of a municipal electricity supply for lighting in 1893. Electric light also alleviated the problems of the deeper-plan office blocks built around confined lightwells, but it was not until the construction of Glasgow's new generating station at Port Dundas in 1898 that sufficient power for electric elevators became readily available. A few of the financial institutions and one or two of the more adventurous property developers had anticipated what was to come, the pace being set by the first really large purpose-built elevator building composed entirely of business chambers for rent, Alexander Petrie's 40–60 St Enoch Square of 1895. Most waited until Port Dundas was safely under way and in 1896–1900 tall elevator blocks mushroomed over the city, while many existing blocks had two to three storeys added to help justify the cost of installation.

In 1896–7 the Pearl Insurance Company built a very large six-storeyed, free Renaissance block at the corner of West George Street and Renfield Street, Glasgow, to designs by James Thomson, quickly followed by an even more enormous and richer block by the same architect, the German Renaissance Liverpool, London & Globe building at the corner of St Vincent Street and Hope Street. It rose still higher to eight storeys in a flurry of columned gables. These insurance buildings were little different from those built wholly for rental, the institution occupying only the prime floor space at the lower levels. They set the scale for those of the home-based institutions, a reflection of the higher land values and the greater demand for office suites in Glasgow, but the Glasgow- and Edinburgh-based institutions tended to be more discriminating in their choice of architect and of style than the London ones.

In 1898 the National Bank held a competition for its new building at the corner of St Vincent and Buchanan Streets in Glasgow, far larger than their Edinburgh head office which still consisted of converted Georgian houses. It was won by J M Dick Peddie and George Washington Browne with a scheme for a vast five-storeyed Baroque palazzo of business chambers enclosing a central banking hall, a row of shops being included on the prime retail frontage to Buchanan Street. In 1906, Peddie and Forbes Smith designed the even taller, French-roofed Scottish Provident Institution at 17–29 St Vincent Place, which rose to a height of six storeys, basement and attic at the front, and a full seven storeys at the rear. It had no shops, but adopted the same formula of a two-storeyed rusticated podium bearing a giant order which encompassed three storeys rather than two. At the Edinburgh Life building (1904) at 122–8 St Vincent Street, John Archibald Campbell followed a similar formula on a shorter frontage with a twin-towered superstructure achieving a height of seven storeys, but at his Northern Building (1908), a little to the east at 84–94 St Vincent Street, the classical conventions of rusticated base and giant order of columns or pilasters were entirely dispensed with in favour of a soaring Portland stone façade, the first in Glasgow, articulated by three canted bay towers breaking a deep cornice and achieving a total height of eight storeys. Both buildings were steel-framed with ground-floor shops and had remarkable rear elevations of white glazed brick with steel casement windows manufactured by Henry Hope of Birmingham, an old-established firm which had originally specialised in conservatories.

All of these buildings had been on large sites previously occupied by several house blocks, but in Glasgow the demand for office accommodation was such that some of the new elevator buildings, particularly those built for rental, were on single house-plots only 30–33 feet (9.14–10.06 metres) wide but over 100 feet (30.48 metres) deep. The best locations for slim buildings of this type were corner sites, as at Burnet & Boston's (1900), 140–2 St Vincent Street, which had a magnificent frontage to Hope Street, or adjacent to a lane, as at Robert Thomson's North British Rubber Company building (1898) at 62 Buchanan Street, and J A Campbell's Britannia Building (also 1898) at 166–8 Buchanan Street. The North British Rubber Company's very showy balconied lane elevation was is in stone, rather than the Britannia's plain brickwork.

These sites posed no particularly difficult problems of construction, lighting or ventilation, but where the building was on a site confined by other buildings on both sides, as at James Salmon Junior's Art Nouveau sculpturesque 'Hatrack' building (1899) at 142a–4 St Vincent Street, considerable ingenuity in planning and construction was required. At the Hatrack the walls were kept to minimum thickness by cantilevering

Figure 29.18 Lion Chambers, 170–2 Hope Street, Glasgow. Crown copyright: RCAHMS, SC 710973).

the floors in steel off a central double row of columns, ventilation, if not much light, being provided by the lightwells formed by its elongated dumb-bell plan. Even more adventurous was Lion Chambers (1904–6; fig. 29.18) at 170 Hope Street by the same architect and his older partner, John Gaff Gillespie, where paper-thin concrete construction by the Yorkshire Hennebique Company enabled eight storeys to be reached on a site only 33 by 46 feet (10.06 by 14.02 metres).

Most of the new office blocks built for rental were, however, on sites previously occupied by large buildings, or were assembled by buying and demolishing two or more houses. Petrie's St Enoch Square block was quickly followed by Mercantile Chambers (1897) on Bothwell Street, again the work of James Salmon junior. It had a rear elevation of close-packed canted bays of steel casements manufactured by Henry Hope, which set the future pattern of lane fenestration in buildings of this size. Its front elevation was remarkable for the originality and quality of its detail, the sculpture being the work of Derwent Wood. Still larger was Duncan McNaughtan's Baltic Chambers (1899) on Wellington Street, a neo-Baroque block which extended around a quadrangle into Cadogan Street and Holm Street.

More remarkable in design than either of these were the office blocks of the Beaux Arts-trained architects, John James Burnet, who had visited the United States in 1896, and J A Campbell, with whom he had broken partnership in 1897. In 1899 Burnet constructed two seven-storeyed blocks, Atlantic Chambers at 43–7 Hope Street, a dumb-bell plan building with the services concentrated at the core, and Waterloo Chambers at 15–23 Waterloo Street (fig. 29.19), which was planned around a galleried staircase atrium on the model of the larger American office blocks of the time. At Atlantic Chambers the elevations were relatively simple with colonnaded eaves galleries at the top, the rear elevation to Cadogan Street following the Chicago pattern of canted bays between pilaster strips. At Waterloo Chambers the elevation was much more elaborate with canted bays set in a salient giant order of Ionic columns and an eaves gallery framed in pylon towerlets.

Considerable ingenuity was required to stretch the conventions of classical architecture to the heights now needed. The problem was often answered by a giant order of columns or pilasters rising above a plainer lower façade, as if one building had been superimposed on another – a solution widely adopted by American Beaux Arts architects at the time – as at three buildings by James Miller, all seven-storeyed: the giant Caledonian Chambers (1901), mainly railway offices, at 75–95 Union Street; 136–48 Queen Street (1902); and the palatial white faience Anchor Shipping offices (1906–7) on St Vincent Place. Frank Burnet, Boston & Carruthers's Gordon Chambers (1903–5) at 87–94 Mitchell Street and J A Campbell's mighty block at the corner of Hope Street (Nos 157–67) and West George Street (Nos 169–75), where the two uppermost floors of its eight storeys were set in deep arched recesses, represented more original examples of the same formula, the latter being the finest of all the turn-of-the-century office blocks. Like most office buildings of that date it was still not regularly steel-framed due to the requirement to provide flues for the coal fires in the principals' rooms.

No commercial block for rent comparable in scale with these was erected in Edinburgh. The nearest equivalents were the Baroque Gresham

749. *Waterloo Chambers, Glasgow*, JOHN JAMES BURNET, A.R.S.A., Architect.
(For Plans, *see* p. 104.)

Figure 29.19 Waterloo Chambers, 15–23 Waterloo Street, Glasgow; perspective view by J J Burnet in *Academy Architecture*, 1899, 115.

House (1908) by T P Marwick at the corner of George and Frederick Streets, and the adjacent Victoria Chambers (1903) at 40–2 Frederick Street by Dunn & Findlay. The only Edinburgh development comparable in grandeur with the Glasgow examples was the comprehensive redevelopment of North Bridge, necessitated by the rebuilding of the bridge itself in 1894–7. A competition won by J N Scott and J A Williamson in 1896 determined the general outline, but on the east side the executed buildings were designed by Sydney Mitchell (southern section, built by the Commercial Bank) and W Hamilton Beattie (Carlton Hotel) in 1898, all modelled stylistically on Burnet's competition design for the North British Hotel with ogee-roofed corner turrets. The west side was entirely developed by J R Findlay of *The Scotsman* newspaper with his son's firm, Dunn & Findlay, as architects in 1899–1902, the magnificently sculptured neo-Baroque north end forming the new headquarters of the newspaper itself with machine shops and loading bays in the basements beneath. The communication of copy between departments in this huge eight- and nine-storeyed structure was achieved by Lamson patent pneumatic tubes, much faster and cheaper than any internal messenger service. At street level, an angled and domed shopping arcade led through to Cockburn Street and a Blois-type staircase reached from the north terrace down to Market Street, where a further six-storeyed and attic block provided new premises for the displaced fruit and vegetable market.

Four further sizeable newspaper buildings were constructed in Glasgow at the turn of the century, though none was as large as that of *The Scotsman*. The first of these was the North British *Daily Mail*'s six-storeyed but somewhat coarsely detailed block (1898) at 102–14 Union Street, designed by George Bell II of Clarke & Bell. An altogether better building was Robert Thomson's *Evening News* (1899) at 65–7 Hope Street, seven-storeyed and attic in similar vein to his North British Rubber building, but with a wider double-arched frontage unified into a tripartite gable over a serpentine balcony at fifth-floor level. Still better, if very much less expensive, was Charles Rennie Mackintosh's *Daily Record* building (1901) in Renfield Lane, a small variant of his earlier *Glasgow Herald* building, stone-faced at the ground and top floors and glazed white brick, patterned in coloured tiles, at the pilastered intermediate floors.

Although D C Thomson & Company's printing and publishing works at Port Dundas, Glasgow, designed by the Dundee civil engineer, Robert Gibson, in 1906, were unremarkable, their headquarters Courier Building at Dundee was one of the finest of the age, designed by the London-Scottish firm of Niven & Wigglesworth after a study tour of the American publishing empire of William Randolph Hearst in 1902. Like Mackintosh's *Daily Record*, its regularly framed five-storeyed and basement structure with boldly arcaded ground floor, pilastered upper floors of tripartite and canted bays, and massive crowning cornice was very much twentieth century rather than fin-de-siècle as at *The Scotsman* and *The Daily Record*. Another equally remarkable multi-storeyed building devoted to printing and publishing was Aird & Coghill's (1898) at 58–9 Cadogan Street, Glasgow, by H E Clifford, which had rippling Art Nouveau façades over a regularly framed fireproof structure with mahogany-faced floors.

The last decade of the nineteenth century and the first of the twentieth also saw the building of some remarkable, and very architectural, warehouses in Glasgow. These were unparalleled in the other Scottish cities, their nearest counterparts being in the English Midlands and the United States. There were two major programmes. The first was in Kingston where the Scottish Co-operative Wholesale Society (SCWS) had established itself from 1872 with warehouses which became progressively taller and more elaborate with each decade, culminating in the building of the colossal Second Empire five-storeyed and attic 95 Morrison Street (1892–7) by Bruce & Hay which, however much they denied it, was probably a re-working of their Municipal Buildings competition design of 1882, domed and rich in statuary representing Light, Liberty and the Four Continents. A marble vestibule and stair led to plain industrial floors of old-fashioned cast-iron columns and timber beams allocated to warehousing, workshops and offices. Of the similar (and several unfinished) warehouses of which it was the nucleus, the only survivor is the French-pavilion roofed 71 Morrison Street (1919–33, but probably designed earlier) by James Ferrigan, a SCWS employee who had previously been a leading draughtsman with Honeyman, Keppie & Mackintosh.

The second major programme was in the city centre and was a municipal initiative undertaken through the provisions of the City Improvement Acts of 1866 and 1897. It centred on Glasgow Cross and Trongate and consisted of five-storeyed and attic flatted warehouses built by the City Improvement Trust for rental, a development made practicable by the introduction of high-pressure hydraulic power mains for heavy goods elevators and cranes in 1895. The most impressive blocks in the southern area (King, Osborne and Parnie Streets) were the subject of a competition held in 1893 and won by John McKissack & Son with a Free Renaissance scheme featuring bold semi-elliptically arched top floors and tall Flemish gables. In 1904 designs by Thomson & Sandilands for the northern area, comprising the east side of Nelson Street, the north side of Trongate and parts of Bell Street, were settled upon, tall quiet Beaux Arts with a magnificent Baroque corner dome on Trongate, a further block (1916) on High Street being built by the City Engineer's Department. The City Improvement Trust scheme was eventually completed after World War I by the quadrant at Glasgow Cross, this representing only the western half of what was intended in the competition won by A Graham Henderson in 1914.

Many of the other high-rise warehouse and flatted factory developments made practicable by Glasgow's hydraulic power main service were also grandly architectural. By far the biggest was Hunter Barr's at 45–57 Queen Street which swallowed up Gibson's National Bank and its associated square. Designed by David Barclay in 1903, Hunter Barr's was six-storeyed, iron- and steel-framed, with arched hollow tile floors and a reticulated façade with over-scaled details. Even more monumental was Stewart & Macdonald's (1900–3) at 134–56 Argyle Street by H K Bromhead, which picked up the giant Corinthian order of William Spence's Macdonald department store on Buchanan Street and superimposed a second giant order on top of it. Stewart & Macdonald had already built another tall elevator clothing factory (1898) in an equally old-fashioned style, Billings-type neo-Jacobean, at the corner of Ingram Street and Montrose Street to designs by James Thomson. The other examples of the genre tended to be slimmer, though equally tall. Among the more interesting were Robertson & Dobbie's quite original Art Nouveau corner block (1908) at 61–5 Glassford Street and H E Clifford's clothing factory (1900) at the corner of Ingram and Albion Streets, an example of the American-inspired mullioned façade. By far the finest example of this very architectural city-centre industrial building-type was J J Burnet's McGeoch ironmongery warehouse (1904–5) at 28 West Campbell Street, a magnificent six-storeyed and basement structure with a regular steel frame and patent concrete floors. Its mullioned grid of windows drew inspiration from the Chicago architect, Louis Sullivan, but were detailed in a characteristically robust Burnetian Baroque.

At the turn of the century, both the Royal Bank and the British Linen Bank undertook a large programme of branch building, mainly to designs by Peddie & Browne. A few, like the Royal's branches (1909) in Ayr and Alloa, Clackmannanshire, High Streets, were refined English Baroque or Palladian, while that in Dundee High Street (1899) was a huge free Renaissance block on the Glasgow scale with ground-floor shops, a domed banking hall to the rear and an elevator to the business chambers above. The great majority were more modest, elegantly François Premier or English neo-Jacobean in style, on corner sites wherever available. They had business chambers, bank house or tenement flats at their upper floors according to local conditions, bank agents being now gradually replaced by managers wholly employed on the business of the bank. In Glasgow, the British Linen's work was in the hands of Salmon Son & Gillespie, who, in addition to transforming the local head office into an elevator building, built richly sculptured Art Nouveau branches (1899) at Govan Cross and 162 Gorbals Street, both with tenement flats above.

Essentially similar to these were the numerous branches of the Glasgow Savings Bank, Salmon Son & Gillespie being responsible for the even more finely sculptured example at Anderston, also of 1899. The bank spread its patronage over several firms, a particular favourite being that of Neil Campbell Duff, architect of the tall domed gusset block (1906) on New City Road, but for the extension of their head office at the corner of Ingram Street and Glassford Street the trustees turned to John James Burnet, son of the architect of the original headquarters building (1865) on Glassford Street. In 1894–7 he added a very American-looking colonnaded top floor to his father's design and built a new banking hall on Ingram Street, a superb domed design with sculpture by George Frampton and an interior rich in marble and mahogany.

The Dundee Savings Bank stuck by a single firm, Johnston & Baxter, which built a Baronial branch at Hilltown and neo-Baroque branches at Lochee (1906) and Albert Street (1914). So did the Aberdeen Savings Bank, which employed William Kelly, winner of the competition for its refined head office palazzo (1895) on Union Terrace. In their branch offices both the Dundee and Aberdeen savings banks preferred to avoid the complications of bank houses, business chambers or tenement flats on the floors above, building smaller structures of one or two storeys exclusively for bank use.

INTER-WAR DEVELOPMENTS

The provision of office space in Glasgow in the period 1880–1910 had been on such a colossal scale that no significant speculative office development took place there between the wars. In Edinburgh, where New Town houses continued to meet the requirement as domestic occupation declined, only one large block of offices was built, the tall Art Deco Lothian House on Lothian Road, designed by Stewart Kaye and including a super cinema, now demolished, by the specialist W R Glen. Opened in 1938, its elevations still conformed to the model of Burnet's Kodak building in London.

The major city blocks of the period were thus mainly built by the financial institutions. In Glasgow, Burnet & Boston's tall Royal Exchange Assurance Building (1911–13) at 91–3 West George Street had been the last to be built in the original Glasgow J J Burnet/J A Campbell Baroque idiom. The style of the new era was firmly set by Alexander Hislop's Phoenix Assurance Building (1912) at 78 St Vincent Street, Glasgow, a slim elegant corner building of six floors of offices over a double-height Roman Doric columned public office at street-level, an American solution to the problems of classical design in tall city blocks which seems to have had its origin in McKim, Meade & White's reconstruction of New York's Merchants' Exchange as the National City Bank in 1904–10. It was severely rectangular, expressing its steel frame; that familiar Glasgow feature, the circled or towered corner, no longer formed part of the architectural vocabulary.

However, it was not the pioneer Hislop but James Miller who became the premier architect of large city blocks in the inter-war years, partly by competition and partly by direct commission. He adopted the new idiom in 1920 at his giant, but uncompleted, McLaren warehouse, an eight-storeyed block at the corner of George Square and Hanover Street. His superb Union Bank of Scotland (1924–7), 110–20 St Vincent Street, Woodhouse Warehouse (1928, subsequently converted to the Prudential Building), 28–36 Renfield Street, J & P Coats office (1928), 155 St Vincent Street, and the Art Deco classical Portland-stone Commercial Bank (1931), 92–8 West George Street, all in Glasgow, conform to the same pattern of double-height giant order base and four to five floors of offices above, those of the banks being set in an upper pilastrade with metal-framed windows and spandrels. A Graham Henderson's Bank of Scotland (1929–31), 235 Sauchiehall Street, was a shorter reversal of the same formula, having an upper pilastrade on a solid two-storeyed base with tall arched ground-floor windows.

In a few instances the provision of upper floors for rent was eliminated, resulting in tall cubic temple-like structures with giant orders of columns screening double-height banking halls and just one or two floors of bank offices above, again a type to be found in the United States. Of these, the first was James Miller's Portland-stone Commercial Bank (1934) on Bothwell Street, Glasgow, followed by the same bank's Union Street branch (1936) in Aberdeen by Jenkins & Marr, both Corinthian columned, and Oldrieve, Bell & Paterson's Ionic Trustee Savings Bank (1938–9) in Hanover Street, Edinburgh, which had a very American blend of refined Greek and early Italian Renaissance detail. On a much more modest scale the same pattern was to be found in many suburban branch banks, the practice of having bank house or tenement flats above the bank premises being universally given up after World War I.

Insurance buildings tended to follow similar patterns to the banks. The earlier examples were all in Glasgow. J J Burnet's North British & Mercantile (1925–7) at 200 St Vincent Street, Glasgow, with five storeys of office floors over a powerfully arcaded business hall floor, was the first and finest of them, its façades being given a subtle entasis which caused no small difficulty in the design of the steelwork. In the same street were, at No 129, the narrow-frontage Legal & General (1927–9) by Burnet's ex-assistant James Napier, and, nearby at No 145, Burnet & Boston's Commercial Union (1931), both of the Hislop-Miller type in having columned ground floors, though very different in detail, the latter being a heavy red sandstone Greek Doric.

All of these buildings were utterly dwarfed in scale by Edward

Grigg Wylie's Scottish Legal Life Building (1927–31) which occupied a complete street block on Bothwell Street, the result of a competition in which Wylie had also offered a more modern elevational treatment. Although a similar American classic to Miller's Bank of Scotland, particularly at the great pilastrade encompassing four floors of metal-framed windows, it differed in eliminating the ground floor colonnade in favour of a rusticated base with a triple-arched entrance loggia flanked by shop windows in architrave frames.

All of these had been designed before 1930 and had been relatively conservative American classic in style. But in 1936 Thomas Waller Marwick rebuilt a block in Edinburgh's George Street as a temporary head office for the National Bank of Scotland in a refined modern idiom with good sculpture, and in 1938–9 Leslie Graham Thomson and Frank Connell built the black and grey marble-faced Caledonian Insurance building at the corner of St Andrew Square and George Street, Edinburgh, six-storeyed with a set-back top floor and a deep-eaved emerald green pantiled roof, refined Scandinavian Art Deco inside and out. A similar scheme by the same architects for a new Edinburgh Stock Exchange was shelved because of World War II. At the National Bank's headquarters (1936–9) on St Andrew Square, however, the inter-war conservatism of the financial institutions had its final triumph; Thomson suffered the indignity of having his half-hearted classical elevations redesigned in an inflated Palazzo Massimi form by the London Beaux Arts-trained architect Arthur Davis.

More modern in concept than any of these were a number of warehouse and industrial office buildings, mainly in Glasgow. The first of them was Campbell, Reid & Wingate's monumental warehouse (1914–21) at 16–20 Bell Street, massively pilastered with metal-framed windows and some very original Greek detailing. It was followed by a number of other tall warehouse blocks in the Wilson Street-Hutcheson Street area which took their style either from Miller's McLaren warehouse or Burnet's Kodak building, the best of them, most notably Links House (1931), being by James Taylor Thomson who had worked in the United States. All of these adhered to elevational formulae developed before World War I, but Sir Owen Williams's all-glass and black vitrolite *Daily Express* building (1936) on Albion Street was entirely new in style, closely following the model of that newspaper's London office.

Outwith Glasgow city centre a few industrial office buildings attempted the horizontal lines of what has come to be known as the International Style, most notably George Boswell's extensions (1934 and 1937) to Templeton's Carpet Factory at Glasgow Green and, outwith the city altogether, William Kerr's simple but sophisticated five-storeyed Paton & Baldwin offices (1936) at Kilncraigs, Alva. The largest industrial headquarters building of the age, Nairn's at Kirkcaldy (1936) by James

Miller, was more conservative, a quadrangle only two storeys high in a modernised English Georgian brick idiom, very simple but of considerable refinement.

POST-WAR DEVELOPMENTS

Prior to World War II the commercial architecture in Glasgow and Edinburgh had in general been of a quality unmatched outside London, and approached only by that of Newcastle in the earlier decades of the nineteenth century, and by that of Manchester and Liverpool in the mid to late nineteenth and early twentieth centuries.

For more than a decade after the war there was little new building in the city centres. But from the later 1950s the insurance companies began an extensive programme of building and rebuilding. The earliest continued to be of decent quality. Gauldie, Hardie, Wright & Needham's Guardian Royal Exchange building (1957) in Albert Square, Dundee, was one of the earliest and best, four-storeyed stone-faced with a recessed portico and a set-back fifth-storey floor with a row of pedimental rectangular bays. In Glasgow, the concrete frame of Harold Bramhill's Royal London House (1957), 54 West Nile Street, rose into a top-floor pergola and was decently clad in sandstone and red granite, while Wright & Kirkwood's large stone-faced office (1958) at 249 West George Street paid tribute to the adjacent Blythswood Square in its rusticated detailing and handsome cornice. In Edinburgh, Gordon & Dey's Scottish Life Building (1960), St Andrew Square, similarly, if less successfully, paid tribute to the David Bryce Junior office previously on the site in its Ionic entrance and the flanking canted bays of its lower floors.

Although, at that date, the banks had still generally respected their old head offices (though not, alas, their branch bank buildings which suffered much ill-advised modernisation in granite), the Edinburgh life insurance companies followed Scottish Life's lead – and indeed that of the Caledonian before the war – and embarked upon a programme of replacement of their old head offices until there remained only the Edinburgh Life, the Scottish Equitable and the Standard Life buildings, the sensitive adaptation and extension of the last (Michael Laird 1964, 1968 and 1975, that of 1968 with Sir Robert Matthew) being a rebuke to the destructiveness of the others. With the single exception of the Scottish Provident Institution by Rowand Anderson, Kininmonth & Paul (South St David Street (1961), grey marble, the adjacent Ivanhoe House (1964), Portland stone, main building (1967–8) on St Andrew Square, sandstone with a glass-topped stair-tower rising from within a granite plinth), the new buildings were not inspiring. Basil Spence, Glover & Ferguson's Scottish Widows (1962) on St Andrew Square, although carefully proportioned, was somewhat bleak, faced in pale grey and black marble to match Thomson's adjacent Caledonian Building, while Mottram Patrick & Dalgleish's Life Association (1966) on George Street was an overworked essay in grey marbles and granite. Both buildings served as head offices only briefly, the latter being demolished after an existence of only 30 years. The other commercial architecture in and around St Andrew Square, Scottish Life's further building on St David Street by Kelly & Surman; the Co-operative Insurance's Edinburgh House (1963 and 1964), St Andrew Street, by H R Stovin-Bradford, 29–31 St Andrew Square by William Nimmo & Partners, and the adjacent Norwich Union by R Seifert & Partners (both 1970) – all equally short-lived – were no more distinguished, and demonstrated the degree to which architects from outwith the city were gaining the largest commissions at the expense of locally-based practitioners.

All of these buildings were at least stone-faced and, although generally higher than their predecessors, respected the existing scale. Outwith the first Edinburgh New Town streets lesser standards were accepted, and there was a depressing trend on the part of both government and business to rent cheap new-build office space rather than commission what was required. Representative of this trend were three commercial developments leased as government offices, Jeffrey Street (1963) by Michael Laird, the St James Centre (1964–70) by Ian Burke & Martin, completed by Hugh Martin & Partners, and Argyle House (1966–9) by Michael Laird & Partners, all of which were rough-faced, mainly pre-cast concrete, although the last at least had an ashlar-faced podium towards Castle Terrace and Lady Lawson Street. The adjoining nine-storeyed West Port House (1967) by Stanley Poole Associates and the tower-block (1964) by Covell Matthews in Torphichen Street, which made such an unfortunate appearance over Atholl Crescent as seen from Walker Street, and in views of the castle as seen from the west, demonstrated the standards of design and material which the city then thought acceptable.

In the 1960s the provision of new office space in central Glasgow was at first concentrated on the sites of redundant churches, theatres, hotels, clubhouses, department stores and public buildings but, as the supply of these became exhausted, older office blocks and warehouses west of Glassford Street were steadily demolished rather than refurbished to achieve higher plot ratio levels, that is, to increase the amount of floor space which could be piled upon the site. Theoretically, this was a maximum of 3.5, but it was frequently exceeded. Speculative office blocks – and shopping developments – of a depressingly low standard of design and material, often by architects remote from Glasgow and with no feeling for the city, were approved by a complaisant city administration desperate for new investment and increased rateable values. Raw concrete, aggregate panels, brick and the cheaper forms of curtain walling now became the norm for elevations; featureless floor space, with much reduced ceiling heights and the most basic provision of elevators, stairs and lavatories, was divided and fitted up according to tenants' needs.

Throughout the decade there was a fashion for two-storeyed podia of shopping development with office blocks and towers above, a type initiated by Michael Laird at 58 Waterloo Street in 1959 where the office tower was of nine storeys and the material mosaic and brick. It was followed by the monster St Andrew House (1961–4), Sauchiehall Street, by Arthur Swift & Partners, with a reinforced concrete-framed tower of 15 storeys clad in aggregate panels forming a mullioned treatment over a long two- to three-storeyed podium. Compared to what was to come later it almost had a certain elegance. Empire House (1962–5, also on a former theatre site) by Covell Matthews was much lower, but its heavy-handed raw concrete and blue engineering brick detailing were even more intrusive. So was the same architects' Alec House (1963) at 94–168 George Street where the two-storeyed podium, stepped with the site, contained an open-deck car park and bore both an office block and a 17-storeyed office tower. It proved to be too far east of the commercial centre and became the David Livingstone Tower of the University of Strathclyde. Charlotte House (1968), 78–82 Queen Street, by John Drummond is a later version of the same type in which an office tower rises above a five-storeyed street frontage conforming in height to its Georgian and early Victorian neighbours. More thoughtful versions of the tower theme were to be found in the work of Derek Stephenson at Heron House (1967–71), between Bothwell Street and St Vincent Street, and at the Bank of Scotland's block (1968) at Lynedoch Place on the Park hilltop, although it could hardly be said that the siting of these buildings was welcome. Seifert's Anderston Centre (1967–73) on Argyle Street, in which the office element was integrated with multi-level shopping and where the towers were domestic, was similar in general concept.

Most of Glasgow's office developments in the 1960s and earlier 1970s were, however, on more restricted sites where the traditional city scale had to be followed. Of these buildings, generally of about six to seven storeys, Leslie Norton's 1–13 Waterloo Street (1963–5, aggregate panels) and Scottish Life House (1968–72, concrete mullions), extending into the retained shell of David Hamilton's Western Club, the Lyon Group's Hellenic House (1970–1, ribbed concrete) on Bath Street and Seifert's even drearier Bowring Building on West George Street may be taken as typical. Much better than any of these as architecture, but unfortunately breaking the traditional building line, was Wylie Shanks's six-storeyed Scottish Union House (1965–7), 174 St Vincent Street, brick with a polygonal bay on a Carrara marble podium. Further west, and in the same banded brick idiom, was Derek Stephenson's colossal Royal

Exchange Assurance House at 320 St Vincent Street, seven-storeyed over a petrol forecourt.

By the later 1970s a reaction had set in against the low standards of material which had hitherto prevailed. A large number of the better older office blocks were refurbished rather than demolished, and a considerable number were the subject of façade retention schemes. In most instances the original interiors were of the simplest kind but in a number of cases important interior work was lost. In both Edinburgh and Glasgow the quality of design and material improved. Tinted glass became fashionable, either banded horizontally with buff sandstone as at Seifert's Hume House (1974–7), West George Street, or vertically in red sandstone or granite as at the huge Amicable House (1973–6), St Vincent Street by King, Main & Ellison. With or without a stone framework, it continued to be so in such megastructures as the Clydesdale Bank (1981), 150 Buchanan Street, by G D Lodge, the Coats Viyella Building (1983–7), St Vincent Street, by Scott, Brownrigg & Turner, 100 Bothwell Street (1987), by the Holmes Partnership, the Britoil building (1983–7) on St Vincent Street by Hugh Martin & Partners – all in Glasgow – and, in Edinburgh, Rutland Court (1991) by Case Design, Apex House (1990) by MAP Architects and Haymarket House by Percy Johnson-Marshall & Associates.

In a number of 1980s and 1990s developments, particularly where there was a need to integrate into the existing townscape, a post-modern treatment was adopted, notably at 58 and 61–9 West Regent Street, Glasgow, by Comprehensive Design (1986) and Scott, Brownrigg & Turner (1988) respectively, at the new Edinburgh House (1993) on St Andrew Street, Edinburgh, by Covell Matthews with Bob Anderson, and, most monumentally of all, at the giant colonnaded pile of Saltire Court (1991), Castle Terrace, Edinburgh, by Campbell & Arnott which, in common with other recent large-scale office developments, featured a great central atrium.

All of these had been on city-centre sites. Since the late 1960s there had been a trend towards building large-scale speculative office development and company headquarters buildings outwith the old business centres, often in what had hitherto been domestic areas. In Edinburgh, J & F Johnston's Finance House (1968), Orchard Brae, was one of the earliest, followed by two large purplish brick office blocks nearby, Orchard Brae House (1974) by Reiach & Hall and the much more thoughtfully detailed slab of 25 Ravelston Terrace (1972) by Roland Wedgwood. Considerably more expensive than any of these was a series of headquarters and computer-centre buildings for the great financial institutions. The prime examples were the Scottish Widows (1976) at Parkside by Basil Spence, Glover & Ferguson, an assemblage of tinted glass hexagons set in stone ramparts and ornamental water, the Royal

Bank Computer Centre (1978) on Dundas Street and Fettes Row by Michael Laird & Partners, long and horizontal in design with a ziggurat-like profile, United Distillers (1984), 33 Ellersly Road by Robert Matthew, Johnson-Marshall & Partners (RMJM), a fine modern suburban palace which marked the end of the old Distillers Company's independent existence, the monumentally columnar tinted-glass Life Association (1990), Dundas Street, by Reiach & Hall, which similarly marked the end of that company's independence, and the Standard Life Computer Centre (1991), Tanfield, by Michael Laird & Partners, a domed low-rise development beautifully integrated into the river bank and the surrounding townscape.

All of these had been on still fairly central Edinburgh sites. Thereafter, the emphasis shifted westwards to peripheral sites near the city by-pass, a trend anticipated in Aberdeen. There, the earlier new large-scale office developments had at first been in the harbour area in the form of tall 11- to 12-storeyed slabs, most notably St Machar's House and Salvesen Tower. But the great oil companies, Shell (1975–8, McInnes, Gardner & Partners), Total (1978) and Chevron (1980, both Jenkins & Marr), all chose peripheral sites, at least partly out of necessity because of the sheer scale of their buildings. A more thoughtful solution to the design of the peripheral megastructure than the Aberdeen examples was James Parr & Partners' General Accident (now Norwich Union) building at Necessity Brae, Perth, where horizontal terraces of offices with roof gardens planned around nine courtyards were skilfully integrated into the contours of Craigie Hill. Its interiors were also much above the norm with coffered ceilings of elm, and tapestry and batik wall panels at the principal spaces.

In Edinburgh, the Bank of Scotland opted for a far less prominent site for its computer centre than the Royal Bank had done, building it on a relatively low-cost site at the industrial park of Sighthill in 1986. In the same year Murray International, with Gareth Hutchison as their architect, began West Office Park at South Gyle, initially with six large, reasonably well-detailed office blocks with central atria, built of red and brown brick with pitched roofs and canted angles. This development was joined by Bamber Gray's excellent GEC Ferranti Building (1988) and further buildings for the Bank of Scotland (1991, David Duncan) and the Royal Bank (1992, Michael Laird).

These developments have since been followed by the pavilions of the West One Business Park (1991) by Forgan & Stewart and by Edinburgh Business Park (begun 1993), an ambitious scheme by the American architect, Richard Meier, and the Edinburgh architects, Campbell & Arnott, in which the Gogar Burn has been fashioned into a series of lochans to provide a much finer setting than was achieved at West Office Park. The new buildings do not all have the consistently high architectural quality envisaged by Meier, who was regrettably not commissioned to design any of them himself, but at least the often illiterate gimmickry of much post-modern design is absent and a considerable number of the buildings are of real quality and elegance. Notable are Page & Park's HSBC G3 Building and Bennetts Associates' Alexander Graham Bell House (both 1999), the latter making particularly effective use of the setting created by Meier.

Concurrently with the construction of Edinburgh Park there have been major changes in the city centre. Charlotte Square and even some relatively recently-constructed office developments in and around St Andrew Square and George Street were suddenly deserted. The grander bank buildings closed and became restaurants. While a few notable redevelopments took place within the New Town – Yorke, Rosenberg & Mardall's stylish block at the corner of Princes Street and Castle Street, and Reiach & Hall's 10 George Street – development now shifted westward to the new financial district on the former railway yards around Festival Square, the Western Approach Road and the new Conference Centre by Terry Farrell. The earliest major developments there were Hugh Martin & Partners' building on the south side of Festival Square, Case Design's glass Baillie Gifford Building (1991), and Michael Laird & Partners' domed Standard Life Block (1994) on the north side of the Western Approach Road. Since that time the south side of the Approach Road has been developed as a high curving terrace which forms the north side of Festival Square, the Clydesdale Bank Plaza (1997–2000), to designs by Cochrane McGregor. Within this same district are the giant Scottish Widows Building (1997) between Morrison Street and Fountainbridge, by Building Design Partnership, Princes Exchange (2000) on Lothian Road, by Percy Johnson-Marshall & Partners, and the complete re-cladding and refitting of the 1960s blocks on Torphichen Street and Canning Street, where a further building has been constructed by RMJM, completing the redevelopment of these Georgian and Victorian street blocks.

In Glasgow, the equivalent of Edinburgh's west-end financial district should have been the area between Broomielaw and Argyle Street, for which the American practice of Kohn, Pedersen & Fox proposed an ambitious, if somewhat dense, scheme embracing the conservation of several Victorian warehouses and office blocks in the early 1990s. Most of this has not been realised, although three impressive blocks, 1 Atlantic Quay (1990) and Atlantic Square (2000), both by Building Design Partnership, and BT Broomielaw (1999), by Fitzroy Robinson, have been built along the waterfront. Within the traditional commercial area there have been a number of major developments, notably the glass tower block of Eagle House by SBT Keppie, 191 West George Street (1998), again largely in glass, by Hugh Martin Partnership, and the post-modern

Alhambra House (1997) by G D Lodge, a more truly Glaswegian essay than most with its circled corners. Further west is Holford's pink Tay House (1991), reaching over the motorway, and the awesomely repetitive Dalian House (1990) on St Vincent Street by Jenkins & Marr. As in Edinburgh, the outdated office blocks of the 1960s and early 1970s are either being demolished or radically reconstructed as virtually new buildings, most notably Holford's reconstruction of Seifert's Anderston Centre as Cadogan Square in 1997.

Not all of these 1990s buildings could be said to be great architecture. But they are much more spacious internally than those of the generation immediately preceding them, and something of the concern for quality, which characterised the business and commercial architecture of the nineteenth and early twentieth centuries, has returned.

NOTE

1. Notes are not given in this chapter as the sources are readily traceable by location, architect or institution in the works listed in the Bibliography. *The Buildings of Scotland* series, and, where relevant for a few of the older buildings, RCAHMS, *Inventory of Edinburgh*, Edinburgh 1951, provide the most detailed information for the buildings which existed at the time of publication, although the RCAHMS volume has also been helpful in respect of the sites of lost buildings in Leith. The Rutland Press *Architectural Guides* provide the best pictorial coverage and often extend to buildings which no longer exist. Bailey, 1996, provides the key to those archives held in public collections or by architects in practice. Since it was published the crucially important Dick Peddie & McKay collection has been taken into the care of RCAHMS and catalogued. Major holdings of architectural drawings are in the possession of the Bank of Scotland (Aberdeen Bank, Caledonian Bank, Central Bank, British Linen Bank, Union Bank of Scotland), The Royal Bank of Scotland (Commercial Bank, National Bank) and the Clydesdale Bank (Aberdeen Town & Country Bank, North of Scotland Bank) although these have been depleted in recent years as a result of the sale of branch bank buildings.

BIBLIOGRAPHY

Allen, N, ed. *Scottish Pioneers of the Greek Revival*, Edinburgh, 1984.
Anon. *Glasgow Delineated: Or a description of the city*, Glasgow, 1821 and later edns.
Anon. *Brief History of the Union Bank of Scotland*, Glasgow, 1910.
Academy Architecture and Architectural Review, London, 1888 – [indexed].
Anderson, J. *The Story of the Commercial Bank of Scotland*, Edinburgh, 1910.
Archaemedia. The Trades House of Glasgow. Unpublished typescript, Glasgow, 2000.
The Architect, London, 1869 -.
Bailey, R M, ed. *Scottish Architects' Papers: A source book*, Edinburgh, 1996.
Beaton, E. *Ross and Cromarty*, Edinburgh, 1992.
Beaton, E. *Sutherland*, Edinburgh, 1995.
Beaton, E. *Caithness*, Edinburgh, 1996.

Booker, J. *Temples of Mammon*, Edinburgh, 1984.
Breeze, D J, ed. *Studies in Scottish Antiquity*, Edinburgh, 1984.
The British Architect, Manchester, 1874 – [indexed].
Brogden, W A. *Aberdeen*, Edinburgh, 1986, 1998.
Brown, R, ed, *The Architectural Outsiders*, London, 1985.
The Builder, London, 1842 – [indexed].
The Builders' Journal, London, 1895–1910 [indexed]. The issue of 28 November 1906 is of special importance for Glasgow.
The Builders' Chronicle, Edinburgh, 1854–56.
Building Industries and Scottish Architect, Glasgow, 1890 – .
The Building News, London, 1854 – [indexed].
Burgher, L. *Orkney*, Edinburgh, 1991.
Cameron, A. *Bank of Scotland, 1695–1995*, Edinburgh, 1995.
Civil Engineer and Architects' Journal, London, 1837–68.
Close, R. *Ayrshire and Arran*, Edinburgh, 1992.
Colvin, H M. *Dictionary of British Architects*, London and New Haven, 3rd edn, 1995.
Crawford, A. *Charles Rennie Mackintosh*, London, 1995.
Cusack, P. Lion Chambers: A Glasgow experiment, *Architectural History*, 28 (1985), 198–206.
Davidson, L, Lowrey, J, eds. *Architectural Heritage*, Vol 3, *The Age of Mackintosh*, Edinburgh, 1992.
Devine, T M. *The Tobacco Lords*, Edinburgh, 1990.
Ewing, A O. *View of the Merchants' House of Glasgow*, Glasgow, 1866.
Fiddes, V, Rowan, A, eds. *Mr David Bryce, 1803–1876*, Edinburgh, 1976.
Finnie, M. *Shetland*, Edinburgh, 1990.
Fleming, R. *Robert Adam and his Circle*, London, 1962.
Forbes, W. *Memoirs of a Banking House*, Edinburgh, 1860.
Fraser, D. *The Making of Buchanan Street*, Glasgow, 1885.
Gifford, J, McWilliam, C, Walker, D. *Buildings of Scotland: Edinburgh*, Harmondsworth, 1984.
Gifford, J. *Buildings of Scotland: Fife*, Harmondsworth, 1988.
Gifford, J. *William Adam, 1689–1748*, Edinburgh, 1989.
Gifford, J. *Buildings of Scotland: Highlands and Islands*, Harmondsworth, 1992.
Gifford, J. *Buildings of Scotland: Dumfries and Galloway*, Harmondsworth, 1996.
The Glasgow Property Circular, Glasgow, 1876–98.
Glasgow Advertiser and Property Circular, Glasgow, 1898 – .
Glendinning, M, McInnes, R, Mackechnie, A. *A History of Scottish Architecture from the Reformation to the Present Day*, Edinburgh, 1996.
Gomme, A, Walker, D. *The Architecture of Glasgow*, London, 1968, 2nd edn, 1987.
Gordon, J F S, ed. *Glasghu Facies: A view of the city of Glasgow ... by John McUre ... MDCCXXXVI* [1736], 2 vols, Glasgow, 1872.
Gow, I. David Rhind, 1808–83. In Brown, 1985, 153–71.
Graham, C (and Morgan, D). *Archibald Simpson Architect of Aberdeen*, Aberdeen, 1990.
Grant, J. *Old and New Edinburgh*, London, 1883.
Gray, A S. *Edwardian Architecture: A biographical dictionary*, London, 1985.
Haynes, N. *Perth and Kinross*, Edinburgh, 2000.
Henderson, T. *The Savings Bank of Glasgow: One hundred years of thrift*, Glasgow, 1938.
Hitchcock, H R. *Early Victorian Architecture in Great Britain*, London and New Haven, 1954.

Howard, D. *Scottish Architecture from the Reformation to the Restoration, 1560–1660*, Edinburgh, 1995.
Howarth, T. *Charles Rennie Mackintosh and the Modern Movement*, London, revised edn 1977.
Hume, J R. *The Industrial Archaeology of Glasgow*, Glasgow, 1974.
Hume, J R. *Dumfries and Galloway*, Edinburgh, 2000.
Jaques, R, McKean, C. *West Lothian*, Edinburgh, 1994.
Johnston, C, Hume, J R. *Glasgow Stations*, Newton Abbot, 1979.
Kaplan, W, ed. *Charles Rennie Mackintosh*, Glasgow and New York, 1996.
Keith, A. *The North of Scotland Bank*, Aberdeen, 1936.
King, D. *The Complete Works of Robert and James Adam*, London, 1991.
Lamb, A C. *Dundee: Its quaint and historic buildings*, Dundee, 1895.
Leith, C. *Alexander Ellis: A fine Victorian architect*, Aberdeen, 1999.
McFadzean, R. *The Life and Work of Alexander Thomson*, London, 1979.
McKean, C, Walker, D. *Dundee*, Edinburgh, 1984, 1993.
McKean, C. *Stirling and the Trossachs*, Edinburgh, 1985, 1994.
McKean, C. *The Scottish Thirties*, Edinburgh, 1987.
McKean, C. *The District of Moray*, Edinburgh, 1987.
McKean, C, Walker, D, Walker, F. *Central Glasgow*, Edinburgh, 1989, 1993, 1999.
McKean, C. *Banff and Buchan*, Edinburgh, 1990.
McKean, C. *Edinburgh*, Edinburgh, 1992.
MacKechnie, A, ed. *David Hamilton, 1768–1843*, Glasgow, 1995.
McKinstry, S. *Rowand Anderson, the Premier Architect of Scotland*, Edinburgh, 1991.
McWilliam, C. *Buildings of Scotland: Lothian*, Harmondsworth, 1978.
Malcolm, C A. *The History of the British Linen Bank*, Edinburgh, 1950.
Maxwell, A H. *Annals of the Scottish Widows Fund Life Association*, Edinburgh, 1914.
Mays, D C. A profile of Sir George Washington Browne. In Davidson, Lowrey, 1992, 52–63.
Miller, R. *The Municipal Buildings of Edinburgh*, Edinburgh, 1895.
Miller, T B Y. *Dundee Past and Present*, Dundee, 1910.
Mudie, R. *Dundee Delineated: Or a history and description of that town*, Dundee, 1822.
Munn, C W. *The Clydesdale Bank: The first one hundred and fifty years*, Glasgow, 1988.
Munn, C W. *The Scottish Provincial Banking Companies*, Edinburgh, 1981.
Munro, N. *History of the Royal Bank of Scotland*, Edinburgh, 1928.
Pagan, J. *Sketch of the History of Glasgow*, Glasgow, 1847.
Peden, A. *The Monklands*, Edinburgh, 1992.
Pride, G L. *Fife*, Edinburgh, 1990, 1999.
Rait, R S. *The History of the Union Bank of Scotland*, Glasgow, 1938.
RCAHMS. *Inventory of Edinburgh*, Edinburgh 1951.
Reid, J M. *The History of the Clydesdale Bank*, Glasgow, 1938.
Richardson, A E. *Monumental Architecture in Great Britain*, London, 1914.
Rock, J. *Thomas Hamilton, Architect*, Edinburgh, 1984.
Rodger, J. *Contemporary Glasgow: An architecture of the 1990s*, Edinburgh, 1999.
Senex (Reid, R). *Glasgow Past and Present*, Glasgow, 1884.
Service, A, ed. *Edwardian Architecture and its Origins*, London, 1975.
Shepherd, I. *Gordon*, Edinburgh, 1994.
Simpson, W D, ed. *A Tribute Offered by the University of Aberdeen to the Memory of William Kelly*, Aberdeen, 1949.
Sinclair, F, ed. *Scotstyle: 150 years of Scottish architects*, Edinburgh, 1984.
Sinclair, F. *Charles Wilson, Architect, 1810–63*, Glasgow, 1995.

Sloan, A, Murray, G. *James Miller, 1860–1947*, Edinburgh, 1993.
Smith, G F. Lanarkshire House, Glasgow: the evolution and regeneration of a 'Merchant City' landmark, *Architectural History*, 42 (1999), 293–306.
Stamp, G. *Alexander Thomson, the Unknown Genius*, Glasgow, 1999.
Stamp, G, McKinstry, S, eds. *Greek Thomson*, Edinburgh, 1994.
Strang, C A. *Borders and Berwick*, Edinburgh, 1994.
Stuart, R. *Views and Notices of Glasgow in Former Times*, Glasgow, 1848.
Swan, A. *Clackmannan and the Ochils*, Edinburgh, 1987 and 2001.
Thomas, J. *Midlothian*, Edinburgh, 1995.
Thomson, D. The Works of the late Charles Wilson. Lecture read before the Glasgow Philosophical Society, 1882 (typescript copy in NMRS).
Walker, D. James Sellars, Architect, Glasgow, *The Scottish Art Review*, new series, 11/1 (1967), 16–19.
Walker, D. Sir John James Burnet; Charles Rennie Mackintosh; The partnership of James Salmon and John Croft Gillespie. In Service, 1975, 192–249.
Walker, D. *Architects and Architecture in Dundee, 1770–1914*, Dundee, 1979.
Walker, D. William Burn and the influence of Sir Robert Smirke and William Wilkins on Scottish Greek Revival design. In Allen, 1984, 3–36.
Walker, D. The architecture of MacGibbon and Ross: The background to the books. In Breeze, 1984, 391–449.
Walker, D. The Honeymans, *Charles Rennie Mackintosh Society Newsletter*, 62 (summer 1993), 7–12; 63 (winter 1993), 5–8; 64 (spring 1994), 5–8.
Walker, D B. *Architects and Architecture on Tayside*, Dundee, 1984.
Walker, F A. *The South Clyde Estuary*, Edinburgh, 1986.
Walker, F A, Sinclair, F. *The North Clyde Estuary*, Edinburgh, 1992.
Walker, F A. *Buildings of Scotland: Argyll and Bute*, Harmondsworth, 2000.
Watson, D. *A Marshall Mackenzie: Architect in Aberdeen*, Aberdeen, 1985.
Williamson, E, Riches, A, Higgs, M. *Buildings of Scotland: Glasgo*w, Harmondsworth, 1990.
Wood, M. All the statelie buildings of Thomas Robertson, *The Book of the Old Edinburgh Club*, 24 (1942), 126–51.
Worsdall, F. *The City that Disappeared*, Glasgow, 1981.
Worsdall, F. *Victorian City*, Glasgow, 1982.
Young, A M, Doak, A M, eds. *Glasgow at a Glance*, Glasgow, 1965 and later edns.
Youngson, A J. *The Making of Classical Edinburgh*, Edinburgh, 1966, 1988.

Index

The index includes names of persons, names of buildings and their locations, building types, parts, methods of construction and architectural styles.

A
abattoirs, 524
    see also killing houses
Abbey Brewery, 544
Abbey Street Works, 540
abbeys, 35
    Abercairnie, 56
    Crossraguel, 35
    Culross, 42, 291, 371
    Dundrennan, 279
    Dunfermline, 43, 282, 291, 448
    Glenluce, 341
    Inchyre, 56
    Melrose, 279, 483
    Pluscarden Abbey, 203
    see also monasteries
Abbotsford, 58–9, 220
Abbotsford Crescent, St Andrews, 76
Abbotshall flax mill, 526
Abercairnie Abbey, 56
Aberchalder Sanatorium, 329
Abercorn, Earl of, 53, 142
Abercrombie, Patrick, 86
Abercrombie, Thomas Graham, 83, 184, 314, 324
Abercromby Street lodging house, Glasgow, 180, 181
Aberdalgie, 603
Aberdeen
    art gallery, 368
    assembly room, 133, 361
    banks, 631, 637–8, 640, 653, 667, 668
    Bishop's Palace, 279
    castles, 388
    cathedral, 279, 281
    chapels, 288
    children's hospital, 332, 333
    clubs, 148
    cotton mill, 537
    custom house, 586
    defences, 388, 398, 401, 402, 409
    dyeworks, 538
    factories, 521, 524, 532, 537
    fish cross, 613
    flax mill, 514, 519–20, 522, 526
    floating docks, 575
    foundries, 258, 530
    harbour office, 586
    head offices, 654, 667
    hospital for incurables, 335
    hospitals, 311, 313, 315, 316, 331, 334
    hotels, 144, 159–60, 163–4, 165, 184
    inns, 133, 140, 143, 144
    ironworks, 529, 532
    lighthouses, 582
    lunatic asylum, 319, 324
    masonic lodge, 133
    mercat cross, 613
    municipal buildings, 251, 258, 261–2, 263
    Music Hall, 361, 373
    night refuge, 182
    offices, 651–2, 655, 657, 674
    oil base, 575
    paper mill, 519
    parks and gardens, 368, 369, 370
    poorhouses, 316, 318
    public housing, 101
    radar station, 410
    railway facilities, 601, 603, 605, 606

Robert Gordon's Institute, 308
schools, 303, 304
seaside pavilion, 378
sewerage, 353
shops, 144, 615, 619
slipways, 574
smokehouses, 580
steam cranes, 580
sugar refinery, 547
tenements and flats, 117
theatre, 362
trades halls, 627, 628, 642
university, 305, 307, 311, 372
villa, 79
water supplies, 343, 344, 526
wool-carding mill, 495
Aberdeen Bank, 631, 637, 640
Aberdeen Daily Free Press office, 655
Aberdeen District Asylum, 324
Aberdeen Grammar School, 304
Aberdeen Hotel, 144, 159
Aberdeen Ironworks, 532
Aberdeen Maternity Hospital, 331
Aberdeen Royal Infirmary, 313, 315, 316
Aberdeen Savings Bank, 667
Aberdeen Town & County Bank, 637, 653
Aberdeen Town House, 261–2, 263
Aberdeen University, 305, 307, 311, 372
Aberdeen Wardhouse, 251
Aberdeenshire
  banks, 632, 639
  bee boles, 203
  bucket mill, 501, 506
  byres, 465, 475
  canals, 598
  castles and palaces, 27, 28, 29, 30, 31, 35–6, 37, 38, 39, 40, 43
  cattle courts, 475
  châteaux, 37, 40, 42, 45
  cheese rooms and presses, 469, 471
  churns, 470
  country houses, 54, 59, 62, 63
  dairy, 473
  docks, 574, 575
  dovecots, 202, 204
  farm bothies, 427
  foundries, 258
  gardens and garden buildings, 211, 215, 216, 371
  gate lodges, 192
  granite quarries, 558
  gun battery, 398
  harbours, 573
  hospitals, 313, 327, 328
  hotels and inns, 139, 140, 172, 176, 177
  kilns, 449, 450, 554
  lighthouses, 582, 583
  mills, 205, 541
  municipal buildings, 253
  paper mills, 539–40
  poultry housing, 478
  prisons, 269
  privies, 204, 205
  radar station, 412, 413
  railway facilities, 534, 604, 606
  rotative beam engine, 520
  salmon fishing facilities, 587
  schools, 296
  smokehouses, 580
  spa, 168, 374
  stables, 194
  threshing machines, 441
  tower-houses, 32, 33, 35–6, 397, 583
  villas, 44
  warehouses, 626–7
  water tower, 540
  whin mills, 456
Aberdeenshire Canal, 598
Aberdour, 211
Abernethy, 495
Abernethy Round Tower, 277
Acharn, 215
Achill workers, 237, 238, 239–40, 242
Achnasheen, 160
Adam brothers, 361
Adam castle style, 54–5
Adam, James, 361, 363–4, 646
Adam, John
  country houses, 53, 54
  Edinburgh Exchange, 630
  Great Inn, Inveraray, 135
  houses, 71, 73
Adam, Matthew, 179
Adam, Robert, 40
  Balbardie, 42
  Charlotte Square, 73, 111
  country houses, 53, 54, 55
  Edinburgh University, 307
  Glasgow Royal Infirmary, 312
  Glasgow trades hall, 630
Adam, William, 71

Chatelherault hunting lodge, 221–2
country houses, 51, 54
Dundee Town House, 629
Edinburgh Royal Infirmary, 311
Hamilton Parish Church, 287
Robert Gordon's Hospital, 303
Royal Bank, Edinburgh, 631
administration buildings, 249–75
see also business chambers; municipal buildings; offices
Advocates' Library, 309
AEG turbine factory, Berlin, 533
aerial gunnery school, 405
Aesthetic Movement, 152, 157, 657
Affleck, 33
agricultural buildings see farm buildings
AI Welders, 535
Aimer, Thomas, 520
air defences, 409–10
air-compressor houses, 562
air-raid shelters, 406
Aird & Coghill, 665
Airdrie, 80–1, 638
airfields, 382, 404, 407, 408
Àiridh nan Sileag, 13
airing see drying facilities; ventilation
Airlie, 40
Aithsting, 485
Albany, Duke of, 382
Albany Hotel, 185
Albert Institute, 308, 368
Albert, Prince, 172, 205
Albion car factory, 517, 534
Albion Court, 647
Albion Street, Glasgow, 666
Alec House, 672
Alexander, Mr, 481
Alexander, Charles, 654
Alexander, Cosmo, 261
Alexander, George, 145
Alexander Graham Bell House, 675
Alexander, James, 263
Alexander, William, 182, 652, 654
Alexander's Mill, 181, 516, 518, 528
Alexandra Hotel
  Glasgow, 156, 166
  Oban, 172, 173
Alexandria, 314, 534
Alford, 541, 606
Alhambra House, 676
Alhambra music hall, 363
Alhambra Theatre, 620

Allan & Mann, 512
Allan Park, Stirling, 76
Allan Ramsay's House, 69
Alley & MacLellan, 533
Alloa
  banks, 639, 666
  baths, 375
  brewery and brewhouse, 545
  collieries, 563, 566
  glassworks kilns, 501
  Great Garden, 212
  half-time school, 528
  houses, 83
  inns, 136
  Kilncraigs Mill, 526, 527, 528, 536
  recreation buildings, 375
Alma Works, 537
almshouses see poorhouses
Alnwickhill, 344
Alt Scherbitz, 322
Altyre, 194, 205
aluminium production, 559
Alva, 512, 669
Alyth, 140
Amicable House, 673
Amisfield, 40, 52, 214, 219
Anchor Mills, 510, 516, 519, 523, 535, 536
Anchor Shipping offices, 662
Anderson, Bob, 673
Anderson, D, & Son, 518
Anderson, R R, 402
Anderson, Sir Robert Rowand
  Beesknowe, 82
  Central Station office and hotel, Glasgow, 162, 165, 656
  Edinburgh Conservative Club, 157
  Edinburgh University buildings, 307
  North British Hotel, Edinburgh, 164
  Scots Renaissance Revival, 80
  Scottish Provident Institution, 670
Anderson shelters, 406
Anderson, William James, 181
Andersonian University, 307, 308
Anderston, 535, 667
Anderston Centre, 672, 676
Anderston Foundry, 537
Andrew Lamb's House, 625, 626
Angel inn, 137–8
Anglo-Norman castles, 28–39
Angus

airfield, 404
assembly rooms, 142, 259
banks, 639
bee boles, 203–4
brewery, 544, 545
Broughty Castle, 402
chapel, 288
châteaux, 37, 40, 42
cheese press, 471
churches, 287
corn exchange, 644
country houses, 54, 57, 59
defences, 397, 398, 402, 410, 412
distilleries, 545
docks, 573, 574
factories, 518, 524, 528, 537, 540
farm bothies, 238
flagstone quarries, 558
foundry, 533
gallery, 37
gardens and garden buildings, 211, 215, 219, 371
goods shed, 605
grain mills, 450
half-time school, 528
harbour facilities, 582, 586
hill bothies, 229, 230
hospitals, 313, 338, 339
hotels and inns, 130–1, 142, 175, 187
kilns, 448, 449, 501, 542, 554
lighthouses and shore stations, 583
lodging house, 182
lunatic asylum, 319, 320
municipal buildings, 253, 259, 267, 319
newspaper office, 655
oil base, 575
paper mill, 540
potato houses, 458
railway facilities, 572, 601
salmon bothies, 226, 227, 587
sanitation, 524
sheep houses, 486
smokehouses, 580
steam engine house chimney, 443
swimming pool, 375
textile mills and factories, 523, 524, 525, 536, 537
threshing machines, 441
tower-houses, 33, 35, 402
towers, 277
trades hall, 631

viewing platform, 38
villas, 77, 79, 80
water supplies, 344, 540
see also Strathmore
Angus, George, 77, 300, 639, 646
Angus Hotel, 185, 186
animal houses, 13–14, 16, 19, 20, 92
see also barmkins; byres; kennels; pig sties; sheep husbandry buildings; stables
animal-powered machinery see horse engines and engine houses
Ann Street, Edinburgh, 73
Annan, 136–7, 166, 187, 253, 626
Annandale Arms, 136, 137
Annandale, Marquess of, 49
Annat House, 76
Anne of Denmark, Queen, 43
Annfield Road, Inverness, 80
Anstruther, 412, 413, 573
anti-aircraft defences, 407, 409–10, 411, 413
anti-invasion defences, 408–9
apartment blocks see tenements and flats
Apex House, 673
apple houses, 219, 371
aqueducts, 344, 599
Arbigland, 54
arboreta, 216
Arbroath
banks, 639
chapel, 288
cistern, 344
corn exchange, 644
docks, 574
factories, 518, 536, 537, 540
foundry, 533
gun battery, 398
harbour facilities, 582, 586
lighthouse shore station, 583
sanitation, 524
smokehouses, 580
swimming pool, 375
trades hall, 631
arcades
business and commerce, 624, 629, 630, 631, 642
country houses, 51, 52
industrial buildings, 530
shops, 154, 156, 620, 621, 664
terraced housing, Moray, 69
archbishop's palace, Glasgow, 30, 35

684 • INDEX

Archer's Hall, 359
architectural source-books *see* pattern books
Arctic Tannery, 541
Ardclach, 391
Ardeer, 556
Ardelve, 577
Ardencaple, 140
Ardersier Manse, 449
Ardgay, 587
Ardgour, 577
Ardkinglas, 62, 83
Ardmore, 545
Ardrossan, 147, 577, 582
Ardtornish House, 197
Argyle Arcade, 620
Argyle House, 671
Argyle Square, 111
Argyle Street, Glasgow, 142, 649, 650, 651
Argyll
    aluminium smelting, 559
    animal housing, 19
    boat building yards, 586
    byre, 467
    byre dwellings, 19–20
    canals, 598, 599, 600
    castles, 29, 30, 35, 390, 397
    church, 288
    coalfields, 560
    Connel Bridge, 603
    country houses, 54, 57, 62, 63
    dams, 559
    defences, 397, 398, 407, 412
    distilleries, 512, 545
    furnaces, 555–6
    gunpowder manufacture, 556, 577
    hand crane, 580
    hay storage, 452
    horse-walk platform, 442
    hotels, 166, 172–3, 175, 183, 184, 187
    houses, 16, 71, 73, 83
    inns, 135, 136, 166
    ironworks, 345, 555, 556
    lades, 345
    laundries, 197
    lighthouses, 583
    longhouses, 93
    lookout tower, 214
    municipal buildings, 267
    newspaper office, 655
    piers, 575, 577
    pump house, 215
    quarries, 497, 558
    radar station, 412
    railway facilities, 577, 603
    roads, 595
    shieling huts, 13
    tenements and flats, 112
    wool stores, 489
Argyll and Bute District Asylum, 321
Argyll Brewery, Edinburgh, 545
Argyll, Duke of, 52, 54, 135, 452
Argyll Hotel
    Dunoon, 166, 187
    Oban, 173
Argyll lodging, 37, 69
Argyll Motor Works, 534
Arisaig, 604
Arkley, Peter, 57
Arkwright, Sir Richard, 512
Armadale, 182
Armadale Castle, 57, 195
Armstrong, John, 155
Armstrong, William, 525
Arniston, 214, 215, 217
Arran, 175, 215, 449, 552, 560
Arrochar, 175
Arrol Brewhouse, 545
Arrol-Johnston car factory, 517, 534
Arrol, Sir William, & Co, 556
Art Deco buildings, 667, 668, 669
art galleries, 308, 360, 363, 364, 366–8, 371
art gallery houses, 78, 366, 367
Art Nouveau style, 81–2, 174, 656, 660, 665, 666
Arthur, J M, 81
Arthur Lodge, 78, 367
Arthur Swift & Partners, 672
Arthur Warehouse, 656
Arts and Crafts architecture, 61, 79
    gate lodges, 192
    hotels, inns and roadhouses, 174, 175, 178, 184, 378
    houses, 79–80, 86
arts centres, 368
    see also art galleries
ash houses, 427
Ashgrove Flax Mill, 519
assembly rooms, 133, 137, 141–8, 259, 361, 632
ASSIST, 124
asylums *see* lunatic asylums; village asylums

athenaea, 360, 361
Athenaeum
    Dundee, 361
    Glasgow, 361
Atholl Arms, 145
Atholl, Duchess of, 455
Atholl, Duke of, 49
Atholl Hydropathic, 170
Atholl Inn, 130
Atkinson, William, 56, 58, 146
Atlantic Chambers, 662
Atlantic Quay, 675
Atlantic Square, 675
Atrium, Kilmacolm, 86
Attadale, 160
Auchenblae, 524
Auchenbowie, 67
Auchencass Castle, 30
Auchentoshan, 545
Auchincruive, 214, 470
Auchindrain, 19–20, 467
Auchinleck, 561, 563
Auchinleck House, 53
Auchmacoy, 202
Auchmedden, 27
Auchterarder, 537
Auchterhouse, 42
Auchterless, 205
Aultmore House, 63
auxiliary hospitals, 336–7
Avalon, 79
aviaries, 215, 371
Aviemore, 176, 177, 186
Aviemore House, 186
Avoch, 578
Ayr
    assembly room, 143
    banks, 632, 639, 666
    defences, 31, 387, 388
    docks, 573
    harbour facilities, 582, 586
    hotels, 143, 163, 185
    houses, 84, 86
    newspaper office, 655
    piers, 575
    poorhouse, 316
    swimming pool, 375
    tannery, 540
    warehouses, 626
    see also Newton-upon-Ayr
Ayr Hospital, 315
Ayrshire
    aerial gunnery school, 405

banks, 137, 632, 639
baths, 375
bell towers, 526
bird housing, 478
boat building yard, 585
byres, 468
castles, 30, 45
churches, 288
coalfields and collieries, 560, 561, 563
cotton mills, 345, 519, 535
country houses, 53, 55, 57, 59, 62, 63, 631
cranes, 580, 582
dairies and dairy product facilities, 469, 470, 471
dams, 343
docks, 574
engineering works, 514, 532
factories, 537, 539, 540, 556
gardens and garden buildings, 214, 219, 221
hay barns, 453
hospitals, 315, 327, 331, 334, 338, 339
hotels, 177–8, 179, 187, 338
houses, 82, 83, 84, 86
inns, 137, 147, 174, 339
ironworks, 556, 557
kilns, 449, 554
lades, 345, 519
lunatic asylum, 321
mercat crosses, 613
municipal buildings, 250, 260, 269, 271
nuclear power station, 346
piers, 575, 577
potato storage, 459
prisons, 269
railway viaducts, 602
salmon fishing facilities, 587
salt extraction, 552
sandstone quarries, 558
seasonal workers' accommodation, 239, 240, 241
smokehouses, 580
steel works, 557
swimming pools, 375
tenements and flats, 112
tollhouses, 595
tower houses, 35
Ayrshire Bank, 632
Ayrshire Central Hospital, 327, 331

Ayton, 482

B
Babbity Bowster, 379
Badenoch Hotel, 186
Baillie Gifford Building, 675
Baillie McMorran's House, 625–6
Baillieknowe, 443
Baird & Thomson, 651
    see also Thomson, James
Baird, John, 617, 646
Baird, John, elder, 649
Baird, John, younger, 655
Bairds and Dalmellington, 561
bakeries, 193, 496, 502
Balbardie, 42
Balbegno, 45
Balblair Inn, 147
Balboughty, 472–3
Balbridie, 10
Balcarres, 215
Balcary, 587
Balcaskie, 49
balconies, 37–8, 69, 102
Baldie, Robert, 330
Baldovan Institution, 325
Balfour Battery, 32
Balfour, James, 57
Balfour, Robert, 73
Balfour Village, 204–5
Balintraid, 575
Ballachulish, 497
Ballantine, John, 470
Ballantrae, 575
Ballantyne & Son, 157
Ballater, 139, 604
Ballindalloch, 449
Ballingall's Park Brewery, 545
Balloch, 38
Balloch Castle, 56
Ballochmyle, 602
Ballochmyle Hospital, 338, 339
Ballyalnach, 442
Balmanno, 61
Balmerino, 227
Balmoral, 216
Balmoral Hotel, Moffat, 136
Baltic Chambers, 662
Baltic Exchange, 361
Baltic Works, 518, 537
Balvenie, 36, 39
Banavie, 600
Banchory, 604

Banff
    brewery, 544
    church, 285
    gun battery, 398
    hospitals, 313
    hotels and inns, 138–9, 145–6, 148, 168
    houses, 37, 71
    lunatic asylum, 321
    mercat cross, 613
    museum, 364
    schools, 300
    warehouses, 626
Banff Hotel, 168
Banffshire
    bee boles, 204
    boat building yards, 585
    Carron Bridge, 529
    castles, palaces and fortified houses, 30, 36, 37–8, 39, 43
    château, 40
    church, 364
    country houses, 51, 195
    fishing stations, 587
    gardens and garden buildings, 214, 215
    harbour lights, 582
    harbours, 571, 573
    hotels and inns, 145–6, 148, 168
    houses, 71, 78
    kiln-barns, 449
    laundry, 195–6
    lighthouses, 582
    lunatic asylum, 321
    military outpost, 397
    swimming pool, 375
    tannery, 540
    warehouses, 626
    see also Banff
Bangour, 324
Bank of Scotland, 638, 639, 640
    Edinburgh, 631, 653, 657, 674
    Glasgow, 637, 657, 668, 672
    see also Exchange Bank of Scotland; National Bank of Scotland; Union Bank of Scotland
Bank of Scotland Chambers, 651
banks, 137, 631–40, 653, 657, 658, 666–7, 668, 670
    Edinburgh, 631–2, 633–7, 642, 653, 657, 658, 660, 664, 668, 669, 674, 675

Glasgow, 632–3, 634, 637, 638,
    642, 646, 651, 653, 657, 658,
    660, 666, 667, 668, 672, 673
  Stirling, 632, 638, 639, 640
  see also names of individual banks
Banner Mill, 537
banqueting house, Edzell Castle, 211, 371
Barbush, 514
Barclay, Andrew, 514, 532
Barclay-Bruntsfield Church, 289
Barclay Curle, 533
Barclay, George, 258
Barclay, Harry, 135
Barclay, Hugh, 644, 649
Barclay, Hugh and David, 289, 294–5, 304, 666
Barclay of Urie, 447
Barham, Samuel, 197
bark mills, 540
barking, salmon nets, 587
barmkins, 28, 33
Barnhill, 318
barns, 70, 193
  as seasonal workers' accommodation, 239, 240
  at shepherds' dwellings, 489
  blackhouses, 92
  byre-dwellings, 425–6
  feed preparation facilities, 456
  grain storage, 439–40, 446
  hay storage, 452
  kilns and, 449–50
  threshing facilities, 438–9, 440, 444–5
  wool storage, 489
  see also girnals; granaries; hay storage
Barnton Quarry, 413
Baronial architecture *see* Scots Baronial architecture
Barony Church, 289
Barony Colliery, 561, 563
Barony Parochial Asylum, 321
Baroque architecture
  Argyll Motor Works, 534
  banks, 653, 660, 666, 667
  clubs, 148
  country houses, 51, 52
  hunting lodge, 221–2
  insurance offices, 640, 658, 667
  Kilncraigs Mill, 527
  offices, 662

Scotsman building, 664
  warehouses, 627, 665, 666
Barossa Place, Perth, 76
Barr & Stroud, 535
Barr, Professor, 165
Barra Head, 583
barracks, 390, 391, 393, 395, 397, 402, 404
  Inverness-shire, 229, 391
Barrie, William, 318–19
Barr's Hotel, 168
Barry, Sir Charles, 148, 168
Barry, Edward M, 152, 153, 156
Barry Mill, 501, 542
Barry's Tavern, 141
bars *see* public houses
Bartholomew's, Geographical Institute, 656
basements *see* cellars
basket-houses, 15–16
bastions, 382, 383, 385–6, 388, 390, 391, 393, 398
Bath House, 168
Bath Street, Glasgow, 651
Bathgate, 182, 535, 559
bathhouses, 211, 215
baths, 347, 374–6
  see also bathhouses; pit-head baths
batteries *see* gun batteries
battery housing, poultry, 478
Battery Point, 368
Bavaria, 19
Bavelaw Castle, 61–2
Baxter, Clark & Paul, 333
Baxter, John, elder, 53, 193
Baxter, John, younger, 192, 630
Baxter Park, 370
Baxter's mill, 528
Bay Hotel
  Gourock, 183
  Stonehaven, 173
Bayle's Tavern, 148
bazaars, 620
Beacon Rock, 399
beam engines and engine houses, 520, 554, 566
beams, 513–16, 650
  see also McConnel patent beams
Beaton, Elizabeth, 439
Beattie, George, & Son, 154, 318
Beattie, George Lennox, 165
Beattie, William Hamilton
  Craiglockhart poorhouse, 318

hotels, 154, 156, 160, 164–5, 170, 664
    Morningside Hydropathic, 171
Beattock, 147, 168
Beauly, 580
Beaux Arts style buildings, 152, 657, 662, 665
Bedford Hotel, 148
bee boles, 203–4
Beech House, 71
beehive buildings, 16
    see also dovecots; shieling huts
Beesknowe, 82
beetling mills, 539
Begg, Ian, 186
Behrens, Peter, 533
Belfast roofs, 518, 533, 584
Belhaven, Lord, 426
Belhaven brewery, 544
Bell, Alexander Graham, 353
Bell family, 37, 45
Bell, George, II, 664
bell-houses see bell towers; bellcotes; town bells and town clocks
Bell, J A, 646
Bell Mill, 513
bell pits, 560
Bell Rock, 583
Bell, Samuel, 361, 630
Bell Street, Glasgow, 650, 665, 669
bell towers, 526
bellcotes, 196, 296, 528, 539
Bellefield Sanatorium, 328
Belleport, 575
Bellevue Hotel
    Dunbar, 174
    Glasgow, 183
bells see bell towers; bellcotes; town bells and town clocks
Bell's Mills, 186
Bell's schools, 298
Bellshill Maternity Hospital, 315
Belmont, Edinburgh, 77
Belvedere, Dunfermline Palace, 43
Ben Wyvis Hotel, 171
Bennetts Associates, 675
Benson and Forsyth, 366
Bentham, Jeremy, 320
Bernera, 32, 391, 393, 397
Berneray, 232
Berriedale, 587–8
berry pickers' accommodation, 239, 241

Berwick, 615
Berwick-on-Tweed, 391
Berwickshire
    banks, 639
    barn or girnal, 440
    barracks, 391
    boathouse, 215
    bone-meal mill, 482
    brewing facilities, 544, 545
    castles, 30, 39, 40, 49
    châteaux, 39, 40
    churches, 287, 297
    country houses, 52, 53, 55, 57, 63, 68
    defences, 31, 383, 401, 410, 412
    farmhouse, 78
    gardens and garden buildings, 212, 215, 216
    grain mills, 450
    hammels, 475
    hotels and inns, 136, 137, 144–5
    killing house, 482
    municipal buildings, 253, 269, 270
    paper mill, 540
    pig housing, 481
    poultry housing, 477
    prison, 269
    root crop storage, 455
    sheep houses, 486
    stables, 194, 431
    threshing machines, 440, 442
Biel, 77
Bield, The, 81
Bielgrange, 71
Bilbohall Hospital, 320
Billings, Robert William, 59, 289, 649
Bilston Glen Colliery, 563
Binns, The, 42, 215
bird housing, 476–9
    see also aviaries; dovecots; pigeon lofts; poultry houses
Bird in the Hand inn, 184
Birkhill, 226
Birkwood, 325
Birnam, 77, 168
Birsay, 17
Bishopbriggs, 344
Bishop's Palace
    Aberdeen, 279
    Dornoch, 271
Black Bull inn
    Banff, 145–6, 148, 168
    Coldstream, 136

Edinburgh, 128–9
Glasgow, 133, 142
Lauder, 137
Moffat, 136
Black, Robert, 632–3, 638, 640, 646
Black Vaults, Leith, 626
Blackford, 544
Blackhills, 458
blackhouses, 17, 92–3, 425
   see also longhouses
Blackie (printer and publisher), 77, 656
Blackness, 31, 35, 495
Blackness Castle, 385, 386, 388, 389, 402
Blackness Foundry, 532
Blackrig, 474
Blackshiels Inn, 130
blacksmiths' workshops see smithies
Blackwater Dam, 559
Blair, 212, 216, 219
Blair Adam, 214
Blair Atholl, 146
Blair Castle, 59
Blair, James, 267
Blair, Patrick, 363
Blairdardie, 600
Blairgowrie, 140, 519, 535, 638
Blairhall, 68
Blairquhan, 57
Blak-Volts, Leith, 626
Blanc, Hippolyte Jean, 83, 324
bleachfields, 196, 502, 505
bleachworks, 345, 495, 520, 539
Blervie, 33
blockhouses, 31, 382–3, 404
   see also pillboxes
Blore, Edward, 58
Bluebell Inn, 127
Blythswood New Town, 75
Blythswoodholm Hotel, 156, 159, 162, 163, 187, 656–7
boarding schools, 303–4
Boarhills, 227
Boar's Head, 127, 135
boat building yards, 585–6
boathouses, 192, 207, 215
Bog of Gight see Gordon Castle
Bog Road, Penicuik, 82
Boghouse of Crawfordjohn, 42, 45
boiler houses, 443, 477, 504, 523, 529, 562
boiler works, 530

boilers see steam power
boiling houses, 207, 456, 580, 587
boiling vats, 502
Bolsham, 443
Bonawe Ironworks, 345, 555, 556
bone-meal mills, 482
Bo'ness, 376, 564, 574, 626
Bo'ness Tolbooth, 253
Bonhill, 495
Bonnington, 516, 547
Boon, 455, 482
boot-making see shoe factories and workshops
booths
   pier-due collecting, 577
   retail trade, 614–16, 627
Borders
   mill kilns, 448
   shepherds' dwellings, 489
   stells, 485
   woollen mills, 518, 520, 521, 535–6
   see also names of individual counties
Bornish, 232
Boroughloch brewery, 544
Borthwick, 35
Borthwick, H, 488
Boswell, Adam, 53
Boswell, George, 669
Boswell, Mitchell & Johnston, 315
Botanic Gardens, 221, 368–9, 371
Botany Mill, 537
bothies, 92, 224–30, 232, 577, 587
   see also farm bothies
Bothwell Castle, 28
Bothwell Chambers, 646
Bothwell Street, Glasgow, 673
bottling plant, Perth, 519
Boturich Castle, 56
Boucher & Cousland, 151
Boucher, James, 79, 649, 654, 656
Boulton & Watt, 350, 514, 520, 524, 561
Bowerhope, 487
Bowhouse, 86
Bowie, Alexander, 76
Bowling, 586
bowling facilities, 214, 372
Bowmore, 288
Bowring Building, 672
Boyd, Thomas, 631
Boyd's Inn, 128
Boyne Castle, 30, 37–8, 43

690 • INDEX

Braco, Lord, 51
Brae, The, 78
Braefoot, 403, 404
Braemar, 40, 172, 397
Braeminzion, 471
Braes of Angus, 397
Braid Estate, 79–80
Braid Hills, 410
Braid, James, 178
Bramah, Joseph, 352
Bramhill, Harold, 670
Brand, Baillie Alexander, 628
Bray, J B, 63
Breadalbane, Earl of, 54, 135, 390
Brechin Round Tower, 277
Brechin Tolbooth, 253
Bremner, James, 572
Brereton, Sir William, 28
Bressay, 404, 443
Bressay Sound *see* Lerwick
breweries, 501, 502, 514, 520, 525, 543–5
    see also maltings
brewhouses, 139, 193, 545
brick construction
    chimneys, 523
    hotels, 177–8
    industrial buildings, 511–12, 544–5, 563
    offices, 672
    railway structures, 601, 602, 604
brick-making workshops, 495
brickworks, 557–8, 563
Bridge of Allan, 77, 168, 170
Bridge of Dee, 140
Bridge of Earn, 139, 168, 338, 339
Bridge of Weir Sanatorium, 328
Bridge Street railway station, Glasgow, 604
bridge-keepers' cottages, 600
Bridgehaugh, 538
Bridgend, 456
bridges, 393, 397, 594–5, 597–8, 603
    Banffshire, 529
    Dunbartonshire, 344
    Dundee, 516
    Glasgow, 600, 601, 603
    see also canal bridges; railway bridges; viaducts
Bridges, James, 632
Brighton Hotel, 153
Britannia Building, 660
British Hotel, Dundee, 143

British Linen Bank, 631, 634–5, 637, 638–9, 658, 666
British Petroleum, 559
Britoil building, 673
Brittany, 20
Britton, John, 59
Broadfoot, Mr, 480
Broadford Mill, 514
Broadford Works, 524
Broadlie, 514
brochs, 10, 191–2
Brodbreidge, John, 628
Brodick, 215
Brodie, 33, 42
Brodie, Dr, 325
Bromhead, H K, 666
Brooke, General Alan, 409
Broom Estate, 84, 86
Broom of Moy, 588
Broomhill Hospital, 336
Broomielaw Quay, 571
Brora, 560, 579
Brothick Mill, 524
Broughton, 63
Broughty Castle, 402
Broughty Ferry, 77, 79, 80, 402, 572
Brown & Carrick, 158
Brown, Mr, 482
Brown Brothers, 532
Brown, James, 158, 650
Brown, William, 524
Browne, George Washington
    banks, 658, 660
    Caledonian Hotel, Edinburgh, 165
    Edinburgh Conservative Club, 157
    houses, 79, 82
    Royal Hospital for Sick Children, 333
    Standard Life building, 657
    see also Peddie & Browne
Browne, W A F, 323
Broxburn, 182
Bruce & Hay, 181, 512, 665
Bruce, Alexander, 51
Bruce, Sir George, 291, 560
Bruce, O Tyndall, 57
Bruce, Robert, 68
Bruce, Robert the, 291
Bruce, Sir William, 43, 48–9, 51, 287, 629
Brucefield, 40, 42
Brucefield House, 51–2
Bruford, Alan, 232, 233

Bruichladdich distillery, 512
Brunswick Hotel, 187
Brunt, The, 76
Brunton House, 78
Bruntsfield *see* Edinburgh
Bruntsfield Hospital, 337
Bruntsfield Links, 371
Bryce, David, 154, 159
  banks, 633, 634–5, 636, 653
  clubs, 150, 151, 157
  country houses, 59, 60
  Dundee Exchange Coffee Room, 643
  Edinburgh music hall, 361, 373
  Fettes College, 304
  houses, 79
  insurance offices, 150, 640
  newspaper offices, 655
Bryce, David, younger, 150, 154, 642, 670
Bryce, John, 75, 646
Bryden, R A, 181
Bryden, Robert, 328
BT Broomielaw, 675
Buccleuch Arms, 168
Buccleuch, Duke of, 137, 147, 485
Buccleuch Mills, 538
Buchan, 412, 413
Buchan, Earl of, 366
Buchan Battery, 32
Buchanan of Arden, John, 56
Buchanan of Ardoch, John, 56
Buchanan Street, Glasgow, 604, 650, 651
Buchanan, William, 366
Buchanhaven, 587
bucket mill, 501, 506
Buckholmside Works, 541
Buckie, 573, 582, 585
Buckquoy, 17, 19
Budapest Ontodei Foundry Museum, 532
Bughties, The, 80
Building Design Partnership, 675
Buith Raw, Edinburgh, 615
bull pens, 476
Bullionfield, 540
Bullough, Sir George, 63
bungalows, 84, 101–2, 104
Burges, William, 642
burgh schools, 294, 300, 303–4
Burghead, 572, 578, 579, 582, 626
Burgie, 33

Burke & Martin, 671
Burke, Ian, 185
Burn, Rev A R G, 228
Burn, James, 73, 631, 637
Burn, William, 317
  asylums, 320
  banks, 636, 638
  country houses, 57, 59
  Edinburgh New Club, 148
  Greenock Custom House, 586
  houses, 77, 168
  inns, 145–6, 147–8, 168
  insurance offices, 640
  schools, 300, 303, 304
  tenements, 111
  Union Street, Dundee, 143
Burnet & Boston, 660, 667, 668
Burnet, Bell & Partners, 86
Burnet, Boston & Carruthers, 662
Burnet, John, 586
Burnet, John, elder
  banks, 653
  Clapperton's warehouse, 647, 648
  Conservative Club, Glasgow, 151
  Glasgow Stock Exchange, 642
  Merchants' House, Glasgow, 657
Burnet, Sir John J
  Barony Church, 289
  Clyde Trust Building, 657
  Glasgow Savings Bank, 667
  hospitals, 333, 334
  hotels, 164, 165, 175, 178
  houses, 75, 80, 83
  McGeoch warehouse, 666
  offices, 662, 668
  Wallace Scott Tailoring Institute, 529
Burnet, Sir John, Tait & Lorne, 183
Burnet, Sir John, Tait & Partners, 186
Burnet, Son & Campbell, 174, 175
Burnett, J Russell, 176
Burns, J & C, 654
Burns-Aitken head office, 654
Burnside Mill, 512
Burntisland
  alumina and red oxide works, 559
  church, 285–6, 287
  coal hoists, 581
  docks, 574
  ferry pier, 577
  hand crane, 580
  harbour lights, 582
  railway facilities, 572, 606

swimming pool, 375
Burray, 579, 627
Burrell Gallery, 366
Burt's Hotel, 140
bus stations, 597
Busby, 605
business and commercial buildings, 624–79
   see also retail trade buildings
business chambers, 646–50, 666
   Dundee, 646, 652, 666
   Edinburgh, 636, 646
   Glasgow, 156, 163, 632, 646, 657, 658, 659, 660, 662
   see also offices
business parks, 674
Bute
   country houses, 54, 162
   grave-markers, 290
   hotels, 166
   hydropathic, 170
   lighthouses, 583
   Marine Biological Research Station, 307
   piers, 575, 577
   potato storage, 459
   seaside pavilion, 378
   tolbooth, 253
   villa, 79
Butt, J, 506
Butter Tron, Edinburgh, 624–5
Butterfield, Irvine, 229
bykes, 446
byre dwellings, 19–21, 92, 425, 465
   see also blackhouses; farmhouses; longhouses
byres, 19, 20, 193, 425–6, 429, 465–8, 476
   Aberdeenshire, 465, 475
   as seasonal workers' accommodation, 239, 240
   as wool stores, 489
   at shepherds' dwellings, 489
   burgh houses, 93
   granaries, 447
   hay lofts, 451
   see also byre dwellings; cattle courts; milking and milk processing facilities

C

Cadder, Thomas, 45
Cadogan Square, Glasgow, 676
Cadzow Castle, 31, 45
Caerlaverock Castle, 28, 30, 31, 39, 44
Café Royal Hotel, 152–3
Caird's Marine Engine works, 532
Cairndhu, 79
Cairndinnis, 445
Cairney Building, 650
Cairnie Point, 587
cairns, 216
Cairns Castle, inn, 146
Caithness
   airfields, 408
   barn, 449
   bee boles, 203
   brewery, 544
   bykes, 446
   byres, 466
   castles, 31, 192
   cemetery, 291–2
   double-skin walled buildings, 17
   dovecot, 203
   farm steadings, 425
   farm workers' housing, 427
   fishing stations, 580, 587–8
   flagstone quarries, 558
   gate lodges, 192
   harbours, 572, 579, 582
   herring curing sheds, 579
   hotels and inns, 147, 165–6
   houses, 587–8
   ice houses, 579, 587–8
   kilns, 449, 450, 554
   lighthouses, 582
   new town, 76
   poultry housing, 477
   prison, 250
   privy, 204
   radar station, 412
   roads, 595
   sheep houses, 483, 485, 486
   smokehouses, 580
   threshing machines, 443
   timber gallery, 37
   tolbooth, 250
   wheelhouses, 10
Calcutta Buildings, 652
Calder, 38, 40
Caldwell Castle, 55
Caledonia Engineering Works, 514, 532
Caledonian Bank, 638, 640
Caledonian Brewery, 545
Caledonian Canal, 395, 598, 599, 600

Caledonian Chambers, 662
Caledonian distillery, 547
Caledonian Hotel
  Aberdeen, 159–60
  Dunoon, 166
  Edinburgh, 160, 165
  Inverness, 137, 185
  Oban, 166
Caledonian Insurance building, 150, 640, 669, 671
Caledonian Ironworks, 532
Caledonian Railway head office, 162, 654, 656
  see also Central Station, Glasgow
Caledonian Station, Dundee, 159
Caledonian Station Hotel, Edinburgh, 153
Caledonian United Services Club, 158
Callander, 140, 168, 170, 174, 604
Cally House, 52
Cally Mains, 453
Calton Hill, 73, 359, 368
Cambus, 547
Cambusnethan, 290
Cameron, Duncan, 166
Cameronbridge, 547
Campbell & Arnott, 673, 674
Campbell, Sir Archibald, 57
Campbell Douglas & Sellars, 332, 653, 655
Campbell Douglas & Stevenson, 644
Campbell, General John, 54
Campbell, Sir John, 390
  see also Breadalbane, Earl of
Campbell, John Archibald, 660, 662
Campbell, Reid & Wingate, 669
Campbell Warehouses, 649
Campbeltown, 398, 545, 547, 577
Campbeltown Town House, 267
Camperdown Dock, Dundee, 574
Camperdown Works, 523, 526, 528
Camphouse, 140
Canada Court, 647
canal bridges, 598, 599–600
canals, 342, 344, 598–600
  see also Forth and Clyde Canal
canalside houses, 95, 600
canalside warehouses, 600
cancer hospitals, 335
Candlemakers' Hall, Edinburgh, 628
Candleriggs, 647
Canisbay Manse, 203
canneries, 580

Canning Street, Edinburgh, 675
Cannon Street Station Hotel, 152
Canongate Tolbooth, 251, 257, 258, 262
canteens, collieries, 563, 565
Cappon, Thomas Martin, 175
car factories, 517, 534
car parks, 597
Cardoness House, 587
Cardowan Colliery, 564
Careston, 287
cargo handling facilities, 578, 580–2
  see also goods movement facilities
Carlekemp, 83
Carlingnose, 403
Carlops, 553
Carlos, King, of Portugal, 163
Carlton Hotel
  Edinburgh, 165, 664
  Forres, 173
Carlton Place, Glasgow, 75
Carluke, 605
Carmichael, J & C, 535
Carmylie Manse, 203–4
Carnegie, Andrew, 309, 370
Carnegie, Sir David, 54
Carnegie, Susan, 319
Carnell, 221
Carnousie, 27
Carnoustie, 528
Caroline Park, 109, 198
carpenters' shops see woodworking workshops
Carr Bridge, 604
carriage houses, 190, 192–4, 597
  see also railway buildings
Carrick, John, 180
Carrington, 288
Carron Bridge, 529
Carron Ironworks, 533, 558
  coal-fired smelting, 556
  foundation, 495, 510
  office and ancillary areas, 527
  sheer legs, 529
  suspended floor, 532
Carronbridge, 606–7
Carruthers, Ballantyne, Cox & Taylor, 86
Carruthers, Frank, 166
Carruthers, William Laidlaw, 79, 80
Carse of Gowrie, 239
Carsluith, 37
Carstairs, 57

Carstairs Junction, 603
Carstairs Street mill, Glasgow, 536
Carswell, James, 647
Carswell, William, 647
cart bays and sheds, 425, 426, 433–5, 447, 448, 500
  see also gig houses
Cartside Mill, 513, 524, 526
Case Design, 673, 675
Cassillis, Earl of, 55
cast-iron see iron construction
castellated architecture
  Armadale Castle, 195
  Bo'ness Tolbooth, 253
  breweries, 544
  cistern, 344
  dairies, 473
  flour mills, 541
  houses, 54–7, 77, 139
  lighthouses, 583
  sluice houses, 344
Castle Douglas, 442
Castle Forbes, 473
Castle Fraser, 37, 40, 59, 194
Castle Grant, 40, 601
Castle Hotel
  Greenlaw, 144–5
  Haddington, 142
Castle Kennedy, 214
Castle Rock walk, Stirling, 368
Castle Semple, 215
Castle Semple Collegiate Church, 291
Castle Street, Edinburgh, 73, 675
Castle Sween, 30
Castle Tioram, 30
Castlecliffe, 79
Castlehill Reservoir, 344
Castlemilk, Dumfriesshire, 59
castles, 27–47, 48, 382
  Aberdeen, 388
  Angus, 402
  Argyll, 29, 30, 35, 390, 397
  Inverness-shire, 28, 30, 390, 391
  see also names of individual castles
Castletown, 579
Cathcart, 529
cathedrals, 278–80
  Aberdeen, 279, 281
  Dunblane, 284, 289
  Edinburgh, 269, 270–1, 282, 285
  Glasgow, 279, 282, 630
  St Andrews, 278–9, 282
Catherine Lodge, 73

Catrine, 112, 345, 519, 526, 535
cattle courts, 426, 448, 468, 474–6, 481
cattle sheds see byres
cattlemen, housing, 427–8
Caulfeild, Major, 131, 397, 594
cellars
  breweries, 545
  châteaux, 42
  houses, 68, 73, 615
  materials storage, 499, 614, 615, 626
  potato storage, 458
  tower-houses, 32, 33
  workshops in, 498, 614
cellular houses, 10, 13, 18–19
cemeteries, 291–2
Central Bank, Perth, 638
Central Chambers, Glasgow, 657
Central Hotel
  Annan, 166
  Edinburgh, 154
Central Station, Glasgow, 162, 604, 654, 656, 658
Central Station Hotel, Glasgow, 162–3, 187, 658
Ceres Weigh House, 260, 269
Chain Home radar stations, 410, 412
Chalmers Hospital, 313
Chalmers, Peter MacGregor, 181, 289
Chambers, Sir William, 53
Chanonry Point, 579
Chapel Works, 523, 524, 536
Chapelcross, 346
chapels, 288
  as family mausolea, 215
  Dundee, 632
  Edinburgh, 288, 368, 628, 630
  Edinburgh District Asylum, 324
  in castles, 28
  in palaces, 36, 43
  in tower-houses, 32, 33
  Stirling Castle, 43
  university, 305
  see also churches
Charing Cross Station Hotel, London, 152, 156
Charlestown, 554
Charlotte House, Glasgow, 672
Charlotte Square, Edinburgh, 73, 111, 182, 675
Charlotte Street, Glasgow, 73
Charteris, Francis, 52–3

INDEX • 695

châteaux, 28, 37–42, 45
Chatelherault, 221–2
cheese rooms, 468–9, 471–2
    see also milking and milk processing facilities
chemical works, 520, 523
    see also petrochemical industries
Chesser, John, 76, 655–6
Chevening, 49
Chevron, 674
Chieftain Forge, 535
children's hospitals, 330, 331–4
    see also orthopaedic hospitals
children's playing fields, 563
chimneys
    bakeries, 502
    boiling houses, 580
    breweries, 544
    châteaux, 42
    girnals, 198
    hotels, 155
    houses, 49, 92, 94, 103
    mills and factories, 521–3, 540
    smithies, 501–2
    steam engine houses, 443
    workshops, 502
Chinese tents, 214
Chirnside Paper Mill, 540
Chirnsyde, James, 129
Christall, William, 311
    see also Crystall, William
Christian Institute lodging house, 181
Church of the Holy Rude, Stirling, 282
Church of the Holy Trinity, Edinburgh, 284
church schools, 296, 297–8, 528
churches, 276–89, 374
    Banffshire, 285, 364
    Edinburgh, 270–1, 282, 284, 285, 286–7, 289
    Fife, 277–9, 280, 282, 284, 285–6, 287, 291
    Glasgow, 262, 289, 630
    Inverness, 368
    Perth, 269, 271, 282, 284, 288
    Perthshire, 130, 285
    Renfrewshire, 288, 291
    see also chapels
Churchill Barriers, 408
churn rooms and churning houses, 470–1
cinemas, 376–8, 667

circular plan forms, 9, 10–16, 18–19, 20–1
    see also shieling huts
cisterns, 344
citadels *see* defences; forts
City & County Buildings, Glasgow, 642
City Chambers, Edinburgh, 630
City Hospital, Edinburgh, 329
City Hotel, Edinburgh, 150
City Mills Hotel, Perth, 186
City of Glasgow Assurance building, 650
City of Glasgow Bank, 633, 642, 646, 653, 655, 658
City of Glasgow District Asylum, 321
City Temperance Hotel, Edinburgh, 150
Clachnaharry, 600
Clackmannan, 536
Clackmannanshire
    banks, 639, 666
    baths, 375
    brewery and brewhouse, 545
    château, 40, 42
    coalfields and collieries, 560, 563, 566
    distilleries, 480, 545, 547
    gardens, 212
    houses, 83, 367
    inns, 130, 136
    kilns, 449, 501
    mill clocks, 526
    mill office, 527
    mills, 512, 528, 535–6
    offices, 669
    paper mill, 540
    recreation buildings, 375
    schools, 303, 304, 528
    tannery, 540
    water tower, 540
    weaving factories, 537
clamps, 500, 553, 558
Clapperton's warehouse, 647, 648–9
Clare Hall, 215
Clarendon Hotel, 154, 156
Clark Kincaid, 532
Clark, R & R, 655–6
Clarke & Bell, 642, 653, 664
Clarke, David *see* David Clarke Associates
classical buildings
    banks, 632, 639

696 • INDEX

business chambers and warehouses, 646
churches, 286, 288, 289
corn exchanges, 644
country houses, 48–9, 53, 57, 63
custom houses, 586
garden buildings, 214
gate lodges, 192
hospitals, 312–13
hotels, 143, 173
houses, 68, 71–2, 73, 76
National Gallery of Scotland, 367
offices, 651, 662, 667
St Andrew's Halls, 373
schools, 298, 300
tenements, 109
textile mills, 536
theatres, 362
tolbooths, 253
tollhouses, 596
see also neo-classical buildings
clay industries, 557–8
clay-pipe making, 501
Claypotts, 37
Clayton, Lieutenant-General Jasper, 131
Clayton, Thomas, 262
cleansing, 348–9
see also sanitation
cleitean, 12, 13, 92
Cleland, James, 78, 374
Clerk, Sir James, 53, 193
Clerk, John, 628
Clerk, Sir John, 51
Clifford, Henry F, 83, 84, 665, 666
climbing huts, 227–30
Cloch, 583
Clock Mill, 526
clock towers, 164, 187, 217–18, 372, 387
mills and factories, 526, 533
clocks, 268, 526, 628
see also town bells and town clocks
Closeburn, 554
clothing factories and workshops, 496, 539, 666
Clouston, Dr Thomas, 323
club huts, mountaineers', 228
clubhouses, 371
clubs, 148–58, 361, 374, 672
see also sports facilities
Clugston, Beatrice, 336
Cluny Hill Hydropathic, 170

Clyde Foundry, 510–11
Clyde Hall, University of Strathclyde, 186
Clyde Navigation Trust, 586
Clyde Street lodging house, 180, 181
Clyde Trust Building, 657
Clyde Yacht Club, 174–5
Clydebank, 117, 526, 533, 582
Clydesdale see Lanarkshire
Clydesdale Bank, 632, 638, 640, 653, 673
Clydesdale Bank Plaza, Edinburgh, 675
Clynelish Farm, 196
Co-operative Insurance, 671
coach houses and carriage houses, 190, 192–4, 597
see also gig houses; stables
coachmen's flats, 597
coal hoists, 581
Coal Wynd flax mill, 526
coalfields and collieries, 109, 560–6
coast defence, 32, 407–8
see also gun batteries
Coastguard, gun battery, 403
Coatbridge, 375, 557
Coatbridge Technical College, 308
Coatbridge Time Capsule, 376
Coats & Clark, 516
Coats, J & P, 521, 529, 668
Coats Viyella Building, 673
Coatts, William, 258
Cobban & Lironi, 185, 187
Cochrane MacGregor, 675
Cochrane Street, Glasgow, 647
Cock of Arran, 552
Cockburn, Lord, 167
Cockburn, Sir Archibald, 440
Cockburn Hotel
Edinburgh, 150
Glasgow, 156
Cockburn Street, Edinburgh, 150, 154, 647, 654
Cockenzie, 571, 626
Cockerell, Charles Robert, 637, 653
Coe & Godwin, 313
coffee houses and coffee rooms, 142, 358, 360–1, 363, 521, 643
Edinburgh, 361, 630
Glasgow, 142, 360–1, 642
see also clubs; exchanges; hotels
Coigach, 227
Coignafearn, 207

Coignet reinforced concrete, 516
Coilsfield House, 631
cold stores *see* ice houses
Coldingham, 410
Coldstream, 136, 544
Coleshill, 49
Colinton, 80
Coll, 13
collieries *see* coalfields and collieries
Collins (printer), 656
Colonies, The, Edinburgh, 97
colony asylums, 322–4
Colquhoun Arms, 175
Coltness Iron Company, 556
Columba Hotel
    Inverness, 173
    Oban, 173
Colville Group, 557
Colville, Sir John, 63
Colvin, Howard, 48
Colzium, 215
combustion-powered machinery, 504
Comelybank Mill, 524, 537
commemorative structures, 215, 289–92
    see also monuments
Commercial Bank, 638, 639
    Aberdeen, 668
    Edinburgh, 631–2, 633–4, 635, 636, 664
    Glasgow, 637, 658, 668
    Stirling, 632
commercial buildings, 624–79
    see also retail trade buildings
Commercial Court, Glasgow, 647
Commercial Hotel, Alyth, 140
commercial hotels, 133
Commercial Union office, 668
Commonside, Renfrewshire, 471
community buildings, 249–420
Comprehensive Design, 673
computer centres, 673–4
Comrie, 563
Con, Alexander, 45
concert halls, 359–60, 373–4, 529
concrete construction
    Central Station, Glasgow, 545
    commercial buildings, 658, 666, 670
    dams, 559
    distilleries, 512
    docks, 573
    estate buildings, 191

industrial buildings, 516, 519
laundry, 197
piers, 577
swimming pools, 375
see also mass concrete construction; precast concrete construction; reinforced concrete construction; shell concrete
confectioners' workshops, 496
Conference Centre, Edinburgh, 675
Conisby, 442
Connachan Lodge, 76
Connel Bridge, 603
Connell, Frank, 669
Conservative Club
    Edinburgh, 157
    Glasgow, 151, 157
conservatories, 60, 219, 221, 371
Considère reinforced concrete, 516
Constitutional Club, London, 157
convalescent homes, 330, 336–7, 563
cooling *see* ice houses; ventilation
Cope, Lieutenant-General Sir John, 131
Corbett, A Cameron, 62
Corgarff, 397
Corkerhill, 606
corn exchanges, 643–4, 646
corn mills *see* grain mills
Corran, 577
Corrienessan, 80
Corstorphine *see* Edinburgh
Cortachy, 40, 471
Cosh, Mary, 191
Cossans farm, 429
Cotswold Manor style, 83
cottage gardens, 80
cottage hospitals, 312, 313
cottage mansions, 76
Cottage, The, Dundee, 80
cottages, 90–107
    lock- and bridge-keepers', 600
    potato pickers', 242
    railway crossing-keepers', 601
    seasonal workers' accommodation, 239
    shooting estates, 207
    see also farm workers' housing; gate lodges
cotton mills, 495, 513–14, 520, 525, 535, 536
    Aberdeen, 537

Ayrshire, 345, 519, 535
Glasgow, 512, 514, 536
Lanarkshire, 514, 521, 524, 525, 535
Perthshire, 345, 512, 513, 514
Renfrewshire, 510, 513, 519, 523, 524, 525, 526, 535, 536
see also lades; textile mills; thread mills; weaving factories
council chambers *see* municipal buildings
country houses, 27–8, 40, 48–66, 83, 221
   as hotels or inns, 139, 140, 182
   Ayrshire, 53, 55, 57, 59, 62, 63, 631
   bird housing, 478
   Bute, 54, 162
   see also estate buildings; gardens and garden buildings; gate lodges; villas
County Buildings, Cupar, 142, 143
County Cinema, Portobello, 377
Coupar Angus, 142, 182
Cour, 63
Courier Building, 664–5
court room, Stirling, 263
Courthill, 442
courts *see* law courts
courtyards, 43, 69, 498, 529, 539, 545, 617
   see also cattle courts
Cousin & Ormiston, 155
Cousin, David, 78, 321
   banks, 637, 639
   corn exchanges, 643–4
   India Buildings, Edinburgh, 646–7
Cousland, 502, 506
Cousland, James, 649
   see also Boucher & Cousland
Coutts, George, 328
Covell Matthews, 671, 672, 673
Cowan, J L, 327
Cowane's Hospital, 497
Cowdenbeath, 376
Cowglen Hospital, 327, 339
Cowie stop-line, 409
Cowlairs, 534
Cox Brothers, 654
Cox's Stack, 512, 523
Coxton Tower, 35
Coylumbridge, 186
Craig
   Aberdeenshire, 45
   West Lothian, 141

Craig Ailey, 77
Craig Castle, 37
Craig House, 323
Craig, James, 369
Craig, John, 311
Craigcrook, 44
Craigdarroch House, 51
Craigend brewery, 544
Craigendoran, 577
Craigends, 59
Craigentinny, 607
Craigie Brewery, 544
Craigie Hall, 79
Craigie, Robert, 53
Craigiehall, 413
Craigiehall House, 49
Craigievar, 37, 38
Craigleith House, 328
Craiglockhart *see* Edinburgh
Craiglockhart Hydropathic, 170
Craigmillar, 37
Craigmillar Brewery, 545
Craignethan Castle, 42, 45, 371
Craignure, 575
Craigston, 37, 38
Craik, William, 54
Crail, 249, 251, 258, 571, 582
Crail Airfield, 408
Crakaig, 481
Craleckan, 555, 556
crames, 614
cranes, 580–2, 600, 605, 606
   see also hand cranes; post cranes; tower cranes; travelling cranes
Cranstonhill, 344
Crathes, 37, 220
Crawford Arcade, 620
Crawford, Earl of, 35
crays *see* pig sties
creameries, 470, 473, 480
creel-houses, 15–16
Crerar & Partners, 186
Crest Hotel, 185
Creswell, John, 261
Crichton Castle, 43, 44
Crichton, Richard, 56, 631
Crichton Royal Asylum, 319, 320, 322, 323
Crieff, 76, 170
Crimond, 313
Crinan Canal, 598, 599, 600
Cringletie, 60
Crockness, 399–401

Crofthead Mill, 526, 536
Cromartie, Earls of, 198
Cromarty, 67, 71, 404, 536, 544
    see also Ross and Cromarty
Cromarty Firth, 404, 407, 575
Cromarty House, 54, 193
Cromwellian fortifications, 31–2, 386–8
Crookston Cottage Homes, 318–19
crop storage and processing, 438–64
    see also barns; girnals; granaries; hay storage; kilns; lofts; root crop storage
Crosbie House, 84
Cross Keys hotel, Peebles, 127
Cross Keys Inn
    Dalkeith, 137
    Kelso, 135, 136
Cross Law, 412, 413
crosses see Eleanor crosses; fish crosses; mercat crosses
crossing-keepers' cottages, 601
Crossraguel Abbey, 35
Crowhill, 441
Crown Hotel, Ayr, 143
Crown Inn
    Coldstream, 136
    Penicuik, 139
Crown Tavern, 141
crowstepped gables, 49, 55, 67, 69, 547, 587
Cruden Bay, 176, 177
Cruden, S, 276, 388
Crystall, William, 263
    see also Christall, William
Cul a' Bhaile, 10, 16
Cullen, 78, 145, 214
Cullen House, 195, 200
Cullen, Lochhead & Brown, 338
Culloden, 456
Culloden House, 54
Culross, 43, 69
Culross Abbey Church, 291
Culross Abbey House, 42, 371
Culross Palace, 27, 262
Culross Town House, 262, 263
Culter Paper Mills, 540
Culzean Castle, 55, 219, 478
Cumberland Barracks, 182
Cumledge, 481
Cummins Engine Works, 535
Cumnock, 343
Cunningham, John, 144

Cunninghame Mansion, 642
Cupar
    corn exchange, 644
    municipal buildings, 142, 143, 250, 269
    Tontine Hotel, 142, 143
    weaving factories, 536
curing sheds, 579
curling houses, 215
Curr Night Refuge, 182
Currie, Edinburgh, 540
curtain walls, 305, 510–11, 672
    castles, 28, 29–30, 385
    Oliver's Fort, 387
Curtis & Davis, 185
custom houses, 586, 626, 630
cycle houses, 83
Cyprus Inn, 139

D

Daily Express buildings, 511, 669
Daily Mail office, 664
Daily Record office, 664, 665
Daily Review office, 655
dairies, 214, 426, 466, 467, 468–73
    see also milking and milk processing facilities
Dairsie Church, 285
dairy farms, 465–8, 480, 481
    see also byres; dairies
Dalem, Pieter Manteau van, 387
Dalian House, Glasgow, 676
Dalkeith
    assembly room, 137
    corn exchange, 644
    garden buildings, 220
    inns, 137
    tannery, 540
    tolbooth, 269
    villa, 79
    water tower, 344
Dalkeith House, 51
Dallas Dhu, 545
Dalmarnock, 537
Dalmellington Ironworks, 557
Dalmeny, 280, 287, 290, 403
Dalmeny House, 57
Dalmore, 79
Dalmuir, 585
Dalnacardoch, 131–2, 391
Dalry, Edinburgh, 495, 539
Dalrymple, Sir John, 55
Dalwhinnie, 131, 545

Dalzell, 221
Dalzell Works, 557
dams, 343–4, 559
Dangerfield Mill, 521
Daniel Stewart's Hospital, 303
Darnaway Castle, 207
Darnley Hospital, 327
Darvel, 537, 539
Daunton, M J, 113
David Clarke Associates, 187
David I, King, 280
David Livingstone Tower, University of Strathclyde, 672
David's Tower, Edinburgh Castle, 384
Davidson's Brewery, 544
Davie of Stirling, 537
Davis, Arthur, 669
Davy's Boorach, 230
'daylight factories', 517, 533
de St Martin, Alexander, 280
Dean Castle, 45
Deane, Major-General Richard, 387, 388
Deanston, 345, 537
defences, 30–2, 381–420, 624
    see also castles; forts; tower-houses
Defoe, Daniel, 131, 133, 626
Delgatie, 33
Denburn Valley, 368, 369
Denholm, 538
Denny, 478–9
Denny, William, & Brothers, 97
Dens Works, 523, 526
dental hospitals, 338
department stores, 164, 187, 619, 620–1, 649, 666
    see also shops
depots, railways and railway buildings, 397, 601
derrick cranes, 582
derricks, 585
Devey, George, 207
Devon Colliery, 566
Devon vernacular, 96
Dewar's whisky bottling plant, 519
Dick Brothers, 521
Dick Place, Edinburgh, 78, 79, 86
Dickson, R & R, 56, 146
Dingwall Tolbooth Tower, 253
Diocletian's Palace, 55
dipping sheds, 488
direction indicators, 596
Dirleton, 407, 410, 412

Dirleton Castle, 28
dispensaries, 312, 328, 329, 331, 335
distilleries, 542, 545, 547
    Clackmannanshire, 480, 545, 547
    concrete construction, 512
    engines, 520, 521, 545
    floors, 516
    head office, 654
    Inverness-shire, 512, 545
    Islay, 512, 545, 547
    kilns, 501
    Moray, 543, 545
    pig-keeping at, 480
    Stirlingshire, 513, 545
    see also maltings; whisky bottling plant
Ditherington Flax Mill, Shrewsbury, 513
docks, 399, 573–5, 578, 581, 582
    see also ports and harbours; quays
dockside warehouses, 578
Doig, C C, 542
Dollan Baths, 375
Dollar, 130
Dollar Academy, 303, 304
Don Brothers, 654
Donald, John, 158
Donaldson's Hospital, 303
Donegal workers, 236–7, 238, 239
donjons, 28, 30
    see also lodgings, in castles
Donnachie, I L, 506
Doolan, Andy, 187
doors
    barns, 439, 444
    cart bays and sheds, 434
    country houses, 49, 54
    gig houses, 431
    granaries, 448
    houses, 14–15, 73, 75, 91, 92
    potato houses, 457
    poultry houses, 477
    shieling huts, 13, 14, 15
    smithies, 502
    stables, 430
    tower-houses, 33
    towers, 35
    workshops, 499
Dop, Adrian, 257
Dornoch, 176, 271
double-skin walls, 10, 12, 16–17
Dougarie, 215
Douglas & Grant's, 532

Douglas & Sellars, 332, 653, 655
    see also Sellars, James
Douglas & Stevenson, 644
Douglas Castle, 54
Douglas, Duke of, 54
Douglas Hotel
    Aberdeen, 184
    Edinburgh, 142, 150
Douglas Tolbooth, 271
Douglas Wood, 410, 412
Doune Castle, 29, 30
Dove Inn, 127
dovecots, 190, 200–3, 204, 211, 478, 479
Dowanhill *see* Glasgow
Downing Point, 403
Dr Bell's schools, 298
Dreadnought Inn, 140
Dreghorn, 288
Dreghorn, Allan, 132, 142, 311, 629
Drem, 450
dressmaking workshops, 496
drinking fountains, 524
Drochduil, 480
Drochil, 37
Drone Hill, 410, 412
drovers' accommodation, 224
Drum, The, 51, 217
Drumbathie Road, Airdrie, 80–1
Drumchapel, Glasgow, 289
Drumlanrig, 214, 215
Drummond, Lord Provost, 629–30
Drummond Castle, 220
Drummond, J G Home, 56
Drummond, John, 672
Drummond, Sir William, 168
Drummonie House, 139
Drumsheugh Baths, 374
Drury Street 13–15, Glasgow, 651
dry docks, 574–5
Drygate lodging house, 180
drying facilities, 196, 468, 502, 523, 536, 538, 587
    see also grass-drying plant; kilns; Richmond drying rack
Duart Castle, 30, 397
Dubh Artach, 583
duck housing, 478–9
Duddingston, 214
Duddingston Burn, 448
Duddingston House, 53
Dudhope, 43, 77
Dudhope Castle, 30, 36

due-collecting booths, 577
Duff House, 37, 51, 146, 215
Duff, Neil C, 181, 667
Duff, William, 51
Duffus Castle, 28
Duke of Gordon's Inn, 146
Duke's Arms, 136, 145, 147
Duke's Inn, 130
Dumbarton, 97, 535, 586
Dumbarton Building Society, 97
Dumbarton Castle, 30, 388
    anti-aircraft battery, 410
    barracks, 393
    defences, 31, 389, 393
Dumfries
    banks, 640
    car factory, 517, 534
    church, 288
    hospitals, 311
    houses, 78
    lunatic asylum, 319, 320, 322, 323
    mercat cross and fish cross, 613
    station hotel, 163
    steeple, 253
    trades hall, 631
    woollen mill, 536
Dumfries, Earl of, 53
Dumfries House, 53
Dumfries Midsteeple, 253
Dumfries Royal Infirmary, 311
Dumfriesshire
    banks, 639, 640
    Bath House, 168
    castles, palaces and fortified houses, 28, 29, 30, 31, 38, 39, 44
    cattle courts, 474
    châteaux, 40
    coalfields, 560
    country houses, 51, 57, 59
    doors, 14–15
    farm machinery, 456
    gardens and garden buildings, 214, 215, 219
    hay barns, 453
    horse engine houses, 442
    hotels, 136, 147, 163, 166, 168
    houses, 14–15
    hydropathic, 170
    inns, 135–7, 139, 187
    lime kilns, 554
    metal mining, 554
    municipal buildings, 253

nuclear power station, 346
potato houses, 458
railway stations, 168, 606–7
sandstone quarries, 558
sheep houses, 485, 486
spas, 168
threshing machines, 441
warehouses, 626
see also Dumfries
Dumphail, 77
Dunaverty Castle, 30
Dunbar
assembly room, 142
boat building yard, 585
brewery, 544
custom house, 586
defences, 382–3, 398
Grieve's House, 70
harbour, 382, 398, 571, 578
hotels and inns, 127, 137, 142, 174, 187
lodging house, 238
municipal buildings, 258, 263
town plan, 69
warehouses, 578, 626
Dunbar, Alexander, 267
Dunbar Castle, 27, 31, 382
Dunbar, John G, 67, 76
Dunbar Town House, 258
Dunbarrow, 441
Dunbartonshire
anti-aircraft battery, 411
aqueduct, 344
boat-building yards, 586
bridges, 344, 601, 603
churches, 287
clock tower, 526
convalescent home, 336
country houses, 56, 57
cranes, 582
custom house, 586
distilleries, 545
factories, 526, 533, 534
farm machinery, 456
forge, 535
hospitals, 314, 336, 339
hotels and inns, 127, 140, 175
houses, 15–16, 77, 79, 80, 86, 97
hydropathic, 169–70
lunatic asylum, 321
piers, 577
potato storage, 459
seasonal workers'

accommodation, 238, 239, 240, 241
shipyards, 582, 585
steam engines, 545
tenements and flats, 117
workshops, 495
see also Dumbarton; Dumbarton Castle
Dunblane Cathedral, 284, 289
Dunblane Hydropathic, 170
Duncan, Andrew, 319
Duncan, David, 674
Dundarg Castle, 29, 30
Dundas Castle, 57
Dundas, James, 57
Dundas, Sir Laurence, 636
Dundee
air-raid shelters, 406
airfield, 404
art galleries and centres, 308, 368
art gallery house, 367
assembly room, 361
banks, 632, 637, 638, 640, 653, 666, 667
baths, 374
bell and clock towers, 526
breweries, 544, 545
bridge, 516
business chambers, 646, 652, 666
chapel, 632
chimneys, 512, 523
churches, 288
cinemas, 376, 377
coffee rooms, 361, 521, 643
cranes, 580
creamery, 473
custom house, 586
defences, 386
dispensary, 329
docks, 573, 574, 578
Dudhope Castle, 30, 36
dyeworks, 510, 518, 538
Eastern Club, 152
exchanges, 361, 521, 643, 644
factories, 514, 521, 523, 528, 533, 534, 541
forge, 535
foundries, 530, 532, 534, 535
gallery, 37
head offices, 654, 664–5
heckling shops, 503
hospitals, 313, 315, 329, 331, 334

hotels and inns, 143, 144, 159, 185, 186
houses, 76–7, 78, 80, 81, 101
laundry, 374
libraries, 308, 361
lodging houses, 179, 181
luckenbooths, 615
lunatic asylums, 319, 320, 325
market, 643
masonic halls, 143, 361
mercat cross, 616
merchants' shelter, 643
mills, 518, 521, 524, 536, 537
Morton's Bond, 513
municipal buildings, 182, 267–8, 629
museums, 308, 363
night refuge, 182
offices, 630, 643, 652, 654, 670
palaces, 36, 43
printing and publishing works, 654, 664–5
public gardens, 368, 370
railway facilities, 159, 604, 605, 606
reading rooms, 361, 643
Sailors' Hall, 362
Sailors' Home, 159
schools, 300, 528
sewerage, 353
shops, 143, 144, 615, 630, 666
slipways, 574
swimming pool, 375
tanneries, 541
taverns, 143
tenements and flats, 109, 114, 117, 652
theatre, 362
trades halls, 361, 630
Trinity House, 362
tron, 614, 627
Union Street, 143
university, 307
warehouses, 578, 626, 627, 646
water supply, 343
weaving sheds and factories, 513, 536, 537
Dundee Advertiser, office and printing works, 654
Dundee Banking Company, 632
Dundee City Arts Centre, 368
Dundee Courier, office, 654
Dundee Exchange Coffee House/Coffee Room, 361, 521, 643
Dundee Foundry, 530, 532
Dundee High School, 300
Dundee Infirmary, lodging house, 179
Dundee Institute of Technology, 308
Dundee Royal Asylum, 319, 320
Dundee Royal Infirmary, 313, 329
Dundee Savings Bank, 667
Dundee Steam Forge, 535
Dunderave Castle, 62
Dundrennan Abbey, 279
Dunecht House, 192
Dunfermline
  churches, 280
  half-time school, 528
  lime kilns, 554
  Pittencrieff Park, 370
  town bells, 257
  villas, 45
  warehouses, 518, 527
  weaving factories, 536, 537
Dunfermline Abbey, 43, 291, 448
Dunfermline Abbey Church, 282
Dunfermline Palace, 37, 42, 43
Dunglass, 214, 384, 602
dungsteads, 481
Dunkeld
  bird housing, 478
  churches, 130
  hermitage, 215
  hotels and inns, 130, 132, 136, 140, 145, 147
Dunkeld House, 49, 130
Dunlop, 471
Dunlop, Sir James, 57
Dunlop, Andrew, 268
Dunlop House, 57
Dunlop Street, Glasgow, 650
Dunmore, 475, 501
Dunmore, Earl of, 57
Dunmore Home Farm, 431
Dunmore Park, 57
Dunmore Pineapple, 219
Dunn & Findlay, 164, 664
Dunn & Watson, 175
Dunn, James Bow, 81, 83, 173–4
Dunninald Castle, 57
Dunning, 285
Dunollie Castle, 30
Dunoon, 166, 184, 187, 575, 577
Dunragit, 470, 473, 480
Dunrobin, 207, 216

Dunrobin Castle, 604
Dunrossness, 450
duns, 10
Duns, Berwickshire, 78, 253, 545, 639
Duns Castle, 57
Duns Town House, 253
Dunsdale Mill, 537
Dunstaffnage Castle, 30, 35
Dunsyre, 222
Duntrune Castle, 30
Dunure, 554
Dunvegan Castle, 30
Dunyvaig Castle, 30
Durness, 412
Dury, Theodore, 390
Dury's Battery, Edinburgh Castle, 390
Duthie Park, Aberdeen, 370
dwellings, 27–245
dyeworks, 502, 510, 518, 538, 559
  see also indigo mill
Dykebar, 324
Dykehead, 468
Dysart, 563, 573

E
Eagle House, 675
Eaglescairnie Mains, 440
Eaglesham, 288
Eaglesham Street boat-building yard, 586
Ear and Skin Dispensary,, 335
ear, nose and throat hospitals, 335
Earl's Palace, Kirkwall, 45
Earlshall, 36, 61, 220, 221, 371
earthworks
  garden landscaping, 214
  see also defences
East Barns, 474
East Bridge Flour Mill, 519
East Gerinish, 478
East Kilbride, 375
East Lothian
  assembly rooms, 142, 259
  barn, 440
  barns, 439
  bleachfield drying house, 502
  boat building yard, 585
  brewery, 544
  byres, 466
  castles, 27, 28, 29, 30, 31, 55, 382, 386
  cattle courts, 474, 475
  chapel as mausoleum, 215

  château, 42
  churches, 280, 282, 285, 288
  collieries, 566
  combine harvesters, 445
  corn exchange, 644
  country houses, 52, 55, 57, 59, 67, 68
  custom houses, 586, 626
  defences, 382–3, 384, 385, 386, 398, 407, 410, 412
  farm stables, 429
  farm steading, 468
  farmhouses, 76
  gardens and garden buildings, 215, 216, 217–18, 219
  grain mills, 450
  granaries, 447, 448
  harbours, 382, 571
  hotels, 127, 142, 170, 174
  houses, 68, 69, 70, 71, 77, 78, 82, 83, 86
  hydropathic, 170
  inns, 127, 137, 187
  kilns, 450, 500, 542
  lodging house, 238
  lunatic asylum, 321
  mercat crosses, 613
  municipal buildings, 257, 258, 259, 263
  nuclear power station, 346
  palaces, 37, 39
  poultry housing, 477
  railway viaduct, 602
  salt extraction, 552
  sawmill, 498
  sheep houses, 486
  summerhouse, 214
  swimming pool, 375
  textile industry, 495
  threshing machines, 440, 441, 442, 443
  town bells, 257
  warehouses, 578, 626
  workshops, 495
  see also Lothians
East Mill, Dundee, 524
East Old Dock, Leith, 573
East Port
  Dundee, 386
  Lennoxlove, 86
East Russell Street lodging house, 180
East Wemyss, 561, 563–4
Easter Buccleuch, 486

INDEX • 705

Easter Eninteer, 471
Easter Ross, 147, 198, 317, 440, 443, 486
  see also Ross and Cromarty
Easterhouse, 344
Eastern Club, Dundee, 152
Eastfield, 477
Eastfield Mill, 538
Easton, Amos & Sons, 153
Edderston Farm, 482
Eden, 40
  see also Mains of Eden
Eden Street, Dundee, 77
Edenmouth, 441
Edgelcliffe, 79
Edina Works, 656
Edinburgh
  abattoir ventilation, 524
  air-raid shelters, 406
  anti-aircraft defences, 410, 413
  arcades and piazzas, 620, 629, 630
  art galleries, 364, 367
  art gallery houses, 367
  assembly rooms, 133, 142–3, 361, 632
  banks, 631–2, 633–7, 642, 653, 657, 658, 660, 664, 668, 669, 674, 675
    see also names of individual banks
  baths, 374
  breweries, 544, 545
  business chambers, 636, 646
  business parks, 674
  cathedral, 269, 270–1, 282, 285
  chapels, 288, 628, 630
  château, 42
  churches, 270–1, 282, 284, 285, 286–7, 289
  cinemas, 376, 667
  clubs, 148, 150, 152, 157, 158, 361, 374
  coffee houses and coffee rooms, 361, 630
  computer centres, 673–4
  concert hall, 373–4
  conference centre, 675
  convalescent home, 336
  country houses, 53, 59
  creamery, 473
  custom house, 630
  defences, 385, 386
    see also Edinburgh Castle
  dental hospital, 338

dispensaries, 312, 328
distilleries, 547
exchanges, 629–30, 641, 642–4, 651, 669
factories, 516, 534
Finance House, 673
foundries, 258, 532
gardens and garden buildings, 214, 368–9
gas lighting, 350
gatehouses, 31
Greyfriars Churchyard, 291
head offices, 669, 670–1, 673
hospitals, 328, 329, 330–1, 332, 333, 334, 335–6, 338, 339
  see also Edinburgh Royal Infirmary
hotels, 141–3, 148, 150, 152–5, 157, 164–5, 182, 183, 185, 186–7, 640, 664
  see also names of individual hotels
houses, 69, 70, 73, 75, 76, 77, 78, 80, 82, 85, 93, 97, 198, 625
hydropathics, 170, 171
inns, 127–9, 132, 147–8, 168, 184
institutions for mentally handicapped, 325
insurance offices, 150, 632, 640, 657–8, 669, 670, 671, 675
libraries, 309
lighthouses, 582
lodging houses, 179, 181
luckenbooths, 615
lunatic asylums, 319, 320, 323–4
maltings, 545
markets, 614, 617, 664
mercat crosses, 217, 629, 630
merchant hall, 628–31, 642
merchants' booths, 615
mills, 186, 448
municipal buildings, 249, 251, 257, 258, 260, 262, 263, 266, 267, 268, 269, 270–1, 630
museums, 363, 364
music halls, 361, 362–3, 373
newspaper offices, 187, 647, 654, 655, 664, 665
night asylum/night refuge, 182
offices, 630, 646, 651, 654–5, 656, 662, 664, 667, 671, 673–5
paper-making, 495, 539
Physicians' Hall, 634

poorhouses, 313, 316, 318, 319
potato house, 458
printing and publishing works, 516, 655–6
railway facilities, 153, 164, 572, 604, 605, 606, 607, 617
roadhouses, 184
rope-making workshops, 495
sandstone quarries, 558
school for mentally deficient children, 325
schools, 298, 300, 303, 304
sewerage, 352, 353
shops, 150, 615, 616, 617, 618, 619, 620, 630–1
sports facilities, 358, 371, 376
stables, 194
steeples, Magdalen Chapel, Edinburgh, 628
suburbs, 76, 77, 79–80
Surgeons' Hall, 634
swimming pool, 375
tannery, 540
taverns, 141, 144
tenements and flats, 108–9, 111–12, 113, 114, 117, 120–2, 615, 630
theatres, 358–9, 362
timber gallery, 37
tower-house, 35, 38
town bells, 257
town plan, 69
trades halls, 359, 627–8, 630
trons, 624–5
underground bunker, 413
universities, 305, 306, 307, 308
villas, 40, 44, 45, 76, 77, 78, 79, 80, 109, 367
warehouses, 616, 646
water supply, 341, 343, 344
workshops, 495, 496
see also Edinburgh New Town
Edinburgh & Leith Bank, 633
Edinburgh Academy, 300, 303
Edinburgh Assembly Rooms, 361, 632
Edinburgh Business Park, 674–5
Edinburgh Castle, 30, 42–3, 388
  barracks, 390, 393, 397
  defences, 31, 384, 389, 390, 391, 393
Edinburgh Charity Workhouse, 316, 319
Edinburgh City Chambers, 630

Edinburgh City Poorhouse, 313
Edinburgh City Travel Inn, 187
Edinburgh Conservative Club, 157
Edinburgh Cooperative Building Company, 97
Edinburgh Courant, office, 655
Edinburgh District Asylum, 324
Edinburgh Eye, Ear and Throat Infirmary, 334, 335
Edinburgh High School, 300
Edinburgh Hilton Hotel, 186
Edinburgh House, 671, 673
Edinburgh Life building, 640, 658, 660, 670
Edinburgh Lodging-house Association, 179
Edinburgh Lying-in Hospital, 330
Edinburgh Museum of Science and Art, 364
Edinburgh New Town, 76, 312, 616
  assembly room, 361
  dispensary, 312
  gardens, 369
  houses, 73–5
  offices, 646, 651, 667, 671
  shops, 646
  statuary, 291
  taverns and hotels, 141
  tenements and flats, 111–12, 115, 122
  see also Castle Street; Charlotte Square
Edinburgh Royal Asylum, 320
Edinburgh Royal Infirmary, 303, 311, 315, 329, 330, 331, 336
Edinburgh School of Art, 308
Edinburgh Southern General Hospital, 339
Edinburgh Stock Exchange, 669
Edinburgh Tolbooth, 263, 266
Edinburgh University, 305, 306, 307
Edinburgh Waverley, 153, 164, 604, 606, 617
Edis, Colonel Robert W, 157
Ednam House, 76
Ednie, John, 82
education buildings, 294–310, 325
  see also schools; universities
Edward, Alexander, 130, 131, 187
Edward, Charles, 644, 654
Edward Street Mill, 518, 537
Edzell, 40, 175, 211, 371
Eglinton Arms, 147

Eglinton, Earl of, 147
Eglinton Engine Works, 516, 530
Egyptian Halls, 650
Eilean Donan Castle, 391
Elcho, 40
Elder & Cannon, 187
Eleanor crosses, 216
electric cranes, 581–2
electric lighting, 162, 172, 345, 350, 524, 659
electricity supply, 345–6
   see also generating houses; power stations
elevator buildings, 152, 153, 154, 155, 156, 157, 162, 659–60, 666
elevators, 525, 658, 659, 665
   see also goods movement facilities; grain elevators; hoists
Elgin, 280
   assembly room, 361
   banks, 640
   hospitals, 312, 320
   houses, 69, 81
   lunatic asylum, 320
   municipal buildings, 266–7
   museum, 364
   station hotel, 160
Elgin Club, 151
Elgin, Earl of, 554
Elgin Railway, 554
Elgin Technical College, 308
Elgin Tolbooth, 267
Elie, 175, 178, 578
Elie Home Farm, 443
Elizabethan style buildings, 173
Ellanbeich, 577, 580
Elliot, Professor and Mrs, 63
Elliot, Archibald, 54, 143
Elliot, Archibald, II, 632, 646
Elliot, James, 54
Ellis & Wilson, 654, 655
Elsie Inglis Hospital, 331
Empire House, 672
employer-provided schools, 296, 298, 528
EMS hospitals, 338–9
endowed schools, 296, 298
engine houses, 443, 444, 521, 560, 566
   see also beam engines and engine houses; horse engines and engine houses; pumping engine houses

engine sheds and depots, railways, 397, 518, 601, 606, 607
engine works, 516, 530, 532, 533, 581, 585
engineering works, 514, 529–35, 606
   see also foundries
English Manorial style, 63
entertainment buildings *see* recreation buildings
Entertrony, 224
Erraid, 583
Erskine Ferry, 578
Erskine of Mar, 449
Eskdalemuir, 485
Esperston, 554
estate buildings, 190–210
   see also animal houses; bird housing; boathouses; coach houses and carriage houses; dovecots; estate workers' houses; gardens and garden buildings; gate lodges; laundries; outbuildings; privies; storehouses
estate hotels, 171–9, 187
   see also shooting lodges
estate workers' houses, 94, 95–6, 207, 219
estate workshops, 498
Etchachan hut, 229
Ettrick, 485
Ettrick Mill, 537
Eugénie, Empress, 150, 159, 163
Evening Citizen office, 658, 659
Evening News office, 664
Eventyr, 86
Ewan, Robert, 170
Ewart, Professor, 82
ewe buchts, 488
Ewing, James, 642
Exchange Bank of Scotland, 635
   see also Bank of Scotland
Exchange Coffee Room
   Dundee, 643
   Edinburgh, 361
exchanges, 624–5, 641–4
   Dundee, 361, 521, 643, 644
   Edinburgh, 629–30, 641, 642–4, 651, 669
   Glasgow, 632, 641–2, 646, 651
   Leith, 624, 644
explosives manufacture, 556
extractive industries, 551–70

eye hospitals, 334
Eyemouth, 31, 383, 401

F

factories, 495, 510–50, 656, 666
   see also mills; workshops
Fairbairn, Alexander, 458
Fairbairn, William, 514, 515, 516, 519
Fairfield Shipbuilding and
      Engineering Company, 530,
      532, 584, 585
Fairlie, 577
Fairlie, Reginald, 309, 338
Fairnilee, 83
Fala, 130
Falcon Foundry, 529
Falcon Hall, 656
Falkirk
   air-raid shelters, 406
   aluminium rolling mill, 559
   bleachfield drying house, 502
   distilleries, 513, 545
   foundries, 533, 556, 605
   grave-markers, 290
   hospital, 315
   houses, 97
   National Bank, 639
   railway station, 605
   Red Lion Hotel, 143
   swimming pool, 375
   see also Carron Ironworks
Falkirk Building Society, 97
Falkirk Royal Hospital, 315
Falkland, 71, 127, 257, 358
Falkland Arms, 127
Falkland House, 57
Falkland Palace, 30, 42
Falkland Town house, 262
fan houses, 562
fanks, 488
farm bothies, 238–9, 240, 427
   see also farm workers' housing
farm buildings, 70, 76, 193, 227,
      423–93
   see also farm workers' housing;
      farmhouses; smithies
farm steadings, 76, 425–8, 453, 468,
      482, 486, 489
farm workers' housing, 95, 236–45,
      427–8, 430, 432, 435, 466, 479
   see also shepherds' dwellings
farmhouses, 70–1, 75–6, 426, 428
   Iceland, 20

   inns and, 139, 187
   pattern books, 78
   seasonal workers'
      accommodation, 239
   servants' quarters, 427
   Taransay, 224
   see also blackhouses; byre
      dwellings; longhouses
Faroe Islands, 17, 20
Farquharson, Colonel Francis, 139,
      644
Farrell, Terry, 675
Faslane, 382
Fast Castle, 30
fenestration *see* windows
Fenton, A, 70, 232
Fenton Barns, 468
Fenton Murray & Wood, 514
Fereneze, 519
Ferguslie Mill School, 528
Ferguslie Mills, 516, 536
Ferguson, Alexander, 51
Ferguson, James D, 481
Fergusson, Robert, 319
fermtouns, 91
ferneries, 219
Fernie, E C, 276
ferries and ferry facilities, 147–8,
      577–8
   see also ports and harbours;
      railway ports; steamer piers
Ferrigan, James, 665
Ferryhill Foundry, 530
Festival Square, Edinburgh, 675
Fetlar, 232, 428
Fettercairn, 175
Fettes College, 304
Fettykill Mill, 540
Fife
   air-raid shelters, 406
   alumina and red oxide works, 559
   assembly room, 142
   banks, 638
   boat building yard, 585
   castles, 29, 30, 31
   chapel, 215
   châteaux, 40, 42
   churches, 277–9, 280, 282, 284,
      285–6, 287, 291
   churns, 470
   cinema, 376
   clay-pipe making premises, 501
   coal hoists, 581

coalfields and collieries, 560, 561, 563–4
corn exchange, 644
country houses, 48–9, 51–2, 55, 57, 61
cranes, 580, 582
defences, 386, 402, 403–4, 407, 408, 409
distilleries, 547
docks, 573, 574
dovecots, 478
engineering works, 532
factories, 527, 528, 536, 537, 538
farm buildings, 425, 432, 455, 481–2
flour mill, 519
forge, 535
foundry, 535
gardens and garden buildings, 49, 61, 211, 220, 221, 370, 371
gate lodge, 61
grain elevators, 582
grain mills, 450
harbour lights, 582
harbours, 571, 573, 578
hospitals, 313–14
hotels, 129, 137, 142, 173, 175, 185
houses, 68, 69, 70, 71, 73, 76, 83, 84, 93, 96, 101
indigo mill, 519
inns, 127, 136
iron mill, 495
kilns, 448, 449–50, 501, 554
lighthouses, 583
lime burning, 553, 554
lunatic asylum, 321
mausoleum, 215
mercat crosses, 613
municipal buildings, 142, 143, 249, 250, 251, 253, 257, 258, 260, 262, 263, 269–70, 271, 272
naval base, 382, 402, 403, 404, 407, 582
offices, 669–70
palaces, 27, 30, 37, 38, 42, 43, 48–9
paper mills, 540
piers, 575, 577
poorhouse, 317
power stations, 346, 523, 566
prisons, 250, 269
radar station, 412
railway facilities, 572, 606
roofs, 227
salmon bothies, 226, 227
salt extraction, 552
sawmill, 524
schools, 304, 528
sports facilities, 358, 375
stables, 432
steam engine house, 443
steeples, 251
stop-line, 409
tenements, 96
textile mills and factories, 518, 526, 536, 537
threshing machines, 442, 443
tower-house, 36, 61
town bells, 257
underground bunker, 413
university, 305
villas, 45, 78, 79
warehouses, 518, 527, 578, 626
water supply, 344
workshops, 495, 502
Fife Arms
    Banff, 145–6
    Braemar, 172
    Turriff, 140
Fife Coal Company, 561, 563
Fife, Duke of, 63
Fife, Earl of, 51, 145
Fife Forge, 535
filter houses, 344
Finance House, Edinburgh, 673
Finavon, 35
Findhorn, 579, 580, 587
Findlater, C, 485
Findlater, Earl and Countess of, 261
Findlay, J R, 664
Findlay, Sir John Ritchie, 364
Findochty, 573
Fingask, 40, 217
Finglen, 16
Finlay, Ian Hamilton, 222
Finnich, 411
Finnieston, 578
Finzean, 501, 506
fire services, 349–50
fire stations, 349
fireclay production, 558
fireplaces, shieling huts, 13
Firth, George, 471
Firth of Clyde
    defences, 404, 405, 410, 411
    electric cranes, 582

ferry facilities, 578
naval base, 382
naval bases, 407
Firth of Forth
    defences, 398–9, 401–4, 405, 407
    piers and ferries, 577
    salt extraction, 552
Firth of Tay, defences, 402
fish crosses, 613
fish farms, 588
Fisherfield Forest, 224
fishermen's and fishworkers' accommodation, 95, 97, 99, 230, 232–3, 587–8
    see also salmon bothies
Fisher's Hotel, 168, 172
fishing industry, 575, 579–80
    see also boat building yards; docks; ports and harbours; salmon fisheries
fishing stations, 229, 230, 232, 580, 587–8
fishing temples, 215
Fitzgerald, Joseph, 257
Fitzroy Robinson, 675
fixed engines, salmon fishing, 587
flagstone quarries, 558
Flatfield, 442
flats see tenements and flats
flax mills, 495, 519, 523, 524, 525, 535, 536
    Aberdeen, 514, 519–20, 522, 526
    Dundee, 521
    Fife, 526
    Perthshire, 519
    Renfrewshire, 514
    steam engines, 520
    see also heckling shops; lint mills; textile mills
flax storage, 499
Fleming, P & R, 453
Fletcher, Andrew, 57
Flisk, 226
Flisk Point, 226
floating docks, 575
Flodden Wall, 385
floor beams, breweries, 514
floorcloth factories, 538
floors
    Central Station, Glasgow, 545
    commercial buildings, 658, 666
    dovecots, 202
    farm buildings, 439, 467, 469, 476
    houses, 94
    industrial buildings, 512–17, 521, 523, 525, 532, 535, 536
    killing houses, 482
    kilns, 541–2
    schools, 297
    stables, 430
    warehouses, 626
    see also beams
Floors Castle, 51, 466
Flotta, 32
flour mills, 519, 520, 541
    see also grain mills
fog houses, 215
Foley, Nelson, 185, 186
follies, 202, 214
football stands and stadia, 372
Forbes, Arthur, 54
Forbes Place, Paisley, 647
Forbes, Sir William, 631
Ford House, 67
Forest Mill, 537
Forfar, 537, 639
Forgan & Stewart, 674
forges, 520, 535
    see also smithies
Formakin, 62, 221
Forres, 69, 170, 173, 313, 640
Fort Augustus, 32, 391, 393, 395, 397
Fort Charlotte, 388–9, 397–8
Fort George, 32, 288, 395–7, 401, 402
Fort William
    aluminium smelting, 559
    fort, 31, 390, 391, 395, 397
    Glen Nevis distillery, 512
    hotel, 166
    see also Inverlochy Fort
fortalices, definition, 27–8
Forter, 40
Forth and Clyde Canal, 574, 598, 599, 600
    see also Port Dundas
Forth Bridge, 603
Forth Hotel, 148
fortifications see defences
fortified houses, 27–47
    see also duns; tower-houses
Fortingall, 80, 96, 175
Fortrose, 580
Fortrose Chapter House, 271, 272
forts, 31, 32, 383–4, 387–9
    Fife, 402
    Inchkeith, 384, 401–2

Inverness-shire, 31–2, 387, 388, 389, 390, 393, 395–7, 401, 402
Perthshire, 215
Shetland, 388–9, 397–8
see also barracks; defences; gun batteries
Fortune's Tavern, 141
foundries, 216, 257–8, 502, 529–30
Angus, 533
Dundee, 530, 532, 534, 535
Edinburgh, 258, 532
Fife, 535
Glasgow, 510–11, 517, 533–4, 649
Inverness, 529, 532
Stirlingshire, 258, 533, 556, 605
see also engineering works; ironworks
Fountainhall, 68
Four Seasons Hotel, 186
Fowke, Captain Francis, 364
Foyers, 559
Frampton, George, 667
Frances Colliery, 563
François Premier style buildings, 162, 658, 666
Fraser department store, 658
Fraser, Douglas, 533
Fraser, R H E, 456
Fraserburgh, 573, 580, 582
Frederick, Empress, 163
Frederick of Prussia, Crown Prince, 172
free Renaissance style buildings *see* Renaissance buildings
French beams, 650
French roofs
breweries, 544
clubs, 152, 156
hotels, 154, 155, 156, 157, 159, 173, 657
Scottish Provident Institution, 660
warehouse, 665
French, Thomas and John, 45
fruit pickers' accommodation, 239, 241
Fryers & Penman, 63, 79
Fulton, James, 471
Fulton, Peter, 173
functional tradition, the, 511
funerary monuments, 289–92
furnaces, 501–2, 555–6
see also ironworks; steam power
further education colleges, 308
Fyrish, 215

Fyvie, palace, 30, 43
Fyvie Castle, 27, 59

G

gables
cottages, 94
grain mills, 541
hay barns, 451–2
houses, 104
poorhouses, 313
potato houses, 458
schools, 298
see also crowstepped gables; tympany gables
Gairloch, 166, 229, 489
Gala Mill, 537
Galashiels
mills, 516, 520, 523, 524, 537
skinworks, 526
tanneries, 541
water supply, 345
Gallanachmore, 407
galleries, 35, 37, 39, 43, 297, 320–1
see also art galleries
Galloway, 445, 446, 452
Galloway, Andrew, 163
Gallowgate, Glasgow, 142
game larders, 207, 482
Gamrie, 364
garages, 194, 500, 597
Garden City developments, 99–100, 101, 118
gardeners' houses, 219
see also estate workers' houses
gardens and garden buildings, 40, 43, 48, 49, 61, 211–23, 368–71
cottages, 80
public housing, 100
sporting hotels, 172
see also bee boles
Gardenstone, Lord, 135
Gardenstone Arms, 127, 135
Gardners', Glasgow, 516
Garlogie, 520
Garrett's Long Shop, 532
Garscadden Pavilion, 372
Garscube, 57
Gartcosh, 557
Gartnavel General Hospital, 315
Garvamore, 229, 391
gas engines, 521
gas lighting, 350, 524, 618
gas works, 350, 524

Gaskell, Philip, 191
gasometers, 205
Gassing Mill, 510
Gasson, Barry, 366
gate lodges, 61, 95–6, 190, 191–2
    see also gatehouses
Gate of Negapatam, 215
Gatehouse of Fleet, 135, 137–8, 587
gatehouses, 28, 30–1, 43, 192
    see also gate lodges
gates
    in gardens, 220
    town defences, 385, 386
Gauldie, Hardie, Wright & Needham, 670
Gayfield House, 76
gazebos, 211–12, 221, 371
GEC Ferranti Building, 674
geese, housing, 478–9
General Accident building, 674
general hospitals, 311–15
generating houses, 190, 205
generating stations see power stations
Geographical Institute, Edinburgh, 656
George, Ernest, 79
George Heriot's Hospital, 303, 311
George Hotel
    Edinburgh, 150
    Perth, 143
George IV Bridge, Edinburgh, 625, 646, 651
George Street
    Edinburgh, 143, 150, 675
    Glasgow, 647
George Watson's Hospital, 303
Georgian architecture
    banks, 638, 640
    corn exchanges, 644
    hospitals, 312
    hotels and inns, 137, 146
    houses, 77
    insurance offices, 640
    Nairn's head office, 669–70
    see also neo-Georgian architecture
German styles, breweries, 545
Gesto, 477
Gheel, 322
GHQ stop-line, 409
Gibbs, A, 104
Gibbs, James, 52, 288, 586
Gibson, David, 122
Gibson, John, 637, 666
Gibson, Miles, 161

Gibson, Robert, 664
Gibson, Thomas, 52, 655
Gibson, Thomas Bowhill, 184
gig houses, 431–2, 469
    see also cart bays and sheds; coach houses and carriage houses
Gight, 45
Gilkes, Gilbert, 520
Gill Pier, 579
Gillespie, James, 61
Gillespie, John Gaff, 661
Gillespie, Kidd & Coia, 289, 315
Gillies's House, 73
Giorgio, Francesco di, 31
girders see iron construction; steel construction
girls' schools, 296, 303, 304
girnals, 197–8, 439–40, 627
    see also barns; granaries; salt girnals
Girnigo, 37
Girouard, Mark, 133
Girvan, 137, 343, 580, 585
Gladstone's Land, 69, 109
Glamis, 37, 40, 59, 211
Glasgow
    air-raid shelters, 406
    aqueducts, 344
    arcades, 620, 642
    art galleries, 366–8, 371
    assembly rooms, 133, 361
    banks, 632–3, 634, 637, 638, 642, 646, 651, 653, 657, 658, 660, 666, 667, 668, 672, 673
        see also names of individual banks
    baths, 347, 374
    boat building yard, 586
    breweries, 545
    bridges, 600, 601, 603
    business chambers, 156, 163, 632, 646, 657, 658, 659, 660, 662
    canal offices, 600
    car factories, 517, 534
    chemical works, 523
    churches, 262, 289, 630
        see also Glasgow Cathedral
    cinemas, 376
    clubs, 148, 151, 152, 154, 156, 157, 158, 374, 672
    coffee rooms, 142, 360–1, 642
    collieries, 564
    concert halls, 373, 374

convalescent home, 336
corn exchange, 644, 646
cranes, 580, 581, 582
custom house, 586
dairies and dairy farms, 466, 467, 468
department stores, 666
dispensaries, 335
docks, 574, 575, 581, 582
engine works, 516, 530, 532, 533, 581, 585
engineering works, 532
exchanges, 632, 641–2, 646, 651
factories, 510, 512, 519, 538, 539, 666
ferry facilities, 578
fish cross, 613
flour mills, 541
forges, 535
foundries, 510–11, 517, 533–4, 649
gardens and garden buildings, 220, 369–71
gas lighting, 350
granaries, 578, 582
harbour office, 586
head offices, 654, 656, 658, 666, 667
Hillington industrial estate, 534
hospitals, 311, 312, 315, 320, 332–3, 334, 335, 337, 339
hospitals for infectious diseases, 325, 327, 329, 339
hotels, 142, 153, 155–7, 159, 160, 161–3, 182, 183, 185, 186, 187, 654, 656–7
  see also names of individual hotels
houses, 73, 75, 76, 79, 632
hydraulic plant, 352, 581
inns, 127, 132–3, 142
insurance offices, 640–1, 650, 658, 659, 660, 667, 668, 670, 672
laundries, 374
libraries, 309
lodging houses, 179–81
lunatic asylums, 319, 320–1
maternity hospitals, 330
mercat cross, 613
merchant hall, 628
Merchants' House, Glasgow, 133, 628, 642, 653, 657
merchants' shelter, 642
municipal buildings, 133, 142,
257, 258, 260, 262, 263, 265–6, 629, 642
museums, 363–4, 366, 371
music halls, 363, 373
newspaper offices, 511, 655, 658, 659, 664, 665, 669
night refuge, 182
offices, 628, 650–1, 658–62, 664, 667, 668, 669, 671–3, 675–6
palaces, 30, 35
pattern shops, 516, 533
pavilions, 369
poorhouses, 316, 318, 320
power stations, 659
printing and publishing works, 511, 656, 664
Professors' Lodgings, 363–4
pumping stations, 581
quays, 571
railway facilities, 534, 601, 603, 606, 607
railway stations, 155, 162, 604, 605, 654, 656, 658
reading rooms, 642
rice mill, 541
rope-making, 495, 519, 521
Sailors' Homes, 158
sandstone quarries, 558
schools, 298–9, 304, 528
sewerage, 352–3
shipbuilding yards, 516, 530, 532, 584, 585
shops, 156, 615, 617, 619, 620, 628, 631, 632, 650, 658, 660, 672
smokehouses, 580
stables, 368
steel works, 557
suburbs, 76
sugar-making, 495, 514, 516, 547
taverns, 141, 142
teacher training college, 297
telephone system, 353
tenements and flats, 109, 111, 112, 113, 114, 115, 117, 119, 122, 124
textile mills and factories, 512, 514, 516, 536, 537
theatres, 362
town bells, 257
trades halls, 630
universities, 305, 306, 307, 308, 372
villas, 79, 83
warehouses, 516, 578, 646, 647–50, 656, 659, 665–6, 668, 669

water supply, 167, 341–3, 344
workshops, 495
Glasgow & Ship Bank, 633, 634
Glasgow & South Western Railway head office, 161, 654, 656
Glasgow and West of Scotland Technical College, 308
Glasgow Architectural Exhibition, 367–8
Glasgow Association for Establishing Lodging-houses for the Working Classes, 179
Glasgow Cancer and Skin Institution, 335
Glasgow Cathedral, 279, 282, 630
Glasgow City Hall, 373, 374
Glasgow Conservative Club, 151, 157
Glasgow Cross, 665, 666
Glasgow Dispensary for Diseases of the Ear, 335
Glasgow Ear, Nose and Throat Hospital, 335
Glasgow Eye Infirmary, 334
Glasgow Green, 369
Glasgow Herald buildings, 368, 655, 656, 659, 664
Glasgow Highland Society, 133
Glasgow Lock Hospital, 337
Glasgow Lying-in Hospital, 330
Glasgow Necropolis, 291
Glasgow, Paisley and Ardrossan Canal, 598, 599
Glasgow Royal Asylum, 319, 320–1
Glasgow Royal Concert Hall, 374
Glasgow Royal Infirmary, 312, 315, 325, 336
Glasgow Savings Bank, 667
Glasgow School of Art, 308
Glasgow Style, 82
Glasgow Tolbooth, 257, 260, 265–6
Glasgow Town Hall, 133, 142, 262, 263, 629, 642
Glasgow Town House, 258
Glasgow University, 305, 306, 307, 372
Glasgow Women's Private Hospital, 337
glass curtain walls, 510–11
glass houses, 218, 219, 371
glass-making workshops, 495
see also glassworks
Glassford Street, Glasgow, 666
glassworks, 501, 520

see also glass-making workshops
Glebe Mill, 538
Glebe sugar refinery, 547
Glen, Kirkcudbrightshire, 68–9
Glen Kinglass, 555, 556
Glen Lednock, 459
Glen Moriston, 391
Glen Nevis distillery, 512
Glen, W R, 667
Glenalmond, 304
Glenburn hydropathic, 170
Glendale, 542
Glendoick House, 53–4
Gleneagles, 178–9, 338, 604
Gleneagles Maltings, 544
Glenesk, 458
Glenfinnan, 604
Glengarnock, 557
Glenhead, 478–9
Glenlockhart Bank, 86
Glenluce Abbey, 341
Glenlyon House *see* Fortingall
Glenlyon, laird of, 16
Glenochil Colliery, 563
Glenogle, 374
Glenshishie, 391
Glenstrae, 16
Glentana Mill, 512
Gogarburn, 325
Golden Lion Hotel, 143
goldsmiths *see* metal-working workshops
golf clubs and courses, 371, 372
Golf Hotel, St Andrews, 173
Golspie, 145
goods lifts *see* hoists
goods movement facilities, 499–500, 525, 543, 585, 604–5
see also cargo handling facilities; cranes; elevator buildings; grain elevators; sheer legs
goods sheds, 605
Gorbals Water Works, 342
Gordon & Dey, 670
Gordon & Dobson, 656
Gordon & Scrymgeour, 182
Gordon, Alexander, 366
Gordon Castle, 37, 39, 192, 219
Gordon Chambers, Glasgow, 662
Gordon, Duke of, 192
Gordon, George, 39
Gordon, James, 627
Gordon, John, 77

Gordon Schools, 296
Gordon Street, Glasgow, 650, 651
Gordon's Inn, 143
Gordonstoun, 194
Gospatric, 280
Gothic architecture
    Armadale Castle, 195
    banks, 640
    Bo'ness Tolbooth, 253
    churches, 281, 285, 286, 288–9
    country houses, 54–7, 58–9, 139
    dairies, 473
    exchanges, 642, 643
    garden buildings, 214
    hotels, 159, 161–2, 163, 173
    Merchants' House steeple, Glasgow, 628
    offices, 162, 654
    'Rogue Gothic' villas, 79
    schools, 298, 304, 528
    sports clubs, 372
    tollhouses, 596
Gothic Revival, 54, 78
Gourdon, 582
Gourlay, Robert, 625
Gourock, 183, 378, 577
Gourock Rope Company, 519
Gourock Rope Works, 521
Govan
    engine works, 532, 535, 585
    ferry facilities, 578
    foundry, 510–11
    offices, 584
    public laundry, 374
    shipbuilding yards, 516, 530, 532, 584, 585
Govan Graving Docks, Glasgow, steam cranes, 581
Govan Press building, 655
Gowans, James, 79, 97
Graham, Angus, 191
Graham family, Kincardine, 40
Graham, James Gillespie, 56–7
    Armadale Castle, 195
    Cairns Castle inn, 146
    Commercial Bank branches, 632, 639
    Duns Town House, 253
    Edinburgh Assembly Rooms, 632
    Edinburgh New Town, 75, 312
    Gray's Hospital, Elgin, 312
    Prestonfield stables, 194
    St Catherine Street, Cupar, 142

Graham, Margaret, 56–7
Graham Square weaving factory, Glasgow, 537
grain elevators, 445, 446, 542, 578, 582
grain mills, 496, 541
    kilns, 448–9, 500–1, 541–2
    on farms, 450, 457
    pig sties at, 480
    power sources, 505
    storage and transport facilities, 499–500, 525
    see also flour mills; granaries
grain storage and handling, 444, 474, 499
    see also barns; girnals; grain elevators; granaries; kilns
grammar schools, 300, 304
granaries, 446–8
    as wool stores, 489
    converted to hotel, 186
    dockside warehouses, 578
    feed preparation in, 456
    Glasgow, 578, 582
    goods movement facilities, 525
    grain mills, 499
    horse engine houses, 442
    location, 425, 426, 433, 447, 474
    seasonal workers' accommodation, 239
    tythe barn, Simprim, 440
    see also girnals
Grand Hotel
    Glasgow, 157, 187
    Lerwick, 154
    St Andrews, 173
    St Pancras, 161, 162
Grandholm Mill, 514, 519–20, 526
Grandholm Works, 521, 522, 538
Grange House, Edinburgh, 59
Grangemouth
    coal hoists, 581
    cranes, 582
    docks, 573, 574, 575
    dockside warehouses, 578
    petrochemical industries, 559
granite quarries, 558
Grant, Sir Alexander, 194
Grant Arms, 172
Granton, 534, 572, 582
Granton Inn, 147–8, 168
Grantown-on-Spey, 172
Grant's Works, 540
Grass Point, 575

grass-drying plant, 445, 455
Grassmarket *see* Edinburgh
graves, 290–2
   see also mausolea
Gray, Lord, 57
Gray, Affleck, 229
Gray, Andrew, 477
Gray, Bamber, 674
Gray, William J, 78
Gray's Hospital, 312, 320
Great Cumbrae, 307, 575
Great Custom House, Cockenzie, 626
Great Eastern Hotel lodging house, 181
Great Garden, Alloa, 212
Great Harbour, Greenock, 574
Great Inn, Inveraray, 135, 136
Great North of Scotland Railway head office, 654
Great Room, Aberdeen Town House, 261–2, 263
Great Western Hotel
   Edinburgh, 154–5
   London, 152
   Oban, 172
Great Western Road, Glasgow, 79
Great Western Terrace, Glasgow, 75
Grecian Building, 650
Greek Revival
   banks, 632, 637, 668
   Dundee Exchange Coffee Room, 643
   houses, 77, 78, 79, 96
   Merchants' House, Glasgow, 642
   offices, 652
   Villafield Works, 656
   see also neo-Greek architecture
Greendyke Street lodging house, 180
Greenknowe, 40
Greenland, 20
Greenlaw, 144–5
Greenlaw Tolbooth Steeple, 269, 270
Greenmarket, Dundee, 627
Greenock
   clubs, 151
   cranes, 581, 582
   custom house, 586
   docks, 574, 575, 582
   dockside warehouses, 578
   engine works, 532
   gun battery, 398
   harbour, 574
   hospitals, 315, 334, 335, 339
   mills, 513
   polychromy, 512
   pumping stations, 581
   railway facilities, 577, 604
   shops, 619
   sugar refineries, 547
   tenements and flats, 115, 117
   water supply, 342, 344
Greenock Club, 151
Greenock Custom House, 586
Greenock Waterworld, 376
Green's of Wakefield, 523
Green's Playhouses, 376, 377
Gresham House, 662, 664
Greyfriars Churchyard, 291
Gribloch, 63
Grieve's House, East Barns, 70
grieves' houses, 70, 428
Griffiths, Moses, 15
Grimsay, 586
Grosvenor Crescent, Glasgow, 183
Grosvenor Hotel, London, 152, 153
Grosvenor Terrace, Glasgow, 75
grottoes, 215, 217, 219, 371
Gruna, 482
guano storage, 482
Guardian Royal Exchange building, 670
guild halls, 359, 627–8
Guisachan, 214, 473
Gullane, 83
gun batteries, 398, 401–4, 405, 407, 411
   see also anti-aircraft defences
gunnery school, 405
gunpowder manufacture, 556, 577
Gurness, 19
Gushetfaulds, 606
Guthrie, 219
Guthrie Castle, 601
gutters' housing, 232–3, 241

H

Hackness, 399–401
Haddington
   assembly rooms, 142, 259
   corn exchange, 644
   defences, 385
   hotels and inns, 127, 142
   houses, 71, 86
   lunatic asylum, 321
   St Martin's Church, 280
   textile industry, 495
   tolbooth bells, 257

Haddington, Earl of, 59, 498
Haddo, 216
Hailes Castle, 30
Half Moon Battery, Edinburgh
    Castle, 384
half-time schools, 298, 528
hall-houses, 28–9
Hall Russell, 532
halls of residence, 29, 307
Hamilton, 258, 287, 483
Hamilton, Dr Alexander, 330
Hamilton, David, 57, 362, 632, 642,
    646
Hamilton, David and James
    banks, 632, 633, 634, 637
    Glenburn hydropathic, 170
    Western Club, Glasgow, 148, 672
Hamilton, Duke and Duchess of, 51
Hamilton House, 70
Hamilton, James see Hamilton, David
    and James
Hamilton, Sir James, of Finnart, 31,
    42, 43, 45
Hamilton, John, 45
Hamilton Palace, 43, 45, 51
Hamilton Parish Church, 287
Hamilton, Thomas, 300, 367, 632, 640,
    656
Hamilton Tolbooth, 258
Hamilton, William, 142, 360
Hamilton's Land, 69
hammels, 475–6
Hammermen's Hall, Edinburgh,
    627–8
hand cranes, 580, 581
hand-loom weaving shops and
    factories, 495, 503, 536–7
Hane, Joachim, 387
Hannayfield, 78
harbour lights, 582
harbour offices, 148, 586
harbours see ports and harbours
Hardwick, Philip, 152
Harland and Wolff, 510
Harley, William, 341, 347, 374, 466,
    467
harness rooms, 430, 432
Harris, 477–8, 498, 580
Harris, John, 48
Harthill, 40, 44
Hartwoodhill, 325
harvest workers' accommodation,
    236–8, 239, 241

see also potato pickers'
    accommodation
Harviestoun House, 367
Haswell, George, 135
'Hatrack' building, 660–1
Hatton, 38, 40, 466, 524
Hawes Inn, 129
Hawick
    corn exchange, 644
    dyeworks, 510, 538
    mills, 519, 521, 526, 538, 541
    railway station, 606
    Royal Bank, 639
    stocking shops, 538
    town house, 253
    water supply, 345
Hawkhead Hospital, 327
Hawkhill, 76
Hay, G, 276
Hay, James, 259
hay storage, 451–3, 455
    at shepherds' dwellings, 489
    at workshops, 500
    cattle court sheds, 474
    feed preparation, 456
    keb houses, 487
    lofts, 193, 430–1, 432, 451, 466
Hay, William, 57
Hayfield, 101
Hayford Mills, 512, 537
Haymarket, Edinburgh, 604, 607
Haymarket House, Edinburgh, 673
head offices, 653–4
    Aberdeen, 654, 667
    Dundee, 654, 664–5
    Edinburgh, 669, 670–1, 673
    Fife, 669–70
    Glasgow, 654, 656, 658, 666, 667
health and welfare buildings, 193,
    311–40
    see also hospitals; hydropathics;
        medical centres
heating, 477, 500–2, 514, 524
    see also metal-working; ovens
Hebrides see Western Isles
heckling shops, 503, 504
    see also flax mills
Heisker, 230
Heiton, Andrew, 140, 163, 170, 182
Helensburgh, 77, 79, 80, 86
Hellenic House, 672
Helmsdale, 147, 579
Hemony, Petrus, 257

718 • INDEX

hemp factories *see* weaving factories
hen houses *see* poultry houses
Henderson, A Graham, 666, 668
Henderson, James, 172
Henderson, John, 148, 304, 361, 646
Henderson, Peter L, 173
Henderson, Richard, 480
Henderson Row, 111
Henderson, William, 640
Henderson, William, & Son, 335
Hennebique, François, 516, 533
    see also Yorkshire Hennebique Company
Henry, A & S, 528
Henry, David, 173
Henry, J McIntyre, 153, 155
Henry Wood Hall, 374
Herbert Hospital, Woolwich, 313, 336
Heriot Row, 75
Heriot Trust, 113, 298
Heriot-Watt College, 308
Heriot-Watt University, 308
Heriot's Hospital, 303, 311
Hermetray, 232
hermitages, 215
Heron House, 672
herring curing sheds, 579
herring gutters' housing, 232–3, 241
Herriot Hill printing works, 656
Hertford, Earl of, 284
Heugh Mills, 448
Hewes & Wren, 519
High Greenan, 83
High Street
    Edinburgh, 109
    Glasgow, 127, 665
    Kirkcaldy, 70
Highland Hotel
    Fort William, 166
    Nairn, 166
    Strathpeffer, 176–7
Highland Railway head office, 654
Highland Society Building, 646
Highlands and Islands
    harbours, 575
    mills, 519
    schools, 294, 297
    sheep houses, 483
    shieling huts, 16
    see also Argyll; Caithness; Inverness-shire; Orkney; Ross and Cromarty; Shetland; Sutherland; Western Isles

Hill, The, 471
hill bothies, 227–30
Hill, David, 631
Hill House, Helensburgh, 77
Hill, Oliver, 63
Hillbank Linen Works, 521
Hillburn Roadhouse, 184
Hillhead halls of residence, Aberdeen, 307
Hillhead radar station, Wick, 412
Hillhead school, Glasgow, 304
Hillhouse, 456
Hillington industrial estate, 534
Hillpark, 84
Hillside, Edinburgh, 73
Hilton Hotel, Edinburgh, 186
Hilton Junction, 606
Hilton of Turnerhall, 441
Hingley, Richard, 9, 10
Hinton House, 83
His Majesty's Theatre, 362
Hislop, Alexander, 667–8
Hoddom, 37, 40
Hogg, James, 224
Hogg, W, 485
hoists, 499, 525
    see also coal hoists; elevators; goods movement facilities
Holford (architect), 676
Holiday Inn
    Edinburgh, 185
    Glasgow, 185, 187
Holmes, John A, 62
Holmes Partnership, 673
Holmwood, 77
Holy Loch, 382
Holy Rude, Church of the, 282
Holy Trinity, Church of the, 284
Holyrood brewery, 544
Holyrood Gardens, 368
Holyrood Palace, 42, 43
Holyrood tennis court, 358
Home, Patrick, of Billie, 53
Home, Patrick, of Wedderburn, 55
Honeyman & Keppie, 659
Honeyman, John, 79, 170–1, 649–50, 654
Honeyman, Keppie & Mackintosh, 665
Hope, Charles, 51
Hope, Henry, 660, 662
Hope Street, Glasgow, 662
Hopeman, 579, 580

Hopetoun, Earl of, 51
Hopetoun House, 49, 51, 211, 215
horse engines and engine houses
    coal mines, 560
    farms, 440, 441–2, 456, 470
    mills, 519
    tanneries, 540
    workshops, 504
horse tramways, 595, 597
horse-walk platforms, 440, 442, 470–1
hosiery factories, 538
Hospital for Consumptives, Isle of Wight, 328
hospitals, 311–15, 325–39, 497
    see also poorhouses
Hotel Beresford, 183
Hotel Buchanan, 183
hotels, 127, 133, 141–8, 150–7, 159–68, 171–9, 182–8, 347, 656
    Aberdeen, 144, 159–60, 163–4, 165, 184
    Aberdeenshire, 139, 172, 176, 177
    Angus, 142, 175
    Argyll, 166, 172–3, 175, 183, 184, 187
    Arran, 175
    Ayr, 143, 163, 185
    Ayrshire, 137, 174, 177–8, 179, 187, 338
    Banff, 138–9, 168
    Berwickshire, 144–5
    Bute, 166
    Caithness, 165–6
    converted country houses, 139, 182
    Dumfriesshire, 136, 147, 163, 166, 168
    Dunbartonshire, 175
    Dundee, 143, 144, 159, 185, 186
    East Lothian, 127, 142, 170, 174
    Edinburgh, 141–3, 148, 150, 152–5, 157, 164–5, 182, 183, 185, 186–7, 640, 664
    Fife, 129, 137, 142, 173, 175, 185
    Glasgow, 142, 153, 155–7, 159, 160, 161–3, 182, 183, 185, 186, 187, 654, 656–7
    Inverness, 137, 144, 160, 163, 173, 185
    Inverness-shire, 166, 176, 177, 186
    Kincardineshire, 173, 175
    Moray, 160, 172, 173
    Nairn, 137, 166, 173
    Peeblesshire, 127, 184

Perth, 129, 142, 143, 163, 186
Perthshire, 140, 146, 147, 168, 172, 175, 178–9, 182, 338
Renfrewshire, 183, 186
Ross and Cromarty, 160, 166, 171, 173, 176–7
Roxburghshire, 140
Shetland, 154
Skye, 160
Stirlingshire, 143, 168
Sutherland, 145, 147, 176, 271
Wigtownshire, 178
see also inns; railway hotels; sporting hotels *and* names of individual hotels
Houldsworth, Henry, 514
Houldsworth's Mill, 512, 514
Hound Point, 403
House of Dun, 37
House of Strathbogie *see* Huntly Castle
House of the Binns *see* Binns, The
Househill Mains, 444
houses, 19, 67–107, 114, 115, 127, 168
    Dumfriesshire, 14–15
    Edinburgh, 69, 70, 73, 75, 76, 77, 78, 80, 82, 85, 93, 97, 198, 625
    fishing stations, 587–8
    Glasgow, 73, 75, 76, 632
    railway stations, 604
    workshops and, 497–8
    see also blackhouses; country houses; farmhouses; longhouses; prehistoric houses; public housing; tenements and flats; villas; workers' housing
Howard, Deborah, 625
Howgate, Midlothian, 139
Hoy, 399–401
HSBC G3 Building, 675
Hughes, T Harold, 372
Humane Society House, 369
Humbie, 86
Hume House, 673
Hume, J R, 506
Hume, John, 510
Hunter Barr's warehouse, 666
Hunter Blair, Sir David, 57
Hunter, J Kennedy, 83, 178
Hunter, Robert, 159
Hunter, Dr William, 363–4
Hunterian Museum, 364

Hunterston, 346, 582
hunting lodges, 40, 45, 221–2
Huntly, 37–8, 253, 296
Huntly Castle, 36, 39, 59
Huntly House, 364
Huntly Town Hall, 253
Hurleycove Tunnel, 215
Hurlford, 468
Hurst, William, 317
Hutcheson Street, Glasgow, 669
Hutchesontown, Glasgow, 111
Hutchison, Gareth, 674
Hutchison, Robert, 142
huts *see* cleitean; gutters' housing; kelp huts; shieling huts
Hyde Park, 540
Hydepark Street lodging house, 180
hydraulic equipment, 351–2, 525, 535, 581, 658
see also water-powered machinery
hydropathics, 168–71, 174, 187
Hyndford Bridge, 140
Hynish, 583

I

ice houses, 198–200, 215, 579, 587–8
ice rinks, 378
Iceland, 16, 17, 20
Imperial Chemical Industries, 556, 559
Imperial Hotel, 159, 160
implement sheds, 434–5
see also cart bays and sheds
Inchgarvie, 403
Inchinnan, 534
Inchkeith, 384, 401–2, 403, 407, 583
Inchture, 146
Inchyre Abbey, 56
India Buildings
  Dundee, 652
  Edinburgh, 646–7, 651
India Tyre Factory, 534
indigo mill, 519
industrial buildings *see* coalfields and collieries; engine works; engineering works; factories; foundries; ironworks; mills; steel works; workshops
industrial estates, 534
industrial schools, 298
infectious diseases, hospitals for, 322, 325–9, 339
infirmaries
  farm buildings, 431
  see also hospitals
Ingleside steel foundry, 535
Inglis, A & J, 584
Inglis, Dr Elsie, 337
Inglis, W Beresford, 183
Inglis's City Hotel, 150
Ingram Court, Glasgow, 647
Ingram Street, Glasgow, 666
Inner Hebrides *see* Western Isles
Innerleithen, 518–19, 521, 655
Innerpeffray library, 308
Innerwick Manse, 78
Innes, 37, 38, 42
inns, 127–41, 187, 596–7
  Aberdeenshire, 133, 139, 140, 143, 144, 172
  Argyll, 135, 136, 166
  Banffshire, 145–6, 148, 168
  Caithness, 147
  Edinburgh, 127–9, 132, 147–8, 168
  Glasgow, 127, 132–3, 142
  Inverness-shire, 131, 146, 229
  Loch Lomond, 166–7
  Midlothian, 129, 130, 135, 137, 139, 140, 146
  Perthshire, 130, 131, 132, 135, 136, 139, 140, 145, 146
  Renfrewshire, 184
  West Lothian, 129, 140–1, 146
  see also hotels; lodging houses; public houses; taverns
Insurance Company of Scotland Building, 640
insurance offices, 640–1, 657–8, 668–9, 670
  Edinburgh, 150, 632, 640, 657–8, 669, 670, 671, 675
  Glasgow, 640–1, 650, 658, 659, 660, 667, 668, 670, 672
  Perth, 674
International Style buildings, 669
Inver, 140
Inveraray
  dwellings, 71, 73, 112
  Great Inn, 135, 136
  hay storage, 452
  lookout tower, 214
  pump house, 215
Inveraray Castle, 54, 57
Inverbervie, 412
Inverbreackie, 577
Invercauld Arms, 172

Inverclyde Royal Hospital, 315
Inveresk, 73, 76
Inverewe, 221
Invergordon, 215, 382, 404, 407, 559, 577
Invergowrie, 540
Inverkeithing, 257, 269
Inverleith *see* Edinburgh
Inverlochy Castle, 28
Inverlochy Fort, 31, 387, 388, 389
    see also Fort William
Invermay, 40
Invermoriston, 391, 536
Inverness
    arts centre, 368
    assembly room, 361
    banks, 638
    churches, 368
    engineering works, 535
    forts, 32, 387, 388, 390, 393, 395–7, 401, 402
    foundries, 529, 532
    head office, 654
    hospitals, 312, 328, 338, 339
    hotels, 137, 144, 160, 163, 173, 185
    houses, 80, 86
    locomotive sheds, 606
    sheer legs, 581
    slipways, 575
    steeples, 251
    tolbooth, 251
    see also Fort George
Inverness Aquadrome, 376
Inverness Castle, 390
Inverness Steeple, 251
Inverness-shire
    aluminium smelting, 559
    barracks, 229, 391
    canals, 598, 599, 600
    castles, 28, 30, 390, 391
    country houses, 54, 63
    dairy, 214, 473
    distilleries, 512, 545
    dry docks, 574
    fishing stations, 580
    Fort George chapel, 288
    forts, 31–2, 387, 388, 389, 395, 397
    hay storage, 452
    hotels, 166, 176, 177, 186
    inns, 131, 146, 229
    lighthouses, 583
    lunatic asylum, 321
    piers, 577
    privies, 205
    railway facilities, 577, 603, 604
    sanatorium, 329
    sawmill, 495
    smithy, 498
    stables, 207
    threshing machines, 443
    weaving factories, 536
    whin mill, 456
    wool stores, 489
Invershin, 580, 587
Inversnaid, 391, 393, 397
Inverteil, 481–2
Inverugie, 40, 43
Inverurie, 327, 328, 534, 540
Ireland, 15, 16, 20
Ireland & Maclaren, 159, 182
iron construction, 558
    banks, 636
    bridges, 595, 600, 601
    Dundee corn exchange roof, 644
    farm buildings, 433, 434
    harbour structures, 402, 577, 582
    horse engine houses, 442
    industrial buildings, 510, 513–16, 518, 530, 532, 534, 537, 543, 547
    lighthouses, 583
    offices, 652
    railway structures, 545, 601, 602–3, 604, 606
    recreation buildings, 375
    St Enoch's Hotel, 162
    shops, 652
    theatres, 362
    viaducts, 602–3
    warehouses, 647, 648, 649–50, 665, 666
    Waverley Market, 617
    workshops, 498
iron mills, 495
iron ore extraction, 555–6
iron-smelting furnaces, 555–6
Ironside, General Sir Edmund, 408, 409
ironworks
    Aberdeen, 529, 532
    Argyll, 345, 555, 556
    Ayrshire, 556, 557
    Lanarkshire, 471, 554, 556, 557, 558
    see also Carron Ironworks;
        foundries; furnaces; smithies
Irvine, 331, 639
Islay

animal housing, 19
castles, 30
church, 288
distilleries, 512, 545
horse-walk platform, 442
longhouses, 93
radar station, 412
Isle of May, 407, 583
isolation hospitals, 325–9
Italianate architecture
    arcade, Edinburgh, 620
    banks, 638, 640, 653, 668
    breweries, 544–5
    Central Station, Glasgow, 656
    clubs, 157
    convalescent home, 336
    corn exchanges, 644
    country houses, 48
    gate lodges, 192
    half-time schools, 528
    hotels and inns, 146, 159, 160, 172, 173
    houses, 75, 77, 79
    hydropathics, 170
    ironworks, 557
    mills, 536, 541
    museums, 364
    offices, 650, 651, 658
    Sailors' Home, Glasgow, 158
    villas, 79
    warehouses, 527, 649
Italianate garden, Drummond, 220
Ivanhoe House, 670
Ivory, James, 268

J

Jacobean style
    banks, 640, 666
    corn exchange, 643
    factory, 666
    hotels, 146, 166, 173
    houses, 57, 77
    hydropathics, 170
    lodging house, 180
    offices, 655, 657–8
    paper mills, 540
    school, 528
'Jacobethan' style, 78, 168
Jaffrey, James, 630
jails *see* prisons
Jamaica Street, Glasgow, 649, 654
James Court, 109
James Watt Dock, 574, 582
Jameson, George, 52, 71
Jamestown, 603
Janet Street, Thurso, 76
Jarlshof, 10, 17, 19, 20
Jedburgh, 78, 140
Jeffrey Street, Edinburgh, 671
Jelfe, Andrews, 391
Jenkins & Marr, 668, 674, 676
Jenner, Charles, 164, 620
Jerdan, John, 86
Jex-Blake, Dr Sophia, 337
Joass, William C, 171
John Knox's House, 69
John Watson's Hospital, 303
Johnson-Marshall, Percy, & Associates, 673
Johnson-Marshall, Percy, & Partners, 675
    see also Robert Matthew Johnson-Marshall
Johnston & Baxter, 667
Johnston, Alexander, 181
Johnston, C S S, 158
Johnston, J & F, 673
Johnston, W & A K, 656
Johnstone, 184
Johnstone Mill, 520, 524, 526
Johnstounburn, 219
joiners' shops *see* woodworking workshops
Jopp, Charles, 153
jougs, 260
Jura, 10, 15, 16, 93
jute mills, 521, 536

K

Kahn, Albert, 534
Kahn reinforced concrete, 517, 533
Kames, 577
Kaye, Stewart, 84, 667
Keathbank Flax Mill, 519
keb houses, 486–7
Kedder, Thomas, 45
keeps, 27, 30
Keil, 183
Keir, 220
Keiss, 572, 579
Keith, 540
Keith Hill, 486
Kellie, 40, 42
Kelly & Surman, 671
Kelly, Dr William, 333
Kelly, William, 667

kelp huts, 232
Kelso, 76, 135, 136, 288, 312, 644
Kelvin Bridge, 344
Kelvingrove, 368, 370
Kelvingrove House, 371
Kelvingrove Park, 97, 369–70, 371
Kelvinhaugh, 578
Kemp, Ebenezer, 532
Kemp's warehouse, 649
Kenmore, 135
Kennedy, David, Earl of Cassillis, 55
Kennedy, George Penrose, 168
kennels, 193, 207, 489
Kennet, 480
Kennetpans, 545
Kennington, Charles, 620
Kennington family, 150
Keppie, Henderson & Gleave, 314
Keppie Henderson & Partners, 315
Kerr, William, 83, 527, 669
Kersmains, 442
Kibble, 220
Kibble, John, 371
Kilbaberton, 40, 45
Kilbagie, 480, 540, 545, 547
Kilbirnie, 343
Kilbirnie Street car factory, 534
Kilbowie Works, 526
Kilchiaran, 412
Kilchurn Castle, 29, 390
Kilcoy, 36
Kilcreggan, 77, 577
Kildonan, 63
Kildrummy Castle, 28
Kiliwhimen, 391, 393
Killearn Hospital, 338, 339
Killermont, 344
Killin, 603
killing houses, 193, 482
    see also abattoirs
Killoch Colliery, 563
Kilmacolm, 86, 171
Kilmarnock, 45, 260, 375, 514, 532, 540
Kilmaurs, 260, 269
Kilmory, 197
Kilmun, 577
Kilmux, 432, 450
kiln-barns, 449–50, 542
Kilncraigs, 669
Kilncraigs Mill, 526, 527, 528, 536
Kilnford, 241
kilns, 16, 511–12
    clay firing, 558

grain drying, 445, 448–50, 500–1, 541–2, 543
    see also kiln-barns; lime kilns; smokehouses
Kilpatrick, 15–16
Kilwinning, 613
Kincardine, 30, 40
Kincardine House, 62
Kincardine Power Station, 523
Kincardineshire
    banks, 640
    châteaux, 37, 45
    defences, 401, 410
    farm buildings, 238, 432, 447, 479
    flax mills, 524
    gardens and garden buildings, 220
    gun battery, 401
    harbour lights, 582
    hotels and inns, 127, 135, 173, 175
    municipal buildings, 271–2
    radar station, 412
    salmon bothies, 587
    steam boilers, 456
    swimming pool, 375
Kinclaven Castle, 30
Kincorth, 101
Kincraig, 407
Kinfauns Castle, 57
King George V Dock, Glasgow, 575, 582
King James VI Hospital, 316
King, Main & Ellison, 673
King Street, Glasgow, 665
Kinghorn
    fort, 402
    gun battery, 403
    inns, 136
    municipal buildings, 253, 262
    St Leonard's Tower, 271, 272
    sawmill, 524
King's Arms
    Ayr, 143
    Girvan, 137
King's College, 305, 372
King's Houses, 131–2
King's Old Buildings, 43
King's Wark, 624
Kingscavil, 287
Kingseat, 324
Kingskettle, 409
Kingston, 512
Kingston Engine Works, 532
Kingussie, 146, 168, 498

Kininmonth & Paul, 670
Kininmonth, James, 481–2
Kininmonth, Sir William, 86
Kinkell, 222
Kinloch Castle, 62, 197
Kinlochleven, 559
Kinmount, 57
Kinnaird, Lord, 146, 179, 457, 476
Kinnaird Castle, 54
Kinnaird Hall, 644
Kinnaird Head, 583
Kinnear, Charles George Hood, 639
Kinneddar, 27
Kinneil, 40, 41, 45, 280, 564
Kinross, 129, 214, 613
Kinross House, 49
Kinross, John, 63, 80, 83, 84, 194, 205
Kinross-Loch Leven, water supply, 344
Kintore, 269
Kintyre, 183
Kippen, 412
Kirkbuddo, 410
Kirkcaldy
   banks, 638
   clay-pipe making premises, 501
   engineering works, 532
   factories, 536, 537, 538
   flour mill, 519
   forge, 535
   foundry, 535
   grain elevators, 582
   head offices, 669–70
   hospital, 313–14
   houses, 70, 96, 101
   kiln, 501
   poorhouse, 317
   stop-line, 409
   tenements, 96
   textile mills, 518, 526
Kirkcaldy and Dysart Poorhouse, 317
Kirkcaldy Cottage Hospital, 313–14
Kirkcudbright, 255–7, 258, 575
Kirkcudbright Tolbooth, 251, 257, 258, 260
Kirkcudbrightshire
   abbey, 279
   country houses, 52, 54
   granite quarries, 558
   harbour, 571
   hay barn, 453
   horse engine houses, 442
   houses, 68–9
   inns, 135
   lead mining, 555
   lighthouse, 583
   salmon fishing, 587
   timber gallery, 37
   tower-house, 33
Kirkintilloch, 238, 241, 287, 336
Kirkland, Alexander, 646, 648
Kirklee Terrace, Glasgow, 75
Kirkliston, 285, 547
Kirkmaiden, 287
Kirknewton, 456
Kirkton Manor, 502
Kirkwall, 45, 578, 586, 639
Kirriemuir, 537, 639, 655
'kit' housing, 104–5
Kittybrewster, 184, 606
Knab, Lerwick, 404
knitting workshops, 495
   see also textile factories and workshops
Knockderry, 339
Knowles, J T, 152
Knox & Hutton, 79
Koch, Robert, 328
Kohn, Pedersen & Fox, 675
Kvaerner Govan *see* Fairfield Shipbuilding and Engineering Company
Kyle of Lochalsh, 160, 452, 577
Kyleakin, 160
Kyles of Bute Hydropathic, 171

L

lace factories, 539
lades, 205, 344–5, 519, 524, 655
Ladhope Mills, 516, 520
Lady Victoria Colliery, 561, 566
Ladybank, 409, 606
Ladykirk, 475
laigh halls, 42, 614
Laigh Moray Hotel, 160
Laigh of Dercullich, 466
laigh shops, 614–15
Laing, John, 67
Laird & Napier, 183
Laird, Michael, 670, 671, 672, 674
Laird, Michael, & Partners, 671, 674, 675
Laird's Inn, 130, 132
Lamb, Andrew, 625, 626
Lamb, Thomas, 159
lambermen's accommodation, 224

lambing houses, 486–8
Lamer Island, 398
Lamlash, 175
Lammerburn, 79
lammiehooses, 483–5
Lanark, 328, 344
Lanark Tolbooth, 262
Lanarkshire
   anti-aircraft battery, 411
   banks, 638
   baths and swimming pools, 375
   bell towers, 526
   canals, 598, 600
   castles, 28, 31, 41, 42, 45, 371
   churches, 287
   coalfields and collieries, 560, 563
   cotton mills, 514, 521, 524, 525, 535
   country houses, 54, 57, 59
   dairies and dairy products, 470, 471, 472
   dams, 343
   engine works, 535
   foundries, 529–30
   gardens and garden buildings, 221–2, 371
   goods shed, 605
   grave slab, 290
   hay storage, 453
   hospitals, 315, 334, 338, 339
   inn, 140
   institutions for mentally handicapped, 325
   ironworks, 471, 554, 556, 557, 558
   lime kilns, 554
   manure storage, 482
   maternity facilities, 315, 334
   metal mining, 554
   mill turbines, 520
   municipal buildings, 258, 262, 271
   palaces, 42, 43, 51
   railway bridges, 603
   sanatorium, 328
   sheep houses, 483
   steel works, 557
   technical college, 308
   threshing machine, 441
   villas, 81
   see also New Lanark
Lancefield Engine Works, 581
Lanfine Hospital, 336
Langholm, 14–15
Lanrick Tree, 215
L'Anse aux Meadows, 20
Lanton, 130, 140
Larachbeg, 197
Larbert, 325
larders, 207
Largo Road, Leven, 84
Largs, 343, 575, 577
Lassodie, 553
Lasswade Cottage, 76
Lauder, 137, 269, 287
Lauder, Sir Thomas Dick, 59
Lauderdale, Duke of, 287
Lauderdale, Earl of, 49, 137, 287
laundries, 195–7, 347, 374, 427, 468
Laurencekirk, 127, 135
Laurenson, J J, 232
Laurieston, Glasgow, 111
lavatories *see* privies
law courts, 249, 258, 260, 262, 263, 642
Law Hospital, 338–9
Lawrence, George, 86
Le Creusot ironworks, 532
Leach, Rhodes & Walker, 185
lead mining, 554–5
Leadhills, 554
leads *see* lades
Leanchoil Hospital, 313
Learmonth Terrace, Edinburgh, 183
leather working *see* shoe factories; tanneries
Leckmelm, 442, 453
Lee Castle, 57
Lee, Sir Richard, 383
Leeming & Leeming, 63
Legal & General, 668
Leiper, William, 79, 82, 512, 658
leisure centres, 375–6
   see also swimming pools
Leitch, Archibald, 533
Leitch, R P, 131
Leith
   banks, 639
   cinema, 377
   cranes, 580, 581, 582
   custom house, 586
   defences, 31, 385–6, 387, 388, 398, 399, 624
   docks, 399, 573, 574, 575, 578
   engineering works, 532
   exchanges, 624, 644
   grain-handling facilities, 578, 582
   harbour, 398
   harbour office, 586
   lighthouses, 583

mills, 512, 519
Navigation School, 158
piazza, 624
pumping station, 581
Sailors' Homes, 158–9
schools, 298
sugar refinery, 516, 547
tenements and flats, 117
tolbooth, 253
tron, 625
warehouses, 578, 625, 626
workshops, 495, 496
Leith Bank, 631
Leith Citadel, 388
Leith Walk, 97
Leng, Sir John, 654
Lennox Castle, 325
Lennoxlove, 86
Lenzie, 321, 336
Leopold, King, of the Belgians, 163
Lerwick
defences, 388–9, 397–8, 404, 409
hotel, 154
piers, 575
Leslie, Bishop, 272
Leslie House
Aberdeenshire, 42
Fife, 48–9
Lessels, John, 153, 157, 652, 655
Lester and Pack, 258
Lethaby, W R, 62
Letham, 412
Lethington, 80
Letterfinlay, 131
Leven, 84, 227
Leverhulme, Lord, 97
Levern, 344
Lewis
blackhouses, 17, 425
byres, 465
custom house, 586
farm steadings, 425
harbour, 573
loom sheds, 498
mills, 450
poultry housing, 477–8
shieling huts, 12, 13, 428
Leys, 440
Libberton Mains, 482
Liberal Clubs, 158
Liberton, Edinburgh, 42
libraries, 308–9
Clare Hall, 215

Dundee, 308, 361
House of Dun, 37
Traquair, 37
universities, 307, 309
Wanlockhead, 554
see also reading rooms
Life Association, 148, 640, 671, 674
lifts *see* elevator buildings; elevators; goods movement facilities; hoists
Lighthouse Centre for Architecture and Design, 368
lighthouses, 582–3, 600
lighting, 350–1, 659
banks, 633, 634, 636
industrial buildings, 518, 524, 539
offices, 650, 651, 652, 657, 659
St Enoch's Hotel, 162
shops, 618, 646
warehouses, 648, 649
workshops, 503–5
see also electric lighting; windows
lily houses, 219
lime burning, 496, 553–4
lime kilns, 553–4, 556
Limekilnburn, 411
Limekilns, 495, 575, 626
Lindsay, Bernard, 624
Lindsay, Ian, 191
Lindsay, Ian G, 86
linen industry *see* flax mills; heckling shops; weaving factories
Linesman radar stations, 412–13
Links House, Glasgow, 669
Linktown Works, 537
Linlithgow
abattoir ventilation, 524
church, 282
houses, 69
municipal buildings, 253, 269
poultry housing, 456
quarry, 287
Linlithgow Palace, 42, 45
linoleum factories, 538
Linsandel House, 79
linseed mills, 519
see also flax mills
lint mills, 495, 496, 499, 505
see also flax mills; textile mills
Linthouse engine works, 516, 532, 585
Lion Chambers, Glasgow, 516, 661
Lipton, Thomas, 619
Lithgow's engine works, 532

Lithgow's shipyard, 585
Little Cumbrae, 583
Little Ferry, Sutherland, 577
Little Sparta, 222
Little, W G, 338
Littlecroft, 86
Liverpool, London & Globe building, 659
livestock housing and products, 465–93
    see also byres; dairies; pig sties; poultry houses; stables
Livingston, 140, 141
local authority housing see public housing
Loch Ailort, 205
Loch Ard, 80
Loch Arkaig, 397
Loch Doon, 405
Loch Doon Castle, 30
Loch Ewe, 407, 410
Loch Fyne, 83, 575
Loch Katrine, 343, 344
Loch Lag, 397
Loch Laggan, 397
Loch Leagh, 397
Loch Lomond, 127, 166–7, 343, 577
Loch Long, 77
Loch Morar, 397
Loch Rannoch, 397
Loch Tay, 409, 459
Loch Vennacher, 343–4
Lochalsh, 489
Lochgilphead, 321
Lochhead, Alfred, 84
Lochhead, William, 649
Lochlogan, 391
Lochmaben, 474
Lochnell, 214
Lochside Brewery, 544, 545
Lock Hospitals, 337
lock-keepers' housing, 95, 600
Lockhart, Sir Charles, 57
Lockhart, N, 537
Lockhart, William, 59
lodberries, 627
Lodge Canongate Kilwinning hall, 359
Lodge, G D, 673, 676
lodges see bothies; fishermen's and fishworkers' accommodation; gate lodges; hunting lodges; shooting lodges; workers' lodges

lodging houses, 179–82, 238
    see also bothies; inns; Sailors' Homes
lodgings
    in castles and tower-houses, 28, 35, 36–7, 43
    in universities, 305, 307
    see also Argyll lodging; donjons
lofts
    byres, 466
    cheese storage, 471
    farm workers' housing, 427, 430
    grain storage, 447
    hay storage, 193, 430–1, 432, 451, 466
    linoleum factories, 538
    materials storage, 499
    potato storage, 458
    seasonal workers' accommodation, 239, 240
    warehouses, 626
    weaving factories, 536
    wool stores, 489
    see also drying facilities; pigeon lofts
Logans of Coatfield, 626
Logie, 194, 374
London & Westminster Bank, 637
London Law Courts, 642
London Road Foundry, 532
Longannet Power Station, 346, 566
Longcroft, 80
longhouses, 19, 20, 90, 93
    see also blackhouses; byre dwellings
Longmorn, 545
Longmuir, A D, 261
Longniddry, 86
lookout towers, 214–15
loom shops and sheds, 498
loose boxes, 193, 239, 240, 431, 432, 476
    see also stables
Lord Milton's House, 73
Lorimer, Sir Robert, 61–2, 80, 83, 221, 371
Lornshill, 449
Lossiemouth, 580
Lothian Coal Company, 561, 566
Lothian House, 667
Lothian Road railway station, 604
Lothians
    coalfields, 560

dovecots, 478
mill kilns, 448
petrochemical industries, 559
seasonal workers'
    accommodation, 238
sheep housing, 483
threshing machines, 442
see also East Lothian; Midlothian;
    West Lothian
Loudon, John Claudius, 77–8, 190
Louise, Princess, 175
Lowe, D, 432
Lower Colmslie, 483
Lower Taes, Flisk, 226
Lowrie, George, 501
lucarnes, 525
luckenbooths, 614, 615
Luffness, 42
Luffness House, 384
Lugar, Robert, 56, 57
Luing, 577
Luma Lightbulb Factory, 510
Lumphanan, 456
lunatic asylums, 319–25
Lundin Links, 173
Lutyens, Edwin Landseer, 175
Lybster, 147, 579
Lyon Group, 672

M

Maam, 452
McAdam, John Loudon, 595
McAlister's boat building yard, 586
McAlpine, Robert, 559
Macaulay, Dr, 57
MacBeth, Alexander, 263
McColl's Hotel, 184
McConnel patent beams, 516, 539,
    649, 650
McConnel, Robert, 649
MacDonald, Lord, 57
Macdonald, D J, 533
Macdonald department store, 666
Macdonald of Sleat, Lord, 195
MacDonald's Sewed Muslin
    warehouses, 649
McDougall, D & J, 651
Macdougall, Leslie Graham, 184
Macduff, 375, 573, 582, 585, 587
McEwan Hall, 307
McEwan maltings, 545
MacFadzean, D, 277
Macfarlane, Walter, 533–4, 649

McGeoch warehouse, 666
MacGibbon & Ross, 150, 154, 255,
    276, 289, 331
MacGibbon, David, 153, 639
    see also Macgibbon & Ross
McGregor, Alexander, 97
MacGregor, C, 232
MacGregor, Don R, 153, 154
Machrahanish, 175, 560
McInnes, Gardner & Partners, 674
Mackay, Hugh, 263
Mackay, Major-General Hugh,
    389–90, 393
McKean, Charles, 84, 86, 184
Mackenzie, A G R, 372
Mackenzie, A Marshall, 63, 172, 173,
    324, 657
Mackenzie, Alexander, 55
Mackenzie, Alexander George
    Robertson, 184
Mackenzie, David, 146
Mackenzie, George, 1st Earl of
    Cromartie, 198
Mackenzie, J Ford, 171
Mackenzie, Osgood, 224
Mackenzie, R W, 61
Mackenzie, Thomas, 160, 364, 638, 640
    see also Matthews & Mackenzie
Mackenzie, Dr William, 334
McKim, Meade & White, 667
McKinnon, William, 529
Mackintosh, Charles Rennie, 42, 82, 83
    Glasgow School of Art, 308
    Hill House, Helensburgh, 77
    newspaper offices, 659, 664, 665
    Scotland Street School, 298–9
Mackison, William, 159, 652
McKissack, John, & Son, 665
Macky, John, 49, 133, 253
McLachlan, John, 642–3
Maclaren & Aitken, 652
McLaren, Mr, 139
Maclaren, David, 159
Maclaren, James, 79, 80, 96, 159, 654
Maclaren, James Marjoribanks, 175
McLaren warehouse, 668, 669
Maclay's Thistle Brewery, 545
Maclean's Windsor Hotel, 157
    see also Windsor Hotel
McLellan, Archibald, 366–7, 647
McLellan Galleries, 367
McLeod, John, 654
McMorran, Baillie, 625–6

McNair, C J, 183
McNaughtan, Duncan, 662
Maconochie Foods, 580
Macrae, Ebenezer, 120
McRae, Ken, 124
Mactaggart & Mickel, 84
Mactaggart, Sir John, 119
McVail, Dr Elizabeth, 240
MacWhannel, Ninian, 337
Madelvic car factory, 534
Madras College, 304
Magdalen Chapel, 628
Magdalen Green, 368
Magdalen Yard, 77
Magdalene's Kirkton, 101
Maidens, 587
Main, A & J, 453
Mains of Eden, 204
Maitland family, 39, 166, 173
Maitland Hotel, 154
Maitland, James Steel, 84, 86, 109
Makdowall, Sir William, 45
Malakof Arch, 215
Mallaig, 577
maltings, 542–3, 544, 545, 547
    engines, 521
    floor construction, 513
    goods movement facilities, 525
    kilns, 501, 541–2, 543
Manderston, 63, 194, 215, 216
Manor House, Inveresk, 73
manor houses, definition, 27
Manor Inn, 130, 140
Manor Powis Colliery, 564
manors, definition, 27
manses, 70
    ancillary buildings, 193, 203–4
    art gallery house, 78
    Old Manse, Cromarty, 67
    patterns, 71, 76, 78
Mansfield, 3rd Earl of, 56
Mansion Hotel, 152
manure storage, 193, 426, 467–8, 476, 481–2
MAP Architects, 673
Mar, Earl of, 369
Mar Lodge, 63
Marcell, Major Lewis, 395
Marchmont, Earl of, 52
Marchmont House, 52
Margaret, Queen, 280
Marine Biological Research Station, 307

marine engine works, 530, 532
Marine Hotel
    Elie, 175, 178
    Nairn, 173
    North Berwick, 170
    Oban, 173
    St Andrews, 173
    Troon, 174
Marischal College, 305, 307
maritime buildings and structures, 571–93
markets and market places, 613, 614, 616–17, 620, 643, 664
    see also mercat crosses
Markinch, 409, 540
Martello Towers, 32, 398–401
Martin, Hugh, & Partners, 671, 673, 675
Martin, Sir Leslie, 374
Marwick, T P, 664
Marwick, Thomas Waller, 669
Maryburgh, 390
    see also Fort William
Maryhill, Glasgow, 124, 574
Mary's Chapel, Edinburgh, 630
masonic halls, 133, 143, 359, 361
Masons' Hotel, 137
masons' yards, 497
mass concrete construction
    harbour structures, 573, 582
    industrial buildings, 512, 516
    railway structures, 601, 603–4
    Waverley Hydropathic, 170
Masson Mill, Derbyshire, 512
Matcham, Frank, 362–3
materials handling facilities *see* goods movement facilities
Maternity, Child Welfare and Special Treatment Centre, 334
maternity facilities, 315, 330–1, 334
Mather's Hotel, 159
Matheson, Donald A, 178
Matthew, Sir Robert, 184, 670
    see also Robert Matthew Johnson-Marshall
Matthews & Lawrie, 160, 654
Matthews & Mackenzie, 159, 163
Matthews, James
    Aberdeen Grammar School, 304
    Aberdeen music hall, 361
    banks, 640, 653
    Marine Hotel, Nairn, 173
    Union Club, Aberdeen, 148

mausolea, 215, 291
  see also graves
Mavisbank, 51
Maxwell, Robert, 258
Maybole, 35, 271, 540, 639
Maybole Castle, 59
Maybury Roadhouse, 184, 379
Meadows
  Dundee, 368
  Edinburgh, 369, 371
Meadows Exhibition, 371
Meadowside Granary, 578, 582
Meadowside shipyard, 584, 585
Meakin, Budgett, 529
Mearnskirk Hospital, 329
Mears and Stinback, 258
Mears, G, 258
Mears, Thomas, 258
measuring houses, 344
Mechanics Institutes, 308
Mechans boat building yard, 586
medical centres, collieries, 563, 565
medieval houses, 19
megacentres, 621
Meier, Richard, 674, 675
Meigle, 140
Meikle, Andrew, 440
Meikle Fardel, 140
Meikle Ferry, Ross and Cromarty, 577
Meikle, James, 440
Meikle, John, 258
Meikleour, 146
Melbourne Place, Edinburgh, 646
Meldrum & Binney, 559
Mellerstain, 212, 214
Melrose, 139, 140, 170, 644
Melrose Abbey, 279, 483
Melsetter House, 62
Melville, 38
Melville Castle, 54
Melville, Earl of, 51
Melville House, 51
Melville, Lord, 291
Melville, Viscount, 54
memorial cairns, 216
Menstrie Mill, 526
mental health care see lunatic asylums
Menzies, 40
Menzies family, 135
Menzies, William, 166
Mercantile Chambers, Glasgow, 662
mercat crosses, 217, 613–14, 616, 624, 629, 630

Merchant City, Glasgow, 75, 539, 646
merchant halls, 628–31, 642
Merchant Maiden Hospital, 303
Merchant Street lodging-house, 179
merchants' booths, 615, 627
Merchants' Exchange, New York, 667
merchants' exchanges see exchanges
Merchant's Hotel, 143, 144, 159
Merchants' House
  Glasgow, 133, 628, 642, 653, 657
  St Andrews, 70
merchants' houses, 221, 575, 627
merchants' shelters, 624, 642, 643
metal construction see iron
  construction; steel construction
metal extraction, 554–7
metal-working workshops, 496, 501–2
  see also forges; foundries;
    ironworks; smithies
Methil, 407, 574, 581
Methodist chapels, 288
Methven, 40
Michael Colliery, 561, 563–4
Mid Calder, 140
middens see manure storage
Middlemore, Thomas, 62
Middlesex Hospital, 335
Middleton's City Temperance Hotel, 150
Midlothian
  assembly room, 137
  barn, 439
  byres, 193, 466
  castles, 44
  collieries, 561, 563, 566
  corn exchange, 644
  country houses, 51, 53, 54, 55, 59, 61–2, 67
  dovecots, 193
  farm bothies, 239
  gardens and garden buildings, 214, 215, 217, 219, 220
  houses and villas, 73, 76, 79, 82, 95
  inns, 129, 130, 135, 137, 139, 140, 146
  lime kilns, 554
  mill school, 528
  municipal buildings, 251, 253, 258, 259, 263, 266, 267, 269
  palaces, 38, 43
  paper mill, 528, 539
  prisons, 269
  sawmills, 435

sheep houses, 485, 486
smithies, 193, 502, 506
stables, 193, 432
steeple, 193
tannery, 540
threshing machines, 443
tower-house, 35
water supplies, 344
whin mills, 456
workshops, 193
see also Lothians
Midmar, 40
Midmar Castle, 59
migrant workers *see* seasonal workers' accommodation
Migvie, 458
Mile End Mill, 536
mileposts, 596
military training areas, 411
milking and milk processing facilities, 13, 468, 469–70
see also byres; cheese rooms; dairies
mill clocks, 526
mill houses, 70
mill kilns, 448–9
Mill of Newmill, 205
mill schools, 528
see also employer-provided schools
Millcraig, 486
Millearne, 56, 473
Miller & Lang offices, 656
Miller, H, 63
Miller, Hugh, 240
Miller, James
banks, 668, 669
country house, 63
Ear, Nose and Throat Hospital, 335
hotels, 163, 177–8
offices, 662, 668, 669–70
Peebles Hydropathic, 178
Queen Street 136–48, Glasgow, 662
warehouses, 668, 669
Miller, Sir James, 63
Miller Street, Glasgow, 647
Miller, T M, 186
Miller, W S, 61
Millers' boat building yard, 585
Millport, 307, 575
mills, 186, 205, 350, 448, 495, 510–50
aluminium rolling mill, Falkirk, 559

see also bark mills; beetling mills; bone-meal mills; bucket mill; flour mills; grain mills; iron mills; jute mills; lades; linseed mills; paper mills; plash mills; sawmills; snuff mills; textile mills; whin mills
Mills and Shepherd, 83
Milne, John, 79
Milne's Court, Edinburgh, 109, 111
Milne's Square, Edinburgh, 109
Milton grain mill, 542
Milton Lockhart, 59
Milwain, W J, 163
miners
convalescent homes, 563
housing, 95, 561
Miners' Welfare Fund, 562–3
mines and mineral works, 551–70
see also coalfields and collieries; metal extraction
Mingary Castle, 30, 397
Minto, 40
Miskolc Ironworks, 532
Mitchell, A MacGregor, 184
Mitchell Hall, Marischal College, 307
Mitchell, Joseph, 160
Mitchell Library, 309
Mitchell Street lodging-house, 179
Mitchell, Sydney
Caledonian United Services Club, 158
Commercial Bank, Glasgow, 658
hospitals, 323, 328
North Bridge, Edinburgh, 664
Ugadale Arms, 175
Mitchell, Sydney, & Wilson, 60, 307
mixing rooms, 457
Moat Pit, 560
mobile cranes, 582
see also travelling cranes
Model Lodging Houses Association, 179
Modern Movement
hospitals, 314–15
hotels, 183, 184
houses, 84, 86
wareroom, Kilncraigs Mill, 527
Moffat
banks, 639, 640
Bath House, 168
hotels, 136, 168
inns, 135–6, 137, 139, 187

732 • INDEX

Moffat Hydropathic, 170
Moffatt, William Lambie, 313, 317
Molleson, Rev, 253
Monach Isles, 230
Monaltrie, 139
Monaltrie Hotel, 139
monasteries, 16, 279, 282, 397, 483
   see also abbeys
Moncreiffe House, 49
Moncreiffe, Sir Thomas, 49
Moncrieff, Sir William, 139
Moncrieffe Arms, 139
Moncur Street lodging house, 180
Monifieth, 130–1, 187
monitoring posts see observation posts
Monkland Canal, 598, 600
Monklands, 556
Monktonhall Colliery, 563, 566
Monmouth and Buccleuch, Duchess of, 51
Monorgan, 40
Monro, J M, 172, 173, 174
Monteath, Dr George, 334
Monteith, Henry, 57
Monteviot, 215
Montgarrie Mill, 541
Montrose
   airfield, 404
   assembly rooms, 259
   banks, 639
   brewery, 544, 545
   cistern, 344
   docks, 573
   goods shed, 605
   gun battery, 398
   hospital, 313
   lunatic asylum, 319, 320
   mills, 523, 524, 536
   municipal buildings, 253, 259, 267, 319
   newspaper office, 655
   oil base, 575
   salmon bothy, 587
   weaving factories, 536
monumentalism, 109, 111, 112, 115, 119
monuments, 285, 289–92, 369
Monymusk Home Farm, 427
Moore, J A, 493
Moorfoot, Midlothian, 485
Moray
   arcades, 69
   assembly room, 361

banks, 640
bee boles, 203
boiling house, 587
castles, 28, 37, 39, 192, 219
cattle courts, 475
châteaux, 37, 38, 42
diocese, 279–80
distilleries, 543, 545
fishing stations, 580, 587, 588
gate lodges, 192
generating house, 205
hand crane, 580
harbours, 572, 578, 579
herring curing sheds, 579
hospitals, 312, 313, 320
hotels, 160, 172, 173
houses, 69, 81, 219, 587, 588
hydropathic, 170
ice houses, 579
kennels, 207
laundries, 196
lighthouses, 582
lunatic asylum, 320
motor house, 194
municipal buildings, 266–7
museum, 364
palace, 43
privies, 204
railway bridges, 601
sheep husbandry, 486
stables, 194
technical college, 308
towers and tower-houses, 33, 35, 43
villas, 77
warehouses, 578, 626
Moray, Colonel, 56
Moray, bishop of, 43
Moray, Earl of, 312, 388
Moray Place, Edinburgh, 75
Morningfield Hospital, 335
Morningside see Edinburgh
Morningside Hydropathic, 171
Morphie, 456
Morphie Farm, 432
Morren's Inn, 143
Morris & Lorimer, 586
Morris & Steedman, 185
Morris, James, 86
Morris, James A, 83
Morris, Roger, 54
Morris, Tom, 176
Morrison Street, Glasgow, 665

Morse, Lieutenant-General, 399
Morton Castle, 29
Morton of Errol, 444
Mortonhall Road, Edinburgh, 80
Morton's Bond, 513
Moseley, W & A, 152
moss houses, 215
Mossman, William, 261
Motherwell, 334, 557
motor houses *see* garages
Mottram Patrick & Dalgleish, 671
Moulin Castle, 30
Moulinearn, 131, 132, 140
Mount Royal Hotel
    Edinburgh, 185
    London, 183
Mount Stuart, 162
mountaineers' bothies, 227–30
Mounthooly, 202
Mouswald Grange, 441
Mowat, John, 258
Moy, 603
Moy House, 204
Much Wenlock Town Hall, 266
Mugdock, 27, 343, 344, 410
Mugdock Castle, 29, 30
Muirkirk, 560
Mull, 575, 583
Mull Castle, 30
Mull of Kintyre, 583
multi-storey housing *see* tenements and flats; tower blocks
municipal buildings, 142, 143, 145, 182, 249–72, 319, 629, 630
    see also law courts; prisons; town bells and town clocks
Munro, James, & Sons, 335
Murchland, William, 468
Murdoch & Rodger, 156
Murdoch, William, 350, 524
Mure of Caldwell, Baron, 55
Murphy, Richard, 368
Murray Arms, 135
Murray International, 674
Murray, James, 52
Murray, Sir James, 45
Murray Royal Asylum, 319, 320
Murray, W H, 230
Murrayfield, 83, 372
Murrays, The, 486
Murrayshall, 554
Murthly, 215, 323
Murthly Castle, 168

Museum of Antiquities, 364
Museum of Scotland, 366
museums, 307, 308, 360, 363–6, 371, 554, 557
Mushet, David, 556
music halls, 361, 362–3, 373
Musselburgh, stop-line, 409
Musselburgh Tolbooth, 251, 253, 258, 259, 263, 266, 267, 269
Muthesius, Hermann, 61, 83
Mylne, John, 45, 48, 286–7, 389
Mylne, Robert
    houses, 48, 52, 71
    roads and bridges, 595
    St Cecilia's Hall, 359–60
    Shore, Leith, 626
Mylne, Thomas, 45
Mylne, William, 595

N

Nairn, 137, 166, 173
Nairn, Michael, 538
Nairn's head office, 669–70
Nairnshire, 29, 391, 444
Naismith, John, 483
Napier House, 181
Napier, James, 668
    see also Laird & Napier
Napier, Lord, 485, 489
Napier Road, Edinburgh, 79
Nasmyth, Alexander, 362
National Bank, 637, 639, 660, 666
National Bank of Scotland, 669
    see also Bank of Scotland
National City Bank, 667
National Gallery of Scotland, 367
National Library of Scotland, 309
National Portrait Gallery, 364
naval bases, 382–3, 401, 404, 407–8
navigation aids, 582–3
Navigation School, 158
navvies' housing, 241
NCR factory, 534
Neave, David, 77, 143, 361, 362
Neilson, James Beaumont, 556
Neilston, 536
Nelson Street, Glasgow, 665
Nelson, Thomas, & Sons, 655
Nelson towers, 215
neo-Baroque architecture *see* Baroque architecture
neo-classical buildings
    Aberdeen assembly room, 361

banks, 640
Geographical Institute, 656
Glasgow Royal Concert Hall, 374
hotels, 143
Hunterian Museum, 364
lodging houses, 180
pattern books, 52
tenements, 111, 115, 116, 122
see also classical buildings
neo-Georgian architecture, 86, 157, 657
see also Georgian architecture
neo-Greek architecture, 143, 144, 152
see also Greek Revival
neo-Italianate architecture *see* Italianate architecture
neo-Jacobean architecture *see* Jacobean style
neo-Romanesque architecture *see* Romanesque architecture
neo-Tudor architecture *see* Tudor style buildings
Neptune Buildings lodging house, 181
Nesfield, William Eden, 79
Nether Abington Farm, 470, 472
Nether Mill, Galashiels, 523
Netherbow Gate, 31
Netherbow Port, Edinburgh, 385, 386
Netherbutton, 410, 412
'New Architecture' designs, 84
New Club
    Edinburgh, 148, 150, 157
    Glasgow, 152, 154
New Craig House, 323
New Inn
    Aberdeen, 133, 143, 144
    Dunbar, 137
New Lanark, 122
    bell towers, 526
    dyeworks, 538
    foundry, 529–30
    mill clock, 526
    mill turbines, 520
    mills, 514, 521, 524, 525, 535
    school, 528
    social facilities, 528–9
    tenements, 112
New Mains, Berwickshire, 477
New Mills, 495
New Ortie, 442
New Town Dispensary, 312
New Towns, 102, 122
Newall, Walter, 78
Newbattle, 129, 135, 140

Newburgh, 226, 409, 502, 536
Newhall, 439
Newhaven, 495, 582
Newington *see* Edinburgh
Newliston, 214
Newmilns, 537, 539
news rooms *see* reading rooms
newspaper offices, 654–5, 664
    Edinburgh, 187, 647, 654, 664, 665
    Glasgow, 511, 655, 658, 659, 664, 665, 669
Newton, Ernest, 174, 178
Newton-upon-Ayr, 250
Newtongrange, 95, 561, 566
Nicholas, Grand Duke, 141
Nicholson, Peter, 71, 75, 147
Nicoll, James, 80, 82–3
Niddry, 33, 37
Nigg, Cromarty Firth, 407
night refuges, 182
Nightingale, Florence, 313, 332
Nimlin, Jock, 228
Nimmo, William, & Partners, 671
Nine Trades Hall, 630
Ninewells Hospital, 315
Nisbet of Dirleton's House, 69
Niven & Wigglesworth, 664–5
Niven, D, 103
Niven, D B, 62
Nobel's Explosives, 556
Noble, Sir Andrew, 62
Noltland, 37
Nordrach-on-Dee, 328–9
Normal Seminary, Glasgow, 297
Norse buildings, 17, 19, 20
North Banks, Papa Stour, 485
North Berwick, 83, 170, 375, 571, 626
North Bovey Manor, 179
North Bridge, Edinburgh, 664
North British & Mercantile insurance offices, 640, 658, 668
North British Hotel
    Edinburgh, 153, 160, 163, 164–5, 187, 664
    Glasgow, 155
North British Rubber Company, 660, 664
North Frederick Street night refuge, 182
North of Scotland Bank, 637–8, 640
North Queensferry, 577, 582
North Ronaldsay, 450
North Row, Peebles, 615

North Scotstarvit, 425
North Street, St Andrews, 73
North Sutor, 404, 407
North Town, Orkney, 410
North Uist, 477–8
North Woodside Road lodging house, 180
Northern Assurance Building, 657
Northern Building, 660
Northern Club, 156
Northern Hotel, 184
Northern Renaissance buildings, 162, 166
Northmavine, 482
Norton, Leslie, 672
Norway, 17, 20
Norwich Union buildings
  Edinburgh, 632, 640, 671
  Perth, 674
Notman, William, 168
nuclear power stations, 346

O

Oakleigh Villa, 77
Oakley, C A, 529
Oban
  defences, 407
  hotels and inns, 166, 172–3, 183, 187
  newspaper office, 655
  railway pier and station, 577
Oban Hills Hydropathic, 171
Oban Times building, 655
obelisks, 216, 523
Obertal, 84
observation posts, 410, 411–12
Ochilvale Mill, 512
O'Dowd, Anne, 242
offices, 516, 584, 625, 628, 630, 643, 646, 647, 650–76
  car parks, 597
  in bank buildings, 632
  in collieries, 562
  in mills and factories, 526–8
  in shipbuilding yards, 584
  see also administration buildings; business chambers; estate buildings; exchanges; harbour offices; insurance offices; newspaper offices
Ogilvie family, 40
Ogilvy, Sir George, 285
Ogilvy, Sir John and Lady, 324–5

oil bases, 575
oil engines and engine houses, 443–4, 521
oil-shale industry, 559–60
Old Aberdeen Tolbooth, 258
Old Assembly Close, Edinburgh, 133
Old Canisbay Manse, 203
Old College, Glasgow, 628
Old Course Hotel, 185
Old Craig House, 323
Old Docks, Leith, 573
Old English style, 82
Old Harbour, Dunbar, 585
Old Kilpatrick, 291
Old Manse, Cromarty, 67
Old Rutherglen Road Mill, 512, 514
Old Somerset House, 631
Old Waverley Hotel, 155
Oldhamstocks, 285
Oldmill Poorhouse, 318
Oldrieve, Bell & Paterson, 668
Oliver's Fort, 387, 388, 390, 395
Ollaberry, 580
orangeries, 219
Orchard Brae House, 673
Ord, Laurence, 127
Orient Boarding House, 181
Orkney
  banks, 639
  barns, 439
  bird housing, 478–9
  bykes, 446
  byres, 465, 466, 476
  churns, 471
  country house, 62
  defences, 32, 382, 398, 399–401, 404, 405, 407–8
  Earl's Palace, 45
  farm steadings, 425
  ferry facilities, 578
  gasometer, 205
  gutters' housing, 232–3
  harbour offices, 586
  herring curing sheds, 579
  horse-engine house and platform, 442
  kilns, 449, 450
  merchants' houses, 627
  mills, 450
  Pictish buildings, 17, 19
  piers, 575
  poultry housing, 476, 478
  privies, 204–5

radar stations, 410, 412
salt extraction, 552
schools, 297
slate roofs, 69
steam power facilities, 443
timber gallery, 37
Viking buildings, 17
warehouses, 578, 627
wind power, 440, 441, 471
Orlit observation posts, 411–12
Ormiston, 71, 495
Orr Ewing, 646
orthopaedic hospitals, 337–8
    see also children's hospitals
Orton House, 196
Osborne Hotel, 154, 157
Osborne Street, Glasgow, 665
Otis, E G, 525
Otter Ferry, 575
Ouderogge, C, 257
outbuildings, 61, 92, 95, 139
    seasonal workers' accommodation, 224, 239
    see also animal houses; bee boles; bird housing; boathouses; cart bays and sheds; cleitean; coach houses and carriage houses; drying facilities; estate buildings; farm steadings; garden buildings; generating houses; hay storage; implement sheds; laundries; privies; storehouses
Outer Hebrides *see* Western Isles
Outer Tolbooth, St Giles Cathedral, 270–1
outhouses *see* outbuildings
oval houses, 18
ovens, 502
Owen, Robert, 112, 528, 529
    see also New Lanark
Ower, Charles and Leslie, 654
ox byres, 429
Oxenfoord, 146
Oxenfoord Castle, 55
Oxenfoord, Viscount, 628

P

Packhouses, 627
Paderewski Hospital, 339
Page & Park, 368, 675
pagoda roofs, 501, 525, 542, 543
Paisley
    assembly rooms, 142, 361
    business chambers, 646
    clubs, 158
    cotton mills, 510, 519, 523, 525, 535, 536
    forge, 535
    hospitals, 314, 327, 334
    hotels and inns, 142, 186
    houses, 83
    lint mills, 495
    mill chimneys, 523
    mills, 186
    offices, 647
    schools, 300, 528
    shawl factories, 539
    shops, 619
    skew bridge, 601
    steam cranes, 580
    tenements and flats, 114, 115, 117
    thread mills, 516, 521
    tolbooth, 260
    warehouses, 646, 647
Paisley Canal *see* Glasgow, Paisley and Ardrossan Canal
Palace Hotel
    Aberdeen, 163, 164, 165
    Edinburgh, 153, 155, 157
    Inverness, 173
Palace of Seton *see* Seton Palace
Palace of Stirling *see* Stirling Castle
palaces, 27–47, 49, 51, 271, 279
Palazzo di Genova, 49
Palladian architecture
    banks, 666
    country houses, 48, 51, 52, 54, 57, 63
    St Mary's Chapel, 359
Palladio, Andrea, 48
Palmer, Sir Thomas, 385
Palnackie, 571
Panmure Arms, 175
Panmure Testimonial, 215
Panmure Works, 528
Pannanich, 139, 168
Pantiles Hotel, 184
Papa Stour, 459, 485
paper mills, 510, 539–40
    heating and ventilation, 502
    Midlothian, 528, 539
    steam engines, 520
    water wheels, 519
paper-making workshops, 495
Papworth, John Buonarotti, 253

paraffin-oil production, 559
parish churches *see* churches
parish schools, 295–7
Park area, Glasgow, 75
Park Brewery, 545
Park House, 330
Park Motor Car Body Works, 517
Park, William, 534
Parker, Charles, 537
Parkhead, 535
parks *see* gardens and garden buildings
Parkside Works, 655
Parliament House, 309
Parnie Street, Glasgow, 665
Parr, James, & Partners, 674
Parsonage of Morebattle, 630
Partickhill *see* Glasgow
Paterson & Broom, 184
Paterson, Dr, 170
Paterson, Alexander Nisbet, 80, 158, 175
Paterson, John, 194, 288, 631
Paterson, Robert, 152, 154
Paton & Baldwin, 527, 669
  see also Kilncraigs Mill
Paton, J T, 375
pattern books
  estate buildings, 190
  farmhouses and farm buildings, 78, 456
  garden buildings, 214
  houses, 52, 59, 71, 77–9, 93, 95
  salmon bothies, 226
pattern shops, 516, 527, 533
Pattiston, 411
pavilion roofs *see* French roofs
pavilions, 371–2
  commercial buildings, 627
  country houses, 49, 51, 52, 54
  garden buildings, 211, 219, 220, 222, 369, 371
  hospitals, 313, 314, 329, 336
  lunatic asylums, 321
  schools, 298
  see also seaside pavilions; tea pavilions
Paxton House, 53
Paxton, Sir Joseph, 370
Paxton, Thomas, 181
Pearl Insurance Company building, 659
peat-flitters' accommodation, 232, 428

Peddie & Browne, 666
  see also Browne, George Washington; Peddie, John More Dick
Peddie & Kinnear
  banks, 639, 640
  Cockburn Street, Edinburgh, 150, 154, 647, 654
  convalescent home, 336
  hotels, 153, 156, 656–7
  houses, 60
  hydropathics, 170, 174
  Leith corn exchange, 644
  offices, 151–2, 650–1
  University Club, Edinburgh, 150
Peddie, John Dick, 157, 178, 635–6, 639, 640
Peddie, John More Dick
  banks, 657, 660
  Central Chambers, Glasgow, 657
  Edinburgh Liberal Club, 157
  Edinburgh School of Art, 308
  hotels, 165, 178
  insurance offices, 657, 658, 660
  see also Peddie & Browne
Peebles, 127, 386, 615
Peebles Hydropathic, 170, 178, 187
Peeblesshire
  brewing plant, 545
  byres, 466
  country houses, 63
  gardens and garden buildings, 211
  hotels and inns, 127, 184
  library, Traquair, 37
  lime burning clamps, 553
  manure storage, 481, 482
  mills, 518–19, 521
  newspaper office, 655
  sheep houses, 485, 486
  smithy, 502
  timber gallery, 37
  water wheels, 519
  weaving factories, 537
Peel, The, 83, 338, 339
peels/peles, definition, 27
penal institutions, 249
  see also prisons
Penicuik, 82, 139, 215, 528, 539
Penicuik House, 53, 139, 193, 215
Penman's Boiler Shop, 532
Pennant, Thomas, 15, 16, 20
Pentland Terrace, Edinburgh, 82

Perpendicular style, Duns Town
    House, 253
Perth
    assembly room, 142
    banks, 638, 639
    churches, 269, 271, 282, 284, 288
    cinemas, 377
    defences, 31, 387, 388
    dyeworks, 518, 538
    hospitals, 313, 316, 336
    hotels, 129, 142, 143, 163, 186
    houses, 76
    insurance offices, 674
    lunatic asylum, 319, 320
    mills, 186, 519, 541
    municipal buildings, 251, 260–1,
        267, 271
    railway facilities, 604, 605, 606
    salmon bothies, 225
    school, 76
    stop-line, 409
    swimming pools, 376
    tenements and flats, 109
    trades halls, 628
    water supply, 345
    whisky bottling plant, 519
Perth Academy, 76
Perth City Mills, 519, 541
Perth Playhouse, 377
Perthshire
    assembly rooms, 146
    banks, 638
    bell towers, 526
    bird housing, 478
    bleachworks, 345
    brewery, 544
    bykes, 446
    byre, 466
    castles, 29, 30
    cathedral, 284, 289
    cattle courts, 475, 476
    châteaux, 40
    churches, 130, 285
    cotton mills, 345, 512, 513, 514,
        521, 524, 535
    country houses, 49, 53–4, 56, 57,
        59, 61
    dairies, 472–3
    dam, 343–4
    farm bothies, 238
    fishing stations, 587
    flax mills, 519
    gardens and garden buildings,
        212, 214, 215, 216, 217, 219, 220
    gate lodge, 61
    granaries, 447
    grass-drying plant, 455
    horse engine house and platform,
        442
    hospitals, 313, 316, 336, 338, 339
    hotels, 140, 146, 147, 168, 172, 175,
        178–9, 182, 338
    houses, 76, 80, 96, 168
    hydropathics, 170, 174
    Innerpeffray library, 308
    inns, 130, 131, 132, 135, 136, 139,
        140, 145, 146
    kilns, 449, 501, 554
    lunatic asylum, 321, 323
    maltings, 544
    military training area, 411
    mill clocks, 526
    mill lades, 345, 519
    palaces, 38, 43
    potato houses, 459
    railway facilities, 178, 603, 604, 605
    roads, 595
    root crop storage, 455
    salmon bothies, 225, 587
    school, 304
    shieling huts, 16
    threshing machines, 443
    villas, 77
    water towers, 526, 540
    weaving shed and factories, 537
    whin mills, 456
    see also Strathmore
Peterculter, 539–40
Peterhead
    Commercial Bank, 639
    docks, 574, 575
    gun battery, 398
    inns, 139
    smokehouses, 580
    spa, 168, 374
    warehouses, 626–7
Petrie, Alexander, 659, 662
Petrie, George, 173
petrochemical industries, 559–60
    see also chemical works
Pevsner, Nikolaus, 4
pheasantries, 215
Philip, Dr Robert, 328
Philiphaugh Saw Mill, 519
Phillips' Harbour, 579, 580

Philorth Castle, 31
Philp, Andrew, 150, 156, 171
Philp's Cockburn Hotel *see* Cockburn Hotel, Edinburgh
Philp's Hotel, 168
Phoenix Assurance Building, 667–8
Physicians' Hall, Edinburgh, 630, 634
piazzas *see* arcades
Pickering, Mrs, 62
Picket Hamilton pillboxes, 408
Pictish buildings, 17, 19
Picturesque, the
    Clydesdale Bank, Glasgow, 653
    country houses, 54, 55, 57
    garden design, 214
    gate lodges, 192
pier buildings, 577
piers, 232, 571, 575, 577–8
    see also ports and harbours; quays
Piershill, 120
Piersland Lodge, 82
pig crays *see* pig sties
pig iron *see* iron
pig sties, 16, 193, 427, 476, 479–81
pigeon lofts, 202
    see also dovecots
Pilkington, F D T, 289
Pilkington, Frederick, 79, 152, 170
pillboxes, 408, 409, 410
    see also blockhouses
Pilmuir, 67, 239
Pilmuir Works, 527, 536
Pilrig Model Buildings Association, 97
Pine Trees Hotel, 182
pineta, 216
Pinkie House, 59
Pirie, J B, 79
pit-head baths, 563, 565
    see also baths
Pitcraigie, 486
Pitfour, 215
Pitkeathly, 139
Pitlochry, 77, 168, 170, 172, 182
Pitmedden, 211, 371
Pitmillie, 227
Pitreavie, 45
Pitsligo, 27, 32, 35–6, 38, 43
Pitt Street Common Lodging House, 181
Pittencrieff Park, 370
Pittenweem, 251, 269–70, 578
planned villages, 93, 94
    see also New Lanark

plantation stells, 486
plash mills, 524, 539
Playfair, James, 54
Playfair Terrace, St Andrews, 76
Playfair, William
    Advocates' Library, 309
    country houses, 59
    Dollar Academy, 303
    Edinburgh University, 307
    houses, 73, 77
    National Gallery of Scotland, 367
    Surgeons' Hall, Edinburgh, 634
Playfield theatre, Edinburgh, 359
playing fields, 563
Pleasance, Edinburgh, 129
pleasances *see* gardens and garden buildings
Plewlands House, 70
Pluscarden Abbey, 203
Pococke, Bishop, 446
Point Hotel, 187
Point of Buckquoy, 17
police boxes, 249
police stations, 249
Polkemmet Colliery, 564
Pollock Halls, 307
Pollokshields *see* Glasgow
Polmadie, 606, 607
Polmaise Colliery, 564
Polonceau truss, 518
polychromy, 512, 523, 535, 541, 545, 547, 653
Pont, Timothy, 27, 39, 40
Ponton, Alexander, 137
poorhouses, 313, 316–19, 320, 321
Port Askaig, 583
Port Dundas
    dockside warehouses, 578
    generating station, 659
    offices, 600
    printing and publishing works, 664
    sugar refinery, 514, 516, 547
Port Errol, 587
Port Glasgow, 521, 532, 574, 585
Port Logan, 582
Port Murray, 86
Port of Ness, 573
Portknockie, 573
Portlethen, 410
Portmahomack, 198, 578, 627
Portmeirion, 86
Portnancon, 580

Portobello, 375, 377, 501, 639
Portobello Baths, 374
Portobello Chocolate Factory, 516
Portormin, 579
Portpatrick, 178, 572, 582
Portree, 579
ports and harbours, 382, 398, 571–83
    see also docks; ferries and ferry facilities; fishing stations; piers
Portsoy, 571, 626
Portugal, 16
Portugal Street lodging house, 180, 181
post cranes, 529, 530
Post House
    Coylumbridge, 186
    Edinburgh, 185, 186
potato houses, 239, 240, 457–9
potato pickers' accommodation, 237, 238, 239–41, 242
potteries, 501, 512
Poulson, John G L, 186
poultry houses, 193, 202, 204, 456, 466, 476–9
Pow of Drummond, 447
power sources, 504–5, 519–21
    see also horse engines and engine houses; steam power; water-powered machinery; water wheels; windmills
power stations, 345–6
    collieries, 562
    Fife, 346, 523, 566
    hydraulic power, 352
    Port Dundas, 659
    shipbuilding yards, 584
Pratt & Keith, 163
precast concrete construction, 512, 671
prefabricated construction, 300, 372, 411, 453
prehistoric houses, 9–11, 20–1
    see also oval houses
Preissnitz, Vincenz, 169, 170
press rooms, 468–9, 471
Preston, 37, 542
Preston Hall, 219
Preston Island, 552
Preston Mill, 500
Prestonfield, 194, 629
Prestongrange, 59, 566
Prestonpans, 70, 552, 613
Prestwick, 375
primary schools, 295–300
Primrose, 449–50

Princes Dock, Glasgow, 574, 581
Princes Exchange, 675
Princes Pier, Greenock, 577
Princes Square, Glasgow, 183, 617, 646
Princes Street, Edinburgh, 675
Princes Street Gardens, 369
Princess Margaret Rose Hospital, 338
printing and publishing works, 511, 516, 517, 654–6, 664–5
    see also newspaper offices
priories, 35
prisons, 249, 253, 268–9
    Fife, 250, 269
    in tower-houses, 32
    Midlothian, 266, 269
    Ross and Cromarty, 269, 272
    Sutherland, 271
private schools, 295, 303, 304
privies, 204–5, 427, 524
    see also sanitation; sewerage
Professors' Lodgings, 363–4
Progress Works, 541
prospect towers, 214–15
Provand's Lordship, 364
Prudential buildings, 658, 668
public baths see baths
public defences see defences
public gardens see gardens and garden buildings
public houses, 378–9
    see also inns; taverns
public housing, 84, 99–101, 102–3, 118–24
public laundry, Glasgow, 374
public libraries, 308–9
public schools see parish schools
public services, 341–57
Public Warehouse, Dundee, 627
pubs see public houses
Pugin, A W N, 288, 317
Pullars Dyeworks, 518, 538
pump houses, 215
pumping engine houses, 562
pumping stations, 581
Purves, Thomas, 485

Q
Quarrier, William, 328
quarries, 287, 496, 497, 551–70
quays, 517, 571
    see also docks; piers; ports and harbours
Quayside Mills, 512

INDEX • 741

Queen Anne Battery, Stirling Castle, 390
Queen Anne style, 80, 158
Queen Court, Glasgow, 647
Queen Street
    Edinburgh, 73
    Glasgow, 662
Queen Street Station, 155, 604
Queens Court, Glasgow, 539
Queen's Crescent, Glasgow, 75
Queens Dock, Glasgow, 574
Queen's Ferry, 577
Queen's Hall, Edinburgh, 374
Queen's Hotel
    Blairgowrie, 140
    Bridge of Allan, 168
    Dundee, 159
    Oban, 173
Queens' Park, Glasgow, 370
Queen's Road, Aberdeen, 79
Queensberry Arms, 136–7
Queensberry, Marquess of, 57
Quennell, C H B, 63
Quittlehead, 456
Quothquhan, 339

R

Raasay, 574
radar stations, 382, 407, 410, 412–13
Rae, George, 76, 79, 137, 173
Raeburn, Robert, 651
ragged schools, 298
Raigmore Hospital, 328, 338, 339
Railton, William, 640
railway bridges, 601, 603
railway buildings, 600–7
    carriage sheds, 397, 601, 606
    engine sheds and depots, 397, 518, 601, 606, 607
    engineering works, 534, 606
    see also railway hotels; railway stations
railway hotels, 153, 160–6, 173, 185, 186
railway piers, 577
railway ports, 572–3
    see also ferries and ferry facilities
railway stations, 218, 577, 601, 604–5
    Dumfriesshire, 168, 606–7
    Dundee, 159, 604, 605
    Edinburgh, 604, 605, 607, 617
    Glasgow, 155, 162, 604, 605, 654, 656

Perthshire, 178, 604, 605
Rait, 76
Rait Castle, 29
Ramsay, Alan, 358
Ramsay Arms, 175
Ramsay, Sir George, 56
ranch houses, 104
Randolph & Elder, 530
Randolph Cliff, 75
Rankeilour Mains, 455
Ratter, 446
Ravelston Terrace, Edinburgh, 673
Ravenscraig, 557
Ravenscraig Castle, 29, 30, 31
Ravenscraig Hospital, 339
Ray, John, 388
reading rooms, 142, 360, 361, 529, 642, 643
    see also libraries
recreation buildings, 358–80
    see also assembly rooms; cinemas; coffee houses and coffee rooms; concert halls; hill bothies; reading rooms; swimming pools; taverns; tennis courts; theatres
rectangular plan form, evolution, 9, 10, 13, 14, 16, 18–21
Red House, Bayswater, 157
Red Lion Hotel, 143
Red Road tower blocks, 122
Redcastle, 227
Redden Farm, 432, 435, 475
Redlands House, 337
Redpath & Brown, 516
Redpoint fishing station, 229
Reed, Henry, 261
reformatory schools, 298
refrigeration *see* ice houses
refuse disposal, 348–9
    see also sanitation
Regent Hotel, 183
Regent Mill, 541
Regent Quay, 547
regional shopping centres, 621
Reiach & Hall, 673, 674, 675
Reid, Mrs, 84
Reid, A & W, 151, 196, 207, 320, 640
Reid, John, 68, 71
Reid, Robert, 142, 320, 631
Reiff, 227
reinforced concrete construction
    bridges, 516, 597

commercial buildings, 661, 672
Daily Express building, 511
hotels, 183, 185
industrial buildings, 512, 516–17, 518–19, 533, 534, 538
lodging house, 181
mine buildings, 563, 564–5
railway coal bunkers, 606
railway stations, 604
silos, 455
Renaissance buildings, 35–45
banks, 640, 666, 668
Burntisland church, 285–6
Central Station, Glasgow, 656
clubs, 152, 157
Curr Night Refuge, 182
Edinburgh Stock Exchange, 642–3
Ferguslie Mill School, 528
fortifications, 32
hotels, 152–3, 154, 159, 162, 164–5, 166
internal stairs, 33
Leith Sailors' Home, 159
offices, 652, 653, 654, 655, 658, 659
warehouses, 648, 649, 665
see also Scots Renaissance architecture
Renfrew, 77
Renfrew Combination Poorhouse, 318–19
Renfrew District Asylum, 324
Renfrew Ferry, 578
Renfrewshire
assembly rooms, 142, 361
business chambers, 646
churches, 288, 291
clubs, 151, 158
cotton mills, 510, 513, 519, 523, 524, 525, 526, 535, 536
country houses, 59, 62
cranes, 580, 581, 582
custom house, 586
dairy farms and equipment, 466, 470, 471
dams, 344
defences, 398, 411
docks, 574, 575, 582
engine works, 532
factories, 512, 521, 534, 539
ferries, 578
flax mills, 514
forge, 535

gardens and garden buildings, 215, 221
harbours, 574
hospitals, 314, 315, 327, 328, 334, 335, 339
hotels and inns, 142, 183, 184, 186
houses, 77, 83, 85, 86
hydropathic, 171
leisure centre, 376
lighthouses, 583
lint mills, 495
lunatic asylum, 324
mill chimneys, 523
mill conversion, 186
mill lades, 519
mill turbine, 520
municipal buildings, 260
offices, 647
poorhouse, 318–19
pumping station, 581
railway facilities, 577, 604
reservoirs, 342
sanatorium, 328
schools, 300, 528
seaside pavilion, 378
seasonal workers' accommodation, 240
shipyard, 585
shops, 619
skew bridge, 601
steel works, 557
sugar refineries, 547
tenements and flats, 114, 115, 117
thread mills, 516, 521
tombs, 291
warehouses, 578, 646, 647
water supplies, 342, 344, 526
see also Paisley
Rennie, John, 572, 573, 578, 595
Reoch, James, 261
research institutes, 307
reservoirs, 342, 343–4
residential clubs see clubs
residential hospitals, 303–4
Restenneth Priory, 277
retail trade buildings, 612–23
see also department stores; shops
Rhifail, 485
Rhind, David
banks, 633–4, 637, 638, 639
Life Association Building, 148, 640
Rhind, John, 207, 313
Rhu, 140

Rhu-na-Haven, 62
Rhum, 20, 62, 197
Ribigill, 488
rice mill, 541
Richardson, H H, 80
Richmond drying rack, 452
Richmond Hotel, 154
Ritchie, Charles, 71
Ritchie, Isa, 232–3
Ritchie, Robert, 443
River Clyde *see* Firth of Clyde
River Forth *see* Firth of Forth
River Tay, 225–6
River Tweed, 225
RMJM, 315, 675
roadhouses, 184–5, 378–9
roads, 93, 131, 393, 397, 594–8
Robert Gordon's Hospital, 303
Robert Gordon's Institute, 308
Robert Gourlay's House, 625
Robert Matthew Johnson-Marshall,
    315, 374, 674, 675
  see also Matthew, Sir Robert
Roberts, James, 185
Robertson & Dobbie, 666
Robertson & Orchar, 513, 514, 518,
    537
Robertson, Rev A E, 228
Robertson, George, 471
Robertson, James, 139
Robertson, John Murray, 80, 643, 652
Robertson stop-line, 409
Robertson, Thomas, 629
Robertson, William, 144, 145, 195, 627
Robertson's boat building yard, 586
Robinson, P, 117
Robroyston Hospital, 329
Rochead, John T
  banks, 637, 646, 651, 653, 657
  Grosvenor Terrace, Glasgow, 75
  Hawick corn exchange, 644
  Sailors' Home, Glasgow, 158
Rococo style, 54, 179
Rodel, 580
Rogerson, William, 337
Romanesque, St Cuthbert's Church,
    Dalmeny, 280
Romanesque architecture
  banks, 631, 632, 633
  Christian Institute, 181
  churches, 281, 288
  Eastern Club, 152
  offices, 653, 658

Romanesque sarcophagus, 290
Romanticism, 54, 55, 59, 62, 79, 214
Romer, Captain John, 391, 393
roof ventilators
  industrial buildings, 524, 538, 540
  kilns, 542
  railway engine sheds, 606
  schools, 296
roofs
  arcades, 620
  cattle courts, 475, 476, 481
  country houses, 49, 54, 67
  dovecots, 200, 202
  Dundee corn exchange, 644
  farm steadings, 453
  fishermen's accommodation, 227,
    230, 232
  horse engine houses, 442
  hotels, 152
  houses, 69, 75, 80, 84, 91, 93, 94,
    95, 103
  industrial buildings, 517–19, 530,
    532, 533, 534, 538, 584
  inns, 128–9, 130, 146
  kilns, 449, 501
  manure storage, 481
  pig sties, 480
  potato houses, 458
  schools, 297
  shieling huts, 13, 16, 230
  workshops, 498, 502–3
  see also Belfast roofs; French roofs;
    pagoda roofs
room gardens, 220–1
root crop storage, 455
root houses, 215
rope-making works and workshops,
    495, 500, 519, 521
Rose Street Foundry, 532
Rose Terrace, Perth, 76
Rosebank, 97
Rosebank Cottages, 97
Rosebank Distillery, 513, 545
Rosebery, Earl of, 57
Rosehearty, 202
Roseneath, 175, 411, 586
Rose's Home, 181
Roslin, 129
Ross & Macbeth, 176, 186
Ross & Mackintosh, 172, 173
Ross, Alexander, 173, 294
Ross and Cromarty
  abattoir ventilation, 524

aluminium smelter, 559
British Linen Bank, 639
country houses, 54
defences, 382, 404, 407
fishing station, 229
gardens and garden buildings, 221, 222
Gate of Negapatam, 215
grain mill, 542
granaries, 198, 578
hand crane, 580
harbours, 578
hay barns, 452, 453
horse-engine house, 442
hotels, 160, 166, 171, 173, 176–7
ice houses, 579
municipal buildings, 251, 253, 269, 271, 272
palace, 36
piers, 577
prisons, 269, 272
railway station, 577
salmon bothie, 227, 587
stables, 193
warehouses, 578, 627
wool stores, 489
see also Cromarty; Easter Ross; Wester Ross
Ross, George, of Pitkerrie, 54
Ross, John, 486
Ross, Thomas *see* MacGibbon & Ross
Rossie Priory, 146
Rosslyn Inn, 129
Rosworm, John, 387
Rosyth
air-raid shelters, 406
naval base, 382, 402, 403, 404, 407, 582
Rothes, 486, 563
Rothes, Earl of, 48
Rothesay, 79, 166, 170, 253, 378, 577
Rothesay Dock, Glasgow, 574, 581
Rothiemay, 449
Rotor bunkers, 413
Rotor Scheme radar stations, 412
rotundas, 214
Round Square, 194
round-houses, 10
see also brochs
Rowallan, 62
Rowan, Professor, 54–5
Roxburgh Castle, 382
Roxburghe, Earl of, 51

Roxburghe Hotel, 182
Roxburghshire
abbey, 279, 483
byre, 466
cattle courts, 474, 475
château, 40
church, 288
commemorative tower, 215
corn exchange, 644
country houses, 51, 58–9
defences, 382
dispensary, 312
dyeworks, 510, 538
flour mill, 541
gardens, 220
grass-drying plant, 455
horse engine houses, 442
hosiery factory, 538
hotels and inns, 130, 135, 136, 139, 140
houses, 73, 78
hydropathic, 170
mills, 519, 521, 526, 538
municipal buildings, 253
potato houses, 457–8
poultry housing, 476
railway station, 606
Redden Farm, 432, 435, 475
Royal Bank, 639
sheep houses, 485
stables, 432
steam engine house, 443
stocking shops, 538
threshing machines, 441
tower-houses, 33
villas, 76
water supply, 345
weaving sheds, 537
Royal Alexandra Infirmary, 314
Royal Bank, 632, 639, 666
Edinburgh, 631, 635–6, 674
Glasgow, 632, 642
Royal Bank Computer Centre, 673–4
Royal Botanic Garden, 368–9
Royal British Hotel
Dundee, 143, 159
Edinburgh, 155
Royal Commonwealth Pool, 375
Royal Edinburgh Asylum, 319, 320
Royal Edinburgh Hospital, 323
Royal Exchange, 632, 642, 646
Royal Exchange Assurance buildings, 667, 672–3

INDEX • 745

Royal Exchange Square, 632, 642, 646
Royal George Hotel, 143
Royal Hospital for Sick Children, 332, 333
Royal Hotel
  Bridge of Allan, 168
  Coupar Angus, 142
  Dundee, 143, 144, 159
  Dunkeld, 147
  Edinburgh, 153, 154, 160, 185
  Fort William, 166
  Glasgow, 155
  Nairn, 137
  Oban, 173
  Rothesay, 166
  St Andrews, 137
  Stirling, 143
  Tain, 173
Royal Infirmary *see* Edinburgh Royal Infirmary
Royal London House, 670
Royal Maternity and Simpson Memorial Hospital, 330–1
Royal Northern Infirmary, 312
Royal Public Dispensary, 312
Royal Samaritan Hospital for Women, 337
Royal Scottish National Institution, 325
Royal Society, 640
Royal Stuart Hotel, 185, 186
Royal Tavern, 141
Royal Victoria Hospital for Consumption, 328
Royalty Theatre, 651
Royston, 109
Royston House, 198
Rubens, 49
Ruchill, 329, 344
rugby stadium, 372
rural schools, 295–7
Rusack's Marine Hotel, 173
Ruskin, John, 289, 512
Russell, James, 648
Russland, 410
rustic style, dairies, 473
Rutherglen, 262, 267
Rutherglen Pottery, 512
Ruthven, 32, 391, 393, 395
Rutland Court, Edinburgh, 673
Rutland House lodging house, 181
Ryder, M L, 483

S

Sailors' Hall, Dundee, 362
Sailors' Homes, 158–9
St Andrew Hotel, 155
St Andrew House, Glasgow, 672
St Andrews
  cathedral, 278–9, 282
  churches, 277–9, 284
  cottage hospital, 313
  defences, 386
  harbour, 571
  hotels, 137, 173, 185
  houses, 70, 73, 76, 79, 83
  Madras College, 304
  swimming pool, 375
  town house, 250
  town plan, 69
  university, 305
St Andrews Castle, 29, 30, 31, 386
St Andrew's Church, Dundee, 288
St Andrew's Church, Glasgow, 262
St Andrew's Halls, 373
St Andrew's House, Edinburgh, 183
St Andrews University, 305
St Ann's, Cromarty, 71
St Cecilia's Hall, 359–60, 373
St Colme, 455, 475
St Cuthbert's Church, 280
St Cyrus, 479
St Enoch Square, 654, 659, 662
St Enoch Station, 604, 656
St Enoch's Hotel, 153, 156, 160, 161–2, 163, 187, 654
St Enoch's Wynd, 182
St Fort, 57
St George's Chambers, 646
St George's Hotel, 127, 142
St George's in the Fields, 289
St Giles Cathedral, 269, 270–1, 282, 285
St James Centre, 671
St John's Church
  Gamrie, 364
  Perth, 269, 271, 282, 284
St Kilda, 12, 13, 92, 93
St Leonard's Tower, 271, 272
St Leonards Works, 527, 528
St Machar's Cathedral, 279, 281
St Machar's House, 674
St Margaret's Hope, 579
St Margaret's Works, 527
St Martin, Alexander de, 280
St Martin's Church, 280

St Mary's, Orkney, 627
St Mary's Chapel, Edinburgh, 359
St Michael's Church
  Dumfries, 288
  Linlithgow, 282
St Monance, 585
St Paul's Church, Perth, 288
St Paul's Episcopal Chapel, Dundee, 632
St Rollox, 523, 606
St Ronan's Standard office, 655
St Roques Works, 533
St Rule's Church, 277–8
St Vincent Street, Glasgow, 651, 660
Salford Twist Mill, 514
Salmon & Gillespie, 82
salmon bothies, 225–7, 587
salmon fisheries, 586–8
Salmon, James, 647
Salmon, James, younger, 660–1, 662
Salmon Son & Gillespie, 666–7
salt extraction, 552
salt girnals, 499
Saltcoats, 552
Saltire Court, 673
Saltoun, 502
Saltoun Hall, 57
saltworks, 499
Salutation Hotel
  Kinross, 129
  Perth, 129, 142
Salvesen Tower, 674
Samalan House, 205
sanatoria, 325–9
  see also hydropathics
Sanday, 442, 443, 450, 478–9
Sandbank, 586
Sanderson & Murray, 526, 541
Sandhaven, 573
Sandside, 203, 204, 449
Sandsting, 485
sandstone quarries, 558
Sango, 412
sanitation
  mills and factories, 524–5
  see also privies; refuse disposal; sewerage
Sanquhar, Orkney, 441
Saracen Foundry, 533–4, 649
Saracen's Head, 132–3, 142
sarcophagus, 290
Saucel Mills, 186
Saughtonhall Exhibition, 371

Saunders, A D, 396
Savoy Croft, 86
sawmill lades, 524
sawmills, 495, 496
  East Lothian, 498
  Fife, 524
  materials handling, 500
  on farms, 435
  power sources, 505
  Selkirkshire, 519
Saxa Vord, 412, 413
SBT Keppie, 315, 675
Scadlaw House, 86
Scandic Crown Hotel, 186
Scandinavia *see* Iceland; Norway; Sweden; Viking buildings
Scapa Flow, 32, 382, 401, 404, 405, 407–8
Scar, Orkney, 443, 450, 478–9
Scarinish, 412
Schaw Convalescent Home, 336–7
Schaw, Marjorie Shanks, 336
Schinkel, Karl Friedrich, 112, 166
School Hill, Portlethen, 410
School of Aerial Gunnery, 405
schoolhouses, 296
schools, 76, 294–305, 325, 528, 563
Scone Palace, 43, 56, 216
Scotch ovens, 502
Scotland Street School, Glasgow, 298–9
Scots Baronial architecture, 58
  Aberdeen Grammar School, 304
  banks, 639, 646, 667
  boarding-houses, 168
  breweries, 544
  business chambers, 646
  Cockburn Street, Edinburgh, 647
  corn exchanges, 644
  country houses, 58–9, 61, 62
  gasometer, 205
  gate lodges, 192
  hotels, 150, 154, 172, 173, 174, 176
  houses, 253
  houses and villas, 78, 79, 83, 96, 168
  hydropathics, 170, 171
  industrial buildings, 527, 528
  Leith Sailors' Home, 158
  poorhouse, 318
  prison, 271
  steam engine house chimney, 443
  tenements, 117

INDEX • 747

warehouses, 646, 649
Scots Renaissance architecture, 80–1, 165, 176, 181, 194, 659
Scotsman buildings, 187, 516, 647, 654, 664, 665
Scotstoun, 534, 584, 585, 586
Scotstoun Mill, 541
Scott, Andrew R, 165
Scott, Brownrigg & Turner, 673
Scott, Sir George Gilbert, 161, 307, 308, 368
Scott, J, 486, 487, 488
Scott, J N, 664
Scott, Mackay Hugh Baillie, 86
Scott Morton & Company, 179
Scott, Peter, 538
Scott, Sir Walter, 58, 59, 79, 146, 167
Scottish Co-operative Wholesale Society, 512, 665
Scottish Dyes, 559
Scottish Equitable building, 657–8, 670
Scottish Exhibition Rooms, 368
Scottish Lands & Building Company offices, 650–1
Scottish Legal Life Building, 657, 669
Scottish Life buildings, Edinburgh, 670, 671
Scottish Life House, Glasgow, 672
Scottish Linoleum Works, 538
Scottish Power headquarters, 529
Scottish Provident Institute, 640
Scottish Provident Institution, 660, 670
Scottish Union House, 672
Scottish vernacular *see* vernacular architecture
Scottish Widows buildings, 151, 670–1, 673, 675
Scrabster, 579
Scroggie, Syd, 229
sculleries, 468, 469, 470
Seafield Baths, 374
Seafield Colliery, 563
Seafield, Earl of, 145
Seafield Works, 514, 528
Seagreens, 479
Seamen's Friend Society, 158
seaplane bases, 404
    see also airfields
seaside pavilions, 378
seasonal dwellings, 224–35
    see also bothies; shielings
seasonal workers' accommodation, 92, 224–7, 230–3, 236–45, 427
    see also shielings
Second Empire style, 154, 170, 665
secondary schools, 295, 304–5
Seifert, R, & Partners, 671, 672, 673, 676
Seil, 577, 580
Selkirk, 83, 280, 537, 538
Selkirk, earl of, 35
Selkirkshire
    clock tower, 526
    hospital, 338, 339
    mills, 516, 519, 520, 523, 524, 537
    potato houses, 457–8
    poultry housing, 476
    sawmill, 519
    sheep husbandry, 485, 486, 487
    skinworks, 526
    tanneries, 541
    water supply, 345
    weaving factories, 537
    wool store, 489
Sellars, James, 152, 154, 158, 373, 651, 657
    see also Douglas & Sellars
Semple, Lord, 291
Sentinel Works pattern shop, 516, 533
servants' quarters
    country houses, 60
    farmhouses, 70–1, 427
    hotels, 154
    inns, 135
    tower-houses, 33
    villas, 83
    see also coachmen's flats; farm workers' housing
Seton Castle, 30, 55
Seton Collegiate Church, 282
Seton Palace, 39
sewerage, 342, 348, 352–3
    see also privies; refuse disposal; sanitation
shaft kilns, 553
Shaftesbury, Lord, 179
Shairp, Alexander, 166, 173
Shandon hydropathic, 169–70
Shapinsay, 204–5
Shaw, Richard Norman, 82, 164, 173, 174
Shaw, William, 45
shawl factories, 539
Shaws Water Mill, 513
Shaws Water Works, 342
sheds

masons' yards, 497
railway engineering workshops, 606
Roseneath, Dunbartonshire, Silvers' motor yacht yard, 586
sheep husbandry, 486
shipbuilding yards, 584–5
stabling, 431
see also animal houses; dipping sheds; implement sheds; railway buildings; storehouses; warehouses
sheep husbandry buildings, 16, 482–9
sheer legs, 529, 530, 581, 582, 585
Shell, 674
shell concrete, 519
Shepherd House, 73
shepherds' dwellings, 224, 428, 486, 488, 489
Sheraton Hotel, 186
Sheriff Courts see law courts
Shetland
　barns, 425, 439
　byre dwellings, 20, 425
　byres, 466
　defences, 388–9, 397–8, 404, 405, 409
　ferry facilities, 578
　fishing stations and lodges, 230, 232
　geese, housing, 478
　gutters' housing, 232
　hand crane, 580
　hotel, 154
　kilns, 449, 450
　lammiehooses, 483–5
　leisure centre, 376
　merchants' booths and houses, 627
　mills, 450
　outhouses, 20
　oval houses, 18
　peat-flitters' accommodation, 232, 428
　piers, 232, 575
　potato houses, 459
　poultry housing, 476, 478
　radar station, 412, 413
　schools, 297
　skeos, 482
　threshing facilities, 443
　Viking buildings, 17, 19, 20
　warehouses, 627
　wheelhouses, 10

Shieldhall, 512, 575
shieling huts, 10, 11–15, 16–17, 18–19, 20, 92, 230, 428
shielings, 224
Shiells, R Thornton, 78, 651
shiels see salmon bothies
Ship Bank
　Glasgow, 632
　see also Glasgow & Ship Bank
shipbuilding yards, 516, 530, 532, 581, 582, 584–5
shoe factories and workshops, 496, 540–1
　see also tanneries
shooting lodges, 63, 205, 207, 227
　see also sporting hotels
shopping arcades, 154, 156, 620, 621, 664
shops, 614, 615–16, 618–22
　Aberdeen, 144, 619
　Dundee, 143, 144, 630, 666
　Edinburgh, 150, 615, 616, 617, 618, 619, 620, 630–1
　Edinburgh New Town, 646
　Glasgow, 156, 615, 617, 619, 620, 628, 631, 632, 650, 658, 660, 672
　in tolbooths, 258
　terraced houses, 69
　workshops and, 498
　see also department stores
Shore, The, Leith, 519, 626
Shortrigg, 441
Shotts, 325, 535, 556
Shotts Ironworks, 471, 556, 557, 558
Sibbald, Dr John, 322
Sighthill, 122
signal boxes, 601, 605–6
Signet Library, 309
silos, 455, 517
silver mining, 554
Silvermills, 540
Silvers' motor yacht yard, 586
silversmiths see metal-working workshops
Sim, J Fraser, 173, 655
Simprim, 440
Simpson, Archibald
　Aberdeen assembly room, 361
　Aberdeen bank buildings, 637–8
　Aberdeen Hotel, 144
　Aberdeen Royal Infirmary, 313
　Duke of Gordon's Inn, 146
　Elgin lunatic asylum, 320

INDEX • 749

Simpson, Sir James Young, 331
Simpson Memorial Pavilion, 331
Sinclair hut, 229
Sinclair, Sir John, 76, 475
Sinclair, General Patrick, 147
Singers works, 526, 533
skeos, 482
Skerryvore, 583
skew bridges, 601
Skimming, David, 135
Skinner, Major-General Lewis, 395
Skinners' Hall, Edinburgh, 628
skinworks, 526
　　see also tanneries
Skirving, Alexander, 656, 657
Skye
　　bothies, 227, 232
　　castles, 30, 57, 195
　　hotels, 160
　　ice houses, 579
　　kilns, 542
　　poultry housing, 477
slate quarries, 558
slaughterhouses see abattoirs; killing houses
Slessor, Mary, 528
Slezer, John, 389, 448, 575
slipways, 574–5, 585–6
sluice houses, 343–4
slum clearance, 114
Smail, Robert, 655
Smailholm, 33
Smart, Iain, 229
smearing and dipping sheds, 488
Smeaton, John, 541, 595
smiddies see smithies
Smirke, Sir Robert, 57
Smith, Archibald, 176
Smith, F & J, 649–50
Smith, George, 361, 643, 646
Smith, J Forbes, 178, 658, 660
Smith, James, 48, 51, 391, 537, 638, 646
Smith, John J, 176
Smith, Roger, 124
Smith, William, 642
smithies, 495, 496, 501–2, 506
　　engineering works, 535
　　Inverness-shire, 498
　　Midlothian, 193, 502, 506
　　on farms, 435
　　tower-houses, 33
　　ventilation, 502–3

windows, 435, 504
　　see also ironworks
smokehouses, 580
　　see also kilns
snuff mills, 456, 495
Soane, Sir John, 366
soap-making workshops, 495, 502
social facilities, mills and factories, 528–9
Society for the Protection of Ancient Buildings, 62
Society of Antiquaries of Scotland, 363, 366
Somerville, Lord, 51
Sorn, 470
source-books see pattern books
South Exchange Court, 646
South Factory, Kirkcaldy, 538
South Gate, Edinburgh, 385
South Queensferry, 70, 129, 257, 582
South Ronaldsay, 32
South St David Street, Edinburgh, 670
South Side railway station, Glasgow, 604
South Sutor, 404, 407
South Uist, 232, 476, 477–8
Southerness, 583
Southwood, 441
Souttar, James, 159
Sparrow Castle, 68
Sparrow, W Shaw, 80
spas, 135–6, 139, 374, 375
　　see also hydropathics
specialist hospitals, 329–38
Spence, Sir Basil, 63
Spence, Glover & Ferguson, 315, 670, 673
Spence, William, 649, 653, 666
Spencer, J F, 532
Speyburn Distillery, 543
Speyside distilleries, 545
spinning mills, 495, 496, 518, 535–6
　　humidification, 524
　　water wheels, 519
　　see also cotton mills; flax mills; jute mills; textile mills; woollen mills
sporting estates, 62, 196–7, 205–7
sporting hotels, 171–9, 187
　　see also shooting lodges
sports facilities, 371–2, 375–6, 563
　　see also swimming pools; tennis courts

750 • INDEX

Spottiswoode, John, Archbishop of St Andrews, 285
Spotts Mains, 455
Spring Garden Ironworks, 529
Springburn, 534
Springfield, 78
Springhill, 77
Spur Inn, 136
Spynie, 43
stables
    as workshops or warehouses, 614
    at shepherds' dwellings, 489
    canal buildings, 600
    Coignafearn, 207
    country houses, 51
    East Lothian, 429
    estate buildings, 190, 192–4, 207
    farm steadings, 425, 426, 427, 428–32, 456
    Glasgow, 368
    granaries and, 447
    Iceland, 20
    inns and hotels, 128, 129, 130, 132, 133, 135, 136, 137, 141, 145, 146, 147, 148, 596
    Ruthven barracks, 393
    seasonal workers' accommodation, 239
    tower-houses, 33
    town houses, 597
    workshops, 500
    see also coach houses and carriage houses; hammels; loose boxes
Stacks, 442
Stafford, Marquis of, 271
Stag Inn, 127
Stair Arms, 146
staircases
    banks and bank houses, 634, 638
    business chambers, 646
    châteaux, 40
    clubs, 148, 157
    country houses, 49, 51, 52, 54, 57, 67
    farmhouses, 427
    gate lodges, 192
    granaries, 448
    hotels, 143, 148
    houses, 68, 69, 75, 80, 93, 96, 97
    industrial buildings, 522, 525, 527, 528, 536, 538, 539
    inns, 128, 135, 137
    North Bridge, Edinburgh, 664

offices, 652, 662
palaces, 36, 39, 44, 45
schools, 298
tenements and flats, 112, 117
tolbooths, 260
towers and tower-houses, 32, 33, 35, 38
trades halls, 628
warehouses, 665
stake net systems, salmon fishing, 587
Standard Life buildings, 640, 657, 670, 675
Standard Life Computer Centre, 674
standing folds, 486
Stanley, 345
    mills, 512, 513, 514, 519, 521, 524, 526
    salmon bothies, 225, 587
Stanley Poole Associates, 671
Star & Garter Hotel, 152
Star Commercial Hotel, 150
Star Hotel
    Edinburgh, 141, 150, 154, 187
    Moffat, 168
    Perth, 143
Starforth, John, 78, 170, 178
Stark, William, 309, 320, 364
Station Hotel
    Aberdeen, 163–4
    Aviemore, 176, 186
    Nairn, 166
    Oban, 166, 173
    Wick, 165–6
statuary, 216, 219, 220, 291
steadings see farm steadings
steam boxes, 585
steam cranes, 580–1, 585
steam power, 471, 496, 504, 520–1
    coal mines, 561
    distilleries, 520, 545
    farms, 442–3, 456, 457, 482
    forges, 535
    tanneries, 540
    workshops, 505
steamer piers, 575, 577
    see also ferries and ferry facilities
steamies, 347, 374
steaming houses, 456
steel construction, 558
    bridges, 595, 597, 600, 601
    commercial buildings, 660–1, 662, 666, 668

industrial buildings, 516, 518,
    532–3, 534, 543, 584–5
lighthouses, 583
mine buildings, 563
North British Hotel, Edinburgh,
    165
Perth Playhouse, 377
railway stations, 604
viaducts, 603
workshops, 498
steel industry, 556–7
Steel, Sir James, 112
steel works, 557
Steele, Sir John, 291
steeples
    Dumfries, 253
    Fife, 251
    Greenlaw Tolbooth, 269, 270
    Inverness, 251
    Kirkcudbright, 255–7
    Magdalen Chapel, 628
    Merchants' House, Glasgow, 628
    Midlothian, 193
    Paisley Tolbooth, 260
    Penicuik House stable block, 193
    Stirling Tolbooth, 253, 257
    Strathmiglo Steeple, 258
stells, 485, 487
Stenton, 71, 77, 82
Stephen, Alexander, 532, 585
Stephens, H, 447, 477
Stephenson, Derek, 672
Steuart, George, 142
Stevens, Alexander, 595
Stevenson, Alan, 583
Stevenson, J J, 157
Stevenson, John James, 78–9, 82
Stevenson, Robert, 572, 583, 595
Stevenston, 561
Steward, James, of Kingarth, 54
Stewart & Lloyds, 557
Stewart & Macdonald, 666
Stewart Hall, 54
Stewart, Earl Patrick, 45
Stewart, Captain Robert, 57
Stewarts of Newhalls, 129
still-houses, 547
Stirling
    arcade, 620
    Argyll lodging, 37, 69
    banks, 632, 638, 639, 640
    bowling club pavilion, 372
    Castle Rock walk, 368

churches, 282
corn exchange, 643
defences, 385, 406, 409, 412
    see also Stirling Castle
foundry, 258
hospitals, 497
hotels, 143
houses, 69, 76
municipal buildings, 253, 257,
    262, 263, 265, 268, 269
prisons, 269
railway station, 604
tenements and flats, 109
theatres, 620
town bells, 257
town plan, 69
Stirling Castle, 42–3, 45, 388
    barracks, 393
    defences, 31, 390, 391, 393
    gallery, 37
    gatehouse, 31
Stirling, John, 56
Stirling Tolbooth, 253, 257
Stirling, William, 143, 168, 361
Stirlingshire
    air-raid shelters, 406
    aluminium rolling mill, 559
    banks, 639
    barn, 440
    barracks, 391
    bird housing, 478–9
    bleachfield drying house, 502
    canal bridge, 600
    canals, 598, 599
    castles, 29, 30
        see also Stirling Castle
    cattle courts, 475
    coal hoists, 581
    coalfields and collieries, 560, 564
    country house, 57, 63, 67
    cranes, 582
    defences, 410, 411
    distilleries, 513, 545
    docks, 573, 574, 575
    dockside warehouses, 578
    dyeworks, 559
    farm stables, 431
    foundries, 533, 556, 605
    gardens and garden buildings,
        215, 219
    goods shed, 605
    grave-markers, 290
    hospitals, 315, 338, 339, 497

hotels, 143, 168
houses, 97
hydropathic, 170
institutions for mentally
    handicapped, 325
kilns, 449, 501, 554
lunatic asylum, 321
mills, 238, 512, 537
National Bank, 639
petrochemical industries, 559
pottery, 501
Red Lion Hotel, 143
reservoirs, 343, 344
seasonal workers'
    accommodation, 238, 240
sheep houses, 483
swimming pool, 375
villas, 77, 80
weaving factories, 537
Stirrat, Frank, 655
Stobhill see Glasgow
Stock Exchange, Glasgow, 642, 651
stock exchanges, 642–3, 651, 669
stocking shops, 538
stone quarries, 558
Stonehaven, 173, 271–2, 375, 401, 640
Stoneleigh, 83
Stoneywood, 495
stop-lines, 409
storage, workshops, 499
storehouses, 16
    see also barns; cheese rooms;
        girnals; granaries; ice houses;
        warehouses
Stormontfield, 345
Stornoway, 586
Stovin-Bradford, H R, 671
Stow, 146
Stow, David, 297
Stracathro Hospital, 338, 339
Strachen, William, 267
Strachur House, 54
Strath Avon, 204
Strathallan Castle, 455
Strathaven, 343
Strathbogie see Huntly Castle
Strathclyde, 93
Strathclyde University, 308
Strathearn House Hydropathic, 170
Strathmiglo Steeple, 258, 269
Strathmore, 429, 433, 474
Strathpeffer, 171, 176–7
Strathy Point, 583

straw barns, 444
Strawfrank, 606–7
Strichen, 253
Stroma, 477
Stromay, 232
Stromness, 575, 586, 627
Stronach, Alexander, 627
Stronsay, 232–3, 443
Strozzi, Piero di, 31, 386
Students' Union, Edinburgh
    University, 307
Sublime, the, 214
suburbs, 76–7, 79–80
Sugar Exchange, 642
sugar refineries, 513, 514, 516, 525,
    547
sugar-making workshops, 495, 502
Sullivan, Louis, 666
summerhouses, 80, 214, 215, 217–18
Summerlee Ironworks, 557
Sun Inn, 129–30, 135, 140, 271
Sun Life building, 658
sundials, 211, 221, 258
Sunlawshill, 474
'Sunlight Cottages', 97
supermarkets, 621
Surgeons' Hall, Edinburgh, 634
Sutherland
    cattle courts, 474
    coalfield, 560
    dipping shed, 488
    dungstead, 481
    Eleanor cross, 216
    ferry pier, 577
    fishing stations, 580, 586, 587
    game larder, 207
    girnals, 439
    herring curing sheds, 579
    hotels, 145, 147, 176, 271
    ice houses, 579, 587
    lighthouses, 583
    lime kilns, 554
    municipal buildings, 271
    radar station, 412
    railway station, 604
    sheep houses, 485
    shieling huts, 16
    smokehouses, 580
    threshing machines, 443
Sutherland Arms, 145
Sutherland, Duke of, 207
Sutherland Hussey, 368
Suttie, E Grant, 59

swan houses, 219
Swanson, Charles W, 179
Swanston Farm, 432, 456
Swanstonhill Sanatorium, 171
Swarbacks Minn, 404, 405
sweat-houses, 16
Sweden, 16
Swift, Arthur, & Partners, 672
swimming pools, 347, 374, 375–6
Symbister, 627

T

Tailefer, John, 628
Tailend, 471
Tailors' Hall, Edinburgh, 359, 628
tailors' workshops, 496, 539
Tain, 173, 251, 253, 269, 524, 639
Tait, T S, 327
Tait, Thomas, 540
tall offices, 656–70
tanks *see* water supply
tanneries, 502, 540–1
    see also shoe factories and workshops; skinworks
Tantallon Castle, 29, 30, 386
Taransay, 224
Tarbet, Loch Lomond, 127, 166–7
Tarbolton & Ochterlony, 185
Tarbolton, Harold Ogle, 324, 331
Tarlair, 375
taverns, 141, 143, 144, 358
    see also inns; public houses
Tay Bridge, railway station, 604
Tay House, 676
Tay Works, 510, 518, 538
Taylor & Sons, 153
Taylor, Frank, 84
Taylor, Harry Ramsay, 656
Taylor, J Herbert, 62
Taylor, James, 183
Taymouth, 214
Taymouth Castle, 54, 182, 215, 473
Tayport, 572
tea pavilions, 207, 214, 370, 371
teacher training colleges, 297
Teacher, William, 654
technical education buildings, 295, 307–8
telecommunications, 341, 353–4
Telfer's Wall, 385
Telford, Thomas
    aqueducts, 599
    ferry piers, 577

harbours, 572
manses, 76
roads and bridges, 145, 147, 595
tollhouses, 596
temples, 214, 215, 371
Templetons, 619–20
Templeton's Carpet Factory, 512, 529, 538, 669
temporary dwellings, 224, 232, 239
    see also shieling huts
tenement factories, 539
tenements and flats, 96, 97, 99, 101, 103, 108–26, 289, 615, 630, 652, 666
Tennant, Robert, 132
Tennant's Stalk, St Rollox chemical works, 523
Tennents' lager brewery, 545
tennis courts, 43, 80, 358
Tentsmuir, 409
Ter Horst, Henrick, 257
terraced houses *see* houses
Tessin, Hans Ewald, 387, 388
textile factories and workshops, 494–5, 496, 497, 502, 538, 539
    see also bleachfields; bleachworks; heckling shops; weaving factories
textile mills, 495, 524, 525
    see also cotton mills; flax mills; lint mills; spinning mills; thread mills; woollen mills
textile warehouses, 647–50
Thalys, 535
Theatre Royal
    Dundee, 362
    Edinburgh, 362
theatres, 358–9, 360, 362–3, 620
Thirlestane Castle, 39, 40, 49
Thirlestane estate, 489
Thistle Brewery
    Alloa, 545
    Edinburgh, 545
Thistle Hall, 143, 159
Thomas, Captain, 13
Thompson, Matthew William, 161
Thoms & Wilkie, 81
Thomson & Sandilands, 665
Thomson, Mrs, 129
Thomson, Alexander, 158, 651
    Bath Street 87–97, 651
    churches, 289
    commercial buildings, 650

Glasgow Architectural Exhibition, 367–8
Great Western Terrace, 75
houses, 77, 79
Macfadzean monograph on, 277
Northern Club, 156
tenement windows, 116
Villafield Works, 656
Washington Hotel, 156
Thomson, D C, & Company, 664–5
Thomson, James
  commercial buildings, 649, 651, 659
  convalescent homes, 336
  Maclean's Windsor Hotel, 157
  Stewart & Macdonald's factory, 666
  turbine, 520
Thomson, James Taylor, 669
Thomson, John, 646
Thomson, Leslie Graham, 669, 671
Thomson, Robert, 660, 664
Thomson, W B, 544
Thomson's Brewery, 544
Thornton, 560, 563
thread mills, 514, 516, 521
  see also cotton mills; spinning mills; textile mills
Threave, 33
threshing facilities, 226, 438–9, 440–5, 452
Thurso, 76, 544, 595
Thurso Castle, 31, 192
Thurston, 475
Tighnabruaich, 577
tile works and workshops, 495, 502, 557–8
Tillicoultry, 526
Tillie & Henderson, 648
Timber Bush, 624, 626
timber construction
  houses, 20–1, 104–5
  piers, 577
  railway stations, 604
  roofs, 517–18
  shieling huts, 16
  viaducts, 602–3
  workshops, 498
timber galleries, 37
timber yards see sawmills
timber-drying kilns, 501
Timex factory, 534
Tinto Firs Hotel, 186

Tiree, 412, 583
Tobacco Exchange, 642
Tod & McGregor, 584
toilets see privies
Tolbooth Hotel, 138–9
Tolbooth Wynd, Leith, 580
tolbooths see municipal buildings
tollhouses, 95, 595–6
  see also due-collecting booths
Tolquhon Castle, 31, 37, 43, 45, 203
Tom-na-Monachan, 182
tombs and tombstones, 290–1
  see also mausolea
Tomintoul, 397
Tontine Building, Glasgow, 642
Tontine coffee room and hotel, Glasgow, 142
Tontine Hotel, Cupar, 142, 143
Tontine inn and assembly room, Paisley, 142
Topham, Edward, 129, 131
Tor House, 79
Torness, 346
Torphichen Street, Edinburgh, 671, 675
Torr Hill, 77
Torrance, 600
Torry, 575, 580
Torry Point, 401, 402
Torsonce Inn, 146
Total, 674
tower blocks, 122–4
tower breweries, 544
tower cranes, 585
Tower Mill, 519
tower silos, 455
tower-houses, 28, 30, 32–7, 48, 385, 397, 402, 583
  tolbooths, 251–3
towers, 27–8, 36, 43, 277, 386
  abbeys, 35
  castles, 28, 29, 30, 31, 35
  châteaux, 40, 42
  country houses, 51, 57, 59, 67
  garden features, 214–15
  hall-houses, 29
  hospitals, 313
  lunatic asylums, 320
  palaces, 35, 43
  poorhouse, 318
  steam engine house chimney, 443
  see also Martello Towers
Towers, James, 330

INDEX • 755

Towie, 45
Town & County Bank, 637, 653
town bells and town clocks, 255–8
town halls and town houses *see* municipal buildings
Town Pier, North Queensferry, 577
Town Yetholm, 73
Town's Hospital, Glasgow, 311, 312, 316, 320
Toxside Farm, 435
Trades Hall
    Aberdeen, 627, 628
    Dundee, 361, 630
trades halls, 627–8, 630, 631
    Aberdeen, 627, 628, 642
    Dundee, 361, 630
    Edinburgh, 359, 627–8, 630
traditional architecture *see* vernacular architecture
train sheds *see* railway buildings
tramcars and tramways, 176, 595, 597, 599
transit sheds *see* warehouses
transport, 594–611
Traprain, Lord, 445
Traquair, 37, 212, 545
Traquair, Phoebe, 333
Traquair, Thomas, 76
travelling cranes, 525, 529, 530, 533
Tregenna Castle Hotel, 179
Tresness, 442
Trinity *see* Edinburgh
Trinity College, Glenalmond, 304
Trinity Hall, Aberdeen, 642
Trinity Hospital, Edinburgh, 368
Trinity House, Dundee, 362
Tron Church, Glasgow, 630
Tron Kirk, Edinburgh, 286–7
Trongate, Glasgow, 142, 665
trons, 613–14, 624–5, 627
Troon, 82, 174, 574, 580
Trossachs Hotel, 167, 168
Truscon, 517, 533
Trustee Savings Bank, 668
tuberculosis hospitals, 328–9
Tudor style buildings
    cistern, 344
    corn exchanges, 644
    country houses, 57
    hotels and inns, 146, 174
    houses, 77, 78
    Trinity Hall, Aberdeen, 642
Tugnet, 580, 587

Tullibody, 540
Tullich, 139
Tullichewan Castle, 56, 57
turbines, 345, 441, 520, 524
Turnberry, 86
Turnberry Hotel, 177–8, 179, 187, 338
Turnbull, Robert, 156
Turnbull's, Hawick, 510
turnip sheds, 451, 455, 475
Turriff, 140, 258
Turtleton Farmhouse, 78
Tweedsmuir, 127
tympany gables, 69
Tyninghame, 59, 216, 498
tyre factories, 534
tythe barn, 439

U

Ugadale Arms, 175
Uig, 13
Uiginish, 477
Uishal, 12
Uist, 230, 232, 476, 477–8
Umpherson & Kerr, 521
underground defences, 411–12, 413
underground passages, 215
Underhoull, 17
Underwood & Partners, 185
Underwood Road, Paisley, 601
Union Bank, 137, 632–3, 636, 638, 640, 653
Union Bank of Scotland, 668, 669
    see also Bank of Scotland
Union Bridge, Glasgow, 603
Union Canal, 598, 599
Union Club, 148
Union Hotel, 144
Union Street
    Dundee, 143
    Glasgow, 649, 650, 651
United Distillers, 674
United Presbyterian Church buildings, 289
universities, 294, 305–7
    Aberdeen, 305, 307, 311, 372
    Edinburgh, 308
    Glasgow, 305, 306, 307, 308, 372
    libraries, 307, 309
University Club, Edinburgh, 150, 152
University College, Dundee, 307
University Gardens, Glasgow, 75
University of Strathclyde, 186, 672
Uphall, 141, 182

Upper City Mills, Perth, 186
Upper Clydesdale *see* Lanarkshire
urban elementary schools, 297–300
Urquhart Castle, 30
Usan, 226
Usher Hall, 373–4

V

Vale of Leven hospital, 314
Valleyfield Paper Mill, 528, 539
Van der Goes, Hugo, 284
vat rooms, 471
Vatersay, 232
ventilation
    barns, 439, 444
    byres, 466, 467
    granaries, 448
    heckling shops, 503
    horse engine houses, 442
    hospitals, 313
    industrial buildings, 523–4, 529, 538, 540, 545
    laundries, 196, 198
    milk houses, 469
    potato houses, 458
    schools, 296
    smokehouses, 580
    stables, 430
    workshops, 502–3
    see also roof ventilators; windows
Verdant Works, 528
vernacular architecture, 61
    dovecots, 200
    estate buildings, 190, 193–4
    Fortingall cottages, 96
    gate lodges, 192
    hotels, 175, 178
    houses, 105
    inns, 139–40, 147
    roofs, 227
    salmon bothies, 226–7
    schools, 294
    stables, 193–4
    workshops, 498
    see also Arts and Crafts architecture; Scots Renaissance architecture
viaducts, 601–4
Victoria Buildings, Glasgow, 646
Victoria Chambers
    Dundee, 652
    Edinburgh, 664
Victoria Dock, Dundee, 574, 578
Victoria Dyeworks, 538
Victoria Hotel
    Forres, 173
    Rothesay, 166
Victoria Linen Works, 537
Victoria Mill, 541
Victoria Park, 370
Victoria, Queen, 79, 131, 143, 163, 172, 205
Victoria Road Calender Works, 528
viewing platforms, 38, 42, 216, 372
viewing shelters, 215
Viking buildings, 17, 19, 20
Villafield Works, 656
village asylums, 322–4
villages, 93, 94, 215
    see also parish schools
villas, 40, 42, 44–5, 71, 73, 76–83, 108, 109, 113, 214
    see also art gallery houses; country houses; houses
Vine, The, 78, 367
Virginia Court, 646
Virginia Mansion, 633
Vogt, Tony, 86
voluntary schools, 297–8

W

Wade, Major-General George, 131, 393, 594
Waghevens, Jacob, 257
wags *see* wheelhouses
waiting rooms, pier buildings, 577
Wakefield, 320
Waldorf Hotel, 184
Wales, 16, 20, 86
Walker, Captain Ewan, 63
Walker, B, 70, 232, 429, 430, 433, 474
Walker, Bruce, 227, 442, 579
Walker, James Campbell, 170
Walker, John, 443
Walkerburn, 519
Walker's Tavern, 141
Wallace Foundry, 534
Wallace Scott Tailoring Institute, 529
walled gardens, 40, 218–20, 371
Wallis, Gilbert & Partners, 534
walls
    blackhouses, 16–17
    byres, 467
    châteaux, 41–2
    granaries, 448

houses, 10, 91, 92, 94
mills and factories, 510–12, 521, 524–5, 530
potato houses, 458
schools, 297
sheep houses, 485
shieling huts, 11–12, 15–17
stables, 430
towers, 35
workshops, 498, 502
Walther, Dr Otto, 328
Wanlockhead, 554
Ward Foundry, 530, 535
Wardhouse, 54
Wardlaw Hall, 79
Wardrop & Reid, 60
Wardrop, Anderson & Browne, 79–80
Wardrop, James Maitland, 60
Ware, Isaac, 53
warehouses, 547, 614, 625–7, 646–50, 656
East Lothian, 578, 626
Edinburgh, 616, 646
Fife, 518, 527, 578, 626
floor construction, 513, 517
Glasgow, 646, 647–50, 656, 659, 665–6, 668, 669
goods movement facilities, 525
see also canalside warehouses; custom houses; dockside warehouses; girnals; goods sheds; granaries
Warrender, 83, 374
wash-houses see laundries
Washington Hotel, 156
Washington Street rice mill, Glasgow, 541
water columns, 606
water-powered machinery, 496, 504, 505, 506
beam engine, 554
bleachworks, 539
cotton mills, 535
farms, 441, 442, 456, 457, 470
printing press, 655
tanneries, 540
see also hydraulic equipment; water wheels
water supply, 167, 341–5, 504
to locomotives, 606–7
water towers, 344, 517, 526, 540
water wheels, 441, 505, 519–20, 526, 530, 545

see also turbines; water-powered machinery; wheelhouses
Waterhouse, Alfred, 174, 658
Waterloo Chambers, 662
Waterloo Place, Edinburgh, 143
Waterloo Street, Glasgow, 672
Waterloo Tavern, 143, 144
Waterloo towers, 215
Watermill Hotel, 186
waterworks see water supply
Watson & Pritchett, 320
Watson, George Lennox, 175
Watson, J A S, 493
Watson Street, Glasgow, 650
Watson, Thomas Lennox, 171, 174, 175, 658
Watt, Christian, 232
wattle construction, 10, 14, 15–16
Waverley Hotel
  Glasgow, 156
  Nairn, 173
Waverley Hydropathic, 170
Waverley Market, 153, 617
Waverley Mill, 518–19, 521
Waverley Station, 153, 164, 604, 606, 617
Waverley Station Hotel see North British Hotel, Edinburgh
Weaver, Lawrence, 83
weaving factories and workshops, 503–4, 513, 524, 536–8
see also hand-loom weaving shops and factories
Webb, John, 631
Weddel & Inglis, 183
Wedderburn Castle, 55
Wedgwood, Roland, 673
Weem, 130, 135
weigh beams see trons
Weir, J & G, 533
Weirs Foundry, 517
Welcombe House, 179
Weldon, Sir Anthony, 368
Well, E P, 516
Wellington Arcade, Glasgow, 620
Wellington Place, Stirling, 76
Wemyss, 495
see also East Wemyss; West Wemyss
Wemyss Bay, 411, 577, 604
Wemyss, Earl of, 52–3
West Bow, Edinburgh, 133
West End Park see Kelvingrove Park

West Gate, Edinburgh, 385
West George Street, Glasgow, 654, 662, 670, 675
West House, Edinburgh Royal Asylum, 320
West Linton, 184
West Lothian
  abattoir ventilation, 524
  Blackness Castle, 385, 386, 388, 389, 402
  Blackness prison, 31, 35
  châteaux, 40, 41, 42, 45
  churches, 280, 282, 285
  cinemas, 376
  collieries, 564
  commemorative tower, 215
  country houses, 49–50, 57
  defences, 403
  distillery, 547
  docks, 574
  forge, 535
  gardens and garden buildings, 211, 214, 215
  granaries, 447
  harbour lights, 582
  houses, 69, 70
  inns, 129, 140–1, 146
  lodging houses, 181–2
  lunatic asylum, 324
  municipal buildings, 253, 257, 269
  palace, 37
  petrochemical industries, 559
  poultry housing, 456
  quarry, 287
  railway bridge, 601
  sarcophagus, 290
  threshing facilities, 442
  tower-houses, 33, 35, 37, 385
  town bells, 257
  villa, 42
  warehouses, 626
  workshops, 495
  see also Lothians
West Nile Street, Glasgow, 650
West Office Park, Edinburgh, 674
West Old Dock, Leith, 573
West One Business Park, 674
West Port, St Andrews, 386
West Port House, Edinburgh, 671
West Port lodging-house, Edinburgh, 179
West Regent Street, Glasgow, 673
West Wemyss, 573

Westburn Foundry, 533
Westburn/Berryards Refinery, 547
Wester Feabuie, 456
Wester Fintray, 450
Wester Gallaberry, 442
Wester Ross, 224, 439, 452
  see also Ross and Cromarty
Western Approach Road, Edinburgh, 675
Western Bank, 632, 636, 639
Western Club, Glasgow, 148, 374, 672
Western Infirmary, Glasgow, 315
Western Isles
  animal houses, 20
  barns, 439
  blackhouses, 17, 92–3, 425
  boat building yard, 586
  byres, 465, 467
  custom house, 586
  double-skin walled buildings, 17
  farm steadings, 425
  fishermen's huts, 230
  hand crane, 580
  harbours, 573
  kilns, 449
  mills, 450
  poultry housing, 476, 477–8
  shieling huts, 13, 428
  weaving sheds, 498, 538
  wheelhouses, 10
  see also names of individual islands
Westerton Arms, 168
Westfield House, 78
Westminster Palace Hotel, 152
Westray, 37, 441, 450, 579
Whaligoe, 579
Whalsay, 627
Whatton Lodge, 83
Wheatsheaf road-house, 184
wheelhouses, 10, 504, 505, 519, 520
  see also water wheels
whin mills, 456
whisky bottling plant, 519
White Hart
  Edinburgh, 132
  Glasgow, 132
White, John, 77
Whitecraigs, 84, 86
Whitefield, George, 133
Whitehall Crescent, Dundee, 159, 652
Whitehorse Inn, 127–8
Whitekirk, 439

Whitelaw, 68
Whitelee, 483
Whitestone Military Range, 411
Whitevale, 374
Whithorn, 71
Whitie, William, 309
Whittingehame, 57, 477
Whittingehame Mains, 445
Whitworth, Robert, 599
Wick
    cemetery, 291–2
    harbour, 572, 579
    herring curing sheds, 579
    hotel, 165–6
    kilns, 450
    lighthouse, 582
    radar station, 412
    roads, 595
    smokehouses, 580
    tolbooth, 250
Wigglesworth, H H, 62
Wight, A, 446
Wight, Andrew, 486
Wigtown, 257, 260
Wigtownshire
    churches, 287
    creamery, 470, 473, 480
    gardens, 214
    Glenluce Abbey, 341
    harbour, 572
    harbour lights, 582
    hotel, 178
    houses, 71
    pig sties, 480
    threshing machines, 445
Wilkins, William, 57
William Black Memorial, 583
Williams, Sir E Owen, 511, 669
Williams-Ellis, Clough, 86
Williamson, J A, 664
Williamsons of Cardrona, 127
Willoughby d'Eresby, Lord, 168
Willowbank Dairy, 466, 467
Willson, Thomas J, 161, 654
Wilson, Charles
    Alexander's Mill, 181, 516
    Buchanan Street 144–52, 650
    Glasgow Royal Asylum, 320–1
    Great Western Hotel, Oban, 172
    houses, 75
    lodging houses, 179
Wilson, John, 334
Wilson, R G, 164

Wilson Street, Glasgow, 669
Wilsontown, 554
Wilton Mill, 526, 538
Wimperis, J T, 172
Winchburgh cutting, 601
wind-powered machinery, 440–1,
    470, 471, 504, 505, 506, 519
winding-engine houses, 561–2, 564
Windmillcroft Quay, 582
windmills, 440–1, 443, 504, 505, 519
windows, 511
    bank buildings, 631, 633, 637, 668,
        669
    byres, 466, 467
    canalside houses, 95
    cheese rooms, 472
    country houses, 52, 54
    farmhouses, 75
    galleries, 37
    heckling shops, 503
    hotels, 174, 183
    houses, 68, 69, 75, 77, 80, 84, 91,
        92, 94, 103
    industrial buildings, 512, 521, 523,
        538, 540, 543
    inns, 129, 130–1, 137
    joiners' shops, 504
    laundries, 196
    milk houses, 469
    offices, 659, 660, 662
    salmon bothy, 587
    shops, 618
    smithies, 435, 504
    stables, 430
    supermarkets and retail
        warehouses, 621
    tenements and flats, 114–17
    tollhouses, 95
    tower-houses, 33
    warehouses, 625, 666, 669
    wood-working premises, 504
    workshops, 435, 502, 503–4
Windsor Hotel, 154
    see also Maclean's Windsor Hotel
Winton, Earl of, 626
Wishart Arch, 386
Wolfe Murray, James, 60
women's hospitals, 337
    see also maternity facilities
Womersley, Peter, 86
Wood, Derwent, 662
wood-turning, power sources, 505
Woodend Hospital, 318

Woodhead, 555
Woodhouse & Morley, 516
Woodhouse Warehouse, 668
Woodside, 456
Woodside Paper Mill, 519
woodworking premises, 496, 499, 501, 504, 506, 584
woodworking workshops, 496
wool stores, 486, 488–9
woollen goods factories, 495
   see also textile factories and workshops
woollen mills, 495, 518, 535–6
   boiler houses, 523
   Borders, 518, 520, 521, 535–6
   drying rooms, 523
   engine houses, 521
   seasonal workers' accommodation, 238
   see also textile mills
workers' housing, 93–9, 112, 113, 117–18, 119, 372
   see also drovers' accommodation; farm workers' housing; lock-keepers' housing; miners, housing; public housing; seasonal workers' accommodation; servants' quarters
workers' lodges, 497
workplaces, 421–679
workshops, 494–509, 527, 533, 614, 665
   farms, 435
   Penicuik House, 193
   railways and railway buildings, 606
   shipbuilding yards, 584
   tower-houses, 33
   windows and lighting, 435

   see also smithies; woodworking premises
Wrech, Robin, 483
Wren, Christopher, 288
Wright & Kirkwood, 670
Wright, C C, 315
Wrights' Hall
   Edinburgh, 630
   Perth, 628
wrought iron *see* iron construction
Wrychtishousis, Edinburgh, 38, 44
Wyatt, James, 634
Wylie & Lochhead, 649, 658
Wylie, Edward Grigg, 534, 668–9
Wylie Shanks, 672
Wylson, James, 647

X

XDO (Extended Defence Officers) Posts, 407

Y

Yarrows shipyard, 584, 585
Yester, 215, 217–18, 219, 288
Yester Castle, 28
Yester Mains, 450
Yetholm Hall, 73
York Buildings, 651
Yorke, F R S, 86
Yorke, Rosenberg & Mardall, 675
Yorkshire Hennebique Company, 661
Young & Meldrum, 159
Young, James, 559
Young's Paraffin Company, 559

Z

Z batteries, 409
Zimmermann, Dr W H, 19